D0087100

SECOND EDITION

DIFFERENTIAL EQUATIONS

SECOND EDITION

DIFFERENTIAL EQUATIONS

DAVID A. SANCHEZ
Lehigh University

RICHARD C. ALLEN, JR.
University of New Mexico

WALTER T. KYNER
University of New Mexico

Addison-Wesley Publishing Company

Reading, Massachusetts ▪ Menlo Park, California ▪ New York
Don Mills, Ontario ▪ Wokingham, England ▪ Amsterdam ▪ Bonn
Sydney ▪ Singapore ▪ Tokyo ▪ Madrid ▪ Bogotá ▪ Santiago ▪ San Juan

Sponsoring Editor: **Thomas N. Taylor** ▪ Production Supervisor: **Marion E. Howe** ▪ Art Consultant: **Loretta Bailey** ▪ Copy Editor: **Lorraine Ferrier** ▪ Illustrator: **Textbook Art Associates** ▪ Manufacturing Supervisor: **Roy Logan** ▪ Cover Designer: **Marshall Henrichs** ▪ Text Designer: **Melinda Grosser**

Library of Congress Cataloging-in-Publication Data

Sanchez, David A.
Differential equations.

Includes index.
1. Differential equations. I. Allen, Richard C.
II. Kyner, Walter T. III. Title.
QA372.S17 1988 515.3′5 87-22949
ISBN 0-201-15407-2

Reprinted with corrections, June 1989

5 6 7 8 9 10 DO 959493

PREFACE

This text is based on a course, successfully given at the University of New Mexico for several years, that is an introduction to differential equations for mathematics majors, engineering students, and majors in the physical sciences. This second edition has incorporated an additional chapter on partial differential equations and Fourier series. This material can be used in a more extensive course on differential equations.

The text differs from more traditional books in that numerical methods are used from the beginning and throughout as a tool to analyze the qualitative behavior of solutions as well as to approximate them. If one examines textbooks that have appeared subsequent to the first edition of this book, one sees that this approach has taken hold. This reflects in part the dramatic availability of personal computers to the undergraduate population.

Nevertheless we emphasize that the book is not intended as an introduction to numerical methods for ordinary differential equations. Our aim is to give the important topics and analytical tools needed to study ordinary and partial differential equations. A course taught with the book should be one with mathematical depth, but attractive to an audience with a large proportion of engineering and physical sciences students.

FEATURES

The following briefly describes the main features of the text.

- **Emphasis on applications.** Applications in areas of mechanics, circuit theory, astronomy, and biology are introduced throughout the text and in the problem sets.
- **Exercises.** Exercises are presented in order of difficulty and matched with the order of the material presented, and the worked examples.
- **Classroom presentation.** Every effort was made to match section lengths with the amount of material needed for a classroom lecture presentation.
- **Algorithmic format.** Upon completion of the discussion of a solution technique, an algorithm is presented to assist the students in using the technique. Examples following the algorithm show its use step-by-step.
- **Numerical techniques.** Numerical methods are introduced throughout the text, beginning with the most elementary ones. Students are asked to compare different algorithms, and to vary step sizes, with the intent of giving them a real appreciation for the use of numerical methods to study the qualitative behavior of solutions of ordinary differential equations.
- **Linear systems.** The emphasis is primarily on two- and three-dimensional systems in which the eigenvalues and eigenvectors can be easily found. The fundamental matrix is computed using both the eigenvector method and the more efficient Laplace transform.
- **Partial differential equations.** An elementary but careful presentation of Fourier series is given using the normal modes of vibration of a taut string as motivation. This is followed by the method of separation of variables used to construct solutions to the classical, second order equations of mathematical physics. A novel addition is a brief study of first order hyperbolic systems using the eigenvalue–eigenvector methods discussed previously; this section can be taught independently of the previous material.
- **Computer programs.** Each elementary numerical algorithm is presented in FORTRAN, but appendixes are included giving the algorithm in BASIC and PASCAL. Later in the text the program RKF45, which uses simultaneously a fourth and fifth order Runge–Kutta method to adjust step size, is introduced and some numerical examples are given. It is not necessary to understand the details of the code to be able to use it effectively.
- **Phase plane.** The phase plane is briefly introduced in the chapter on second order linear equations, as an aid to analyze the behavior of solutions, and later in the chapter on nonlinear systems.
- **Theoretical considerations.** Where such concepts as existence and uniqueness of solutions, direction fields, linear independence, and fundamental sets of solutions are introduced, students are given prior examples and exercises to help underpin their understanding of the concept.

- **Conservative systems.** The chapter on nonlinear systems includes a discussion of one degree of freedom conservative systems analyzed using energy methods. If a brief discussion of nonlinear systems is needed, which does not require phase plane analysis, this section can be taught independently.

ORGANIZATION

We have written a text that offers a great deal of flexibility in the choice of material to be covered. A basic course in ordinary differential equations for engineers and physical sciences majors would consist of Chapters 1, 2, 3 (Runge–Kutta methods only), 4, 5 and 7 (selected topics). Where Laplace transform methods are not required to be taught, Chapter 4 can be omitted, and Chapter 6 or further topics in nonlinear systems from Chapter 7 can be taught.

A course in linear systems, including both ordinary and partial differential equations, would consist of Chapters 1, 2, 4, 5, 6, and 9. A one-quarter separate course in partial differential equations, or as part of an engineering mathematics course, could be constructed from the material in Chapters 6 and 9.

The book contains more material than can be used in a one-semester course. But whether the course desired is a basic introductory course, or one emphasizing linear or nonlinear systems, or introducing partial differential equations, the authors believe the material presented in the text can be suitably arranged to meet any need.

CHANGES IN THE SECOND EDITION

We have made extensive changes in the second edition of *Differential Equations.* It should be emphasized that many of the changes came as a result of classroom experience, both here at New Mexico and at other institutions, as well as from reviewers' comments. We believe these changes will greatly improve the presentation of the material to the student reader as well as to the instructor. Some examples of some of the changes in the second edition are as follows:

- There is a 30% increase in the number of exercises, and exercises are ordered by level of difficulty and to match the material presented and the worked examples, which have also been increased.
- A new chapter (Chapter 9) on Fourier series and partial differential equations has been added. The final section of this chapter discusses first order hyperbolic systems and transmission line equations using eigenvalue–eigenvector methods—a first in introductory texts, to our knowledge.

- Much of the theoretical material has been introduced only after students have had some hands on experiences solving the relevant differential equations.
- Numerical methods based on the trapezoidal rule and extrapolation have been used in Chapter 1 to solve first order linear equations whose solutions cannot be obtained by quadrature.
- A discussion of higher order linear equations with application to beam problems has been added to Chapter 2.
- The discussion of linear independence and fundamental sets of solutions in Chapter 2 has been greatly simplified and better motivated, as has the discussion of the phase plane.
- The emphasis of the Heaviside formulas in the discussion of the Laplace transform (Chapter 4) has been reduced, and partial fractions expansions are used more.
- More emphasis has been given to the Problem/Solution format in worked examples.
- Material has been reordered and some sections split to make for better classroom presentation and reader absorption.
- An appendix has been added in which the elementary numerical techniques are written in BASIC.

ACKNOWLEDGMENTS

We would like to give our thanks to the following reviewers of this current edition for their valuable help and suggestions:

Christopher L. Cox, Clemson University
Charles K. Cook, University of South Carolina, Sumter
Zita Divis, Ohio State University
Juan A. Gatica, University of Iowa
Harry Hochstadt, Polytechnic University of New York
James A. Hummel, University of Maryland
Joseph Johnson, Rutgers University
Kenneth M. Larson, Brigham Young University
Gerald D. Ludden, Michigan State University
Walter J. Neath, California State University, Chico
Merle D. Roach, University of Alabama in Huntsville
John T. Scheick, Ohio State University
David R. Scribner, University of Maine at Farmington
Monty J. Strauss, Texas Tech University
William F. Trench, Trinity University
Fred Van Vleck, University of Kansas
Joseph. J. Wolcin, U. S. Coast Guard Academy

We would like to express our special appreciation to Steven G. Krantz of Washington University in St. Louis for his insightful assistance throughout the entire development of this edition.

The reviewers of the first edition gave valuable advice and suggestions. Our thanks to:

W.J. Kammerer, Georgia Institute of Technology
F.H. Mathis, Baylor University
R.E. Plant, University of California, Davis
L.F. Shampine, Sandia National Laboratories
R.E. Showalter, The University of Texas at Austin
A.P. Stone, University of New Mexico
D.D. Warner, Clemson University
K. Schmitt, University of Utah
T. Bowman, University of Florida

We wish to thank Gary Bauerschmidt, Scott Dumas, David Rutschman, and Ferenc Varadi for their technical assistance and Jackie Damrau for her expert typing. Finally, we thank the staff of Addison-Wesley for their support and advice during the writing of this text, and for their assistance in the technical preparation of this book.

Albuquerque, New Mexico

D.A.S.
R.C.A.
W.T.K.

CONTENTS

1 FIRST ORDER DIFFERENTIAL EQUATIONS 1

2 LINEAR SECOND ORDER DIFFERENTIAL EQUATIONS 127

3 ELEMENTARY NUMERICAL METHODS 243

4 THE LAPLACE TRANSFORM 271

7 NONLINEAR DIFFERENTIAL EQUATIONS 471

8 MORE ON NUMERICAL METHODS 559

9 FOURIER SERIES AND PARTIAL DIFFERENTIAL EQUATIONS 587

1

FIRST ORDER DIFFERENTIAL EQUATIONS

1.1 GENERAL REMARKS

An *ordinary differential equation*[1] is an equation expressing a relationship among derivatives of an unknown function of a *single variable.* Such equations often result from the mathematical expression of scientific laws connecting physical quantities and their rates of change. For example, Newton's law of cooling states that the rate of change of temperature of a body is proportional to the difference in temperature between a cooling body and its surroundings. If $T(t)$ represents the temperature of the body at time t and S is the constant temperature of the surroundings, Newton's law leads to the differential equation

$$\frac{dT}{dt} = -k(T - S), \tag{1.1.1}$$

where k is a constant of proportionality. We know from experience that the temperature of the body will change monotonically until we cannot detect a difference between its temperature and that of its surroundings. But we shall need some mathematics to answer questions such as: How rapidly does $T(t)$ approach S? or How does $T(t)$ depend on the proportionality constant k and on the temperature at the time we started our observations?

1. It is not clear why "ordinary" ever became standard terminology in a subject that motivated the invention of the calculus, that has such a wide applicability to science and technology, and that contains so many fascinating ideas and methods.

EXAMPLE 1 Show that if $T(t) = S + Ce^{-kt}$ is substituted into (1.1.1) the expression becomes an identity in t. Write C in terms of S and the temperature of the body at $t = 0$.

SOLUTION We substitute $T(t)$ into (1.1.1) to obtain

$$\frac{d}{dt}T(t) = \frac{d}{dt}[S + Ce^{-kt}] = -kCe^{-kt} = -k[T(t) - S],$$

an identity in t.

If we set $t = 0$ in the formula for $T(t)$, then

$$T(0) = S + C,$$

or

$$C = S - T(0). \qquad \blacksquare$$

The function $T(t) = S + (T(0) - S)e^{-kt}$ is said to satisfy the differential equation (1.1.1). We see that $T(t)$ approaches S exponentially as t increases and that the proportionality constant k appears in the argument of the exponential function. It is called a rate constant.

Techniques for constructing solutions to differential equations will be presented in subsequent sections, but we need only the tools of calculus to test if a function is or is not a solution to a differential equation. To do this we differentiate the function as many times as needed and substitute the function and its derivatives into the differential equation. This is illustrated in the following examples.

EXAMPLE 2 Show that $y(t) = 6 \sin 2t + 7 \cos 3t$ is a solution to the differential equation $y'' + 4y = -35 \cos 2t$.

SOLUTION From

$$y'' = -24 \sin 2t - 63 \cos 2t,$$

$$4y = 24 \sin 2t + 28 \cos 2t,$$

we have

$$y'' + 4y = -35 \cos 2t,$$

as required. $\blacksquare$

EXAMPLE 3 Show that $y(t) = 6 \sin t + 7 \cos 3t$ is not a solution to the differential equation $y'' + 4y = -5 \cos 2t$.

SOLUTION From

$$y'' = -6 \sin t - 63 \cos 3t,$$
$$4y = 24 \sin t + 28 \cos 3t,$$

we have

$$y'' + 4y = 18 \sin t - 35 \cos 3t \neq -5 \cos 2t$$

on any t-interval of nonzero length. This illustrates the fact that to be a solution a function must satisfy the differential equation in a nontrivial way. ■

Our goal in studying differential equations is to determine both the qualitative and quantitative properties of those functions that satisfy them; such functions are called *solutions.* This can sometimes be done by representing the solutions as elementary functions, e.g., polynomials or trigonometric functions. More often, geometric arguments are needed to determine the qualitative properties, and numerical methods are needed to determine the quantitative properties. Numerical techniques are especially important when an explicit representation of the solution cannot be found.

Three differential equations that arise quite often in applications and that will be discussed later in the book are:

1. $\dfrac{dy}{dt} = k(t)F(y)$, the growth equation;
2. $L\dfrac{d^2Q}{dt^2} + R\dfrac{dQ}{dt} + \dfrac{Q}{C} = E(t)$, the LCR oscillator equation; and
3. $\dfrac{d^2\theta}{dt^2} + \dfrac{g}{l}\sin\theta = 0$, the pendulum equation.

In all three equations t is called the *independent variable,* and the unknown functions y, Q, and θ, whose derivatives explicitly appear, are called the *dependent variables.* A solution is a function $y(t)$, $Q(t)$, or $\theta(t)$ that satisfies the equation on an open t-interval. Hence the solutions y, Q, and θ are functions of t but, in contrast with the explicit notation $k(t)$ and $E(t)$, this functional dependence is to be understood from the context.

Our study is made easier by grouping together those differential equations that have a significant common property. The most important property is that of *order.* The *order of a differential equation* is the order of the highest derivative of the dependent variable which appears in the equation. For example, the oscillator and pendulum equations are second order, while the growth equation is first order. In the remainder of this chapter, first order differential equations will be studied, and in later chapters higher order equations will be discussed.

A *partial differential equation* is a relationship involving an unknown function of *at least two* variables and one or more of its derivatives. For instance, a partial differential equation that describes the temperature $u(x, t)$ in a thin heated wire as a function of position x and time t is

$$\frac{\partial u}{\partial t} = k\frac{\partial^2 u}{\partial x^2} \qquad \text{or} \qquad u_t = ku_{xx},$$

where k is a constant dependent on the physical properties of the wire. An introduction to some of the partial differential equations of mathematical physics is given in Chapter 9.

EXERCISES 1.1

In Exercises 1–6 determine which of the given equations are ordinary differential equations and which are partial differential equations. Identify the dependent and independent variables.

1. $y'' + 3y' + 2y = \sin t$

2. $\left(\frac{dy}{dt}\right)^2 + 4y = t^2$

3. $u_{xx} - 9u_{tt} = 3 \sin x \cos t$

4. $\frac{\partial u}{\partial x} + t^2\frac{\partial u}{\partial t} = 0$

5. $\frac{d^3x}{dt^3} - 4\frac{dx}{dt} + 7x = 4e^{-t}$

6. $\varphi'' + (\cos t)\varphi = 0$

In Exercises 7–12 determine the order of each of the given differential equations.

7. $y' + 4y = 0$

8. $\frac{d^2y}{dt^2} + 4y^2 = 7$

9. $y'' + \epsilon y'(y^2 - 1) + 4y = 3 \sin t$

10. $(p(t)y')' + q(t)y = 0$

11. $y'y'' + 4yy' = 0$

12. $(y')^2 + y^2 = 4$

In Exercises 13–20 verify by direct substitution that the given function $y(t)$ is a solution.

13. $\frac{dy}{dt} = 10(y - t), \qquad y(t) = t + 0.1$

14. $\frac{dy}{dt} = 4ty^2, \qquad y(t) = \frac{1}{(1 - 2t^2)}$

15. $\frac{dy}{dt} = -2y + e^t, \qquad y(t) = Ce^{-2t} + \frac{1}{3}e^t,$ C any constant

16. $\frac{dy}{dt} = \frac{t(3 - 2y)}{(t^2 - 1)}, \qquad y(t) = \frac{1}{(t^2 - 1)} + \frac{3}{2}$

17. $\dfrac{d^2y}{dt^2} - 4y = 0, \qquad y(t) = 3e^{2t} - \dfrac{1}{4}e^{-2t}$

18. $\dfrac{d^2y}{dt^2} + 5\dfrac{dy}{dt} + 6y = 10e^{2t}, \qquad y(t) = Ae^{-2t} + Be^{-3t} + \dfrac{1}{2}e^{2t}$, A, B any constants

19. $t\dfrac{d^2y}{dt^2} - \dfrac{dy}{dt} + 4t^3y = 0, \qquad y(t) = 4\sin t^2 - 3\cos t^2, \qquad t > 0$

20. $\dfrac{d^2y}{dt^2} - \left(\dfrac{dy}{dt}\right)^3 - \dfrac{dy}{dt} = 0, \qquad y(t) = \sin^{-1}e^t + C$, C any constant

---

1.2 LINEAR FIRST ORDER DIFFERENTIAL EQUATIONS

Some important concepts and techniques of the theory of differential equations can be introduced by studying linear first order differential equations. It is convenient to start with homogeneous equations. Inhomogeneous equations and some of their applications will be discussed in the following section.

Definition

A first order differential equation is linear if it can be written in the form

$$\frac{dy}{dt} + p(t)y = g(t). \tag{1.2.1}$$

If $g(t) = 0$, the differential equation is said to be *homogeneous;* otherwise, the equation is *inhomogeneous* and $g(t)$ is called the *inhomogeneous term.*

Observe that y and its derivative occur to the first degree only and that the coefficient of y depends only on t. We shall assume in the discussions to follow that $p(t)$ and $g(t)$ are continuous functions of t.

The differential equations

$$\frac{dy}{dt} + t^2y = 0 \tag{1.2.2}$$

and

$$\frac{dy}{dt} + y = \sin t \tag{1.2.3}$$

are linear, while the equations

$$\frac{dy}{dt} + t^2y^2 = 0 \quad \text{and} \quad y\frac{dy}{dt} + y = \sin t$$

are not. Equation (1.2.2) is homogeneous and (1.2.3) is inhomogeneous.

HOMOGENEOUS LINEAR FIRST ORDER DIFFERENTIAL EQUATIONS

We shall begin our study of techniques for constructing solutions to differential equations by considering the homogeneous equation

$$\frac{dy}{dt} + p(t)y = 0 \quad \text{or} \quad \frac{dy}{dt} = -p(t)y. \tag{1.2.4}$$

If $p(t) = k$, a nonzero constant, then we have a differential equation,

$$\frac{dy}{dt} = -ky,$$

that reminds us of the calculus rule for differentiating the exponential function, $y(t) = e^{-kt}$, namely,

$$\frac{d}{dt}e^{-kt} = -ke^{-kt}. \tag{1.2.5}$$

This formula is used in the following example.

EXAMPLE 1 Find a solution to the differential equation,

$$\frac{dy}{dt} + ky = 0, \tag{1.2.6}$$

such that $y(0) = C$, an arbitrary constant.

SOLUTION If we multiply equation (1.2.5) by a constant A, we obtain

$$A\frac{d}{dt}e^{-kt} = \frac{d}{dt}(Ae^{-kt}) = -k(Ae^{-kt}).$$

Therefore $y(t) = Ae^{-kt}$ is a solution to (1.2.6) for all t. To satisfy the constraint $y(0) = C$, we take $t = 0$ and compute $y(0) = Ae^0 = A$. Hence $A = C$ and $y(t) = Ce^{-kt}$, a function that satisfies the given differential equation and the constraint. ■

The differential equation $dy/dt = -ky$ models the decay of radioactive substances, for it is observed that the rate of disintegration is proportional to the amount of radioactive material. If the amount of radioactive material at time t is denoted by $y(t)$, then from Example 1 we have

$$y(t) = y_0 e^{-kt},$$

where y_0 is the amount present at $t = 0$. A graph of the logarithm of $y(t)$ versus time is a straight line since

$$\ln y(t) = \ln (y_0 e^{-kt}) = \ln y_0 - kt. \tag{1.2.7}$$

(Recall that $\ln (ab) = \ln a + \ln b$ and $\ln e^w = w$.) The slope of the line, the decay constant k, can be found from experiments. It determines the half-life $t_{1/2}$ of a radioactive substance, the time it takes for half the substance to decay. To compute the half-life, we set $y(t_{1/2}) = \frac{1}{2}y_0$, then

$$\ln \tfrac{1}{2}y_0 = \ln y_0 - kt_{1/2}.$$

Since

$$\ln \frac{1}{2} y_0 = \ln \frac{1}{2} + \ln y_0 = -\ln 2 + \ln y_0,$$

we have $kt_{1/2} = \ln 2$, or

$$t_{1/2} = \frac{1}{k} \ln 2.$$

For example, the radioactive carbon C^{14} decays to ordinary carbon with a decay constant 1.244×10^{-4} per year. Therefore the half-life of C^{14} is 5570 years ($1.244 \times 10^{-4} \times 5570 \cong \ln 2$).

An important application of formula (1.2.7) was made by an American chemist, Willard Libby, who developed a method called *radioactive dating* for determining the approximate age of fossils.

Carbon dating

The carbon in living matter contains a tiny proportion of the radioactive isotope C^{14}, which arises from cosmic ray bombardment in the upper atmosphere. The isotope oxidizes to carbon dioxide and then is mixed completely with nonradioactive carbon by atmospheric winds. This mixture of carbon dioxide is absorbed by plants and by the animals that eat the plants and is fixed in their tissues.

When a plant or animal dies the amount of radioactive carbon decreases due to the decomposition of radiocarbon into ordinary carbon and beta-rays. Since the dead plant or animal cannot absorb any new radiocarbon into its tissues, the age of its remains can be found by

1. comparing the proportion of C^{14} with the proportion assumed to have been present before death, and

2. calculating the number of years that have elapsed since the carbon was extracted from the atmosphere.

Many objects of historical interest have been dated by this method. Several are noted on the plot of the logarithm of C^{14} concentration (expressed in radioactive decay counts per minute per gram of carbon) against time that is given in Fig. 1.1.

Figure 1.1 **Carbon dating of ten objects. (From *LIFE: The Science of Biology*, 2nd ed., by W. K. Purves and G. H. Orians. Copyright © 1987, Sinauer Associates, Inc., Publishers, Sunderland, Mass. Reprinted by permission of the publisher.)**

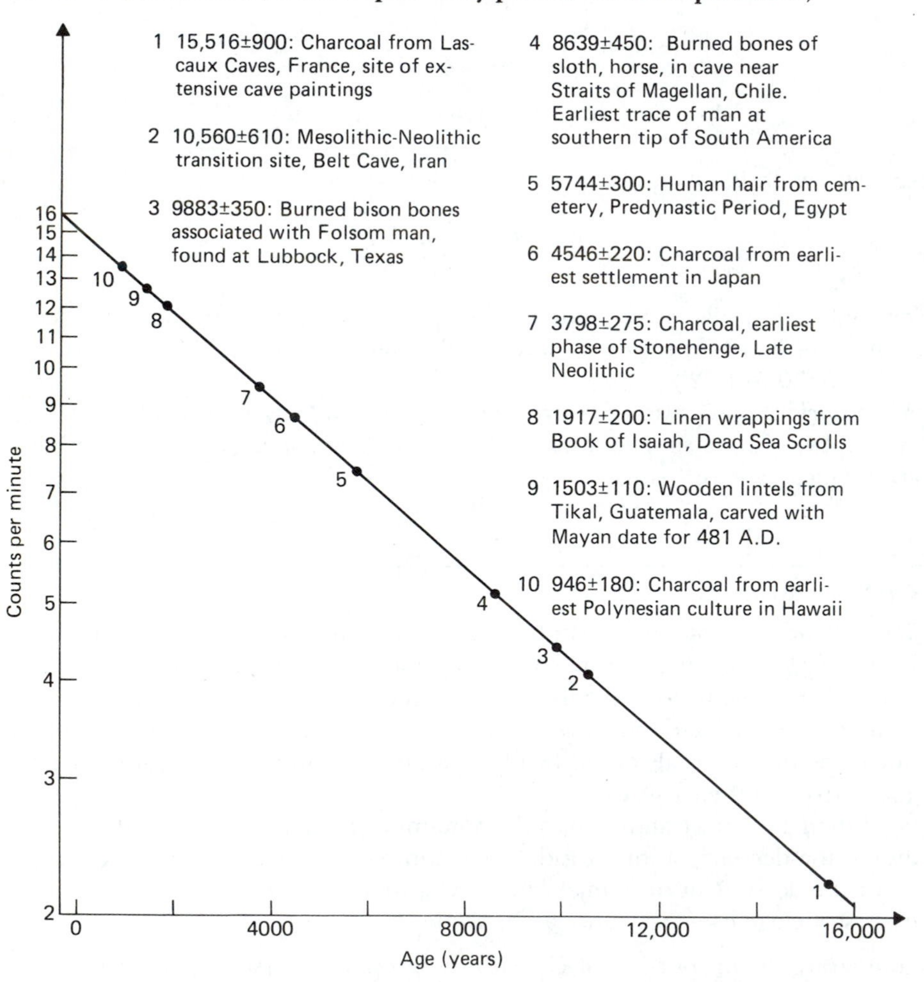

EXAMPLE 2 In carbon extracted from living tissue there are approximately 6×10^{10} atoms of C^{14} per gram of carbon. A wood beam in the Egyptian tomb of the First Dynasty was found to contain approximately 3.33×10^{10} atoms of C^{14} per gram of carbon. Approximately how old is the tomb?

SOLUTION We substitute $y(t) = 3.33 \times 10^{10}$, $y_0 = 6 \times 10^{10}$, $k = 1.244 \times 10^{-4}$ into (1.2.7) to obtain the approximate age of the tomb:

$$t \cong \frac{1}{1.244 \times 10^{-4}} [\ln (6 \times 10^{10}) - \ln (3.33 \times 10^{10})]$$

$$\cong 4730 \text{ years.}$$

■

Now we turn to the differential equation

$$\frac{dy}{dt} + p(t)y = 0. \tag{1.2.8}$$

It can be solved by several methods. The simplest is based on the observation that since $y = Ce^{-kt}$ is a solution to

$$\frac{dy}{dt} + ky = 0,$$

then we might be able to select a function $P(t)$ so that

$$y(t) = Ce^{-P(t)} \tag{1.2.9}$$

is also a solution where C is an arbitrary constant. Let us try. Substituting (1.2.9) into the differential equation we obtain

$$\begin{aligned}\frac{d}{dt} y(t) + p(t)y(t) &= \frac{d}{dt}[Ce^{-P(t)}] + p(t)Ce^{-P(t)} \\ &= -Ce^{-P(t)} \frac{dP(t)}{dt} + p(t)Ce^{-P(t)} \\ &= Ce^{-P(t)} \left[-\frac{dP(t)}{dt} + p(t) \right] = 0.\end{aligned}$$

If $C \neq 0$ ($e^{-P(t)}$ is always positive), we must have

$$-\frac{dP(t)}{dt} + p(t) = 0 \quad \text{or} \quad \frac{dP(t)}{dt} = p(t).$$

Therefore if $P(t)$ is an antiderivative of $p(t)$, then $y(t) = Ce^{-P(t)}$ is a solution to (1.2.8).

EXAMPLE 3 Solve $dy/dt + (\cos t)y = 0$.

SOLUTION Here $p(t) = \cos t$ and $P(t) = \sin t$ is an antiderivative. Using (1.2.9), we get a solution

$$y(t) = Ce^{-\sin t}. \qquad \blacksquare$$

The function $P(t)$ in (1.2.9) is an antiderivative of the coefficient function $p(t)$. Any other antiderivative is of the form $P(t) + c$, where c is a constant. Therefore the solution can have the form

$$y(t) = Ce^{-P(t)-c} = Be^{-P(t)}$$

with $B = Ce^{-c}$. In other words, the arbitrary additive constant of integration c is absorbed into the arbitrary multiplicative constant of the solution of the linear homogeneous differential equation (1.2.8). In Example 3, we could have taken $P(t) = \sin t + 10$; then

$$y(t) = Ce^{-\sin t - 10} = Ce^{-10}e^{-\sin t} = Be^{-\sin t}$$

with $B = Ce^{-10}$. Both solutions have the same degree of arbitrariness and are equivalent.

The presence of an arbitrary multiplicative constant in our trial solution (1.2.9) illustrates an important property of linear homogeneous differential equations. If $y(t)$ is a solution of

$$\frac{dy}{dt} + p(t)y = 0,$$

then $u(t) = Cy(t)$ is also a solution for any constant C. This last statement is verified by noting that

$$\frac{d}{dt}u(t) + p(t)u(t) = \frac{d}{dt}Cy(t) + p(t)Cy(t)$$

$$= C\left[\frac{d}{dt}y(t) + p(t)y(t)\right] = 0,$$

since $y(t)$ is a solution. Thus $u(t)$ is also a solution.

We shall now show that any solution of (1.2.8) is a constant times $e^{-P(t)}$. It is not obvious how to prove this statement, so we invoke a mathematical adage: "When stuck, reformulate." With this in mind, we pose an equivalent problem: Show that if $u(t)$ is any solution of (1.2.8) then

$$\frac{d}{dt}\{u(t)e^{P(t)}\} = 0.$$

The equivalence of the two statements is clear because the last relation holds if and only if $u(t)e^{P(t)} = C$, a constant. Since $e^{P(t)} > 0$, the last equation is equivalent to $u(t)$

$= Ce^{-P(t)}$, the desired conclusion. We now proceed to solve the equivalent problem. By assumption, $u(t)$ is a solution to (1.2.8); therefore,

$$\frac{d}{dt}\{u(t)e^{P(t)}\} = e^{P(t)}\frac{d}{dt}u(t) + u(t)e^{P(t)}\frac{d}{dt}P(t)$$

$$= e^{P(t)}\left[\frac{d}{dt}u(t) + p(t)u(t)\right] = 0.$$

The above discussion is the basis of the following definition.

Definition

A *general solution* of the homogeneous linear differential equation

$$\frac{dy}{dt} + p(t)y = 0 \tag{1.2.10}$$

is

$$y(t) = Ce^{-P(t)}, \tag{1.2.11}$$

where $P(t)$ is any antiderivative of $p(t)$ and C is an arbitrary constant.

EXAMPLE 4 Find a solution to $dy/dt + 2ty = 0$ such that $y(10) = 3$.

SOLUTION To solve this problem we first find a general solution of the differential equation. Since $p(t) = 2t$, an antiderivative is $P(t) = t^2$, and so $y(t) = Ce^{-t^2}$ is a general solution. To satisfy the condition $y(10) = 3$, we let $t = 10$ and $y = 3$ to obtain $3 = Ce^{-100}$. Hence, $C = 3e^{100}$ and the solution is

$$y(t) = 3e^{100}e^{-t^2} = 3e^{(100-t^2)}.$$ ■

A general solution (1.2.11) to the differential equation (1.2.10) determines a one-parameter family of curves called *integral curves,* namely, for each C, the set of points $\{t, Ce^{-P(t)}\}$ in the ty-plane. The constraint

$$y(t_0) = y_0 \tag{1.2.12}$$

specifies the integral curve that passes through the point (t_0, y_0). In many physical problems, t denotes the time and $y_0 = y(t_0)$ denotes the initial state of the system being investigated. Because of this, the differential equation (1.2.10) together with the constraint (1.2.12) constitute an *initial value problem.* Example 4 is such a problem. A sketch of several integral curves, including the one passing through the point (10, 3), is given in Fig. 1.2.

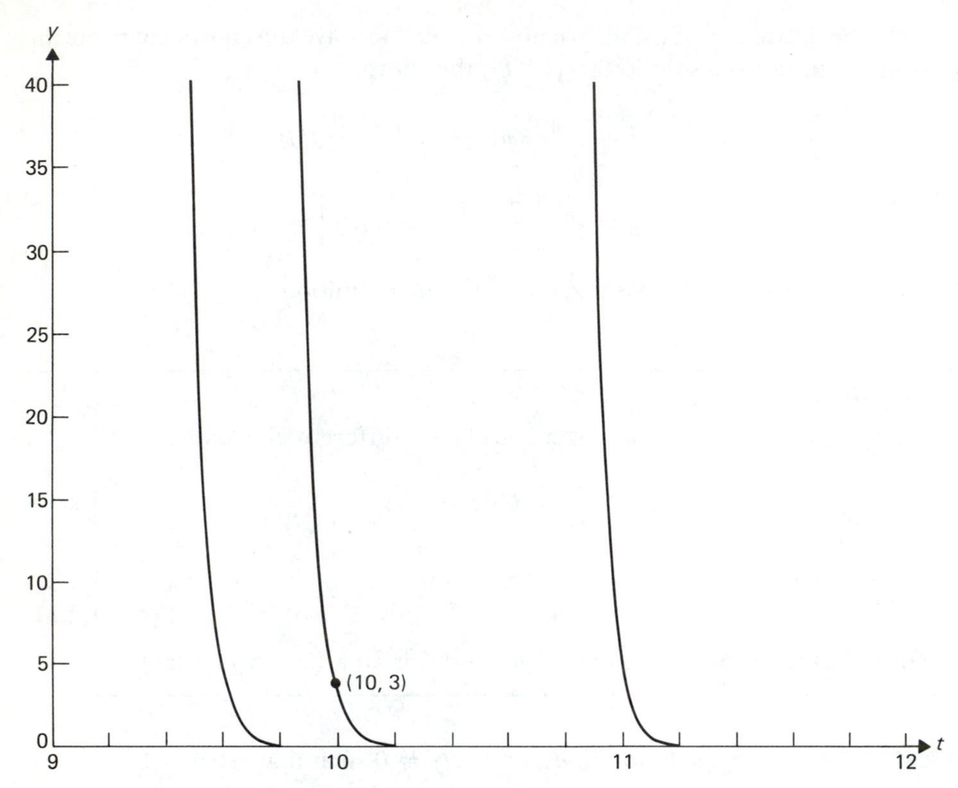

Figure 1.2 **Several integral curves of $dy/dt + 2ty = 0$.**

It should be noted that a solution and its corresponding integral curve need not be defined for all t; singularities in the coefficient function $p(t)$ may limit the interval of definition. This is illustrated in the following initial value problem.

EXAMPLE 5 Solve

$$\frac{dy}{dt} = -(\sec^2 t)y, \qquad y\left(\frac{9\pi}{4}\right) = -3$$

and find the maximal interval of validity of the solution.

SOLUTION Since an antiderivative of $\sec^2 t$ is $\tan t$, a general solution is $y(t) = Ce^{-\tan t}$. The constant C is determined by setting

$$-3 = Ce^{-\tan(9\pi/4)} = Ce^{-1}.$$

Therefore $C = -3e$, and $y(t) = -3e^{1-\tan t}$ is the solution to the differential equation such that

$$y\left(\frac{9\pi}{4}\right) = -3.$$

The tangent function is singular at odd integral multiples of $\pi/2$. The closest singularities to $9\pi/4$ are $3\pi/2$ and $5\pi/2$. Therefore the maximal interval of validity of the solution is

$$\frac{3\pi}{2} < t < \frac{5\pi}{2}.$$

Notice that this is also the maximal interval of validity of the differential equation,

$$\frac{dy}{dt} = -(\sec^2 t)y,$$

that contains $t_0 = 9\pi/4$. ■

EXAMPLE 6 A 1000-liter tank contains a mixture of water and chlorine. In order to reduce the concentration of chlorine in the tank, fresh water is pumped in at a rate of 6 liters per second. The fluid is well stirred and pumped out at a rate of 8 liters per second. If the initial concentration of chlorine is 0.02 grams per liter, find the amount of chlorine in the tank as a function of t and the interval of validity of the mathematical model.

SOLUTION Let $Q(t)$ denote the amount of chlorine and $V(t)$ denote the volume of fluid in the tank at time t (see Fig. 1.3). Then

$$\frac{dQ}{dt}(t) = \text{the rate of change of the amount of chlorine in the tank}$$

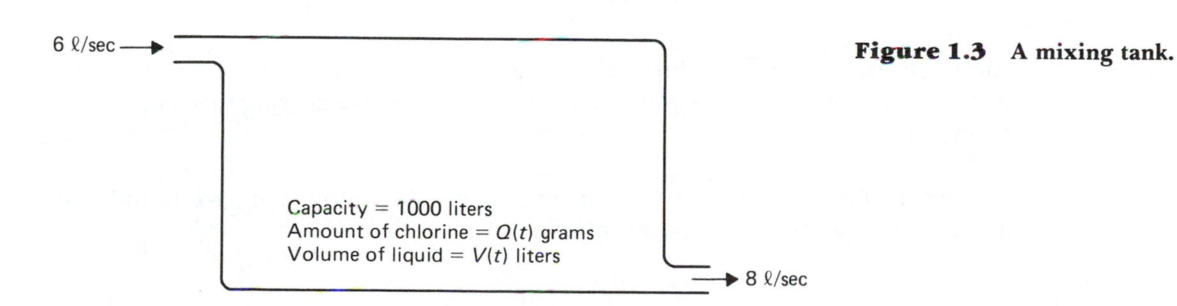

Figure 1.3 A mixing tank.

and

$$8\,\frac{Q(t)}{V(t)} = \text{the rate the fluid is pumped out times the concentration of the chlorine}$$
$$= \text{the outflow rate of chlorine.}$$

Since the rate of change of the amount of chlorine in the tank plus the outflow rate of chlorine must sum to zero, the governing differential equation is

$$\frac{dQ}{dt} + \frac{8Q}{V(t)} = 0,$$

where

$$V(t) = (\text{the initial volume}) - (\text{the net loss of fluid in liters per second}) \times (\text{the elapsed time in seconds})$$

or

$$V(t) = 1000 - (8 - 6)t = 1000 - 2t.$$

The complete mathematical model is an initial value problem:

$$\frac{dQ}{dt} + \frac{8Q}{1000 - 2t} = 0,$$

$$Q(0) = (1000 \text{ liters}) \times (0.02 \text{ grams/liter}) = 20 \text{ grams}.$$

It is easy to solve. We first set $p(t) = 8/(1000 - 2t)$ and then integrate to get $P(t) = -4\ln(500 - t)$. Then

$$Q(t) = Ce^{-P(t)} = Ce^{4\ln(500-t)} = C(500 - t)^4$$

is a general solution to the differential equation. To determine C, set $t = 0$ and $Q(0) = 20$. It follows that $C = 20(500)^{-4}$. Therefore the amount of chlorine in the tank at time t is

$$Q(t) = 20\left[\frac{500 - t}{500}\right]^4 \qquad 0 < t < 500.$$

Although the mathematical interval of validity of the solution is $0 < t < 500$, the process must be stopped well before it reaches 500 or the stirring mechanism will be ineffective. ■

Let us summarize what has been covered in this section. We have found that a homogeneous linear first order differential equation,

$$\frac{dy}{dt} + p(t)y = 0,$$

has a general solution

$$y(t) = Ce^{-P(t)},$$

where $P(t)$ is an antiderivative of the coefficient function $p(t)$. Furthermore, if an initial condition, $y(t_0) = y_0$, is imposed, the constant C can be selected so that the initial condition is satisfied.

EXERCISES 1.2

In each of the Exercises 1–8, find a general solution to the given differential equation.

1. $\frac{dy}{dt} + y = 0$

2. $\frac{dy}{dt} - 4y = 0$

3. $\frac{dy}{dt} + t^2y = 0$

4. $\frac{dy}{dt} - \frac{1}{t}y = 0,\ 0 < t$

5. $\frac{dy}{dt} - \frac{1}{\sqrt{1 - t^2}}y = 0,\ -1 < t < 1$

6. $\frac{dy}{dt} = \frac{3t - 2}{t^2 - 2t - 8}y,\ -2 < t < 4$

7. $\frac{dy}{dt} = \frac{2t + 1}{t^2 - 2t + 1}y,\ t < 1$

8. $\frac{dy}{dt} - \ln(t)y = 0,\ 0 < t$

In each of the Exercises 9–14, find a general solution and the solution satisfying the given initial conditions. Determine the maximal interval of validity of the solution.

9. $\frac{dy}{dt} + 2y = 0,\ y(0) = 4$

10. $\frac{dy}{dt} = \frac{1}{2}y,\ y(2) = -1$

11. $\frac{dy}{dt} = \frac{t}{t^2 + 4t - 5}y,\ y(0) = 1$

12. $\frac{dy}{dt} + t\ln(2t)y = 0,\ y(\tfrac{1}{2}) = 0$

13. $\frac{dy}{dt} + \sqrt{1 + 2t}\,y = 0,\ y(4) = 10$

14. $\frac{dy}{dt} = \frac{1}{t(1 - t)}y,\ y(\tfrac{1}{4}) = 1$

15. A certain radioactive material decays at a rate proportional to the amount present. Initially there are 100 milligrams of the material present, and after two hours the material has lost 10% of its original mass.
    - **a)** Find an expression for the mass remaining at any time t.
    - **b)** Find the mass of the material after six hours.
    - **c)** Find the half-life of the material.

16. A radioactive material decays at a rate proportional to the amount present. If 80% of the material at time $t = 100$ remains when $t = 200$, what percentage of the original amount remains when $t = 400$?

17. A bacterial culture grows at a rate proportional to the amount present. After one hour 400 strands of the bacteria are observed, and after two hours 500 strands of the bacteria are observed. How long will it take for the initial number of strands in the culture to double?

18. The population of Idaho in 1920 was 431,866, and in 1960 it was 667,191. Assuming a rate of growth proportional to the present size, when will the population of Idaho double its 1920 size?

19. A tank contains 200 gallons of a toxic solution containing 20 lb of pollutant. Fresh water is poured into the tank at a rate of 8 gal/min and leaves at the same rate. Assuming instant stirring

a) find an expression for the amount of pollutant in the tank at any time t, and

b) find at what time t the percent concentration [(lb of pollutant)/200 $\times$ 100] reaches a predetermined non-toxic level of 0.1%.

20. A tank with a 200 gallon capacity contains 100 gallons of a toxic solution containing 10 lb of pollutant. Fresh water is poured into the tank at a rate of 8 gal/min and leaves the tank at the rate 6 gal/min. Assuming instant stirring find the amount of pollutant at the time of overflow.

21. A pint of ice cream at a temperature of 5° F is removed from the refrigerator and placed in a room where the temperature is 65° F. If after 20 minutes the temperature of the ice cream is 10° F, how long will it take the ice cream to reach a temperature of 32° F?

22. A metal bar at a temperature of 200° F is moved to a room where the temperature is 72° F. After five minutes the temperature of the bar is 150° F. How long will it take the bar to reach a temperature of 100° F?

23. Exponential growth of a population is usually valid only for a short period of time, after which it greatly overestimates the actual size of the population. A possible modification would be to model the growth by the equation

$$\frac{dy}{dt} = k(t)y.$$

Suggest some possible nonnegative functions $k(t)$, and compare the growth curves with those obtained when $k(t) \equiv 1$. For convenience let $y(0) = 1$.

1.3 INHOMOGENEOUS LINEAR FIRST ORDER DIFFERENTIAL EQUATIONS

Our next topic is the study of inhomogeneous linear first order differential equations,

$$\frac{dy}{dt} + p(t)y = g(t). \tag{1.3.1}$$

The inhomogeneous term $g(t)$ can represent an applied force and is frequently referred to as a forcing function. After a brief treatment of equations with $p(t) = 0$, we shall discuss two solution methods for inhomogeneous equations, the integrating factor method and the method of undetermined coefficients.

The simplest type of inhomogeneous differential equation is

$$\frac{dy}{dt} = g(t), \tag{1.3.2}$$

where $g(t)$ is a continuous function. A solution to (1.3.2) is a function that is an antiderivative of $g(t)$; we write

$$y(t) = \int^t g(s)\,ds + c. \tag{1.3.3}$$

The additive constant can be chosen so that the solution takes on a prescribed initial value.

EXAMPLE 1 Solve $dy/dt = 4te^{2t}$, $y(0) = 6$.

SOLUTION Set[2]

$$y(t) = \int^t 4se^{2s}\,ds + c = (2t - 1)e^{2t} + c.$$

If $y = 6$ at $t = 0$, then

$$6 = (-1)e^0 + c.$$

Therefore $c = 7$, and the solution to the initial value problem is

$$y(t) = (2t - 1)e^{2t} + 7.$$

It is defined for all t. ■

The following problem illustrates the use of inhomogeneous differential equations in mechanics. If we throw a ball straight up, how long will it take to reach its maximum altitude?

To convert this into a mathematical statement we need to select a coordinate system (Fig. 1.4), determine which physical forces are relevant, and invoke Newton's law, which states that the time rate of change of the ball's momentum equals the forces acting on it. Let $x(t)$ denote the height of the ball at time t and $v(t) = dx(t)/dt$ its velocity. We shall neglect the curvature of the earth and assume that the downward pull of gravity is the only relevant force. It follows that the governing differential equation is

$$\frac{d}{dt}(mv) = -mg \quad \text{or} \quad \frac{dv}{dt} = -g,$$

2. This integral can be evaluated by integrating by parts or by consulting a table of integrals.

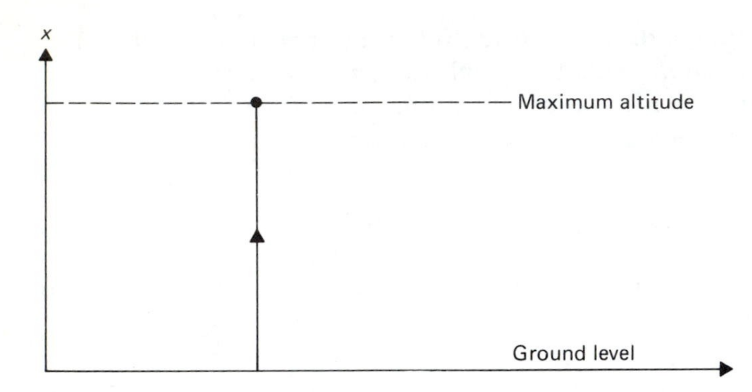

Figure 1.4 A mechanics problem.

where m is the mass of the ball, mv its momentum, and $g = 32$ ft/sec^2 is the gravitational acceleration.

If the ball is thrown upward at time zero with velocity $v = 60$ ft/sec, then the physical problem is represented by an initial value problem,

$$\frac{dv}{dt} = -g, \qquad v(0) = 60.$$

After integrating both sides of the differential equation from 0 to t, we have

$$\int_0^t \frac{d}{ds} v(s)\, ds = -\int_0^t g\, ds$$

or

$$v(t) - v(0) = -(t - 0)g.$$

Therefore

$$v(t) = 60 - 32t \text{ ft/sec}, \qquad 0 < t.$$

The velocity must be zero at the time the maximum altitude is reached. Thus

$$t_{\text{max}} = \frac{60}{32} = 1.875 \text{ sec.}$$

If we want to find the maximum altitude, we can integrate

$$\frac{dx(t)}{dt} = v(t) = 60 - 32t$$

to obtain

$$x(t) - x(0) = 60t - 32\frac{t^2}{2} = 60t - 16t^2.$$

Taking $x(0) = 0$, and $t = t_{\max}$, we have that the maximum altitude of the ball is

$$x_{\max} = 60t_{\max} - 16(t_{\max})^2 = 56.25 \text{ ft.}$$

Our solution to the thrown ball problem employs the *definite integral representation* of the solution to an initial value problem of the type

$$\frac{dy}{dt} = g(t), \qquad y(t_0) = y_0. \tag{1.3.4}$$

Instead of using the expression (1.3.3), we write

$$y(t) = y_0 + \int_{t_0}^{t} g(s)\, ds \tag{1.3.5}$$

and verify that

$$\frac{dy(t)}{dt} = \frac{d}{dt}\left\{y_0 + \int_{t_0}^{t} g(s)\, ds\right\} = 0 + g(t),$$

and $y(t_0) = y_0$.

EXAMPLE 2 Solve $dy/dt = \ln t$, $y(1) = 10$.

SOLUTION An antiderivative of $\ln t$ is

$$\int \ln t\, dt = t \ln t - t.$$

Therefore we set

$$y(t) = 10 + \int_{1}^{t} \ln s\, ds = 10 + (s \ln s - s)\big|_1^t.$$

Thus

$$y(t) = 10 + (t \ln t - t) - (1 \ln 1 - 1)$$

or

$$y(t) = 11 + t \ln t - t.$$

This solution is defined on the interval $0 < t < \infty$. ■

THE INTEGRATING FACTOR METHOD

Let us now consider the inhomogeneous differential equation,

$$\frac{dy}{dt} + ry = g(t), \tag{1.3.6}$$

where r is a nonzero constant. We start with this simpler equation to demonstrate the method and will discuss the general equation (1.3.1) later.

Many years ago it was observed that the rule for differentiating a product of an exponential and a solution to (1.3.6) could be used to generate a much simpler equation. Since

$$\begin{aligned}\frac{d}{dt}\{e^{rt}y(t)\} &= e^{rt}\frac{dy(t)}{dt} + re^{rt}y(t)\\ &= e^{rt}\left\{\frac{dy(t)}{dt} + ry(t)\right\},\end{aligned}$$

one can multiply both sides of (1.3.6) by e^{rt} to obtain

$$e^{rt}\left\{\frac{dy(t)}{dt} + ry(t)\right\} = e^{rt}g(t).$$

It follows that

$$\frac{d}{dt}\{e^{rt}y(t)\} = e^{rt}g(t).$$

This equation is easy to solve. We have first, by integrating,

$$e^{rt}y(t) = \int^t e^{rs}g(s)\,ds + C,$$

and then

$$y(t) = e^{-rt}\int^t e^{rs}g(s)\,ds + Ce^{-rt}. \qquad \textbf{(1.3.7)}$$

Even though an antiderivative of $e^{rt}g(t)$ may be difficult to construct, we have obtained an explicit formula for a solution to the inhomogeneous differential equation (1.3.6).

Because multiplying the differential equation by e^{rt} converts it into an equation that can be solved by integrating, the function e^{rt} is called an *integrating factor.*

The solution that we have obtained has a form that is characteristic of solutions to inhomogeneous linear differential equations of any order. Namely, $y(t)$ is the sum of two terms, a solution to the inhomogeneous equation (1.3.6), and a general solution to the corresponding homogeneous equation,

$$\frac{dy}{dt} + ry = 0. \qquad \textbf{(1.3.8)}$$

We write

$$y(t) = y_p(t) + y_h(t)$$

with

$$y_p(t) = e^{-rt} \int^t e^{rs} g(s)\, ds$$

and

$$y_h(t) = Ce^{-rt}.$$

The solution $y_p(t)$ is called a *particular solution* to the inhomogeneous equation (1.3.6), and $y_h(t)$, a general solution to the homogeneous equation (1.3.8), is called a *complementary solution.* If $y(t)$ is of the form (1.3.7), it is called a *general solution* to the inhomogeneous equation (1.3.6).

EXAMPLE 3 Find a general solution to

$$\frac{dy}{dt} + y = 3t - 5.$$

SOLUTION Using e^t as an integrating factor, we have

$$\frac{d}{dt}\{e^t y\} = e^t(3t - 5).$$

An antiderivative of $e^t(3t - 5)$ is $3te^t - 8e^t$. Therefore,

$$e^t y(t) = 3te^t - 8e^t + C,$$

and

$$y(t) = 3t - 8 + Ce^{-t}$$

is a general solution. It is valid for all t. ■

The definite integral representation (1.3.5) to the inhomogeneous initial value problem (1.3.4) can be used as a model for a corresponding representation for the solution to the initial value problem,

$$\frac{dy}{dt} + ry = g(t), \qquad y(t_0) = y_0. \tag{1.3.9}$$

We set

$$\frac{d}{dt}\{e^{rt}y(t)\} = e^{rt}g(t)$$

and integrate from t_0 to t. Thus

$$e^{rt}y(t) - e^{rt_0}y(t_0) = \int_{t_0}^{t} e^{rs}g(s)\, ds,$$

and, after replacing $y(t_0)$ by y_0, we solve for $y(t)$ to obtain

$$y(t) = e^{-rt}\int_{t_0}^{t} e^{rs}g(s)\,ds + y_0e^{-r(t-t_0)}. \tag{1.3.10}$$

In this representation,

$$y_p(t) = e^{-rt}\int_{t_0}^{t} e^{rs}g(s)\,ds$$

is a particular solution to the inhomogeneous equation that vanishes at $t = t_0$, and

$$y_h(t) = y_0e^{-r(t-t_0)}$$

is a solution to the corresponding homogeneous equation that takes on the initial value $y(t_0) = y_0$. The verification of these statements will be left to the reader.

EXAMPLE 4 Solve the initial value problem

$$\frac{dy}{dt} + 2y = \sin t, \qquad y(\pi) = -7.$$

SOLUTION We write

$$\frac{d}{dt}\{e^{2t}y(t)\} = e^{2t}\sin t$$

and integrate from π to t. Thus

$$\begin{aligned}
e^{2t}y(t) + 7e^{2\pi} &= \int_{\pi}^{t} e^{2s}\sin s\,ds \\
&= \frac{e^{2s}}{5}(2\sin s - \cos s)\Big|_{\pi}^{t} \\
&= \frac{e^{2t}}{5}(2\sin t - \cos t) - \frac{e^{2\pi}}{5}(2\sin \pi - \cos \pi) \\
&= \frac{e^{2t}}{5}(2\sin t - \cos t) - \frac{e^{2\pi}}{5},
\end{aligned}$$

and

$$y(t) = \frac{1}{5}(2\sin t - \cos t) - \frac{1}{5}e^{2(\pi-t)} - 7e^{-2(t-\pi)}.$$

In this problem, the particular solution

$$y_p(t) = \frac{1}{5}(2\sin t - \cos t) - \frac{1}{5}e^{2(\pi-t)}$$

vanishes at $t = \pi$ and the complementary solution,

$$y_h(t) = -7e^{-2(t-\pi)}$$

takes on the initial value -7 at $t = \pi$. ■

Let us now consider an inhomogeneous equation with a continuous time-dependent coefficient function $p(t)$:

$$\frac{dy}{dt} + p(t)y = g(t). \tag{1.3.11}$$

Since e^{rt} is an integrating factor for the constant coefficient equation (1.3.6), it is reasonable to take $e^{P(t)}$ as a trial integrating factor for (1.3.11). The function $P(t)$ will be selected so that

$$\frac{d}{dt}\{e^{P(t)}y(t)\} = e^{P(t)}g(t). \tag{1.3.12}$$

By the product rule,

$$\begin{aligned}\frac{d}{dt}\{e^{P(t)}y(t)\} &= e^{P(t)}\frac{dy(t)}{dt} + e^{P(t)}\frac{dP(t)}{dt}y(t) \\ &= e^{P(t)}\left\{\frac{dy(t)}{dt} + \frac{dP(t)}{dt}y(t)\right\}.\end{aligned}$$

Equation (1.3.12) will be satisfied if we require that

$$\frac{dP(t)}{dt} = p(t). \tag{1.3.13}$$

Taking $P(t)$ as an antiderivative of $p(t)$ and integrating (1.3.12), we obtain

$$e^{P(t)}y(t) = \int^t e^{P(s)}g(s)\,ds + C.$$

Therefore

$$y(t) = e^{-P(t)}\int^t e^{P(s)}g(s)\,ds + Ce^{-P(t)}. \tag{1.3.14}$$

We see that this expression for $y(t)$ has the characteristic form of a solution to an inhomogeneous linear differential equation. It is composed of two terms: The first is a particular solution of the inhomogeneous equation (1.3.11), and the second is a general solution to the corresponding homogeneous equation,

$$\frac{dy}{dt} + p(t)y = 0. \tag{1.3.15}$$

This last statement is easy to verify, but before doing so, let us use this approach in the following example.

EXAMPLE 5 Solve

$$\frac{dy}{dt} + \frac{2}{t}y = 8t - 6, \qquad 0 < t.$$

SOLUTION Since $p(t) = 2/t$, $P(t) = \ln t^2$ is an antiderivative and $e^{P(t)} = e^{\ln t^2} = t^2$ is an integrating factor. Hence

$$\frac{d}{dt}\{t^2y(t)\} = t^2(8t - 6),$$

and integrating both sides gives

$$t^2y(t) = 8\frac{t^4}{4} - 6\frac{t^3}{3} + C,$$

where C is a constant of integration. It follows that

$$y(t) = 2(t^2 - t) + \frac{C}{t^2}, \qquad 0 < t.$$

Observe that the solution is the sum of two functions, one of which, $2(t^2 - t)$, is a *particular solution* to the inhomogeneous equation

$$\frac{dy}{dt} + \frac{2}{t}y = 8t - 6,$$

and the other, Ct^{-2}, is a *general solution* to the homogeneous equation

$$\frac{dy}{dt} + \frac{2}{t}y = 0.$$

We suggest that the reader verify these statements. ■

To conclude our discussion we return to the expression (1.3.14) for the solution to the inhomogeneous differential equation. We must show that the first term,

$$y_p(t) = e^{-P(t)} \int^t e^{P(s)}g(s)\, ds, \tag{1.3.16}$$

satisfies the inhomogeneous equation. Differentiating $y_p(t)$, we obtain

$$\begin{aligned}
\frac{d}{dt} y_p(t) &= \left(\frac{d}{dt} e^{-P(t)}\right) \int^t e^{P(s)} g(s)\, ds + e^{-P(t)} \left(\frac{d}{dt} \int^t e^{P(s)} g(s)\, ds\right) \\
&= -p(t) e^{-P(t)} \int^t e^{P(s)} g(s)\, ds + e^{-P(t)} e^{P(t)} g(t) \\
&= -p(t) y_p(t) + g(t),
\end{aligned}$$

and therefore $y_p(t)$ is a particular solution to the inhomogeneous differential equation. The expression (1.3.14) is a sum of a particular solution and a general solution $Ce^{-P(t)}$ to the homogeneous equation. The constant C can be determined if initial conditions are given.

The formula (1.3.14) is called a general solution to the linear inhomogeneous differential equation (1.3.11) since any solution can be so represented. A justification of this statement is based on two observations.

1. If $z(t)$ is any solution to the inhomogeneous equation (1.3.11), then $z(t) - y_p(t)$ is a solution to the corresponding homogeneous equation (1.3.15).
2. Any solution to (1.3.15) is a constant times $e^{-P(t)}$.

If (1) and (2) are valid, then

$$z(t) - y_p(t) = Ce^{-P(t)} \quad \text{or} \quad z(t) = y_p(t) + Ce^{-P(t)},$$

as required.

Statement (2) was proved in the previous section. Let us verify (1). We are given that

$$z'(t) + p(t)z(t) = g(t) \quad \text{and} \quad y_p'(t) + p(t)y_p(t) = g(t).$$

Subtracting the two equations gives

$$[z(t) - y_p(t)]' + p(t)[z(t) - y_p(t)] = 0.$$

Thus we have established that $z(t) - y_p(t)$ is a solution to the homogeneous equation (1.3.15), the key step in establishing that (1.3.14) is a general solution to the inhomogeneous equation (1.3.11).

We now summarize all of this in the following algorithm:

The integrating factor algorithm

1. Write the equation in the form

$$\frac{dy}{dt} + p(t)y = g(t).$$

2. Find $P(t)$, an antiderivative of $p(t)$, and the integrating factor $e^{P(t)}$.
3. Rewrite the differential equation as

$$\frac{d}{dt}\{e^{P(t)}y(t)\} = e^{P(t)}g(t).$$

4. Integrate to obtain

$$e^{P(t)}y(t) = \int^t e^{P(s)}g(s)\,ds + C.$$

5. Divide by $e^{P(t)}$ to obtain a general solution $y(t)$:

$$y(t) = e^{-P(t)}\int^t e^{P(s)}g(s)\,ds + Ce^{-P(t)}.$$

6. If initial data are given, substitute into the above equation and solve for C.

EXAMPLE 6 Find the general solution of $dy/dt - 2y = 4t$.

SOLUTION We follow the algorithm step by step.

1. In the required form already.
2. Because $p(t) = -2$, let

$$P(t) = \int^t -2\,ds = -2t,$$

and the integrating factor is e^{-2t}.

3. Hence

$$\frac{d}{dt}[e^{-2t}y(t)] = 4te^{-2t}.$$

4. Therefore

$$e^{-2t}y(t) = \int^t 4se^{-2s}\,ds = (-2t-1)e^{-2t} + C.$$

5. A general solution is

$$y(t) = -2t - 1 + Ce^{2t}.$$ ■

EXAMPLE 7 Solve

$$\frac{1}{t}\frac{dy}{dt} + \frac{2}{1+t^2}y = 3, \qquad y(1) = 4.$$

SOLUTION Employing the above algorithm step by step we can write:

1. $\dfrac{dy}{dt} + \dfrac{2t}{1 + t^2}y = 3t$
2. $p(t) = \dfrac{2t}{1 + t^2}$, so let

$$P(t) = \int \frac{2t}{1 + t^2}\, dt = \ln(1 + t^2)$$

and an integrating factor is

$$e^{\ln(1+t^2)} = 1 + t^2.$$

3. Hence

$$\frac{d}{dt}[(1 + t^2)y(t)] = (1 + t^2)3t.$$

4. Therefore

$$(1 + t^2)y(t) = \int^t (1 + s^2)3s\, ds + C = \frac{3}{4}t^4 + \frac{3}{2}t^2 + C.$$

5. A general solution is

$$y(t) = \frac{3}{4}\left(\frac{t^4 + 2t^2}{1 + t^2}\right) + \frac{C}{1 + t^2}.$$

6. To determine C set $t = 1$, $y(1) = 4$ to obtain

$$4 = \frac{3}{4} \cdot \frac{3}{2} + \frac{C}{2},$$

and so $C = \frac{23}{4}$. The solution to the initial value problem is

$$y(t) = \frac{3}{4}\left(\frac{t^4 + 2t^2}{1 + t^2}\right) + \frac{23}{4}\left(\frac{1}{1 + t^2}\right).$$

A variant of the algorithm that is efficient for initial value problems is to integrate the formulas of step 3 from the initial time ($t = 1$) to the generic time ($t = t$) and then solve for $y(t)$. In this example, we obtain

$$(1 + t^2)y(t) - 2y(1) = \int_1^t (1 + s^2)s\, ds = \frac{3}{4}t^4 + \frac{3}{2}t^2 - \frac{9}{4}.$$

Since $y(1) = 4$, the solution is

$$y(t) = \frac{3}{4}\left(\frac{t^4 + 2t^2 - 3}{1 + t^2}\right) + \frac{8}{(1 + t^2)}\,.$$

It is clearly equivalent to the solution of step 6. ■

We generalize this trick (see 1.3.10) to obtain the solution of the initial value problem

$$\frac{dy}{dt} + p(t)y = g(t), \qquad y(t_0) = y_0$$

as

$$y(t) = e^{-P(t)} \int_{t_0}^{t} e^{P(s)} g(s)\, ds + y_0 e^{-P(t)}, \tag{1.3.17}$$

where

$$P(t) = \int_{t_0}^{t} p(s)\, ds.$$

The main topic of this section is the integrating factor method that transforms the problem of solving an inhomogeneous linear first order differential equation into the problem of computing two antiderivatives. The resulting formula (1.3.14) is the sum of a particular solution of the inhomogeneous equation (1.3.11) and a general solution of the corresponding homogeneous equation (1.3.15). If initial data is given, the arbitrary constant in (1.3.14) can be chosen so that the constraint is satisfied. An alternative representation (1.3.17) of the solution to an initial value problem was derived. It will be used in the next section where some numerical methods are introduced.

EXERCISES 1.3

In each of the Exercises 1–12, find a general solution of the given differential equation.

1. $\frac{dy}{dt} + y = 0$ **2.** $\frac{dy}{dt} - 4y = 0$ **3.** $\frac{dy}{dt} + t^2 y = 0$ **4.** $\frac{dy}{dt} - \frac{1}{t} y = 0$

5. $\frac{dy}{dt}$ [illegible] $3y = 0$ **6.** $\frac{dy}{dt} + 2y = t$ **7.** $\frac{dy}{dt} - y = \sin 2t$ **8.** $\frac{dy}{dt} + 2ty = 4t$

9. $t^2 \frac{dy}{dt} + ty + 1 = 0$

10. $t \frac{dy}{dt} - 4y = t^3 + 3t^2$

11. $\frac{dy}{dt} + (\cot t)y = 3 \sin t \cos t$

12. $t \frac{dy}{dt} + \frac{1}{\ln(t)} y = 1$

In each of the Exercises 13–27, find a general solution and the solution satisfying the given initial conditions.

13. $\frac{dy}{dt} + 2y = 0,\ y(0) = 4$

14. $\frac{dy}{dt} - \frac{1}{2}y = 0,\ y(2) = -1$

15. $\frac{dy}{dt} = (\sec t)y,\ y\left(\frac{\pi}{4}\right) = \frac{\sqrt{2}}{2}$

16. $\frac{dy}{dt} + |t|y = 0,\ y(0) = 4$

17. $\frac{dy}{dt} + y = 2t + 5,\ y(0) = 4$

18. $\frac{dy}{dt} + 4y = 6 \sin 2t,\ y(0) = -\frac{3}{5}$

19. $\frac{dy}{dt} - 2y + e^{2t} = 0,\ y(0) = 2$

20. $\frac{dy}{dt} - 2y = t^2e^{2t},\ y(0) = 2$

21. $\frac{dy}{dt} - 2ty - t = 0,\ y(0) = 1$

22. $e^y \frac{dy}{dt} + e^y = 4 \sin t,\ y(0) = 0$ (*Hint:* Let $u = e^y$.)

23. $(t^2 + 1)\frac{dy}{dt} - 2ty = t^2 + 1,\ y(1) = \pi$

24. $\frac{dy}{dt} - (\tan t)y = e^{\sin t},\ 0 < t < \pi/2,\ y(1) = 0$

25. $\frac{dy}{dt} + 2ty = e^{-t^2},\ y(0) = 1$

26. $\frac{dy}{dt} - 2y = t^2,\ y(0) = 2$

27. $\frac{dy}{dt} + \frac{2}{t}y = \frac{\cos t}{t^2},\ y(\pi) = 0,\ t > 0$

An alternate method for finding the general solution of the inhomogeneous linear differential equation

$$\frac{dy}{dt} + p(t)y = q(t)$$

is to use the method of *variation of parameters.* Assume the solution is of the form $y(t) = v(t)y_h(t)$ where $y_h(t) = \exp[-P(t)]$ is a solution of the homogeneous equation. Substitute $y(t)$ into the inhomogeneous equation and solve the resulting differential equation for $v(t)$. In Exercises 28–33 use the method of variation of parameters to find the general solution of the given differential equation.

28. $\frac{dy}{dt} + 3y = 6t;\ y_h(t) = e^{-3t}$

29. $\frac{dy}{dt} - y = 4e^{2t};\ y_h(t) = e^t$

30. $\frac{dy}{dt} - \frac{2}{t}y = \sin t;\ y_h(t) = t^2$

31. $\frac{dy}{dt} - \frac{t}{t^2 + 1}y = 1;\ y_h(t) = (t^2 + 1)^{1/2}$

32. $\dfrac{dy}{dt} - (\tan t)y = 1;\ y_h(t) = \dfrac{1}{\cos t}$

33. $\dfrac{dy}{dt} + \dfrac{1}{t}y = 2 + \dfrac{1}{t^2};\ y_h(t) = \dfrac{1}{t}$

The equation $dy/dt + p(t)y = q(t)y^n$ is known as Bernoulli's equation. It is linear when $n = 0$ or 1. Show that it can be reduced to a linear equation for any other value of n by the change of variables $z = y^{1-n}$. Apply this method to solve the equations in Exercises 34–37.

34. $\dfrac{dy}{dt} - y = ty^2$

35. $\dfrac{dy}{dt} - \dfrac{t}{2}y = ty^5$

36. $\dfrac{dy}{dt} + \dfrac{y}{t} = t^3y^3$

37. $2\dfrac{dy}{dt} - \dfrac{1}{t}y + (\cos t)y^3 = 0$

38. Newton's law of cooling states that the rate of change of temperature of a body is proportional to the difference of temperature between a cooling body and its surroundings. If $T(t)$ represents the temperature of a body in a room whose ambient temperature $S(t)$ varies with the time of day, then

$$\frac{dT}{dt} = -k(T - S(t)),$$

where k is a constant of proportionality. In a room whose temperature is fixed at 70° F, it would take 10 minutes for the body to cool to 80° F from an initial temperature of 100° F. Use this information to find k. If $S(t) = 70(1 + 5 \sin wt)$, $w = 2\pi/1440$ radians per minute, and $T(0) = 100$, find $T(t)$.

39. A ball is dropped with zero initial velocity from a building and encounters air resistance proportional to its velocity. Let its steady state ($t \to \infty$) velocity be 24 ft/sec.

a) Find its velocity at the end of three seconds and how far it has dropped.

b) If the building is 222 feet high, estimate at what time the ball hits the ground.

40. A simple model for the growth of a population is exponential growth where, if $P(t)$ is the size of a population (always nonnegative) at time t, then

$$P'(t) = rP(t), \qquad r > 0, \qquad P(0) = P_0,$$

where P_0 is the initial size. If the population is being harvested, say, by a predator or disease, at a constant rate $H > 0$, then a possible model is

$$P'(t) = rP(t) - H, \qquad P(0) = P_0.$$

a) If $r = 0.01$, $P_0 = 4000$, show that the population will expire in finite time, that is, $P(T) = 0$ for some $T > 0$, if $H > 40$, but will increase without bound if $H < 40$.

b) If $H = 50$, find the time T at which the population expires.

c) What is the solution if $H = 40$?

d) For the general case, show that $H = rP_0$ is the critical harvest rate above which the population expires in finite time.

1.4 THE METHOD OF UNDETERMINED COEFFICIENTS AND SOME NUMERICAL INTEGRATION RULES

The integrating factor method can be used to construct a particular solution to an inhomogeneous linear first order differential equation,

$$\frac{dy}{dt} + p(t)y = g(t), \tag{1.4.1}$$

if the antiderivatives,

$$P(t) = \int^t p(s)\, ds \quad \text{and} \quad \int^t e^{P(s)} g(s)\, ds,$$

can be evaluated. In this section we shall consider a simpler way of constructing particular solutions for problems where the forcing function $g(t)$ is a polynomial, an exponential, a sinusoidal function, or a sum of products of such functions. Then we shall study some elementary numerical techniques that can be employed if $g(t)$ is such that our analytical techniques are inadequate to produce a solution to the differential equation.

THE METHOD OF UNDETERMINED COEFFICIENTS

The method of undetermined coefficients rests on the observation that for some problems considerable effort is needed to evaluate the antiderivative

$$\int^t e^{rs} g(s)\, ds,$$

and yet the particular solution that results from this is so simple that a trial and error approach seems reasonable. In the preceding section we found that the differential equation (Example 3)

$$\frac{dy}{dt} + y = 3t - 5$$

has a particular solution, $y_p(t) = 3t - 8$, and the equation (Example 4)

$$\frac{dy}{dt} + 2y = \sin t$$

has a particular solution, $y_p(t) = \frac{1}{5}(2 \sin t - \cos t)$. A judicious guess is that polynomial forcing functions result in polynomial solutions and sinusoidal forcing functions result in sinusoidal solutions.

As we shall see, the labor of evaluating antiderivatives can be avoided if the differential equation is of the form

$$\frac{dy}{dt} + ky = g(t),$$

where k is a nonzero constant and $g(t)$ is

1. a polynomial,
2. an exponential e^{rt},
3. a product of an exponential and a polynomial,
4. a sum of trigonometric functions $\sin mt$, $\cos mt$,
5. a sum of products $e^{rt} \sin mt$, $e^{rt} \cos mt$, or
6. a sum of terms $P(t) \sin mt + Q(t) \cos mt$ where $P(t)$ and $Q(t)$ are polynomials.

The basic idea is to try a particular solution $y_p(t)$ of the most general form as $g(t)$, with undetermined coefficients. One substitutes the trial solution into the differential equation and determines the coefficients by comparison, if possible. Once having found $y_p(t)$, we know a general solution is of the form

$$y(t) = y_p(t) + Ce^{-kt}.$$

This technique, called the *method of undetermined coefficients,* is inherited from the theory of higher order constant coefficient linear differential equations, but it also works for first order equations. We give a few examples to show its advantages.

EXAMPLE 1 Find a general solution of $y' + 2y = 4e^{3t}$.

SOLUTION Since the right-hand side is a multiple of e^{3t}, try a trial solution $y_p(t) = Ae^{3t}$. Substitute to obtain

$$3Ae^{3t} + 2Ae^{3t} = 4e^{3t},$$

which after cancelling e^{3t} gives $5A = 4$ or $A = \frac{4}{5}$. Consequently, $y_p(t) = \frac{4}{5}e^{3t}$, and a general solution is

$$y(t) = \frac{4}{5}e^{3t} + Ce^{-2t}.$$ ■

EXAMPLE 2 Find a general solution of $y' - 4y = 8t^2$.

SOLUTION Since the right-hand side is a polynomial of degree 2, try a trial solution

$$y_p(t) = At^2 + Bt + C,$$

the most general polynomial of degree 2. Substitute to obtain

$$(2At + B) - 4(At^2 + Bt + C) = 8t^2,$$

and compare powers of t on both sides to obtain the equations

$$-4A = 8, \qquad 2A - 4B = 0, \qquad B - 4C = 0.$$

These can be solved successively to obtain

$$A = -2, \qquad B = -1, \qquad C = -\frac{1}{4};$$

hence

$$y_p(t) = -2t^2 - t - \frac{1}{4},$$

and a general solution is

$$y(t) = -2t^2 - t - \frac{1}{4} + Ce^{4t}.$$

This example points out the effectiveness of the technique, since using integrating factors would involve two integrations by parts. ■

EXAMPLE 3 Find a periodic solution of $y' + y = 4 \cos 2t$.

SOLUTION A trial solution should be of the form

$$y_p(t) = A \sin 2t + B \cos 2t.$$

We substitute the trial solution into the differential equation to obtain

$$(2A \cos 2t - 2B \sin 2t) + (A \sin 2t + B \cos 2t) = 4 \cos 2t,$$

or

$$(2A + B) \cos 2t + (-2B + A) \sin 2t = 4 \cos 2t + 0 \sin 2t.$$

Since the coefficients of $\cos 2t$ and $\sin 2t$ on both sides must be equal, we have

$$2A + B = 4, \qquad -2B + A = 0.$$

Solving these gives $B = \frac{4}{5}$, $A = \frac{8}{5}$, so

$$y_p(t) = \frac{8}{5} \sin 2t + \frac{4}{5} \cos 2t.$$

We see that $y_p(t)$ has period π; that is,

$$y_p(t + \pi) = y_p(t), \quad \text{for all } t.$$

■

EXAMPLE 4 Find a general solution of $y' - 2y = 6e^{2t}$.

SOLUTION A first guess of a trial solution would be $y_p(t) = Ae^{2t}$. Substituting it into the differential equation we obtain

$$2Ae^{2t} - 2(Ae^{2t}) = 6e^{2t},$$

or $6e^{2t} = 0$, which is nonsense. Our trial solution failed because the forcing term $6e^{2t}$ is a solution of the homogeneous equation $y' - 2y = 0$; so a trial solution Ae^{2t} won't work.

Our second guess, $y_p(t) = Ate^{2t}$, may seem arbitrary, but it is based on a study of solutions obtained by the integrating factor method. Substituting into the differential equation gives

$$(Ae^{2t} + 2Ate^{2t}) - 2(Ate^{2t}) = 6e^{2t},$$

and consequently $A = 6$. Therefore $y_p(t) = 6te^{2t}$, and a general solution is

$$y(t) = 6te^{2t} + Ce^{2t}.$$ ■

Although the method described works only for linear equations in which the homogeneous equation has constant coefficients, for those cases it is applicable for a large class of forcing terms. For that reason it is worth adding to one's mathematical toolbox.

NUMERICAL INTEGRATION RULES

The method of undetermined coefficients works well for a restricted class of differential equations. The integrating factor method can, in theory, be applied to any inhomogeneous linear first order differential equation, but the solution is represented in terms of integrals that might not be possible to evaluate. For example,

$$y(t) = e^{-t} \int_1^t e^s \frac{\sin s}{s}\, ds$$

is the solution to the initial value problem,

$$\frac{dy}{dt} + y = \frac{\sin t}{t}, \qquad y(1) = 0,$$

but $y(t)$ cannot be expressed in terms of elementary functions. If the value of the solution is needed at $t = 2$, then numerical techniques must be employed.

Numerical techniques have been part of mathematics from the beginning of its recorded history. The computation of areas, a problem of interest to the Egyptians thousands of years ago, here takes the form of the numerical evaluation of

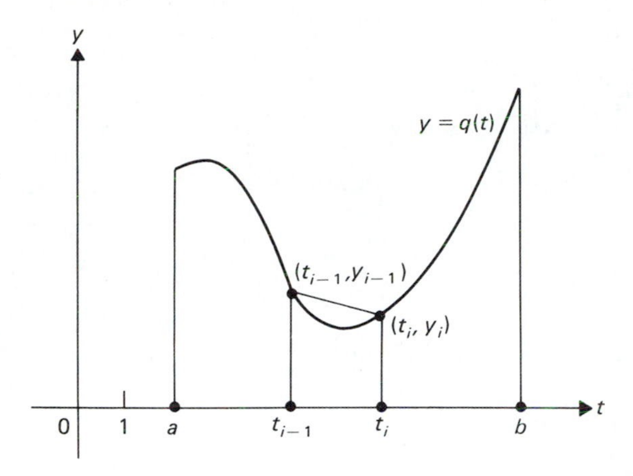

Figure 1.5 **The trapezoid approximation:** $h[q(t_{i-1}) + q(t_i)]/2$.

definite integrals. There are many methods available; one of the simplest is the *trapezoidal rule:*

$$\int_a^b q(t)\,dt \cong T(h) = \frac{h}{2}[q(t_0) + 2q(t_1) + 2q(t_2) + \cdots + 2q(t_{n-1}) + q(t_n)], \tag{1.4.2}$$

where the interval of integration $a \le t \le b$ is divided into n equal subintervals, each of length $h = (b - a)/n$.[3] The endpoints of the subintervals are $t_0 = a$, $t_1 = a + h$, $t_2 = a + 2h, \ldots, t_{n-1} = a + (n - 1)h$, $t_n = b$.

The trapezoidal rule can be derived by setting

$$\int_a^b q(t)\,dt = \int_{t_0}^{t_1} q(t)\,dt + \int_{t_1}^{t_2} q(t)\,dt + \cdots + \int_{t_{n-1}}^{t_n} q(t)\,dt$$

and approximating the integrals on the subintervals by areas of trapezoids; for example,

$$\int_{t_0}^{t_1} q(t)\,dt \cong \frac{h}{2}[q(t_0) + q(t_1)],$$

(see Fig. 1.5). It can be shown [11] that if $q(t)$ is twice differentiable, then the error of the approximation is very nearly proportional to h^2:

$$\int_a^b q(t)\,dt - T(h) \cong k_1 h^2. \tag{1.4.3}$$

The constant k_1 depends on the smoothness of $q(t)$.

3. $\cong$ means *approximately equal to.*

EXAMPLE 5 Approximate

$$y(2) = e^{-2} \int_1^2 e^s \frac{\sin s}{s}\, ds$$

by the trapezoidal rule with $n = 2, 4, 6, 8, 10, \ldots, 18, 20$.

SOLUTION With $h = 1/n$, we set

$$T(h) = \frac{h}{2}\left[q(t_0) + 2\sum_{j=1}^{n-1} q(t_j) + q(t_n) \right]$$

and then evaluate $T(h)$. The results are in Table 1.1. ■

One strategy for picking h, the interval width, is to begin with a fairly large width and then halve it and compare the values of T until they agree to the desired accuracy. In Example 5, $h = \frac{1}{8}$ and $T = 0.396$ are acceptable if three significant figures are wanted in the approximate value of $y(2)$.

Another numerical method for evaluating definite integrals, called *Simpson's rule,* approximates an integral on a subinterval by the area under the arc of a parabola (see Fig. 1.6). The result is

$$\int_a^b q(t)\, dt \cong S(h) = \frac{h}{3}\,[q(t_0) + 4q(t_1) + 2q(t_2) + 4q(t_3) + 2q(t_4) + \cdots + 2q(t_{n-2}) + 4q(t_{n-1}) + q(t_n)]. \quad \textbf{(1.4.4)}$$

Table 1.1 Trapezoidal and Simpson approximations

n	$h = 1/n$	$T(h)$	$S(h)$	$[S(h) - T(h)]/h^2$
2	0.500000	0.392723	0.396262	1.41560×10^{-2}
4	0.250000	0.395491	0.396413	1.47624×10^{-2}
6	0.166667	0.396008	0.396421	1.48723×10^{-2}
8	0.125000	0.396189	0.396423	1.49136×10^{-2}
10	0.100000	0.396274	0.396423	1.49280×10^{-2}
12	8.33333×10^{-2}	0.396319	0.396423	1.49302×10^{-4}
14	7.14286×10^{-2}	0.396347	0.396423	1.49478×10^{-2}
16	6.25000×10^{-2}	0.396365	0.396423	1.49460×10^{-2}
18	5.55556×10^{-2}	0.396377	0.396423	1.49378×10^{-2}
20	5.00000×10^{-2}	0.396386	0.396423	1.49488×10^{-2}

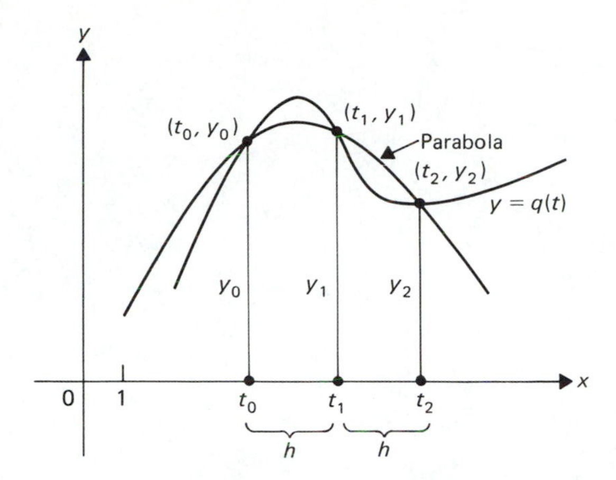

Figure 1.6 The parabolic approximation: $h[q(t_0) + 4q(t_1) + q(t_2)]/3$.

An even number of subintervals must be used. If $q(t)$ is four times differentiable, then the error of the approximation (see [11]) is very nearly proportional to h^4:

$$\int_a^b q(t)\,dt - S(h) \cong k_2 h^4. \tag{1.4.5}$$

The constant k_2 depends on the smoothness of $q(t)$.

In general, Simpson's rule gives a more accurate result than the trapezoidal rule for the same small value of h. This is illustrated in the following example.

EXAMPLE 6 Approximate

$$y(2) = e^{-2} \int_1^2 e^s \frac{\sin s}{s}\,ds$$

by Simpson's rule with $n = 2, 4, 6, 8, 10, \ldots, 18, 20$. Use the results of Example 5 and plot $[S(h) - T(h)]/h^2$ versus h.

SOLUTION We use

$$q(t) = e^{(t-2)} \frac{\sin t}{t}$$

in the preceding formula and evaluate $S(h)$. The results are in Table 1.1.

From the error estimates (1.4.3) and (1.4.5), we have

$$\frac{S(h) - T(h)}{h^2} \cong k_1 - k_2 h^2.$$

The graph in Fig. 1.7 is consistent with this result. ■

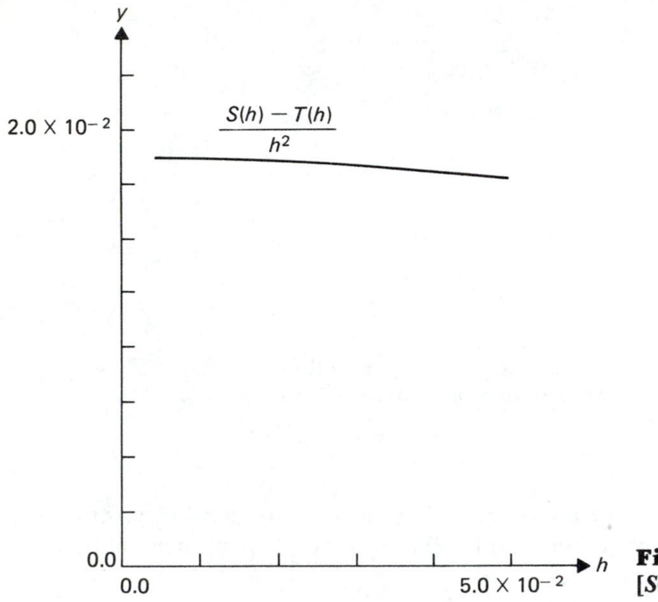

Figure 1.7 Graph of $[S(h) - T(h)]/h^2 \cong k_1 - k_2h^2$.

An extrapolation idea can be applied to a numerical integration method such as the trapezoidal rule or Simpson's rule if the dependence of the error on h, the subinterval width, is known. The trapezoidal rule [see (1.4.3)] has an error approximately proportional to h^2, so if we set $I = \int_a^b q(t)\,dt$ and use two widths h and $2h$, we have

$$I - T(h) \cong kh^2,$$
$$I - T(2h) \cong k(2h)^2.$$

Therefore

$$(I - T(2h)) \cong 4(I - T(h)),$$

or

$$I \cong T(h) + \tfrac{1}{3}[T(h) - T(2h)]. \qquad \textbf{(1.4.6)}$$

The addition of the term, $\frac{1}{3}[T(h) - T(2h)]$ gives an approximation to I that is, in general, much better than the trapezoidal rule. This is illustrated in the following example.

EXAMPLE 7 Show that

$$y(t) = e^{-t^2} \int_0^t s^2 e^{s^2}\, ds$$

is the solution to the initial value problem,

$$\frac{dy}{dt} + 2ty = t^2, \qquad y(0) = 0.$$

Use the extrapolation rule (1.4.6) with $h = \frac{1}{8}$ and then with $h = \frac{1}{16}$ to find approximate values of $y(1)$. Compare the results with those obtained by Simpson's rule with $h = \frac{1}{8}$ and $h = \frac{1}{16}$.

SOLUTION

$$\begin{aligned}\frac{d}{dt}y(t) &= \frac{d}{dt}\left(e^{-t^2}\int_0^t s^2e^{s^2}\,ds\right)\\ &= -2t\left(e^{-t^2}\int_0^t s^2e^{s^2}\,ds\right) + t^2\\ &= -2ty(t) + t^2.\end{aligned}$$

This computation plus the fact that $y(0) = 0$ shows that $y(t)$ is the solution to the initial value problem. Using the formula (1.4.2), we find that

$$T\left(\frac{1}{4}\right) = 0.251436,$$

$$T\left(\frac{1}{8}\right) = 0.236146,$$

$$T\left(\frac{1}{16}\right) = 0.232261;$$

so by formula (1.4.6)

$$y(1) \cong T\left(\frac{1}{8}\right) + \frac{1}{3}[T\left(\frac{1}{8}\right) - T\left(\frac{1}{4}\right)] = 0.231049,$$

$$y(1) \cong T\left(\frac{1}{16}\right) + \frac{1}{3}[T\left(\frac{1}{16}\right) - T\left(\frac{1}{8}\right)] = 0.230966.$$

On the other hand, Simpson's rule gives

$$y(1) \cong S\left(\frac{1}{8}\right) = 0.231049,$$

$$y(1) \cong S\left(\frac{1}{16}\right) = 0.230966. \qquad \blacksquare$$

The agreement between the extrapolation rule and Simpson's rule in Example 7 is no accident. It is easy to show that they are equivalent.

It should be noted that the numerical methods of this section are applicable only to problems for which one can evaluate $P(t)$, an antiderivative of the coefficient function $p(t)$ in the differential equation (1.4.1). Otherwise, more general methods such as those introduced in Section 1.7 and in Chapter 3 should be used.

EXERCISES 1.4

In Exercises 1–9 use the method of undetermined coefficients or trial solutions to find a general solution of the given differential equation.

1. $\dfrac{dy}{dt} + 2y = 6t$; try $y_p(t) = At + B$.

2. $\dfrac{dy}{dt} - 6y = 8e^{2t}$; try $y_p(t) = Ae^{2t}$.

3. $\dfrac{dy}{dt} + 4y = 12$; try $y_p(t) = A$.

4. $\dfrac{dy}{dt} + y = -t^2 + 1$; try $y_p(t) = At^2 + Bt + C$.

5. $\dfrac{dy}{dt} - 2y = 3 \sin t$; try $y_p(t) = A \sin t + B \cos t$.

6. $\dfrac{dy}{dt} + 4y = 6e^{-4t}$; try $y_p(t) = Ae^{-4t}$. Why didn't it work? Now try $y_p(t) = Ate^{-4t}$.

7. $\dfrac{dy}{dt} - y = te^{2t}$; try $y_p(t) = (At + B)e^{2t}$.

8. $\dfrac{dy}{dt} - y = te^{t}$; try $y_p(t) = (At + B)e^{t}$. Why didn't it work? Now try $y_p(t) = (At^2 + Bt)e^{t}$.

9. $\dfrac{dy}{dt} + 2y = 3t \sin t$; try $y_p(t) = (At + B) \sin t + (Ct + D) \cos t$.

In each of the Exercises 10–14, find a general solution of the given differential equation.

10. $\dfrac{dy}{dt} + y = 3t + 2$

11. $\dfrac{dy}{dt} - 4y = 4t^2 - 3t + 1$

12. $\dfrac{dy}{dt} - 3y = 3e^{-t} - 7$

13. $\dfrac{dy}{dt} + 2y = t + 6$

14. $\dfrac{dy}{dt} - y = 3 \sin 2t + 2 \cos 2t$

In each of the Exercises 15–20, find a general solution and the solution satisfying the given initial conditions.

15. $\dfrac{dy}{dt} + 2y = te^t,\ y(0) = 4$

16. $\dfrac{dy}{dt} - \dfrac{1}{2}y = 2\cos t,\ y(\pi) = -1$

17. $\dfrac{dy}{dt} + y = 2t^2 + 15,\ y(0) = 4$

18. $\dfrac{dy}{dt} + 4y = 6\sin^2 2t,\ y(0) = 4$. (*Hint:* Express $\sin^2 2t$ in terms of $\cos 4t$.)

19. $\dfrac{dy}{dt} - 2y + 7e^{2t} = 0,\ y(0) = 2$

20. $\dfrac{dy}{dt} - 2y = te^{2t},\ y(0) = 12$

Consider the linear equation

$$y' = -ky + \phi(t), \qquad k > 0, \qquad y(0) = y_0,$$

where $\phi(t)$ is periodic with period T, that is, $\phi(t + T) = \phi(t)$ for all t. The solution can be written in the form

$$y(t) = y_0(t) + y_p(t),$$

where $y_0(t) \to 0$ as $t \to \infty$ and is called a *transient solution* and $y_p(t)$ is periodic with period T and is called the *steady state solution.* Find the steady state solution of the equations in Exercises 21–24.

21. $y' = -y + 4\cos t$

22. $y' = -2y + 1 + \sin 2t$

23. $y' = -y + 2\cos^2 t$. (*Hint:* Express $\cos^2 t$ in terms of $\cos 2t$.)

24. $y' = -2y - 5\sin t,\ y(0) = 3$. Find a time T for which $|y(t) - y_p(t)| < 10^{-5}$ for $t > T$.

25. Show that the extrapolation rule (1.4.6) and Simpson's rule are equivalent.

26. Show that

$$y(t) = \int_0^t e^{-(t-s)} g(s)\, ds$$

is the solution to the initial value problem, $y' + y = g(t),\ y(0) = 0$.

In Exercises 27–32, the forcing function $g(t)$ is given. Use integration by parts to evaluate the integral of Exercise 26 and compute $y(1)$. Then use the trapezoidal rule (1.4.2) to compute approximate values of $y(1)$ for $h = \frac{1}{4}, \frac{1}{6}, \frac{1}{8}, \frac{1}{10}, \frac{1}{12}, \frac{1}{14}$, and $\frac{1}{16}$.

Plot the error (the difference between the exact and approximate value of $y(1)$) versus h. We suggest that you use a programmable calculator or a computer for these exercises.

27. $g(t) = 2t + 1$ **28.** $g(t) = 4t - 5$ **29.** $g(t) = \cos 3t$

30. $g(t) = \sin 3t$ **31.** $g(t) = |t - \frac{1}{2}|$ **32.** $g(t) = |t^2 - \frac{1}{4}|$

Use the extrapolation rule (1.4.6) with $h = \frac{1}{4}$ and $h = \frac{1}{8}$ to compute approximate values of $y(1)$.

33. $g(t) = \cos 3t$ **34.** $g(t) = \sin 3t$

35. Show that the solution of

$$\frac{dy}{dt} + 2ty = 1, \qquad y(0) = 0$$

is

$$y(t) = e^{-t^2} \int_0^t e^{s^2}\, ds.$$

Use Simpson's rule with four and eight subdivisions to find $y(0.5)$.

36. Find the solution of

$$\frac{dy}{dt} - 2y = \sin t^2, \qquad y(0) = 4,$$

and use Simpson's rule with four and eight subdivisions to find $y(0.8)$.

1.5 SEPARABLE EQUATIONS

Although it is not possible to construct explicit formulas for a general first order differential equation,

$$\frac{dy}{dt} = f(t, y),$$

there are classes of equations that can be solved, for example, the linear equations,

$$\frac{dy}{dt} = -p(t)y + g(t),$$

where $f(t, y) = -p(t)y + g(t)$. Another important class of first order differential equations for which solutions can be constructed consists of those that can be written as

$$\frac{dy}{dt} = q(t)h(y); \tag{1.5.1}$$

so $f(t, y) = q(t)h(y)$, the product of a function of t and a function of y. They are called *separable equations* because the key step in their solution is the separation of the independent and dependent variables. Separable equations serve as mathematical models for a variety of interesting problems, e.g., the determination of the velocity of an object falling in a gravitational field, the description of the growth of populations, and the analysis of certain nonlinear electric circuits. In this section we shall give a description of a technique for solving separable equations.

Let us start with a typical example of a separable equation.

EXAMPLE 1 Solve

$$\frac{dy}{dt} = -e^t y^2$$

SOLUTION By inspection, we find that $y(t) \equiv 0$ is one solution to the differential equation. If $y(t)$ is not zero, we can write the equation as

$$-\frac{1}{y(t)^2}\frac{dy(t)}{dt} = e^t$$

so that the variables are separated. The next step is to take the indefinite integral of both sides to get

$$\frac{1}{y(t)} = e^t + C$$

where C is an arbitrary constant. Solving for $y(t)$ yields

$$y(t) = \frac{1}{e^t + C}$$

with the restriction that $e^t + C$ cannot be zero. ■

Let us now consider

$$\frac{dy}{dt} = q(t)h(y).$$

The first step, just as in the preceding example, is so simple that it is frequently overlooked; namely, one should seek constant solutions. If $y(t) = a$ is a constant solution on an interval, then its derivative is zero. Therefore, by substitution we get

$$0 = q(t)h(a),$$

and, unless $q(t) = 0$ on the interval, which is unlikely in problems of physical interest, we must have $h(a) = 0$. Conversely, if $h(a) = 0$, then $y(t) = a$ is a constant solution. Hence, the zeros of h are constant solutions, and, in general, they are the only constant solutions.

Now assume that $y(t)$ is a nonconstant solution of (1.5.1) and that $h(y(t))$ is nonzero on an interval I. We divide both sides of Eq. (1.5.1) by $h(y(t))$ to obtain the *separated form* of the differential equation:

$$\frac{1}{h(y(t))}\frac{d}{dt}y(t) = q(t). \qquad \textbf{(1.5.2)}$$

If $Q(t)$ is an antiderivative of $q(t)$ and $H(y)$ is an antiderivative of $1/h(y)$, then we can integrate both sides of (1.5.2) to obtain

$$H(y(t)) = Q(t) + C. \qquad \textbf{(1.5.3)}$$

The constants of integration associated with H and Q have been combined in the single constant C. Note that the construction of the composite function $H(y(t))$ from an antiderivative of $1/h(y)$ is based on the chain rule. To see this, we differentiate $H(y(t))$:

$$\frac{d}{dt}H(y(t)) = \frac{d}{dy}H(y)\,|_{y=y(t)}\,\frac{d}{dt}y(t) = \frac{1}{h(y(t))}\frac{d}{dt}y(t),$$

which is the left side of Eq. (1.5.2).

Equation (1.5.3) is not an explicit solution of the differential equation (1.5.1) but is a functional relationship between the nonconstant solution $y(t)$ and the independent variable t. It is called an *implicit solution.* One would like to take an implicit solution and solve for $y(t)$ and hence obtain an *explicit solution.* In theory, this can be done; however, it is not always possible to represent an explicit solution in closed form, i.e., in terms of familiar functions.[4] This is illustrated in Example 6 of this section.

EXAMPLE 2 Solve the initial value problem

$$\frac{dy}{dt} = 3t^2e^{-y}, \qquad y(0) = 1.$$

4. It is not easy to give a satisfactory definition of a closed form solution. We have waffled by appealing to a reader-dependent concept, *familiar function.*

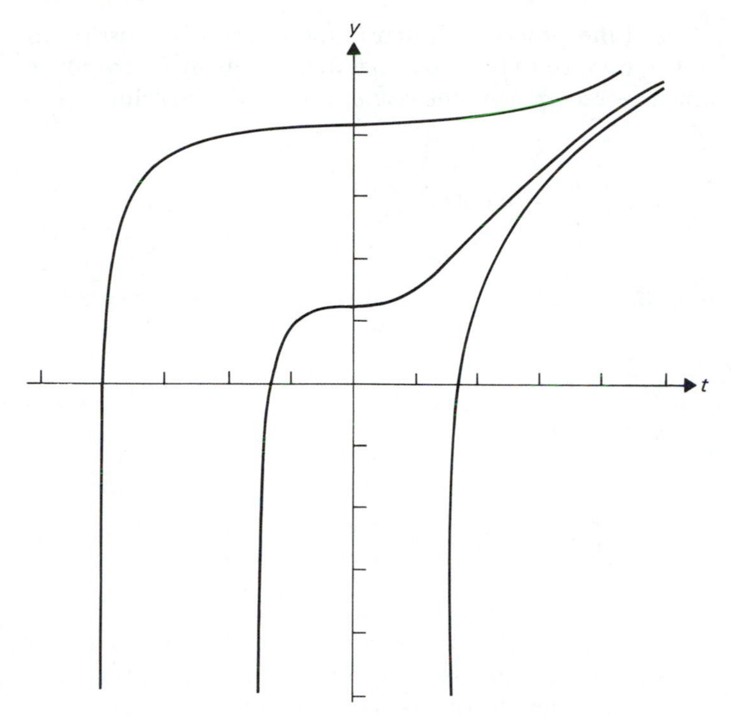

Figure 1.8 Several integral curves of $dy/dt = 3t^2e^{-y}$.

SOLUTION Since $h(y) = e^{-y}$ is never zero, there are no constant solutions and we separate variables to obtain

$$e^{y(t)}\frac{d}{dt}y(t) = 3t^2.$$

Integrating both sides gives the implicit solution

$$e^{y(t)} = t^3 + C. \tag{1.5.4}$$

The explicit solution is found by taking the natural logarithm of both sides of the above equation yielding

$$y(t) = \ln(t^3 + C), \tag{1.5.5}$$

which is defined on any interval where $t^3 + C$ is positive. To solve the initial value problem, C must be chosen so that $y(0) = 1$. Substituting $y = 1$ and $t = 0$ into (1.5.4) gives $C = e$. Therefore, the explicit solution to the initial value problem is

$$y(t) = \ln(t^3 + e).$$

The interval of definition of the solution is determined by the initial data. The fact that $\ln u$ is defined only for $u > 0$ requires that $t^3 + e > 0$ or equivalently, $t > -\exp(\frac{1}{3}) \cong -1.396$. Several integral curves for various initial values for the differential equation in Example 2 are sketched in Fig. 1.8. ∎

In Example 2 we followed the procedure given in the previous discussion to obtain the solution. However, we wish to point out that there is a formal procedure to obtain the solution, which we recommend the reader follow when solving separable equations. Given

$$\frac{dy}{dt} = q(t)h(y),$$

first find constant solutions, if any, by finding zeros of $h(y)$. Then write the equation in the form

$$\frac{dy}{h(y)} = q(t)\, dt, \tag{1.5.6}$$

and integrate to obtain

$$\int \frac{dy}{h(y)} = \int q(t)\, dt. \tag{1.5.7}$$

If $H(y)$ is an antiderivative of $1/h(y)$ and $Q(t)$ is an antiderivative of $q(t)$ then we have $H(y) = Q(t) + C$, which defines the implicit solution $y = y(t)$ obtained previously. As before, the constant C incorporates all constants of integration.

The step (1.5.6) of writing the differential equation in the form given justifies naming the solution technique, *separation of variables.* Although some mathematical justification can be given for the step, it is much simpler to think of it as a device to obtain the correct expression (1.5.3) for the implicit solution.

EXAMPLE 3 Solve

$$\frac{dy}{dt} = -\frac{y^2}{t}, \qquad t \neq 0.$$

SOLUTION First we note that since $h(y) = -y^2$, $y(t) = 0$ is a constant solution defined on any interval that does not contain $t = 0$. To find the nonconstant solutions we separate variables to obtain

$$-\frac{dy}{y^2} = \frac{1}{t}\, dt.$$

Integrating both sides gives us the implicit solution

$$\frac{1}{y(t)} = \ln|t| + C,$$

which we can solve to obtain the solution

$$y(t) = \frac{1}{\ln|t| + C},$$

where C is to be determined from given initial conditions. For example, if it is required that $y(1) = 2$, then

$$2 = \frac{1}{\ln 1 + C} = \frac{1}{C}.$$

Hence $C = \frac{1}{2}$ and the solution is

$$y(t) = \frac{1}{\ln|t| + \frac{1}{2}}.$$

But $y(t)$ is defined only for an interval I that contains the initial point $t = 1$ and for which $\ln|t| + \frac{1}{2} \neq 0$. Since $\ln(e^{-1/2}) = -\frac{1}{2}$, the correct expression for the solution is

$$y(t) = \frac{1}{\ln t + \frac{1}{2}}, \qquad t > e^{-1/2} \cong 0.6065,$$

where $\ln|t|$ has been replaced by $\ln t$ since t is positive. Note that the same expression for $y(t)$ would satisfy the initial conditions $y(e^{-5/2}) \cong y(0.0821) \cong -\frac{1}{2}$, but in this case $y(t)$ would be defined for $0 < t < e^{-1/2}$ instead. Several integral curves for Example 3 are sketched in Fig. 1.9. ■

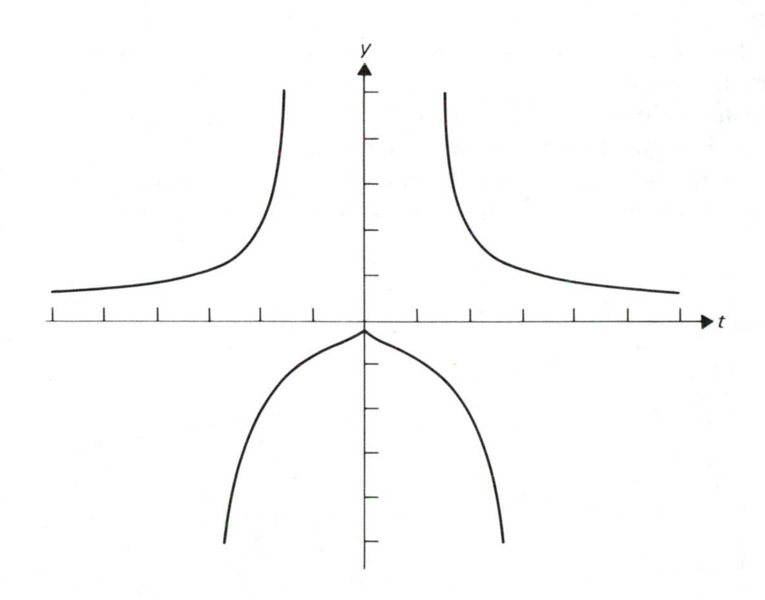

Figure 1.9 Several integral curves of $dy/dt = -y^2/t$.

EXAMPLE 4 Solve

$$\frac{dy}{dt} = \frac{t^4}{y^4}.$$

SOLUTION Since $h(y) = 1/y^4$, there are no constant solutions. We separate variables to obtain

$$y^4\, dy = t^4\, dt$$

and integrate to obtain an implicit solution,

$$[y(t)]^5 = t^5 + C.$$

Note that the factor $\frac{1}{5}$ which occurred in the integration is incorporated in the constant C. An explicit solution is

$$y(t) = (t^5 + C)^{1/5}, \tag{1.5.8}$$

where the constant C can be specified if initial conditions are given. Note that $y'(t) = t^4(t^5 + C)^{-4/5}$.

What is the interval of definition of the solution? This question requires a careful answer. The function $y(t)$ given by (1.5.8) is defined and continuous for all t, but if $C \neq 0$, it is not differentiable at the point $t = b$, where $b^5 + C = 0$. Therefore, one solution will be defined on the interval $b < t < \infty$ and a second solution on the interval $-\infty < t < b$. This is not just a technical nicety; a solution is a *continuously differentiable* function that satisfies the differential equation on an interval. It is important that $(t, y(t))$, for t in the interval, be in the domain of definition of the differential equation. In this example, $y(b) = 0$ but since $f(t, y) = t^4/y^4$, the differential equation is not defined for $y = 0$, and the function (1.5.8) is not a solution on any interval containing $t = b = (-C)^{1/5}$. Several integral curves for Example 4 are sketched in Fig. 1.10 for various values of C.

On the other hand, if $C = 0$, $y(t) = t$ can be considered as a solution for all t if we agree to redefine the differential equation at $t = 0$, $y = 0$ by setting

$$\frac{dy}{dt} = \lim_{t \to 0} \left(\frac{t}{y(t)}\right)^4,$$

if the limit exists. ■

EXAMPLE 5 Solve

$$\frac{dy}{dt} = y^3.$$

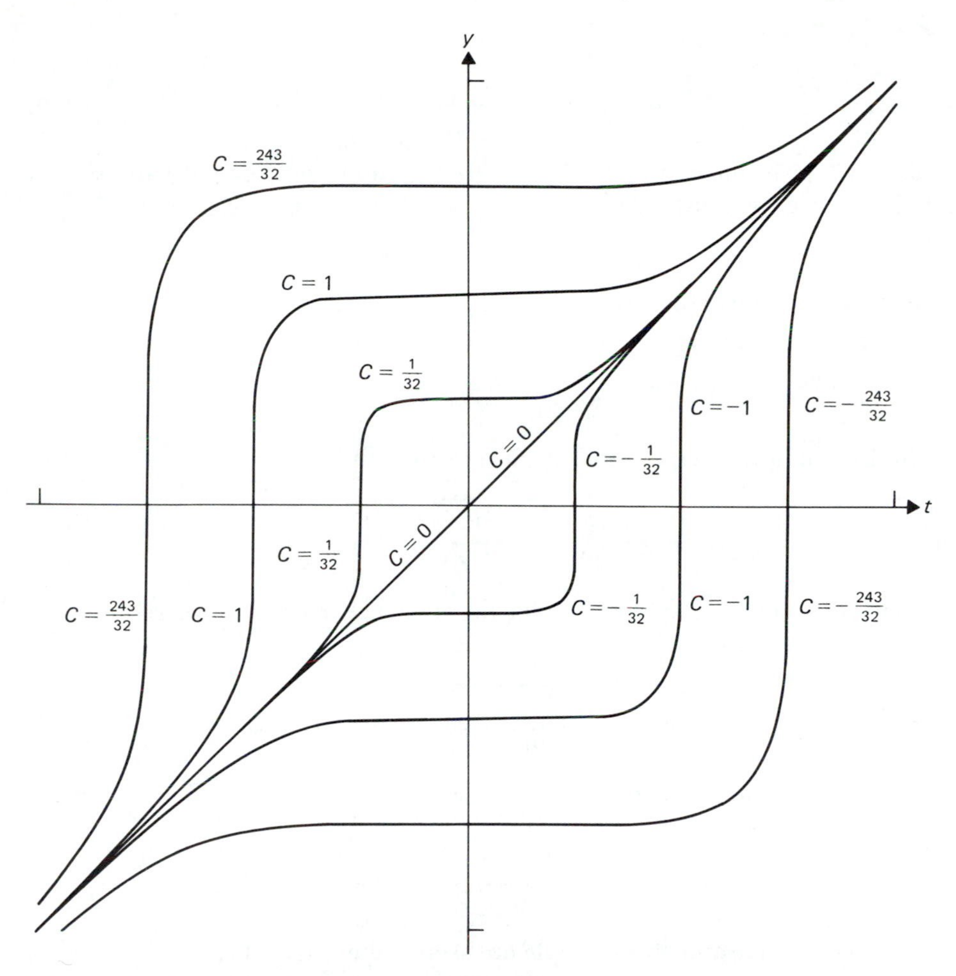

Figure 1.10 Several integral curves of $dy/dt = t^4/y^4$.

SOLUTION Here $q(t) \equiv 1$ and $h(y) = y^3$ so $y(t) = 0$ is the only constant solution. We separate variables:

$$\frac{dy}{y^3} = dt$$

and integrate to obtain the implicit solution

$$-\frac{1}{2y^2} = t + C$$

or

$$y^2 = -\frac{1}{2(t + C)}. \tag{1.5.9}$$

No value of C gives the constant solution $y(t) = 0$, and since the left-hand side of (1.5.9) is positive, we must have

$$t + C < 0 \quad \text{or} \quad t < -C.$$

Equation (1.5.9) represents a pair of explicit solutions:

$$y_1(t) = \sqrt{\frac{-1}{2(t + C)}}, \qquad t < -C$$

if the initial conditions require that $y(t)$ be positive, and

$$y_2(t) = -\sqrt{\frac{-1}{2(t + C)}}, \qquad t < -C$$

if the initial conditions require that $y(t)$ be negative. For instance, if we require that $y(0) = \frac{1}{2}$, the solution is

$$y_1(t) = \sqrt{\frac{-1}{2(t - 2)}}, \qquad t < 2.$$

However, if we require that $y(-1) = -2$, the solution is

$$y_2(t) = -\sqrt{\frac{-1}{2(t + \frac{7}{8})}}, \qquad t < -\frac{7}{8}.$$

Several integral curves for this example are sketched in Fig. 1.11. ■

EXAMPLE 6 Solve

$$\frac{dy}{dt} = \frac{t^2}{1 + y^5}.$$

SOLUTION In the last four examples it has been possible to solve explicitly for the solution $y(t)$. This is not the case here since separating variables gives

$$[1 + y^5]\, dy = t^2\, dt,$$

and integrating gives the implicit solution

$$y + \frac{1}{6}y^6 = \frac{1}{3}t^3 + C.$$

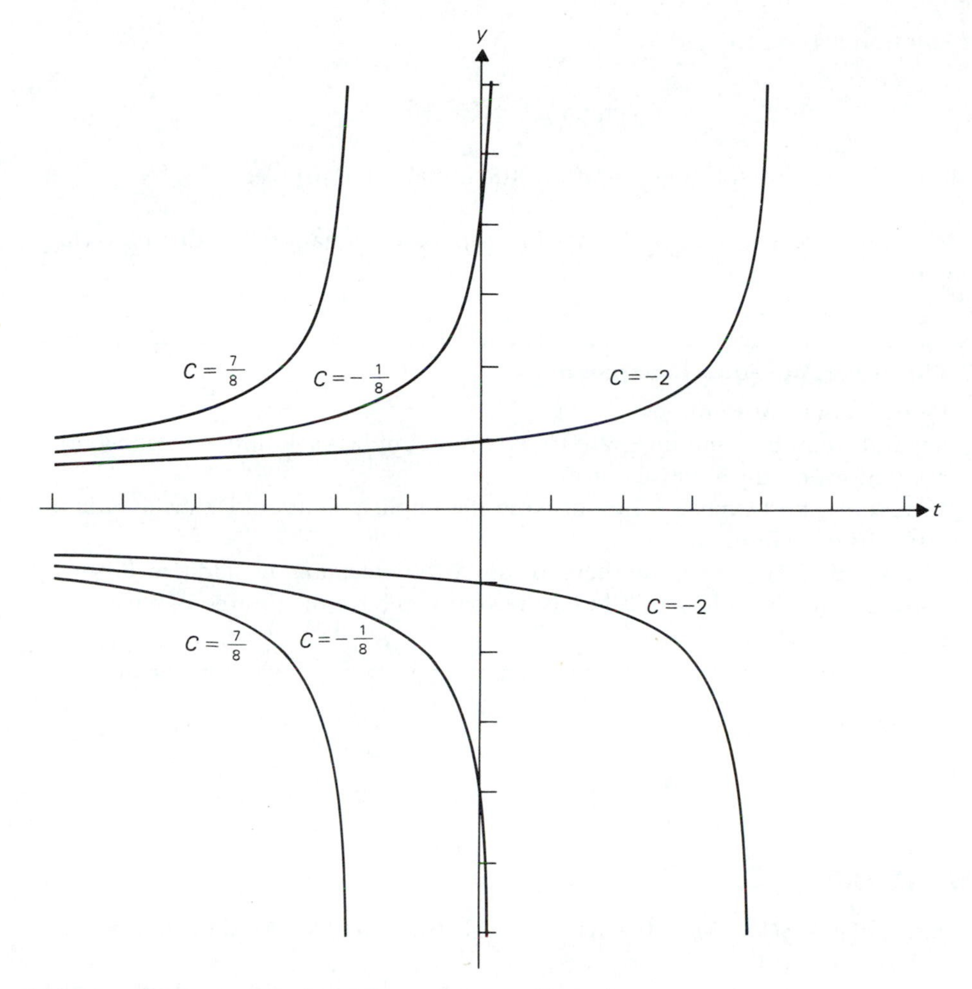

Figure 1.11 Several integral curves of $dy/dt = y^3$.

The last expression cannot be solved for $y = y(t)$, and therefore for each value of C it defines an implicit solution.

However, given initial conditions, the constant C can be specifically determined. For instance, if it is required that $y(-1) = 2$, then letting $t = -1$ and $y = 2$ gives

$$2 + \frac{1}{6}2^6 = \frac{1}{3}(-1)^3 + C.$$

The solution is $C = 13$, and

$$y + \frac{1}{6}y^6 = \frac{1}{3}t^3 + 13$$

implicitly defines the solution $y = y(t)$ of the initial value problem. ■

We summarize the steps for solving separable equations in the following algorithm.

The separable equation algorithm

1. Find all constant solutions.
2. Separate variables and integrate to obtain an implicit solution with possible restrictions on the additive constant.
3. If possible, find explicit solutions from the implicit solution. Determine their intervals of definition.
4. If initial data are given, use them in step 2 to determine the additive constant and again in step 3 to ensure that you have the desired explicit solution.

EXAMPLE 7 Solve

$$\frac{dy}{dt} = y(1 - y), \qquad y(0) = 2.$$

SOLUTION

1. Since $h(y) = y(1 - y) = 0$ when $y = 0, 1$, there are two constant solutions, $y(t) = 0$ and $y(t) = 1$.
2. To construct the nonconstant solutions, we separate variables and use partial fractions to obtain

$$\frac{1}{y(1 - y)}\,dy = \left[\frac{1}{y} + \frac{1}{1 - y}\right] dy = dt.$$

An implicit solution obtained by integration is

$$\ln|y| - \ln|1 - y| = t + C$$

or

$$\ln\left|\frac{y}{1 - y}\right| = t + C.$$

Since $-\infty < \ln |u| < \infty$, this implies that C is unrestricted. Because we are solving an initial value problem we go to step 4 to determine C, then return to step 3 to try to find the specific explicit solution. This usually makes calculations easier.

4. To solve the initial value problem, set $t = 0$ and $y = 2$:

$$\ln \left| \frac{2}{1 - 2} \right| = \ln 2 = 0 + C,$$

and so $C = \ln 2$. Therefore the implicit solution is

$$\ln \left| \frac{y}{1 - y} \right| = t + \ln 2.$$

Since $y(0) = 2$, $1 - y(t) < 0$ (why?). The implicit solution becomes

$$\ln \left[\frac{y}{y - 1} \right] = t + \ln 2,$$

and exponentiating both sides of the equation gives

$$\frac{y}{y - 1} = 2e^t.$$

3. The above expression can be solved to find the explicit solution $y(t)$:

$$y(t) = \frac{-2e^t}{1 - 2e^t} = \frac{2e^t}{2e^t - 1}.$$

Note that the expression is valid for all $t > \ln \frac{1}{2}$ and $y(t)$ is always larger than 1 for $t \geq 0$. If we multiply the numerator and denominator by $\exp(-t)$, we obtain

$$y(t) = \frac{2}{2 - e^{-t}},$$

and it is seen that $y(t)$ approaches the constant solution $y(t) = 1$ as t approaches infinity. Several integral curves for Example 7 are sketched in Fig. 1.12. ■

EXAMPLE 8 In Section 1.3, we discussed the motion of a ball thrown upward with initial velocity v_0 and subject only to the force of gravity. If it is also subject to air resistance proportional to its velocity $v(t)$, then its motion is described by the initial value problem

$$\frac{dv}{dt} = -g - kv, \qquad v(0) = v_0,$$

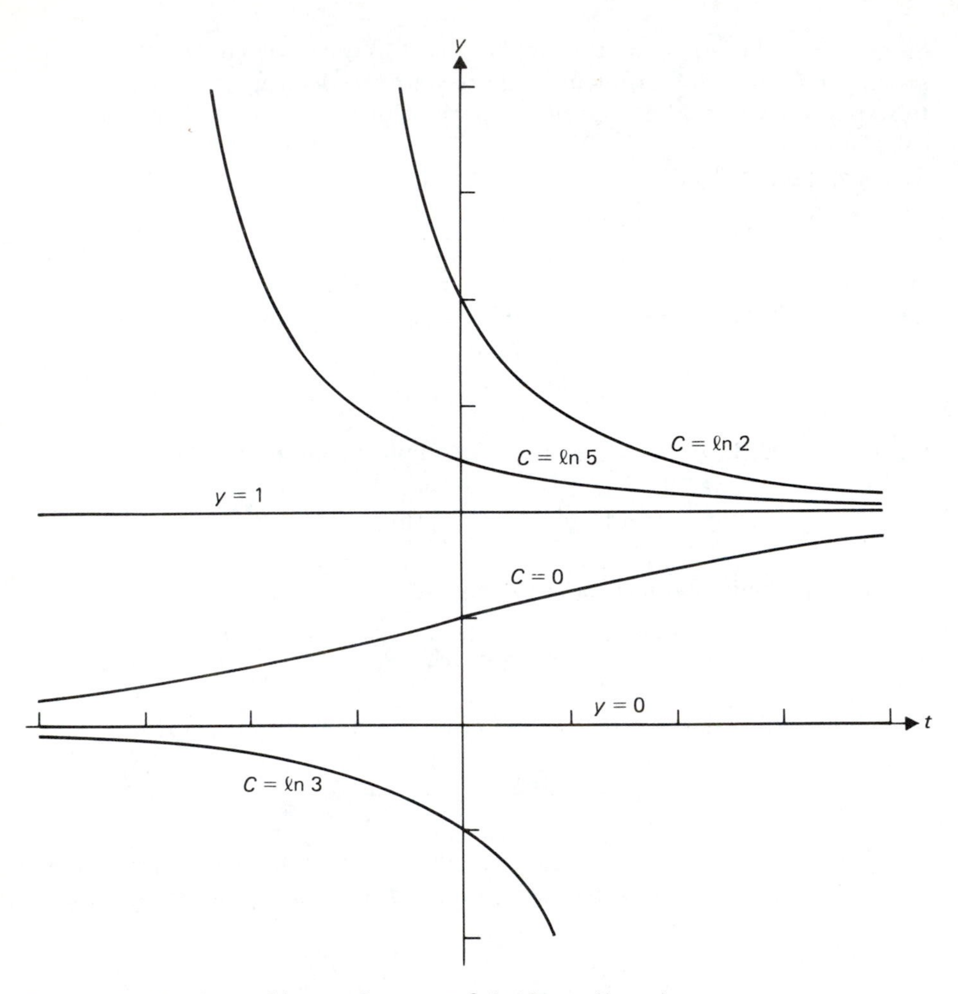

Figure 1.12 Several integral curves of $dy/dt = y(1 - y)$.

where $k > 0$ is the constant proportionality. Show that the solution of the initial value problem is

$$v(t) = \left(\frac{g}{k} + v_0\right) e^{-kt} - \frac{g}{k}.$$

SOLUTION

1. A constant solution is $v(t) = -g/k$. It is valid for this initial value problem only if $v_0 = -g/k$. But v_0 is positive since the ball was thrown upward, so there is no constant solution.

2. Since $v_0 \neq -g/k$, we separate variables to obtain

$$\frac{dv}{g + kv} = -dt.$$

An implicit solution is

$$\frac{1}{k} \ln |g + kv(t)| = -t + C.$$

To determine the constant C, we go to step 4.

4. At $t = 0$, $v = v_0$:

$$\frac{1}{k} \ln |g + kv_0| = C.$$

Therefore our implicit solution is

$$\frac{1}{k} \ln |g + kv(t)| = -t + \frac{1}{k} \ln |g + kv_0|,$$

or

$$\ln \left| \frac{g + kv(t)}{g + kv_0} \right| = -kt.$$

The function $[g + kv(t)]/[g + kv_0]$ is positive for all values of t since it is equal to unity at $t = 0$ and cannot change sign without violating the restriction that there is only one (constant) solution equal to $-k/g$. Therefore the preceding equation can be replaced by

$$\ln \left(\frac{g + kv(t)}{g + kv_0} \right) = -kt.$$

3. We can now solve for $v(t)$. First exponentiate both sides

$$\frac{g + kv(t)}{g + kv_0} = e^{-kt};$$

then we have

$$v(t) = \left(\frac{g}{k} + v_0 \right) e^{-kt} - \frac{g}{k},$$

as required. Note that

$$\lim_{t \to \infty} v(t) = -\frac{g}{k},$$

is the limiting or terminal velocity of the ball. ■

EXERCISES

1.5

Solve each of the following differential equations.

1. $\dfrac{dy}{dt} = \dfrac{t^2}{y}$

2. $\dfrac{dy}{dt} = \dfrac{t^2\sqrt{y}}{1+t^3}$

3. $\dfrac{dy}{dt} = e^{t+y}$

4. $\dfrac{dy}{dt} = \dfrac{t-ty^2}{y+t^2y}$

5. $t\dfrac{dy}{dt} = y^2 - 3y + 2$

6. $\dfrac{dy}{dt} = \dfrac{t+1}{y^4+1}$

7. $\dfrac{dy}{dt} = \dfrac{te^t}{2y}$

8. $\dfrac{dy}{dt} = \dfrac{1}{\tan t \cos^2 y}$

9. $t^2y\dfrac{dy}{dt} = y - 1$

10. $\dfrac{dy}{dt} = \dfrac{\ln t \cos y}{t \sin 2y}$

Solve each of the following initial value problems.

11. $\dfrac{dy}{dt} = \dfrac{\sin t}{y}$, $y(0) = -1$; for what values of t is $y(t)$ defined?

12. $\dfrac{dy}{dt} = t^2e^{-y}$, $y(0) = 2$

13. $\dfrac{dy}{dt} = 1 - y^2$, $y(0) = 0$

14. $\dfrac{dy}{dt} = \dfrac{y^2+y}{t}$, $y(1) = -1$

15. $\dfrac{dy}{dt} = \dfrac{3t^2+2}{2(y-1)}$, $y(0) = -1$. What other initial condition does the implicit solution satisfy at $t = 0$?

16. $\dfrac{dy}{dt} = \sqrt{1-y^2}$, $y\left(\dfrac{\pi}{2}\right) = 0$

17. $t^2\dfrac{dy}{dt} - \dfrac{1}{2}\cos 2y = \dfrac{1}{2}$, $y(1) = \dfrac{\pi}{4}$. What is $\lim_{t\to\infty} y(t)$?

18. $\dfrac{dy}{dt} = ty^4\sqrt{1+3t^2}$, $y(1) = -1$

19. $\dfrac{dy}{dt} = -3y^{4/3}\sin t$, $y\left(\dfrac{\pi}{2}\right) = \dfrac{1}{8}$. For what values of t is $y(t)$ defined? What if $y(\pi/2) = 8$?

20. $\dfrac{dy}{dt} + \dfrac{4y}{1-t^2} = 0$, $y(2) = 9$. What is $\lim_{t\to\pm\infty} y(t)$?

21. Refer to Example 8.

a) Show that the ball reaches its maximum height when

$$t = t_{max} = \frac{1}{k} \ln\left(1 + \frac{k}{g} v_0\right).$$

b) If $g = 32$ ft/sec^2, $v_0 = 128$ ft/sec, and $t_{max} = 3.5$ sec, then estimate the maximum height attained by the ball. (*Hint:* To find k you may need the approximation $\ln(1 + x) \cong x - \frac{1}{2}x^2 + \frac{1}{3}x^3$ with error less than 0.016 for $|x| < \frac{1}{2}$.)

22. If a falling object is subject to gravity and an opposing force $f(v)$ of air resistance, then its velocity satisfies the initial value problem

$$\frac{dv}{dt} = g - f(v), \qquad v(0) = v_0.$$

a) If $f(v) = kv^2$, $k > 0$, that is, if the air resistance is proportional to the square of the velocity, show that $v(t)$ approaches $\sqrt{g/k}$ as t approaches infinity. This limiting or free-fall velocity is independent of v_0 but you may assume $v_0 = 0$.

b) If the limiting velocity of a human in free fall is 50 m/sec, estimate at what time a parachutist should open his parachute if he falls from an altitude of 2000 meters and the parachute must be opened when he reaches an altitude of 500 meters. Assume $v_0 = 0$ and $g = 9.8$ m/sec^2.

A family of curves that intersects each member of a given family of curves $F(x,y) = C$, where C is a parameter, at an angle of $\pi/2$ is called the *orthogonal trajectories* of F. One can find them by differentiating $F(x,y)$ implicitly with respect to x to obtain

$$\frac{\partial F}{\partial x} + \frac{\partial F}{\partial y}\frac{dy}{dx} = 0$$

and then replacing dy/dx with $-1/(dy/dx)$ and solving the resulting differential equation. In Exercises 23–26 find the orthogonal trajectories of the given family. Graph a few members of each family so as to show the orthogonality.

23. The family of parabolas $x^2/y = C$

24. The family of circles $x^2 + y^2 - C^2 = 0$

25. The family of ellipses $x^2 + 4y^2 - C = 0$

26. The family of exponentials $e^{2x}y - C = 0$

27. Water escapes from a tank through a small hole in the bottom; the rate of escape is given by

$$\frac{dV}{dt}\,(\text{ft}^3/\text{sec}) = -4.8 \cdot \text{area of hole } (\text{ft}^2) \cdot \sqrt{h}\,,$$

where $V(t)$ is the volume of water in the tank after t sec and the depth of the water is h ft. If by reasons of symmetry one can express the volume as $V(h)$, one obtains the differential equation

$$V'(h)\frac{dh}{dt} = -4.8 \cdot \text{area of hole (ft}^2) \cdot \sqrt{h}\,.$$

This equation can be solved to determine the time T when the tank will be empty: $h(T) = 0$.

a) Find the time required to empty a cylindrical tank of radius 4 ft and height 12 ft, if water is leaking from the bottom through a square hole with a 1-inch side.

b) Find the time required to empty a conical funnel 12 inches high and with base diameter 3 inches, if the apex hole has a diameter of $\frac{1}{2}$ inch.

c) Using the fact that 1 ft $\cong$ 0.305 m, modify the constant -4.8 to find the metric equivalent of the differential equation.

1.6 DIRECTION FIELDS AND EXISTENCE AND UNIQUENESS THEOREMS

This section is an introduction to the geometric theory of first order differential equations and a statement of conditions that ensure existence and uniqueness of solutions. Suppose that $f(t, y)$ is defined and continuous on an open rectangle R in the plane. Then

$$\frac{dy}{dt} = f(t,y) \tag{1.6.1}$$

is called the first order differential equation associated with $f(t,y)$. For example, the differential equation associated with $f(t,y) = -p(t)y + g(t)$ is

$$\frac{dy}{dt} = -p(t)y + g(t).$$

Definition

A *solution* to the differential equation (1.6.1) is a differentiable function $y(t)$ defined on a t-interval I, $t_1 < t < t_2$, with the property that substitution of $y(t)$ into the differential equation yields the identity,

$$\frac{dy}{dt}(t) \equiv f(t,y(t)), \qquad t_1 < t < t_2.$$

The graph of a solution is called an *integral curve* of the differential equation.

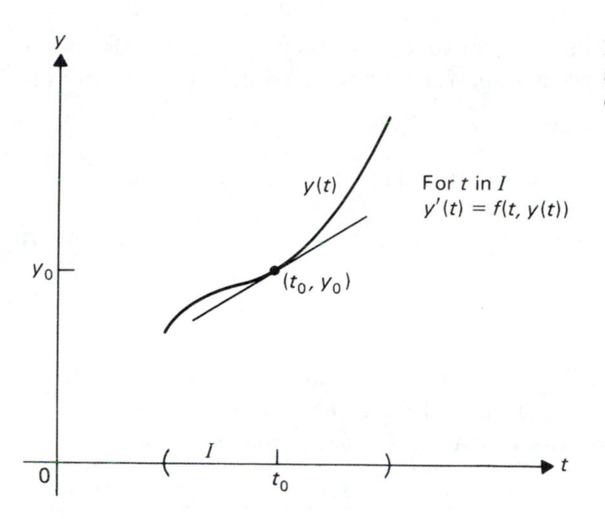

Figure 1.13 The slope of a tangent line to an integral curve at (t, y) is $f(t, y)$.

Geometrically, this means that the tangent line to the integral curve $y = y(t)$ has slope $f(t,y(t))$ for each t in I. This is illustrated in Fig. 1.13.

EXAMPLE 1 Solve $dy/dt = -2y + 2t^2$ and show that the slope of an integral curve passing through the point $t = -1$, $y = \frac{5}{2}$, is given by $f(-1, \frac{5}{2}) = -3$. Find the equation of the tangent line of the integral curve at the given point.

SOLUTION Using the method of undetermined coefficients, we take $y_p(t) = At^2 + Bt + C$ and find that a general solution to the differential equation is

$$y(t) = \frac{1}{2} - t + t^2 + Ce^{-2t}.$$

If an integral curve passes through $t = -1$, $y = \frac{5}{2}$, then $y(-1) = \frac{5}{2}$, $C = 0$, and

$$y(t) = \frac{1}{2} - t + t^2.$$

From

$$\frac{dy}{dt}(t) = -1 + 2t,$$

we obtain

$$\frac{dy}{dt}(-1) = -3 = f\left(-1, \frac{5}{2}\right),$$

as required.

The equation of the tangent line through the points $(-1, \frac{5}{2})$ has slope $f(-1, \frac{5}{2}) = -3$. By means of the point-slope formula for a straight line, we obtain that

$$y - \frac{5}{2} = -3(t + 1)$$

as the equation of the tangent line at the prescribed point. ■

Definition

The differential equation (1.6.1) assigns to each point (t, y) in R, its domain of definition, a slope equal to $f(t, y)$ that can be represented as a line segment. The totality of such slopes is called the *direction field* of the differential equation.

The direction field of the differential equation of Example 1, $dy/dt = -2y + 2t^2$, for the rectangle, $R: -2 < t < 2, -6 < y < 6$ and several of its integral curves are sketched in Fig. 1.14. The direction field of a differential equation displays the geometric characteristics of its solutions. For example, it is clear from the direction field in Fig. 1.14 that a typical integral curve approaches a parabola, the integral curve of the solution to Example 1.

The sketching of a direction field can be laborious. It is sometimes helpful to sketch curves of the form $f(t, y) =$ constant. These are called *isoclines.* However, most computing centers have graphics programs that can provide accurate and attractive sketches of direction fields. The direction fields used in this text are generated by such a program.

EXAMPLE 2 Sketch several isoclines of the differential equation

$$\frac{dy}{dt} = t^2 - y^2.$$

SOLUTION The isoclines are a set of hyperbolas, $t^2 - y^2 =$ constant. They are sketched in Fig. 1.15. ■

The differential equation of the preceding example cannot be solved by elementary methods even though its direction field is quite regular. How can we be sure that a solution even exists? To be more precise, is it always true that the differential equation (1.6.1) has at least one solution whose graph contains (t_0, y_0), an arbitrary point in the domain of definition of $f(t, y)$? The answer to this is given by the following theorem.

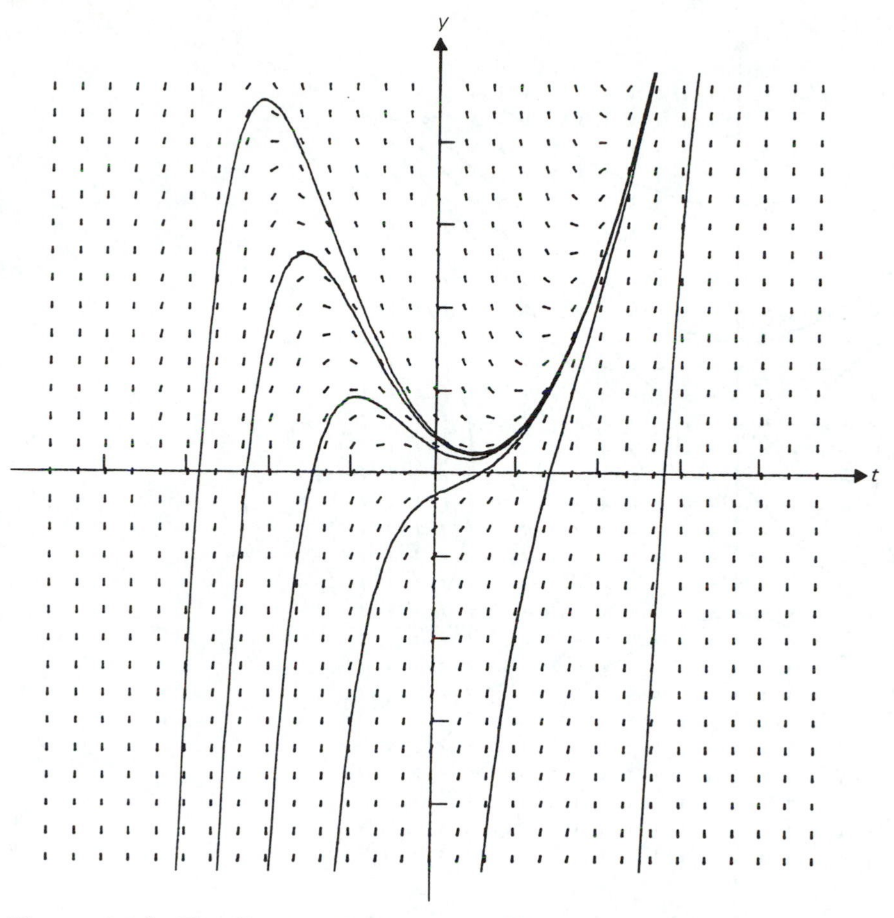

Figure 1.14 The direction field and several integral curves of $dy/dt = -2y + 2t^2$.

The basic existence theorem

Let $f(t, y)$ be a real-valued continuous function defined on a rectangle R in the ty-plane and let (t_0, y_0) be an arbitrary point in R. Then there exists a solution $y(t)$ of the differential equation

$$\frac{dy}{dt} = f(t, y)$$

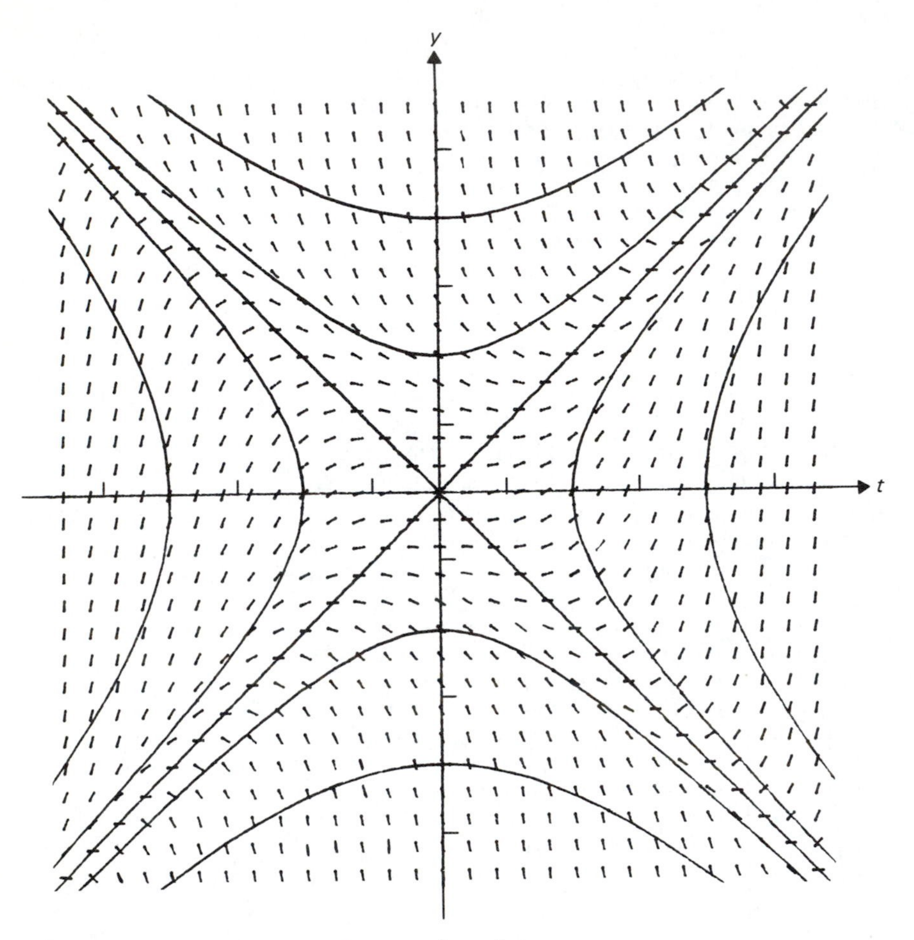

Figure 1.15 Isoclines of $dy/dt = t^2 - y^2$.

which is defined on some interval I: $t_1 < t < t_2$, containing t_0 and which satisfies the condition

$$y(t_0) = y_0.$$

The geometric meaning of the basic existence theorem is shown in Fig. 1.13.

Since the graph of $y(t)$ contains the point (t_0, y_0), we say that the solution *passes through the point* (t_0, y_0). As we saw earlier, a differential equation together with an *initial condition* $y(t_0) = y_0$ is called an *initial value problem,* and the basic existence theorem states that under very reasonable conditions on $f(t, y)$, it will always have a solution. The condition $y(t_0) = y_0$ is frequently a mathematical translation of the statement "at the start of an experiment the time is t_0 and the initial state of the system of interest is y_0." For instance, if the differential equation describes the motion of a particle, then y_0 could represent the initial displacement from a fixed reference point. Or if the differential equation describes the growth of a population, then y_0 could be the size of the population at time t_0.

Note furthermore that the basic existence theorem is a *local* theorem in the sense that it does not specify the length of the interval I on which the solution is defined. The theorem asserts only that some interval containing t_0 exists and that

Figure 1.16 **Several integral curves of $dy/dt = t^2 - y^2$.**

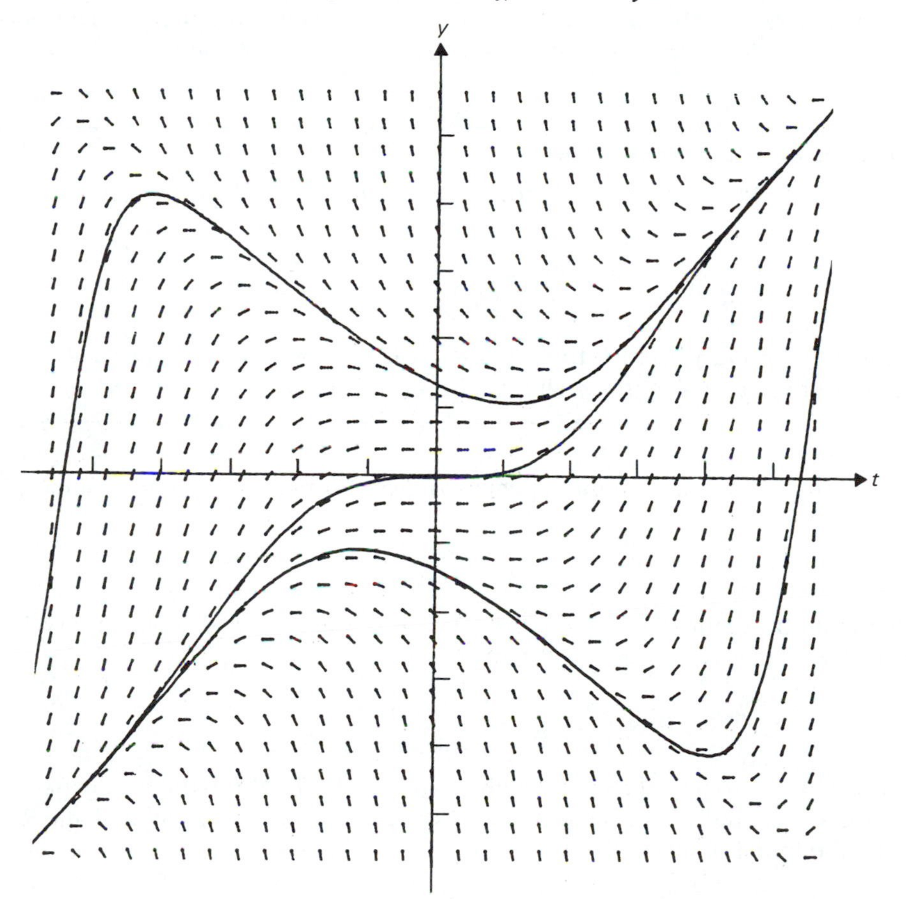

the solution is defined for all t in that interval. As the above examples show, solutions may be defined on finite intervals, half intervals, or the whole real line; this depends in part on the choice of initial conditions. The proof of the basic existence theorem is difficult and will be discussed in Appendix 1 along with related theorems.

Although the above theorem asserts that under very simple conditions there are solutions to the differential equation (1.6.1), it does not suggest a method for constructing one. Furthermore, it does not address the question of uniqueness; i.e., can there be more than one solution passing through the point (t_0, y_0)? This is an important question. For any physical system modeled by an initial value problem, nonuniqueness would mean lack of predictability, and consequently the model should probably be scrapped.

Continuity of the function f alone, as required in the basic existence theorem, is not sufficient to guarantee that only one solution of a given differential equation passes through a given point (t_0, y_0). The addition of one simple condition, that $\partial f/\partial y$ be continuous in the rectangle R, is enough to assure uniqueness. This is summarized in the following theorem, which is also proved in Appendix 1.

The basic uniqueness theorem

Let $f(t, y)$ and $\partial f(t, y)/\partial y$ be continuous in R and let $y_1(t)$ and $y_2(t)$ be any two solutions of

$$\frac{dy}{dt} = f(t, y)$$

with $(t, y_1(t))$ and $(t, y_2(t))$ in R for $t_1 < t < t_2$. If $y_1(t_0) = y_2(t_0)$ for some t_0, $t_1 < t_0 < t_2$, then $y_1(t) = y_2(t)$ for all t, $t_1 < t < t_2$.

A spectacular example of nonuniqueness is given by the initial value problem

$$\frac{dy}{dt} = y^{1/3}, \qquad y(0) = 0,$$

which has an infinite number of solutions, all of which are defined on the interval $-\infty < t < \infty$. They are, for any constant $c \geq 0$,

$$y(t) = \begin{cases} 0 & \text{for } t \leq c, \\ \left[\frac{2}{3}(t - c)\right]^{3/2} & \text{for } t > c. \end{cases}$$

A graph of $y(t)$ for some values of c is given in Fig. 1.17. The basic uniqueness theorem does not apply since $\partial f/\partial y = \frac{1}{3}y^{-2/3}$ is not continuous in any domain R containing the initial point $(0, 0)$.

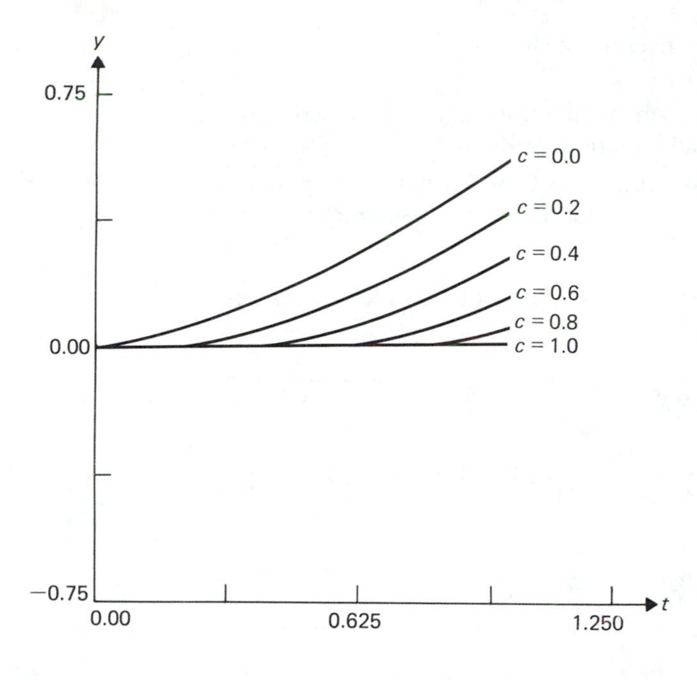

Figure 1.17 Some solutions of $dy/dt = y^{1/3}$, $y(0) = 0$.

EXERCISES 1.6

For each of the differential equations in Exercises 1–10, verify by direct substitution that the given function $y(t)$ is a solution. State the interval I on which it is defined.

1. $\dfrac{dy}{dt} = -\dfrac{1}{(t+2)^2}$, $y(t) = \dfrac{1}{t+2} + 4$

2. $\dfrac{dy}{dt} = \ln t$, $y(t) = t \ln t - t + 8$

3. $\dfrac{dy}{dt} = 10(y - t)$, $y(t) = t + 0.1$

4. $\dfrac{dy}{dt} = 2ty$, $y(t) = Ke^{t^2}$, K an arbitrary constant

5. $\dfrac{dy}{dt} = -2y + e^t$, $y(t) = \dfrac{1}{3}e^t + \dfrac{2}{3}e^{-2t}$

6. $\dfrac{dy}{dt} = 4ty^2$, $y(t) = \dfrac{1}{(1 - 2t^2)}$

7. $\dfrac{dy}{dt} = \dfrac{3}{4}y^{5/3}$, $y(t) = \left(1 - \dfrac{t}{2}\right)^{-3/2}$

8. $\dfrac{dy}{dt} = \dfrac{t(3 - 2y)}{(t^2 - 1)}$, $y(t) = \dfrac{1}{(t^2 - 1)} + \dfrac{3}{2}$

9. $\dfrac{dy}{dt} = y\left(1 - \dfrac{1}{2}y\right)$, $y(t) = \dfrac{2}{1 + e^{-t}}$

10. $\dfrac{dy}{dt} = \dfrac{-t}{t+1}y$, $y(t) = K(t + 1)e^{-t}$, K an arbitrary constant

11. Show that a solution to $dy/dt = e^{t-y}$ is given implicitly by $e^y = e^t + c$, where c is an arbitrary constant.

12. Show by implicit differentiation that a pair of functions $(x(t), y(t))$ satisfying $x^2 + y^2 = c^2$ also satisfy the differential equation $dy/dx = -x/y$.

13. Use the basic theorems to deduce that equations of the form $y' = g(t)y + h(t)$, with $g(t)$, $h(t)$ continuous on some interval I, always possess unique solutions on I.

In Exercises 14–19 find points (t_0, y_0) where there might not be a unique solution satisfying $y(t_0) = y_0$ for the given differential equations.

14. $\dfrac{dy}{dt} = -\ln|y - 1|$ **15.** $\dfrac{dy}{dt} = \dfrac{1}{t}\sec y$ **16.** $\dfrac{dy}{dt} = \sqrt{(y-1)(y-2)}$

17. $\dfrac{dy}{dt} = \dfrac{y^2}{y - t}$ **18.** $\dfrac{dy}{dt} = \dfrac{t}{t+1}y$ **19.** $\dfrac{dy}{dt} = ty^{1/2}$

20. Show that the functions

$$y_1(t) = 0, \qquad -\infty < t < \infty,$$

and

$$y_2(t) = (t - t_0)^3, \qquad -\infty < t < \infty,$$

are both solutions of the initial value problem

$$\frac{dy}{dt} = 3y^{2/3}, \qquad y(t_0) = 0.$$

Graph both solutions for $t_0 = -1, 0, 1$ and explain why the basic uniqueness theorem does not apply.

21. Consider the problem $dy/dt = \sqrt{|1 - y^2|}$, $y(0) = 1$. Verify that

a) $y(t) = 1$ is a solution on any interval containing $t = 0$;

b) $y(t) = \cosh t$ is a solution on any interval $0 \le t \le b$ for any $b > 0$;

c) $y(t) = \cos t$ is a solution on a suitable interval. What is the largest interval containing $t = 0$ on which $\cos t$ is a solution?

For each of the differential equations in Exercises 22–27 sketch the direction field and graph the given integral curves in the specified rectangle R.

22. $\dfrac{dy}{dt} = 2y$, $R = \{0 < t < 1, -1 < y < 1\}$, $y(t) = Ke^{2t}$, $K = 0, \pm\dfrac{1}{4}, \pm\dfrac{1}{2}$

23. $\dfrac{dy}{dt} = y - t$, $R = \{0 < t < 1, -1 < y < 2\}$, $y(t) = Ke^t + t + 1$, $K = -1, -\dfrac{1}{2}, -\dfrac{1}{4}, 0$

24. $\dfrac{dy}{dt} = \dfrac{1}{y^2}$, $R = \{0 < t < 2, -2 < y < 2\}$, $y(t) = [3(t - c)]^{1/3}$, $c =$ $0, \pm\dfrac{1}{2}, \pm 1$

25. $\dfrac{dy}{dt} = \tan t$, $R = \{-1 < t < 1, -4 < y < 4\}$, $y(t) = K - \ln(\cos t)$, $K =$ $0, \pm 1, \pm 2$

26. $\dfrac{dy}{dt} = \dfrac{t}{y}$, $R = \{-1 < t < 1, 0 < y < 1\}$, $y(t) = \sqrt{t^2 + c}$, $c = 0, \pm\dfrac{1}{16}, \pm\dfrac{1}{4}$

27. $\dfrac{dy}{dt} = \dfrac{2y}{t}$, $R = \{0 < t < 1, -2 < y < 2\}$, $y(t) = Kt^2$, $K = 0, \pm\dfrac{1}{2}, \pm 1, \pm\dfrac{3}{2}$

28. A less stringent condition, which ensures uniqueness of solutions of $y' = f(t, y)$ in R, is that $f(t, y)$ be continuous in R and there exist a constant K such that

$$|f(t, y_1) - f(t, y_2)| \leq K|y_1 - y_2|$$

for any points (t, y_1), (t, y_2) in R. This condition is called a *Lipschitz condition.*

a) For the function $f(t, y) = |y|$ with $R = \{-\infty < t < \infty, -\infty < y < \infty\}$ show that $\partial f/\partial y$ is not continuous at $y = 0$ but f does satisfy the above condition.

b) Show that the functions

$$y_1(t) \equiv 0, \qquad y_2(t) = Ce^{-t}, \qquad C < 0$$

and

$$y_3(t) = Ce^{t}, \qquad C > 0$$

for $-\infty < t < \infty$ are all possible solutions of $y' = |y|$. Show also that only one of these solutions will pass through any point in R as defined in (a).

c) If $R = \{0 < t < 2, -\infty < y < \infty\}$ and if $f(t, y) = \sin|y|$, what is a possible choice of K?

1.7 SOME ELEMENTARY NUMERICAL METHODS

In the previous sections we have concentrated on methods for finding explicit solutions for certain differential equations, that is, solutions expressible in terms of elementary functions. We saw in Section 1.4 that some differential equations have solutions that cannot be written down in this manner, e.g.,

$$\frac{dy}{dt} + y = \frac{\sin t}{t}.$$

In fact, many of the differential equations encountered in mathematical models of physical phenomena cannot be solved explicitly, so we are led to study methods for obtaining approximations to solutions. Such solutions are usually called *numerical solutions.*

In this section we will study some fundamental ideas that serve as the basis for numerical methods and present two examples of such methods. In Chapters 3 and 8, numerical methods will be discussed in more detail.

We begin our discussion of numerical methods by considering the initial value problem

$$\frac{dy}{dt} = f(t, y), \qquad y(t_0) = y_0, \tag{1.7.1}$$

defined on some interval $t_0 \leq t \leq b$. The numerical method to be studied will generate a table of approximate values for $y(t)$, the solution to (1.7.1). For now, we suppose that the entries in the table are for equally spaced values of the independent variable t. If we choose an integer n and let the *step size* be

$$h = (b - t_0)/n,$$

our task is to find approximations to $y(t)$ at the equally spaced points $t_i = t_0 + ih$, $i = 1, 2, \ldots, n$. In what follows, the expression $y(t_i)$ will always represent the actual solution of (1.7.1) evaluated at $t = t_i$, whereas y_i will represent an approximation to that value. The reader should keep this notational distinction in mind.

EULER METHOD

The simplest example of a numerical method for obtaining an approximate solution to (1.7.1) is the *Euler method.* In a sense, the Euler method produces a solution that follows the direction field of the differential equation. Consider a ty coordinate system and construct the vertical lines $t = t_0, t = t_1, \ldots, t_n$ as in Fig. 1.18.

The exact solution to (1.7.1) passes through the point (t_0, y_0) and has the slope $f(t_0, y_0)$ there. Let us draw a straight line through (t_0, y_0) with slope $f(t_0, y_0)$. This line intersects the vertical line $t = t_1$ at some point with $y = y_1$. Next, we draw a straight line through (t_1, y_1) with slope $f(t_1, y_1)$ that will intersect the line $t = t_2$ at some point with $y = y_2$. Continuing in this fashion, we generate a set of points (t_i, y_i), $i = 1, 2, \ldots, n$, such that the y_i values are approximations to $y(t_i)$.

To derive an analytical expression for the above, proceed as follows. The slope of the direction field at (t_0, y_0) is $f(t_0, y_0)$ and, by using the point-slope form for the equation of a straight line, we get the equation of the first line segment:

$$\frac{y - y_0}{t - t_0} = f(t_0, y_0).$$

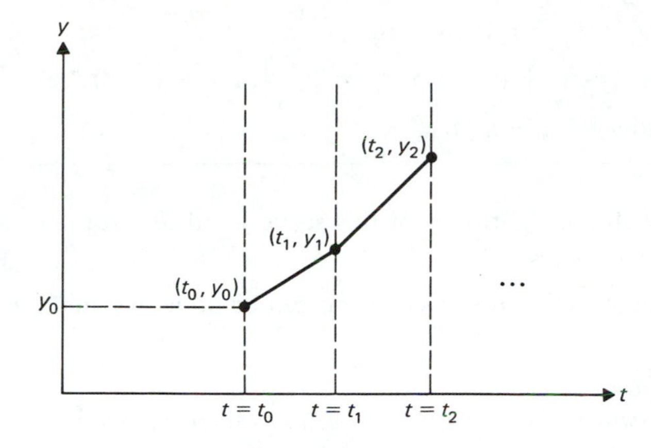

Figure 1.18 The Euler solution.

Hence, letting $t = t_1$, we obtain the following expression for y_1, the approximation to $y(t_1)$:

$$y_1 = y_0 + (t_1 - t_0)\, f(t_0, y_0),$$

or, since $h = t_1 - t_0$,

$$y_1 = y_0 + hf(t_0, y_0).$$

The slope at (t_1, y_1) is $f(t_1, y_1)$, and so the equation of the second line segment is

$$\frac{y - y_1}{t - t_1} = f(t_1, y_1).$$

Letting $t = t_2$ and remembering that $t_2 - t_1 = h$, we obtain the expression for y_2, the approximation to $y(t_2)$:

$$y_2 = y_1 + hf(t_1, y_1).$$

In general, if we are at the point (t_i, y_i), we calculate the slope $f(t_i, y_i)$, then use the equation of the ith line segment to obtain the approximation y_{i+1} to the value $y(t_{i+1})$ of the actual solution:

$$y_{i+1} = y_i + hf(t_i, y_i).$$

This is summarized in the following algorithm.

The Euler algorithm

An approximate solution to the initial value problem

$$\frac{dy}{dt} = f(t, y), \qquad y(t_0) = y_0$$

at the equally spaced points $t_0, t_1, t_2, \ldots, t_n$ is given by

$$y_{i+1} = y_i + hf(t_i, y_i), \quad t_{i+1} = t_0 + (i + 1)h, \quad i = 0, \ldots, n - 1, \tag{1.7.2}$$

where h, t_0, and y_0 are given and $h = (t_n - t_0)/n$.

To familiarize the reader with the algorithms of this section and to test their accuracy, we will initially choose examples

1. for which the step size is large so that the calculations can be done by hand or with a calculator, and
2. that can be solved explicitly.

Having done this we will then switch to a computer program, which will allow us to reduce greatly the step size and to consider problems for which no exact solutions can be found.

EXAMPLE 1 Find an approximate solution to the initial value problem

$$y' = y, \qquad y(0) = 1$$

at the point $t = 1$ by using the Euler algorithm with step size $h = \frac{1}{4}$.

SOLUTION Here $f(t, y) = y$, so (1.7.2) simplifies to

$$y_{i+1} = y_i + hy_i = (1 + h)y_i,$$

or, since $h = \frac{1}{4}$,

$$y_{i+1} = \frac{5}{4}y_i.$$

Starting then with $t_0 = 0$, $y_0 = 1$, we compute

$$y_1 = \frac{5}{4}y_0 = \frac{5}{4} \cdot 1 = \frac{5}{4},$$

$$t_1 = t_0 + 1 \cdot \frac{1}{4} = 0 + \frac{1}{4} = \frac{1}{4},$$

$$y_2 = \frac{5}{4}y_1 = \frac{5}{4} \cdot \frac{5}{4} = \frac{25}{16},$$

$$t_2 = t_0 + 2 \cdot \frac{1}{4} = 0 + \frac{1}{2} = \frac{1}{2},$$

$$y_3 = \frac{5}{4}y_2 = \frac{5}{4} \cdot \frac{25}{16} = \frac{125}{64},$$

$$t_3 = t_0 + 3 \cdot \frac{1}{4} = 0 + \frac{3}{4} = \frac{3}{4},$$

$$y_4 = \frac{5}{4} y_3 = \frac{5}{4} \cdot \frac{125}{64} = \frac{625}{256},$$

$$t_4 = t_0 + 4 \cdot \frac{1}{4} = 0 + 1.$$

Therefore the approximation at t_4 is $y_4 = 625/256 = 2.441$ (correct to three decimal places). The actual solution is $y(t) = e^t$ and $y(1) = 2.718$ (again correct to three decimal places). The absolute error in the numerical solution is $|y_4 - y(1)| = 0.277$. The two solutions are represented graphically in Fig. 1.19, where the approximation points y_0, y_1, y_2, y_3, and y_4 are joined by straight line segments. ■

A point worth noting here is that the first segment of the Euler approximation is part of the tangent line to the exact solution. This is not true in general for subsequent segments. The ith segment is tangent to a solution of the differential equation going through the point (t_i, y_i), but not necessarily tangent to the solution of interest, namely, the one passing through the initial point (t_0, y_0). Consequently, it is possible and usual for the "Euler solution" to drift away from the true solution after many steps of the algorithm.

With $h = \frac{1}{4}$ in Example 1, Euler's method gave an approximation that had an absolute error of 0.277. Our intuition tells us that we would probably make a

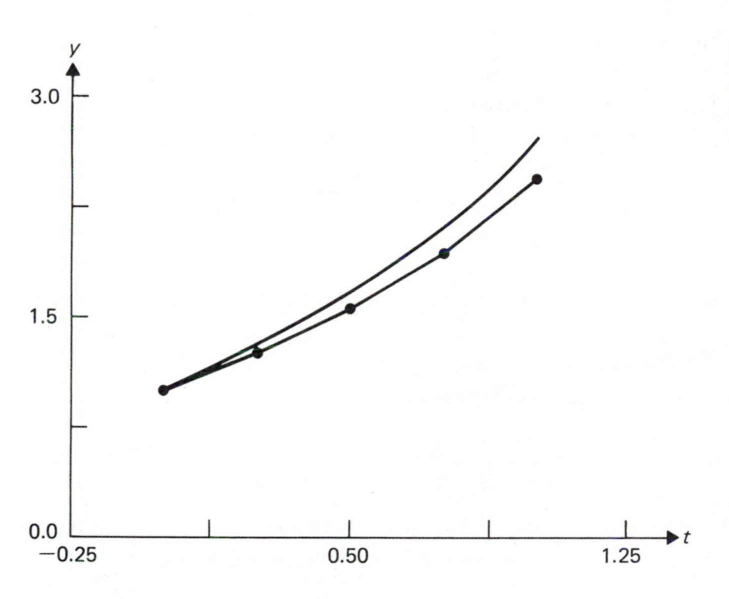

Figure 1.19 Plot of the Euler solution and the exact solution to $dy/dt = y$, $y(0) = 1$.

smaller error by using a smaller value of h. Computing with $h = \frac{1}{8}$ and $t_0 = 0$ and $y_0 = 1$ as before, we get

$$y_1 = (1 + h)y_0 = \frac{9}{8} \cdot 1 = \frac{9}{8},$$

$$t_1 = t_0 + 1 \cdot \frac{1}{8} = \frac{1}{8},$$

$$y_2 = \frac{9}{8} y_1 = \frac{9}{8} \cdot \frac{9}{8} = \left(\frac{9}{8}\right)^2,$$

$$t_2 = t_0 + 2 \cdot \frac{1}{8} = \frac{1}{4},$$

$$\vdots$$

$$y_8 = \frac{9}{8} y_7 = \frac{9}{8}\left(\frac{9}{8}\right)^7 = \left(\frac{9}{8}\right)^8,$$

$$t_8 = t_0 + 8 \cdot \frac{1}{8} = 1.$$

The approximation is $y_8 = 2.566$, which has an absolute error

$$|y_8 - y(1)| = 0.152$$

(correct to three decimal places). This result is slightly better. If we are willing to do more work, the result can be improved even more as shown in Table 1.2.

Note that if we make the step size half as big, the absolute error will also be approximately half as big. This suggests that the error in the numerical approximation produced by Euler's method is roughly proportional to the step size h. We shall see later that this is true for most problems.

A Fortran program that implements Euler's method is given in Fig. 1.20. It is set up to solve the initial value problem

$$\frac{dy}{dt} = ty^{1/3}, \qquad y(1) = 1. \tag{1.7.3}$$

The solution is computed and printed at 10 equally spaced points in the interval $1 \leq t \leq 2$, so $N = 10$, and the final time is $\text{TF} = 2.0$. This results in a step size $H = (2.0 - 1.0)/10 = 0.1$. The output of the program is reproduced in Table 1.3.

Equation (1.7.3) is separable and the solution is

$$y(t) = \left(\frac{t^2 + 2}{3}\right)^{3/2},$$

so $y(2) \cong 2.828$. With 10 steps, the above program produced 2.724, a value in error by about 0.104. As in Example 1, the step size h could be decreased to obtain a better result.

Table 1.2 The Euler method approximation and errors for $dy/dt = y$, $y(0) = 1$

n	$h = 1/n$	y_n	$\lvert y_n - y(1)\rvert$
16	0.063	2.638	0.080
32	0.031	2.677	0.041
64	0.016	2.697	0.021
128	0.008	2.708	0.010
256	0.004	2.713	0.005
512	0.002	2.716	0.002
1024	0.001	2.717	0.001

```
C PROGRAM USES THE EULER METHOD TO SOLVE THE
C INITIAL VALUE PROBLEM
C
C     DY/DT = F(T,Y), Y(T0) = Y0
C
C DEFINE FUNCTION F(T,Y) AND INITIAL CONDITIONS.
C
      F(T,Y) = T*Y**(1.0/3.0)
      T0=1.0
      Y0=1.0
C
C SET FINAL T AND NUMBER OF STEPS N TO BE TAKEN.
C
      TF=2.0
      N=10
C
C EULER LOOP
C
      T=T0
      Y=Y0
      H=(TF - T0)/FLOAT(N)
      WRITE(6,2)T,Y
    2 FORMAT(1X,F6.2,F10.5)
      DO 1 I=1,N
         Y=Y+H*F(T,Y)
         T=T+H
         WRITE(6,2)T,Y
    1 CONTINUE
      STOP
      END
```

Figure 1.20 Euler program to solve $dy/dt = ty^{1/3}$, $y(1) = 1$.

Table 1.3 The Euler method approximation for $dy/dt = ty^{1/3}$, $y(1) = 1$, $h = 0.1$

t	y	t	y
1.00	1.000000	1.60	1.821700
1.10	1.099999	1.70	2.017109
1.20	1.213550	1.80	2.231903
1.30	1.341546	1.90	2.467134
1.40	1.484922	2.00	2.723868
1.50	1.644643		

IMPROVED EULER METHOD

The above discussions clearly indicate that Euler's method is not very practical for realistic computation. One must take a large number of steps to achieve even modest accuracy for most problems. How then can we generate a better approximation? One approach is the following. First note that in going from t_0 to t_1, Euler's method uses only information about the slope of the solution at t_0, namely $f(t_0, y_0)$. The fact that $f(t, y)$ may change in the interval is ignored. A more accurate slope value to use might be the average of the slopes at (t_0, y_0) and (t_1, y_1). However, $f(t_1, y_1)$, the slope at (t_1, y_1), cannot be calculated until we know y_1. If this value is approximated by Euler's method, we obtain the following approximation to $y(t_1)$:

$$y_1 = y_0 + \frac{h(s_1 + s_2)}{2},$$

where

$$s_1 = f(t_0, y_0),$$

$$s_2 = f(t_0 + h, y_0 + hs_1).$$

Note that s_2 is the value of f at t_1 and at a value of y given by Euler's method. This idea easily extends to the other line segments and leads to the following algorithm.

The improved Euler algorithm

An approximate solution to the initial value problem

$$\frac{dy}{dt} = f(t, y), \qquad y(t_0) = y_0$$

at the equally spaced points $t_0, t_i, \ldots, t_n$ is given by

$$s_1 = f(t_i, y_i),$$

$$s_2 = f(t_i + h, y_i + hs_1),$$

$$y_{i+1} = y_i + h(s_1 + s_2)/2,$$
$$t_{i+1} = t_0 + (i + 1)h, \; i = 0, \ldots, n - 1,$$

where n, t_0, and y_0 are given, and $h = (t_n - t_0)/n$.

EXAMPLE 2 Find an approximate solution to the initial value problem in Example 1 at the point $t = 1$ by using the improved Euler algorithm with $h = \frac{1}{4}$.

SOLUTION Starting with $t_0 = 0$ and $y_0 = 1$, we compute

$$s_1 = y_0 = 1.000$$

$$s_2 = y_0 + \frac{1}{4} s_1 = 1.250$$

$$y_1 = y_0 + \frac{1}{8}(s_1 + s_2) = 1.281$$

$$t_1 = t_0 + \frac{1}{4} = 0.250$$

$$s_1 = y_1 = 1.281$$

$$s_2 = y_1 + \frac{1}{4} s_1 = 1.601$$

$$y_2 = y_1 + \frac{1}{8}(s_1 + s_2) = 1.641$$

$$t_2 = t_0 + 2 \cdot \frac{1}{4} = 0.500$$

$$s_1 = y_2 = 1.641$$

$$s_2 = y_2 + \frac{1}{4} s_1 = 2.051$$

$$y_3 = y_2 + \frac{1}{8}(s_1 + s_2) = 2.103$$

$$t_3 = t_0 + 3 \cdot \frac{1}{4} = 0.750$$

$$s_1 = y_3 = 2.103$$

$$s_2 = y_3 + \frac{1}{4} s_1 = 2.629$$

$$y_4 = y_3 + \frac{1}{8}(s_1 + s_2) = 2.695$$

$$t_4 = t_0 + 4 \cdot \frac{1}{4} = 1.000$$

In this case the absolute error to three decimal places is $|y_4 - y(1)| = 0.023$, which is approximately 10% of the error produced by using the Euler algorithm. ■

A Fortran program implementing the improved Euler algorithm is given in Fig. 1.21. Again, it is set up to solve the Eq. (1.7.3). Its output is presented in Table 1.4.

For the improved Euler algorithm, $y_{10} \cong 2.828$, which differs from the true solution by 0.0008. This is a considerable improvement over the Euler prediction, $y_{10} \cong 2.724$.

```
C PROGRAM USES THE IMPROVED EULER METHOD TO SOLVE
C THE INITIAL VALUE PROBLEM
C
C     DY/DT = F(T,Y), Y(T0) = Y0
C
C DEFINE FUNCTION F(T,Y) AND INITIAL CONDITIONS.
C
      F(T,Y)=T*Y**(1.0/3.0)
      T0=1.0
      Y0=1.0
C
C SET FINAL T AND NUMBER OF STEPS N TO BE TAKEN.
C
      TF=2.0
      N=10
C
C IMPROVED EULER LOOP
C
      T=T0
      Y=Y0
      H=(TF-T0)/FLOAT(N)
      WRITE(6,2)T,Y
    2 FORMAT(1X,F6.2,F10.5)
      DO 1 I=1,N
         S1=F(T,Y)
         S2=F(T+H,Y+H*S1)
         Y=Y+H*(S1+S2)/2.0
         T=T+H
         WRITE(6,2)T,Y
    1 CONTINUE
      STOP
      END
```

Figure 1.21 Improved Euler program to solve $dy/dt = ty^{1/3}$, $y(1) = 1$.

Table 1.4 The improved Euler method approximation for $dy/dt = ty^{1/3}, y(1) = 1, h = 0.1$

t	y	t	y
1.00	1.000000	1.60	1.873597
1.10	1.106775	1.70	2.080564
1.20	1.227788	1.80	2.307834
1.30	1.363985	1.90	2.556483
1.40	1.516345	2.00	2.827600
1.50	1.685873		

EXAMPLE 3 Approximate the solution to the initial value problem

$$\frac{dy}{dt} = 1 + y^2, \qquad y(0) = 0,$$

at the point $t = 1$ by using the improved Euler method with $h = 0.1$, 0.01, and 0.001. The actual solution to this problem is $y(t) = \tan t$.

SOLUTION Using the improved Euler program with

```
F(T,Y) = 1.0 + Y**2
      T0 = 0.0
      Y0 = 0.0
      TF = 1.0
N = 10, N = 100, and N = 1000
```

we obtain Table 1.5. These data suggest that the error in the improved Euler algorithm might be proportional to h^2, rather than to h, as in the Euler algorithm. For most problems, this is the case. ■

Table 1.5 The improved Euler method approximation for $dy/dt = 1 + y^2, y(0) = 0, h = 0.1, 0.01, 0.001$

n	h	y_n	$\lvert y_n - \tan(1)\rvert$
10	10^{-1}	1.553790	$0.4 \cdot 10^{-2}$
100	10^{-2}	1.557369	$0.4 \cdot 10^{-4}$
1000	10^{-3}	1.557407	$0.4 \cdot 10^{-6}$

EXERCISES 1.7

In Exercises 1–6, perform the indicated computations by hand, retaining only four significant digits at each step of the calculations.

1. Consider the initial value problem

$$\frac{dy}{dt} = -ty^2, \quad y(0) = 2.$$

a) Use the Euler algorithm to compute an approximate solution at $t = 1.0$. Use the values $n = 2, 4, 8$, which correspond to step sizes of $h = \frac{1}{2}, \frac{1}{4}, \frac{1}{8}$.
b) Repeat part (a) using the improved Euler algorithm.
c) Determine the actual solution and compare the value of $y(1)$ with your results in (a) and (b).

Repeat Exercise 1 for the initial value problems

2. $\dfrac{dy}{dt} + 2y = t,\ y(0) = 1$

3. $\dfrac{dy}{dt} = t + 1,\ y(0) = 1$

4. $\dfrac{dy}{dt} = t^3e^{-2y},\ y(0) = 0$

5. $\dfrac{dy}{dt} = \dfrac{e^t}{y},\ y(0) = 1$

6. Use the improved Euler algorithm with $h = 1$ to obtain an approximate solution at $t = 1$ for Exercise 3 above with the initial condition $y(0) = -\frac{1}{4}$. Why is the answer so accurate?

7. Use the Euler and improved Euler programs to obtain an approximate solution at $t = 0.5$ in Exercises 1–5. What is the smallest value of n (or largest value of h) that will produce three significant digits of accuracy in each case?

8. Consider the initial value problem

$$\frac{dy}{dt} = y + t^2, \qquad y(0) = 1.$$

a) Find the solution $y(t)$ and evaluate it for $t = 0.2, 0.4, \ldots, 1.0$.
b) Using the improved Euler program with a step size of $h = 0.2$, find approximate values for the solution at the t values in part (a).
c) Repeat part (b) by using $h = 0.1$.
d) Compare the results of part (c) with those of part (b) and the exact values. The differences in the results for $h = 0.2$ and $h = 0.1$ tend to indicate whether a smaller step size must be used for this range of t values. (A good rule of thumb is to use the solution corresponding to the smaller step size if the two solutions agree to the required accuracy for all t values of interest. If they do not agree, reduce h and repeat. This

procedure gives an *indication* but not a proof of the accuracy of the result.)

9. Modify the programs in the text so that the values of t and y are printed out at every pth step, where p is to be specified. (If many integration steps are used, one normally does not want to print out t and y at every step.)

10. Consider the problem

$$\frac{dy}{dt} = y^2, \qquad y(0) = 1.$$

a) Find the solution $y(t)$ by using the methods in Section 1.5.

b) Use the Euler algorithm to approximate the solution at $t = 0.5$ by using h values of 0.05 and 0.025. Compare with the actual solution.

c) Reduce h until you have achieved three significant figures of accuracy.

11. Repeat Exercise 10 using the improved Euler method. What is the largest value of h that will produce three significant figures of accuracy?

The initial value problems of Exercises 12–15 cannot be solved by analytical methods. Use the improved Euler program to approximate the solution at $t = 1$ to three significant figures (see Exercise 8.d).

12. $\dfrac{dy}{dt} = y^2 - t^2,\ y(0) = \dfrac{1}{2}$

13. $\dfrac{dy}{dt} = y^{3/2} + t,\ y(0) = 1$

14. $\dfrac{dy}{dt} = \sin y + e^t,\ y(0) = 0$

15. $\dfrac{dy}{dt} = y(2 - y) - t,\ y(0) = \dfrac{3}{2}$

16. Dawson's integral

$$y(t) = e^{-t^2} \int_0^t e^{x^2}\, dx$$

is the solution to the initial value problem

$$\frac{dy}{dt} = -2ty + 1,\ y(0) = 0.$$

Approximate $y(0.5)$ to four significant figures by using the improved Euler program. The actual result (correct to five decimal places) is $y(0.5) = 0.42444$. Compare your results with those obtained in Exercise 35 of Section 1.4, if you worked it.

Apply the improved Euler program to each of the initial value problems of Exercises 17–25 to approximate the solution at the indicated value of t correct to three significant figures. Solve the differential equation analytically and compare with your numerical solution.

17. $\dfrac{dy}{dt} = \dfrac{1}{1 + t^2},\ y(0) = 0;\ t = 1$

18. $\dfrac{dy}{dt} = 2y,\ y(0) = 1;\ t = 0.5$

19. $\dfrac{dy}{dt} = \dfrac{2y}{t}$, $y(1) = 1$; $t = 2$

20. $\dfrac{dy}{dt} = y - t$, $y(0) = 2$; $t = 1$

21. $\dfrac{dy}{dt} = t^3e^{-y}$, $y(1) = 0$; $t = 2$

22. $\dfrac{dy}{dt} = y - e^{-t}$, $y(0) = 1$; $t = 1$

23. $\dfrac{dy}{dt} = e^{y+t}$, $y(0) = 1$; $t = 0.25$

24. $\dfrac{dy}{dt} = \dfrac{2t}{y + t^2y}$, $y(0) = -2$; $t = 1$

25. $\dfrac{dy}{dt} = 4y + 1 - t$, $y(0) = 1$; $t = 0.5$

26. Approximate the solution to the initial value problem

$$\frac{dy}{dt} = 1 + y^2, \qquad y(0) = 0$$

at $t = 0.1, 0.2, \ldots, 1.0$ by using the improved Euler program with $h = 0.1$. Plot the difference between your numerical solution and the actual solution $y(t) = \tan t$. What do you think would happen if you tried to approximate the solution near $t = \pi/2$?

27. Verify that the error in the Euler algorithm appears to be proportional to h and the error in the improved Euler algorithm proportional to h^2 by using the initial value problem

$$\frac{dy}{dt} = -y, \qquad y(0) = 1$$

and examining the numerical solution at $t = 1$. Use the h values: $\frac{1}{2}, \frac{1}{4}, \frac{1}{8}, \frac{1}{16}, \frac{1}{32}$, and $\frac{1}{64}$.

28. Estimate the time required to empty a conical funnel 12 inches high with a 3-inch base diameter if the apex hole has a diameter of $\frac{1}{2}$ inch. Use the improved Euler program with TF = 5.0, modified with a command to STOP if the height is not positive. (See Exercise 27 of Section 1.5.)

29. Extrapolation methods such as the one employed in Section 1.4 can be used with both the Euler and improved in Euler methods. Let $E(h)$ and $IE(h)$ denote the approximations to $y(T)$ produced by the Euler and improved Euler methods. Assume that

$$y(T) - E(h) \cong k_1h,$$

$$y(T) - IE(h) \cong k_2h^2,$$

and derive the extrapolation formulas

$$y(T) \cong E(h) + [E(h) - E(2h)], \qquad \textbf{(1.7.4)}$$

$$y(T) \cong IE(h) + \tfrac{1}{3}\,[IE(h) - IE(2h)]. \qquad \textbf{(1.7.5)}$$

30. Use The Euler extrapolation formula (1.7.4) of Exercise 29 in Exercise 10(b) and compare the result with the exact solution.

31. Use the improved Euler extrapolation formula (1.7.5) in Exercise 11(b) and compare the result with the exact solution.

1.8 IMPLICIT SOLUTIONS OF FIRST ORDER EQUATIONS

EXACT DIFFERENTIAL EQUATIONS

There is an important class of differential equations called *exact differential equations,* which have *implicit solutions.* The separable equations that we studied in the preceding section are of that type but so are many others. In this section we shall discuss a criterion for a differential equation to be exact and a method for constructing its implicit solution.

It is sometimes convenient to represent a first order differential equation of the form

$$\frac{dy}{dt} = -\frac{M(t, y)}{N(t, y)}$$

in differential notation as

$$M(t, y)\, dt + N(t, y)\, dy = 0$$

with the understanding that t remains the independent variable and y the dependent variable. For instance, the differential equation

$$\frac{dy}{dt} = \frac{e^y}{t}, \qquad t > 0,$$

can be written

$$e^y\, dt - t dy = 0, \qquad t > 0.$$

Our goal in this section is to construct implicit solutions for first order differential equations that are not necessarily separable. We shall start with an equation,

$$F(t, y) = C, \tag{1.8.1}$$

construct a differential equation from it, derive a constraint that must be satisfied by the differential equation, and then use the constraint as part of a procedure for constructing implicit solutions.

Let us assume that $F(t, y)$ is a differentiable function and that there is a constant C and a differentiable function $y(t)$ such that $F(t, y(t)) \equiv C$ on some open

t-interval. After differentiating this equation with respect to t we obtain

$$\frac{dF}{dt}(t, y(t)) \equiv F_t(t, y(t)) + F_y(t, y(t))\frac{dy}{dt}(t) \equiv 0. \tag{1.8.2}$$

Thus if $F_y(t, y) \neq 0$, we have a first order differential equation

$$\frac{dy}{dt} = -\frac{F_t(t, y)}{F_y(t, y)} \tag{1.8.3}$$

that can be written

$$dF = F_t(t, y)\,dt + F_y(t, y)\,dy = 0.$$

By construction, it has an implicit solution $F(t, y) = C$.

The graphs in the ty-plane determined by equation (1.8.1) are called the level curves of the function $F(t, y)$. The vector with components $(F_t(t, y), F_y(t, y))$ is normal to the level curve at (t, y), and (1.8.2) shows that the vector with components $(1, dy/dt)$ is tangent to the curve. It follows that the slopes of the tangent line to the level curves are determined by the differential equation (1.8.3) (see Fig. 1.22).

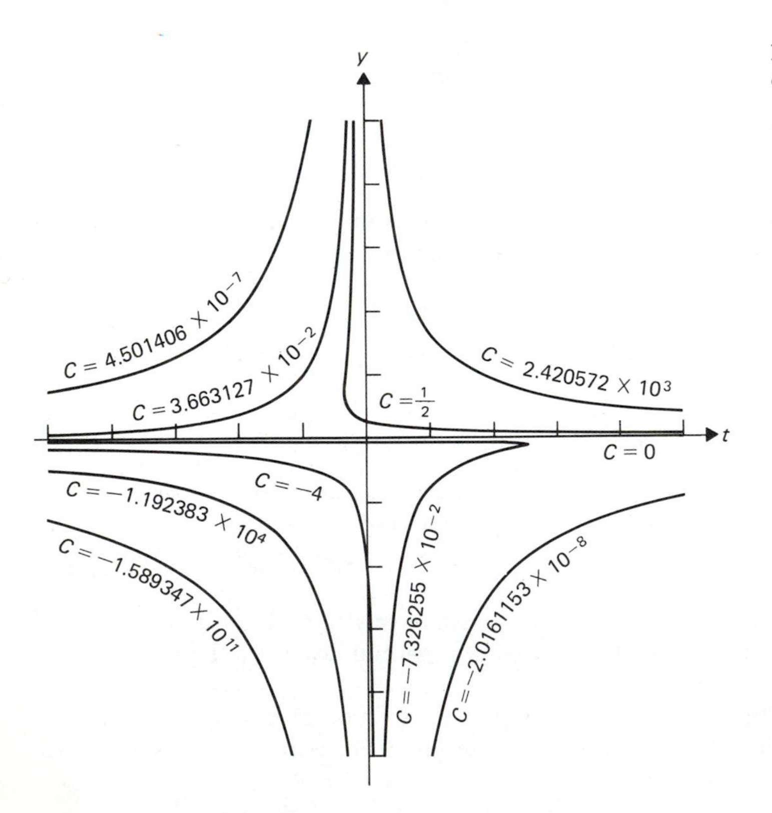

Figure 1.22 Several level curves of $ye^{ty} = C$.

EXAMPLE 1 Derive a first order differential equation from $ye^{ty} = C$ and sketch several level curves.

SOLUTION Set $F(t, y) = ye^{ty}$. Then $F_t(t, y) = y^2e^{ty}$, $F_y(t, y) = (1 + ty)e^{ty}$, so

$$dF = y^2e^{ty}\,dt + (1 + ty)e^{ty}\,dy = 0$$

and

$$\frac{dy}{dt} = -\frac{y^2}{1 + ty}.$$

Sketches of several level curves are in Fig. 1.22. ■

We note that this differential equation has an implicit solution but is not separable. It is an example of an important class of equations, *exact differential equations.*

Definition

A differential equation

$$M(t, y)\,dt + N(t, y)\,dy = 0 \tag{1.8.4}$$

is said to be *exact* if there is a differentiable function $F(t, y)$ such that

$$dF = M(t, y)\,dt + N(t, y)\,dy. \tag{1.8.5}$$

Hence,

$$F_t(t, y) = M(t, y), \qquad F_y(t, y) = N(t, y). \tag{1.8.6}$$

The level curves of $F(t, y)$ are called *integral curves* and

$$F(t, y) = \text{constant} \tag{1.8.7}$$

is an *implicit solution* or *integral* of the differential equation (1.8.4).

Some differential equations are so simple that exactness can be determined by inspection; e.g., the left-hand side of

$$\cos y\,dt - t\sin y\,dy = 0$$

is the differential of $F(t, y) = t\cos y$ and therefore $t\cos y = C$ is an implicit solution. For more complicated equations we need a test for exactness and a method for constructing an implicit solution. If the equation (1.8.4) is exact and if (1.8.5) is satisfied and all of the second derivatives of $F(t, y)$ are continuous, then it is known from calculus that the mixed second partial derivatives of $F(t, y)$ are equal:

$$\frac{\partial^2 F}{\partial y\,\partial t}(t, y) = \frac{\partial^2 F}{\partial t\,\partial y}(t, y)$$

or, from (1.8.6),

$$\frac{\partial M}{\partial y}(t, y) = \frac{\partial N}{\partial t}(t, y).$$

Clearly, the converse should hold, and it does, if the region R is nice, say, a rectangle,[5] and the coefficient functions are continuously differentiable in R. This is the content of the following theorem.

Theorem

Let $M(t, y)$ and $N(t, y)$ be continuously differentiable in a rectangle R. The differential equation

$$M(t, y)\, dt + N(t, y)\, dy = 0$$

is exact if and only if

$$\frac{\partial M}{\partial y}(t, y) = \frac{\partial N}{\partial t}(t, y) \text{ in } R. \tag{1.8.8}$$

Proof. If the equation is exact, then $\partial M/\partial y = \partial N/\partial t$ follows from the equality of the mixed partial derivatives of smooth functions.

If $\partial M/\partial y = \partial N/\partial t$, we shall show that the differential equation is exact by constructing a suitable function $F(t, y)$. To do this, pick a convenient value of the independent variable $t = a$, and integrate $M(t, y)$ with respect to t while holding y fixed:

$$F(t, y) = \int_a^t M(s, y)\, ds + W(y), \tag{1.8.9}$$

where $W(y)$ is an arbitrary differentiable function of y corresponding to an integration constant, and (s, y) is in R for $a \leq s \leq t$. By construction,

$$\frac{\partial}{\partial y}\int_a^t M(s, y)\, ds + W'(y) = N(t, y)$$

or

$$W'(y) = N(t, y) - \frac{\partial}{\partial y}\int_a^t M(s, y)\, ds. \tag{1.8.10}$$

5. A precise statement would be that R is a *simply connected region,* of which rectangles or discs are examples. For the definition of this concept the interested reader should consult an advanced calculus text.

But is the right side of (1.8.10) a function of y alone? It is if its derivative with respect to t is zero. Let us check:

$$\begin{aligned}\frac{\partial}{\partial t}\left[N(t, y) - \frac{\partial}{\partial y}\int_a^t M(s, y)\, ds\right] &= \frac{\partial N}{\partial t}(t, y) - \frac{\partial^2}{\partial t\, \partial y}\int_a^t M(s, y)\, ds \\ &= \frac{\partial N}{\partial t}(t, y) - \frac{\partial}{\partial y}\left[\frac{\partial}{\partial t}\int_a^t M(s, y)\, ds\right] \\ &= \frac{\partial N}{\partial t}(t, y) - \frac{\partial}{\partial y}[M(t, y)] = 0,\end{aligned}$$

since we have assumed that $\partial M/\partial y = \partial N/\partial t$ in R. The interchanging of the order of differentiation is permitted because of the assumed smoothness of $M(t, y)$.

Now set

$$W(y) = \int_b^y N(t, z)\, dz - \int_b^y \frac{\partial}{\partial z}\int_a^t M(s, z)\, ds\, dz, \tag{1.8.11}$$

where (s, z) is in R for $b \leq z \leq y$, $a \leq s \leq t$. By virtue of the relation $\partial M/\partial y = \partial N/\partial t$, the differential of the function $F(t, y)$ defined by (1.8.9), where $W(y)$ is given by (1.8.11), equals $M(t, y)\, dt + N(t, y)\, dy$. Therefore $F(t, y) = C$ is an implicit solution of the differential equation, which completes the proof of the theorem. □

An advantage of the proof just given is that it is constructive inasmuch as it gives us a procedure for finding a solution. This is illustrated in the following example.

EXAMPLE 2 Construct an implicit solution $F(t, y) = C$, to

$$(6ty + 4)\, dt + (3t^2 + 10y)\, dy = 0,$$

and sketch several of its level curves.

SOLUTION Let $M(t, y) = 6ty + 4$, $N(t, y) = 3t^2 + 10y$. Since

$$\frac{\partial M}{\partial y}(t, y) = 6t = \frac{\partial N}{\partial t}(t, y),$$

the condition (1.8.8) for exactness is satisfied. This result, together with the smoothness of the coefficient functions $M(t, y)$ and $N(t, y)$ insures that there exists an implicit solution. To construct it, set

$$F(t, y) = \int M(t, y)\, dt + W(y) = 3t^2y + 4t + W(y).$$

Note that when we find the antiderivative $F(t, y)$ of $M(t, y)$ with respect to t for fixed y, the arbitrary constant of integration becomes an arbitrary differentiable function of y. The correctness of this can be seen by differentiating

$$\frac{\partial F}{\partial t}(t, y) = \frac{\partial}{\partial t}(3t^2y + 4t + W(y)) = 6ty + 4 = M(t, y).$$

From the fact that

$$N(t, y) = 3t^2 + 10y = \frac{\partial F}{\partial y}(t, y) = 3t^2 + W'(y),$$

we have

$$W'(y) = 10y \quad \text{or} \quad W(y) = 5y^2.$$

Hence,

$$F(t, y) = 3t^2y + 4t + 5y^2,$$

and

$$3t^2y + 4t + 5y^2 = C$$

is an implicit solution to the differential equation. Several of its level curves are sketched in Fig. 1.23.

We could have started instead with $N(t, y)$ and constructed an antiderivative with an additive function $R(t)$. To do this, set

$$F(t, y) = \int N(t, y)\, dy + R(t) = 3t^2y + 5y^2 + R(t).$$

Note that in this case the arbitrary constant of integration becomes an arbitrary differentiable function of t since we are antidifferentiating $N(t, y)$ with respect to y for fixed t. From

$$M(t, y) = 6ty + 4 = \frac{\partial F}{\partial t}(t, y) = 6ty + R'(t),$$

we have

$$R'(t) = 4 \quad \text{and} \quad R(t) = 4t.$$

Hence,

$$F(t, y) = 3t^2y + 5y^2 + 4t,$$

as before. Note that $W(y)$ and $R(t)$ are determined up to an additive constant that is immaterial in the construction of an implicit solution $F(t, y) = \text{constant}$. ■

INTEGRATING FACTORS

It is sometimes possible to construct an implicit solution to a nonexact differential equation by multiplying it by an *integrating factor* to convert it into an exact differential equation. The finding of an integrating factor usually rests on a clever obser-

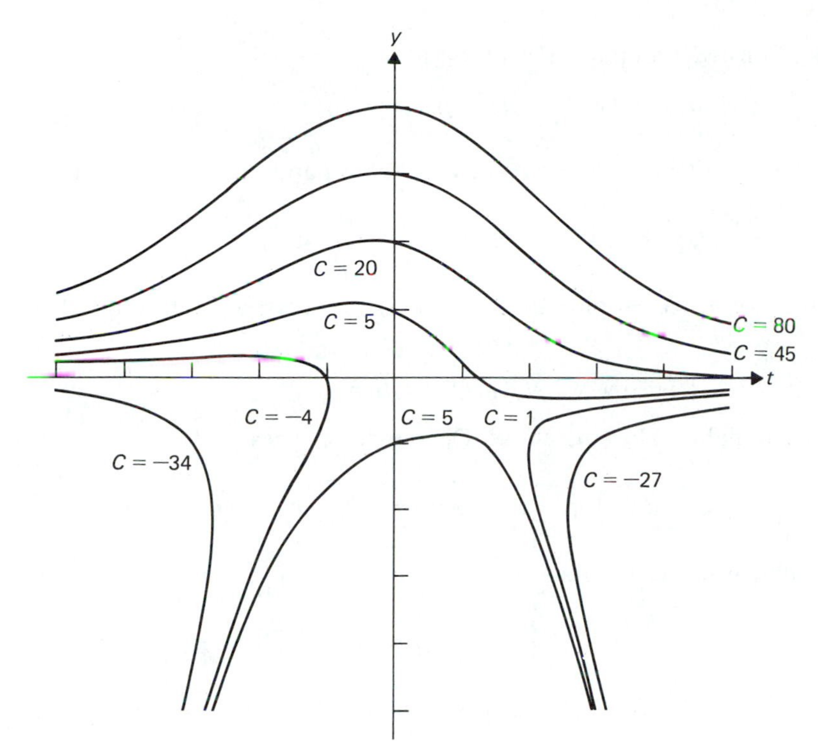

Figure 1.23 Several level curves of $3t^2y + 4t + 5y^2 = C$.

vation that an expression, such as $t\,dy - y\,dt$, reminds one of the differential

$$d\left(\frac{y}{t}\right) = \frac{t\,dy - y\,dt}{t^2}.$$

On the other hand, it is also true that

$$d\left(-\frac{t}{y}\right) = \frac{t\,dy - y\,dt}{y^2}$$

and

$$d\left(\ln\frac{y}{t}\right) = \frac{t\,dy - y\,dt}{ty}.$$

This suggests that $1/t^2$, $1/y^2$, or $1/ty$ are all possible integrating factors. Which, if any, will lead to an exact differential equation must be determined by further analysis.

This is illustrated in the following example.

EXAMPLE 3 Construct an implicit solution to

$$(ty - 1)y\,dt + (ty + 1)t\,dy = 0.$$

SOLUTION Here, $M(t, y) = ty^2 - y$, $N(t, y) = t^2y + t$ and

$$\frac{\partial M}{\partial y}(t, y) = 2ty - 1 \neq 2ty + 1 = \frac{\partial N}{\partial t}(t, y),$$

so the equation is not exact. Hoping for inspiration, we rearrange the terms. The arrangement

$$t\,dy - y\,dt + ty^2\,dt + t^2y\,dy = 0$$

suggests that $d(y/t)$ might be lurking about. Dividing by t^2 gives

$$d\left(\frac{y}{t}\right) + \frac{y^2}{t}\,dt + y\,dy = 0,$$

which unfortunately is not exact since

$$\frac{\partial}{\partial y}\left(\frac{y^2}{t}\right) = \frac{2y}{t} \neq \frac{\partial}{\partial t}(y) = 0.$$

Dividing by y^2 gives the same result, but after dividing the equation by ty, we obtain

$$\frac{t\,dy - y\,dt}{ty} + (y\,dt + t\,dy) = 0.$$

The second term is the differential of ty; therefore

$$\ln\frac{y}{t} + ty = C$$

is an implicit solution in any region where $0 < y/t < \infty$.

This example supports the widely held opinion that the finding of integrating factors is an arcane art that may lead to madness. ■

HOMOGENEOUS DIFFERENTIAL EQUATIONS

It is also possible to transform certain equations that are not exact into exact equations by a clever substitution. We shall illustrate the technique in the following example.

EXAMPLE 4 Show that the differential equation $(t - y)\,dt + 2t\,dy = 0$ is not exact. Investigate the effect of the substitution, $y = tu$. Find an implicit solution whose integral curve contains the point $t = 4$, $y = 4$. Sketch several of its integral curves.

SOLUTION Since $M(t, y) = t - y$, $N(t, y) = 2t$, then

$$\frac{\partial M}{\partial y} = -1 \neq 2 = \frac{\partial N}{\partial t},$$

so that the equation is not exact. Next, set $y = tu$, then $dy = u\,dt + t\,du$. The differential equation becomes

$$(t - tu)\,dt + 2t(u\,dt + t\,du) = 0,$$

or if $t \neq 0$,

$$(1 + u)\,dt + 2t\,du = 0.$$

This separable equation has an implicit solution,

$$\tfrac{1}{2}\ln|t| + \ln|1 + u| = C$$

or

$$\ln|t^{1/2}(1 + u)| = C.$$

Now make the substitution $u = y/t$ to obtain

$$\ln\left|t^{1/2}\left(1 + \frac{y}{t}\right)\right| = C.$$

If $t = 4$ and $y = 4$, then $C = \ln 4$, and

$$\ln\left|t^{1/2}\left(1 + \frac{y}{t}\right)\right| = \ln 4$$

is an implicit solution to the original equation whose integral curve contains the point $t = 4$, $y = 4$. Sketches of several of the level curves including the one through (4, 4) are in Fig. 1.24. ■

There is an important class of differential equations that are not exact but can be transformed into separable (and therefore exact) equations by the substitution $y = tu$ that was used in Example 4. These equations are called *homogeneous equations.*

Definition

A differential equation

$$M(t, y)\,dt + N(t, y)\,dy = 0$$

is said to be *homogeneous* if

$$M(\alpha t, \alpha y) = \alpha^{\gamma} M(t, y), \qquad N(\alpha t, \alpha y) = \alpha^{\gamma} N(t, y)$$

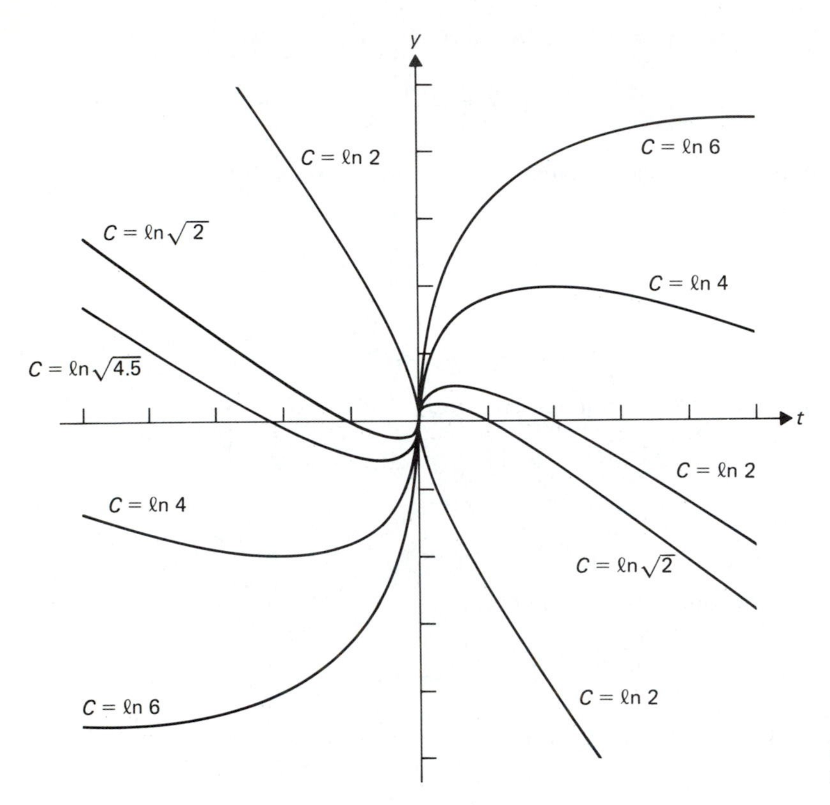

Figure 1.24 Several level curves of $\ln|t^{1/2}(1 + y/t)| = C$.

for all suitably restricted t, y, and α. The coefficient functions $M(t, y)$ and $N(t, y)$ are said to be *homogeneous of degree* γ. A differential equation

$$\frac{dy}{dt} = f(t, y)$$

is said to be homogeneous if $f(\alpha t, \alpha y) = f(t, y)$, that is, if the function $f(t, y)$ is homogeneous of degree zero. The two definitions are consistent for if $M(t, y)$ and $N(t, y)$ are homogeneous of the same degree then $f(t, y) = -M(t, y)/N(t, y)$ is homogeneous of degree zero.

EXAMPLE 5 Show that the homogeneous differential equation

$$2y^3 dt + (2t^3 - 3ty^2)\, dy = 0$$

is not exact but that an implicit solution can be constructed with the aid of the substitution $y = tu$.

SOLUTION The equation is not exact since

$$\frac{\partial}{\partial y} M(t, y) = 6y^2 \neq 6t^2 - 3y^2 = \frac{\partial}{\partial t} N(t, y).$$

As in Example 4, we set $y = tu$, $dy = u\,dt + t\,du$, and the differential equation becomes

$$2t^3u^3\,dt + (2t^3 - 3t^3u^2)(u\,dt + t\,du) = 0,$$

or if $t \neq 0$,

$$(2u - u^3)\,dt + (2 - 3u^2)t\,du = 0.$$

This separable equation has an implicit solution,

$$\ln |t(2u - u^3)| = C.$$

Finally, we make the substitution $u = y/t$ to obtain an implicit solution of the original differential equation,

$$\ln \left| 2y - \frac{y^3}{t^2} \right| = C.$$

Sketches of several of the level curves are in Fig. 1.25. ■

It is sometimes possible to make a preliminary transformation to produce a homogeneous differential equation that is then transformed into a separable differential equation. This is illustrated by a classic pursuit problem. The version given here is based on the one in [5].

EXAMPLE 6 A rabbit starts at the origin and runs up the y-axis with speed a. At the same time a dog starts at the point $(c, 0)$ and pursues the rabbit. The dog runs with speed b and always points toward the rabbit. What is the path of the dog? If b is greater than a, when will the dog catch the rabbit?

SOLUTION As shown in Fig. 1.26, the tangent line to the path of the dog intersects the y-axis at $(0, at)$, the location of the rabbit. If θ denotes the acute angle between the tangent line and the x-axis, then the components of the dog's velocity are

$$\frac{dx}{dt} = -b \cos\theta = -b\frac{x}{r}, \qquad \frac{dy}{dt} = b \sin\theta = \frac{b(at - y)}{r},$$

where $r^2 = x^2 + (at - y)^2$. The slope of the tangent line is

$$\frac{dy}{dx} = \frac{b(at - y)}{-bx} = \frac{y - at}{x}.$$

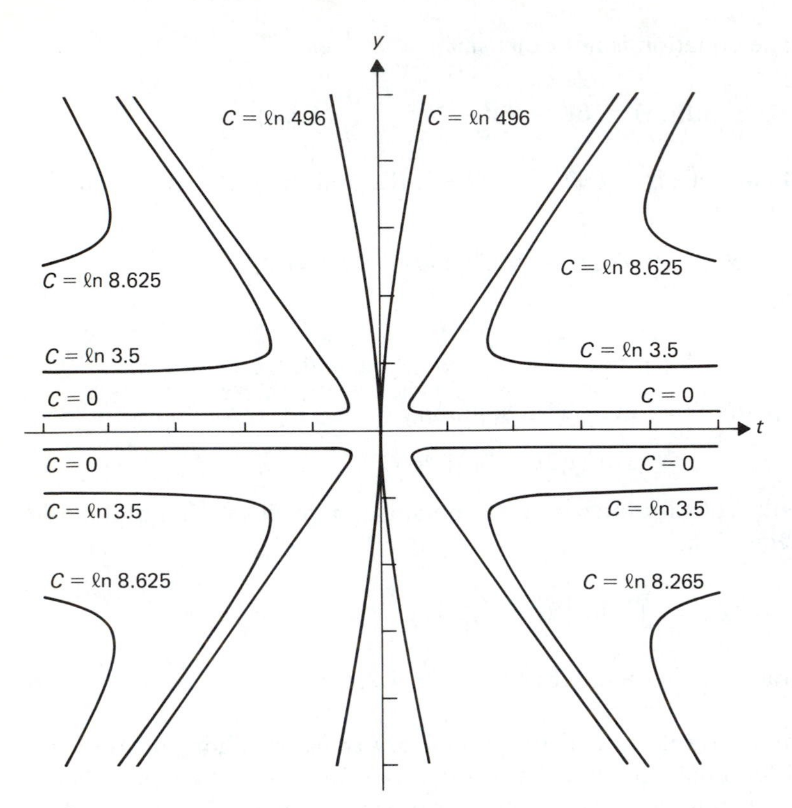

Figure 1.25 Several level curves of $\ln|2y - y^3/t^2| = C$.

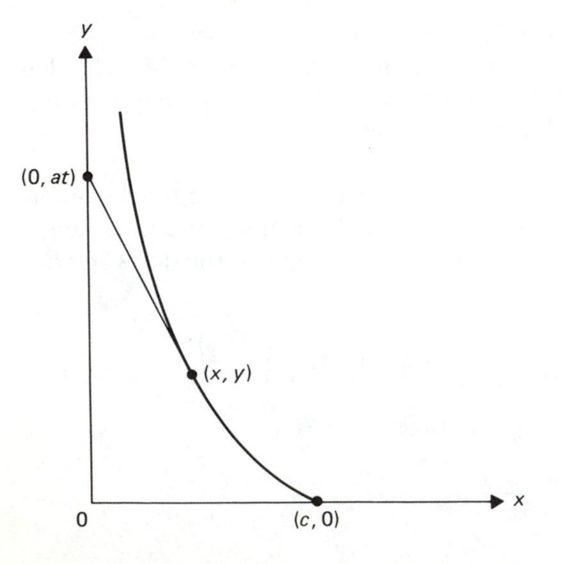

Figure 1.26 A pursuit problem.

A transformation that will result in a homogeneous equation is $z = y - at$. Then

$$\frac{dx}{dt} = -b\frac{x}{r}, \qquad \frac{dz}{dt} = \frac{dy}{dt} - a = -\frac{bz}{r} - a,$$

with $r^2 = x^2 + z^2$. The differential equation

$$\frac{dz}{dx} = \frac{-b(z/r) - a}{-b(x/r)} = \frac{bz + ar}{bx}$$

is homogeneous, and an implicit solution can be constructed by setting $z = xu$. We have

$$\frac{dz}{dx} = x\frac{du}{dx} + u = \frac{bxu + ax(1 + u^2)^{1/2}}{bx}$$

or

$$\frac{du}{(1 + u^2)^{1/2}} = k\frac{dx}{x},$$

where $k = a/b$. From a table of integrals, we can find that an antiderivative of $(1 + u^2)^{-1/2}$ is $\ln(u + \sqrt{u^2 + 1})$. Therefore

$$\ln(u + \sqrt{u^2 + 1}) = \ln x^k + C.$$

At $t = 0$, $x = c$, $y = 0$, so $z = 0$ and $u = 0$. Therefore $C = -\ln c^k$, so

$$\ln(u + \sqrt{u^2 + 1}) = \ln\left(\frac{x}{c}\right)^k,$$

$$u + \sqrt{u^2 + 1} = \left(\frac{x}{c}\right)^k.$$

To solve for u, we write

$$\sqrt{u^2 + 1} = \left(\frac{x}{c}\right)^k - u,$$

square both sides of the equation, and obtain

$$u^2 + 1 = \left(\frac{x}{c}\right)^{2k} - 2\left(\frac{x}{c}\right)^k u + u^2.$$

Thus

$$u = \frac{1}{2}\left[\left(\frac{x}{c}\right)^k - \left(\frac{c}{x}\right)^k\right].$$

The next step is to substitute $u = z/x$ and then $z = y - at$. We have

$$\frac{y - at}{x} = \frac{1}{2}\left[\left(\frac{x}{c}\right)^k - \left(\frac{c}{x}\right)^k\right].$$

But

$$\frac{dy}{dx} = \frac{y - at}{x},$$

so

$$\frac{dy}{dx} = \frac{1}{2}\left[\left(\frac{x}{c}\right)^k - \left(\frac{c}{x}\right)^k\right].$$

Recalling that at $t = 0$, $x = c$, and $y = 0$, we integrate and obtain the path of the dog:

1. If $a = b$, then $k = 1$, and

$$y = \frac{1}{2}\left[\frac{x^2}{2c} - c \ln x\right] + \frac{1}{2}\left[c \ln c - \frac{c}{2}\right].$$

2. If $a \neq b$, then $k \neq 1$, and

$$y = \frac{c}{2}\left[\frac{1}{1 + k}\left(\frac{x}{c}\right)^{k+1} - \frac{1}{1 - k}\left(\frac{x}{c}\right)^{1-k}\right] + \frac{ck}{1 - k^2}.$$

If b, the speed of the dog, is greater than a, the speed of the rabbit, then $0 < k < 1$ and we can let x tend to zero to obtain that the terminal position of the rabbit is $x = 0$, $y = ck/(1 - k^2)$. The time needed to reach this position is t_f, where

$$t_f = \frac{ck}{a(1 - k^2)} = \frac{cb}{b^2 - a^2}. \qquad \blacksquare$$

Implicit solutions can be found for separable equations, homogeneous equations, exact equations, and equations that can be converted to exact equations by multiplying by an integrating factor. The slope of a tangent line of a level curve of an implicit solution is equal to dy/dt, and it can be computed at a point (t, y) from the differential equation, $dy/dt = f(t, y)$. Therefore the integral curves of the differential equations are also level curves of the implicit solution. This was illustrated by a classical problem in differential equations, the pursuit of a rabbit by a faster dog.

EXERCISES 1.8

Show that the following differential equations are exact and find an implicit solution to each of them.

1. $(2ty + y^2)\,dt + (t^2 + 2ty)\,dy = 0$
2. $\left(\frac{1}{y} - \frac{y}{t^2}\right) dt + \left(\frac{1}{t} - \frac{t}{y^2}\right) dy = 0$
3. $(2ty - y \sin t - \sin t)\,dt + (t^2 + \cos t + \cos y)\,dy = 0$
4. $(\cosh t \cosh y + \sinh t)\,dt + (\sinh t \sinh y + \cosh y)\,dy = 0$
5. $(y + \sin y \sec^2 t + 2t)\,dt + (t + \cos y \tan t - 2y)\,dy = 0$
6. $[y \cos(t + y) - ty \sin(t + y)]\,dt + [t \cos(t + y) - ty \sin(t + y)]\,dy = 0$
7. $\left[\frac{y}{t} - 2t \ln|y| + \frac{1}{t}\right] dt + \left[\ln|t| - \frac{t^2}{y} + \frac{1}{y}\right] dy = 0$
8. $(ye^{ty} + 2ty + 2te^{t^2})\,dt + (te^{ty} + t^2 - 2ye^{y^2})\,dy = 0$

Implicit solutions to the pair of differential equations

$$M(x, y)dx + N(x, y)dy = 0,$$
$$N(x, y)dx - M(x, y)dy = 0$$

form an orthogonal family of curves since the product of their slopes is -1. For both equations to be exact,

$$\frac{\partial M}{\partial y}(x, y) = \frac{\partial N}{\partial x}(x, y)$$

and

$$\frac{\partial N}{\partial y}(x, y) = -\frac{\partial M}{\partial x}(x, y) .$$

These equations are called *Cauchy-Riemann equations;* they play a fundamental role in the theory of two-dimensional flow of incompressible irrotational fluids. Verify that the following pairs of equations are exact and find implicit solutions.

9. $e^x \cos y\,dx - e^x \sin y\,dy = 0,\ e^x \sin y\,dx + e^x \cos y\,dy = 0$
10. $(2x + 2)\,dx - 2y\,dy = 0,\ 2y\,dx + (2x + 2)\,dy = 0$
11. $\sin x \cosh y\,dx - \cos x \sinh y\,dy = 0,\ \cos x \sinh y\,dx + \sin x \cosh y\,dy = 0$
12. $(x^2 - y^2)\,dx - 2xy\,dy = 0,\ 2xy\,dx + (x^2 - y^2)\,dy = 0$

Multiply the following differential equations by the given integrating factor and show that you obtain an exact equation. Find an implicit solution for each of them.

13. $(2t^3y - y^3)\,dt - 2t^4\,dy = 0,\ Z(t, y) = t^{-2}y^{-3}$

14. $(2t^3y - y^3)\,dt + (ty^2 - t^4)\,dy = 0,\ Z(t, y) = t^{-2}y^{-2}$

15. $(2y^2 + ty)\,dt - (t^2 - 2ty)\,dy = 0,\ Z(t, y) = -t^{-1}y^{-2}$

16. $(3yt^2 - 2y^3)\,dt - (3t^3 - 2ty^2)\,dy = 0,\ Z(t, y) = y^{-2}(t^2 + y^2)^{-1}$

17. Which of the following functions are homogeneous?

a) $\dfrac{3t + 2y}{t - y}$ **b)** $\dfrac{y + t}{4t}$ **c)** $\dfrac{2t + y^2}{ty}$

d) $\dfrac{ty^2 + t^3}{t^3 + t^2y + y^3}$ **e)** $\dfrac{6t^{1/3}y^{2/3} + ty}{4t - y}$ **f)** $\dfrac{3 \ln 2^y}{t + \sqrt{ty}}$

A differential equation $dy/dt = f(t, y)$, where $f(t, y)$ is a homogeneous function of degree zero, is called a *homogeneous differential equation.* It can be solved by the substitution

$$y = tu, \qquad \frac{dy}{dt} = u + t\frac{du}{dt},$$

which transforms it to a separable equation. Use this technique to solve the following equations in Exercises 18–23.

18. $\dfrac{dy}{dt} = \dfrac{y^2 + ty + t^2}{t^2},\ y(1) = 0$

19. $\dfrac{dy}{dt} = \dfrac{2t + y}{y}$

20. $\dfrac{dy}{dt} = \dfrac{t^2 + y^2}{ty},\ y(1) = 4$

21. $\dfrac{dy}{dt} = \dfrac{2t - y}{t + 4y}$

22. $\dfrac{dy}{dt} = \dfrac{2t \sin(y/t) + 3y \cos(y/t)}{3t \cos(y/t)},\ y(8) = 4\pi$

23. $\dfrac{dy}{dt} = \dfrac{t + y}{t - y},\ y(1) = 0$

Equations of the form

$$\frac{dy}{dt} = \frac{At + By + C}{Dt + Ey + F}$$

can sometimes be reduced to homogeneous equations by translation of the ty coordinates. Show that if the pair of simultaneous equations

$$At + By + C = 0, \qquad Dt + Ey + F = 0$$

has a unique solution $t = h$, $y = k$, then the change of variables $t = x + h$, $y = w + k$ gives the transformed differential equation

$$\frac{dw}{dx} = \frac{Ax + Bw}{Dx + Ew},$$

which is homogeneous. Use this idea to solve the following equations in Exercises 24–26:

24. $\dfrac{dy}{dt} = \dfrac{3t + 4y - 5}{2t + 2}$ **25.** $\dfrac{dy}{dt} = \dfrac{t + 2y}{2t + 3y - 1}$ **26.** $\dfrac{dy}{dt} = \dfrac{5y - 2t - 8}{4y - t - 7}$

27. A man and his dog are separated by a river whose current has speed a. When the man calls to the dog, the dog dives into the river and swims toward the man with constant speed b. Let the y-axis and the line $x = c$ denote the banks of the river, and orient the y-axis so that the velocity of the current is positive. At $t = 0$, the man is at $(0, 0)$ and the dog is at $(c, 0)$. If b is greater than a, what will be the path of the dog and when will the dog reach the man?

28. If the speed of the current in Exercise 27 is equal to the speed of the dog, where will the dog land?

29. The man in Exercise 27 walks downstream with speed a while calling to his dog. What will be the path of the dog if $b > a$? $b = a$? $b < a$?

1.9 A MATHEMATICAL MODEL OF AN ELECTRIC-CIRCUIT PROBLEM

In the following sections we shall study some aspects of the process of describing a physical problem in mathematical terms. This process is called *mathematical modeling.* It is not a logical exercise, like that of solving a linear differential equation, nor is it an action requiring divine guidance. The process is a combination of physical insight, mathematical reasoning, and common sense. We shall first illustrate it with the problem of finding the flux variation in an iron-core induction coil.

The experimental equipment consists of the coil, a multitester (voltmeter and ohmmeter), a variable resistor, a battery, and a switch. The main properties of the iron-core induction coil used here are that the flux is a function of the current and that the voltage drop across the coil due to inductance is the time rate of change of the flux. Since the flux cannot be measured directly, we shall have to deduce it from measurements of the voltage drop across the coil. If the coil, variable resistor, and battery are connected in a loop, i.e., in a series circuit, then the current in each circuit element is the same and the total resistance of the circuit is the sum of the individual resistances of the coil and the resistor. By Ohm's law, the voltage drop across each circuit element is equal to its resistance times the current flow through it. Therefore the total voltage drop due to resistance is

$$V_{\text{res}} = R \cdot I,$$

where I denotes the current and R the total resistance of the circuit. The induction coil opposes any change in the current with the result that there is a voltage drop across it,

$$V_{\text{ind}} = \frac{d\phi}{dt},$$

where ϕ denotes the magnetic flux of the current. In accordance with Kirchhoff's law, the sum of the voltage drops across the coil and the resistor must equal the voltage of the battery:

$$V_{\text{ind}} + V_{\text{res}} = E$$

or

$$\frac{d\phi}{dt} + RI = E. \qquad \textbf{(1.9.1)}$$

We now have part of our mathematical model, an ordinary differential equation. The model is not complete since a relation between the magnetic flux ϕ and the current I is not yet formulated. With such a relation, the differential equation (1.9.1) would become a first order equation in one of the unknowns ϕ or I. An acceptable flux–current relation must be simple and stable. Without launching into an extensive study of the theory of electricity and magnetism, let us try to formulate an acceptable relation. It is known that if the current is zero, there is no flux, and if the current is small, the flux is small. A simple relation consistent with these observations is that of proportionality. Hence, suppose that $\phi = LI$, where L, the proportionality constant, is called the *inductance of the coil.* The variable resistor must be adjusted so that the current is small enough for this linear flux–current relation to be valid. The differential equation for the series circuit is then

$$L\frac{dI}{dt} + RI = E. \qquad \textbf{(1.9.2)}$$

The constants R and L are not known and must be estimated from experimental data. Figure 1.27 is a diagram of the simple RL circuit just described.

EXAMPLE 1 Use the integrating factor method to construct a general solution to the differential equation (1.9.2) and then find a solution that is zero at $t = t_0$.

SOLUTION We first divide by L to get

$$\frac{dI}{dt} + \frac{R}{L}I = \frac{E}{L}.$$

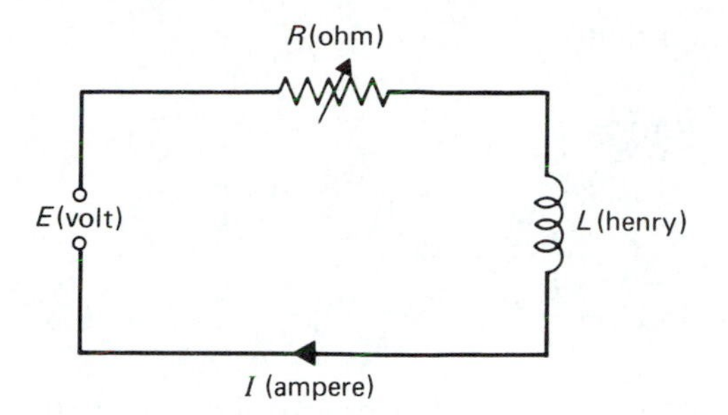

Figure 1.27 A simple *RL* circuit.

Then we choose $e^{Rt/L}$ as an integrating factor and obtain

$$\frac{d}{dt}\{e^{Rt/L}\, I(t)\} = \frac{E}{L}\, e^{Rt/L}.$$

We take antiderivatives and find that

$$e^{Rt/L}\, I(t) = \frac{E}{R}\, e^{Rt/L} + K,$$

where K is a constant of integration. Thus

$$I(t) = \frac{E}{R} + Ke^{-Rt/L}$$

is a general solution. If $I(t_0) = 0$, then

$$I(t) = \frac{E}{R}(1 - e^{-R(t-t_0)/L}).$$

■

The estimation of constants (also called *parameters*) is an essential aspect of mathematical modeling. If the parameters of a model cannot be estimated, the model is of little value. In the physical sciences, this restriction is so obvious, it requires little discussion. In the biological and social sciences, where there seems to be a great temptation to construct elaborate and untestable mathematical models, this restriction deserves strong emphasis.

Let us return to the study of the flux variation of the coil. The circuit elements are connected in series with the switch open, and the ohmmeter measures R, the resistance of the coil and variable resistor. Once this is done, the switch is closed ($t = 0$). Since $I(0) = 0$, the solution to the differential equation (1.9.2) is

$$I(t) = \frac{E}{R}(1 - e^{-Rt/L}) \quad \text{and} \quad \phi(t) = LI(t) = \frac{EL}{R}(1 - e^{-Rt/L}). \qquad \textbf{(1.9.3)}$$

It follows from (1.9.1) and the last relation that

$$\frac{d\phi}{dt}(t) = E - RI(t) = Ee^{-Rt/L}; \tag{1.9.4}$$

taking logarithms of both sides gives

$$\ln \frac{d\phi}{dt}(t) = \ln E - \frac{Rt}{L}. \tag{1.9.5}$$

This last equation shows that a plot of $\ln(d\phi/dt)$ against t gives a straight line with slope $-R/L$. Values of $(t, d\phi/dt)$ can be found by measuring the voltage drop across the coil with a voltmeter. Since R has been measured earlier, we can estimate the remaining circuit parameter L, the inductance, by estimating the slope of the line (1.9.5). Actual measurements will not lie exactly on the line because of experimental errors or slight deviations from our mathematical model. By plotting some values of $\ln(d\phi/dt)$ against t, the slope R/L can be obtained graphically, or its value can be estimated by linear regression. This last procedure consists of performing a least-squares fit of a straight line to the data and is explained below.

The (simulated) measurements of $d\phi/dt$ in Table 1.6 will be used to illustrate the method of estimating R/L. The data were generated by adding random numbers to values of $d\phi/dt$ computed from (1.9.4) with $R = 10$ and $L = 30$.

Linear regression is a computational procedure for finding the equation of a line that best describes a given set of data by using the *principle of least squares.* To apply the idea here, let $y = \ln(d\phi/dt)$ and let the line be represented by the equation

$$y = \alpha + \beta t.$$

The data $\{(t_i, y_i), i = 1, 2, \ldots, 10\}$ from Table 1.6 do not lie on any line (see Fig. 1.28), so we try to choose a line in such a way that the sum of the squares of the deviations $[y_j - (\alpha + \beta t_j)], j = 1, \ldots, 10$, of the data from the line is as small as

Table 1.6 **Measurements of the rate of change of the flux, $d\phi/dt$**

t	$d\phi(t)/dt$	$\ln d\phi(t)/dt$	t	$d\phi(t)/dt$	$\ln d\phi(t)/dt$
1.2	1.467	0.383	7.2	0.197	−1.625
2.4	0.661	−0.414	8.4	0.0855	−2.459
3.6	0.416	−0.877	9.6	0.0744	−2.598
4.8	0.361	−1.019	10.8	0.0607	−2.802
6.0	0.152	−1.884	12.0	0.0329	−3.414

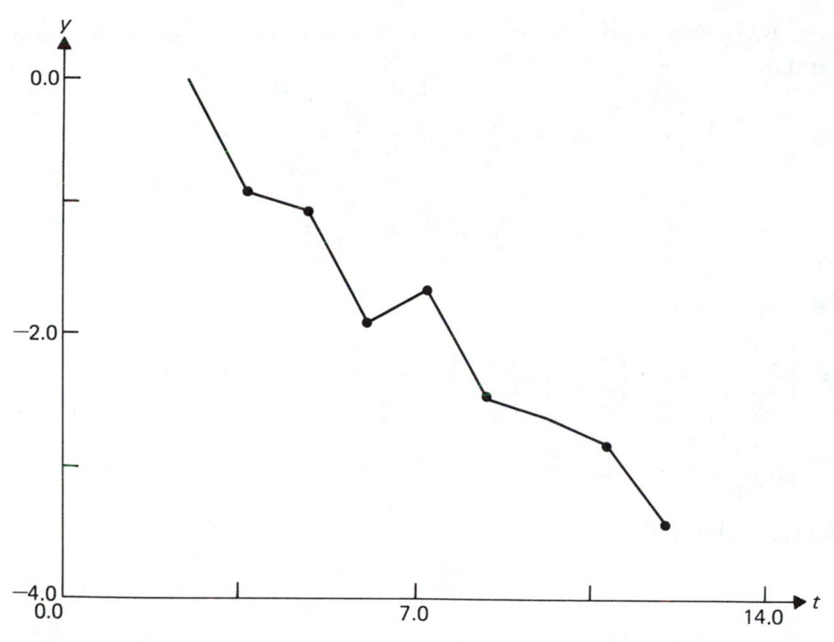

Figure 1.28 A plot of the data, $y = \ln(d\phi/dt)$ versus t.

possible. This leads us to the problem of choosing values of the parameters α and β so that

$$F(\alpha, \beta) = \sum_{j=1}^{10} [y_j - (\alpha + \beta t_j)]^2 \tag{1.9.6}$$

is a minimum; hence the term, principle of least squares. If $F(\tilde{\alpha}, \tilde{\beta})$ is the minimum of $F(\alpha, \beta)$, then

$$y = \tilde{\alpha} + \tilde{\beta} t$$

is the equation of the best-fitting line. From calculus, we know that $\tilde{\alpha}$ and $\tilde{\beta}$ must satisfy

$$0 = \frac{\partial F}{\partial \alpha}(\tilde{\alpha}, \tilde{\beta}) = -2 \sum_{j=1}^{10} [y_j - (\tilde{\alpha} + \tilde{\beta} t_j)],$$

$$0 = \frac{\partial F}{\partial \beta}(\tilde{\alpha}, \tilde{\beta}) = -2 \sum_{j=1}^{10} t_j [y_j - (\tilde{\alpha} + \tilde{\beta} t_j)].$$

Expanding the two expressions above leads to the following pair of linear equations for $\tilde{\alpha}$ and $\tilde{\beta}$:

$$10\tilde{\alpha} + \left(\sum_{j=1}^{10} t_j\right)\tilde{\beta} = \sum_{j=1}^{10} y_j,$$

$$\left(\sum_{j=1}^{10} t_j\right)\tilde{\alpha} + \left(\sum_{j=1}^{10} t_j^2\right)\tilde{\beta} = \sum_{j=1}^{10} y_j t_j.$$

Solving for $\tilde{\alpha}$ and $\tilde{\beta}$ gives

$$\tilde{\beta} = \left[\sum_{j=1}^{10} y_j t_j - \frac{1}{10}\left(\sum_{j=1}^{10} t_j\right)\left(\sum_{j=1}^{10} y_j\right)\right] \div \left[\sum_{j=1}^{10} t_j^2 - \frac{1}{10}\left(\sum_{j=1}^{10} t_j\right)^2\right],$$
$$\tilde{\alpha} = \frac{1}{10}\left(\sum_{j=1}^{10} y_j - \tilde{\beta}\sum_{j=1}^{10} t_j\right). \tag{1.9.7}$$

From the data in Table 1.6,

$$\sum_{j=1}^{10} t_j = 66.00, \qquad \sum_{j=1}^{10} y_j = -16.709,$$
$$\sum_{j=1}^{10} t_j^2 = 554.4, \qquad \sum_{j=1}^{10} y_j t_j = -148.41,$$

and so $\tilde{\beta} = -0.321$ and $\tilde{\alpha} = 0.448$. Since we know that $y = \tilde{\alpha} + \tilde{\beta}t$ approximates $\ln(d\phi/dt) = \ln E - (R/L)t$, we set

$$\ln E \cong \tilde{\alpha} = 0.448 \quad \text{or} \quad E \cong 1.565$$

and

$$-\frac{R}{L} \cong \tilde{\beta} = -0.321 \quad \text{or} \quad L \cong \frac{R}{0.321}.$$

If $R \cong 9.963$ is the measured value of the total resistance, then $L \cong 31.04$.

We have constructed a mathematical model for the flux variation in an iron-core induction coil and have shown that the parameters of the model can be estimated from measured data. In the next section, a model for population growth is described.

EXERCISES 1.9

1. Solve (1.9.2) when $L = 5$ henrys, $R = 15$ ohms, and $E = 60$ volts if $I(0) = 0$. At what time will $|I(t) - 4| < 10^{-4}$ amperes?

2. Solve (1.9.2) when $L = 4$ henrys, $R = 8$ ohms, $I(0) = 4$ amperes, and $E(t) = 0$ for $0 \leq t < 1$, $E(t) = 6$ volts for $t > 1$. Do this by solving the initial value problem at $t = 0$, then find the general solution for $t > 1$, and finally match the two solutions at $t = 1$, so that $I(t)$ is continuous there. Is $I'(t)$ continuous at $t = 1$?

3. Use (1.9.5) and the following data to estimate E and L given that $R = 10$ ohms.

t	$d\phi(t)/dt$	$\ln d\phi(t)/dt$
1.5	6.5500	1.8795
2.5	5.3938	1.6852
3.5	4.1274	1.4172
4.5	3.2588	1.1814
5.5	2.5301	0.9283
6.5	1.9233	0.6540

4. Use (1.9.5) and the following data to estimate E and L given that $R = 15$ ohms.

t	$d\phi(t)/dt$	$\ln d\phi(t)/dt$
1.1	4.6497	1.5368
2.2	3.5896	1.2780
3.3	2.8026	1.0305
4.4	2.0912	0.7377
5.5	1.6430	0.4965
6.6	1.2632	0.2336

5. The nationwide average Scholastic Aptitude Test scores in mathematics for selected years from 1963 are given below (range of possible scores is from 200 to 800):

Year	Score
1963	502
1967	492
1970	488
1974	480
1977	470

a) Letting $t = 0$ correspond to 1963, $t = 4$ to 1967, etc., construct a least-squares fit of a straight line to the data. Plot the data and the line.

b) Use the straight line to predict the average SAT mathematics scores for 1979 and 1980 (the actual scores were 467 and 466, respectively).

6. For the data of Table 1.6 plot $(t_j, \ln(d\phi/dt)_j)$ and then with a straightedge draw a best-fitting line. From it estimate E and L given that $R = 10$ ohms. This is called the *eyeballing method* of linear regression.

7. Compute the *residuals,*

$$r_j = (d\phi/dt)_j - E\exp(-Rt_j/L),$$

first with the least-square values of E and L, then with the values you obtained in the preceding problem. Compare the corresponding values of F, $\partial F/\partial\alpha$, and $\partial F/\partial\beta$, recalling that F should be a minimum and $\partial F/\partial\alpha$, $\partial F/\partial\beta$ should be equal to zero.

1.10 A MODEL OF POPULATION GROWTH

In the construction of a mathematical model for population growth, the basic difficulty is that there are no physical principles (like Ohm's law or Newton's laws of motion) on which to anchor the construction. Nevertheless, the model should reflect biological reality and this can be verified by an analysis of the qualitative behavior of solutions and, it is hoped, by comparison with biological data. Finally, the model should be simple enough so that the above analysis is possible—a model with 50 parameters is of little value either mathematically or biologically.

In this section, we will create a model for the growth of a single population. Let $x(t)$ be the size of the total population at time t and let $\dot{x}(t) = dx(t)/dt$ be its rate of growth. Then the per capita rate of growth of the population at time t is represented by

$$\frac{\dot{x}(t)}{x(t)},$$

since it is the rate of growth divided by the total population at time t. To obtain a differential equation, there must be some assumption made about the dependency of $\dot{x}/x$ on the time t, the total population $x(t)$, and possibly on some biological parameters. The simplest assumption would be that the per capita rate of growth is a positive constant, which leads to the equation

$$\frac{\dot{x}(t)}{x(t)} = r \quad \text{or} \quad \dot{x} = rx, \qquad r > 0.$$

This is the equation for exponential growth, and if x_0 is the initial size of the population at time $t = 0$, then its solution is $x(t) = x_0e^{rt}$. In some sense, this model is too simple, since one would expect that, as the population size increases, the per capita rate of growth will decrease (due to overcrowding, competition for food, resources, etc.). At best, one would expect pure exponential growth when there is unlimited space and resources. Therefore the per capita rate of growth should depend on the population size, and so the model should be of the form

$$\frac{\dot{x}(t)}{x(t)} = f(x(t)) \quad \text{or} \quad \dot{x} = xf(x).$$

Now the task is to select a plausible function $f(x)$. A possible set of principles which might guide the selection of $f(x)$ are:

1. For small values of $x = x(t)$, the function $f(x)$ should be positive, reflecting the assumption that when the population is small, its growth is *locally* exponential.
2. For large values of $x = x(t)$, the function $f(x)$ should be negative, meaning that too large a population inhibits the rate of growth.
3. For all values of $x = x(t)$, the function $f(x)$ should be decreasing since the per capita growth rate should decrease as the population increases.

These principles are surely not the only ones that could be devised, but they will suffice for the model we wish to describe and they are biologically plausible.

The next problem is that of selecting a candidate for $f(x)$. Certainly, there is no unique function satisfying all the conditions (1), (2), and (3); however, we must adhere to our principle of simplicity. Too complicated a function $f(x)$ will make the analysis of the ensuing differential equation virtually impossible. The simplest function satisfying conditions (1), (2), and (3) is a linear one:

$$f(x) = r - mx, \quad r, m > 0.$$

We see that

a) $f(x) > 0$ for $x < r/m$,
b) $f(x) < 0$ for $x > r/m$,
c) $f'(x) = -m < 0$.

Hence, the biological guidelines are satisfied. The mathematical model for the population growth is given by the differential equation

$$\dot{x} = x(r - mx), \qquad x(0) = x_0, \quad r, m, x_0 > 0,$$

where $x(t)$ is the total population at time t and x_0 is the initial population. Letting $K = r/m$, we can write it as

$$\dot{x} = rx\left(1 - \frac{x}{K}\right), \qquad x(0) = x_0, \quad r, K, x_0 > 0. \tag{1.10.1}$$

This differential equation is called the *logistic equation* of population growth and has been used successfully to model the growth of yeast cells, fruit flies, the population of Sweden, the height of sunflower plants, and the Pacific Halibut fishery, to name a few examples.

An examination of (1.10.1) shows immediately that $x(t) = K$ is a constant solution (so is $x(t) = 0$ but it has no biological interest) and that

$$\dot{x} = rx\left(1 - \frac{x}{K}\right)\begin{cases} >0 & \text{for} \quad x < K, \\ <0 & \text{for} \quad x > K. \end{cases}$$

This means that if the initial value x_0 satisfies $0 < x_0 < K$, the solution $x(t)$ satisfying $x(0) = x_0$ is a strictly increasing function of t. As long as $0 < x(t) < K$ the population will increase. But since $x(t)$ can never cross the line $x = K$ (why?), and the direction field in $0 \leq t < \infty$, $0 < x < K$ always points upward, $x(t)$ must become asymptotic to $x = K$ as $t \to \infty$. By a similar line of reasoning one can conclude that any solution with initial value $x(0) = x_0 > K$ is a strictly decreasing function of t, which becomes asymptotic to $x = K$ as $t \to \infty$. Without solving the equation (which can be done—see Exercise 1a), it can be concluded that every solution $x(t)$, for which $x(0) = x_0 \neq K$ and $x_0 > 0$, approaches the value K as $t \to \infty$. For this reason the constant K is called the *carrying capacity* of the population and represents the natural limit of its size (Fig. 1.29).

If $x(t)$ is small and positive, then for a short period of time the population grows like the solution of $\dot{x} = rx$, since the term $-rx^2/K$ will be negligible. Hence

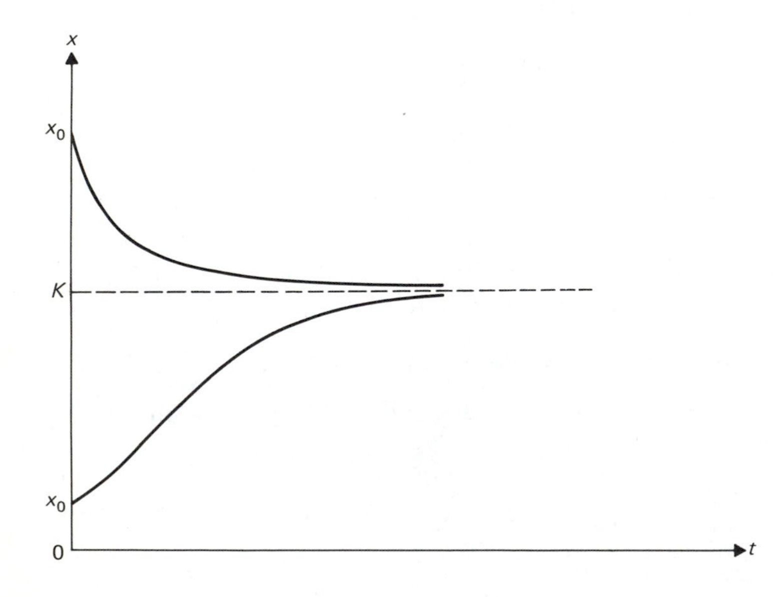

Figure 1.29 Graph of solutions to the logistic equation.

for small values of x the growth looks exponential, and for that reason the parameter r is called the *intrinsic growth rate* of the population.

EXAMPLE 1 Approximate census figures for the United States (with the population in millions) are in Table 1.7. A reasonable fit to the data is given by

$$x(t) = \frac{265}{1 + 69\, e^{-0.03t}} \tag{1.10.2}$$

A plot of this solution to the logistic equation and the data are given in Fig. 1.30. Find the values of the carrying capacity and the intrinsic growth rate of this model.

SOLUTION The carrying capacity is the limiting value of the population:

$$K = \lim_{t\to\infty} x(t) = 265.$$

From (1.10.1), we see that the intrinsic growth rate,

$$r = \frac{\dot{x}}{x\left(1 - \frac{x}{K}\right)}.$$

From

$$\frac{\dot{x}(t)}{x(t)} = \frac{69(0.03)\, e^{-0.03t}}{1 + 69\, e^{-0.03t}}$$

Table 1.7 Population growth in the United States

Year	Population (in millions)	Year	Population (in millions)
1790	4	1890	63
1800	5	1900	76
1810	7	1910	92
1820	10	1920	106
1830	13	1930	123
1840	17	1940	132
1850	23	1950	152
1860	31	1960	180
1870	39	1970	204
1880	50	1980	226

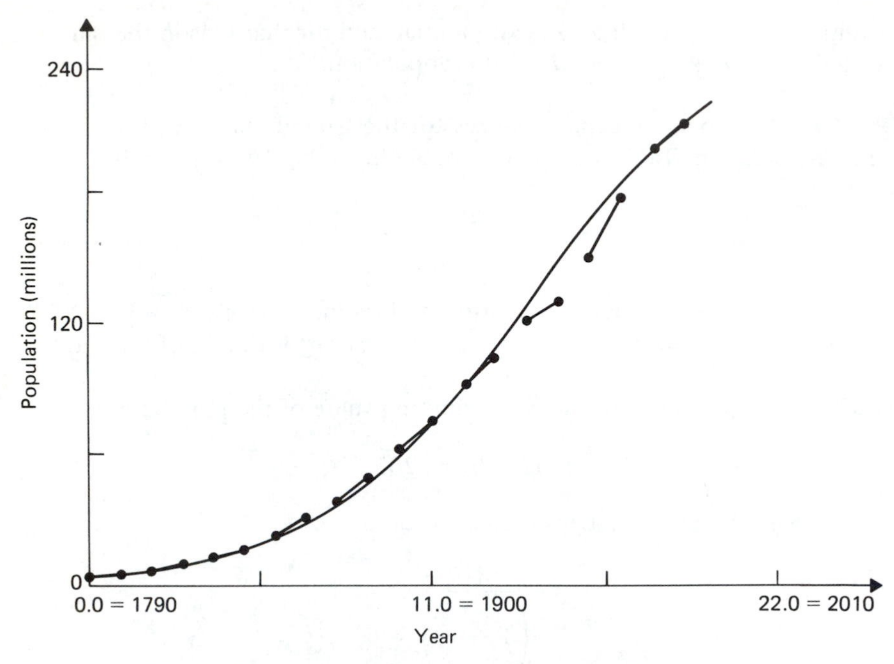

Figure 1.30 A plot of the census data and the modeling solution to the logistic equation.

and

$$1 - \frac{x(t)}{K} = 1 - \frac{1}{1 + 69\, e^{-0.03t}} = \frac{69\, e^{-0.03t}}{1 + 69\, e^{-0.03t}}$$

we find that $r = 0.03$. ■

Our mathematical results are biologically plausible since one would expect that a small population will initially grow exponentially. Then, as the competition for food and the effects of overcrowding begin to be felt, its growth rate will diminish. One also would expect that there is a natural limit, K, to the size of the population, which is, in part, determined by the resources available. Of course, if it is believed or observed that the growth of a population under study behaves differently, then this would call for a different set of assumptions and, consequently, a different $f(x)$.

EXERCISES 1.10

1. a) Using the method of separation of variables, show that the solution of the logistic equation with initial data $x(0) = x_0$ is given by

$$x(t) = \frac{Kx_0}{x_0 + (K - x_0)e^{-rt}}.$$

b) Show that $\lim_{t\to\infty} x(t) = K$ and that $x(t)$ is increasing if $0 < x_0 < K$ and decreasing if $x_0 > K$.

2. The function

$$x(t) = \frac{665}{1 + \exp[4.1896 - 0.5355t]}$$

was used by R. Pearl in 1925 to model the growth of a yeast *(Saccharomyces)* population. Here t is measured in hours and $x(t)$ is a number proportional to the quantity of yeast cells.

a) Find the logistic differential equation and the initial conditions that $x(t)$ satisfies.

b) Find a value T for which $K - x(t) < 1$ whenever $t > T$.

c) Graph $x(t)$, $0 \le t \le 16$.

3. The function

$$P(t) = 1.535 + x(t),$$

where $x(t)$ is the solution of the logistic equation

$$\dot{x} = 0.023x\left(1 - \frac{x}{6.336}\right), \qquad x(0) = 0.767,$$

is an excellent fit to the population of Sweden from 1800 to 1920. Here t is measured in years, $x(t)$ is population size in millions, and $t = 0$ corresponds to the year 1800.

Solve the logistic equation and use it to predict the population of Sweden in 1930 (6.142), 1940 (6.372), 1950 (7.041), 1960 (7.495), and 1970 (8.040). The figures in parentheses are the actual populations.

4. A model for the harvesting of a population with logistic growth is

$$\dot{x} = rx\left(1 - \frac{x}{K}\right) - H,$$

where H can be regarded as a constant harvesting rate. Such harvesting could occur as the result of hunting, fishing, or a disease.

a) Show that if $H = 0$, the maximum rate of growth of $x(t)$ is $H_c = rK/4$ and that if $H > H_c$, the population expires in finite time (meaning $x(T) = 0$ for some $T > 0$).

b) Show that if $0 < H < H_c$, the equation has two constant solutions $x(t) = K_1$ and $x(t) = K_2$ with $0 < K_1 < K_2 < K$. Furthermore, all solutions $x(t)$

with $x(0) = x_0$ and $x_0 > K_1$ satisfy $\lim_{t\to\infty} x(t) = K_2$. The effect of harvesting is to reduce the carrying capacity of the population.

5. The sandhill crane *(Grus canadensis)* has a suggested carrying capacity $K = 194.6$ in thousands of birds and an intrinsic growth rate $r = 0.0987$. The sandhill crane is hunted because of its destructive foraging in grain fields.

a) Using the results of Exercise 4, find the critical harvest rate (birds per year) H_c.

b) What is the effect on the carrying capacity if 3000 birds per year ($H = 3.0$) are harvested?

c) Assuming the population is near equilibrium, that is, $x(0) = K = 194.6$, and the harvesting rate is 13,000 birds per year ($H = 13.0$), use the improved Euler method with step size $h = 0.5$ to estimate when the population expires. (This will occur prior to $T = 25$ years.)

d) The same as (c), but harvesting 20,000 birds per year.

6. By writing the expression for the solution of the harvested logistic equation in the form

$$\int_K^0 \frac{1}{rx(1 - x/K) - H}\, dx = \int_0^T dt, \quad H > H_c,$$

one can solve for the time T necessary to reach extinction. Do this and, using the data from Exercise 5, compare your computed results in 5(c) and 5(d) with the actual answer.

7. A simpler model for harvesting is based on exponential growth. If the per capita growth rate is a constant $r = b - d$, representing the difference between birth rates and death rates, then one can model the effect of a constant harvesting rate h by using the model for growth $\frac{dx}{dt} = (b - d - h)x$.

Consequently hx is the size of the harvest per unit time.

a) For whales it is estimated that $b = 0.14$ and $d = 0.09$ where t is measured in years. What catch per year will maintain the population at a constant size of 5000 whales? If $h = 0.6$ and harvesting takes place for one year, how long will it take for the population to return to its original size?

b) For buffaloes $b = 0.081$ and $d = 0.05$ where t is measured in years. If there were 40 million buffaloes, what is the maximum annual harvest required to keep the herd at that size?

8. The relative error $= (x_{\text{data}} - x(t))/x(t)$. Compute and plot the relative error versus t for the data and model of Example 1.

9. Another model for population growth that has been applied to the study of animal tumors is the Gompertz equation

$$\frac{dx}{dt} = rx \ln\left(\frac{K}{x}\right),$$

where r, K are positive constants.

a) Using an argument similar to that for the logistic equation, show that any population satisfying $x(0) = x_0 > 0$ will satisfy $\lim_{t\to\infty} x(t) = K$.

b) Letting $y = \ln x$, find the exact solution satisfying $x(0) = x_0 > K$.

10. Below are some other models of population growth that have been suggested by various researchers. Discuss their underlying biological assumptions and qualitative features and compare them to those of the logistic model.

a) $\dot{x} = rx\left(\frac{K}{x} - 1\right)$

b) $\dot{x} = \frac{rx(K - x)}{K + \epsilon x}$, where $\epsilon > 0$ is small

c) $\dot{x} = rx[1 - (x/K)^{\alpha}]$; consider two cases: $0 < \alpha < 1$ and $\alpha > 1$.

11. The logistic equation where t is measured in years was used successfully to model the growth of the population of the United States in the first half of the century. From the expression in Exercise 1, use the following population values to determine the constants x_0, r, and K:

1790: 3,929,000, 1850: 23,192,000, 1910: 91,972,000.

a) Use the resulting expression to calculate the values of the population for the years 1920, 1930, 1940, and 1950. The census figures for those years are

1920: 105,711,000, 1930: 123,124,000, 1940: 131,669,000

1950: 150,699,000 (rounded to nearest thousand).

Compute the percent relative error for each year.

b) The census figures for the years 1960, 1970, and 1980 are

1960: 179,300,000, 1970: 204,000,000, 1980: 226,500,000.

Use the population values for the years 1920, 1930, and 1940 to determine the values of x_0, r, and K, and calculate the values of the population for the years 1960, 1970, and 1980. Compute the percent relative error.

c) Compare your model with the model of Example 1.

1.11 A SIMPLE MIXING PROBLEM AND FEEDBACK CONTROL

In this section we shall study the problem of adjusting the concentration of salt to a desired level in a brine mixing tank. Simple extensions of this model will apply to a number of problems in dilution or contamination. For instance, a dam or reservoir into which pollutants are flowing via feeder streams can be modeled as a mixing tank.

Let us suppose that at $t = 0$ the tank contains V liters of brine with an initial salt concentration of c_0 grams per liter. Brine is pumped into the top of the tank at a rate of r liters per second and its salt concentration is s grams per liter. The tank contains a mixing device which insures that the concentration of salt c in the tank is uniform throughout the tank. Finally, the mixture in the tank is pumped out through the bottom at the same rate r as the incoming solution.

Our task is to adjust s, the incoming concentration of salt, so that the concentration of salt in the tank attains (and remains at) a predetermined concentration k. Our mathematical model will show that unless the initial concentration c_0 and the incoming concentration s both equal k, our task cannot be accomplished in finite time. To overcome this difficulty, we must introduce the notion of a feedback control law which governs s.

The fundamental principle in this and other mixing problems is that the rate of change of the amount of salt in the tank is equal to the rate of change of the amount of incoming salt minus the rate of change of the amount of outgoing salt. Since the amount of salt in the tank is c (grams/liter) $\times$ V (liters),

$$\frac{d}{dt}(cV) = (\text{Rate in}) - (\text{Rate out}).$$

The data given imply that

$$\begin{aligned} \text{Rate in} &= s\ (\text{grams/liter}) \times r\ (\text{liters/second}), \\ \text{Rate out} &= c\ (\text{grams/liter}) \times r\ (\text{liters/second}), \end{aligned}$$

and so the governing equation is

$$\frac{d}{dt}(cV) = sr - cr. \tag{1.11.1}$$

Because the volume of the tank is constant, the equation (1.11.1) becomes

$$\frac{dc}{dt} = ps - pc, \qquad p = \frac{r}{V}, \tag{1.11.2}$$

with initial condition $c(0) = c_0$.

EXAMPLE 1 A tank contains a homogeneous solution of 4 kilograms of salt and 400 liters of water. Starting at $t = 0$ a brine solution containing 0.25 kilograms of salt per liter enters at a rate of 8 liters per minute. A mixing device maintains homogeneity, and the uniform solution leaves the tank at the same rate. Find the concentration and the amount of salt at any time.

SOLUTION The entering and exiting rates of the solution are equal and therefore the volume of liquid in the tank is constant. Since

$$\text{Rate in} = 0.25 \text{ kg/liter} \times 8 \text{ liter/min} = 2 \text{ kg/min},$$

$$\text{Rate out} = c \text{ kg/liter} \times 8 \text{ liter/min} = 8c \text{ kg/min},$$

the differential equation governing the concentration $c(t)$ is

$$400\frac{dc}{dt} = 2 - 8c \quad \text{or} \quad \frac{dc}{dt} = \frac{1}{200} - \frac{1}{50}c,$$

with initial conditions $c(0) = 4/400$ kg/liter $= 0.01$ kg/liter. The solution to this differential equation is

$$c(t) = (0.25 - 0.24e^{-t/50}) \text{ kg/liter}.$$

Hence the amount of salt in the tank is

$$400\,c(t) = (100 - 96e^{-t/50}) \text{ kg}.$$

■

EXAMPLE 2 Suppose that in Example 1 the solution is leaving the tank at a rate of 6 liters per minute. What is the differential equation governing the concentration at any time?

SOLUTION In this case the volume of the tank is not constant but is increasing 2 liters per minute. Therefore $V = 400 + 2t$. Now

$$\text{Rate in} = 0.25 \text{ kg/liter} \times 8 \text{ liter/min} = 2 \text{ kg/min},$$

$$\text{Rate out} = c \text{ kg/liter} \times 6 \text{ liter/min} = 6c \text{ kg/min},$$

and the differential equation governing the concentration $c(t)$ is

$$\frac{d}{dt}[(400 + 2t)c] = 2 - 6c, \; c(0) = 0.01.$$

By performing the differentiation we have

$$2c + (400 + 2t)\frac{dc}{dt} = 2 - 6c$$

or

$$\frac{dc}{dt} = \frac{2 - 8c}{400 + 2t} = \frac{1 - 4c}{200 + t},$$

which can be solved by separation of variables. ■

Equation (1.11.2) is linear with constant coefficients and it is easily solved to obtain

$$c(t) = s + (c_0 - s)e^{-pt}. \quad \textbf{(1.11.3)}$$

As t increases, $c(t)$ approaches s in the limit, and therefore we see that we must take $s = k$ to solve our problem of attaining a concentration k. Hence the inflow concentration must equal the desired concentration.

From equation (1.11.3) it is seen that if $s = k$ and the initial concentration $c_0 \neq k$, an infinite amount of time is required before $c(t) = k$. If this is not desirable, the design of our system must be improved, and to do this we will employ a fundamental concept in modern engineering, *feedback control,* the control of a process through monitoring. In this problem, this will be accomplished by changing the incoming concentration s, where the change is governed by the existing concentration in the tank. (See Fig. 1.31.)

The concept of feedback control is neither esoteric nor new. The household thermostat is a feedback control device that turns the furnace on or off depending

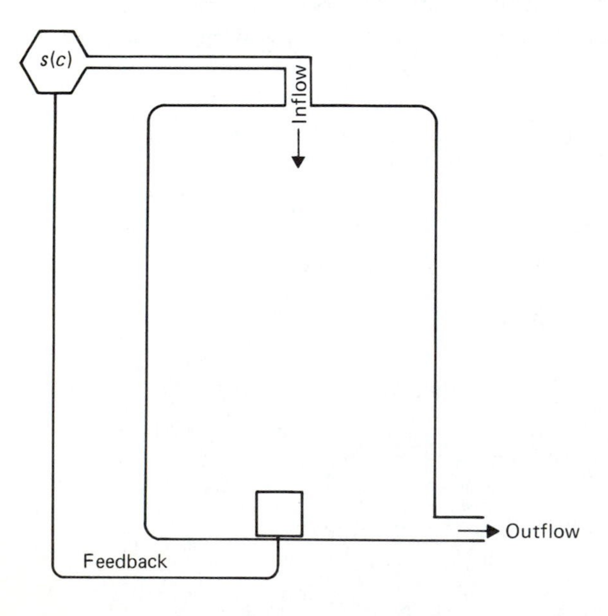

Figure 1.31 **A mixing tank with feedback control.**

on the measured temperature. An autopilot mechanism, which measures the actual heading of an airplane and adjusts the control surfaces so as to maintain a fixed heading, is another example. A much older feedback device, not understood until recently, is the control of respiration by measuring the hydrogen-ion concentration in the brain.

Returning to our problem, suppose that the concentration c of salt in the tank can be measured at any instant. This could be accomplished, for example, by a pressure-sensitive device on the bottom of the tank since the mass of fluid is equal to the mass of dissolved salt plus the mass of the water. Suppose furthermore that instrumentation can be devised so that the measurements obtained can be used to control the incoming concentration s. Then the simplest example of a *feedback control law* is the following one:

$$s = s(c) = \begin{cases} 0 & \text{if} \quad c > k, \\ k & \text{if} \quad c = k, \\ z & \text{if} \quad c < k, \end{cases}$$

where z is some convenient value greater than k. For instance, the first equation says to pump in fresh water if the concentration is greater than k. The last equation says to pump in a heavier concentration if $c(t)$ is less than k.

With this control law, the differential equation (1.11.2) now becomes

$$\frac{dc}{dt} = ps(c) - pc, \tag{1.11.4}$$

and the form of the solution depends on the initial concentration $c(0) = c_0$.

Case 1. If $c_0 > k$, then $s(c) = 0$ and the solution of (1.11.4) is

$$c(t) = c_0 e^{-pt}.$$

The desired concentration will be attained at $t = t^*$, where $c(t^*) = k$. Hence $t^* = (1/p) \ln(c_0/k)$.

Case 2. If $c_0 = k$, then $s(c) = k$, $c(t) = k$, and $t^* = 0$.

Case 3. If $c_0 < k$, then $s(c) = z$ and from (1.11.3) the solution is

$$c(t) = z + (c_0 - z)e^{-pt},$$

where $z > k > c_0$. The desired concentration will be obtained when $c(t^*) = k$, and this gives

$$t^* = \frac{1}{p} \ln\left(\frac{z - c_0}{z - k}\right).$$

On the face of it, our simple control law has solved the problem of attaining k in finite time.

We continue our discussion of the mixing problem and begin with the following observation. The simple feedback model has a serious drawback—we have assumed instantaneous mixing of the solution in the tank and the incoming brine. In reality, once equilibrium k is attained, the input mechanism will oscillate (or *chatter,* in the control terminology) as it attempts to follow the fluctuating feedback directions.

Therefore the model should be designed to allow for errors in the adjustments of the input concentration, as well as errors in the measurements of the brine concentration in the tank. One way to do this is to weaken the requirement that the concentration c attain a specific value k and instead require that it be close to c. Hence we require that c be in the interval

$$I_\epsilon = \{c\colon |c - k| \leq \epsilon\},$$

where $\epsilon > 0$ is some acceptable measure of error.

This last requirement will not eliminate chatter since it is still possible for the concentration c to drift back and forth across the boundary of I_ϵ. A simple way to eliminate chatter is to design a feedback law for the input concentration s with a time delay $\tau > 0$ as follows:

1. $$s = s(c) = \begin{cases} 0 & \text{if} \quad c > k + \epsilon, \\ k & \text{if} \quad k - \epsilon \leq c \leq k + \epsilon, \\ z & \text{if} \quad c < k - \epsilon, \text{ where } z > k. \end{cases}$$

2. Change values of $s(c)$ exactly τ seconds after the measurement of the tank concentration c indicates that the boundary of I_ϵ has been reached.

To guard against chatter, no new measurements (and hence no new instructions) are accepted during the delay period.

As before, the form of the solution will depend on the value c_0, the initial concentration in the tank.

Case 1. $c_0 > k + \epsilon$. In this case $s(c) = 0$ and $c(t) = c_0e^{-pt}$, and this will remain so until $c(t)$ reaches $k + \epsilon$. This will occur at the time t^* which satisfies the relation

$$c_0e^{-pt^*} = k + \epsilon.$$

However, because of the delay, $s(c)$ will not switch from 0 to k until $t = t^* + \tau$. At the switching time,

$$c(t^* + \tau) = c_0e^{-p(t^*+\tau)} = c_0e^{-pt^*} e^{-p\tau},$$

which implies (from the previous equation) that

$$c(t^* + \tau) = (k + \epsilon)e^{-p\tau}.$$

In the last relation it is seen that if the delay time τ is too large, the right-hand side could be very small. It could then occur that $c(t^* + \tau)$ will be less than $k - \epsilon$, the lower boundary of I_ϵ. Therefore, in the design of our system τ must be restricted, so that

$$k - \epsilon < (k + \epsilon)e^{-p\tau}.$$

When $s(c)$ switches to k, the differential equation and initial conditions are

$$\frac{dc}{dt} = pk - pc, \qquad c(t^* + \tau) = (k + \epsilon)e^{-p\tau}.$$

The solution is

$$c(t) = k + [(k + \epsilon)e^{-p\tau} - k]e^{-p(t-t^*-\tau)}, \; t \geq t^* + \tau,$$

and the restriction on τ and a simple estimate give

$$c(t) > k + [(k - \epsilon) - k]e^{-p(t-t^*-\tau)} \geq k - \epsilon$$

for $t \geq t^* + \tau$. Furthermore,

$$c(t) - k = (k + \epsilon)e^{-p(t-t^*)} - ke^{-p(t-t^*-\tau)} \leq (k + \epsilon)e^{-p(t-t^*)} - ke^{-p(t-t^*)} \leq \epsilon;$$

hence $|c(t) - k| \leq \epsilon$ for $t \geq t^* + \tau$ as desired.

Case 2. $k - \epsilon \leq c_0 \leq k + \epsilon$. Now $s(c) = k$, and for $t > 0$ the solution of (1.11.2) is

$$c(t) = k + [c_0 - k]e^{-pt}.$$

We then have the estimates

$$\begin{aligned} c(t) &= k(1 - e^{-pt}) + c_0 e^{-pt} \leq k(1 - e^{-pt}) + (k + \epsilon)e^{-pt} \\ &= k + \epsilon e^{-pt} \leq k + \epsilon \end{aligned}$$

and

$$c(t) \geq k(1 - e^{-pt}) + (k - \epsilon)e^{-pt} = k - \epsilon e^{-pt} \geq k - \epsilon.$$

Therefore $|c(t) - k| \leq \epsilon$, as required.

Case 3. $c_0 < k - \epsilon$. In this case, $s(c) = z > k > c_0$, and the concentration will increase and reach $k - \epsilon$, the lower boundary of I_ϵ at a time $t^* > 0$. At time $t^* + \tau$, $s(c)$ will switch to k, and to ensure that $x(t^* + \tau)$ is less than $k + \epsilon$ (the upper boundary of I_ϵ), a further restriction on the delay time τ is needed. We ask the reader (see Exercise 14) to show that this restriction is

$$z + (k - \epsilon - z)e^{-p\tau} < k + \epsilon,$$

and that with this requirement $|c(t) - k| \leq \epsilon$ for $t \geq t_* + \tau$. Note that the last restriction can be written as

$$(z - k) - \epsilon < [(z - k) + \epsilon]e^{-p\tau},$$

which is analogous to the requirement on τ in Case 1.

EXAMPLE 3 Take $\epsilon = \frac{1}{100}$, $\tau = 60$, $c_0 = \frac{1}{12}$, $k = \frac{1}{20}$, $p = \frac{1}{1200}$, and determine $c(t)$.

SOLUTION Since $c_0 > k$, this is a case 1 problem. The switching time, $t^* + \tau$, can be found from

$$c(t^* + \tau) = (k + \epsilon)e^{-p\tau} = .0.06e^{-0.05} \cong 0.05707$$

and

$$\frac{1}{12}\exp(-(t^* + 60)/1200) \cong 0.05707.$$

Thus $t^* + \tau \cong 454$ seconds.

The concentration is given by

$$c(t) = \frac{1}{12}\exp(-t/1200), \qquad 0 \leq t \leq t^* + \tau,$$

$$c(t) \cong 0.05 + [0.05707 - 0.05]\exp[-(t - t^* - \tau)/1200], \qquad t > t^* + \tau,$$

or

$$c(t) \cong 0.05 + 0.00707\exp[-(t - 454)/1200], \qquad t > 454.$$ ■

EXERCISES 1.11

1. A tank contains a homogeneous solution of 5 kilograms of salt and 500 liters of water. Starting at $t = 0$, fresh water is poured into the tank at a rate of 4 liters/min. A mixing device maintains homogeneity. The uniform solution leaves the tank at 4 liters/min.
   a) What is the differential equation governing the amount of salt in the tank at any time?
   b) In how many minutes will the concentration of salt reach a 0.1% level?
2. In Exercise 1, change the exit rate to 2 liters/min and assume the tank has a 1000-liter capacity. What is the amount and concentration of salt in the tank at the instant of overflow? (*Hint:* Find V as a function of time and use it in the differential equation for c.)

3. A tank contains a homogeneous solution of 5 kilograms of salt and 1000 liters of water. A brine solution containing 0.5 kilograms of salt per liter enters at 5 liters per minute. The uniform solution leaves the tank at the same rate. When will the tank contain 100 kilograms of salt?

4. A tank contains a homogeneous solution of 2 kilograms of salt and 50 liters of water. Pure salt is fed into the tank at the rate of 1 kilogram per minute. The uniform solution leaves the tank at the rate of 2 liters per minute. Find the amount and concentration of the salt after 25 minutes.

5. In Exercise 3 suppose the exit rate is changed to 3 liters/min and the tank has a 2000-liter capacity. What is the amount and concentration of salt after 250 minutes? At overflow?

6. For the simple feedback control strategy given, suppose that the tank contains 100 liters of brine and the inflow and exit rates are 4 liters/min. If the desired concentration is $k = 12$ grams/liter, find the time t^* at which it is achieved if:

a) The initial concentration in the tank is $c_0 = 20$ grams/liter,

b) The initial concentration in the tank is $c_0 = 4$ grams/liter and we use an incoming concentration of $z_1 = 16$, $z_2 = 13$, and $z_3 = 12.1$ grams/liter.

7. Two tanks initially contain 200 liters of fresh water each. Starting at $t = 0$, brine containing 5 kg/liter of salt is added to the first tank at a rate of 2 liters/min. The uniform solution from the first tank is transferred to the second tank at a rate of 2 liters/min. The uniform solution in the second tank is removed at the same rate. Find the concentration in each tank at any time.

8. A 1000-liter tank contains a homogeneous solution of 5 kilograms of salt and 500 liters of water. In order to attain a concentration of 0.02 ± 0.0006 kg/liter, brine is added at a rate of 4 liters/min. The uniform solution leaves at the same rate. The concentration of the incoming brine is governed by the following feedback law:

$$s = s(c) = \begin{cases} 0 & \text{if } c > 0.026, \\ 0.02 & \text{if } 0.0194 \leq c \leq 0.026, \\ 0.04 & \text{if } c < 0.0194. \end{cases}$$

Let t^* be the time at which the concentration is equal to 0.0194. Find t^* and $c(t^*)$. Find the formula for $c(t)$ with $0 \leq t \leq t^*$ and with $t^* < t$. For this problem $\tau = 0$; that is, there is no delay.

9. Assume that the tank of Exercise 8 initially contains 15 kilograms of salt and 500 liters of water. Let t^* be the time at which the concentration is equal to 0.026. Find the formula for $c(t)$ with $0 \leq t \leq t^*$ and with $t^* < t$.

10. Suppose in Exercise 8 a delay of $\tau = 2$ min is introduced in the control and the initial conditions are 5 kilograms of salt dissolved in 1000 liters of water.

Find $c(t)$ on the intervals $0 \leq t \leq t^* + \tau$ and $t^* + \tau < t$, where t^* is the time at which the concentration is equal to 0.0194.

11. With a delay of $\tau = 2$ min in the control of Exercise 8 and the initial conditions of 30 kilograms of salt and 1000 liters of water, find $c(t)$ on the invervals $0 \leq t \leq t^* + \tau$ and $t^* + \tau < t$, where t^* is the time at which the concentration is equal to 0.026.

12. With no delay and with the initial concentration in the 1000-liter tank denoted by c_0, find the formulas for $c(t)$ if the control law is

$$s(c) = \begin{cases} 0 & \text{if} \quad c > k, \\ z & \text{if} \quad c \leq k. \end{cases}$$

Show that t^* is a decreasing function of z. Therefore to minimize t^*, take z as large as possible. Assume V equals 500 liters.

13. Solve Exercise 12 with a delay τ.

14. Derive the restriction on the delay time τ given in the text for case 3 and show that

$$|c(t) - k| \leq \epsilon \quad \text{for} \quad t \geq t^* + \tau.$$

15. Sewage treatment affords another example of mixing problems. Polluted water with a concentration c_1 of pounds of pollutant per gallon is pumped at a rate of r_1 gal/min into a tank holding V_0 gallons of clean water. The tank is pumped out at a rate of r_2 gal/min with $r_2 \leq r_1$. When the concentration $c(t)$ of pollutant in the tank reaches a manageable level αc_1, $0 < \alpha < 1$, the tank is bypassed and then is aerated to reduce the concentration further to a safe level for discharge.

a) Write down the differential equation and initial conditions governing the concentration $c(t)$.

b) If $r_1 = 400$, $r_2 = 200$, and $V_0 = 10{,}000$, find the solution $c(t)$.

c) With the data given in (b) estimate the time it takes for the concentration to reach a level of $\frac{1}{3} c_1$.

16. Another example similar to a mixing problem is a simplified model of the heart. If the volume of blood in the heart is $v(t)$, the inflow is $I(t)$, and the outflow is $F(t)$ then

$$\frac{dv}{dt} = I(t) - F(t).$$

However, it is more sensible to use pressure as the dependent variable, so let

$$p(t) = K[v(t) - v_0], \qquad F(t) = p(t)/R$$

where K, v_0, and R are constants.

a) Find the differential equation satisfied by $p(t)$.

b) The systolic stage, when blood is pumped out, lasts from $t = t_0$ to $t = T$ and $I(t) \equiv 0$. Find $p(t)$ given the initial condition $p(t_0) = p_0$, $t_0 \leq t \leq T$.

c) The diastolic stage, when blood is pumped in, lasts from $t = 0$ to $t = t_0$, and $I(t) \equiv I_0$, a constant. Find $p(t)$ satisfying the endpoint condition $p(t_0) = p_0$, $0 \leq t \leq t_0$.

d) Finally, since the heart is cyclic, it is required that $p(0) = p(T)$, hence the terminal value of the solution found in (b) must match the initial value of the solution found in (c). Find the value of R that guarantees this.

SUMMARY

The concept of a differential equation and its solution is introduced, but the main topic of the chapter is the first order differential equation. The basic existence and uniqueness theorem for solutions of this equation is given as well as the geometric concepts of integral curves and direction fields. Two numerical techniques, Euler's algorithm and the improved Euler algorithm, for approximating solutions are presented.

Although a number of types of differential equations are analyzed, such as separable equations and exact equations, the most important equation discussed is the first order linear equation

$$\frac{dy}{dt} + p(t)y = q(t).$$

This equation, possibly with associated initial conditions $y(t_0) = y_0$, has widespread applicability in the physical and life sciences, and the structure of its solutions generalizes to the discussion of higher order linear equations in the chapters following.

Specific applications, discussed at length, of first order linear and nonlinear differential equations are to electric circuit problems, mixing problems with feedback control, and models of population growth. In the latter a discussion of the assumptions leading to the growth model described by the logistic differential equation is given.

MISCELLANEOUS EXERCISES

1.1. The Riccati differential equation (Count J. Riccati, 1676–1754) is a first order nonlinear differential equation of the form

$$\frac{dy}{dt} = p(t)y + q(t)y^2 + r(t).$$

(If $r(t) = 0$, it is a special case of Bernoulli's equation [see Exercise 1.3] and can be solved exactly.) If $y_1(t)$ is a solution let

$$y(t) = y_1(t) + \frac{1}{x(t)},$$

and by substitution into the differential equation, show that $x(t)$ satisfies a linear differential equation. Hence, knowing a particular solution, we can find the general solution.

1.2. Use the result of the previous exercise to find the solution of the following initial value problems, given a particular solution $y_1(t)$.

a) $\dfrac{dy}{dt} = y^2 - 1,\ y(0) = 3;\ y_1(t) = 1$

b) $\dfrac{dy}{dt} = y + \dfrac{2}{t^3}y^2 - t^2,\ y(1) = 4;\ y_1(t) = t^2$

c) $\dfrac{dy}{dt} = -y^2 + \dfrac{15}{4t^2},\ y(1) = 4;\ y_1(t) = \dfrac{a}{t}$. What are possible values of a?

1.3. The Riccati differential equation occurs in the design of a feedback-control law for the linear-regulator problem with quadratic cost. A one-dimensional version of that problem is the following: A system with output $x(t)$ and input or control $u(t)$ is governed by the linear differential equation

$$\frac{dx}{dt} = a(t)x + b(t)u, \qquad x(t_0) = x_0, \quad t_0 \leq t \leq t_1.$$

Choose a control $u = u(t)$ so as to minimize the cost or performance measure

$$C[u] = \frac{1}{2}kx(t_1)^2 + \frac{1}{2}\int_{t_0}^{t_1}[w_1(t)x(t)^2 + w_2(t)u(t)^2]dt,$$

where $w_1(t)$ and $w_2(t)$ are given weighting functions and k is a nonnegative constant.

The theory of this problem shows that the optimal control $u(t)$, which minimizes $C[u]$, is realized through the feedback control law:

$$u(t) = p^*(t)x(t), \qquad p^*(t) = -\frac{b(t)}{w_2(t)}p(t),$$

where $p(t)$ satisfies the Riccati differential equation

$$\frac{dp}{dt} = -2a(t)p + \frac{b(t)^2}{w_2(t)}p^2 - w_1(t), \qquad p(t_1) = k.$$

If one can find $p(t)$, the problem is completely solved. Hence, we find the solution of

$$\frac{dx}{dt} = [a(t) + b(t)p^*(t)]x, \qquad x(t_0) = x_0,$$

and then evaluate

$$C[u] = \frac{1}{2} kx(t_1)^2 + \frac{1}{2} \int_{t_0}^{t_1} [w_1(t) + w_2(t)p^*(t)^2]x(t)^2 \, dt,$$

which will be a minimum.

Given a linear regulator problem

$$\frac{dx}{dt} = x + u, \qquad x(0) = 1,$$

$$C[u] = \frac{1}{2}kx(1)^2 + \frac{1}{2} \int_0^1 [3x(t)^2 + u(t)^2] \, dt.$$

a) Solve the problem completely for the case $k = 3$; i.e., find the feedback control law, the optimal output $x(t)$ and the value of $C[u]$.

b) Find the feedback control law for the case $k = 0$.

c) For the case $k = 2.75$, find the feedback control law, then use the improved Euler method to approximate $x(t)$ on $0 \leq t \leq 1$. With the numerical values you can use a numerical quadrature formula (e.g., Simpson's rule) to estimate the minimum cost $C[u]$.

1.4. Given the first order linear equation

$$\frac{dx}{dt} + ax = \phi(t),$$

where $a \neq 0$ and $\phi(t)$ is a continuous periodic function of period T, show that it can have at most one solution of period T. (*Hint:* You could assume it has two solutions and find the differential equation satisfied by their difference.) Show by example that if $a = 0$, there could exist infinitely many periodic solutions or none.

1.5. It began to snow sometime in the evening. At midnight the snowplow started out and by 1:00 A.M. it had plowed two miles down the main highway. By 2:00 A.M. it had plowed only one more mile. Assuming it snows at a constant rate and the plow removes snow at a constant rate, when did it start to snow? (An almost classic problem devised by Prof. R. P. Agnew.)

1.6. The fact that $y^2 < y^2 + t^2$ for $t > 0$ implies that the solution of

$$\frac{dy}{dt} = y^2 + t^2, \qquad y(0) = y_0 > 0$$

grows faster than the solution of

$$\frac{dy}{dt} = y^2, \qquad y(0) = y_0.$$

Use this fact to show that the solution of the first differential equation becomes infinite for a finite value of $t > 0$. (A little analysis of this kind prior to doing any numerical work pays off!)

1.7. Following the idea developed in the previous problem, show that the solution of

$$\frac{dy}{dt} = y^{1/2} + t^2, \qquad y(0) = 1,$$

is bounded for $t > 0$ and satisfies

$$1 + t + \frac{t^3}{3} < y(t) < 3e^t - t^2 - 2t - 2.$$

1.8. Show that the solution of

$$\frac{dy}{dt} = y^2 \cos t, \qquad y(0) = y_0 \neq 0,$$

is defined for $-\infty < t < \infty$ if $|y_0|^{-1} > 1$, whereas it is defined only on a finite interval if $|y_0|^{-1} < 1$.

1.9. Find the exact solution for $\epsilon > 0$ of the equation

$$\epsilon \frac{dy}{dt} + y = 1 + t, \qquad y(0) = 0,$$

and sketch its graph for $t \geq 0$ and $\epsilon = 10^{-1}, 10^{-2}$. For each value of ϵ, determine the interval $0 \leq t \leq \epsilon$ for which the solution $y(t)$ satisfies

$$(1 + t) - y(t) \geq 2\epsilon.$$

This narrow region is an example of a *boundary layer*—a region of sudden transition. The boundary layer phenomenon occurs in many problems of fluid dynamics.

Another class of equations that occur frequently in applications are called *integral equations.* They can be roughly characterized as equations in which integrals of the unknown function $y(t)$ appear. In some cases integral equations can be converted to ordinary differential equations by successive differentiations. Use the fundamental theorem of calculus,

$$\frac{d}{dt} \int_0^t f(s)ds = f(t),$$

to convert the integral equations of Exercises 1.10–1.13 into ordinary differential equations.

1.10. $y(t) = \int_0^t e^s y(s)ds$

1.11. $y(t) = e^{2t} + \int_1^t sy(s)ds$

1.12. $y(t) = \sin t + \int_0^t (t - s)y(s)ds$

1.13. $y(t) = \int_0^t \cos(t - s)y(s)ds$

REFERENCES

There are many introductory texts on ordinary differential equations. A few studies the reader will find worth examining are:

1. W.E. Boyce and R.C. DiPrima, *Elementary Differential Equations,* 4th ed., Wiley, New York, 1986.
2. F. Brauer and J.A. Nohel, *Ordinary Differential Equations with Applications,* Harper & Row, New York, 1986.
3. M. Braun, *Differential Equations and Their Applications,* Springer-Verlag, New York, 1975.
4. W.R. Derrick and S.I. Grossman, *Elementary Differential Equations with Applications,* 2nd ed., Addison-Wesley, Reading, Mass., 1981.
5. S.L. Ross, *Introduction to Ordinary Differential Equations,* 3rd ed., Wiley, New York, 1974.

For further study of the modeling of populations and ecological systems a classic study is

6. E.C. Pielou, *An Introduction to Mathematical Ecology,* Wiley-Interscience, New York, 1969.

Other interesting discussions of population models as well as the modeling of other biological systems are

7. J.M. Smith, *Mathematical Ideas in Biology,* Cambridge Press, Cambridge, 1968.
8. J.M. Smith, *Models in Ecology,* Cambridge Press, Cambridge, 1974.
9. S. Levin (Ed.), *Studies in Mathematical Biology,* Parts I and II, Mathematical Association of America Studies in Mathematics, Volumes 15 and 16, 1978. This is a collection of articles, some of which are quite advanced in level; it gives an idea of the scope of problems in biology being modeled mathematically.
10. D.S. Jones and B.D. Sleeman, *Differential Equations and Mathematical Biology,* Allen and Unwin, London, 1983.

A good reference for numerical methods is

11. C.F. Gerald, *Applied Numerical Analysis,* 2nd ed., Addison-Wesley, Reading, Mass., 1978.

2

LINEAR SECOND ORDER DIFFERENTIAL EQUATIONS

2.1 INTRODUCTION

Second order ordinary differential equations play a central role in the theory of differential equations as well as in their applications to engineering, physics, and other sciences. Their role in the theory is due, in part, to their ability to model concepts, such as linear independence (which occurs in the study of equations of order two or higher, but not in equations of order one) and to their usefulness in studying such topics as the phase plane analysis of nonlinear oscillators (in which the topology of the two-dimensional plane is essential). Their role in applied mathematics is due to the great variety and importance of phenomena which can be described by second order differential equations, such as LCR circuits and mechanical systems with one degree of freedom, and because they are an essential tool in solving many of the equations of mathematical physics.

Although we shall emphasize linear second order differential equations in this chapter, a brief introduction to higher order linear equations is included. Nonlinear differential equations are discussed in Chapter 7.

2.2 EXAMPLES FROM MECHANICS AND CIRCUIT THEORY

The analysis of simple mechanical systems, which give rise to oscillatory or damped oscillatory motion, occurs in almost every introduction to differential equations because it is based on easily understood concepts of mechanics, namely Newton's and Hooke's laws. Furthermore, it illustrates the effects of damping and the phenomenon of resonance, which are of fundamental importance in the design of vibration absorbing devices. More important, the physical description, and its consequent mathematical translation, goes hand in hand with physical reality, as evidenced by the widespread use of the model and variations of it in engineering, structural analysis, and applied mechanics.

The result of translating the physical description of a simple oscillatory mechanical system into a mathematical model is a linear second order differential equation with constant coefficients. In the opinion of the authors, a thorough understanding of the qualitative behavior of solutions of this equation as parameters vary, or external driving forces are added, is *the most important topic* to be learned from an introductory course in differential equations. We now proceed to construct a mathematical model that gives us a differential equation.

A MASS–SPRING SYSTEM

Consider a light spring of length l suspended from a fixed support. If a mass m is attached to the bottom of the spring, it will be extended by a length Δl (see Fig. 2.1a). This extension is due to a gravitational force mg, where g is the gravitational constant. By Hooke's law of linear elasticity, if a spring is compressed or extended from its natural length, it will exert a restoring force proportional to the displacement.[1] Since the restoring force must equal the gravitational force, we have the relation,

$$k\Delta l = mg,$$

where k, the constant of proportionality, is called the *spring constant.* It is a measure of the stiffness of the spring.

Now suppose that at an initial time t_0 the mass is vertically displaced y_0 units from its equilibrium or rest position and that it has a velocity v_0 upon release. If $y = y(t)$ is the distance of the mass (measured downwards) at time t from equilibrium, then the spring extension at time t is $\Delta l + y(t)$. From Newton's second law of motion,

Rate of change in downward momentum = Net downward force,

we have

$$\frac{d}{dt}\left[m\frac{dy}{dt}(t)\right] = mg - k[\Delta l + y(t)].$$

1. Robert Hooke (1635–1703), an English physicist and inventor, published his law of linear elasticity in 1676.

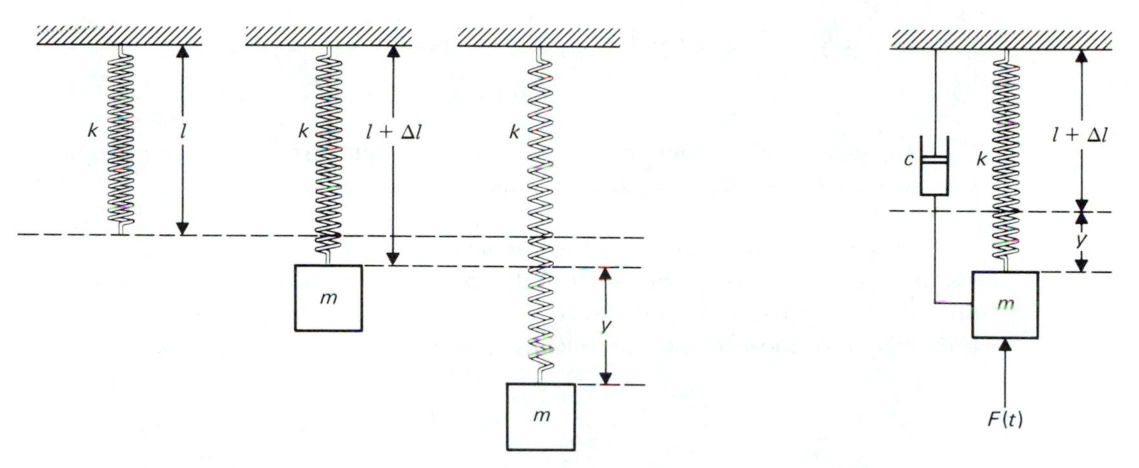

Figure 2.1 (a) A mass–spring system with no damping or forcing. (b) A mass–spring system with damping and forcing.

Since $k\Delta l = mg$, this becomes (letting $y = y(t)$)

$$m\frac{d^2y}{dt^2} = -ky \quad \text{or} \quad m\frac{d^2y}{dt^2} + ky = 0 \tag{2.2.1}$$

a linear second order differential equation. In addition we are given the initial conditions of displacement and velocity,

$$y(t_0) = y_0, \qquad \frac{dy}{dt}(t_0) = v_0. \tag{2.2.2}$$

A typical problem is to solve the constant coefficient second order linear differential equation (2.2.1) subject to the initial conditions (2.2.2). Its solution gives the displacement of the mass from equilibrium at any time $t \geq t_0$.

EXAMPLE 1 A one kilogram mass stretches a spring by 20 centimeters. It is then stretched an additional 5 centimeters and released. If $y(t) = A\cos\omega t$ describes the subsequent motion, find its amplitude A and angular frequency ω.

SOLUTION Since $m = 1$ kg, $\Delta l = 20$ cm, and $g = 980$ cm/sec^2, we have from the force balance equation, $k\Delta l =$ mg, $k = 980/20 = 49$ kg/sec^2. The governing differential equation is

$$\frac{d^2y}{dt^2} + 49y = 0.$$

Substituting $y = A \cos \omega t$ and $d^2y(t)/dt^2 = -A\omega^2 \cos \omega t$, we obtain

$$(-\omega^2 + 49)A \cos \omega t = 0.$$

Hence, $-\omega^2 + 49 = 0$ and $\omega = 7 \text{ sec}^{-1}$ is the angular frequency. The condition $y(0) = 5$ gives $A = 5$ cm as the amplitude of the vibration. ■

However, this description of the physical model is quite simple, and we might wish to take into account the effect of damping due, for instance, to air resistance on the mass. A plausible model for damping is to assume that it resists the motion with a force proportional to the velocity. Therefore the damping force F_d is

$$F_d = -c \frac{dy}{dt}(t),$$

where the positive constant c is called the *damping coefficient.* Such damping is called *viscous damping* and is often modeled by the addition of a piston-like device called a *dashpot* (see Fig. 2.1b).

There could also be a time dependent external force $F(t)$ acting on the mass. This could be the result of a movable, rather than fixed, support, or the mass could be attached to another mass–spring system, as in a shock or vibration absorbing mechanism. If the forces F_d and $F(t)$ are included in our model, we have from Newton's second law,

$$\frac{d}{dt}\left(m \frac{dy}{dt}\right) = \text{Net downward force} = -ky - c\frac{dy}{dt} + F(t)$$

or

$$m \frac{d^2y}{dt^2} + c\frac{dy}{dt} + ky = F(t). \tag{2.2.3}$$

The solution $y = y(t)$ and its derivative $v = y'(t)$ will give the displacement from equilibrium and velocity of the mass at time t subject to the initial conditions

$$y(t_0) = y_0, \qquad y'(t_0) = v_0. \tag{2.2.4}$$

A principal objective of this chapter is to analyze and solve, whenever possible, the initial value problem (2.2.3) and (2.2.4).

EXAMPLE 2 Show that $y(t) = (-B \cos t + A \sin t)/c$ is a 2π-periodic solution to the mass–spring equation,

$$y'' + cy' + y = B \sin t + A \cos t,$$

where A and B are constants and $c \neq 0$.

SOLUTION From

$$y(t) = (-B\cos t + A\sin t)/c,$$
$$y'(t) = (B\sin t + A\cos t)/c,$$
$$y''(t) = -(-B\cos t + A\sin t)/c,$$

we have

$$y''(t) + cy'(t) + y(t) = B\sin t + A\cos t.$$

Next we observe that for any t,

$$\begin{aligned} y(t + 2\pi) &= (-B\cos(t + 2\pi) + A\sin(t + 2\pi))/c \\ &= (-B\cos t + A\sin t)/c = y(t), \end{aligned}$$

the condition that $y(t)$ have period equal to 2π. ■

The next example from circuit theory shows that the differential equation (2.2.3) is not solely a mathematical model for a mass–spring system. A useful mathematical model described by, say, a differential equation and/or initial value problem will often represent a number of distinct physical situations. One can then study the mathematical model in the abstract and, with an appropriate choice of the parameters, interpret the quantitative and qualitative results obtained in terms of the physical reality under study.

AN ELECTRICAL CIRCUIT

Suppose three circuit elements are connected in series, as shown in Fig. 2.2; the symbols for the resistor, capacitor, and inductor are identified by the letters R, C, and L, respectively. The imposed electromotive force, which could be a battery or a signal generator, is represented by a circular symbol and denoted by $E(t)$. One can assume the presence of voltmeters or oscilloscopes that would monitor the process without affecting it.

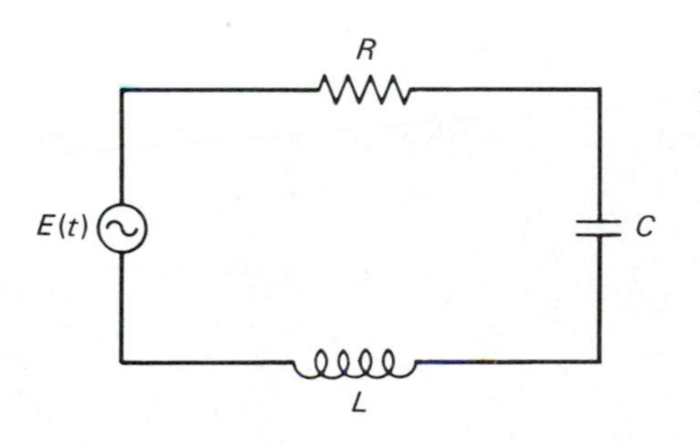

Figure 2.2 An *LCR* circuit.

Let I denote the current flowing through the circuit; then the voltage drop across the resistor is proportional to the current,

$$V_R = RI,$$

where R denotes the magnitude of the resistor. On the other hand, if current is flowing through an inductor, the voltage drop across the inductor is proportional to the rate of change of the current with time,

$$V_L = L\,dI/dt,$$

where L denotes the magnitude of the inductor. Note that there is no voltage drop due to inductance if the current is constant. Since the inductor consists of a coil of wire, there is always resistance present but, for convenience, it is lumped together with the resistance R of the resistor.

A perfect capacitor does not allow current to flow through it but, instead, it stores electric charges. As the amount of charge Q changes, current flows into and out of the capacitor. The voltage drop across the capacitor is proportional to its charge:

$$V_C = Q/C,$$

where C denotes the magnitude of the capacitor.

The three equations above are related through a fundamental principle of circuit theory, namely, that the sum of the voltage drops in a closed loop is equal to the imposed electromotive force, hence,

$$V_L + V_R + V_C = E(t).$$

Substituting the above equations for V_L, V_R, and V_C, we have the differential equation

$$L\frac{dI}{dt} + RI + \frac{Q}{C} = E(t). \tag{2.2.5}$$

A second differential equation is obtained by noting that the current is the time rate of change of the charge:

$$\frac{dQ}{dt} = I. \tag{2.2.6}$$

These two first order equations are coupled and are equivalent to a single second order equation constructed by setting

$$\frac{dI}{dt} = \frac{d^2Q}{dt^2}, \qquad I = \frac{dQ}{dt}$$

in (2.2.5) to obtain

$$L\frac{d^2Q}{dt^2} + R\frac{dQ}{dt} + \frac{Q}{C} = E(t). \tag{2.2.7}$$

Equation (2.2.7) is a linear second order differential equation with constant coefficients whose solution is $Q(t)$, the charge.

Sometimes it is desirable to model the circuit by a differential equation with I as the dependent variable. This can be accomplished by differentiating (2.2.5) to obtain

$$L\frac{d^2I}{dt^2} + R\frac{dI}{dt} + \frac{I}{C} = \frac{dE(t)}{dt}. \tag{2.2.8}$$

However, if the impressed electromotive force $E(t)$ is rapidly changing, e.g., as in a square wave or in a saw-toothed wave, equation (2.2.7) is preferable to (2.2.8) since the right-hand side function, *the forcing function,* is smoother.

The pair of differential equations (2.2.5) and (2.2.6) or the equivalent second order differential equation (2.2.7) can be solved for $Q(t)$ and $I(t)$. More information is needed to specify the solution functions completely. It is sufficient, and for this problem natural, to specify the charge and current at a convenient time, say, $Q = Q_0$ and $I = I_0$ at $t = t_0$, to give the initial conditions

$$Q(t_0) = Q_0, \qquad I(t_0) = Q'(t_0) = I_0. \tag{2.2.9}$$

The reader can easily see that there is no essential *mathematical* difference between the initial value problem (2.2.3), (2.2.4) for the mass–spring system and the initial value problem (2.2.7), (2.2.9) for an LCR circuit.

EXAMPLE 3 Show that $Q(t) = Q_0e^{-4000t}\cos(3000t)$ is a solution to the LCR differential equation (2.2.7) if $L = 1$ henry, $C = 4 \times 10^{-8}$ farads, and $R = 8 \times 10^3$ ohms and $E(t) = 0$. Find a value of t, t^*, so that $|Q(t)| \leq |Q_0|/2$ for $t \geq t^*$.

SOLUTION The governing differential equation is

$$\frac{d^2Q}{dt^2} + 8 \times 10^3\frac{dQ}{dt} + 25 \times 10^6 Q = 0.$$

To check that the given $Q(t)$ is a solution, substitute to obtain

$$\begin{aligned}
&\frac{d^2}{dt^2}[Q_0e^{-4000t}\cos(3000t)] \\
&+ 8 \times 10^3\frac{d}{dt}[Q_0e^{-4000t}\cos(3000t)] \\
&+ 25 \times 10^6[Q_0e^{-4000t}\cos(3000t)] = Q_0 \times 10^6 e^{-4000t}\{(16 - 9)\cos(3000t) \\
&\qquad + 24\sin(3000t) - 32\cos(3000t) \\
&\qquad - 24\sin(3000t) + 25\cos(3000t)\} \\
&\qquad = 0.
\end{aligned}$$

Since

$$|Q(t)| \leq |Q_0| e^{-4000t},$$

take t^* so that

$$\exp(-4000t^*) = \frac{1}{2}, \quad \text{or} \quad t^* = 2.5 \times 10^{-4} \ln 2.$$

Then $t \geq t^*$ implies that $|Q(t)| \leq |Q_0|/2$, as required. ■

In the next six sections we shall mathematically analyze the second order linear differential equation and the corresponding initial value problem. The solution techniques apply mainly to the case of constant coefficients, but much of the theory applies to the case of time-dependent coefficients as well. In the latter sections we shall return to the two examples above and interpret the mathematical results in terms of the physical reality.

EXERCISES 2.2

In Exercises 1–4 the MKS system of units is being used with $g = 9.8$ m/sec^2.

1. a) A mass of 10 kg attached to a spring stretches it 0.5 m from its natural length. Find the spring constant k.

b) The mass–spring system is set in motion by displacing the 10 kg mass 0.1 m above the equilibrium position. Find the differential equation and the initial conditions satisfied by the system assuming no air resistance.

2. a) A mass of 10 kg attached to a spring stretches it 2 m from its natural length. Find the spring constant k.

b) The mass–spring system is set in motion by starting the 10 kg mass from equilibrium with an initial velocity of 4 m/sec in the downward direction. Find the differential equation and the initial conditions satisfied by the system assuming no air resistance.

3. A mass of 4 kg attached to a spring stretches it 0.2 m from its natural length. The damping coefficient is 10 kg/sec and the mass–spring system is set in motion from equilibrium with a downward initial velocity of 2 m/sec. Find the differential equation and the initial conditions satisfied by the system.

4. A mass of $\frac{1}{2}$ kg is attached to a spring having a spring constant of 5 kg/sec^2. The mass–spring system is set in motion by initially displacing it 2 m in the downward direction. If the damping coefficient is 4 kg/sec find the differential equation and the initial conditions satisfied by the system.

5. Suppose the configuration shown in Fig. 2.1(b) is modified so that the dashpot is permanently anchored to the floor, and the vertical support is oscillating with a displacement of $A \cos wt$. What is the differential equation satisfied by the system?

6. An LCR circuit has $R = 120$ ohms, $C = 10^{-2}$ farads, $L = 20$ henrys, and an imposed voltage of $E(t) = 10$ volts. There is no initial charge on the capacitor but there is an initial current of 2 amperes at $t = 0$. Find the differential equation and the initial conditions satisfied by the charge.

7. An LCR circuit has $R = 10$ ohms, $C = 10^{-2}$ farads, $L = 1$ henry, and an imposed voltage of $E(t) = 8 \sin t$ volts. Assuming no initial current or charge at $t = 0$, find the differential equation and the initial conditions satisfied by the charge.

8. Suppose the LCR circuit of Fig. 2.2 is modified so that the voltage source is replaced by a switch. At $t = 0$ a charge of $\frac{1}{10}$ coulomb is placed on the capacitor and the switch is closed. If $R = 4$ ohms, $C = 2 \times 10^{-2}$ farads and $L = \frac{1}{5}$ henry, find the differential equation and the initial conditions satisfied by the charge.

9. A disc of mass m is suspended from its center by a wire. If the disc is rotated the stiffness of the wire provides a restoring force proportional to the angular displacement θ. Find the differential equation satisfied by θ using the fact that the torque equals $I(d^2\theta/dt^2)$, where I is the moment of inertia about the axis of rotation.

10. **a)** Show that a simple harmonic motion

$$y(t) = A \cos(\omega t - \varphi)$$

where A = amplitude, ω = frequency and φ = phase shift, satisfies the second order linear differential equation

$$y'' + \omega^2 y = 0,$$

with initial conditions $y(0) = A \cos \varphi$, $y'(0) = \omega A \sin \varphi$.

b) Show that a damped harmonic motion described by

$$y(t) = Ae^{-kt} \cos(\omega t - \varphi), \qquad k > 0,$$

satisfies the second order linear differential equation

$$y'' + 2ky' + (\omega^2 + k^2)y = 0,$$

with initial conditions $y(0) = A \cos \varphi$, $y'(0) = A\sqrt{\omega^2 + k^2} \sin(\varphi - \theta)$ where $\tan \theta = k/\omega$.

2.3 GENERAL THEORY—INTRODUCTION

Before studying techniques for finding solutions to linear second order differential equations, we discuss some basic properties of such equations and their solutions. We assume our equations have the form

$$a(t)\frac{d^2y}{dt^2} + b(t)\frac{dy}{dt} + c(t)y = f(t), \tag{2.3.1}$$

where $a(t)$, $b(t)$, $c(t)$, and $f(t)$ are continuous functions on some open interval. In order to insure that the equation is always of second order, the coefficient $a(t)$ must never be zero in the interval. If the forcing function $f(t)$ is zero, the differential equation

$$a(t)\frac{d^2y}{dt^2} + b(t)\frac{dy}{dt} + c(t)y = 0 \tag{2.3.2}$$

is called *homogeneous;*[2] otherwise, it is referred to as *inhomogeneous.*

The essential property of linear equations is, of course, *linearity.* Namely, if $y_1(t)$ and $y_2(t)$ are solutions of the homogeneous equation (2.3.2), then

$$y(t) = c_1y_1(t) + c_2y_2(t)$$

is also a solution for any choice of the constants c_1 and c_2. This follows easily from the linearity of differentiation:

$$\begin{aligned}
a(t)\frac{d^2}{dt^2}[c_1y_1(t) + c_2y_2(t)] &+ b(t)\frac{d}{dt}[c_1y_1(t) + c_2y_2(t)] + c(t)[c_1y_1(t) + c_2y_2(t)] \\
&= c_1\left[a(t)\frac{d^2}{dt^2}y_1(t) + b(t)\frac{d}{dt}y_1(t) + c(t)y_1(t)\right. \\
&\quad + c_2\left[a(t)\frac{d^2}{dt^2}y_2(t) + b(t)\frac{d}{dt}y_2(t) + c(t)y_2(t)\right].
\end{aligned}$$

Each term in brackets is zero because $y_1(t)$ and $y_2(t)$ are solutions. Therefore $y(t)$ is a solution to the homogeneous equation (2.3.2).

By analogy with linear first order equations, we find that if $y_p(t)$ is a solution to the inhomogeneous equation (2.3.1) and $w(t)$ is a solution to the homogeneous equation (2.3.2), then

$$y(t) = y_p(t) + Kw(t)$$

is a one-parameter family of solutions to the inhomogeneous equation (2.3.1). As usual, the constant (or parameter) K is arbitrary. To see this, substitute the

2. That unfortunate usage again! This should not be confused with homogeneous first order equations discussed in Section 1.8.

assumed solution into the differential equation (2.3.1) to obtain

$$a(t)\frac{d^2}{dt^2}[y_p(t) + Kw(t)] + b(t)\frac{d}{dt}[y_p(t) + Kw(t)] + c(t)[y_p(t) + Kw(t)]$$
$$= \left[a(t)\frac{d^2}{dt^2}y_p(t) + b(t)\frac{d}{dt}y_p(t) + c(t)y_p(t)\right]$$
$$+ K\left[a(t)\frac{d^2}{dt^2}w(t) + b(t)\frac{d}{dt}w(t) + c(t)w(t)\right].$$

The first bracketed term is equal to $f(t)$ since $y_p(t)$ is a solution to the inhomogeneous equation (2.3.1), and the second bracketed term is equal to zero since $w(t)$ is a solution to the homogeneous equation (2.3.2). Thus $y(t)$ is a solution to the inhomogeneous equation (2.3.1).

EXAMPLE 1 Show that $y(t) = \frac{1}{2}t^4 + Kt^{-2}$ is a solution to the inhomogeneous differential equation

$$t^2\frac{d^2y}{dt^2} + t\frac{dy}{dt} - 4y = 6t^4 \tag{2.3.3}$$

by verifying that $y_p(t) = \frac{1}{2}t^4$ is a solution to the inhomogeneous equation and that $w(t) = t^{-2}$ is a solution to the corresponding homogeneous equation.

SOLUTION Compute the derivatives

$$\frac{dy_p}{dt}(t) = 2t^3, \qquad \frac{dw}{dt}(t) = -2t^{-3},$$

$$\frac{d^2y_p}{dt^2}(t) = 6t^2, \qquad \frac{d^2w}{dt^2}(t) = 6t^{-4},$$

and substitute into the appropriate differential equation to obtain

$$t^2(6t^2) + t(2t^3) - 4\left(\frac{1}{2}t^4\right) = 6t^4$$

$$t^2(6t^{-4}) + t(-2t^{-3}) - 4(t^{-2}) = 0.$$

This shows that $y_p(t) = \frac{1}{2}t^4$ is a solution to the inhomogeneous equation and $w(t) = t^{-2}$ is a solution to the corresponding homogeneous equation. Therefore $y(t) = \frac{1}{2}t^4 + Kt^{-2}$ is a one-parameter family of solutions to (2.3.3). ■

Given a one-parameter family of solutions $y(t) = y_p(t) + Kw(t)$ to equation (2.3.1), do we have all possible solutions? In other words, if $z(t)$ is a solution, can K be selected so that $z(t) = y_p(t) + Kw(t)$? The answer to this question is no. As an example we refer to the differential equation in Example 1 above. It is easy to check that $z(t) = \frac{1}{2}t^4 + 3t^2$ is a solution to equation (2.3.3) and we have already shown that $y(t) = \frac{1}{2}t^4 + Kt^{-2}$ is a one-parameter family of solutions. Clearly there is no choice for K for which

$$\tfrac{1}{2}t^4 + 3t^2 = \tfrac{1}{2}t^4 + Kt^{-2}$$

on any nonempty interval.

Upon reflection, the attempt to represent all solutions to a second order differential equation by a one-parameter family is unreasonable. As was pointed out in the preceding section, linear second order differential equations can be used to model mechanical or electrical problems whose solution is determined by specifying two quantities, e.g., the displacement and velocity or the charge and current at a given time. One should therefore expect that a linear second order differential equation would have a two-parameter family of solutions. For example,

$$y(t) = \tfrac{1}{2}t^4 + c_1t^2 + c_2t^{-2}$$

is a two-parameter family of solutions to the differential equation of Example 1. The following theorem is a mathematical justification for two-parameter families of solutions.

The existence and uniqueness theorem

Let $a(t)$, $b(t)$, $c(t)$, and $f(t)$ be continuous functions on an interval $t_1 < t < t_2$, with $a(t)$ nonzero. Then there exists a pair of solutions, $y_1(t)$ and $y_2(t)$, of the homogeneous differential equation (2.3.2) and a solution $y_p(t)$ of the inhomogeneous differential equation (2.3.1), defined on the interval, such that the two-parameter family

$$y(t) = y_p(t) + c_1y_1(t) + c_2y_2(t) \tag{2.3.4}$$

contains all the solutions of the inhomogeneous differential equation (2.3.1). Furthermore, for any t_0 in the interval and any numbers r and s, there is a unique choice of c_1 and c_2 such that the solution (2.3.4) satisfies the *initial conditions*

$$y(t_0) = r, \qquad \frac{dy}{dt}(t_0) = s. \tag{2.3.5}$$

The solution (2.3.4) is called a *general solution.*

A proof of this theorem will be found in Appendix 1. In the case of a homogeneous equation, where $f(t) = 0$, the term $y_p(t)$ will not appear in the expression (2.3.4).

The task of solving a differential equation subject to constraints of the type (2.3.5) is called an *initial value problem.*

EXAMPLE 2 Solve the initial value problem

$$t^2 \frac{d^2y}{dt^2} + t\frac{dy}{dt} - 4y = 6t^4,$$

$$y(2) = 3, \qquad \frac{dy}{dt}(2) = 13,$$

given that a two-parameter family of solutions to the differential equation is

$$y(t) = \frac{1}{2}t^4 + c_1t^2 + c_2t^{-2}.$$

SOLUTION First compute the derivative of $y(t)$ to get

$$\frac{dy}{dt}(t) = 2t^3 + 2c_1t - 2c_2t^{-3}.$$

It is required that at $t = 2$, $y = 3$ and $dy/dt = 13$. Therefore parameters c_1 and c_2 must satisfy the algebraic equations obtained by substituting these data into the expressions for $y(t)$ and $dy(t)/dt$:

$$3 = 8 + 4c_1 + \frac{1}{4}c_2, \qquad 13 = 16 + 4c_1 - \frac{1}{4}c_2.$$

Solving, we get $c_1 = -1$, $c_2 = -4$, and the unique solution to this initial value problem is

$$y(t) = \frac{1}{2}t^4 - t^2 - 4t^{-2}.$$

■

EXERCISES 2.3

For each of the linear second order differential equations in Exercises 1–7, show that each of the given functions is a solution. What initial conditions does each satisfy at $t = t_0$?

1. $y'' + 3y' + 2y = 0$; $y_1(t) = 4e^{-t}$, $y_2(t) = -3e^{-2t}$; $t_0 = 0$
2. $y'' + 9y = 0$; $y(t) = 5\cos 3t - 2\sin 3t$; $t_0 = \pi/6$
3. $y'' - 2y' + y = t$; $y_1(t) = t + 2$, $y_2(t) = t + 2 + 3e^t - te^t$; $t_0 = 0$
4. $y'' + 4y = 3e^{2t}$; $y_1(t) = \frac{3}{8}e^{2t}$, $y_2(t) = \frac{3}{8}e^{2t} + 4\cos(2t - \pi/4)$; $t_0 = \pi/2$
5. $t^2y'' - 3ty' + 4y = 0$; $y_1(t) = t^2$, $y_2(t) = t^2 \ln t$; $t_0 = 1$
6. $y'' - 2y' + 2y = 3\sin t - \cos t$; $y_1(t) = \sin t + \cos t$, $y_2(t) = \sin t + \cos t - 6e^t \sin t$; $t_0 = 0$
7. $t^2y'' + ty' + (t^2 - \frac{1}{4})y = 0$; $y_1(t) = (\cos t)/\sqrt{t}$, $y_2(t) = (\sin t)/\sqrt{t}$; $t_0 = \pi$

For the following differential equations from Exercises 1–6, the given function represents a two-parameter family of solutions. Use it to solve the given initial value problem.

8. From Exercise 1: $y(t) = Ae^{-t} + Be^{-2t}$; $y(0) = 1$, $y'(0) = 2$

9. From Exercise 2: $y(t) = A \cos 3t + B \sin 3t$; $y(0) = 0$, $y'(0) = 1$

10. From Exercise 2: $y(t) = K \cos(3t - \phi)$, where K and ϕ are constants, $K > 0$, $-\pi/2 \leq \phi \leq \pi/2$; $y(0) = 0$, $y'(0) = 1$

11. From Exercise 3: $y(t) = t + 2 + Ae^t + Bte^t$; $y(1) = 3$, $y'(1) = 2$

12. From Exercise 4: $y(t) = \frac{3}{8}e^{2t} + A \cos 2t + B \sin 2t$; $y(\pi/4) = 0$, $y'(\pi/4) = 1$

13. From Exercise 5: $y(t) = At^2 + Bt^2 \ln t$; $y(2) = 4$, $y'(2) = 8$

14. From Exercise 6: $y(t) = \sin t + \cos t + Ke^t \cos(t - \phi)$, where K and ϕ are constants, $K > 0$, $-\pi < \phi \leq \pi$; $y(\pi/2) = 1$, $y'(\pi/2) = 3$

2.4 CONSTANT COEFFICIENT HOMOGENEOUS EQUATIONS

PRELIMINARY REMARKS

Linear differential equations with constant coefficients are easy to solve and provide a rich class of examples to illustrate the general theory of linear differential equations. They can model a surprisingly large number of interesting physical processes. In this section, we shall study the homogeneous second order linear differential equation

$$a\frac{d^2y}{dt^2} + b\frac{dy}{dt} + cy = 0, \qquad \textbf{(2.4.1)}$$

where a, b, and c are constants with $a \neq 0$. The inhomogeneous equation with constant coefficients will be studied in Sections 2.7 and 2.8.

The concept of a general solution to a linear second order differential equation was introduced in Theorem 1, the existence and uniqueness theorem, presented in the previous section. Because of the importance of general solutions in our study of linear homogeneous differential equations, we restate it here.

Definition

A two-parameter family of solutions to the linear homogeneous equation (2.4.1),

$$y(t) = c_1y_1(t) + c_2y_2(t), \qquad \textbf{(2.4.2)}$$

is said to be a *general solution* if the family contains all the solutions to the differential equation.

It is a consequence of the existence and uniqueness theorem that general solutions always exist and that if we are presented with a solution $z(t)$, it is always possible to pick the parameters (constants) c_1 and c_2 so that

$$z(t) = c_1y_1(t) + c_2y_2(t).$$

This is illustrated in the following example.

EXAMPLE 1 Show that $z(t) = \cosh t$ is a solution to

$$y'' - y = 0 \tag{2.4.3}$$

and that

$$y(t) = c_1e^t + c_2e^{-t} \tag{2.4.4}$$

is a general solution.

SOLUTION To show that $z(t)$ is a solution we set

$$\cosh t = \frac{e^t + e^{-t}}{2}, \qquad (\cosh t)' = \frac{e^t - e^{-t}}{2} = \sinh t,$$

$$(\cosh t)'' = (\sinh t)' = \frac{e^t + e^{-t}}{2} = \cosh t.$$

Therefore

$$z''(t) - z(t) = (\cosh t)'' - \cosh t = \cosh t - \cosh t = 0,$$

as required.

Next we check that (2.4.4) is a solution by computing

$$\begin{aligned} y''(t) - y(t) &= (c_1e^t + c_2e^{-t})'' - (c_1e^t + c_2e^{-t}) \\ &= (c_1e^t + c_2e^{-t}) - (c_1e^t + c_2e^{-t}) = 0. \end{aligned}$$

To show that the family of solutions (2.4.4) contains all the solutions to (2.4.3) we must demonstrate that given arbitrary initial conditions there is a unique choice of c_1 and c_2 so that they are satisfied. Let t_0, r, and s be arbitrary numbers and require that $y(t_0) = r$, $y'(t_0) = s$. This leads to two linear algebraic equations that determine c_1 and c_2:

$$r = y(t_0) = c_1e^{t_0} + c_2e^{-t_0},$$

$$s = y'(t_0) = c_1e^{t_0} - c_2e^{-t_0}.$$

By Cramer's rule for solving linear algebraic equations,[3]

$$c_1 = \frac{\begin{vmatrix} r & e^{-t_0} \\ s & -e^{-t_0} \end{vmatrix}}{\begin{vmatrix} e^{t_0} & e^{-t_0} \\ e^{t_0} & -e^{-t_0} \end{vmatrix}} = \frac{-(r+s)e^{-t_0}}{-2} = (r+s)e^{-t_0},$$

$$c_2 = \frac{\begin{vmatrix} e^{t_0} & r \\ e^{t_0} & s \end{vmatrix}}{\begin{vmatrix} e^{t_0} & e^{-t_0} \\ e^{t_0} & -e^{-t_0} \end{vmatrix}} = \frac{(s-r)e^{t_0}}{-2} = \frac{(r-s)}{2}\, e^{t_0}.$$

By the existence and uniqueness theorem and the preceding computation of c_1 and c_2, if $z(t)$ is any solution to (2.4.3) and $z(t_0) = r$, $z'(t_0) = s$, then $z(t)$ has a unique representation,

$$z(t) = \frac{1}{2}(r+s)e^{t-t_0} + \frac{1}{2}(r-s)e^{-(t-t_0)}.$$

In this example $z(t) = \cosh t$, so if we take $t_0 = 0$, $z(0) = 1$, $z'(0) = 0$, then $c_1 = \frac{1}{2}$, $c_2 = \frac{1}{2}$. Therefore

$$z(t) = \frac{1}{2}e^{t} + \frac{1}{2}e^{-t} = \cosh t,$$

as required. ■

This example motivates the following theorem that will be proven in Section 2.6.

Determinant test theorem

The two-parameter family of solutions to (2.4.1) is a general solution if and only if the determinant

$$\begin{vmatrix} y_1(t) & y_2(t) \\ y_1'(t) & y_2'(t) \end{vmatrix} \neq 0 \tag{2.4.5}$$

for all t.

---

3. Recall that $\begin{vmatrix} a & b \\ c & d \end{vmatrix} = ad - bc$.

SOLUTION TECHNIQUES

Recalling that exponential functions are solutions to linear first order differential equations with constant coefficients, it is natural to ask if they also satisfy second order equations. Let $y(t) = \exp(\lambda t)$; substitute it into (2.4.1) to obtain

$$a\lambda^2 e^{\lambda t} + b\lambda e^{\lambda t} + ce^{\lambda t} = e^{\lambda t}(a\lambda^2 + b\lambda + c) = 0.$$

Since $\exp(\lambda t)$ can never be zero, this equality can only be satisfied if λ is a root of the quadratic equation

$$a\lambda^2 + b\lambda + c = 0. \tag{2.4.6}$$

It is no accident that this equation is of second degree in λ; each time $\exp(\lambda t)$ is differentiated, another factor of λ appears. The two roots of equation (2.4.6) are given by the quadratic formula

$$\lambda_1 = \frac{-b + \sqrt{b^2 - 4ac}}{2a}, \qquad \lambda_2 = \frac{-b - \sqrt{b^2 - 4ac}}{2a}. \tag{2.4.7}$$

Definition

The quadratic equation (2.4.6) is called the *characteristic equation* of the differential equation (2.4.1). Its roots are called *characteristic roots.*

The characteristic roots can be classified as:

Type I. Distinct Real with $b^2 - 4ac > 0$,

Type II. Equal Real with $b^2 - 4ac = 0$,

Type III. Distinct Complex with $b^2 - 4ac < 0$.

The nature of the solution to the differential equation (2.4.1) depends on the sign of the *discriminant,* $b^2 - 4ac$. We consider real characteristic roots with $b^2 - 4ac \geq 0$ in this section. Complex characteristic roots with $b^2 - 4ac < 0$ are considered in the next section.

TYPE I. DISTINCT REAL CHARACTERISTIC ROOTS

If the roots of the characteristic equation (2.4.6) are real and distinct, there are two solutions

$$y_1(t) = e^{\lambda_1 t} \quad \text{and} \quad y_2(t) = e^{\lambda_2 t}, \tag{2.4.8}$$

where λ_1 and λ_2 are defined in (2.4.7) with $b^2 - 4ac > 0$. A general solution to the differential equation (2.4.1) is

$$y(t) = c_1 e^{\lambda_1 t} + c_2 e^{\lambda_2 t}, \tag{2.4.9}$$

since the determinant (see the determinant test theorem),

$$\begin{vmatrix} e^{\lambda_1 t} & e^{\lambda_2 t} \\ \lambda_1 e^{\lambda_1 t} & \lambda_2 e^{\lambda_2 t} \end{vmatrix} = e^{(\lambda_1 + \lambda_2)t}(\lambda_2 - \lambda_1) \neq 0.$$

The constants c_1 and c_2 can be chosen so that initial data are satisfied.

EXAMPLE 2 Find a general solution to

$$y'' + y' - 6y = 0,$$

and then choose c_1 and c_2 so that $y(0) = 5$, $y'(0) = 0$.

SOLUTION Let $y(t) = e^{\lambda t}$ and substitute to obtain

$$e^{\lambda t}(\lambda^2 + \lambda - 6) = 0.$$

The characteristic equation, $\lambda^2 + \lambda - 6 = 0$, is easily factored,

$$(\lambda - 2)(\lambda + 3) = 0,$$

giving the characteristic roots $\lambda_1 = 2$ and $\lambda_2 = -3$. The solutions corresponding to these roots are

$$y_1(t) = e^{2t} \quad \text{and} \quad y_2(t) = e^{-3t}$$

and a general solution is

$$y(t) = c_1 e^{2t} + c_2 e^{-3t}$$

with c_1 and c_2 arbitrary constants. To choose c_1 and c_2 so that the initial conditions are satisfied, we form

$$y'(t) = 2c_1 e^{2t} - 3c_2 e^{-3t}$$

and evaluate the expressions for $y(t)$ and $y'(t)$ at $t = 0$. Set $y(0) = 5$ and $y'(0) = 0$ to get

$$5 = c_1 + c_2,$$
$$0 = 2c_1 - 3c_2.$$

This system of linear equations in c_1 and c_2 can be solved using Cramer's rule to obtain

$$c_1 = \frac{\begin{vmatrix} 5 & 1 \\ 0 & -3 \end{vmatrix}}{\begin{vmatrix} 1 & 1 \\ 2 & -3 \end{vmatrix}} = \frac{-15}{-5} = 3,$$

$$c_2 = \frac{\begin{vmatrix} 1 & 5 \\ 2 & 0 \end{vmatrix}}{\begin{vmatrix} 1 & 1 \\ 2 & -3 \end{vmatrix}} = \frac{-10}{-5} = 2.$$

Therefore the solution to the initial value problem is

$$y(t) = 3e^{2t} + 2e^{-3t}. \qquad \blacksquare$$

EXAMPLE 3 Solve the initial value problem

$$y'' + 6y' + 4y = 0, \qquad y(0) = 1, \qquad y'(0) = -3.$$

SOLUTION The characteristic equation

$$\lambda^2 + 6\lambda + 4 = 0$$

has roots $\lambda_1 = -3 + \sqrt{5}$, $\lambda_2 = -3 - \sqrt{5}$. Therefore a general solution is

$$y(t) = c_1 \exp[(-3 + \sqrt{5})t] + c_2 \exp[(-3 - \sqrt{5})t].$$

To determine c_1 and c_2, form

$$y'(t) = (-3 + \sqrt{5})c_1 \exp[(-3 + \sqrt{5})t] + (-3 - \sqrt{5})c_2 \exp[(-3 - \sqrt{5})t],$$

evaluate the expressions for $y(t)$ and $y'(t)$ at $t = 0$, and set $y(0) = 1$, and $y'(0) = -3$ to get

$$1 = c_1 + c_2, \qquad -3 = (-3 + \sqrt{5})c_1 + (-3 - \sqrt{5})c_2.$$

By Cramer's rule,

$$c_1 = \frac{\begin{vmatrix} 1 & 1 \\ -3 & (-3 - \sqrt{5}) \end{vmatrix}}{\begin{vmatrix} 1 & 1 \\ (-3 + \sqrt{5}) & (-3 - \sqrt{5}) \end{vmatrix}} = \frac{-\sqrt{5}}{-2\sqrt{5}} = \frac{1}{2},$$

$$c_2 = \frac{\begin{vmatrix} 1 & 1 \\ (-3 + \sqrt{5}) & -3 \end{vmatrix}}{\begin{vmatrix} 1 & 1 \\ (-3 + \sqrt{5}) & (-3 - \sqrt{5}) \end{vmatrix}} = \frac{-\sqrt{5}}{-2\sqrt{5}} = \frac{1}{2},$$

and so the solution to the initial value problem is

$$y(t) = \frac{1}{2}e^{(-3+\sqrt{5})t} + \frac{1}{2}e^{(-3-\sqrt{5})t} = e^{-3t}\left[\frac{e^{\sqrt{5}t} + e^{-\sqrt{5}t}}{2}\right],$$

or, in a more compact form,

$$y(t) = e^{-3t}\cosh(\sqrt{5}t). \qquad \blacksquare$$

TYPE II. EQUAL REAL CHARACTERISTIC ROOTS

In this case the discriminant, $b^2 - 4ac$, is zero. Since $c = b^2/4a$, the characteristic equation (2.4.6) is equivalent to

$$a\lambda^2 + b\lambda + \frac{b^2}{4a} = a\left(\lambda + \frac{b}{2a}\right)^2 = 0.$$

Thus there is only one exponential solution, $y(t) = c_1e^{\lambda t}$, where $\lambda = -b/2a$ and c_1 is an arbitrary constant. Another solution is needed if we are to solve arbitrary initial value problems. This second solution can be constructed by a mathematical method called *reduction of order.*[4] It is closely related to the variation of parameters method that will be employed in Section 2.8.

We take as a trial solution

$$y(t) = u(t)e^{\lambda t}, \qquad \lambda = \frac{-b}{2a}$$

where $u(t)$ is a nonconstant function that is to be determined. After substituting the expression for $y(t)$ into the differential equation (2.4.1), we obtain, after dividing by $e^{\lambda t}$,

$$a\frac{d^2u}{dt^2}(t) + (2a\lambda + b)\frac{du}{dt}(t) + (a\lambda^2 + b\lambda + c)u(t) = 0.$$

Since

$$\lambda = -b/2a, \qquad a \neq 0, \qquad 2a\lambda + b = 0, \quad \text{and} \quad a\lambda^2 + b\lambda + c = 0,$$

we have

$$\frac{d^2u}{dt^2}(t) = 0,$$

4. If one solution of a homogeneous linear second ordinary differential equation is known, a second solution that is not a constant multiple of the first can be constructed by the reduction of order procedure. In general, a first order differential equation must be solved, hence the "reduction of order" terminology.

a differential equation whose solutions are a linear combination of $u_1(t) = 1$ and $u_2(t) = t$. By construction

$$y_1(t) = u_1(t)e^{\lambda t} = e^{\lambda t},$$

$$y_2(t) = u_2(t)e^{\lambda t} = te^{\lambda t}, \qquad \lambda = \frac{-b}{2a},$$

are two solutions to the differential equation (2.4.1). This fact can be verified by substitution.

A general solution is

$$y(t) = c_1 e^{\lambda t} + c_2 t e^{\lambda t}, \qquad \lambda = \frac{-b}{2a}, \tag{2.4.10}$$

since the determinant

$$\begin{vmatrix} e^{\lambda t} & te^{\lambda t} \\ \lambda e^{\lambda t} & (1 + \lambda t)e^{\lambda t} \end{vmatrix} = e^{2\lambda t}(1 + \lambda t - \lambda t) = e^{2\lambda t} \neq 0$$

for all t.

EXAMPLE 4 Find a general solution to $y'' + 6y' + 9y = 0$.

SOLUTION The characteristic equation,

$$\lambda^2 + 6\lambda + 9 = (\lambda + 3)^2 = 0$$

has only one root, $\lambda = -3$. One solution to the differential equation is $y_1(t) = e^{-3t}$, and using the reduction of order argument, we find the second solution $y_2(t) = te^{-3t}$. It should be noted that we do not have to repeat the derivation each time we solve a double root problem. A general solution is

$$y(t) = c_1 e^{-3t} + c_2 t e^{-3t}.$$

Let us check our result by substituting it into the differential equation:

$$\begin{aligned} y'' + 6y' + 9y = {} & [9c_1 e^{-3t} + (-6 + 9t)c_2 e^{-3t}] \\ & + 6[-3c_1 e^{-3t} + (1 - 3t)c_2 e^{-3t}] \\ & + 9[c_1 e^{-3t} + c_2 t e^{-3t}] = 0. \end{aligned}$$

■

As the next example illustrates, the two-parameter family (2.4.10) can be used to solve initial value problems.

EXAMPLE 5 Solve $4y'' + 4y' + y = 0$, $y(2) = -6$, $y'(2) = 4$.

SOLUTION The characteristic equation

$$4\lambda^2 + 4\lambda + 1 = (2\lambda + 1)^2 = 0$$

has only one root, $\lambda = -\frac{1}{2}$. Therefore one solution is $y_1(t) = e^{-(1/2)t}$, and the reduction of order procedure yields the second, $y_2(t) = te^{-(1/2)t}$. A general solution is then

$$y(t) = c_1 e^{-(1/2)t} + c_2 t e^{-(1/2)t}.$$

To satisfy the initial data, we form

$$y'(t) = c_1 \left(-\frac{1}{2} e^{-(1/2)t}\right) + c_2 \left(1 - \frac{1}{2} t\right) e^{-(1/2)t},$$

evaluate $y(t)$ and $y'(t)$ at $t = 2$, and set $y(2) = -6$, $y'(2) = 4$ to get the conditions

$$-6 = c_1 e^{-1} + 2c_2 e^{-1}$$

$$4 = c_1 \left(-\frac{1}{2} e^{-1}\right).$$

Solving for c_1 and c_2 gives $c_1 = -8e^1$, $c_2 = e^1$, and the desired solution

$$y(t) = -8e^{[1-(1/2)t]} + te^{[1-(1/2)t]}.$$ ■

The solution techniques of this section are summarized in the following statement.

Solution to the constant coefficient homogeneous case: real roots

1. Solve the characteristic equation, $a\lambda^2 + b\lambda + c = 0$, to obtain the characteristic roots

$$\lambda_1 = \frac{-b + \sqrt{b^2 - 4ac}}{2a}, \qquad \lambda_2 = \frac{-b - \sqrt{b^2 - 4ac}}{2a}$$

If $\lambda_1 \neq \lambda_2$, let

$$y_1(t) = e^{\lambda_1 t}, \qquad y_2(t) = e^{\lambda_2 t}.$$

If $\lambda_1 = \lambda_2 = \lambda = -b/2a$, let

$$y_1(t) = e^{\lambda t}, \qquad y_2(t) = te^{\lambda t}.$$

In either case, a general solution is

$$y(t) = c_1 y_1(t) + c_2 y_2(t).$$

2. If initial data $y(t_0) = r$ and $y'(t_0) = s$ are given, select c_1 and c_2 so that

$$r = c_1 y_1(t_0) + c_2 y_2(t_0),$$
$$s = c_1 y_1'(t_0) + c_2 y_2'(t_0).$$

EXERCISES 2.4

For each of the following constant coefficient linear differential equations, the characteristic equation has real roots. Write down the general solution in each case and, if initial data are given, the solution satisfying that data.

1. $y'' - 9y' + 20y = 0$
2. $y'' - 4y = 0$
3. $y'' - ry' + 4y = 0; \quad r \geq 4$
4. $y'' + 2\sqrt{5}\, y' + 5y = 0$
5. $y'' - y' - 30y = 0$
6. $y'' - 6y' = 0$
7. $y'' = 0$
8. $y'' + 2y' + y = 0$
9. $y'' - 5y = 0; \quad y(0) = 0, \quad y'(0) = 1$
10. $y'' - 7y' = 0; \quad y(0) = 1, \quad y'(0) = 1$
11. $y'' + y' + \frac{1}{4} y = 0; \quad y(2) = 1, \quad y'(2) = 1$
12. $y'' + 32y' + 256y = 0; \quad y(-1) = 2, \quad y'(-1) = 1$
13. $y'' - 4y' - 12y = 0; \quad y(0) = 4, \quad y'(0) = -4$
14. $y'' - 2y' - 3y = 0; \quad y(0) = 0, \quad y'(0) = 0$
15. $y'' - 6y' + 9y = 0; \quad y(1) = 0, \quad y'(1) = 3$
16. $y'' = 0, \quad y(-11) = 4, \quad y'(-11) = -3$
17. $y'' + \frac{1}{2}y' + \frac{1}{16}y = 0; \quad y(4) = 4, \quad y'(4) = 0$
18. $y'' + \sqrt{10}\, y' + \frac{5}{2}y = 0; \quad y(-2) = 10, \quad y'(-2) = 0$
19. $y'' - \frac{1}{16} y = 0; \quad y(0) = 2, \quad y'(0) = -1$
20. $y'' + 3y' + 2y = 0; \quad y(1) = 1, \quad y'(1) = 3$
21. Show that the relation $y(t) = c_1 e^{kt} + c_2 e^{-kt}$ can be written in the form

$$y(t) = k_1 \cosh kt + k_2 \sinh kt,$$

where $k_1 = c_1 + c_2$, $k_2 = c_1 - c_2$. Recall that

$$\sinh x = \frac{e^x - e^{-x}}{2}, \qquad \cosh x = \frac{e^x + e^{-x}}{2}.$$

Exercise 21 shows that if the roots of the characteristic equation are $\lambda_1 = -\lambda_2 = k$, then the general solution can always be written in the form

$$y(t) = c_1 \cosh kt + c_2 \sinh kt.$$

Express the solution of the following initial value problems in this form.

22. $y'' - 16y = 0; \quad y(0) = 3, \quad y'(0) = 2$
23. $y'' - \frac{1}{9}y = 0; \quad y(0) = 1, \quad y'(0) = -1$
24. $y'' - 2y = 0; \quad y(0) = 0, \quad y'(0) = 1$

2.5 CONSTANT COEFFICIENT HOMOGENEOUS EQUATIONS WITH COMPLEX CHARACTERISTIC ROOTS

To complete our discussion of the constant coefficient, linear differential equation

$$a\frac{d^2y}{dt^2} + b\frac{dy}{dt} + cy = 0, \tag{2.5.1}$$

we now consider Type III, where the roots of the characteristic equation

$$a\lambda^2 + b\lambda + c = 0$$

are complex numbers. The characteristic roots can be written in the form

$$\begin{aligned} \lambda &= p + iq = -\frac{b}{2a} + i\frac{\sqrt{4ac - b^2}}{2a}, \\ \lambda^* &= p - iq = -\frac{b}{2a} - i\frac{\sqrt{4ac - b^2}}{2a}, \end{aligned} \tag{2.5.2}$$

since $b^2 - 4ac < 0$. As usual, λ^* denotes the complex conjugate of λ, that is, i is replaced by $-i$ (in some texts, the notation $\bar{\lambda}$ is used for the complex conjugate of λ). There are two complex-valued exponential solutions, $e^{\lambda t}$ and $e^{\lambda^* t}$. Because

$$\begin{vmatrix} e^{\lambda t} & e^{\lambda^* t} \\ \lambda e^{\lambda t} & \lambda^* e^{\lambda^* t} \end{vmatrix} = e^{(\lambda+\lambda^*)t}(\lambda^* - \lambda) = e^{(\lambda+\lambda^*)t}\frac{i\sqrt{4ac - b^2}}{a} \neq 0$$

for all t, a general solution to (2.5.1) is

$$y(t) = c_1 e^{\lambda t} + c_2 e^{\lambda^* t} \tag{2.5.3}$$

where c_1 and c_2 are arbitrary real or complex constants.

The verification that the complex-valued exponentials are solutions to the differential equation (2.5.1) rests on the fact that the computational rules for complex-valued exponentials parallel those for real-valued exponentials.[5] For example,

$$\frac{d}{dt} e^{\lambda t} = \lambda e^{\lambda t}$$

is valid if λ is a real or complex constant.

For many applications of differential equations, particularly in electrical engineering, complex-valued solutions are employed. However, real-valued solutions are also needed. They can be derived from the complex-valued solutions as follows. The solution (2.5.3) can be written as

$$y(t) = c_1 e^{(p+iq)t} + c_2 e^{(p-iq)t} = c_1 e^{pt} e^{iqt} + c_2 e^{pt} e^{-iqt}.$$

Factoring out e^{pt}, we have

$$y(t) = e^{pt}(c_1 e^{iqt} + c_2 e^{-iqt}).$$

Next, we employ Euler's formulas (see Appendix 5),

$$e^{iqt} = \cos qt + i \sin qt, \qquad e^{-iqt} = \cos qt - i \sin qt,$$

and obtain

$$y(t) = e^{pt}[(c_1 + c_2) \cos qt + i(c_1 - c_2) \sin qt].$$

If we let

$$k_1 = c_1 + c_2, \qquad k_2 = i(c_1 - c_2),$$

then we have a linear combination of two real functions,

$$y(t) = k_1 e^{pt} \cos qt + k_2 e^{pt} \sin qt, \tag{2.5.4}$$

where the coefficients k_1 and k_2 may be real or complex. It can be verified by substitution that

$$y_1(t) = e^{pt} \cos qt \quad \text{and} \quad y_2(t) = e^{pt} \sin qt$$

are real solutions of the differential equation (2.5.1). Furthermore, (2.5.4) is a general solution to (2.5.1) since

$$\begin{aligned}
&\begin{vmatrix} e^{pt} \cos qt & e^{pt} \sin qt \\ e^{pt}[p \cos qt - q \sin qt] & e^{pt}[p \sin qt + q \cos qt] \end{vmatrix} \\
&\qquad = e^{2pt} [\cos qt(p \sin qt + q \cos qt)] - e^{2pt} [\sin qt(p \cos qt - q \sin qt)] \\
&\qquad = e^{2pt}(q \cos^2 qt + q \sin^2 qt) \\
&\qquad = q e^{2pt} \neq 0,
\end{aligned}$$

for all t because $q = (4ac - b^2)^{1/2}/2a \neq 0$.

5. Review material on the calculus of complex-valued functions is presented in Appendix 5.

EXAMPLE 1 Find two-parameter families of real- and complex-valued solutions of $y'' + 4y' + 5y = 0$.

SOLUTION The characteristic equation, $\lambda^2 + 4\lambda + 5 = 0$, has the distinct complex roots $\lambda = -2 + i$ and $\lambda^* = -2 - i$. One two-parameter family of complex-valued solutions is

$$y(t) = c_1 \exp[(-2 + i)t] + c_2 \exp[(-2 - i)t].$$

Since $\lambda = p + iq = -2 + i$, $p = -2$, $q = 1$, and a two-parameter family of real-valued solutions is

$$y(t) = k_1 e^{-2t} \cos t + k_2 e^{-2t} \sin t. \qquad \blacksquare$$

EXAMPLE 2 Solve $y'' + 6y' + 13y = 0, \quad y(\pi/2) = -2, \quad y'(\pi/2) = 8.$

SOLUTION The characteristic equation $\lambda^2 + 6\lambda + 13 = 0$ has complex distinct roots $\lambda = -3 + 2i$ and $\lambda^* = -3 - 2i$, so (2.5.4) becomes

$$y(t) = k_1 e^{-3t} \cos 2t + k_2 e^{-3t} \sin 2t.$$

Differentiating, we obtain

$$y'(t) = -3(k_1 e^{-3t} \cos 2t + k_2 e^{-3t} \sin 2t) + 2e^{-3t}(-k_1 \sin 2t + k_2 \cos 2t)$$

or

$$y'(t) = -3y(t) + 2e^{-3t}(- k_1 \sin 2t + k_2 \cos 2t).$$

(The reader will find the trick of writing $y'(t)$ in the form above very useful in the computing of the values of k_1 and k_2.) Substituting the initial data gives

$$-2 = -k_1 e^{-3\pi/2}, \qquad 8 = 6 - 2k_2 e^{-3\pi/2},$$

and hence $k_1 = 2e^{3\pi/2}$, $k_2 = -e^{3\pi/2}$. The solution to the initial value problem is

$$\begin{aligned} y(t) &= 2e^{3\pi/2}e^{-3t} \cos 2t - e^{3\pi/2}e^{-3t} \sin 2t \\ &= 2e^{-3(t-\pi/2)} \cos 2t - e^{-3(t-\pi/2)} \sin 2t. \end{aligned} \qquad \blacksquare$$

In many applications it is sometimes useful to replace the real two-parameter family of solutions by *amplitude–phase angle* solutions

$$y(t) = Ae^{pt} \cos(qt - \phi), \tag{2.5.5}$$

where the two parameters are A, the amplitude of the solution, and ϕ, the phase angle. For instance, if $p = 0$, the value A immediately tells us the minimum and maximum values of the solution, namely $\pm A$. A relationship between the parameters (k_1, k_2) and (A, ϕ) is easy to derive. Since it is assumed that not both k_1 and k_2 are zero, rewrite (2.5.4) as

$$y(t) = \sqrt{k_1^2 + k_2^2}\, e^{pt} \left[\frac{k_1}{\sqrt{k_1^2 + k_2^2}} \cos qt + \frac{k_2}{\sqrt{k_1^2 + k_2^2}} \sin qt \right].$$

The numbers $\alpha = k_1/\sqrt{k_1^2 + k_2^2}$ and $\beta = k_2/\sqrt{k_1^2 + k_2^2}$ satisfy conditions

$$|\alpha| \leq 1, \qquad |\beta| \leq 1, \qquad \alpha^2 + \beta^2 = 1.$$

Therefore, letting

$$A = \sqrt{k_1^2 + k_2^2}, \qquad \cos\phi = \alpha = \frac{k_1}{A}, \qquad \sin\phi = \beta = \frac{k_2}{A}, \tag{2.5.6}$$

we can write the solution $y(t)$ as

$$\begin{aligned} y(t) &= Ae^{pt}(\cos\phi \cos qt + \sin\phi \sin qt) \\ &= Ae^{pt}\cos(qt - \phi). \end{aligned}$$

EXAMPLE 3 Solve $y'' - 2y' + 26y = 0$, $y(0) = 3/\sqrt{2}$, $y'(0) = -12/\sqrt{2}$ and write the solution in the amplitude–phase angle form (2.5.5).

SOLUTION The characteristic equation is $\lambda^2 - 2\lambda + 26 = 0$, with roots $\lambda = 1 + 5i$, $\lambda^* = 1 - 5i$, and therefore

$$y(t) = k_1 e^t \cos 5t + k_2 e^t \sin 5t,$$

is a general solution. Substituting the initial data gives the equations

$$\frac{3}{\sqrt{2}} = k_1, \qquad \frac{-12}{\sqrt{2}} = \frac{3}{\sqrt{2}} + 5k_2.$$

So $k_1 = 3/\sqrt{2}$, $k_2 = -3/\sqrt{2}$, and the solution is

$$y(t) = \frac{3}{\sqrt{2}} e^t (\cos 5t - \sin 5t).$$

To express the solutions in the amplitude–phase angle form, use equations (2.5.6) to obtain

$$k_1 = \frac{3}{\sqrt{2}} = A\cos\phi, \qquad k_2 = \frac{-3}{\sqrt{2}} = A\sin\phi;$$

and

$$A = \sqrt{k_1^2 + k_2^2} = \sqrt{\frac{9}{2} + \frac{9}{2}} = 3$$

and

$$\phi = \arctan\left(\frac{k_2}{k_1}\right) = \arctan(-1) = -\frac{\pi}{4}.$$

The amplitude–phase angle form of the solution is

$$y(t) = 3e^t \cos\left(5t + \frac{\pi}{4}\right).$$

Note that the phase angle is measured in radians. ■

The solution techniques of this section are summarized in the following statement.

Solution to the constant coefficient homogeneous case: complex roots

1. Solve the characteristic equation, $a\lambda^2 + b\lambda + c = 0$, to obtain the characteristic roots

$$\lambda = p + iq, \qquad \lambda^* = p - iq,$$

where

$$p = -\frac{b}{2a}, \qquad q = \frac{\sqrt{4ac - b^2}}{2a}.$$

2. There are three types of general solutions:

a) complex,

$$y(t) = c_1 e^{(p+iq)t} + c_2 e^{(p-iq)t},$$

b) real,

$$y(t) = k_1 e^{pt} \cos qt + k_2 e^{pt} \sin qt,$$

c) amplitude–phase angle,

$$y(t) = Ae^{pt} \cos(qt - \phi).$$

3. If initial data $y(t_0) = r$, $y'(t_0) = s$, are given:

a) select c_1 and c_2 so that

$$r = c_1 e^{(p+iq)t_0} + c_2 e^{(p-iq)t_0},$$

$$s = (p + iq)c_1 e^{(p+iq)t_0} + (p - iq)c_2 e^{(p-iq)t_0};$$

b) select k_1 and k_2 so that

$$r = k_1 e^{pt_0} \cos qt_0 + k_2 e^{pt_0} \sin qt_0,$$

$$s = pr - qk_1 e^{pt_0} \sin qt_0 + qk_2 e^{pt_0} \cos qt_0;$$

c) select A and ϕ so that

$$r = Ae^{pt_0} \cos (qt_0 - \phi),$$
$$s = pr - qAe^{pt_0} \sin (qt_0 - \phi).$$

EXERCISES 2.5

Write down the general solution of the following linear differential equations and, if initial data are given, the solution satisfying that data. Give only real solutions.

1. $y'' + 3y' + \frac{13}{4}y = 0$

2. $y'' + 9y = 0$

3. $y'' - 2y' + 2y = 0$

4. $y'' - y' + 3y = 0$

5. $y'' + \frac{49}{4}y = 0$

6. $y'' + 4y' + 5y = 0; \quad y(0) = 0, \quad y'(0) = 1$

7. $y'' + y = 0; \quad y(0) = 2, \quad y'(0) = -6$

8. $y'' + 3y' + \frac{5}{2}y = 0; \quad y(0) = 2, \quad y'(0) = -6$

9. $y'' + 4y = 0; \quad y(\pi/8) = 1, \quad y'(\pi/8) = 2$

10. $y'' - 4y' + 7y = 0; \quad y(1) = 0, \quad y'(1) = \sqrt{3}$

Express the solution to each of the following initial value problems in the form

$$y(t) = Ae^{pt} \cos (qt - \phi),$$

where A is the amplitude of the solution and ϕ is the phase angle.

11. $y'' + y = 0; \quad y(0) = 1, \quad y'(0) = 1$

12. $y'' + 2y' + 5y = 0; \quad y(0) = 0, \quad y'(0) = 4$

13. $y'' + 4y = 0; \quad y(\pi/4) = 1, \quad y'(\pi/4) = -2$

14. $y'' + 6y' + 13y = 0; \quad y(\pi/2) = 4, \quad y'(\pi/2) = 0$

15. If the solution of an initial value problem is given in the amplitude–phase angle form

$$y(t) = Ae^{pt} \cos (qt - \phi),$$

and the initial data $y(0)$ and $y'(0)$ are given, show that

$$A^2 = y(0)^2 + \left[\frac{y'(0) - py(0)}{q}\right]^2,$$

$$\cos\phi = \frac{y(0)}{A},$$

$$\sin\phi = \frac{y'(0) - py(0)}{qA}.$$

Use the result of Exercise 15 to find the solutions of the following initial value problems.

16. $y'' + 25y = 0; \quad y(0) = 1, \quad y'(0) = 5$

17. $y'' + 2y' + 10y = 0; \quad y(0) = 4, \quad y'(0) = 5$

18. $y'' - y' + \frac{13}{4}y = 0; \quad y(0) = -2, \quad y'(0) = 2$

19. The amplitude–phase angle form of the general solution of $y'' + w^2y = 0$ is $y(t) = A\cos(wt - \phi)$. Show that the solution determines an ellipse

$$y^2 + \frac{(y')^2}{w^2} = K^2$$

in the variables (y, y').

Find the ellipse for the following initial value problems.

20. $y'' + y = 0; \quad y(0) = 4, \quad y'(0) = 3$

21. $y'' + 4y = 0; \quad y(0) = 1, \quad y'(0) = -1$

22. $y'' + \frac{25}{16}y = 0; \quad y(0) = 0, \quad y'(0) = 7$

23. The series for e^z,

$$e^z = 1 + z + \frac{z^2}{2!} + \cdots + \frac{z^n}{n!} + \cdots = \sum_{n=0}^{\infty} \frac{z^n}{n!},$$

converges for all real or complex z. By letting $z = i\theta$ and rearranging the series into its real and imaginary parts, prove Euler's formula

$$e^{i\theta} = \cos\theta + i\sin\theta.$$

(*Hint:* See Appendix 5 for a sketch of a proof.)

24. If initial data for a differential equation are given at more than one point, the differential equation together with the data is called a *boundary value problem.* For example, the differential equation

$$y'' - y' - 2y = 0$$

together with the conditions

$$y(0) = 0, \qquad y(1) = 1$$

is called a *two-point boundary value problem* because a solution $y(t)$ of the differential equation is sought whose graph passes through the two points (0, 0) and (1, 1). Find $y(t)$.

Solve the following two-point boundary value problems.

25. $y'' + y = 0; \quad y(0) = 1, \quad y(\pi/2) = -1$

26. $y'' + 2y' - 3y = 0; \quad y(0) = 0, \quad y(1) = 1$

27. $y'' + 2y' + y = 0; \quad y(0) = 2, \quad y'(2) = -2$

28. $y'' + y = 0; \quad y(0) = 1, \quad y(\pi) = \alpha$. For what values of α will there be one solution, many solutions, or no solution?

29. $y'' + \omega^2 y = 0; \quad y(0) = 0, \quad y(\pi) = 0$. For what values of ω will there be one solution, many solutions, or no solution?

2.6 THEORY OF HOMOGENEOUS SECOND ORDER LINEAR DIFFERENTIAL EQUATIONS

Linear second order differential equations with constant coefficients are emphasized in this chapter since they are important in applications and are easy to solve. Furthermore, as we study them we employ concepts such as general solutions and the determinant test theorem, Section 2.4, that are needed in the study of higher order linear differential equations (Section 2.12), linear differential equations with variable coefficients (Chapter 6) and systems of linear differential equations (Chapter 5). In this section we shall formulate and prove several results about solutions to linear homogeneous second order differential equations including a proof of the determinant test theorem that was stated, but not proved, in Section 2.4.

We shall now consider linear homogeneous second order differential equations,

$$a(t)\frac{d^2y}{dt^2} + b(t)\frac{dy}{dt} + c(t)y = 0, \tag{2.6.1}$$

where $a(t)$, $b(t)$, and $c(t)$ are continuous functions on an interval, $t_1 < t < t_2$, and $a(t)$ is nonzero. If the coefficients are not constant, there is no general method for constructing solutions and it is important to have criteria for those solutions that can be constructed to insure that initial value problems can be solved.

We begin our discussion with the following definition.

Definition

A two-parameter family of solutions to the linear homogeneous equation (2.6.1) defined on the interval, $t_1 < t < t_2$,

$$y(t) = c_1 y_1(t) + c_2 y_2(t), \tag{2.6.2}$$

is said to be a *general solution* if the family contains all the solutions to the differential equation.

This definition of general solution differs from that given earlier for solutions to differential equations with constant coefficients in that the interval of definition of the coefficients is included, a detail that is clearly not needed if the coefficients are constant.

We saw in the two preceding sections that the requirement that any solution be represented by (2.6.2) leads to a restriction on $y_1(t)$ and $y_2(t)$, namely,

$$\begin{vmatrix} y_1(t) & y_2(t) \\ y_1'(t) & y_2'(t) \end{vmatrix} \neq 0.$$

This type of determinant occurs frequently in the theory of differential equations.

Definition

$$W(y_1, y_2)(t) = \begin{vmatrix} y_1(t) & y_2(t) \\ y_1'(t) & y_2'(t) \end{vmatrix} \tag{2.6.3}$$

is called the Wronskian of the two functions $y_1(t)$ and $y_2(t)$.[6]

Definition

Two solutions $y_1(t)$, $y_2(t)$ of the homogeneous differential equation (2.6.1) form a *fundamental set of solutions* if their Wronskian, $W(y_1, y_2)(t)$, is nonzero on the interval $t_1 < t < t_2$.

EXAMPLE 1 Show that the functions $y_1(t) = e^{2t}$, $y_2(t) = te^{2t}$ form a fundamental set of solutions for the differential equation,

$$y'' - 4y' + 4y = 0.$$

6. Named after the Polish mathematician H. Wronski (1778–1853).

SOLUTION We first verify that $y_1(t)$ and $y_2(t)$ are, in fact, solutions to the differential equation:

$$(e^{2t})'' - 4(e^{2t})' + 4(e^{2t}) = (4 - 8 + 4)e^{2t} = 0,$$

$$(te^{2t})'' - 4(te^{2t})' + 4(te^{2t}) = [(4 + 4t) - 4(1 + 2t) + 4t]e^{2t} = 0.$$

Next, we calculate the Wronskian,

$$W(e^{2t}, te^{2t}) = \begin{vmatrix} e^{2t} & te^{2t} \\ 2e^{2t} & (1 + 2t)e^{2t} \end{vmatrix} = e^{4t} \neq 0.$$

Because the Wronskian is nonzero on the interval $-\infty < t < \infty$, $y_1(t) = e^{2t}$, $y_2(t) = te^{2t}$ form a fundamental set of solutions to the differential equation for all t. ■

A geometric interpretation of the Wronskian $W(y_1, y_2)(t)$ can be given by forming two vectors,

$$\mathbf{Y}_1(t) = y_1(t)\mathbf{e}_1 + y_1'(t)\mathbf{e}_2,$$
$$\mathbf{Y}_2(t) = y_2(t)\mathbf{e}_1 + y_2'(t)\mathbf{e}_2,$$

where $\mathbf{e}_1$ and $\mathbf{e}_2$ are orthonormal vectors, and then taking their cross-product,

$$\begin{aligned}\mathbf{Y}_1(t) \times \mathbf{Y}_2(t) &= [y_1(t)y_2'(t) - y_2(t)y_1'(t)](\mathbf{e}_1 \times \mathbf{e}_2) \\ &= W(y_1, y_2)(t)(\mathbf{e}_1 \times \mathbf{e}_2).\end{aligned}$$

Since the length of the vector $\mathbf{Y}_1(t) \times \mathbf{Y}_2(t)$ is proportional to the sine of the angle between $\mathbf{Y}_1(t)$ and $\mathbf{Y}_2(t)$ (see Fig. 2.3), the same is true for the Wronskian of $y_1(t)$ and $y_2(t)$. If the Wronskian is zero at a point, then the vectors are parallel at that point. If the Wronskian is zero on an interval, then the vectors are parallel on the interval. But if the vectors are parallel on the interval, then for some constant a,

$$\mathbf{Y}_1(t) = a\mathbf{Y}_2(t),$$

for all t in the interval. This vector equation is equivalent to two scalar equations,

$$y_1(t) = ay_2(t) \quad \text{and} \quad y_1'(t) = ay_2'(t).$$

If the Wronskian is nonzero on an interval, then the vectors are not parallel on the interval and the sine of the angle between them is nonzero. Therefore $y_1(t)$ is not a multiple of $y_2(t)$, and the solutions $y_1(t)$ and $y_2(t)$ form a fundamental set of solutions. The concept of fundamental set of solutions is useful because we can construct a general solution from a fundamental set of solutions.

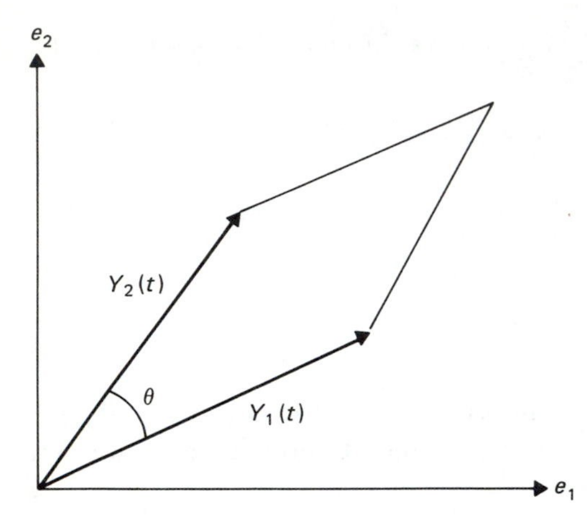

Figure 2.3 The area of the parallelogram equals $\|Y_1(t)\| \cdot \|Y_2(t)\| \cdot |\sin\theta|$.

The determinant test theorem

A two-parameter family of solutions to (2.6.1),

$$y(t) = c_1y_1(t) + c_2y_2(t), \tag{2.6.4}$$

is a general solution on the interval $t_1 < t < t_2$ if and only if $W(y_1, y_2)(t)$ is nonzero on the interval.

Proof. If $W(y_1, y_2) \neq 0$ on the interval $t_1 < t < t_2$, and $z(t)$ is any solution of (2.6.1) on the interval, we shall show that c_1 and c_2 can be picked so that

$$z(t) = c_1y_1(t) + c_2y_2(t), \qquad t_1 < t < t_2.$$

To do this we pick t_0 in the interval and evaluate $z(t)$ and $z'(t)$ at t_0 to get $z(t_0) = r$ and $z'(t_0) = s$. Then, as we have done before, we set

$$r = c_1y_1(t_0) + c_2y_2(t_0), \qquad s = c_1y_1'(t_0) + c_2y_2'(t_0) \tag{2.6.5}$$

and solve for c_1 and c_2,

$$c_1 = \frac{ry_2'(t_0) - sy_2(t_0)}{W(y_1, y_2)(t_0)}, \qquad c_2 = \frac{sy_1(t_0) - ry_1'(t_0)}{W(y_1, y_2)(t_0)}.$$

Since the values of $z_1(t)$ and $z_1'(t)$ at t_0 uniquely characterize $z(t)$ and also determine c_1 and c_2, we conclude that $z(t)$ is a member of the family of solutions (2.6.4).

However, if for some t_0,

$$W(y_1, y_2)(t_0) = y_1(t_0)y_2'(t_0) - y_2(t_0)y_1'(t_0) = 0,$$

we shall show that the initial data cannot be arbitrary and, therefore, $y_1(t)$ and $y_2(t)$ cannot form a general solution to (2.6.1). If both $y_1(t_0)$ and $y_2'(t_0)$ are zero, r must be a multiple of $y_2(t_0)$ and s must be a multiple of $y_2'(t_0)$. If they are not both zero, we multiply the first equation of (2.6.3) by $y_1'(t_0)$, the second by $y_1(t_0)$, and subtract one equation from the other. We then have

$$ry_1'(t_0) - sy_1(t_0) = c_2(y_1'(t_0)y_2(t_0) - y_1(t_0)y_2'(t_0)) = 0,$$

a restriction on the initial data r and s. Therefore the initial data cannot be arbitrary and the theorem is proved. □

The Wronskian of two solutions to a homogeneous linear second order differential equation,

$$a(t)\frac{d^2y}{dt^2} + b(t)\frac{dy}{dt} + c(t)y = 0,$$

has the remarkable property that it is either identically zero or it is never zero on $t_1 < t < t_2$. This is a consequence of Abel's formula,[7]

$$W(y_1, y_2)(t) = W(y_1, y_2)(t_0)\exp\left[-\int_{t_0}^{t}\frac{b(u)}{a(u)}\,du\right]. \qquad \textbf{(2.6.6)}$$

Abel's formula can be derived by adding the equations

$$-y_2[ay_1'' + by_1' + cy_1] = 0,$$
$$y_1[ay_2'' + by_2' + cy_2] = 0$$

to obtain

$$a(y_1y_2'' - y_2y_1'') + b(y_1y_2' - y_2y_1') = 0.$$

Since

$$\frac{d}{dt}W(y_1, y_2)(t) = \frac{d}{dt}(y_1y_2' - y_2y_1') = y_1y_2'' - y_2y_1'',$$

we have

$$a(t)\frac{dW}{dt}(y_1, y_2)(t) + b(t)W(y_1, y_2)(t) = 0, \qquad \textbf{(2.6.7)}$$

a linear first order differential equation whose solution is (2.6.6).

7. N. H. Abel (1802–1829) was a Norwegian mathematician who made major contributions in both algebra and analysis.

EXAMPLE 2 Find a general solution to

$$t^2 \frac{d^2y}{dt^2} + t\frac{dy}{dt} - 4y = 0, \qquad t > 0.$$

Compute the corresponding Wronskian directly and compare it to Abel's formula.

SOLUTION This is an example of a special type of differential equation called *Euler's equation.* It can be solved by setting $y(t) = t^r$ and then deriving a quadratic equation for the undetermined exponent r. (See Exercise 14 at the end of this section.) Substitution gives

$$\begin{aligned} t^2 \frac{d^2}{dt^2}(t^r) + t\frac{d}{dt}(t^r) - 4t^r &= t^2[r(r-1)t^{r-2}] + t(rt^{r-1}) - 4t^r \\ &= t^r[r(r-1) + r - 4] = t^r(r^2 - 4) = 0. \end{aligned}$$

Since $t > 0$, $r^2 - 4 = 0$, and therefore $r = 2$ or $r = -2$. This implies that $y_1(t) = t^2$ and $y_2(t) = t^{-2}$ are both solutions and a two-parameter family of solutions is

$$y(t) = c_1t^2 + c_2t^{-2}.$$

This is a general solution on any interval not containing $t = 0$ since the Wronskian

$$W(y_1, y_2)(t) = \begin{vmatrix} t^2 & t^{-2} \\ 2t & -2t^{-3} \end{vmatrix} = -4t^{-1} \neq 0.$$

Finally, from Abel's formula

$$\begin{aligned} W(y_1, y_2)(t) &= W(y_1, y_2)(t_0)\exp\left[-\int_{t_0}^{t} \frac{du}{u}\right] \\ &= W(y_1, y_2)(t_0)e^{-\ln(t/t_0)} = t_0W(y_1, y_2)(t_0)/t. \end{aligned}$$

Abel's formula shows that $W(y_1, y_2)(t)$ is inversely proportional to t but the constant of proportionality depends on the choice of $y_1(t)$, $y_2(t)$, and t_0. ■

The theory of fundamental solution sets of homogeneous second order linear differential equations seems overly elaborate for we can usually tell by inspection if two solutions do or do not form a fundamental system. Nevertheless, this theory serves as a model for higher order equations and for systems of first order equations. A significant concept needed for the general theory is that of linear independence of functions.

Definition

Two functions $f_1(t)$ and $f_2(t)$ are *linearly dependent* on an interval $t_1 < t < t_2$ if there exist constants c_1 and c_2, not both zero, such that the linear combination

$$c_1 f_1(t) + c_2 f_2(t) = 0, \qquad t_1 < t < t_2.$$

The functions $f_1(t)$ and $f_2(t)$ are *linearly independent* if they are not linearly dependent.

EXAMPLE 3 Show that $f_1(t) = 3t^2 - 6$, $f_2(t) = 7t^2 + 3t$ are linearly independent on any interval, $t_1 < t < t_2$.

SOLUTION If

$$c_1(3t^2 - 6) + c_2(7t^2 + 3t) = 0, \qquad t_1 < t < t_2,$$

then

$$(3c_1 + 7c_2)t^2 + 3c_2 t - 6c_1 = 0, \qquad t_1 < t < t_2.$$

But a quadratic polynomial can have at most two zeros. Therefore its coefficients must be zero:

$$3c_1 + 7c_2 = 0, \qquad 3c_2 = 0, \qquad -6c_1 = 0.$$

It follows that both c_1 and c_2 are zero and $f_1(t) = 3t^2 - 6$, $f_2(t) = 7t^2 + 3t$ are linearly independent on the interval $t_1 < t < t_2$. ■

There is a close relationship between the concepts of linear independence and fundamental solution sets.

Theorem

If $y_1(t)$ and $y_2(t)$ are solutions to the homogeneous linear second order differential equations (2.6.1) on the interval $t_1 < t < t_2$, then they form a fundamental set of solutions if and only if they are linearly independent.

Proof. Suppose that $y_1(t)$ and $y_2(t)$ form a fundamental set of solutions and that a linear combination, $c_1 y_1(t) + c_2 y_2(t) \equiv 0$ on a subinterval; then $c_1 y_1'(t) + c_2 y_2'(t) \equiv 0$. If t_0 is any point in the subinterval, then the linear homogeneous algebraic equations

$$c_1 y_1(t_0) + c_1 y_2(t_0) = 0, \qquad c_1 y_1'(t_0) + c_2 y_2'(t_0) = 0$$

have a solution $c_1 = 0$ and $c_2 = 0$. But since the coefficient determinant, $y_1(t_0)y_2'(t_0) - y_2(t_0)y_1'(t_0) = W(y_1, y_2)(t_0) \neq 0$, the solution is unique. Therefore $y_1(t)$ and $y_2(t)$ are linearly independent.

If $y_1(t)$ and $y_2(t)$ are solutions but do not form a fundamental set, then by the determinant test theorem and Abel's formula, the Wronskian must be identically zero on $t_1 < t < t_2$. Take any t_0 in the interval and set $c_1 = y_2(t_0)$, $c_2 = -y_1(t_0)$. Then by linearity,

$$y(t) = y_2(t_0)y_1(t) - y_1(t_0)y_2(t)$$

is a solution to the differential equation (2.6.1) such that

$$y(t_0) = 0, \qquad y'(t_0) = W(y_1, y_2)(t_0) = 0.$$

It follows that $y(t)$ is identically zero on the interval and that $y_1(t)$ and $y_2(t)$ are linearly dependent. □

The following theorem is a consequence of preceding theorems.

Theorem

Two solutions $y_1(t)$ and $y_2(t)$ to (2.6.1) are linearly independent on $t_1 < t < t_2$ if and only if their Wronskian $W(y_1, y_2)(t)$ is nonzero on the interval.

It is important to note that linearly independent differentiable functions can have a Wronskian that is identically zero. But if so, these functions cannot be solutions to the same homogeneous second order linear differential equation. This is illustrated by the following example.

EXAMPLE 4 Show that $f_1(t) = t^3$, $f_2(t) = |t|t^2$ are linearly independent on $-2 < t < 2$ and that $W(f_1, f_2)(t) = 0$ on the interval.

SOLUTION We have $f_1(t) = t^3$, $f_1'(t) = 3t^2$,

$$f_2(t) = \begin{cases} t^3 & \text{if} \quad t \geq 0 \\ -t^3 & \text{if} \quad t < 0, \end{cases} \qquad f_2'(t) = \begin{cases} 3t^2 & \text{if} \quad t \geq 0 \\ -3t^2 & \text{if} \quad t < 0 \end{cases}$$

or $f_2(t) = |t|t^2$, $f_2'(t) = 3|t|t$. It follows that

$$W(f_1, f_2)(t) = \begin{vmatrix} t^3 & |t|t^2 \\ 3t^2 & 3|t|t \end{vmatrix} = 3|t|t^4 - 3|t|t^4 = 0,$$

for all t. Nevertheless, the functions are linearly independent on $-2 < t < 2$, for

$$c_1t^3 + c_2|t|t^2 = 0$$

is equivalent to

$$(c_1 + c_2)t^3 = 0, \qquad 0 \le t < 2$$

$$(c_1 - c_2)t^3 = 0, \qquad -2 < t < 0.$$

Hence $c_1 = c_2 = 0$ as required for linear independence. ■

The results proved in this section can be summarized as follows: Given the homogeneous second order linear differential equation,

$$a(t)y'' + b(t)y' + c(t)y = 0,$$

where $a(t)$, $b(t)$, $c(t)$ are continuous on the interval $a < t < b$, and two solutions, $y_1(t)$ and $y_2(t)$ defined on $a < t < b$, then the following statements are equivalent:

1. the solutions $y_1(t)$ and $y_2(t)$ are linearly independent;
2. the Wronskian of $y_1(t)$ and $y_2(t)$, $W(y_1, y_2)(t)$, is nonzero, $a < t < b$;
3. the solutions $y_1(t)$ and $y_2(t)$ form a fundamental solution set.

In this case we say that the solution,

$$y(t) = c_1y_1(t) + c_2y_2(t),$$

where c_1 and c_2 are arbitrary constants, is a general solution to the differential equation.

EXERCISES 2.6

In Exercises 1–7, compute the Wronskian $W(y_1, y_2)(t)$ of the following pairs of functions and determine for what values of t it is nonzero.

1. $y_1(t) = e^{2t}, \quad y_2(t) = e^{-2t}$

2. $y_1(t) = e^t, \quad y_2(t) = te^t$

3. $y_1(t) = \sin \omega t, \quad y_2(t) = \cos \omega t$

4. $y_1(t) = t^3, \quad y_2(t) = |t|^3, \quad t \neq 0$

5. $y_1(t) = t^2, \quad y_2(t) = t^2 \ln |t|, \quad t \neq 0$

6. $y_1(t) = \cos t, \quad y_2(t) = \cos(t - \pi)$

7. $y_1(t) = \dfrac{\sin t}{\sqrt{t}}, \quad y_2(t) = \dfrac{\cos t}{\sqrt{t}}, \quad t < 0$

Using the methods developed in Sections 2.4 and 2.5 for finding general solutions, find two linearly independent solutions for each of the differential equations in Exercises 8–13.

8. $y'' + 5y' + 6y = 0$

9. $y'' - 10y' + 25y = 0$

10. $y'' + 2y' + 5y = 0$

11. $y'' - 2y' + 5y = 0$ **12.** $y'' + 3y' - 10y = 0$ **13.** $y'' + 9y = 0$

14. The *Euler differential equation* of second order is an equation of the form

$$t^2y'' + bty' + cy = 0, \qquad t > 0,$$

where b and c are constants.

a) Assume a solution of the form $y(t) = t^r$ and show that $y(t)$ is a solution if and only if r is a root of the equation

$$r^2 + (b - 1)r + c = 0.$$

b) There are three cases. In each case compute the Wronskian $W(y_1, y_2)(t)$ to show that the given function $y(t) = Ay_1(t) + By_2(t)$ is a general solution.
Case 1: distinct roots $r = \alpha_1$, $r = \alpha_2$; then

$$y(t) = At^{\alpha_1} + Bt^{\alpha_2}.$$

Case 2: a double root $r = \alpha$; then

$$y(t) = At^{\alpha} + Bt^{\alpha} \ln t.$$

Case 3: complex roots $r = \alpha + i\beta$, $r^* = \alpha - i\beta$; then

$$y(t) = At^{\alpha} \cos(\beta \ln t) + Bt^{\alpha} \sin(\beta \ln t).$$

Note that the identity $t = \exp(\ln t)$ implies $t^{i\beta} = \exp(i\beta \ln t)$.

Use the results of Exercise 14 to find a general solution of the following Euler equations in Exercises 15–20. Does the solution approach zero as $t \to \infty$?

15. $t^2y'' + 6ty' + 6y = 0$ **16.** $t^2y'' + 6ty' - 6y = 0$ **17.** $t^2y'' + 7ty' + 9y = 0$

18. $t^2y'' - 7ty' + 16y = 0$ **19.** $t^2y'' + 3ty' + 5y = 0$ **20.** $t^2y'' - 2ty' + \frac{25}{4}y = 0$

21. Show that

a) $y_1(t) = t^2$ and $y_2(t) = t^3$ cannot both be solutions to the differential equation

$$y'' + b(t)y' + c(t)y = 0,$$

where $b(t)$ and $c(t)$ are continuous on an interval containing $t = 0$;

b) $y(t) = c_1 + c_2 \cos t$ cannot be a general solution of the differential equation

$$y'' + by' + cy = 0,$$

where b and c are constants.

2.7 INHOMOGENEOUS EQUATIONS—THE METHOD OF UNDETERMINED COEFFICIENTS

We shall now study inhomogeneous linear second order differential equations with constant coefficients,

$$a\frac{d^2y}{dt^2} + b\frac{dy}{dt} + cy = g(t), \tag{2.7.1}$$

with $a \neq 0$ and $g(t)$ a continuous function on an interval $t_1 < t < t_2$. According to the existence and uniqueness theorem, Section 2.3, and the determinant test theorem, Section 2.6, a general solution to (2.7.1) can be written as a sum,

$$y(t) = y_p(t) + c_1y_1(t) + c_2y_2(t), \tag{2.7.2}$$

where

1. $y_p(t)$ is a *particular solution*[8] to the inhomogeneous equation (2.7.1);
2. $y_1(t)$ and $y_2(t)$ are a fundamental set of solutions to the corresponding homogeneous equation,

$$a\frac{d^2y}{dt^2} + b\frac{dy}{dt} + cy = 0; \tag{2.7.3}$$

3. if t_0, $y(t_0)$, and $y'(t_0)$ are given, then the constants c_1 and c_2 are the unique solutions to the linear algebraic equations,

$$\begin{aligned} y(t_0) - y_p(t_0) &= c_1y_1(t_0) + c_2y_2(t_0) \\ y'(t_0) - y_p'(t_0) &= c_1y_1'(t_0) + c_2y_2'(t_0). \end{aligned} \tag{2.7.4}$$

Each equation of (2.7.4) determines a straight line in the c_1, c_2 plane. Since there is a unique solution to the two algebraic equations, the lines must intersect at a point. Therefore the lines are not parallel and the determinant of the coefficients of the equations,

$$\begin{vmatrix} y_1(t_0) & y_2(t_0) \\ y_1'(t_0) & y_2'(t_0) \end{vmatrix} = W(y_1, y_2)(t_0) \neq 0.$$

From this we conclude that $y_1(t)$ and $y_2(t)$ form a fundamental set of solutions to (2.7.3).

The task of constructing a general solution to (2.7.1) can be split into finding a fundamental solution set to (2.7.3) and a particular solution to (2.7.1). The first task was studied in Sections 2.4 and 2.5. We shall now study the second task,

8. It is standard to refer to parameter-free solutions to inhomogeneous equations as *particular solutions.*

namely, the construction of particular solutions to the inhomogeneous linear second order differential equation (2.7.1). Two methods will be presented, the *method of undetermined coefficients* and the *variation of parameters method.* In this section, we shall study the simpler of the two, the method of undetermined coefficients.

An inspection of problems solved by various methods reveals that constant coefficient differential equations with exponential, polynomial, or sinusoidal forcing functions usually have exponential, polynomial, or sinusoidal particular solutions. As we saw in Section 1.4, we can construct particular solutions by deducing the type of solution from the forcing function and then determining its exact structure by substituting a trial solution into the differential equation. This is called the *method of undetermined coefficients,* or the *method of judicious guessing.* A comparison of forcing functions and trial (guessed) solutions is given in Table 2.1.

Let us use the method to find the solutions to

$$y'' + y' - 6y = g(t)$$

with four different choices of $g(t)$.

EXAMPLE 1 a) Find a particular solution to

$$y'' + y' - 6y = 5e^{4t}.$$

SOLUTION Since the derivative of an exponential is again an exponential, and the forcing function is an exponential, let $y_p(t) = Ae^{4t}$, an exponential of the same type. The undetermined coefficient A will be found by substituting the trial solution into the differential equation

$$(Ae^{4t})'' + (Ae^{4t})' - 6Ae^{4t} = Ae^{4t}(16 + 4 - 6) = 14Ae^{4t} = 5e^{4t}.$$

Thus, $A = \frac{5}{14}$ and the particular solution is $y_p(t) = \frac{5}{14}e^{4t}$.

Table 2.1 Forcing functions and trial solutions

$g(t)$	Trial solution
ae^{rt}	Ae^{rt}
$a \sin \omega t$ or $a \cos \omega t$	$A \sin \omega t + B \cos \omega t$
at^n, where n is a positive integer	$P(t)$, a general polynomial of degree n
$at^n e^{rt}$, where n is a positive integer	$P(t)e^{rt}$, with $P(t)$ a general polynomial of degree n
$t^n(a \sin \omega t + b \cos \omega t)$, where n is a positive integer	$P_1(t) \sin \omega t + P_2(t) \cos \omega t$, with $P_1(t)$, $P_2(t)$ general polynomials of degree n
$e^{rt}(a \sin \omega t + b \cos \omega t)$	$e^{rt}(A \sin \omega t + B \cos \omega t)$

b) Find a general solution to

$$y'' + y' - 6y = 5e^{2t}.$$

SOLUTION With confidence based on the previous example we try $y_p(t) = Ae^{2t}$. But

$$(Ae^{2t})'' + (Ae^{2t})' - 6Ae^{2t} = Ae^{2t}(4 + 2 - 6) = 0 \neq 5e^{2t},$$

so our guess fails. Fortunately there is an obvious explanation for the failure and an easy modification to prevent such failures. It failed because the trial solution, $y_p(t) = Ae^{2t}$, is a solution to the corresponding homogeneous equation and its substitution gave us zero. To guard against this difficulty, one should always solve the homogeneous equation first. Its solutions will be needed to construct a general solution or to satisfy initial data, so this is not wasteful. Then check the trial solution to see if it is also a solution to the homogeneous equation. If it is not, proceed; otherwise, the method must be modified. The modification is suggested by the form of the solution constructed by the reduction of order method of Section 2.4:

$$y_p(t) = Ate^{2t}.$$

Substituting gives us

$$\begin{aligned}(Ate^{2t})'' + (Ate^{2t})' - 6Ate^{2t} &= A(4te^{2t} + 4e^{2t}) + A(2te^{2t} + e^{2t}) - 6Ate^{2t} \\ &= Ate^{2t}(4 + 2 - 6) + Ae^{2t}(4 + 1) = 5e^{2t}.\end{aligned}$$

Clearly, $A = 1$, and since $y_1(t) = e^{2t}$ and $y_2(t) = e^{-3t}$ form a fundamental solution set of the homogeneous equation,

$$y(t) = te^{2t} + c_1e^{2t} + c_2e^{-3t}$$

is a general solution.

c) Find a general solution to

$$y'' + y' - 6y = 5t.$$

SOLUTION Since the forcing term is a polynomial, this suggests a trial solution $y_p(t) = At + B$, which, when substituted into the differential equation, gives

$$(At + B)'' + (At + B)' - 6(At + B) = (A - 6B) - 6At = 5t.$$

Since two polynomials are equal if and only if the coefficients of like powers of t are equal, A and B must satisfy

$$A - 6B = 0, \qquad -6A = 5.$$

Hence $A = -\frac{5}{6}$, $B = -\frac{5}{36}$, and a general solution is

$$y(t) = -\frac{5}{6}t - \frac{5}{36} + c_1e^{2t} + c_2e^{-3t}.$$

The reader should note that had the guess been simply $y(t) = At$, its substitution would have yielded

$$(At)'' + (At)' - 6At = A - 6At = 5t,$$

an equation with no solution. Since $g(t) = 5t$ is a polynomial of degree one, the trial solution must be $y_p(t) = At + B$, the most general polynomial of degree one.

d) Find a general solution to

$$y'' + y' - 6y = 5 \cos t.$$

SOLUTION The derivative of $\cos t$ is $-\sin t$ and so we would expect that both $\cos t$ and $\sin t$ are needed in the trial solution. Choose $y_p(t) = A \cos t + B \sin t$ and substitute to obtain

$$\begin{aligned}(A \cos t + B \sin t)'' + (A \cos t + B \sin t)' - 6(A \cos t + B \sin t)&\\ = (-7A + B) \cos t + (-A - 7B) \sin t&\\ = 5 \cos t = 5 \cos t + 0 \sin t.&\end{aligned}$$

Equating coefficients of $\cos t$ and $\sin t$ we get

$$-7A + B = 5, \qquad -A - 7B = 0,$$

which gives $A = -\frac{7}{10}$, $B = \frac{1}{10}$, and a general solution is

$$y(t) = -\frac{7}{10} \cos t + \frac{1}{10} \sin t + c_1e^{2t} + c_2e^{-3t}.$$

■

The next problem shows that occasionally a trial solution requires a factor of t^2 or even t^3.

EXAMPLE 2 Find a general solution to

$$y'' + 6y' + 9y = 4e^{-3t}.$$

SOLUTION First solve the homogeneous equation

$$y'' + 6y' + 9y = 0$$

to obtain $y_1(t) = e^{-3t}$, $y_2(t) = te^{-3t}$. Examination of the inhomogeneous term $4e^{-3t}$ would normally suggest a trial solution of the form Ae^{-3t}. But since both Ae^{-3t} and Ate^{-3t} are solutions to the homogeneous equation, we multiply the trial solution by the smallest integer power of t so that it is no longer a solution. We shall find that

t^2 will do, and so we let $y_p(t) = At^2e^{-3t}$. Substitution, some careful differentiation, and collection of terms gives

$$(At^2e^{-3t})'' + 6(At^2e^{-3t})' + 9At^2e^{-3t} = 2Ae^{-3t} = 4e^{-3t}.$$

Hence, $A = 2$ and a general solution is

$$y(t) = 2t^2e^{-3t} + c_1e^{-3t} + c_2te^{-3t}.$$ ■

The reader should notice that if the forcing function were $4te^{-3t}$, one might normally try a solution of the form $y_p(t) = (At + B)e^{-3t}$. However, for the above example both Ate^{-3t} and Be^{-3t} are solutions of the homogeneous equation, so the trial solution would be instead

$$\begin{aligned} y(t) &= t^2(At + B)e^{-3t} \\ &= At^3e^{-3t} + Bt^2e^{-3t}. \end{aligned}$$

If the forcing function $g(t)$ is a sum of exponential, polynomial, and sinusoidal functions, then the problem can be split into simpler problems whose solutions can be added to give a solution to the original problem. This procedure is illustrated in the next example.

EXAMPLE 3 Find a general solution to

$$y'' + y = 6e^t + 6\sin t.$$

SOLUTION We begin by noticing that $y_1(t) = \cos t$ and $y_2(t) = \sin t$ are two linearly independent solutions of the homogeneous equation. The first equation to be solved is $y'' + y = 6e^t$. A trial solution is $y_p(t) = Ae^t$ and

$$(Ae^t)'' + Ae^t = 2Ae^t = 6e^t.$$

Therefore $A = 3$ and $y_p(t) = 3e^t$. The second equation to be solved is $y'' + y = 6\sin t$. Since $6\sin t$ is a solution to the homogeneous equation, let

$$y_p(t) = t(B\cos t + C\sin t).$$

Then

$$\begin{aligned} (Bt\cos t + Ct\sin t)'' + (Bt\cos t + Ct\sin t) &= -2B\sin t + 2C\cos t \\ &= 6\sin t + 0\cos t; \end{aligned}$$

hence $B = -3$, $C = 0$, and $y_p(t) = -3t\cos t$. Now add the two particular solutions and a complementary solution to get a general solution

$$y(t) = 3e^t - 3t\cos t + c_1\cos t + c_2\sin t.$$ ■

The following are guidelines for the method of undetermined coefficients. Of course, the reader should have carefully gone over the preceding examples before attempting any problems.

Guidelines for the method of undetermined coefficients

Given the *constant coefficient* linear differential equation

$$ay'' + by' + cy = g(t),$$

where $g(t)$ is an exponential, a simple sinusoidal function, a polynomial, or a product of these functions:

1. Solve the homogeneous equation for a pair of linearly independent solutions $y_1(t)$ and $y_2(t)$.
2. If $g(t)$ is *not* a solution of the homogeneous equation, take a trial solution of the same type as $g(t)$ according to the suggestions given in Table 2.1.
3. If $g(t)$ is a solution of the homogeneous equation, take a trial solution of the same type as $g(t)$ multiplied by the lowest power of t for which no term of the trial solution is a solution of the homogeneous equation.
4. Substitute the trial solution into the differential equation and solve for the undetermined coefficients so that it is a particular solution $y_p(t)$.
5. Set $y(t) = y_p(t) + c_1y_1(t) + c_2y_2(t)$, where the constants c_1 and c_2 can be determined if initial conditions are given.
6. If $g(t)$ is a sum of forcing functions of the type described above, split the problem into simpler parts. Find a particular solution for each part, then add the particular solutions to obtain $y_p(t)$.

EXERCISES 2.7

Using the method of undetermined coefficients, find a general solution of each of the following inhomogeneous differential equations.

1. $y'' - y' - 2y = t$
2. $y'' - 4y = 6e^t$
3. $y'' - 2y' + y = 4 \sin t$
4. $y'' + 5y' + 4y = 2e^{-t}$
5. $y'' = t^2 + t + 1$
6. $y'' + 2y' + 5y = 10t^2 - e^{-2t}$
7. $y'' + 4y = t - 2 \sin 2t$
8. $y'' + y = t \cos 2t$
9. $y'' + y' - 4y = 2 \sinh t$
10. $y'' + y = 8 \cos^3 t$

Solve the following initial value problems.

11. Exercise 1; $y(0) = 0, \quad y'(0) = 1$
12. Exercise 2; $y(0) = 1, \quad y'(0) = 2$

13. Exercise 3; $y(0) = 1$, $y'(0) = 0$

14. Exercise 4; $y(0) = 1$, $y'(0) = 1$

15. Exercise 7; $y(\pi) = 0$, $y'(\pi) = 1$

16. Exercise 10; $y(\pi/6) = 0$; $y'(\pi/6) = 0$

17. An alternative method of solving the Euler differential equation

$$t^2y'' + bty' + cy = 0, \quad t > 0,$$

is to make the change of independent variable $t = e^z$.
Show that

$$\frac{dy}{dt} = \frac{1}{t}\frac{dy}{dz}, \qquad \frac{d^2y}{dt^2} = \frac{1}{t^2}\frac{d^2y}{dz^2} - \frac{1}{t^2}\frac{dy}{dz},$$

and that the substitution gives the constant coefficient differential equation

$$\frac{d^2y}{dz^2} + (b-1)\frac{dy}{dz} + cy = 0$$

with solutions $y(z) = y(\ln t)$.

The change of variables in Exercise 17 transforms an inhomogeneous Euler equation

$$t^2y'' + bty' + by = P(t),$$

where $P(t)$ is a linear combination of powers of t, to the equation

$$\frac{d^2y}{dz^2} + (b-1)\frac{dy}{dz} + cy = P(e^z).$$

The last equation can be solved by the method of undetermined coefficients. Use this trick to find a general solution of

18. $t^2y'' - 2y = 4t^3$

19. $t^2y'' + 5ty' - 6y = 4/t^2 - 12$

20. $t^2y'' + 5ty' + y = 3/t + 5t^2$

21. $t^2y'' + 5ty' + 4y = 6/t^2$

22. $t^2y'' + ty' + 9y = At^k$, k any real number.

2.8 INHOMOGENEOUS EQUATIONS—THE VARIATION OF PARAMETERS METHOD

In the previous section we discussed a method for determining particular solutions to inhomogeneous differential equations with constant coefficients in those cases where the inhomogeneous term was of a suitable form. In this section we determine a *general method* for finding a particular solution of the inhomogeneous second order linear differential equation.

$$a(t)\frac{d^2y}{dt^2} + b(t)\frac{dy}{dt} + c(t)y = g(t), \tag{2.8.1}$$

where $a(t)$, $b(t)$, $c(t)$, and $g(t)$ are continuous on some interval and $a(t)$ does not vanish. It requires that the corresponding homogeneous equation

$$a(t)\frac{d^2y}{dt^2} + b(t)\frac{dy}{dt} + c(t)y = 0 \tag{2.8.2}$$

be solved for a linearly independent pair of solutions $y_1(t)$ and $y_2(t)$. This is easy to do if the coefficients in the differential equation are constants; otherwise, it is difficult, if not impossible. In any case, suppose that this has been done. Then building on our experience with the reduction of order method of Section 2.4, we seek functions $u_1(t)$ and $u_2(t)$ so that

$$y_p(t) = u_1(t)y_1(t) + u_2(t)y_2(t) \tag{2.8.3}$$

is a solution to equation (2.8.1).

To derive appropriate equations for $u_1(t)$ and $u_2(t)$, we first form the first and second derivatives of $y_p(t)$,

$$\begin{aligned} y_p'(t) &= u_1(t)y_1'(t) + u_2(t)y_2'(t) + s(t), \\ y_p''(t) &= u_1(t)y_1''(t) + u_2(t)y_2''(t) + s'(t) + u_1'(t)y_1'(t) + u_2'(t)y_2'(t), \end{aligned}$$

where we set

$$s(t) = u_1'(t)y_1(t) + u_2'(t)y_2(t). \tag{2.8.4}$$

Then substituting into the differential equation (2.8.1) and simplifying, we get

$$\begin{aligned} a(t)y_p''(t) + b(t)y_p'(t) + c(t)y_p(t) = {} & u_1(t)[a(t)y_1''(t) + b(t)y_1'(t) + c(t)y_1(t)] \\ & + u_2(t)[a(t)y_2''(t) + b(t)y_2'(t) + c(t)y_2(t)] \\ & + a(t)[u_1'(t)y_1'(t) + u_2'(t)y_2'(t)] + a(t)s'(t) \\ & + b(t)s(t) = g(t). \end{aligned}$$

The coefficients in the brackets following $u_1(t)$ and $u_2(t)$ are both zero since $y_1(t)$ and $y_2(t)$ are given solutions to the homogeneous equation (2.8.2). Hence if $y_p(t)$ is to be a solution, then $u_1'(t)$ and $u_2'(t)$ must be selected so that

$$a(t)[u_1'(t)y_1'(t) + u_2'(t)y_2'(t)] + a(t)s'(t) + b(t)s(t) = g(t).$$

Now is the time to be clever. This expression can be simplified if

$$a(t)s'(t) + b(t)s(t) = 0, \tag{2.8.5}$$

the same first ordinary differential equation satisfied by the Wronskian $W(y_1, y_2)(t)$ (see (2.6.7)). Hence $s(t) = QW(y_1, y_2)(t)$ where Q is a constant at our disposal. Using the definition of $s(t)$ from (2.8.4), it follows that $u_1'(t)$ and $u_2'(t)$ must satisfy the two linear algebraic equations

$$u_1'(t)y_1(t) + u_2'(t)y_2(t) = QW(y_1, y_2)(t),$$
$$u_1'(t)y_1'(t) + u_2'(t)y_2'(t) = g(t)/a(t).$$

We shall take $Q = 0$ for it is easy to show that the effect of a nonzero Q is to add a solution of the homogeneous equation to the particular solution produced by this method.

Once the equations

$$\begin{aligned} u_1'(t)y_1(t) + u_2'(t)y_2(t) &= 0, \\ u_1'(t)y_1'(t) + u_2'(t)y_2'(t) &= g(t)/a(t) \end{aligned} \qquad \textbf{(2.8.6)}$$

are solved for $u_1'(t)$ and $u_2'(t)$, one can, in theory, find their antiderivatives $u_1(t)$ and $u_2(t)$ and construct a particular solution

$$y_p(t) = u_1(t)y_1(t) + u_2(t)y_2(t) \qquad \textbf{(2.8.7)}$$

to the inhomogeneous equation (2.8.1). The *constraint equations* (2.8.6) have a unique solution because the determinant of the coefficients of $u_1'(t)$ and $u_2'(t)$ is $W(y_1, y_2)(t)$, which is nonzero since $y_1(t)$ and $y_2(t)$ are required to be linearly independent.

EXAMPLE 1 Solve the initial value problem $y'' + y = 4 \cos t$, $y(\pi/2) = 2\pi$, $y'(\pi/2) = -3$.

SOLUTION First solve the homogeneous equation for two linearly independent solutions, $y_1(t) = \cos t$ and $y_2(t) = \sin t$. Then set

$$y_p(t) = u_1(t) \cos t + u_2(t) \sin t. \qquad \textbf{(2.8.8)}$$

With $a(t) = 1$ and $g(t) = 4 \cos t$, the equations (2.8.6) become

$$u_1'(t) \cos t + u_2'(t) \sin t = 0,$$
$$-u_1'(t) \sin t + u_2'(t) \cos t = 4 \cos t.$$

Since the Wronskian $W(y_1, y_2)(t) = \cos^2 t + \sin^2 t = 1$, these can always be solved for any t to get

$$u_1'(t) = \begin{vmatrix} 0 & \sin t \\ 4 \cos t & \cos t \end{vmatrix} = -4 \cos t \sin t = -2 \sin 2t,$$

$$u_2'(t) = \begin{vmatrix} \cos t & 0 \\ -\sin t & 4 \cos t \end{vmatrix} = 4 \cos^2 t = 2[1 + \cos 2t],$$

where we have used the trigonometric identities

$$2 \cos t \sin t = \sin 2t \quad \text{and} \quad 2 \cos^2 t = 1 + \cos 2t.$$

To find $u_1(t)$ and $u_2(t)$, we antidifferentiate to get

$$u_1(t) = \int -2 \sin 2t \, dt = \cos 2t$$

$$u_2(t) = \int 2(1 + \cos 2t) \, dt = 2t + \sin 2t,$$

where the constants of integration have been omitted since we are looking for a particular solution of the inhomogeneous problem. From (2.8.8), we have

$$y_p(t) = \cos 2t \cos t + [2t + \sin 2t] \sin t,$$

which after simplification becomes

$$\begin{aligned} y_p(t) &= [\cos 2t \cos t + \sin 2t \sin t] + 2t \sin t \\ &= \cos (2t - t) + 2t \sin t = \cos t + 2t \sin t. \end{aligned}$$

The solution $y_p(t)$ is a *particular* solution of the inhomogeneous problem since it does not depend on arbitrary parameters. A *general* solution is obtained by adding to the particular solution a general solution of the homogeneous equation, sometimes called the *complementary* solution. Therefore

$$y(t) = \cos t + 2t \sin t + c_1 \cos t + c_2 \sin t$$

is a two-parameter family of solutions to the inhomogeneous equation.

To find c_1 and c_2 that satisfy the initial conditions, substitute the data into $y(t)$ and $y'(t)$:

$$2\pi = 0 + \pi + 0 + c_2 \quad \text{and} \quad -3 = 1 + 0 - c_1 + 0,$$

and solve to get $c_1 = 4$ and $c_2 = \pi$. The solution to the initial value problem is therefore

$$\begin{aligned} y(t) &= \cos t + 2t \sin t + 4 \cos t + \pi \sin t \\ &= 2t \sin t + 5 \cos t + \pi \sin t. \end{aligned}$$

We see that in the expression for $y(t)$, the first term is the particular solution, and the remainder is a solution of the homogeneous equation. A direct substitution shows that if $y_p(t) = 2t \sin t$, then

$$y_p'' + y_p = (4 \cos t - 2t \sin t) + 2t \sin t = 4 \cos t,$$

so $y_p(t)$ is a particular solution. As this example shows, the variation of parameters algorithm may produce a term in the particular solution that is a solution to the homogeneous equation. Such a term can always be absorbed in the complementary solution, e.g.,

$$y(t) = \cos t + 2t \sin t + c_1 \cos t + c_2 \sin t$$

can be written as

$$y(t) = 2t \sin t + k_1 \cos t + k_2 \sin t,$$

where $k_1 = c_1 + 1$, $k_2 = c_2$. ∎

EXAMPLE 2 Find a general solution to

$$t^2 y'' + ty' - 4y = 4t^6, \qquad t > 0.$$

SOLUTION Earlier it was shown that $y_1(t) = t^2$ and $y_2(t) = t^{-2}$ are linearly independent solutions to the corresponding homogeneous differential equation. Let

$$y_p(t) = u_1(t)t^2 + u_2(t)t^{-2};$$

then, since $a(t) = t^2$ and $g(t) = 4t^6$, the constraint equations (2.8.6) are

$$u_1'(t)t^2 + u_2'(t)t^{-2} = 0,$$
$$t^2[2u_1'(t)t - 2u_2'(t)t^{-3}] = 4t^6,$$

with Wronskian

$$W(y_1, y_2)(t) = \begin{vmatrix} t^2 & t^{-2} \\ 2t & -2t^{-3} \end{vmatrix} = -4t^{-1}.$$

Some manipulation gives

$$u_1'(t) = \frac{\begin{vmatrix} 0 & t^{-2} \\ 4t^4 & -2t^{-3} \end{vmatrix}}{-4t^{-1}} = t^3,$$

and

$$u_2'(t) = \frac{\begin{vmatrix} t^2 & 0 \\ 2t & 4t^4 \end{vmatrix}}{-4t^{-1}} = -t^7,$$

and so

$$u_1(t) = \frac{t^4}{4}, \qquad u_2(t) = -\frac{t^8}{8}.$$

A particular solution to the inhomogeneous differential equation is then

$$y_p(t) = \frac{t^4}{4}t^2 + \left(-\frac{t^8}{8}\right)t^{-2} = \frac{t^6}{8},$$

and a general solution

$$y(t) = \frac{t^6}{8} + c_1 t^2 + c_2 t^{-2}$$

is obtained by adding a general solution of the homogeneous equation to $y_p(t)$. ∎

EXAMPLE 3 Find the solution of

$$y'' + y' - 6y = g(t), \qquad y(t_0) = A, \qquad y'(t_0) = B.$$

SOLUTION Two linearly independent solutions of the corresponding homogeneous equation are $y_1(t) = e^{2t}$, $y_2(t) = e^{-3t}$, and their Wronskian is

$$W(y_1, y_2)(t) = \begin{vmatrix} e^{2t} & e^{-3t} \\ 2e^{2t} & -3e^{-3t} \end{vmatrix} = -5e^{-t}.$$

Therefore

$$y_p(t) = u_1(t)e^{2t} + u_2(t)e^{-3t}$$

where $u_1(t)$ and $u_2(t)$ satisfy the constraint equations (2.8.6):

$$u_1'(t)e^{2t} + u_2'(t)e^{-3t} = 0,$$

$$2u_1'(t)e^{2t} - 3u_2'(t)e^{-3t} = g(t).$$

Solving this system for $u_1'(t)$ and $u_2'(t)$ gives

$$u_1'(t) = \frac{1}{5}\,g(t)e^{-2t}, \qquad u_2'(t) = -\frac{1}{5}\,g(t)e^{3t}.$$

Since $g(t)$ is not specified and the initial data are given at $t = t_0$, it is convenient (but not necessary) to use the antiderivatives

$$u_1(t) = \frac{1}{5}\int_{t_0}^{t} g(s)e^{-2s}\,ds,$$

$$u_2(t) = -\frac{1}{5}\int_{t_0}^{t} g(s)e^{3s}\,ds,$$

which gives the particular solution

$$y_p(t) = \left(\frac{1}{5}\int_{t_0}^{t} g(s)e^{-2s}\,ds\right)e^{2t} + \left(-\frac{1}{5}\int_{t_0}^{t} g(s)e^{3s}\,ds\right)e^{-3t}.$$

The general solution is the particular solution plus the complementary solution:

$$y(t) = y_p(t) + c_1e^{2t} + c_2e^{-3t},$$

where the constants c_1 and c_2 are chosen so as to satisfy the initial conditions,

$$y(t_0) = A, \quad y'(t_0) = B.$$

For instance, if $g(t) = 5t$ and $t_0 = 0$, then

$$u_1(t) = \frac{1}{5}\int_0^t 5se^{-2s}\,ds = -\frac{1}{4}e^{-2t}(2t + 1) + \frac{1}{4},$$

$$u_2(t) = -\frac{1}{5}\int_0^t 5se^{3s}\,ds = -\frac{1}{9}e^{3t}(3t - 1) - \frac{1}{9},$$

and

$$y_p(t) = u_1(t)e^{2t} + u_2(t)e^{-3t} = -\frac{5}{6}t - \frac{5}{36} + \frac{1}{4}e^{2t} - \frac{1}{9}e^{-3t}.$$

Note that the choice of antiderivative insures that $y_p(0) = y_p'(0) = 0$. (Why?)

On the other hand, if $g(t) = \sec t$ and $t_0 = 1$, then

$$u_1(t) = \frac{1}{5}\int_1^t \sec s\, e^{-2s}\, ds$$

$$u_2(t) = -\frac{1}{5}\int_1^t \sec s\, e^{3s}\, ds,$$

and neither integral can be evaluated by elementary means. Nevertheless, we can write

$$y_p(t) = \frac{1}{5}e^{2t}\int_1^t \sec s\, e^{-2s}\, ds - \frac{1}{5}e^{-3t}\int_1^t \sec s\, e^{3s}\, ds,$$

and a general solution is

$$y(t) = y_p(t) + c_1e^{2t} + c_2e^{-3t}.$$

If the initial conditions are, for instance,

$$y(1) = 0 \quad \text{and} \quad y'(1) = 2,$$

then

$$y(1) = y_p(1) + c_1e^2 + c_2e^{-3} = 0 + c_1e^2 + c_2e^{-3} = 0,$$

$$y'(1) = y_p'(1) + 2c_1e^2 - 3c_2e^{-3} = 0 + 2c_1e^2 - 3c_2e^{-3} = 2,$$

which can be solved to get $c_1 = \frac{2}{5}e^{-2}$ and $c_2 = -\frac{2}{5}e^{3}$. The advantage of the choice of antiderivatives is clear in this example. If a different antiderivative (i.e., with a different lower limit than $t_0 = 1$) had been used, then $y_p(1)$ and $y_p'(1)$ would have to be calculated by some numerical technique in order to find c_1 and c_2. ■

The method is summarized below.

The variation of parameters method

1. Solve the homogeneous equation for a linearly independent pair of solutions $y_1(t)$, $y_2(t)$.
2. Solve the pair of linear equations

$$u_1'(t)y_1(t) + u_2'(t)y_2(t) = 0,$$

$$a(t)[u_1'(t)y_1'(t) + u_2'(t)y_2'(t)] = g(t)$$

for $u_1'(t)$ and $u_2'(t)$; then find antiderivatives $u_1(t)$ and $u_2(t)$.

3. A particular solution of the inhomogeneous equation is

$$y_p(t) = u_1(t)y_1(t) + u_2(t)y_2(t)$$

with

$$y_p'(t) = u_1(t)y_1'(t) + u_2(t)y_2'(t).$$

A general solution is

$$y(t) = y_p(t) + c_1y_1(t) + c_2y_2(t),$$

where c_1 and c_2 are arbitrary constants.

4. If initial data $y(t_0) = r$ and $y'(t_0) = s$ are given, select c_1 and c_2 so that

$$r = y_p(t_0) + c_1y_1(t_0) + c_2y_2(t_0),$$
$$s = y_p'(t_0) + c_1y_1'(t_0) + c_2y_2'(t_0).$$

In this case it is convenient to use the antiderivatives

$$u_1(t) = \int_{t_0}^{t} u_1'(s)\,ds \quad \text{and} \quad u_2(t) = \int_{t_0}^{t} u_2'(s)\,ds$$

in step 2, so that $y_p(t_0) = y_p'(t_0) = 0$.

EXERCISES 2.8

Using the variation of parameters method, find a general solution of the inhomogeneous differential equations in Exercises 1–9.

1. $y'' - y = 2e^{3t}$

2. $y'' + 9y = 3t + 2$

3. $y'' - 7y' + 12y = 5e^{4t}$

4. $y'' - 4y = 3e^{-2t} - t^2$

5. $y'' + 4y = 4\cos^3 2t$. (*Hint:* A useful identity is $\cos^3\theta = \frac{3}{4}\cos\theta + \frac{1}{4}\cos 3\theta$.)

6. $y'' + y = \sec t$

7. $y'' + 6y' + 9y = \dfrac{e^{-3t}}{t^2}$

8. $y'' - 4y = \dfrac{1}{\cosh 2t}$

9. $y'' - 4y' + 3y = \dfrac{1}{1 + e^{-t}}$

Using the method of variation of parameters, find a general solution of the inhomogeneous Euler differential equations (see Exercise 18 of Section 2.6) in Exercises 10–13.

10. $t^2y'' - 3ty' + 4y = t + 2$

11. $t^2y'' - 2ty' + 2y = te^t$

12. $t^2y'' + 7ty' + 5y = 10 - 4/t$

13. $t^2y'' - 4ty' + 6y = t^2e^t + \ln t$

Solve the initial value problems in Exercises 14–19.

14. Exercise 1: $y(0) = 1, \quad y'(0) = 2$
15. Exercise 2: $y(0) = 0, \quad y'(0) = 1$
16. Exercise 6: $y(0) = 2, \quad y'(0) = -1$
17. Exercise 7: $y(1) = 1, \quad y'(1) = 0$
18. Exercise 11: $y(1) = 0, \quad y'(1) = 1$
19. Exercise 12: $y(1) = 1, \quad y'(1) = 0$

20. Show that a general solution $y(t)$ of the linear inhomogeneous differential equation

$$a(t)y'' + b(t)y' + c(t)y = g(t)$$

can be written as a sum $y(t) = y_p(t) + y_h(t)$, where $y_p(t)$ is a particular solution of the inhomogeneous equation and $y_h(t)$ is a general solution of the homogeneous equation. (*Hint:* What differential equation is satisfied by $y(t) - y_p(t)$?)

21. The solution of

$$y'' + w^2y = f(t); \qquad y(0) = y'(0) = 0$$

can be expressed in the form

$$y(t) = u_1(t) \sin wt + u_2(t) \cos wt,$$

where $u_1(t)$ and $u_2(t)$ are found by the variation of parameters algorithm. By combining $\sin wt$ and $\cos wt$ with the integral expressions for $u_1(t)$ and $u_2(t)$ show that

$$y(t) = \frac{1}{w} \int_0^t \sin[w(t - s)]f(s)\, ds.$$

For the inhomogeneous differential equations given in Exercises 22–24, the functions $y_1(t)$ and $y_2(t)$ are linearly independent solutions of the corresponding homogeneous equation. Use them and the variation of parameters method to find a general solution of the inhomogeneous equation.

22. $ty'' - (2t + 1)y' + (t + 1)y = 4t^2e^t; \qquad y_1(t) = e^t, \quad y_2(t) = t^2e^t$

23. $y'' - \dfrac{2}{t}y' + \left(1 + \dfrac{2}{t^2}\right)y = te^t; \qquad y_1(t) = t \cos t, \quad y_2(t) = t \sin t$

24. $ty'' - y' + 4t^3y = 8t^4; \qquad y_1(t) = \sin t^2, \quad y_2(t) = \cos t^2$

2.9 UNFORCED OSCILLATIONS OF ELECTRICAL AND MECHANICAL SYSTEMS

Oscillations of mechanical and electrical systems are so varied and so commonplace that most of us do not appreciate their shared character; the systems are attempting to return to equilibrium states in the presence of isolated or persistent disturbances. After a tennis racquet strikes a tennis ball, the racquet vibrates and energy is dissipated in the frame and handle until the racquet attains its original configuration and is ready for the next stroke. A clock is fed energy in such a way that it oscillates with a fixed period and, if it is well designed and carefully constructed, its periodic oscillation can be maintained with remarkable accuracy for years. The oscillatory circulation of blood in our bodies results from the pumping of an electro-mechanical system, the heart. Although some oscillatory phenomena are intrinsically nonlinear, the gross behavior of many can be modeled by linear second order differential equations with constant coefficients and can be analyzed with the techniques that we have studied. In this section we shall investigate oscillatory systems that are not subject to external forces. Forced oscillatory systems are the topic of the next section.

Before starting this discussion, let us consider an important physical example, which is illustrated in Fig. 2.4. The figure depicts a mass supported on a level surface; the mass is attached to a rigid wall by a light spring. If the mass is displaced from its equilibrium position ($y = 0$), the spring either stretches or contracts. The spring then attempts to return to its original length by applying a restoring force to the mass that is proportional to the displacement (Hooke's law). Other forces acting on the mass are a frictional force and, perhaps, an external force. The sum of these forces, by Newton's second law, equals the time rate of change of the momentum of the system. Hence,

$$m\frac{dy}{dt} = p, \qquad \frac{dp}{dt} = -ky + f\left(y, \frac{dy}{dt}\right) + F(t), \tag{2.9.1}$$

where y denotes the displacement from equilibrium, p is the momentum, m is the mass, k is the Hooke's constant, $f(y, dy/dt)$ is the friction force, and $F(t)$ is the external force. The first equation defines the momentum, and the second models the forces acting to change the momentum.

The frictional force $f(y, dy/dt)$ is difficult to model. The simplest useful model assumes that the dissipation of energy due to friction is proportional to the velocity

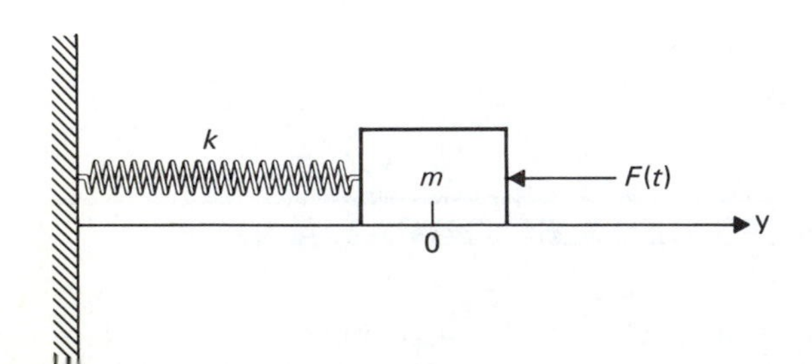

Figure 2.4 A mass–spring system.

and is independent of position, and therefore

$$f\left(y, \frac{dy}{dt}\right) = -c\frac{dy}{dt},$$

where c is a positive constant. The minus sign in this formula is due to the resistance of the surface to the motion of the mass. This model of friction is called *viscous damping*. The differential equations describing the mass–spring system with viscous damping are

$$m\frac{dy}{dt} = p, \qquad \frac{dp}{dt} = -ky - c\frac{dy}{dt} + F(t), \tag{2.9.2}$$

which implies

$$m\frac{d^2y}{dt^2} = \frac{dp}{dt} = -ky - c\frac{dy}{dt} + F(t)$$

or

$$m\frac{d^2y}{dt^2} + c\frac{dy}{dt} + ky = F(t). \tag{2.9.3}$$

This equation is called the *oscillator equation*. If $c = 0$, the oscillator is said to be *undamped*. Clearly, Eqs. (2.9.2) and (2.9.3) are equivalent: A solution to the pair of first order equations is a solution to the second order system, and conversely. We found it convenient to use second order equations in the discussion of solution techniques. We shall now find it convenient to use first order systems in the discussion of the geometry of solution curves. By *first order systems* we mean a set of first order differential equations.

The *oscillator equation* of a closely related electrical example, the *LCR* circuit,

$$L\frac{d^2Q}{dt^2} + R\frac{dQ}{dt} + \frac{Q}{C} = E(t), \tag{2.9.4}$$

was derived in Section 2.2 from the two first order equations,

$$\frac{dQ}{dt} = I, \qquad L\frac{dI}{dt} + RI + \frac{Q}{C} = E(t). \tag{2.9.5}$$

As the reader can see, the differential equation describing the mass–spring system when the mass is sliding on a level surface is the same as when the mass is suspended from a vertical support (page 128). In the first case the term $c\,dy/dt$ describes the effect of friction, while in the second case it describes the effect of air resistance. Along with the *LCR* circuit, all these systems are described by the oscillator equation, and this results in an economy of analysis but also in a transfer of intuition from one discipline to another. For example, electrical-network analogies are used in the analysis of stresses in structures, in underwater acoustics, and in thermodynamics.

With the *LCR* circuit and the mass–spring systems as motivating examples, let us study the homogeneous oscillator equation,

$$m\frac{d^2y}{dt^2} + c\frac{dy}{dt} + ky = 0 \tag{2.9.6}$$

and the equivalent first order system,

$$\frac{dy}{dt} = \frac{p}{m}, \qquad \frac{dp}{dt} = -ky - \frac{cp}{m}. \tag{2.9.7}$$

We already know that there will be three types of solutions whose features are determined by the roots of the corresponding characteristic equation,

$$m\lambda^2 + c\lambda + k = 0, \tag{2.9.8}$$

and, in particular, by the sign of the discriminant, $c^2 - 4mk$. What we have not done is study the effect of varying the parameters of the differential equation (2.9.6) on the geometry of its solutions curves. In doing this, new concepts such as the phase plane of a differential equation will be employed.

The simplest and perhaps the most striking type of unforced oscillation is one in which a configuration or state of the system is taken on again and again. This type of motion is called periodic.

Definition

The *state* of a mass–spring system at time t is the ordered pair of real numbers, $(y(t), p(t))$, where $y(t)$ is the position of the mass and $p(t)$ is its momentum. The oscillations of the system are said to be *periodic* if there exists a positive number T such that for all t,

$$y(t) = y(t + T) \quad \text{and} \quad p(t) = p(t + T).$$

The least positive number T such that the above equations are valid is called the *period* of the system.

The corresponding definitions for an *LCR* circuit use the charge and current to determine the state of the system.

Unforced periodic motion can occur only if there is no loss of energy. This means no frictional forces can be acting on the system. While this is physically obvious, a mathematical proof permits the introduction of energy concepts that are also used in a study of nonlinear oscillations in Chapter 7.

Definition

The *total energy* of the mass–spring system illustrated in Fig. 2.4. is represented by the function,

$$E(y, p) = \frac{1}{2}\frac{p^2}{m} + \frac{1}{2}ky^2. \tag{2.9.9}$$

The total energy is the sum of the *kinetic energy*, $\frac{1}{2}\frac{p^2}{m}$, and the *potential energy*, $\frac{1}{2}ky^2$.

The time rate of change of the total energy can be computed by substituting a solution $(y(t), p(t))$ into (2.9.9) and differentiating:

$$\frac{d}{dt}E(y(t), p(t)) = \frac{p(t)}{m}\frac{dp}{dt}(t) + ky(t)\frac{dy}{dt}(t).$$

This expression can be simplified with the aid of the system of first order equations (2.9.7). Thus,

$$\begin{aligned}\frac{d}{dt}E(y(t), p(t)) &= \frac{p(t)}{m}\left(-ky(t) - c\frac{p(t)}{m}\right) + ky(t)\left(\frac{p(t)}{m}\right)\\ &= -c\frac{p(t)^2}{m^2}.\end{aligned} \tag{2.9.10}$$

From this we can conclude that if friction is present $(c > 0)$, then the total energy is a monotonic decreasing function of time unless the system is at rest. If no frictional forces are acting $(c = 0)$, the total energy is constant.

One way to present a graphical description of the time history of a solution to an oscillator equation is to display two graphs, one of y versus t and the other of p versus t. Another way is to plot the set of points $\{(y(t), p(t))\}$ for t in a convenient interval. The resulting curve is called an *integral curve* of the differential equation; the y, p plane is called the *phase plane*. We shall investigate the effect of changing the amount of viscous damping on both the solutions to the differential equation and on the graphical description of the motion. We have four possibilities:

Type 1.	Undamped, $c = 0$.
Type 2.	Underdamped, $c > 0$, $c^2 - 4km < 0$.
Type 3.	Critically damped, $c > 0$, $c^2 - 4km = 0$.
Type 4.	Overdamped, $c > 0$, $c^2 - 4km > 0$.

Let us now consider the case of an undamped spring–mass system.

EXAMPLE 1 Solve the undamped oscillator equation,

$$y'' + ky = 0, \quad \text{or} \quad \frac{dy}{dt} = p, \qquad \frac{dp}{dt} = -ky,$$

with initial conditions $y(0) = 2$, $y'(0) = 0$ and (a) $k = 4$, (b) $k = \frac{1}{4}$. Compute the total energy and sketch the graphs of y versus t, p versus t, and p versus y. Show that the oscillations are periodic and compute the period T.

SOLUTION The amplitude–phase angle form of the solution is

$$y(t) = A\cos(\sqrt{k}\,t - \phi).$$

Since $m = 1$, we have that $p = y'$. Therefore

$$p(t) = -\sqrt{k}\,A\sin(\sqrt{k}\,t - \phi).$$

The trigonometric functions $y(t)$ and $p(t)$ are periodic with period $2\pi/\sqrt{k}$. The total energy is given by

$$\begin{aligned} E(y(t), p(t)) &= \frac{1}{2}kA^2\sin^2(\sqrt{k}\,t - \phi) + \frac{1}{2}kA^2\cos^2(\sqrt{k}\,t - \phi) \\ &= \frac{1}{2}kA^2. \end{aligned}$$

The numerical values of the amplitude and phase can be computed from the initial conditions: $A = 2$, $\phi = 0$. It follows that

a) $y(t) = 2\cos 2t, \qquad p(t) = -4\sin 2t, \qquad E = 8, \qquad T = \pi,$

b) $y(t) = 2\cos\dfrac{t}{2}, \qquad p(t) = -\sin\dfrac{t}{2}, \qquad E = \dfrac{1}{2}, \qquad T = 4\pi.$

The graphs are in Figs. 2.5 through 2.7.

The motion starts at $t = 0$ with $y = 2$ and $p = 0$. Since the stretched spring pulls the mass, y decreases and p becomes negative. The mass passes the equilibrium position, $y = 0$, at its maximum speed. At this time, the spring force vanishes and the integral curve in the phase plane takes on its minimum value. As the mass continues to move to the left, it is pulled by the stretched spring, its speed decreases, and in the phase plane, the integral curve rises toward the p-axis. At its maximum distance from the equilibrium position, $y = -2$ and $p = 0$. The mass is then pulled to the right and returns to its starting point. The motion is repeated. Since the motion is symmetric with respect to the equilibrium position, the integral curve in the phase plane is symmetric with respect to the p axis. The elliptical integral curve is a level curve of the energy function, $E(y, p) =$ constant. This follows from the computation, $E(y(t), p(t)) = \frac{1}{2}kA^2$. ■

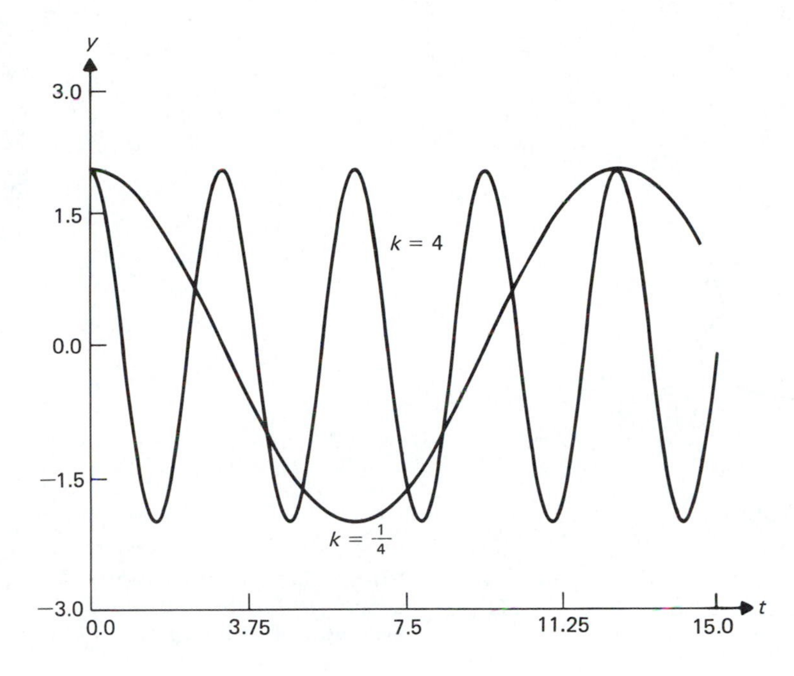

Figure 2.5 Graphs of $y(t)$ for the undamped oscillator of Example 1.

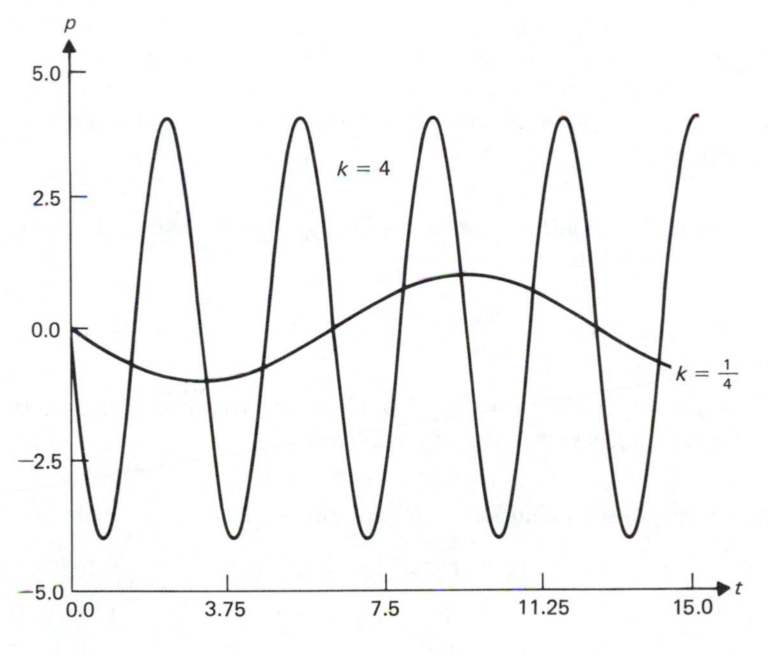

Figure 2.6 Graphs of $p(t)$ for the undamped oscillator of Example 1.

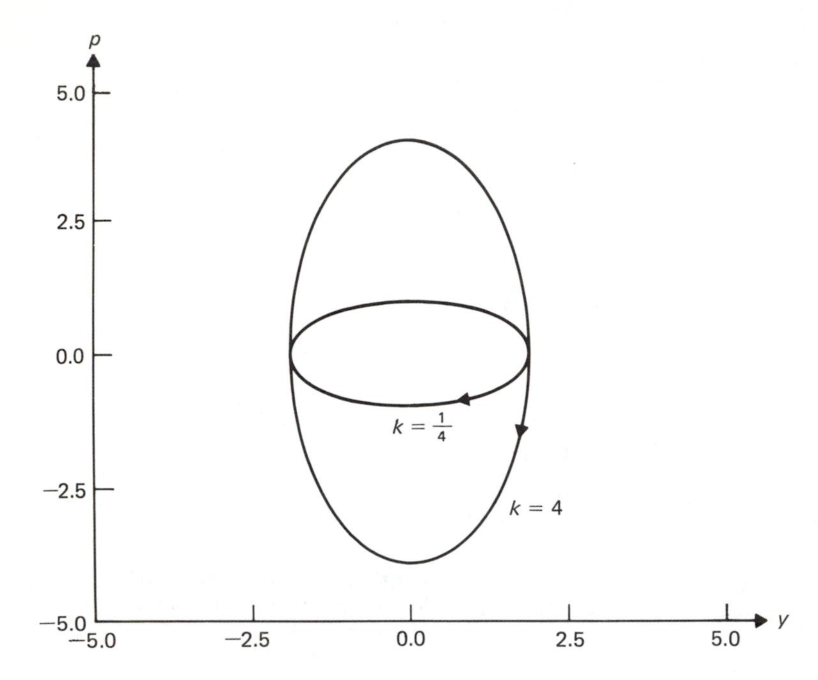

Figure 2.7 Graphs of $p(t)$ versus $y(t)$ for the undamped oscillator of Example 1.

If a small amount of viscous damping is present, the energy and the maximum excursions of the mass decrease as time increases. There is a transfer of energy between the kinetic and potential energies with the kinetic energy taking on its local maxima at $y = 0$ and the potential energy taking on its local maxima at $p = 0$. The mass oscillates with exponentially decreasing amplitude. This is illustrated in the following example.

EXAMPLE 2 Describe the motions of a mass–spring system governed by the lightly damped oscillator equation,

$$4y'' + 4y' + 17y = 0, \quad \text{or} \quad \frac{dy}{dt} = \frac{p}{4}, \qquad \frac{dp}{dt} = -p - 17y,$$

with initial conditions $y(0) = 4$, $p(0) = 4y'(0) = 0$. Compute the total energy and sketch the graphs of y versus t, p versus t, and p versus y.

SOLUTION The corresponding characteristic equation,

$$4\lambda^2 + 4\lambda + 17 = 0,$$

has roots, $\lambda = -\frac{1}{2} \pm 2i$. From this we deduce that a general solution to the differential equation in amplitude–phase angle form (see Section 2.5) is

$$y(t) = Ae^{-t/2}\cos(2t - \phi),$$

$$p(t) = 4y'(t) = -4Ae^{-t/2}\left[\frac{1}{2}\cos(2t - \phi) + 2\sin(2t - \phi)\right].$$

The initial data are $t = 0, y = 4, p = 0$. Substituting these values into the preceding equations, we obtain

$$4 = A\cos\phi, \qquad 0 = \frac{1}{2}\cos\phi - 2\sin\phi.$$

Therefore $\tan\phi = \frac{1}{4}$ and

$$\cos\phi = \frac{4}{\sqrt{17}}, \qquad \sin\phi = \frac{1}{\sqrt{17}}, \qquad A = \sqrt{17}.$$

With ϕ defined by these trigonometric equations, we have that the solution to the initial value problem is

$$y(t) = \sqrt{17}\, e^{-t/2}\cos(2t - \phi),$$

$$p(t) = -4\sqrt{17}\, e^{-t/2}\left[\frac{1}{2}\cos(2t - \phi) + 2\sin(2t - \phi)\right].$$

An alternative representation of the solution is

$$y(t) = e^{-t/2}(4\cos 2t + \sin 2t),$$

$$p(t) = -34e^{-t/2}\sin 2t.$$

The total energy,

$$E = \frac{1}{2}\frac{p^2}{m} + \frac{1}{2}ky^2,$$

is

$$E(y(t), p(t)) = \frac{289}{2}e^{-t}\left[(\sin 2t)^2\right] + \left[(4\cos 2t + \sin 2t)^2\right]\frac{17}{2}e^{-t}.$$

As the mass oscillates its amplitude decreases exponentially. Its motion is said to be underdamped. The term $e^{-t/2}$ is called the damping factor. The maximum excursions of the mass are characterized by the zeros of $p(t)$:

$$t = 0, \frac{\pi}{2}, \pi, \frac{3\pi}{2}, \ldots.$$

The corresponding monotonically decreasing values of y (in absolute value) and E are

$$y = 4, -4e^{-\pi/4}, 4e^{-\pi/2}, -4e^{-3\pi/4}, \ldots,$$
$$E = 136, 136e^{-\pi/2}, 136e^{-\pi}, 136e^{-3\pi/2}, \ldots.$$

Because of the damping, the motion is not periodic, and its mathematical representation is the product of an exponential function and a periodic trigonometric function. The period of the trigonometric factor is called the *quasi-period* of the system. It is the time between successive maximum values of solution. In this problem, the quasi-period is equal to π.

Graphs of y versus t, p versus t, and p versus y are in Figs. 2.8 through 2.10. The integral curve in the phase plane, Fig. 2.10, can be thought of as being generated by a particle moving on an ellipse that is slowly shrinking with increasing time. The points on the integral curve with $p = 0$ are local maxima of the potential energy and those with $y = 0$ are local maxima of the kinetic energy. ■

As the viscous damping increases, the exponential damping factor decreases and the quasi-period increases. At critical damping the quasi-period is infinite and the characteristic equation (2.9.8) has a double characteristic root, $\lambda = -c/2m$.

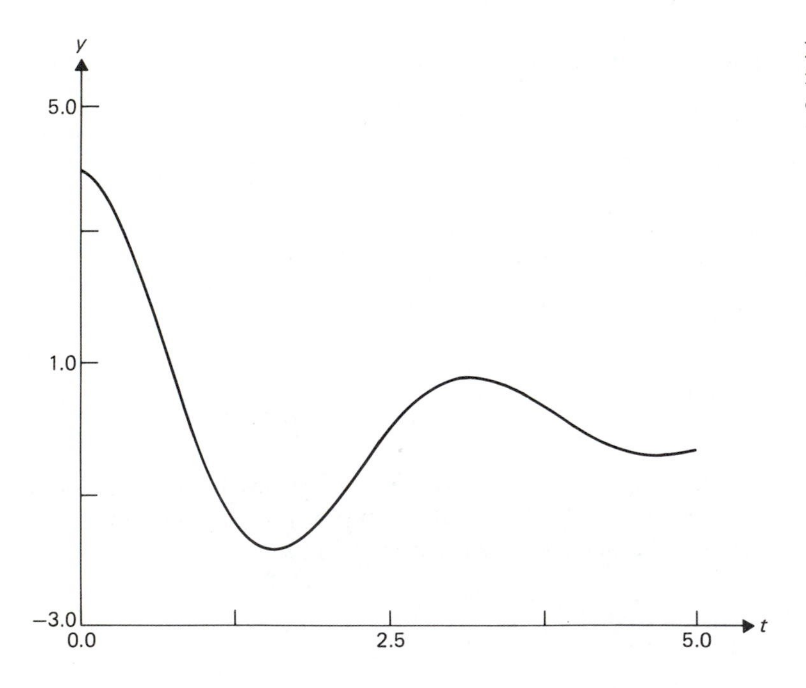

Figure 2.8 Graph of $y(t)$ for the underdamped oscillator of Example 2.

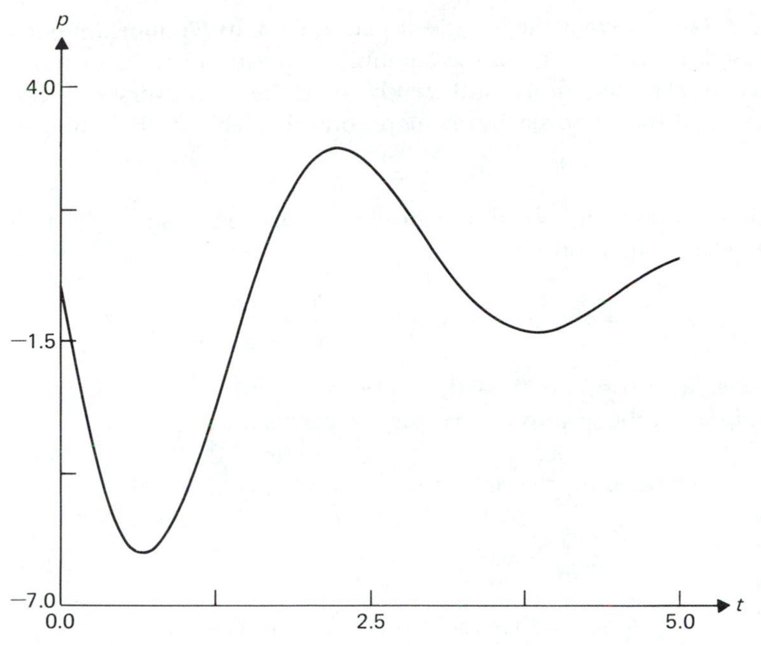

Figure 2.9 Graph of $p(t)$ for the underdamped oscillator of Example 2.

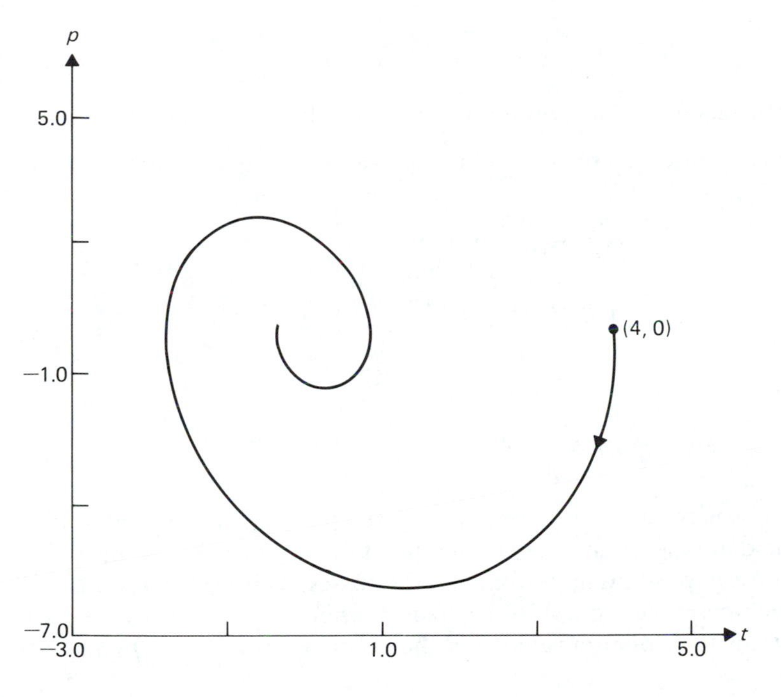

Figure 2.10 Graph of $p(t)$ versus $y(t)$ for the underdamped oscillator of Example 2.

In a critically damped system if the mass is displaced from its equilibrium position and then released, its distance from the equilibrium position decreases monotonically. But if its initial momentum is sufficiently large, the mass overshoots the equilibrium position and then approaches it monotonically. This is illustrated in the next example.

EXAMPLE 3 Describe the motion of a critically damped mass–spring system governed by the oscillator equation,

$$2y'' + y' + \frac{1}{8}y = 0, \quad \text{or} \quad \frac{dy}{dt} = \frac{p}{2}, \qquad \frac{dp}{dt} = \frac{-p}{2}\frac{-y}{8},$$

with initial conditions, (a) $y(0) = 4$, $p(0) = 0$, (b) $y(0) = 4$, $p(0) = -10$. Compute the total energy and sketch the graphs of y versus t, p versus t, and p versus y.

SOLUTION The corresponding characteristic equation,

$$2\lambda^2 + \lambda + \frac{1}{8} = 0,$$

has a double root, $\lambda = -\frac{1}{4}$. A general solution to the differential equation is

$$y(t) = e^{-t/4}(c_1 t + c_2),$$

$$p(t) = 2e^{-t/4}\left(c_1 - \frac{1}{4}c_2 - \frac{1}{4}c_1 t\right).$$

The graphs for both sets of initial conditions are in Figs. 2.11 through 2.13.

a) From the initial conditions, $y(0) = 4$, $p(0) = 0$, we deduce that $c_1 = 1$, $c_2 = 4$, and

$$y(t) = e^{-t/4}(t + 4),$$

$$p(t) = -\frac{1}{2}te^{-t/4}.$$

The total energy,

$$E(y(t), p(t)) = \frac{1}{16}e^{-t/2}(t^2 + (t + 4)^2).$$

Once the mass is released, it moves to the left with increasing speed until $t = 4$. Then the damping dominates and its speed decreases as its distance from the equilibrium position monotonically decreases. This information is displayed on the phase plane graph of p versus y, where as y decreases, p decreases, attains its minimum value, and then increases until $y = 0$ and $p = 0$.

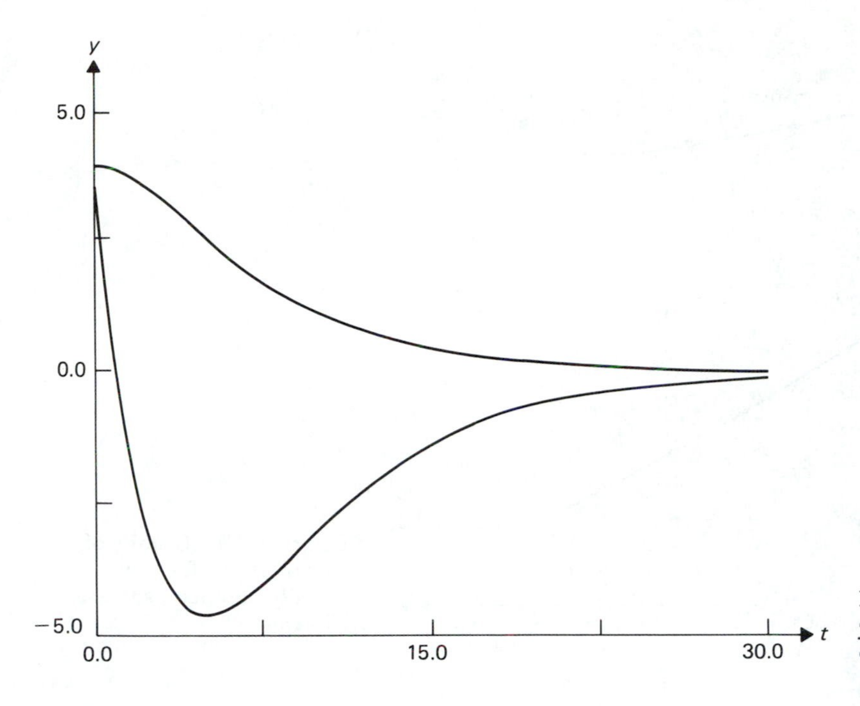

Figure 2.11 Graphs of $y(t)$ for the critically damped oscillator of Example 3.

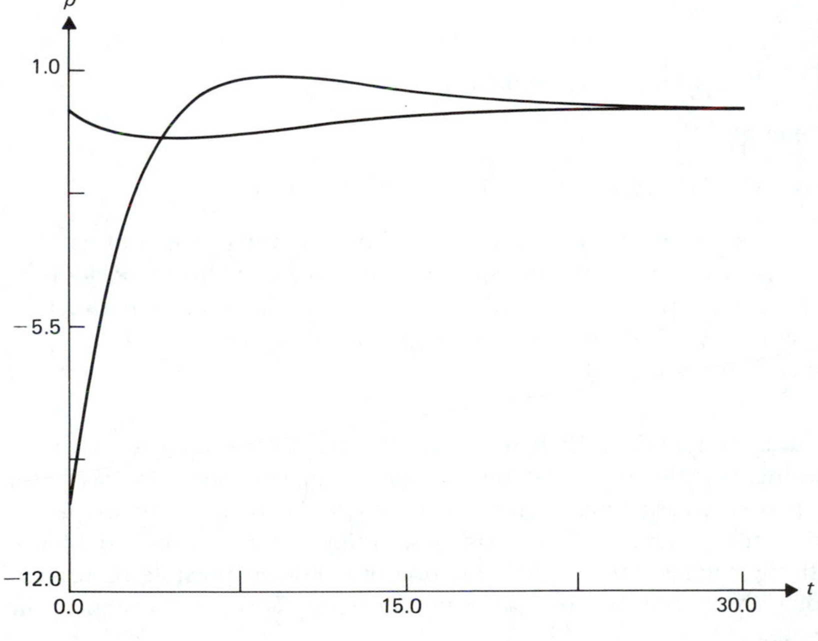

Figure 2.12 Graphs of $p(t)$ for the critically damped oscillator of Example 3.

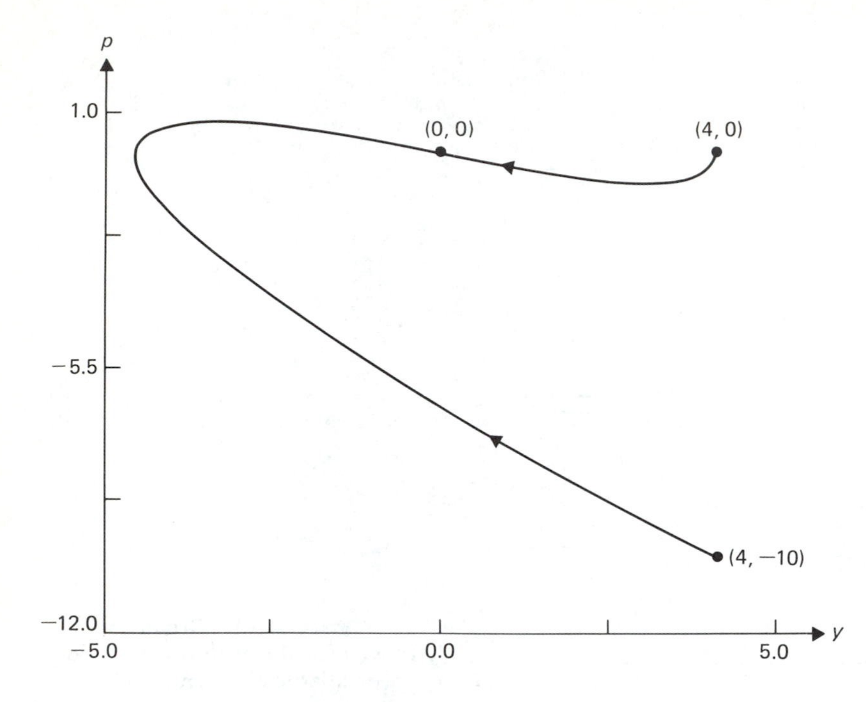

Figure 2.13 Graphs of $p(t)$ versus $y(t)$ for the critically damped oscillator of Example 3.

b) From the initial conditions, $y(0) = 4$, $p(0) = -10$, we deduce that $c_1 = -4$, $c_2 = 4$, and

$$y(t) = 4e^{-t/4}(1 - t),$$
$$p(t) = 2e^{-t/4}(t - 5).$$

The total energy,

$$E(y(t), p(t)) = e^{-t/2}[(t - 5)^2 + (1 - t)^2].$$

At $t = 0$, the mass is moving to the left with decreasing speed but its initial speed is sufficient for the mass to overshoot the equilibrium position and then return slowly to it. In the phase plane graph the distance between the curve and the y-axis decreases monotonically as time increases. The overshoot is clearly displayed. ■

If the viscous damping is sufficiently large, the characteristic equation of the system has distinct negative roots and any solution to the corresponding oscillator equation can be written as a linear combination of two damped exponentials. The system is said to be overdamped. The gross description of the motion is similar to that of a critically damped system. At most one overshoot is possible before the mass monotonically approaches its equilibrium position. This is illustrated in the following example.

EXAMPLE 4 Solve

$$y'' + 3y' + 2y = 0, \quad \text{or} \quad \frac{dy}{dt} = p, \qquad \frac{dp}{dt} = -3p - 2y,$$

with initial conditions (a) $y(0) = 4$, $p(0) = 0$, (b) $y(0) = 4$, $p(0) = -12$. Compute the total energy and sketch y versus t, p versus t, and p versus y.

SOLUTION The characteristic equation,

$$\lambda^2 + 3\lambda + 2 = 0,$$

has characteristic roots $\lambda_1 = -1$ and $\lambda_2 = -2$. A general solution to the overdamped oscillator equation is

$$y(t) = c_1 e^{-t} + c_2 e^{-2t},$$

$$p(t) = -c_1 e^{-t} - 2c_2 e^{-2t}.$$

a) From the initial conditions $y(0) = 4$, $p(0) = 0$, we obtain $c_1 = 8$, $c_2 = -4$. Therefore

$$y(t) = 4(2e^{-t} - e^{-2t}),$$

$$p(t) = -8(e^{-t} - e^{-2t}),$$

$$E(y(t), p(t)) = 32(e^{-t} - e^{-2t})^2 + 16(2e^{-t} - e^{-2t})^2.$$

b) From the initial conditions, $y(0) = 4$, $p(0) = -12$, we obtain $c_1 = -4$, $c_2 = 8$. Therefore

$$y(t) = 4(-e^{-t} + 2e^{-2t}),$$

$$p(t) = 4(e^{-t} - 4e^{-2t}),$$

$$E(y(t), p(t)) = 8(e^{-t} - 4e^{-2t})^2 + 16(-e^{-t} + 2e^{-2t})^2.$$

The graphs are in Figs. 2.14 through 2.16. ■

A SERIES ELECTRICAL CIRCUIT

The charge Q and the current I of the series LCR electric circuit shown in Fig. 2.17 is governed by a first order system of differential equations (see Section 2.2),

$$\frac{dQ}{dt} = I, \qquad L\frac{dI}{dt} + RI + \frac{Q}{C} = 0,$$

where L denotes the inductance, R the resistance, and C the capacitance of the circuit. An equivalent second order equation is

$$L\frac{d^2Q}{dt^2} + R\frac{dQ}{dt} + \frac{Q}{C} = 0.$$

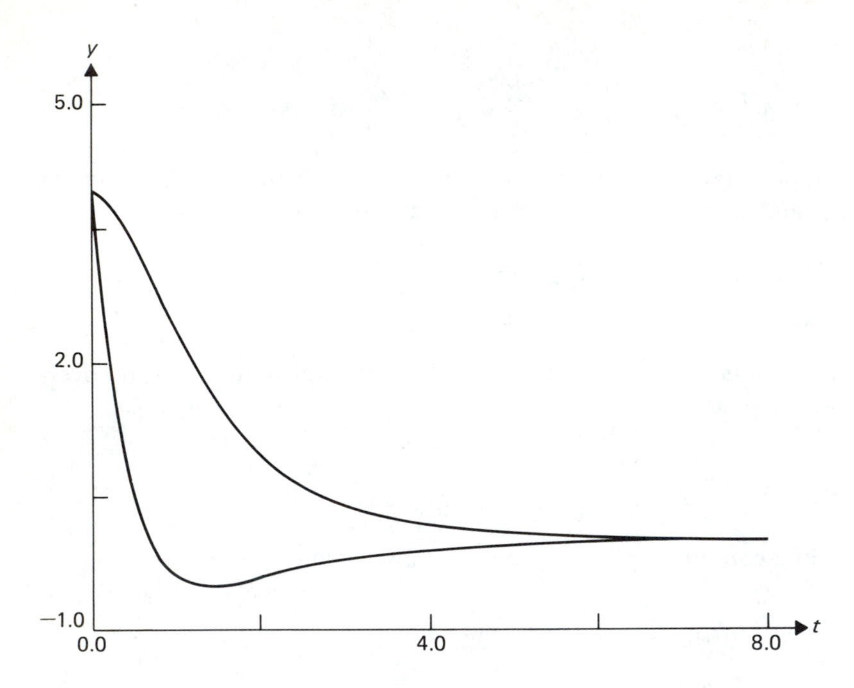

Figure 2.14 Graphs of $y(t)$ for the overdamped oscillator of Example 4.

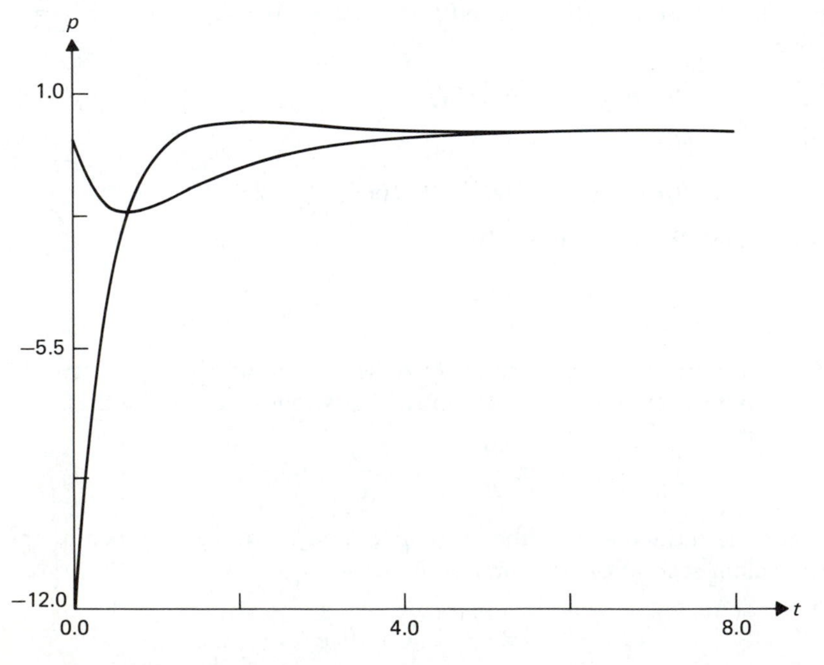

Figure 2.15 Graphs of $p(t)$ for the overdamped oscillator of Example 4.

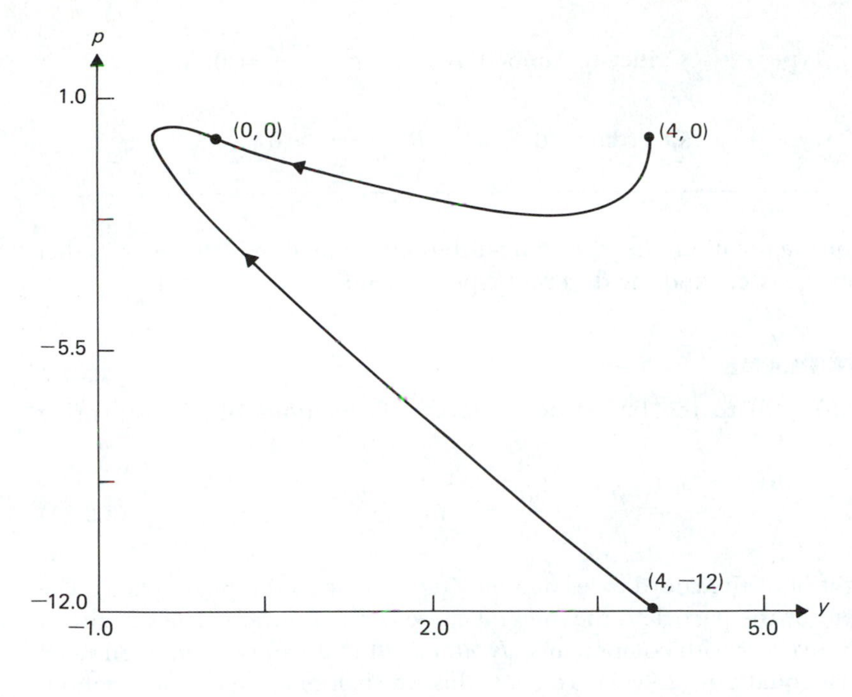

Figure 2.16 Graphs of $p(t)$ versus $y(t)$ for the overdamped oscillator of Example 4.

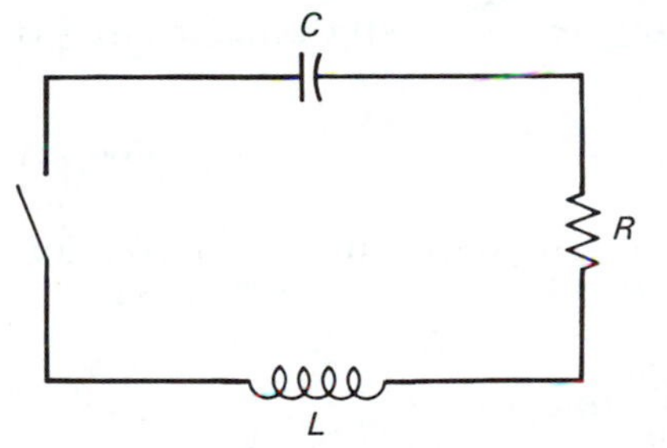

Figure 2.17 An *LCR* circuit.

The roots of its characteristic equation,

$$L\lambda^2 + R\lambda + \frac{1}{C} = 0,$$

determine the nature of the flow of electrons. We again have four possibilities:

Type 1. Undamped, $R = 0$.

Type 2. Underdamped, $R > 0$, $R^2 - \frac{4L}{C} < 0$.

Type 3. Critically damped, $R > 0$, $R^2 - \dfrac{4L}{C} = 0$.

Type 4. Overdamped, $R > 0$, $R^2 - \dfrac{4L}{C} > 0$.

Except for the notation, the governing differential equation is the same as that of a mass–spring system and the different types of solutions considered.

THE PHASE PLANE

A solution $(y(t), p(t))$ to the first order differential equations of a mass–spring system,

$$\frac{dy}{dt} = \frac{p}{m}, \qquad \frac{dp}{dt} = -ky - \frac{cp}{m}, \tag{2.9.11}$$

can be thought of as the coordinates of a moving particle in the phase plane. The path traced out by the particle is the integral curve of the solution. The velocity of the particle is a vector with components $(dy/dt, dp/dt)$ that can be computed from the differential equations (2.9.11). To show this we shall employ vector notation. If the unit vectors $\mathbf{e}_1$ and $\mathbf{e}_2$ are chosen as coordinate vectors, the *solution vector* is defined by

$$\mathbf{Y}(t) = y(t)\mathbf{e}_1 + p(t)\mathbf{e}_2. \tag{2.9.12}$$

It extends from the origin to the integral curve. The derivative of the solution vector,

$$\frac{d\mathbf{Y}}{dt}(t) = \frac{dy}{dt}(t)\mathbf{e}_1 + \frac{dp}{dt}(t)\mathbf{e}_2,$$

is tangent to the integral curve at each point. This is a basic result of vector calculus that can be established by studying the vector

$$\frac{\Delta\mathbf{Y}}{\Delta t} = \frac{\mathbf{Y}(t + \Delta t) - \mathbf{Y}(t)}{\Delta t}$$

as Δt approaches zero.

For an integral curve passing through any point (y, p), the velocity vector is given by

$$\frac{d\mathbf{Y}}{dt} = \frac{dy}{dt}\mathbf{e}_1 + \frac{dp}{dt}\mathbf{e}_2,$$

where $\mathbf{e}_1$ and $\mathbf{e}_2$ are unit coordinate vectors. The components of the velocity vector *at any point* in the phase plane of the system (2.9.11) are

$$\frac{dy}{dt} = \frac{p}{m}, \qquad \frac{dp}{dt} = -ky - \frac{cp}{m};$$

hence

$$\frac{d\mathbf{Y}}{dt} = \frac{p}{m}\mathbf{e}_1 + \left[-ky - \frac{cp}{m}\right]\mathbf{e}_2. \tag{2.9.13}$$

For instance, at the point $(y, p) = (2, 1)$ the velocity vector is

$$\frac{1}{m}\mathbf{e}_1 + \left[-2k - \frac{c}{m}\right]\mathbf{e}_2.$$

It is a tangent vector to the integral curve passing through (2, 1).

We see that it is not necessary to solve the differential equation in order to know the direction of the flow of the points in the phase plane. Given a system of differential equations, one can graph the flow by selecting various points $\{(y, p)\}$ and then computing the corresponding velocity vector. This is done by substituting the values of y and p into the right-hand sides of the differential equations to obtain the components of the vector. Thus the concept of a *direction field* of a single first order differential equation is extended to a *vector field* of a system of differential equations. To sketch the vector field for the above example, take the vector (2.9.13) and compute it for some points $\{(y, p)\}$. Then at each point (y, p) place a small arrow in the direction of the vector (compare this procedure with the one used to construct direction fields in Chapter 1).

The vector fields and several integral curves of the following oscillator equations are sketched in Figs. 2.18 through 2.21.

$$\frac{dy}{dt} = p, \qquad \frac{dp}{dt} = -4y, \qquad \text{undamped, see Example 1;} \tag{2.9.14}$$

$$\frac{dy}{dt} = \frac{p}{4}, \qquad \frac{dp}{dt} = -17y - p, \qquad \text{underdamped, see Example 2;} \tag{2.9.15}$$

$$\frac{dy}{dt} = \frac{p}{2}, \qquad \frac{dp}{dt} = -\frac{y}{8} - \frac{p}{2}, \qquad \text{critically damped, see Example 3;} \tag{2.9.16}$$

$$\frac{dy}{dt} = p, \qquad \frac{dp}{dt} = -2y - 3p, \qquad \text{overdamped, see Example 4.} \tag{2.9.17}$$

From a vector field much can be learned about the qualitative behavior of all the solutions to a system of differential equations in a region of the phase plane.

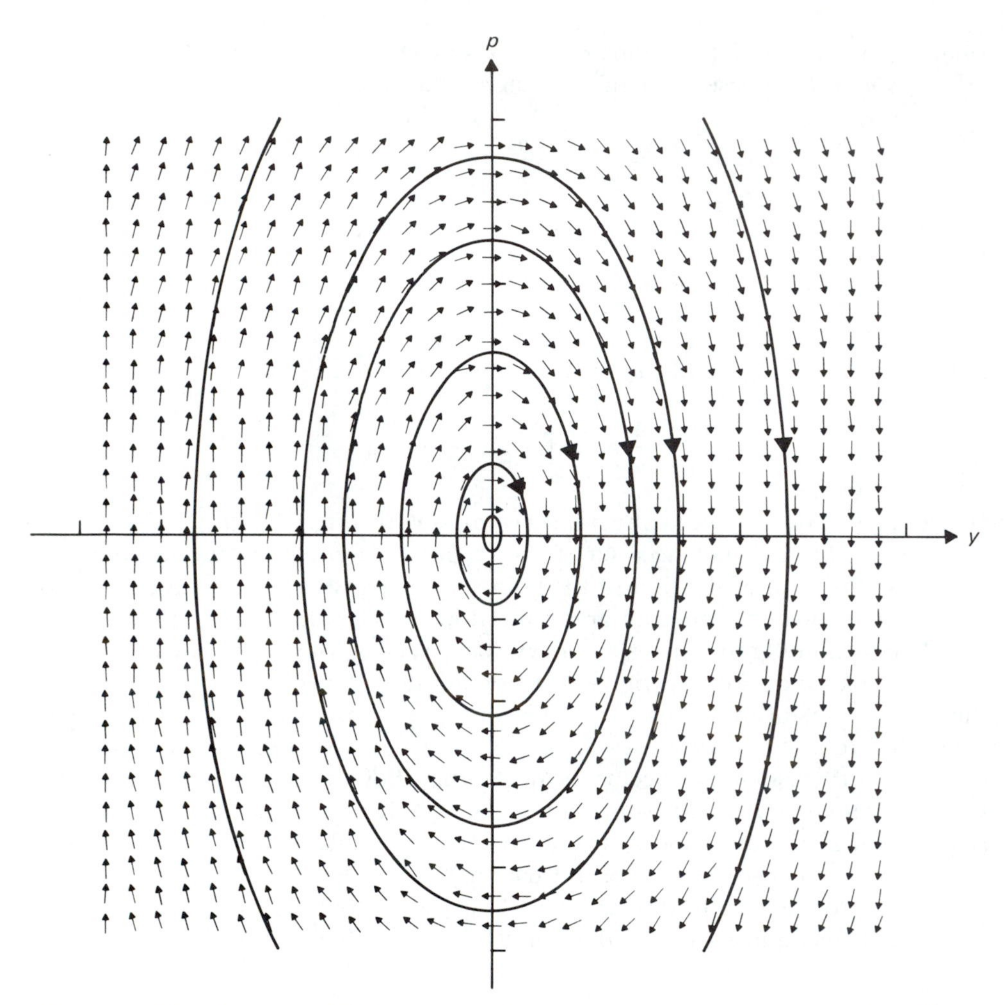

Figure 2.18 The vector field and several integral curves of the undamped oscillator of Example 1.

An inspection of the vector field of the undamped system (2.9.14) shows that all the integral curves are ellipses, the level curves of the energy function $E(y, p)$. The integral curves of the underdamped system (2.9.15) spiral into the origin as time increases. Although the vector fields of critically and overdamped systems (2.9.16–17) differ in detail, they show that the motion is nonoscillatory and that the integral curves approach the origin, the equilibrium state of the mass–spring system, as time increases.

The unforced mass–spring system and series *LCR* circuit are rather simple physical systems, but they deserve thorough study for they are the building blocks

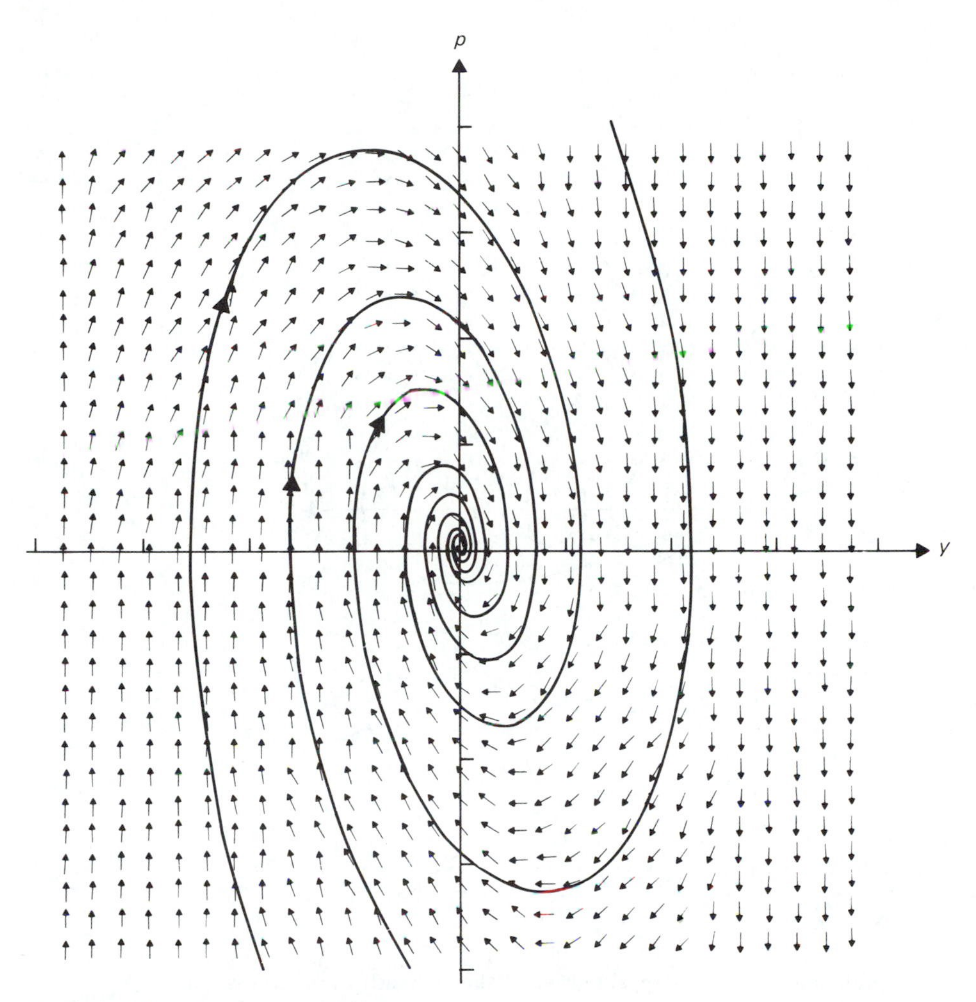

Figure 2.19 The vector field and several integral curves of the underdamped oscillator of Example 2.

of more complicated systems. We shall examine the forced oscillator equation in the next section.

EXERCISES 2.9

Letting $p = dy/dt$, write equivalent first order systems for the second order differential equations in Exercises 1–12 and sketch their vector fields in the yp-plane. Find a real general solution of the system by solving the corresponding

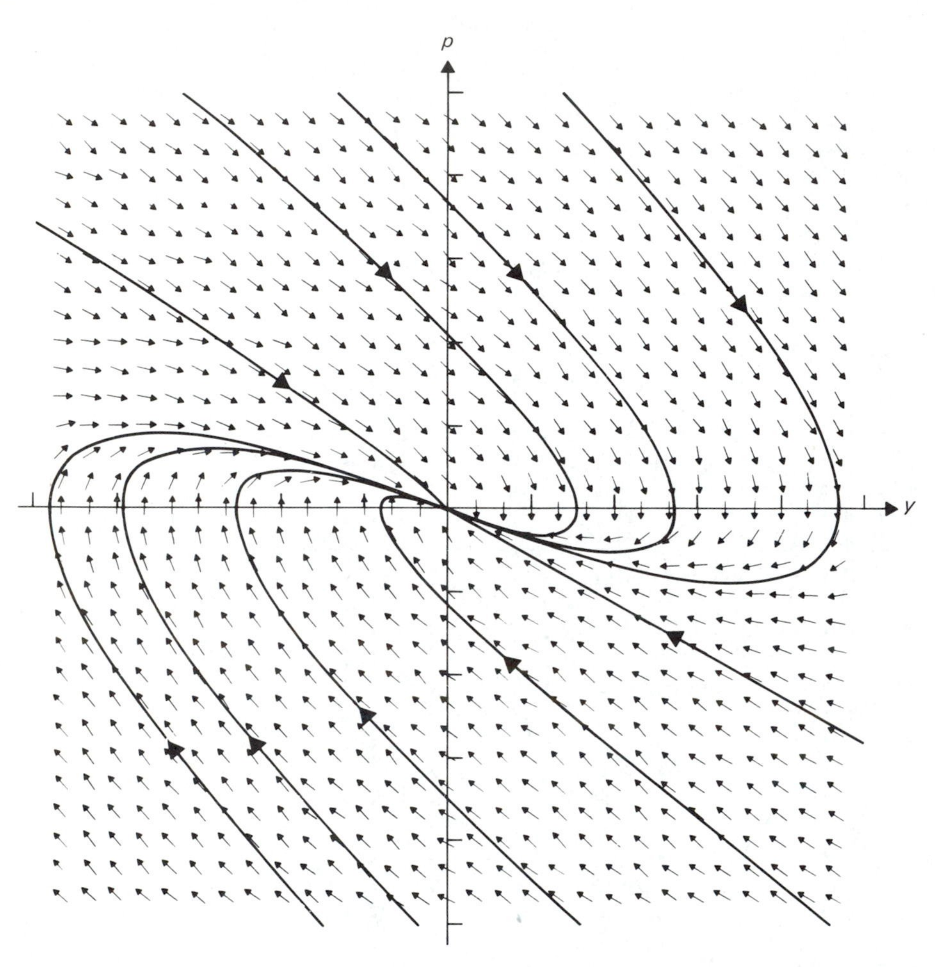

Figure 2.20 **The vector field and several integral curves of the critically damped oscillator of Example 3.**

second order equation. Find a solution such that $y(0) = 1$, $p(0) = y'(0) = 0$ and plot it on the same paper used for the vector field of the equation. Classify the equations as not damped, underdamped, critically damped, or overdamped.

1. $y'' + 4y = 0$
2. $y'' + 9y = 0$
3. $y'' + y = 0$
4. $y'' + 4y' + 5y = 0$
5. $y'' + 4y' + 13y = 0$
6. $y'' + 6y' + 13y = 0$
7. $y'' + 4y' + 4y = 0$
8. $y'' + 6y' + 9y = 0$
9. $y'' + 2y' + y = 0$
10. $y'' + 5y' + 6y = 0$
11. $y'' + 3y' + 2y = 0$
12. $y'' + 5y' + 4y = 0$

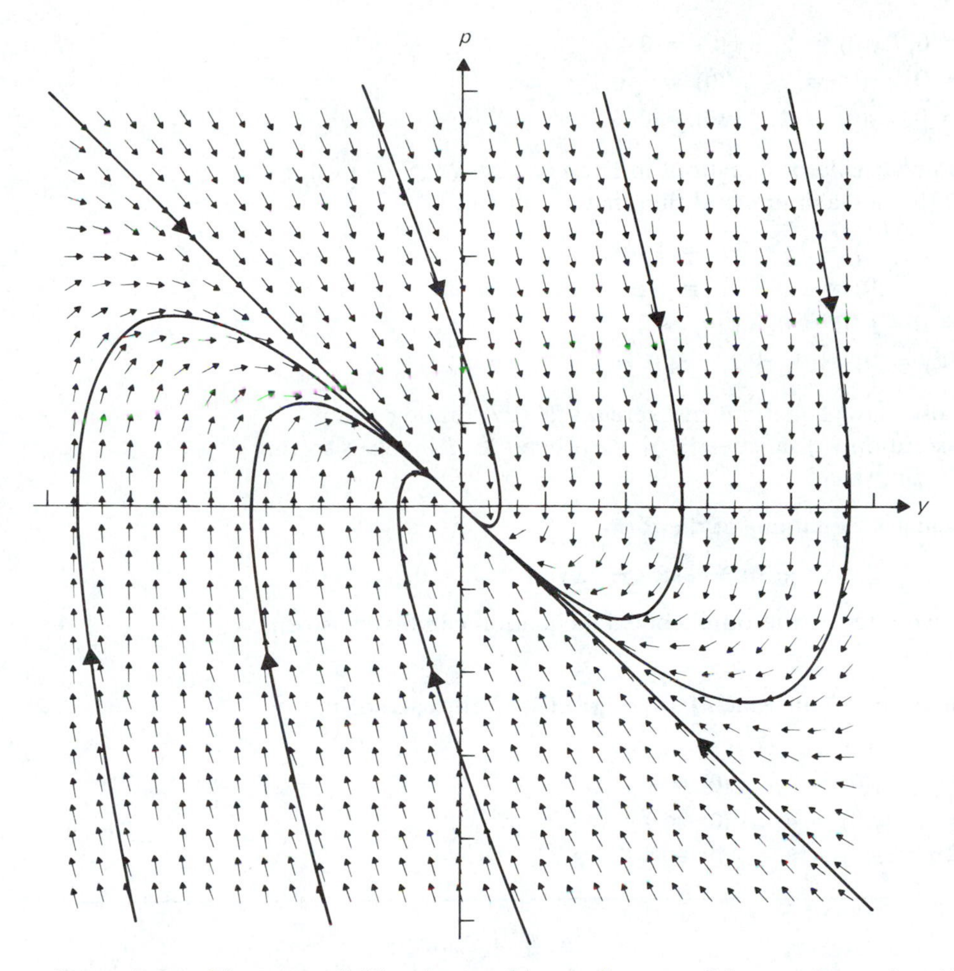

Figure 2.21 The vector field and several integral curves of the overdamped oscillator of Example 4.

For undamped oscillator equations in Exercises 13–16, find the amplitude and period of the motion and find the equation of the ellipse in the phase plane which the integral curve describes.

13. $y'' + y = 0, \quad y(0) = 2, \quad y'(0) = 1$

14. $9y'' + 4y = 0, \quad y(0) = 4, \quad y'(0) = 2$

15. $y'' + 4y = 0, \quad y(0) = 0, \quad y'(0) = 8$

16. $y'' + 10^{-4}y = 0, \quad y(0) = 1, \quad y'(0) = 1$

For the underdamped oscillator equations in Exercises 17–19, find the damping factor and period of the motion. Find a time T for which the motion has magnitude $|y(t)| < 10^{-4}$ for all $t > T$.

17. $y'' + 2y' + 2y = 0, \quad y(0) = 2, \quad y'(0) = 0$
18. $y'' + \frac{1}{10}y' + 2y = 0, \quad y(0) = 2, \quad y'(0) = -0.1$
19. $y'' + \frac{14}{5}y' + 2y = 0, \quad y(0) = 2, \quad y'(0) = 0$

For the critically damped oscillator equations in Exercises 20–23, find a value $T \geq 0$ for which $|y(T)|$ is a maximum and find that maximum.

20. $y'' + 2y' + y = 0, \quad y(0) = 1, \quad y'(0) = 1$
21. $y'' + 2y' + y = 0, \quad y(0) = 1, \quad y'(0) = -3$
22. $y'' + \frac{1}{5}y' + \frac{1}{100}y = 0, \quad y(0) = 0, \quad y'(0) = 8$
23. $y'' + 200y' + 10^4y = 0, \quad y(0) = 2, \quad y'(0) = 1$

For the overdamped oscillator equations in Exercises 24–27, finding a time $T \geq 0$ for which the solution $y(t)$ satisfies $|y(t)| < \epsilon$ when $t \geq T$ involves solving a transcendental equation.

24. Show that for oscillator equations of the form

$$y'' + (n + 1)ay' + na^2y = 0, \; a > 0,$$

where n is a positive integer, the transcendental equation can be transformed into an algebraic one.

Let $\epsilon = 10^{-2}$ and find a time T for which $|y(t)| < \epsilon$ if $t > T$ for each of the following.

25. $y'' + 3y' + 2y = 0, \quad y(0) = 1, \quad y'(0) = 1$
26. $y'' + 6y' + 8y = 0, \quad y(0) = 4, \quad y'(0) = 0$
27. $y'' + 4y' + 3y = 0, \quad y(0) = 3, \quad y'(0) = -5$

2.10 FORCED OSCILLATIONS OF ELECTRICAL AND MECHANICAL SYSTEMS

The forced vibrations of an oscillatory system depend on the parameters of the system and the nature of the forcing function. For example, if a signal generator is attached to a series *LCR* circuit as in Fig. 2.22, then the flow of electrons depends on the inductance, the resistance, and the capacitance of the circuit as well as the amplitude and frequency of the signal. The governing differential equations are

$$\frac{dQ}{dt} = I, \qquad L\frac{dI}{dt} + RI + \frac{Q}{C} = P\cos\omega t,$$

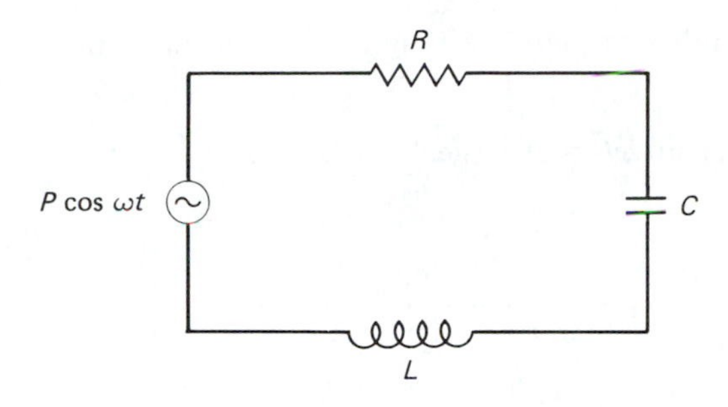

Figure 2.22 An *LCR* circuit.

or the equivalent second order equation,

$$L\frac{d^2Q}{dt^2} + R\frac{dQ}{dt} + \frac{Q}{C} = P\cos\omega t.$$

Let us assume that the corresponding homogeneous equation,

$$L\frac{d^2Q}{dt^2} + R\frac{dQ}{dt} + \frac{Q}{C} = 0,$$

is underdamped. Then a general solution to the inhomogeneous equation is

$$Q(t) = c_1e^{-st}\cos qt + c_2e^{-st}\sin qt + A\cos(\omega t - \phi),$$

where c_1 and c_2 are constants that depend on initial conditions,

$$s = \frac{R}{2L}, \qquad q = \left(\frac{1}{LC} - \frac{R^2}{4L^2}\right)^{1/2},$$

and A and ϕ are constants that depend on L, R, C, P, and ω. In this section we examine the effects of changing the input angular frequency ω on the solution, in particular on A, the amplitude of the periodic component. It is quite easy to construct a series *LCR* circuit with a signal generator in a laboratory and to monitor the current as the frequency is changed. However, the magnitude of the current can increase sharply and therefore the experimenter should be prepared to replace charred components of the circuit.

EXAMPLE 1 Describe the effect of changing the input angular frequency on the current in the circuit modeled by the differential equation,

$$4Q'' + 4Q' + 17Q = 16\cos\omega t.$$

Assume that $Q(0) = 0$, $I(0) = Q'(0) = 0$. Plot $I(t)$ versus t for $\omega = 1, 2, 3$.

SOLUTION Using the method of undetermined coefficients we find that a particular solution is

$$Q(t) = M\cos\omega t + N\sin\omega t = A\cos(\omega t - \phi),$$

where

$$M = \frac{16(17 - 4\omega^2)}{(17 - 4\omega^2)^2 + 16\omega^2}, \qquad N = \frac{64\omega}{(17 - 4\omega^2)^2 + 16\omega^2},$$

$$A = (M^2 + N^2)^{1/2} = 16[(17 - 4\omega^2)^2 + 16\omega^2]^{-1/2},$$

$$\phi = \arctan\left(\frac{N}{M}\right).$$

A general solution to the differential equation is

$$Q(t) = c_1 e^{-t/2}\cos 2t + c_2 e^{-t/2}\sin 2t + A\cos(\omega t - \phi).$$

After imposing the initial conditions, $Q(0) = 0$, $Q'(0) = 0$, we have

$$Q(t) = -Ae^{-t/2}\left[\cos\phi\cos 2t + \left(\frac{1}{4}\cos\phi + \frac{\omega}{2}\sin\phi\right)\sin 2t\right] + A\cos(\omega t - \phi).$$

The current is

$$I(t) = Ae^{-t/2}\left[(-\omega\sin\phi)\cos 2t + \left(\frac{17}{8}\cos\phi + \frac{\omega}{4}\sin\phi\right)\sin 2t\right]$$
$$- \omega A\sin(\omega t - \phi).$$

The primary effect of changing the input angular frequency is on A, the amplitude of the particular solution. It is easy to verify that the maximum value of A is attained at $\omega = \sqrt{15}/2$. The plots of the integral curves in the t I-plane are in Fig. 2.23. A striking property of the solution is that the effect of the initial conditions disappears as time increases. The solution is the sum of a periodic function that has the same period as the forcing function and an exponentially damped function that is a solution to the corresponding homogeneous differential equation. ■

We now examine the *forced-oscillator equation* written in the form

$$m\frac{d^2y}{dt^2} + c\frac{dy}{dt} + ky = F(t) \qquad \textbf{(2.10.1)}$$

or as an equivalent system

$$m\frac{dy}{dt} = p, \qquad \frac{dp}{dt} = -ky - \frac{cp}{m} + F(t). \qquad \textbf{(2.10.2)}$$

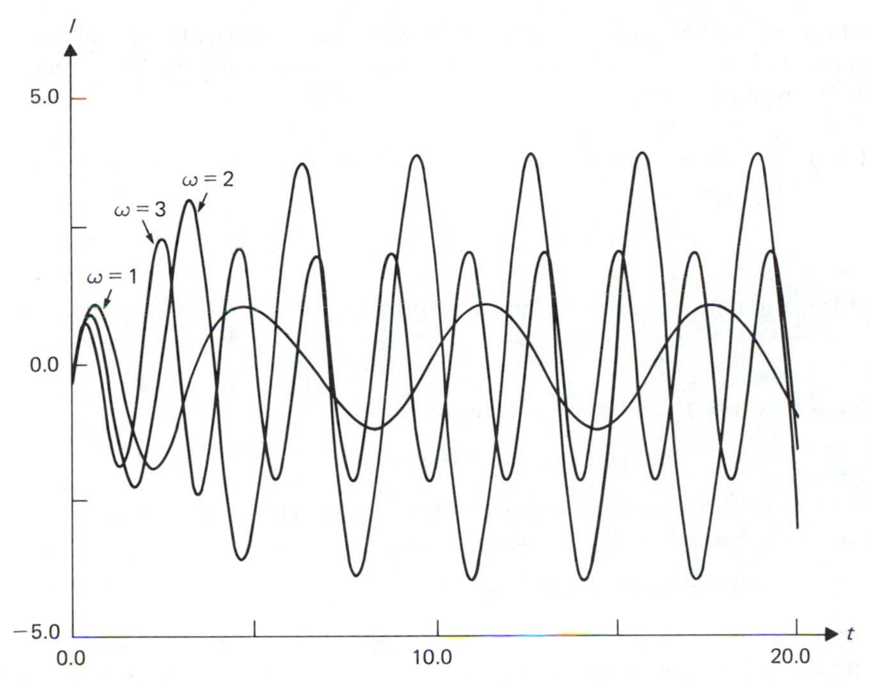

Figure 2.23 Plots of $I(t)$ for $\omega = 1, 2, 3$. The maximum value of the amplitude is attained at $\omega = \sqrt{15}/2$.

It is assumed that the quantities m, c, and k are positive constants. In this section we shall restrict our attention to equations with periodic forcing functions, i.e., functions such that

$$F(t + T) = F(t)$$

for all t, where T, the period, is a positive number.

It can be proved that the damped-oscillator equation ($c > 0$) with a periodic forcing function has a unique periodic solution and that this solution has the same period as the period of the forcing function [3, pp. 100–101]. Therefore a general solution can be written as

$$y(t) = c_1y_1(t) + c_2y_2(t) + w(t), \tag{2.10.3}$$

where $y_1(t)$ and $y_2(t)$ are linearly independent solutions of the corresponding homogeneous differential equation and $w(t)$ is the unique periodic solution of the inhomogeneous differential equation. Since c is positive, both $y_1(t)$ and $y_2(t)$ tend to zero as t increases and we have $y(t) \cong w(t)$ for large t. For this reason, $c_1y_1(t) + c_2y_2(t)$ is called the *transient solution* and $w(t)$ the *steady state solution.* Clearly, the initial conditions, which are incorporated into the constants c_1 and c_2, will have little effect on the solution when t is large.

EXAMPLE 2 Find the steady state solution in amplitude–phase angle form and the transient solution to

$$y'' + 2y' + 2y = 4 \cos t - 3 \sin t.$$

SOLUTION A general solution of the corresponding homogeneous equation is

$$y_h(t) = c_1 e^{-t}\cos t + c_2 e^{-t}\sin t.$$

A particular solution to the inhomogeneous equation,

$$w(t) = 2 \cos t + \sin t,$$

can be constructed by the method of undetermined coefficients. Therefore a general solution to the forced oscillator equation is

$$y(t) = 2 \cos t + \sin t + c_1 e^{-t}\cos t + c_2 e^{-t}\sin t.$$

It is clear that $w(t)$ is the steady state solution and $y_h(t)$ is the transient solution.

To write the steady state solution in amplitude–phase angle form, we set

$$w(t) = 2 \cos t + \sin t = A \cos(t - \phi) = A \cos t \cos \phi + A \sin t \sin \phi.$$

Therefore

$$A \cos \phi = 2, \qquad A \sin \phi = 1,$$

and $A = \sqrt{5}$, $\phi = \arctan \frac{1}{2} \cong 0.464$ radians, $w(t) = \sqrt{5} \cos(t - \phi)$. ■

In order to exhibit the dependence of the solution on the period of the forcing function, we make the further restriction that $F(t)$ be a finite sum of sinusoidal functions:

$$F(t) = P_0 + \sum_{n=1}^{N} [P_n \cos(n\omega t) + R_n \sin(n\omega t)]. \tag{2.10.4}$$

The period of $F(t)$ is $T = 2\pi/\omega$. To verify this we notice that

$$\cos\left[n\omega\left(t + \frac{2\pi}{\omega}\right)\right] = \cos(n\omega t + n2\pi) = \cos(n\omega t),$$

$$\sin\left[n\omega\left(t + \frac{2\pi}{\omega}\right)\right] = \sin(n\omega t + n2\pi) = \sin(n\omega t),$$

which implies that $F(t + 2\pi/\omega) = F(t)$ for all t. The unique periodic solution $w(t)$ can now be found by the method of undetermined coefficients.

Let us first solve the case where $P_0 = 0$ and $N = 1$:

$$m\frac{d^2y}{dt^2} + c\frac{dy}{dt} + ky = P_1 \cos \omega t + R_1 \sin \omega t. \tag{2.10.5}$$

Set $w_1(t) = A_1 \cos \omega t + B_1 \sin \omega t$ as the trial solution, and after substituting this for y in (2.10.5) one obtains the following equations for A_1 and B_1:

$$(k - m\omega^2)A_1 + c\omega B_1 = P_1, \qquad -c\omega A_1 + (k - m\omega^2)B_1 = R_1.$$

Hence,

$$A_1 = \frac{P_1(k - m\omega^2) - R_1 c\omega}{\Delta}, \qquad B_1 = \frac{R_1(k - m\omega^2) + P_1 c\omega}{\Delta},$$

where $\Delta = (k - m\omega^2)^2 + c^2\omega^2$, and the solution is

$$w_1(t) = \frac{P_1}{\Delta}[(k - m\omega^2)\cos \omega t + c\omega \sin \omega t] + \frac{R_1}{\Delta}[-c\omega \cos \omega t + (k - m\omega^2)\sin \omega t].$$

The next step is to write $w_1(t)$ in the amplitude–phase angle form introduced in Section 2.5. We define $Q(\omega)$, the *phase displacement,* by

$$\cos Q(\omega) = \Delta^{-1/2}(k - m\omega^2),$$
$$\sin Q(\omega) = \Delta^{-1/2}c\omega.$$

Then with the aid of the trigonometric identities,

$$\cos Q \cos \omega t + \sin Q \sin \omega t = \cos(\omega t - Q),$$
$$-\sin Q \cos \omega t + \cos Q \sin \omega t = \sin(\omega t - Q),$$

we have

$$w_1(t) = K(\omega)\{P_1 \cos[\omega t - Q(\omega)] + R_1 \sin[\omega t - Q(\omega)]\}, \tag{2.10.6}$$

where $K(\omega)$, the *amplification factor,* is defined by $K(\omega) = \Delta^{-1/2}$ or

$$K(\omega) = [(k - m\omega^2)^2 + c^2\omega^2]^{-1/2}. \tag{2.10.7}$$

The *phase displacement* $Q(\omega)$ can be computed by

$$Q(\omega) = \arctan\left(\frac{c\omega}{k - m\omega^2}\right) \tag{2.10.8}$$

where $Q(\omega)$ is in the first quadrant if $k - m\omega^2$ is positive and in the second quadrant if $k - m\omega^2$ is negative.

The solutions corresponding to the *higher harmonics* of the forcing function, $P_n \cos(n\omega t) + R_n \sin(n\omega t)$, are found by replacing ω by $n\omega$ and P_1, R_1 by P_n, R_n in (2.10.5) and (2.10.6):

$$w_n(t) = K(n\omega)\{P_n \cos[n\omega t - Q(n\omega)] + R_n \sin[n\omega t - Q(n\omega)]\}.$$

The constant term in the forcing function produces a constant solution, $w_0(t) = P_0/k = w_0$. Because of the linearity of the differential equation, the periodic solution corresponding to the periodic forcing solution $F(t)$ is the sum of the $w_n(t)$:

$$w(t) = w_0 + \sum_{n=1}^{N} w_n(t)$$

or

$$w(t) = \frac{P_0}{k} + \sum_{n=1}^{N} K(n\omega)\{P_n \cos[n\omega t - Q(n\omega)] + R_n \sin[n\omega t - Q(n\omega)]\}. \quad \textbf{(2.10.9)}$$

EXAMPLE 3 Find the steady state solution, the transient solution, the phase displacements, and the amplification factors for

$$4y'' + 8y' + 3y = 65 \cos t + 10 \sin \frac{t}{2}.$$

SOLUTION As in Example 2, we solve the corresponding homogeneous equation to get the transient solution,

$$y_h(t) = c_1 e^{-t/2} + c_2 e^{-3t/2}.$$

By employing the method of undetermined coefficients, we construct the steady state solution,

$$w(t) = -\cos t + 8 \sin t - 2 \cos \frac{t}{2} + \sin \frac{t}{2}.$$

The formulas for the phase displacements and the amplification factors are

$$\tan Q(\omega) = \frac{8\omega}{3 - 4\omega^2}, \qquad K(\omega) = [(3 - 4\omega^2)^2 + 64\omega^2]^{-1/2},$$

with $\omega = 1$ because of the forcing term $65 \cos t$ and $\omega = \frac{1}{2}$ because of the forcing term $10 \sin t/2$. Therefore

$$K(1) = (65)^{-1/2}, \qquad Q(1) = \arctan(-8) \cong 1.695 \text{ radians},$$

$$K\left(\frac{1}{2}\right) = (20)^{-1/2}, \qquad Q\left(\frac{1}{2}\right) = \arctan(2) \cong 1.107 \text{ radians}.$$

Note that $Q(1)$ is in the second quadrant and $Q(\frac{1}{2})$ is in the first quadrant.

In this example $\omega_1 = \frac{1}{2}$, $\omega_2 = 2(\frac{1}{2}) = 1$, $P_1 = 0$, $R_1 = 10$, $P_2 = 65$, $R_2 = 0$ and the steady state solution can be written

$$w(t) = 10K\left(\frac{1}{2}\right)\sin\left[\frac{t}{2} - Q\left(\frac{1}{2}\right)\right] + 65K(1)\cos[t - Q(1)]. \qquad ■$$

The amplification factor $K(\omega)$ plays a key role in the analysis of the input–output relationship of oscillator equations. The phase displacement $Q(\omega)$ is of secondary importance in applications. As is seen from Eq. (2.10.9), the amplitude of the nth harmonic of the input function (2.10.4) is multiplied by $K(n\omega)$. The function $K(\omega)$ depends on m, c, and k, the parameters of the oscillator equation, and its graph gives important information on how $w(t)$, the periodic response of the system, depends on the forcing function $F(t)$. However, a collection of graphs of $K(\omega)$, also called *resonance curves,* would be overwhelming for many choices of the three parameters. This can be avoided by a preliminary study of the amplification factor as a function of the four variables ω, m, c, and k.

First, let us determine $\overline{\omega}$, the *maximum response frequency,* or *resonance frequency.* It is the value of ω at which the amplification factor $K(\omega)$ takes on its maximum. Knowing it would enable us to solve the problem of choosing the input frequency so as to maximize the amplitude of the output. From the expression

$$K(\omega) = [(k - m\omega^2)^2 + c^2\omega^2]^{-1/2} = [m^2\omega^4 - (2km - c^2)\omega^2 + k^2]^{-1/2}$$

it follows that

$$K'(\omega) = -K^3(\omega)[2m^2\omega^3 - (2km - c^2)\omega].$$

Then $K'(\omega) = 0$ implies that

$$\overline{\omega}^2 = \frac{2km - c^2}{2m^2} > 0,$$

the condition for $K(\omega)$ to have a maximum for positive ω. If $2km - c^2 \le 0$ or equivalently, $c \ge \sqrt{2km}$, the expression for $K'(\omega)$ is negative, hence $K(\omega)$ is a strictly decreasing function of ω. Sketches of several resonance curves are shown in Fig. 2.24.

Next set $\omega_0 = (k/m)^{1/2}$, the frequency of the undamped oscillator, and write

$$\begin{aligned} K(\omega) &= \left[m^2\left(\omega^4 - 2\,\frac{2km - c^2}{2m^2}\,\omega^2 + \frac{k^2}{m^2}\right)\right]^{-1/2} \\ &= m^{-1}[\omega^4 - 2\overline{\omega}^2\omega^2 + \omega_0^4]^{-1/2}. \end{aligned}$$

We started with three parameters m, c, and k and still have three parameters m, $\overline{\omega}$, and ω_0. Has anything been gained? The answer depends on the use of the physical system, e.g., the oscillator can serve as a simplified model for microphone diaphragms, seismographs, and oscillator circuits in audio amplifiers. If the maximum

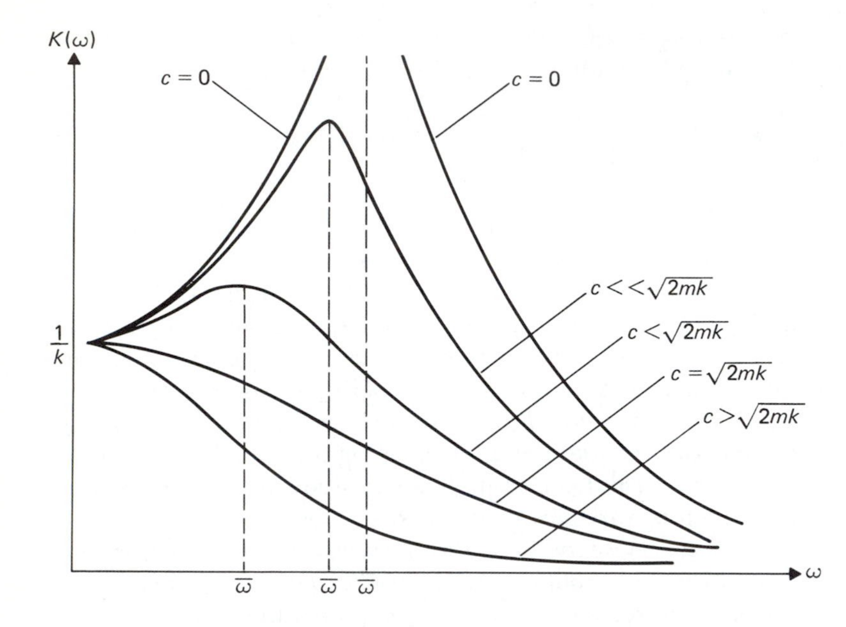

Figure 2.24 Sketches of several resonance curves.

response frequency $\overline{\omega}$ is fixed, then ω_0 and m can be selected to reduce distortion for a range of frequencies centered at $\overline{\omega}$ and simultaneously to maximize the sensitivity of the device. It may also be advantageous to fix ω_0 and study $K(\omega)$ for different values of ω, to insure, for instance, that the system will not be damaged at or near the maximum response frequency.

EXAMPLE 4 Find $\overline{\omega}$, the maximum response frequency, ω_0, the frequency of the undamped oscillator, $K(\omega)$, the amplification factor, and plot $K(\omega)$ versus ω for the oscillator equation,

$$2y'' + 3y' + 3y = F(t).$$

SOLUTION

$$\omega_0^2 = \frac{k}{m} = \frac{3}{2}, \qquad \overline{\omega}^2 = \frac{2km - c^2}{2m^2} = \frac{3}{8}$$

$$K(\omega) = \frac{1}{2}\left[\omega^4 - \frac{3}{4}\omega^2 + \frac{9}{4}\right]^{-1/2} = \frac{1}{2}\left[\left(\omega^2 - \frac{3}{8}\right)^2 + \frac{135}{64}\right]^{-1/2}.$$

Therefore the maximum response frequency is $(\frac{3}{8})^{1/2}$ and the undamped frequency is $(\frac{3}{2})^{1/2}$. A plot of $K(\omega)$ versus ω is given in Fig. 2.25. ■

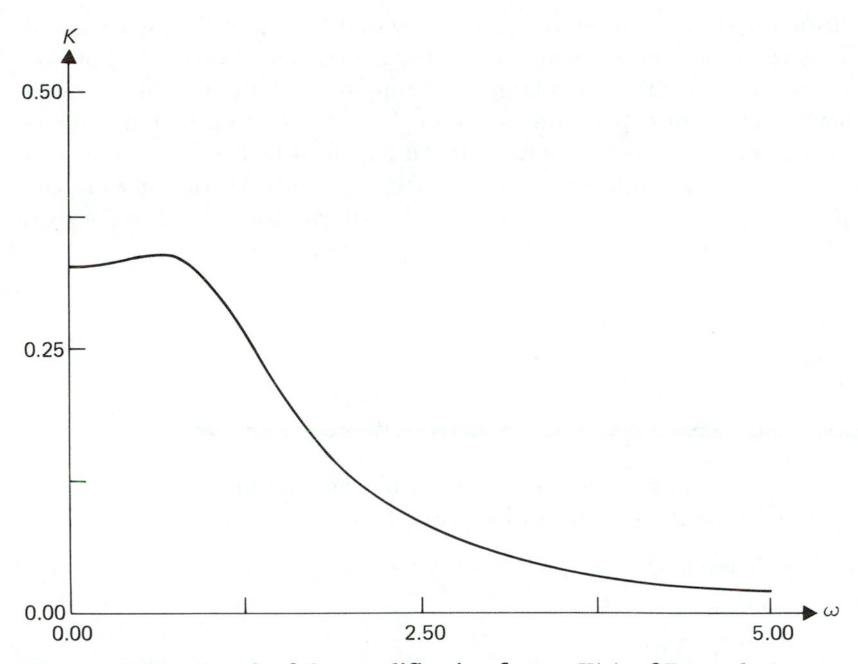

Figure 2.25 Graph of the amplification factor, $K(\omega)$, of Example 4.

Finally, we wish to discuss briefly the undamped case (when $c = 0$). Again consider the situation where $P_0 = 0$ and $N = 1$:

$$m\frac{d^2y}{dt^2} + ky = P_1 \cos \omega t + R_1 \sin \omega t.$$

The analysis is the same as before and (2.10.7) implies that the amplification factor is

$$K(\omega) = [(k - m\omega^2)^2]^{-1/2} = |k - m\omega^2|^{-1}.$$

But $K(\omega)$ is not defined for $\omega = (k/m)^{1/2} = \omega_0$, the frequency of the undamped oscillator, which is in keeping with the fact that when $\omega = \omega_0$, the forcing function is a solution of the homogeneous equation. In the case where $\omega = \omega_0$, the trial solution must be of the form

$$\omega_1(t) = A_1 t \cos \omega_0 t + B_1 t \sin \omega_0 t, \qquad \omega_0 = \sqrt{k/m}.$$

This is the case of *pure resonance,* where solutions will be oscillatory but their amplitudes become unbounded as t increases.

In a realistic physical situation, the probability that the input frequency ω will be exactly ω_0 is zero, but the problem is that for ω in a neighborhood of ω_0 the amplification factor $K(\omega)$ will be very large, resulting in oscillations of huge amplitude. This may or may not explain the apocryphal tales of certain famous singers shattering wine glasses with high C notes, but it is a serious factor in the design of towers or bridges which are subject to sustained strong winds. Movies taken of the collapse of the Tacoma Narrows Bridge in 1940 due to gale force winds are a vivid example of the destructive force of near-resonance excitations.

EXERCISES 2.10

Find the steady state solution, the transient solution, the phase displacement, and the amplification factor for the differential equations in Exercises 1–6.

1. $y'' + 4y' + 13y = 6 \cos 2t - 5 \sin 2t$
2. $y'' + 3y' + 2y = \cos 3t + 2 \sin 3t$
3. $y'' + 4y' + 5y = 7 \cos t + 8 \sin t$
4. $4y'' + 12y' + 9y = 3 \cos t$
5. $3y'' + 13y' + 4y = 4 \cos 2t - \sin 2t$
6. $y'' + cy' + y = A \cos t, \quad c = 2, 0.2, 0.002$

Find the amplification factor and the phase displacement for the undamped forced oscillators in Exercises 7–10.

7. $3y'' + 12y = 5 \cos 3t$
8. $4y'' + 36y = 3 \sin 2t$
9. $y'' + \pi^2 y = 2 \sin \dfrac{22}{7} t$
10. $y'' + y = A \cos (1 + \epsilon)t, \quad \epsilon = 1, 0.1, 0.001$
11. Graph the resonance curves for the forced-oscillator equation and in the first two cases find the maximum response frequency:

$$2y'' + cy' + 9y = A \cos \omega t + B \sin \omega t$$

with $c = 1, 5, 8$.

12. Find an *approximate* solution of the form

$$y(t) = A \sin \epsilon t \sin 2t$$

to the initial value problem

$$y'' + (2 + \epsilon)^2 y = \epsilon F \cos 2t, \quad y(0) = 0, y'(0) = 0.$$

(*Hint:* Find the exact solution and use the trigonometric identity

$$\cos(2 + \epsilon)t = \cos 2t \cos \epsilon t - \sin 2t \sin \epsilon t$$

and the approximation $\cos \epsilon t \cong 1$. The term $A \sin \epsilon t$ can be thought of as a slowly varying amplitude for the vibration $\sin 2t$. This illustrates the *phenomenon of beats.* Take $A = 2$, $\epsilon = \frac{1}{16}$ and sketch the approximate solution on the interval $0 \leq t \leq 32\pi$).

13. For the undamped forced oscillator in the case of pure resonance

$$y'' + \omega^2 y = P_1 \cos \omega t + R_1 \sin \omega t$$

show that a particular solution is

$$w(t) = \frac{t}{2\omega}\left[P_1 \cos\left(\omega t - \frac{\pi}{2}\right) + R_1 \sin\left(\omega t - \frac{\pi}{2}\right)\right].$$

14. A system is described by the forced-oscillator equation with initial data as

$$y'' + 4y = \frac{8}{\pi}\cos 2t, \qquad y(0) = 0,\ y'(0) = 0.$$

It becomes overloaded when $|y(t)| \geq 12$. Show that this will first occur somewhere in the interval $6\pi < t < 25\pi/4$.

2.11 ELEMENTARY NUMERICAL METHODS FOR SECOND ORDER EQUATIONS

The Euler and improved Euler algorithms for generating numerical solutions to single first order differential equations will be extended in this section to algorithms that can be applied to a pair of first order equations. The algorithms will then be applied to coupled first order equations that are equivalent to linear second order equations.

The differential equation

$$a(t)\frac{d^2y}{dt^2} + b(t)\frac{dy}{dt} + c(t)y = q(t) \qquad \textbf{(2.11.1)}$$

with initial data $y(t_0) = r$, $(dy/dt)(t_0) = s$ can be written as a first order system by introducing a new variable, $z = dy/dt$. Then, since

$$\frac{dz}{dt} = \frac{d^2y}{dt^2},$$

the system

$$\frac{dy}{dt} = z, \qquad \frac{dz}{dt} = \frac{-c(t)y - b(t)z + q(t)}{a(t)} \tag{2.11.2}$$

with initial data $y(t_0) = r$, $z(t_0) = s$ is equivalent to (2.11.1). For instance,

$$y'' + 4y' + 3y = 4 \sin 2t, \qquad y(0) = 1, \qquad y'(0) = 1$$

is equivalent to

$$y' = z, \qquad y(0) = 1,$$

$$z' = -3y - 4z + 4 \sin 2t, \qquad z(0) = 1.$$

Similarly,

$$t^2y'' - 2y = e^{-t}, \qquad y(1) = 0, \qquad y'(1) = 4$$

is equivalent to

$$y' = z, \qquad y(1) = 0,$$

$$z' = \frac{2y + e^{-t}}{t^2} = \frac{2}{t^2}y + \frac{e^{-t}}{t^2}, \qquad z(1) = 4.$$

The system (2.11.2) is a special case of the general pair of coupled first order equations

$$\frac{dy}{dt} = F(y, z, t), \qquad \frac{dz}{dt} = G(y, z, t) \tag{2.11.3}$$

with initial data $y(t_0) = r$, $z(t_0) = s$. Assuming that $F(y, z, t)$, $G(y, z, t)$ are continuous functions defined on a domain in yzt space, we now extend the previously developed numerical algorithms to this system. These will then be applied to the system of linear differential equations (2.11.2).

We can associate with the two equations of (2.11.3) a direction field in the three-dimensional yzt-space. At a point with coordinates (y, z, t), the direction field is determined by the triple $(dy, dz, dt) = (F(y, z, t)dt, G(y, z, t)dt, dt)$. The simplest numerical method for obtaining an approximate solution to the system (2.11.3) is Euler's method, which produces a solution by following, in discrete steps, the direction field in the yzt-space. Since $F(y, z, t)$ and $G(y, z, t)$ are the slopes of a solution pair $y(t)$ and $z(t)$, Euler's method for a single equation extends to the algorithm

$$y_{i+1} = y_i + hF(y_i, z_i, t_i),$$

$$z_{i+1} = z_i + hG(y_i, z_i, t_i),$$

$$t_{i+1} = t_i + h, \quad i = 0, 1, 2, \ldots, n - 1,$$

with the initial data $y_0 = y(t_0)$, $z_0 = z(t_0)$. For the linear system (2.11.2),

$$F(y, z, t) = z, \qquad \textbf{(2.11.4)}$$
$$G(y, z, t) = \frac{-c(t)y - b(t)z + q(t)}{a(t)}.$$

This is summarized in the following algorithm.

The Euler algorithm for linear second order differential equations

Given the initial value problem

$$a(t)y'' + b(t)y' + c(t)y = q(t), \qquad y(t_0) = y_0, \qquad y'(t_0) = z_0,$$

write it as a system

$$y' = z, \qquad z' = G(y, z, t), \qquad y(t_0) = y_0, \qquad z(t_0) = z_0, \qquad \textbf{(2.11.5)}$$

with

$$G(y, z, t) = \frac{-c(t)y - b(t)z + q(t)}{a(t)}.$$

An approximate solution to the initial value problem (2.11.5) at the equally spaced points $t_0, t_1, \ldots, t_n$ is given by

$$y_{i+1} = y_i + hz_i,$$
$$z_{i+1} = z_i + hG(y_i, z_i, t_i), \qquad \textbf{(2.11.6)}$$
$$t_{i+1} = t_i + h, \quad i = 0, 1, 2, \ldots, n - 1,$$

where h, t_0, y_0, and z_0 are given.

EXAMPLE 1 Use the Euler algorithm with $h = 0.1$ to find an approximate solution to the initial value problem

$$y'' + 2y' + y = -2t, \qquad y(0) = 4, \qquad y'(0) = 0,$$

at the points $t_i = ih$, where $i = 0, 1, \ldots, 10$.

Find the exact solution and then compare the exact and approximate values of y and y' at the prescribed times.

SOLUTION The exact solution to the initial value problem is

$$y(t) = 2te^{-t} + 4 - 2t,$$
$$y'(t) = 2(1 - t)e^{-t} - 2.$$

Set $y' = z$, $z' = -y - 2z - 2t$, $y(0) = 4$, $z(0) = 0$. Then

$$G(y, z, t) = -(y' + 2z + 2t),$$

and (2.11.6) becomes

$$y_{i+1} = y_i + hz_i,$$
$$z_{i+1} = z_i - h(y_i + 2z_i + 2t_i),$$
$$t_{i+1} = t_i + h.$$

Starting with $t_0 = 0$, $y_0 = 4$, $z_0 = 0$, and using $h = 0.1$, we obtain Table 2.2 with $\Delta y = y_i - y(t_i)$, $\Delta z = z_i - z(t_i)$.

Table 2.2 **The Euler method approximation and errors for $y'' + 2y' + y = -2t$, $y(0) = 4$, $y'(0) = 0$, with $h = 0.1$**

t_i	y_i	z_i	Δy_i	Δz_i
0.0	4.0000	0.0000	0.0000	0.0000
0.10	4.0000	−0.4000	0.0190	−0.0287
0.20	3.9600	−0.7400	0.0325	−0.0500
0.30	3.8860	−1.0280	0.0415	−0.0651
0.40	3.7832	−1.2710	0.0469	−0.0754
0.50	3.6561	−1.4751	0.0496	−0.0816
0.60	3.5086	−1.6457	0.0500	−0.0848
0.70	3.3440	−1.7874	0.0488	−0.0854
0.80	3.1653	−1.9043	0.0463	−0.0841
0.90	2.9748	−2.0000	0.0430	−0.0813
1.00	2.7748	−2.0775	0.0391	−0.0775

It may be surprising that the errors Δy_i and Δz_i do not increase with t_i. This is a special circumstance; both the exact and approximate solutions approach $y(t_i) = 4 - 2t_i$, $z(t_i) = -2$ as t_i tends to infinity. The proof of this requires a study of difference equations and will not be given here. ■

It is also easy to extend the improved Euler method. For the general system (2.11.3), set the Euler approximation

$$\tilde{y}_i = y_i + hF(y_i, z_i, t_i),$$
$$\tilde{z}_i = z_i + hG(y_i, z_i, t_i),$$

and define

$$y_{i+1} = y_i + \frac{h}{2}[F(y_i, z_i, t_i) + F(\tilde{y}_i, \tilde{z}_i, t_i + h)],$$

$$z_{i+1} = z_i + \frac{h}{2}[G(y_i, z_i, t_i) + G(\tilde{y}_i, \tilde{z}_i, t_i + h)],$$

$$t_{i+1} = t_i + h, \quad i = 0, 1, 2, \ldots, n - 1,$$

with the initial data $y_0 = y(t_0)$, $z_0 = z(t_0)$. As previously, $F(y, z, t)$ and $G(y, z, t)$ are given by (2.11.4) for the linear system (2.11.2). This is summarized in the following algorithm.

The improved Euler algorithm for linear second order differential equations

Given the initial value problem

$$a(t)y'' + b(t)y' + c(t)y = q(t), \qquad y(t_0) = y_0, y'(t_0) = z_0,$$

write it as a system

$$y' = z, \qquad z' = G(y, z, t), \qquad y(t_0) = y_0, \qquad z(t_0) = z_0, \tag{2.11.7}$$

with

$$G(y, z, t) = \frac{-c(t)y - b(t)z + q(t)}{a(t)}.$$

An approximate solution to the initial value problem (2.11.7) at the equally spaced points $t_0, t_1, \ldots, t_n$ is given by

$$\begin{aligned}
\tilde{y}_i &= y_i + hz_i, \qquad \tilde{z}_i = z_i + hG(y_i, z_i, t_i), \\
y_{i+1} &= y_i + \frac{h}{2}(z_i + \tilde{z}_i), \\
z_{i+1} &= z_i + \frac{h}{2}[G(y_i, z_i, t_i) + G(\tilde{y}_i, \tilde{z}_i, t_i + h)], \\
t_{i+1} &= t_i + h, \quad i = 0, 1, 2, \ldots, n - 1,
\end{aligned} \tag{2.11.8}$$

where h, t_0, y_0, and z_0 are given.

EXAMPLE 2 Use the improved Euler algorithm with $h = 0.1$ to construct an approximate solution to the initial value problem,

$$y'' + 4y = 2.5 \sin 3t, \qquad y(0) = 0, \qquad y'(0) = 0$$

at the points $t_i = ih$, $i = 0, 1, \ldots, 25$. Plot the exact solution and a continuous graph of the approximate solution in the zy phase plane.

SOLUTION With $G(y, z, t) = -4y + 2.5 \sin 3t$, algorithm (2.11.8) becomes

$$\tilde{y}_i = y_i + hz_i, \qquad \tilde{z}_i = z_i + h(-4y_i + 2.5 \sin 3t_i),$$

$$y_{i+1} = y_i + \frac{h}{2}(z_i + \tilde{z}_i)$$

$$z_{i+1} = z_i + \frac{h}{2}[(-4y_i + 2.5 \sin 3t_i) + (-4\tilde{y}_i + 2.5 \sin 3(t_i + h))]$$

$$t_{i+1} = t_i + h.$$

Table 2.3 is obtained by letting $h = 0.1$ and starting with $y_0 = 0$, $z_0 = 0$. The graphs are in Fig. 2.26. ■

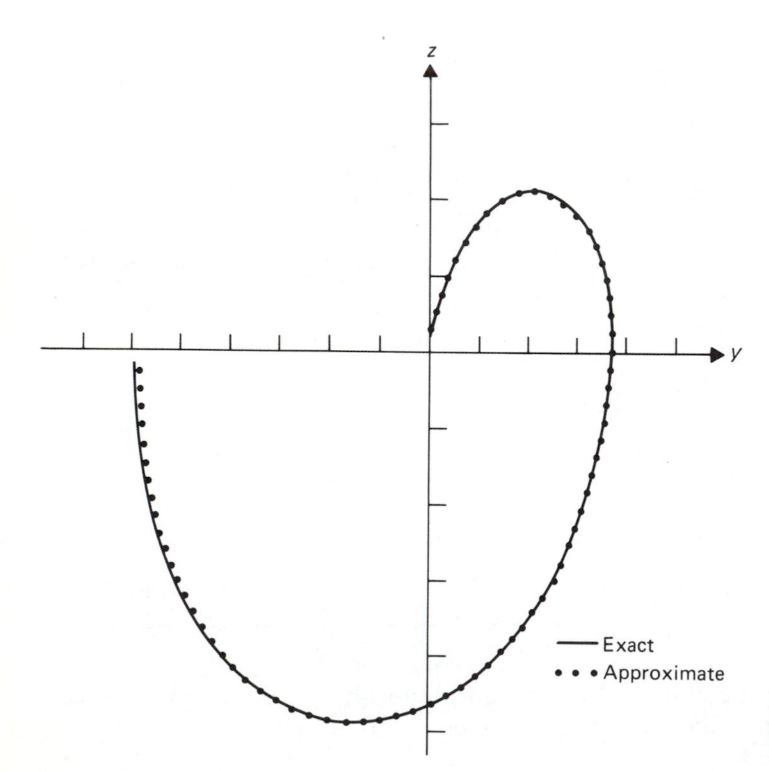

Figure 2.26 Plots of the exact and approximate solutions of Example 2.

Table 2.3 The improved Euler method approximation and errors for $y'' + 4y = 2.5 \sin 3t$, $y(0) = 0$, $y'(0) = 0$, with $h = 0.1$

t_i	y_i	z_0	Δy_i	Δz_i
0	0	0	0	0
0.10	0	0.0369	0.0012	0.0002
0.20	0.0074	0.1437	0.0024	−0.0001
0.30	0.0287	0.3064	0.0031	−0.0008
0.40	0.0685	0.5032	0.0035	−0.0017
0.50	0.1291	0.7069	0.0032	−0.0026
0.60	0.2097	0.8876	0.0024	−0.0032
0.70	0.3064	1.0156	0.0010	−0.0033
0.80	0.4127	1.0650	−0.0007	−0.0027
0.90	0.5193	1.0165	−0.0027	−0.0012
1.00	0.6160	0.8595	−0.0045	0.0013
1.10	0.6913	0.5938	−0.0064	0.0046
1.20	0.7349	0.2304	−0.0071	0.0087
1.30	0.7377	−0.2095	−0.0072	0.0130
1.40	0.6934	−0.6953	−0.0064	0.0174
1.50	0.5991	−1.1899	−0.0045	0.0211
1.60	0.4560	−1.6525	−0.0017	0.0238
1.70	0.2691	−2.0421	0.0021	0.0249
1.80	0.0480	−2.3212	0.0065	0.0240
1.90	−0.1948	−2.4594	0.0112	0.0209
2.00	−0.4437	−2.4361	0.0158	0.0153
2.10	−0.6819	−2.2427	0.0198	0.0075
2.20	−0.8923	−1.8840	0.0229	−0.0023
2.30	−1.0590	−1.3782	0.0245	−0.0137
2.40	−1.1684	−0.7555	0.0244	−0.0258
2.50	−1.2107	−0.0566	0.0225	−0.0379

The sample program given in Chapter 1 for the Euler and the improved Euler algorithms for first order equations can be easily modified to compute solutions of second order linear equations. See Fig. 2.27 for a Fortran program to do the improved Euler method.

```
C PROGRAM USES THE IMPROVED EULER METHOD TO SOLVE THE
C INITIAL VALUE PROBLEM
C
C    A(T)*Y"(T)+B(T)*Y'(T)+C(T)*Y(T)=Q(T)
C    Y(T0)=Y0, Y'(T0)=Z0
C
C DEFINE FUNCTIONS A(T), B(T), C(T), Q(T), G(Y,Z,T)
C AND INITIAL CONDITIONS
C
      A(T)=1.0
      B(T)=0.0
      C(T)=-T
      Q(T)=0.0
      G(Y,Z,T)=(-C(T)*Y-B(T)*Z+Q(T))/A(T)
      T0=0.0
      Y0=0.35503
      Z0=1.0
C
C SET FINAL TIME AND NUMBER OF STEPS TO BE TAKEN
C
      TF=1.0
      N=10
C
C BASIC LOOP
C
      T=T0
      Y=Y0
      Z=Z0
      H=(TF-T0)/FLOAT(N)
      WRITE(6,1)T,Y,Z
    1 FORMAT(1X,F6.2,2F10.5)
      DO 2 I=1,N
         YBAR=Y+H*Z
         GYZ=G(Y,Z,T)
         ZBAR=Z+H*GYZ
         Y=Y+H*(Z+ZBAR)/2.0
         Z=Z+H*(GYZ+G(YBAR,ZBAR,T+H))/2.0
         T=T+H
         WRITE(6,1)T,Y,Z
    2 CONTINUE
      STOP
      END
```

Figure 2.27 Improved Euler program to solve $y'' - ty = 0$, $y(0) = 0.35503$, $y'(0) = 1$.

EXERCISES 2.11

Use the Euler algorithm in Exercises 1–5 to find an approximate solution to each initial value problem at the t-values 0.25, 0.50, 0.75, 1.00 with $h = \frac{1}{4}$. Do your calculations by hand, carrying four significant digits in all results. Find the exact solution in each case and compare it with your numerical approximation at each t-value.

1. $y'' - y = 0, \quad y(0) = 1, \quad y'(0) = 0$
2. $y'' + 3y' + 2y = 1, \quad y(0) = 0, \quad y'(0) = 1$
3. $y'' - y' - 2y = t^2, \quad y(0) = 1, \quad y'(0) = 0$
4. $y'' - y' - 2y = t + 1, \quad y(0) = 0, \quad y'(0) = 0$
5. $y'' + 4y = e^{-t}, \quad y(0) = 0, \quad y'(0) = 0$

Use the improved Euler algorithm in Exercises 6–10 to find an approximate solution to each initial value problem at the t-values 0.25, 0.50, 0.75, 1.00. Do your calculations by hand, carrying four significant digits in all results. Find the exact solution in each case and compare it with your numerical approximation at each t-value. Use $h = \frac{1}{4}$.

6. $y'' - y = 0, \quad y(0) = 1, \quad y'(0) = 0$
7. $y'' + 3y' + 2y = 1, \quad y(0) = 0, \quad y'(0) = 1$
8. $y'' - y' - 2y = t^2, \quad y(0) = 1, \quad y'(0) = 0$
9. $y'' - y' - 2y = t + 1, \quad y(0) = 0, \quad y'(0) = 0$
10. $y'' + 4y = e^{-t}, \quad y(0) = 0, \quad y'(0) = 0$
11. Compare the results of Exercises 1 and 6 with the exact solution by plotting all three on the same graph. Repeat for Exercises 4 and 9.

The Fortran program given in Fig. 2.27 implements the improved Euler algorithm in the text to approximate the solution of the initial value problem

$$y'' - ty = 0, \quad y(0) = 0.35503, \quad y'(0) = 1$$

at $t = 0.1, 0.2, \ldots, 1.0$ using the step size $h = 0.1$. Modify the program to find approximate solutions to the initial value problems in Exercises 12–16 at $t = 0.1, 0.2, \ldots, 1.0$ using the step size $h = 0.1$. Find the exact solution in each case and plot the error of your numerical approximation for $0 \leq t \leq 1$.

12. $y'' - y' = \sin t; \quad y(0) = 1, \quad y'(0) = 0$
13. $y'' + y = 4 \cos t; \quad y(0) = 1, \quad y'(0) = 1$
14. $y'' + 3y' - 4y = e^{-4t} + te^{-t}; \quad y(0) = 0, \quad y'(0) = 0$

15. $y'' + 3y' + 2y = \dfrac{1}{e^t + 1}; \quad y(0) = 0, \quad y'(0) = 1$

16. $y'' + 4ty' + (4t^2 + 2)y = 0; \quad y(0) = 0, \quad y'(0) = 1 \; [y(t) = te^{-t^2}]$

17. Use the improved Euler program to tabulate the approximate solution of the initial value problem

$$y'' + ty = 0, \quad y(0) = 0.35503, \quad y'(0) = 1,$$

at $t = 0.1, 0.2, \ldots, 3.0$. Estimate the first zero of $y(t)$ (i.e., the first value of t where $y(t) = 0$) from your output. Discuss what procedure you might follow if you wanted to estimate the first zero to an accuracy of two or three decimal places.

18. Consider the initial value problem

$$y'' + (t^2 - 1)y = 0, \qquad y(0) = 1, \qquad y'(0) = 0.$$

a) Show that the solution is symmetric about $t = 0$, i.e., $y(t) = y(-t)$.

b) By considering the h-values 0.4, 0.2, 0.1, 0.05, . . . , use the improved Euler program to approximate $y(0.4)$ correct to four decimal places.

In Exercises 19–22 use the improved Euler program to approximate the solution to each initial value problem at the indicated t-values correct to two decimal places. In each case the inhomogeneous term, $f(t)$, is discontinuous over the t-interval of interest. To handle problems of this nature numerically, the integration must be stopped and restarted at each point of discontinuity of $f(t)$. For example, if $f(t)$ were defined by

$$f(t) = \begin{cases} 1 & \text{if } 0 \le t < 1, \\ 0 & \text{if } 1 \le t < 2, \end{cases}$$

the integration should be stopped and restarted at $t = 1$, and in the interval $0 \le t \le 1$ use $f(t) = 1$. In the interval $1 \le t \le 2$ use $f(t) = 0$. Why is this stopping and restarting important?

19. $y'' + 4y = f(t), \quad y(0) = 0, \quad y'(0) = 0;$

$$f(t) = \begin{cases} 1 & \text{if } 0 \le t < \frac{1}{2} \\ 0 & \text{if } \frac{1}{2} \le t \end{cases}; \qquad t = 0.25, 0.5, 0.75, 1.0$$

20. $y'' + 3y' + 2y = f(t), \quad y(0) = 0, \quad y'(0) = 1;$

$$f(t) = \begin{cases} 0 & \text{if } 0 \le t < \frac{3}{4} \\ 1 & \text{if } \frac{3}{4} \le t \end{cases}; \qquad t = 0.25, 0.5, 0.75, 1.0$$

21. $y'' - y' - 2y = f(t), \quad y(0) = 1, \quad y'(0) = 0;$

$$f(t) = \begin{cases} 0 & \text{if } 0 \le t < 0.1 \\ \frac{1}{2} & \text{if } 0.1 \le t < 0.7 \\ 1 & \text{if } 0.7 \le t \end{cases}; \qquad t = 0.2, 0.4, 0.6, 0.8, 1.0$$

22. $y'' + y = f(t), \quad y(0) = 1, \quad y'(0) = 0;$

$$f(t) = \begin{cases} 0 & \text{if } 0 \le t < 1 \\ 4 & \text{if } 1 \le t < 2\,; \\ 0 & \text{if } 2 \le t \end{cases} \qquad t = 0.25, 0.5, 0.75, \ldots, 3.0$$

2.12 HIGHER ORDER EQUATIONS

The theory of higher order linear differential equations is modeled after the second order theory. The definitions as well as the theorems and their proofs are straightforward extensions of those presented earlier in this chapter. We shall study inhomogeneous equations,

$$a_0(t)\frac{d^ny}{dt^n} + a_1(t)\frac{d^{n-1}y}{dt^{n-1}} + \cdots + a_{n-1}(t)\frac{dy}{dt} + a_n(t)y = g(t), \qquad \textbf{(2.12.1)}$$

and the corresponding homogeneous equations

$$a_0(t)\frac{d^ny}{dt^n} + a_1(t)\frac{d^{n-1}y}{dt^{n-1}} + \cdots + a_{n-1}(t)\frac{dy}{dt} + a_n(t)y = 0, \qquad \textbf{(2.12.2)}$$

where $a_0(t)$, $a_1(t)$, . . . , $a_n(t)$, and $g(t)$ are continuous functions on an interval $t_1 < t < t_2$ and $a_0(t)$ does not vanish.

The following terminology will be needed.

Definitions

1. The functions $f_1(t), f_2(t), \ldots, f_k(t)$ are said to be *linearly dependent* on the interval $t_1 < t < t_2$ if there exist constants $c_1, c_2, \ldots, c_k$ not all zero such that

$$c_1f_1(t) + c_2f_2(t) + \cdots + c_kf_k(t) \equiv 0, \qquad t_1 < t < t_2.$$

Otherwise, they are said to be *linearly independent.*

2. A linear combination of n solutions to the homogeneous differential equation (2.12.2),

$$y(t) = c_1y_1(t) + c_2y_2(t) + \cdots + c_ny_n(t), \qquad \textbf{(2.12.3)}$$

is called a *general solution* if the coefficients $c_1, c_2, \ldots, c_n$ can be chosen so that $y(t_0), y'(t_0), \ldots, y^{(n-1)}(t_0)$ take on arbitrary prescribed values

$$y(t_0) = r_0, \qquad y'(t_0) = r_1, \ldots, y^{(n-1)}(t_0) = r_{n-1}$$

for any t_0 in the interval $t_1 < t < t_2$. If $y(t)$ is a general solution, then $y_1(t)$, $y_2(t)$, . . . , $y_n(t)$ are said to form a *fundamental set of solutions.*

3. The *Wronskian* of n solutions is the nth order determinant

$$W(y_1, y_2, \ldots, y_n)(t) = \begin{vmatrix} y_1(t) & y_2(t) & \cdots\ y_n(t) \\ y_1'(t) & y_2'(t) & \cdots\ y_n'(t) \\ \vdots & \vdots & \vdots \\ y_1^{(n-1)}(t) & y_2^{(n-1)}(t) & \cdots\ y_n^{(n-1)}(t) \end{vmatrix} \quad \textbf{(2.12.4)}$$

Abel's formula for n solutions to (2.12.2),

$$W(y_1, y_2, \ldots, y_n)(t) = W(y_1, y_2, \ldots, y_n)(t_0) \exp\left[-\int_{t_0}^{t} \frac{a_1(u)}{a_0(u)}\,du\right], \quad \textbf{(2.12.5)}$$

is valid on the interval $t_1 < t < t_2$. The proof requires rather tedious computations with determinants but is not difficult.

EXAMPLE 1 Show that the functions $y_1(t) = t$, $y_2(t) = t^2$, $y_3(t) = t^3$ are linearly independent on any nonempty interval, have a nonzero Wronskian if $t \neq 0$, and form a fundamental set of solutions to the Cauchy–Euler differential equation,

$$t^3y''' - 3t^2y'' + 6ty' - 6y = 0, \qquad t \neq 0.$$

Take $t_0 = 1$ and verify Abel's formula for $t > 0$.

SOLUTION The functions are linearly independent since the equation

$$c_1t + c_2t^2 + c_3t^3 = 0$$

has at most three roots unless $c_1 = c_2 = c_3 = 0$. Their Wronskian

$$W(y_1, y_2, y_3)(t) = \begin{vmatrix} t & t^2 & t^3 \\ 1 & 2t & 3t^2 \\ 0 & 2 & 6t \end{vmatrix} = 2t^3 \neq 0 \quad \text{if} \quad t \neq 0.$$

Since $W(y_1, y_2, y_3)(t) = 2t^3$, $W(y_1, y_2, y_3)(1) = 2$, and

$$\exp\left[-\int_1^t \frac{-3u^2}{u^3}\,du\right] = \exp\,[\ln t^3 - 3\ln 1] = t^3,$$

Abel's formula is verified.

To verify that the three functions are solutions to the differential equation, we set $y(t) = t^m$, substitute, simplify, and obtain the equation

$$m^3 - 6m^2 + 11m - 6 = (m-1)(m-2)(m-3) = 0.$$

Therefore, if $m = 1, 2$, or 3, $y(t) = t^m$ is a solution. To show that $\{t, t^2, t^3\}$ form a fundamental set of solutions, we will show that constants c_1, c_2, and c_3 can be selected so that the solution,

$$y(t) = c_1 t + c_2 t^2 + c_3 t^3,$$

satisfies the constraints

$$y(t_0) = r_0, \qquad y'(t_0) = r_1, \qquad y''(t_0) = r_2, \qquad t_0 \neq 0,$$

for any $t_0 \neq 0$, r_0, r_1, and r_2. This leads to the set of equations

$$\begin{aligned} r_0 &= c_1 t_0 + c_2 t_0^2 + c_3 t_0^3, \\ r_1 &= c_1 + 2c_2 t_0 + 3c_3 t_0^2, \\ r_2 &= 0 + 2c_2 + 6c_3 t_0, \end{aligned}$$

which has a unique solution since the coefficient determinant, $W = 2t_0^3 \neq 0$. ■

As in the second order case, the concepts of linear independence, fundamental set of solutions, and nonvanishing of the Wronskian are closely related.

Theorem

Let $y_1(t), y_2(t), \ldots, y_n(t)$ be solutions to the homogeneous linear differential equations (2.12.2). Then the following statements are equivalent:

1. The solutions are linearly independent.
2. They form a fundamental set of solutions.
3. Their Wronskian is nonzero.

Corollary

Let $y(t)$ be any solution to (2.12.2); then there is a unique set of constants d_1, $d_2, \ldots, d_n$ such that

$$y(t) = d_1 y_1(t) + d_2 y_2(t) + \cdots + d_n y_n(t).$$

The proofs of these results are left to the reader.

EXAMPLE 2 Show that $y_1(t) = e^t$, $y_2(t) = t^2$, $y_3(t) = t$ form a fundamental set of solutions to

$$(t^2 - 2t + 2)y''' - t^2 y'' + 2ty' - 2y = 0, \qquad -\infty < t < \infty.$$

SOLUTION The functions are solutions since

$$(t^2 - 2t + 2)(e^t)''' - t^2(e^t)'' + 2t(e^t)' - 2(e^t) = e^t(t^2 - 2t + 2 - t^2 + 2t - 2) = 0,$$

$$(t^2 - 2t + 2)(t^2)''' - t^2(t^2)'' + 2t(t^2)' - 2(t^2) = 0 - 2t^2 + 4t^2 - 2t^2 = 0,$$

$$(t^2 - 2t + 2)(t)''' - t^2(t)'' + 2t(t)' - 2(t) = 2t - 2t = 0.$$

To prove that these solutions form a fundamental set, it is sufficient to show that their Wronskian is nonzero:

$$\begin{vmatrix} e^t & t^2 & t \\ e^t & 2t & 1 \\ e^t & 2 & 0 \end{vmatrix} = -(t^2 - 2t + 2)e^t \neq 0. \qquad \blacksquare$$

As before, any solution to the inhomogeneous equation (2.12.1) is called a *particular solution.* If $y_p(t)$ is a particular solution and if $y_1(t), y_2(t), \ldots, y_n(t)$ form a fundamental set of solutions to the homogeneous equation (2.12.2), then a *general solution* to (2.12.1) can be written:

$$y(t) = y_p(t) + c_1y_1(t) + \cdots + c_ny_n(t).$$

This means that if $v(t)$ is any solution, it is possible to find constants $d_1, d_2, \ldots, d_n$ so that

$$v(t) = y_p(t) + d_1y_1(t) + \cdots + d_ny_n(t).$$

The existence of both particular solutions and fundamental sets of solutions follows from the linear existence theorem that is proved in Appendix 1.

Solutions to the nth order equation with constant coefficients can be found by the methods developed for second order equations. Let us consider the differential equations

$$a_0\frac{d^ny}{dt^n} + a_1\frac{d^{n-1}y}{dt^{n-1}} + \cdots + a_{n-1}\frac{dy}{dt} + a_ny = 0, \qquad \textbf{(2.12.6)}$$

where $a_0, a_1, \ldots, a_n$ are constants and a_0 is nonzero. We set $y(t) = e^{\lambda t}$, substitute into (2.12.6), and, after dividing by $e^{\lambda t}$, obtain

$$P(\lambda) = a_0\lambda^n + a_1\lambda^{n-1} + \cdots + a_{n-1}\lambda + a_n = 0. \qquad \textbf{(2.12.7)}$$

The polynomial $P(\lambda)$ is called the *characteristic polynomial* of (2.12.6), and its zeros are called the *characteristic roots* of the *characteristic equation.*

If $\lambda_1, \lambda_2, \ldots, \lambda_k$ are the distinct characteristic roots with respective multiplicities $m_1, m_2, \ldots, m_k$, then

$$P(\lambda) = a_0(\lambda - \lambda_1)^{m_1}(\lambda - \lambda_2)^{m_2}\cdots(\lambda - \lambda_k)^{m_k}. \qquad \textbf{(2.12.8)}$$

If there are n distinct roots, then we have n distinct solutions,

$$y_1(t) = e^{\lambda_1 t}, \qquad y_2(t) = e^{\lambda_2 t}, \ldots, y_n(t) = e^{\lambda_n t}, \qquad \textbf{(2.12.9)}$$

and it can be shown that these solutions are linearly independent. For example, if $n = 3$, the Wronskian,

$$W(e^{\lambda_1 t}, e^{\lambda_2 t}, e^{\lambda_3 t}) = \begin{vmatrix} e^{\lambda_1 t} & e^{\lambda_2 t} & e^{\lambda_3 t} \\ \lambda_1 e^{\lambda_1 t} & \lambda_2 e^{\lambda_2 t} & \lambda_3 e^{\lambda_3 t} \\ \lambda_1^2 e^{\lambda_1 t} & \lambda_2^2 e^{\lambda_2 t} & \lambda_3^2 e^{\lambda_3 t} \end{vmatrix}$$

$$= e^{(\lambda_1+\lambda_2+\lambda_3)t} \begin{vmatrix} 1 & 1 & 1 \\ \lambda_1 & \lambda_2 & \lambda_3 \\ \lambda_1^2 & \lambda_2^2 & \lambda_3^2 \end{vmatrix}$$

$$= e^{(\lambda_1+\lambda_2+\lambda_3)t}[(\lambda_2\lambda_3^2 - \lambda_3\lambda_2^2) - (\lambda_1\lambda_3^2 - \lambda_3\lambda_1^2) + (\lambda_1\lambda_2^2 - \lambda_2\lambda_1^2)]$$

$$= e^{(\lambda_1+\lambda_2+\lambda_3)t}(\lambda_3 - \lambda_2)(\lambda_2\lambda_3 + \lambda_1^2 - \lambda_1\lambda_3 - \lambda_1\lambda_2)$$

$$= e^{(\lambda_1+\lambda_2+\lambda_3)t}(\lambda_3 - \lambda_2)(\lambda_2 - \lambda_1)(\lambda_3 - \lambda_1) \neq 0$$

since the characteristic roots λ_1, λ_2, λ_3 are distinct.

EXAMPLE 3 Find three linearly independent solutions to $y''' - 10y'' + 31y' - 30y = 0$.

SOLUTION The characteristic polynomial is

$$P(\lambda) = \lambda^3 - 10\lambda^2 + 31\lambda - 30 = (\lambda - 2)(\lambda - 3)(\lambda - 5).$$

Thus we see that the characteristic roots are $\lambda_1 = 2$, $\lambda_2 = 3$, $\lambda_3 = 5$, and the corresponding solutions are $y_1(t) = e^{2t}$, $y_2(t) = e^{3t}$, $y_3(t) = e^{5t}$. They are linearly independent since their Wronskian equals $e^{(2+3+5)t}(5 - 3)(3 - 2)(5 - 2) = 6e^{10t} \neq 0$. ■

The characteristic roots may be complex; if so, they will form conjugate pairs, $\lambda = p \pm iq$, because the characteristic polynomial has real coefficients. The complex-valued solutions, $\exp[(p + iq)t]$ and $\exp[(p - iq)t]$, can be replaced by real-valued solutions, $e^{pt} \cos qt$ and $e^{pt} \sin qt$, if desired.

EXAMPLE 4 Find a general solution to

$$y''' - 2y'' + 2y' = 0.$$

SOLUTION The characteristic equation,

$$\lambda^3 - 2\lambda^2 + 2\lambda = \lambda(\lambda^2 - 2\lambda + 2) = 0,$$

has roots, $\lambda = 0$, $\lambda = 1 \pm i$. Therefore

$$y_1(t) = 1\ (=e^{0t}), \qquad y_2(t) = e^{(1+i)t}, \qquad y_3(t) = e^{(1-i)t}$$

form a fundamental set of solutions since the characteristic roots are distinct. The set of real solutions,

$$y_1(t) = 1, \qquad y_2(t) = e^t\cos t, \qquad y_3(t) = e^t\sin t,$$

has a nonzero Wronskian,

$$W(t) = \begin{vmatrix} 1 & e^t\cos t & e^t\sin t \\ 0 & e^t\cos t - e^t\sin t & e^t\sin t + e^t\cos t \\ 0 & -2e^t\sin t & 2e^t\cos t \end{vmatrix} = 2e^{2t},$$

and hence also forms a fundamental set. ■

If λ is a characteristic root with multiplicity m, then it can be shown that

$$y_1(t) = e^{\lambda t}, \qquad y_2(t) = te^{\lambda t}, \ldots, y_m(t) = t^{m-1}e^{\lambda t}$$

are linearly independent solutions. If $\lambda = p + iq$ is a complex root with multiplicity m, then its complex conjugate $\lambda^* = p - iq$ has the same multiplicity. It follows that there are $2m$ linearly independent real solutions,

$$t^k e^{pt}\cos qt, \qquad t^k e^{pt}\sin qt, \qquad k = 0, 1, \ldots, m - 1.$$

Therefore, if all the characteristic roots and their multiplicities are known, a general solution can be constructed.

EXAMPLE 5 Find a fundamental set of solutions to

$$\frac{d^4y}{dt^4} + 4\frac{d^3y}{dt^3} + 14\frac{d^2y}{dt^2} + 20\frac{dy}{dt} + 25y = 0.$$

SOLUTION The characteristic equation

$$P(\lambda) = \lambda^4 + 4\lambda^3 + 14\lambda^2 + 20\lambda + 25 = (\lambda^2 + 2\lambda + 5)^2 = 0$$

has roots $\lambda = -1 + 2i$, $\lambda^* = -1 - 2i$, each with multiplicity 2. Therefore

$$e^{(-1+2i)t}, \qquad e^{(-1-2i)t}, \qquad te^{(-1+2i)t}, \qquad te^{(-1-2i)t}$$

form a complex-valued fundamental set of solutions. The corresponding real set is

$$e^{-t}\cos 2t, \qquad e^{-t}\sin 2t, \qquad te^{-t}\cos 2t, \qquad te^{-t}\sin 2t.$$ ■

The algebraic theory for homogeneous nth order constant coefficient linear differential equations is quite tidy, but when we attempt to solve problems other than those designed to have easily determined characteristic roots, we encounter difficulties. Unfortunately, there are no formulas for exhibiting solutions to polynomial equations of degree greater than four. Numerical methods can be used to generate approximate roots and, if the roots are not too close together, the errors due to the approximation can be estimated. It is also possible to construct accurate approximate solutions by numerical methods—a topic for subsequent chapters.

The method of undetermined coefficients and the variation of parameters method can be used to find general solutions to inhomogeneous nth order differential equations. The following examples illustrate how this can be done.

EXAMPLE 6 Find a general solution to

$$y''' - 2y'' - y' + 2y = 24e^{3t}$$

by the method of undetermined coefficients and the variation of parameters method.

SOLUTION Let us first solve the homogeneous equation,

$$y''' - 2y'' - y' + 2y = 0.$$

Its characteristic equation,

$$\lambda^3 - 2\lambda^2 - \lambda + 2 = (\lambda - 2)(\lambda^2 - 1) = 0,$$

has roots $\lambda_1 = 2, \lambda_2 = 1, \lambda_3 = -1$. Therefore a fundamental set of solutions is

$$y_1(t) = e^{2t}, \qquad y_2(t) = e^t, \qquad y_3(t) = e^{-t}.$$

To construct a general solution by the variation of parameters method, set

$$y(t) = u_1(t)y_1(t) + u_2(t)y_2(t) + u_3(t)y_3(t)$$

and determine $u_1'(t)$, $u_2'(t)$ and $u_3'(t)$ by solving the algebraic equations,

$$\begin{aligned} u_1'(t)y_1(t) + u_2'(t)y_2(t) + u_3'(t)y_3(t) &= 0, \\ u_1'(t)y_1'(t) + u_2'(t)y_2'(t) + u_3'(t)y_3'(t) &= 0, \\ u_1'(t)y_1''(t) + u_2'(t)y_2''(t) + u_3'(t)y_3''(t) &= 24e^{3t}. \end{aligned}$$

For this problem,

$$\begin{aligned} u_1'(t)e^{2t} + u_2'(t)e^t + u_2'(t)e^{-t} &= 0, \\ 2u_1'(t)e^{2t} + u_2'(t)e^t - u_3'(t)e^{-t} &= 0, \\ 4u_1'(t)e^{2t} + u_2'(t)e^t + u_3'(t)e^{-t} &= 24e^{3t}. \end{aligned}$$

Hence,

$$u_1'(t) = 8e^t, \qquad u_2'(t) = -12e^{2t}, \qquad u_3'(t) = 4e^{4t},$$

$$u_1(t) = 8e^t + c_1, \qquad u_2(t) = -6e^{2t} + c_2, \qquad u_3(t) = e^{4t} + c_3,$$

and

$$y(t) = (8e^t + c_1)e^{2t} + (-6e^{2t} + c_2)e^t + (e^{4t} + c_3)e^{-t}.$$

After simplifying, we have a general solution,

$$y(t) = 3e^{3t} + c_1e^{2t} + c_2e^{t} + c_3e^{-t}.$$

To construct a particular solution by the method of undetermined coefficients, set $y_p(t) = Ae^{3t}$, substitute into the differential equation, and obtain

$$A(27 - 18 - 3 + 2)e^{3t} = 24e^{3t}.$$

Therefore $A = 3$ and $y_p(t) = 3e^{3t}$.

The variation of parameters method is not efficient for this problem but, as we saw with second order equations, has greater applicability than the method of undetermined coefficients. ■

EXAMPLE 7 Find a general solution to

$$y''' + 4y'' + y' + 4y = 12t^3 + t^2 + 5. \tag{2.12.10}$$

SOLUTION The characteristic equation of the corresponding homogeneous differential equation,

$$\lambda^3 + 4\lambda^2 + \lambda + 4 = (\lambda + 4)(\lambda^2 + 1) = 0,$$

has roots $\lambda = -4$, $\lambda = \pm i$. Therefore $y_1(t) = e^{-4t}$, $y_2(t) = \cos t$, $y_3(t) = \sin t$ form a fundamental set of solutions.

A particular solution to (2.12.10) can be found by the method of undetermined coefficients. Set

$$y_p(t) = At^3 + Bt^2 + Ct + D,$$

substitute, and simplify. Then

$$(4A)t^3 + (3A + 4B)t^2 + (24A + 2B + 4C)t + (6A + 8B + C + 4D) = 12t^3 + t^2 + 5.$$

The coefficients of the powers of t must be equal:

$$4A = 12, \quad 3A + 4B = 1, \quad 24A + 2B + 4C = 0,$$
$$6A + 8B + C + 4D = 5.$$

Therefore $A = 3$, $B = -2$, $C = -17$, $D = 5$. A general solution to equation (2.12.10) is

$$y(t) = 3t^3 - 2t^2 - 17t + 5 + c_1e^{-4t} + c_2\cos t + c_3\sin t.$$ ■

An important application of fourth order differential equations is the study of the bending of elastic rods. A thin elastic rod is known to deform in such a way that its curvature is proportional to the sum of the moments of all the forces acting on the sides of the rod. If the vertical displacement of the axis of the rod is denoted by $y(x)$ at the point x, then

$$\text{curvature} = \frac{y''(x)}{[1 + y'(x)^2]^{3/2}} = \alpha M,$$

where M is the moment and α, the constant of proportionality, depends on the elastic properties of the rod.

For many problems, the slope $y'(x)$ is small and the term $y'(x)^2$ can be neglected. If this is done, we have the linear equation

$$\frac{d^2y}{dx^2} = \alpha M.$$

The moment M is related to the external force F by

$$\frac{d^2M}{dx^2} = F.$$

Eliminating M, we obtain the fourth order elastic rod equation,

$$\frac{d^4y}{dx^4} = \alpha F.$$

If the rod is resting on an elastic foundation, there is a restraining force $-ky$, and the appropriate equation is

$$\frac{d^4y}{dx^4} + ky = \alpha F(x),$$

where k is the elastic constant of the foundation.

Typically, there are constraints at the ends of the rod. The constraint $y = 0$ means the end cannot deflect vertically, $y' = 0$ means the slope at the end is zero, and $y'' = 0$ means that the end is free to rotate in the plane of deformation. If an end is *clamped,* $y = 0$ and $y' = 0$. If an end is *pinned,* $y = 0$ and $y'' = 0$. These conditions are illustrated in Fig. 2.28.

Since the constraints are imposed at two distinct points, the mathematical model for the bending of an elastic rod acted on by external forces is a *linear two-point boundary value problem.* A solution procedure for such problems is illustrated in the following example.

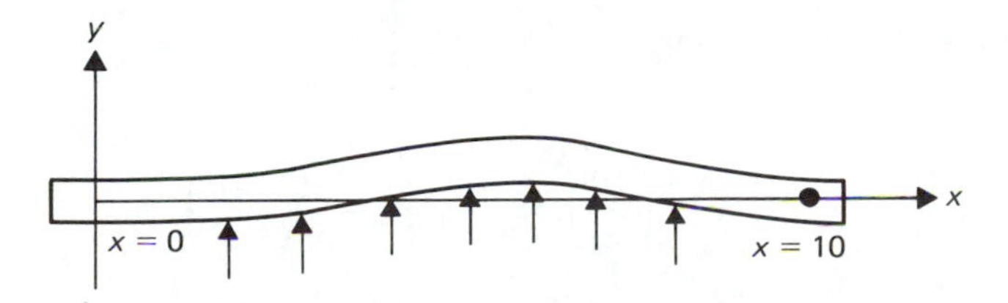

Figure 2.28 An elastic rod with clamped and pinned constraints.

EXAMPLE 8 A load $F(x) = 1000 \sin\left(\frac{\pi x}{10}\right)$ is applied to a rod that is 10 feet long. One end of the rod, $x = 0$, is clamped; the other end, $x = 10$, is pinned. Find the deflection if α, the elastic constant of the rod, is equal to 0.01.

SOLUTION The corresponding two-point boundary value problem is

$$y^{(4)} = (0.01)(1000) \sin\left(\frac{\pi x}{10}\right),$$

$$y(0) = 0, \qquad y'(0) = 0, \qquad y(10) = 0, \qquad y''(10) = 0.$$

We first seek a particular solution and let

$$y_p(x) = A \sin\left(\frac{\pi x}{10}\right).$$

After substituting we find that

$$A = 10\left(\frac{10}{\pi}\right)^4.$$

Clearly, 1, x, x^2, and x^3 are linearly independent solutions to the homogeneous differential equation, $y^{(4)} = 0$. Therefore a general solution is

$$y(x) = 10\left(\frac{10}{\pi}\right)^4 \sin\left(\frac{\pi}{10}x\right) + c_1 + c_2 x + c_3 x^2 + c_4 x^3,$$

and

$$y'(x) = 10\left(\frac{10}{\pi}\right)^3 \cos\left(\frac{\pi}{10}x\right) + 0 + c_2 + 2c_3 x + 3c_4 x^2,$$

$$y''(x) = -10\left(\frac{10}{\pi}\right)^2 \sin\left(\frac{\pi}{10}x\right) + 0 + 0 + 2c_3 + 6c_4 x.$$

From the boundary conditions, we have

$$y(0) = 0 = c_1,$$

$$y'(0) = 0 = 10\left(\frac{10}{\pi}\right)^3 + c_2,$$

$$y(10) = 0 = 10c_2 + 100c_3 + 1000c_4,$$

$$y''(10) = 0 = 2c_3 + 60c_4.$$

Thus,

$$c_1 = 0, \qquad c_2 = -10\left(\frac{10}{\pi}\right)^3, \qquad c_3 = \frac{3}{2}\left(\frac{10}{\pi}\right)^3, \qquad c_4 = -\frac{1}{20}\left(\frac{10}{\pi}\right)^3,$$

and

$$y(x) = \left(\frac{10}{\pi}\right)^3 \left\{10\left(\frac{10}{\pi}\right)\sin\left(\frac{\pi}{10}x\right) - 10x + \frac{3}{2}x^2 - \frac{1}{20}x^3\right\}. \quad \blacksquare$$

EXERCISES 2.12

In Exercises 1–8 determine whether the given sets of functions are linearly independent on $(-\infty, \infty)$. Use the Wronskian where it is appropriate.

1. $1, t, e^t$ **2.** $1, e^{-2t}, te^{-2t}$ **3.** $e^{2t}, \sin t, \cos t$

4. $1, t^3, |t|^3$ **5.** $t + 1, t^2 - t, 2t^2 - 3$ **6.** $1, t, t^2, e^{4t}$

7. $1, e^t, e^{-t}, \cosh t$ **8.** $1, \sin 2t, \cos 2t, \sin^2 t$

In Exercises 9–17 show that the given sets of functions are a fundamental set of solutions of a homogeneous linear differential equation and find that equation.

9. $1, t, e^{-4t}$ **10.** e^{-3t}, e^t, e^{2t} **11.** $e^{4t}, te^{4t}, t^2e^{4t}$

12. $1, \sin 2t, \cos 2t$ **13.** $1, t, t^2$ **14.** $t^{-1}, t, t^3; t \neq 0$

15. $1, t, e^t, e^{-t}$ **16.** $1, e^{-t}, \sin t, \cos t$ **17.** $e^{-2t}, e^{2t}, e^{3t}, te^{3t}$

In Exercises 18–25 find the general solution of the given differential equation.

18. $y^{(4)} - y'' = 0$

19. $y''' + y' = 0$

20. $y^{(4)} + 8y'' + 16y = 0$

21. $y''' + 2y'' - y' + y = 0$. (*Hint:* The characteristic polynomial does not have rational roots.)

22. $y''' + y'' + \frac{1}{3}y' + \frac{1}{27}y = 0$

23. $y''' - 8y'' + 21y' - 18y = 0$

24. $y''' - 3y' + y = 0$. (*Hint:* The characteristic polynomial does not have real roots.)

25. $y^{(4)} + 7y'' + 18y' + 10y = 0$

In Exercises 26–33 find the solution of the given initial value problem.

26. $y''' - y' = 0, \quad y(0) = 0, \quad y'(0) = 0, \quad y''(0) = 2$

27. $y^{(4)} - y = 0, \quad y(0) = y'(0) = 0, \quad y''(0) = 2, \quad y'''(0) = -6$

28. $y''' + y'' + 4y' + 4y = 0, \quad y(0) = y'(0) = 1, \quad y''(0) = 2$

29. $y^{(4)} + 2y'' + y = 0, \quad y(0) = y''(0) = 1, \quad y'(0) = y'''(0) = 2$

30. $y^{(4)} + 5y'' + 4y = 0, \quad y(0) = y'''(0) = 0, \quad y'(0) = y''(0) = 3$

31. $y^{(5)} - y^{(3)} = 0, \quad y(0) = 2, \quad y'(0) = \frac{7}{2}, \quad y''(0) = 2, \quad y'''(0) = 3, \quad y^{(4)}(0) = -1$

32. $t^3y^{(3)} - t^2y'' - 2ty' + 6y = 0, \quad y(1) = 4, \quad y'(1) = 9, \quad y''(1) = 24$. (*Hint:* Let $y = t^r$.)

33. $t^3y''' - t^2y'' - 7ty' + 16y = 0, \quad y(1) = 0, \quad y'(1) = y''(1) = -6$. (*Hint:* Let $y = t^r$.)

In Exercises 34–41 find a particular solution of the given differential equation.

34. $y^{(4)} + y = 3e^{2t}$

35. $y''' + 2y' = 2 \sin t$

36. $y^{(4)} - y'' = 6t + 1$

37. $y''' + y'' - y' - y = 8e^{-t}$

38. $y''' - y = 5 \cos 2t$

39. $y^{(4)} + 3y'' - 4y = -4t^4 + \frac{1}{2}t$

40. $y''' - 4y' = \cos^2 t$

41. $y''' + y = 4 \cosh t$

42. A load $F = 200 \cos((\pi x)/10)$ is applied to a rod that is 10 feet long. One end of the rod, $x = 0$, is pinned; the other end, $x = 10$, is clamped. Find the deflection of the rod if α, the elastic constant of the rod, equals 0.01. Determine the location and magnitude of the maximum deflection.

43. A beam that is clamped at both ends has a length of 6 feet and is subjected to a load $F(x) = 20(6 - x)^2$. Find the deflection of the beam and determine the magnitude and location of the maximum deflection if α, the elastic constant of the rod, equals $\frac{1}{10}$.

44. A beam that is pinned at both ends (also called a simply supported beam) has length of 20 feet and elastic constant $\alpha = 0.05$. If it is subjected to a uniform load, $F(x) = 200$, find the deflection of the rod and determine the location of the maximum deflection.

45. A beam that is clamped at one end and completely free at the other is called a cantilevered beam. In this case the boundary conditions at the free end $x = L$ are $y''(L) = y'''(L) = 0$. If a cantilevered beam has length of 10 feet and its elastic constant $\alpha = \frac{1}{20}$, find its deflection if it is subjected to a uniform load $F(x) = 200$. Determine the location and magnitude of the maximum deflection.

46. A rod subjected to an axial force T and a load F satisfies the differential equation

$$y^{(4)} - \alpha T y'' = \alpha F.$$

A beam of length 4 feet is pinned at both ends and has an elastic constant $\alpha = \frac{1}{400}$. It is subjected to an axial force $T = 100$ and a uniform load $F(x) = 400$. Find the deflection of the rod.

47. Repeat Exercise 46 for the case in which the rod is clamped at both ends.

SUMMARY

Undoubtedly the most important differential equation in applied mathematics, engineering, and the physical sciences is the second order linear differential equation

$$a(t)\frac{d^2y}{dt^2} + b(t)\frac{dy}{dt} + c(t)y = g(t).$$

It can be used as a model of *LCR* electrical circuits, mass–spring systems, and forced oscillations as well as for approximating the solutions of nonlinear phenomena such as the pendulum equation.

The structure of the set of solutions of linear differential equations is basically an algebraic one, analogous to the notion of a vector space. In this chapter is discussed the concept of a linearly independent set of solutions of the second order linear equation, how one obtains from them the general solution of the homogeneous equation ($g(t) \equiv 0$), and how one uses the general solution to solve the equation with initial conditions

$$y(t_0) = r,$$

$$\frac{dy}{dt}(t_0) = s.$$

For the case where $a(t)$, $b(t)$, and $c(t)$ are constants a complete description of the general solution can be obtained by finding the roots of an associated quadratic polynomial. The various cases, depending on the nature of the roots, are analyzed, and the behavior of each type of solution is studied by examining the oscillator equation describing an *LCR* circuit or a mass–spring system.

For the inhomogeneous equation, where $g(t)$ is present, a general expression for the solution is developed from the general solution of the homogeneous equation. This is known as the variation of parameters method. However, for the constant coefficient equation a much simpler method, called undetermined coefficients, is applicable for a large class of functions $g(t)$.

The theory of the nth order differential equations is modeled after the second order theory. The definitions, theorems, and solution techniques that are presented are straightforward extensions of those presented in the earlier sections.

Two-dimensional versions of the Euler algorithm and the improved Euler algorithm are given for constructing approximate solutions numerically.

MISCELLANEOUS EXERCISES

Find general solutions to Exercises 2.1–2.13 by either the method of undetermined coefficients or the variation of parameters method. When applicable, the method of undetermined coefficients is more efficient.

2.1. $y'' - y' - 2y = t\cos t + 3\sin t$

2.2. $y'' - 4y = 6\cosh 2t + 3t^3$

2.3. $y'' + 2y' + y = 4t^5e^{-t}$

2.4. $y'' - 4y' + 4y = (t^2 + 2t)e^{2t}$

2.5. $2y'' + 18y = 72\csc 3t$

2.6. $y'' - 9y = 18t^{10}e^{3t}$

2.7. $y'' - 4y' + 4y = 6e^{2t}/(1 + t^2)$

2.8. $y''' - 3y'' = 6t^3 + 2t - 1$

2.9. $y''' - 2y'' - y' + 2y = 4e^{2t} - 7e^{3t}$

2.10. $t^2y'' + 5ty' - 6y = 4t^3 - (4/t^2)$

2.11. $t^2y'' + 5ty' + y = 6e^t + (6/t^2)$

2.12. $2y''' + 9y'' + 12y' + 5y = 3t^2e^t$

2.13. $3y''' + 10y'' + 15y' + 4y = 8t^2e^{-t/3}$

2.14. Show that the general solution of

$$\epsilon y'' + y' + y = 0, \quad 0 < \epsilon \ll 1, \quad t > 0,$$

will approach the general solution of $y' + y = 0$ as $\epsilon \to 0$. Will this occur for $t < 0$? (*Hint:* In your analysis you may find the expansion

$$(1 - u)^{1/2} = 1 - \frac{1}{2}u - \frac{1}{8}u^2 + \ldots, \; |u| < 1$$

useful.)

2.15. The general solution of $y'' + 3y' + 2y = f(t)$ is of the form

$$y(t) = Ae^{-t} + Be^{-2t} + y_p(t),$$

where $y_p(t)$ is obtained from the variation of parameters formula. If $f(t)$ is bounded for $t \geq 0$ (say, $|f(t)| \leq M$), show that

$$|y(t)| \leq |A| + |B| + \frac{3}{2}M, \; t \geq 0.$$

2.16. Generalize the result of the previous problem to the case

$$y'' + ay' + by = f(t),$$

where the associated characteristic polynomial has roots with negative real parts. Specifically, show that if $f(t)$ is bounded for $t \geq 0$, then so is the general solution. This is an example of the engineering dictum "bounded input gives bounded output."

2.17. Given the homogeneous equation

$$y'' + a(t)y = 0$$

and one nontrivial solution $y_1(t)$, show that

$$y_2(t) = y_1(t) \int \frac{1}{[y_1(t)]^2}\, dt$$

is also a solution. Furthermore, show that $y_1(t)$ and $y_2(t)$ are a pair of linearly independent solutions. (This is a special case of the method of *reduction of order* to be discussed in Chapter 6.)

Use the result of the previous problem and the one given solution to find the general solution of the following differential equations:

2.18. $y'' + 9y = 0, \quad y_1(t) = \cos 3t$

2.19. $y'' + \frac{1}{4}t^{-2}y = 0, \quad y_1(t) = t^{1/2}$

2.20. $y'' - (2\sec^2 t)y = 0, \quad y_1(t) = \tan t$

2.21. a) Show that the general second order equation $y'' + a(t)y' + b(t)y = 0$ becomes the first order Riccati equation $du/dt = u^2 - a(t)u + b(t)$ under the transformation $y(t) = \exp[-\int u(t)\, dt]$.

b) Conversely, show that the Riccati equation $du/dt = u^2 + p(t)u = q(t)$ becomes the second order equation $y'' - p(t)y' + q(t)y = 0$ under the transformation $u(t) = -y'(t)/y(t)$.

Using the results of the previous exercise, solve the following Riccati equations by converting them to second order linear equations.

2.22. $\dfrac{du}{dt} = -u^2 + 2u - 2$

2.23. $\dfrac{du}{dt} = u^2 + 1$

2.24. $\dfrac{du}{dt} = u^2 - \dfrac{3}{t}u + \dfrac{1}{t^2}$

Using the results of Exercise 2.21, find a solution of the following second order linear equations by converting them to Riccati equations and solving.

2.25. $y'' + (\cos^2 t + \cos t - 1)y = 0$

2.26. $y'' + \left(\dfrac{3}{t} - 1\right)y' - \dfrac{2}{t}y = 0$

2.27. Given a system governed by the equation $y'' + 2by' + y = \sin 2t$, where $t > 0$, $b > 0$, determine the smallest value of b which insures that the maximum amplitude of the steady-state solution is less than $\frac{1}{4}$.

2.28. Using the ideas developed in Exercise 17 of Section 2.7 show that

a) the general solution of $t^2y'' + ty' + y = 4e^t$ is

$$y(t) = A\sin(\ln t) + B\cos(\ln t) + 4\sum_{n=0}^{\infty}\frac{t^n}{n!(n^2+1)};$$

b) the general solution of $t^2y'' - 5ty' + 9y = 3\sin t$ is

$$y(t) = At^3 + Bt^3\ln t + \frac{3}{4}t - \frac{1}{2}t^3(\ln t)^2 + \frac{3}{4}\sum_{n=2}^{\infty}\frac{(-1)^n t^{2n+1}}{(2n+1)!(n-1)^2}.$$

The following method uses second order linear equations to solve systems of first order linear equations. Given

$$\frac{dx}{dt} = ax + by, \qquad x(0) = A,$$

$$\frac{dy}{dt} = cx + dy, \qquad y(0) = B,$$

differentiate the first equation and then successively substitute for y and dy/dt:

$$\frac{d^2x}{dt^2} = a\frac{dx}{dt} + b\frac{dy}{dt} = a\frac{dx}{dt} + b(cx + dy)$$

$$= a\frac{dx}{dt} + bcx + bd\left(\frac{1}{b}\frac{dx}{dt} - \frac{a}{b}x\right).$$

We obtain a second order equation in x whose initial conditions are

$$x(0) = A, \qquad x'(0) = ax(0) + by(0) = aA + bB.$$

Find $x(t)$, then substitute in the first differential equation to find $y(t)$. Use this procedure to find solutions $x(t)$, $y(t)$ of the following:

2.29. $\dfrac{dx}{dt} = 2x + 3y,\ \ x(0) = 0;\qquad \dfrac{dy}{dt} = 3x + 4y,\ \ y(0) = 4$

2.30. $\dfrac{dx}{dt} = x - 3y,\ \ x(0) = 2;\qquad \dfrac{dy}{dt} = x - y,\ \ y(0) = -1$

2.31. The variation of parameters solution to the initial value problem,

$$y'' + y = 1/(t+1), \qquad y(0) = 0, \qquad y'(0) = 0$$

is

$$y(t) = u_1(t)\cos t + u_2(t)\sin t$$

where

$$u_1(t) = -\int_0^t \frac{\sin z}{z+1}\,dz,$$

$$u_2(t) = \int_0^t \frac{\cos z}{z+1}\,dz.$$

Use Simpson's rule to approximate $u_1(6)$, $u_2(6)$ and then compute an approximation to $y(6)$ with 60 subintervals of width $h = 0.1$. Solve the same initial value problem using the improved Euler method with a stepsize picked so that the approximate value of $y(6)$ constructed by the improved Euler method agrees to three significant figures with the value constructed using Simpson's rule.

REFERENCES

In Chapter 1 we gave some references to introductory texts in ordinary differential equations. Additional references with an emphasis on applications and/or some historical background are

1. P.D. Ritger and N.D. Rose, *Differential Equations with Applications,* McGraw-Hill, New York, 1968.
2. G.F. Simmons, *Differential Equations with Applications and Historical Notes,* McGraw-Hill, New York, 1972.

A more theoretically inclined treatment is given in

3. O. Plaat, *Ordinary Differential Equations,* Holden Day, San Francisco, 1971.

For a discussion of the broader aspects of mathematics applied to the physical world, the reader might enjoy

4. R. Haberman, *Mathematical Models,* Prentice-Hall, Englewood Cliffs, N.J., 1977.
5. B. Noble, *Applications of Undergraduate Mathematics in Engineering,* Macmillan, New York, 1967.
6. G. Polya, *Mathematical Methods in Science,* Mathematical Association of America, 1977.
7. G. Strang, *Introduction to Applied Mathematics,* Wellesley-Cambridge Press, Wellesley, Mass., 1986.

3

ELEMENTARY NUMERICAL METHODS

3.1 INTRODUCTION

Why do we need numerical methods? It should be clear from what has been done so far that only a few differential equations can be solved completely in terms of elementary functions. For example, in Chapter 1 we found that first order equations are usually solvable if they are linear and sometimes solvable if they are not (e.g., separable or exact equations). Equations of order two and higher are generally solvable only if they are linear with constant coefficients. In fact, very few differential equations that arise in applications can be solved exactly, and so one must resort to numerical or approximate methods to find solutions.

3.2 A GENERAL ONE-STEP METHOD

In Chapter 1 two simple numerical methods were introduced: the Euler and the improved Euler algorithms. These methods are adequate only for very simple problems. In this chapter methods will be developed that apply to a wider class of problems and that lay the groundwork for the highly efficient computer program RKF45, which will be discussed in Chapter 8. Before we begin, however, let us briefly review the notion of a numerical solution and introduce some of the notation and definitions that will be used.

To start with, consider the first order initial value problem

$$\frac{dy}{dt} = f(t, y), \quad \alpha \leq t \leq \beta, \tag{3.2.1}$$

$$y(a) = y_0, \quad \alpha < a < \beta. \tag{3.2.2}$$

The methods we develop for solving the problem (3.2.1, 3.2.2) can easily be extended to systems of first order differential equations and to higher order differential equations. The reader has already seen an example of this in Chapter 2, where the Euler and improved Euler methods were extended to second order differential equations. These methods are called step-by-step methods, and they generate a sequence of t-points, $t_1, t_2, \ldots$, at which the solution to the differential equation is approximated. The *approximate* solution at the point $t = t_i$, denoted by y_i, is computed from approximations at prior t-points. As in previous sections, the quantity $y(t_i)$ will always be used to denote the *exact* solution to the problem (3.2.1, 3.2.2) at the point $t = t_i$.

Suppose we want to approximate the solution to the initial value problem (3.2.1, 3.2.2) on the interval $[a, b]$ where $\alpha < a$, $b < \beta$. Let the t-points be equally spaced; so, for some positive integer n and $h = (b - a)/n$, $t_i = a + ih$, $i = 0, 1, \ldots, n$. If $a < b$, h is positive and we are integrating forward. If $a > b$, h is negative and we are integrating backward. The latter case could occur if we were solving for the initial point of an integral curve given the terminal point. For ease of exposition, we assume $h > 0$. A general method for solving (3.2.1, 3.2.2) can be written in the form

$$y_{i+1} = y_i + h\theta(t_i, y_i) \tag{3.2.3}$$

with

$$y_0 = y(t_0). \tag{3.2.4}$$

Referring to Section 1.7, one sees that Euler's method has

$$\theta(t_i, y_i) = f(t_i, y_i),$$

while for the improved Euler method

$$\theta(t_i, y_i) = \frac{f(t_i, y_i) + f(t_i + h, y_i + hf(t_i, y_i))}{2}.$$

We seek an accurate algorithm (3.2.3, 3.2.4). Since the approximate solution generated by the algorithm is to be close to $y(t)$, the exact solution of the initial value problem (3.2.1, 3.2.2), we require that the algorithm be such that it is almost satisfied by $y(t)$, i.e.,

$$y(t_{i+1}) = y(t_i) + h\theta(t_i, y(t_i)) + h\tau_i \tag{3.2.5}$$

with τ_i "small." The quantity $h\tau_i$ is called the *local* (truncation) error. The method (3.2.3) is said to be of *order p* if for all t_i, $a \leq t_i \leq b$, and for all sufficiently small h, there are constants C and p such that

$$|\tau_i| \leq Ch^p.$$

This can be interpreted as meaning that $|\tau_i|$ goes to zero no slower than Ch^p as h approaches zero. The order p condition can be abbreviated by using the big-oh notation[1]

$$\tau_i = O(h^p)$$

as h approaches zero. The constant C, in general, depends on the solution $y(t)$, its derivatives and the length of the interval over which the solution is to be found, but is independent of h.

The importance of local error and of the order of a method are easily explained. Suppose the exact value $y(t_i)$ of the solution is known (as it is, for instance, at the initial point $t = a$). Then $y_i = y(t_i)$ and the next step of the method gives

$$y_{i+1} = y_i + h\theta(t_i, y_i) = y(t_i) + h\theta(t_i, y(t_i))$$

as an approximation to $y(t_{i+1})$. In fact, relation (3.2.5) implies that $h\tau_i$, the local error, is equal to $y(t_{i+1}) - y_{i+1}$. If the method is of order p, then

$$|y(t_{i+1}) - y_{i+1}| = |h\tau_i| \leq Ch^{p+1}.$$

The order of a method is taken as p even though the order of the local error is $p + 1$ because local errors accumulate and, in general, the actual error of the approximate solution at a fixed point t is $O(h^p)$.

EXAMPLE 1 Find the order p of Euler's method.

SOLUTION Suppose $y(t)$ is a solution of $y' = f(t, y)$ and $y''(t)$ is continuous for $a \leq t \leq b$. Then Taylor's formula with remainder implies

$$\begin{aligned} y(t_{i+1}) &= y(t_i + h) = y(t_i) + hy'(t_i) + \frac{h^2}{2} y''(\zeta_i) \\ &= y(t_i) + hf(t_i, y(t_i)) + h\left[\frac{h}{2} y''(\zeta_i)\right], \end{aligned}$$

1. The big-oh notation is a useful one for error estimating. We say $g(h) = O(h^p)$ as $h \to 0$ if $\lim_{h\to 0} |g(h)|/|h^p|$ is bounded. For instance, $\sin(h) = O(h)$ and $\cos(h^2) - 1 = O(h^4)$ as $h \to 0$.

where $t_i \le \zeta_i \le t_{i+1}$, and we have used the fact that $y' = f(t, y)$. Letting

$$\tau_i = \frac{h}{2} y''(\zeta_i)$$

and observing that the continuity of $y''(t)$ on $[a, b]$ implies that it is bounded there, we see that

$$|\tau_i| = \left| \frac{h}{2} y''(\zeta_i) \right| \le Ch,$$

where C could be chosen to be the maximum value of $|y''(t)/2|$ on $a \le t \le b$. Hence Euler's method has order $p = 1$. ■

The order of a method may also be viewed as a measure of how fast the error in the computed solution goes to zero at a fixed point t as more and more steps are taken, i.e., as $h \to 0$. Recall that in the examples in Chapter 1 the error in the Euler algorithm was proportional to h, which is consistent with the method being of order $p = 1$. It was also indicated that the error in the improved Euler algorithm was proportional to h^2. As will be seen shortly, this algorithm has order $p = 2$. Clearly, the larger the value of p, the faster the error goes to zero as $h \to 0$. In the next section we shall describe a class of methods known as Taylor series methods, which can be used to develop algorithms of very high orders.

3.3 TAYLOR SERIES METHODS

Perhaps the simplest methods of order p are based on the Taylor series expansion of the solution $y(t)$ of the differential equation (3.2.1). If $y^{(p+1)}(t)$ is continuous on $[a, b]$, then Taylor's formula gives

$$\begin{aligned} y(t_{i+1}) &= y(t_i + h) \\ &= y(t_i) + y'(t_i)h + \cdots + y^{(p)}(t_i)\frac{h^p}{p!} + y^{(p+1)}(\zeta_i)\frac{h^{p+1}}{(p+1)!} \qquad \textbf{(3.3.1)} \\ &= y(t_i) + h\left[y'(t_i) + \cdots + y^{(p)}(t_i)\frac{h^{p-1}}{p!} \right] + y^{(p+1)}(\zeta_i)\frac{h^{p+1}}{(p+1)!}, \end{aligned}$$

where $t_i \le \zeta_i \le t_{i+1}$. The continuity of $y^{(p+1)}(t)$ implies that it is bounded on $a \le t \le b$ and therefore

$$y^{(p+1)}(\zeta_i)\frac{h^{p+1}}{(p+1)!} = O(h^{p+1}) = hO(h^p).$$

Since $y' = f(t, y)$, Eq. (3.3.1) can be simplified somewhat. To evaluate the higher derivatives of $y(t)$, we use the chain rule and the fact that $dy/dt = f(t, y)$ to get

$$\begin{aligned} y''(t) &\stackrel{d}{=} f^{(1)}(t, y) = \frac{d}{dt} f(t, y) = \frac{\partial}{\partial t} f(t, y) \frac{dt}{dt} + \frac{\partial}{\partial y} f(t, y) \frac{dy}{dt} \\ &= f_t(t, y) + f_y(t, y) f(t, y) = f_t + f_y f, \\ y'''(t) &\stackrel{d}{=} f^{(2)} = \frac{d}{dt} f^{(1)} = \frac{\partial}{\partial t} f^{(1)} \frac{dt}{dt} + \frac{\partial}{\partial y} f^{(1)} \frac{dy}{dt} \\ &= f_t^{(1)} + f_y^{(1)} f = f_{tt} + f_{yt} f + f_y f_t + [f_{ty} + f_{yy} f + f_y f_y] f \\ &= f^2 f_{yy} + 2 f f_{ty} + f_{tt} + f_t f_y + f f_y^2, \end{aligned}$$

etc.[2] In general, all the higher derivatives can be found from the recurrence relation

$$\begin{aligned} y^{(n+1)}(t) &\stackrel{d}{=} f^{(n)} = \frac{d}{dt} f^{(n-1)}(t) \\ &= f_t^{(n-1)} + f_y^{(n-1)} f, \quad n = 1, 2, \ldots, \end{aligned} \tag{3.3.2}$$

where $f^{(0)}(t, y) = f(t, y)$. The quantity $f^{(n)}(t, y)$ is called the nth *total derivative* of $f(t, y)$. By using (3.3.2), the expansion (3.3.1) can now be put into the form

$$y(t_{i+1}) = y(t_i) + h \left\{ f(t_i, y(t_i)) + \cdots + f^{(p-1)}(t_i, y(t_i)) \frac{h^{p-1}}{p!} \right\} + hO(h^p). \tag{3.3.3}$$

Comparison of (3.2.5) and (3.3.3) shows that if a method of order p is desired, we can let

$$\theta(t_i, y(t_i)) = f(t_i, y(t_i)) + \cdots + f^{(p-1)}(t_i, y(t_i)) \frac{h^{p-1}}{p!} \tag{3.3.4}$$

to obtain a family of methods called *Taylor series methods.* These are given in the following algorithm.

The Taylor series algorithm

To obtain an approximate solution of order p to the initial value problem (3.2.1, 3.2.2) on $[a, b]$, let $h = (b - a)/n$ for some positive integer n and

2. The symbol $\stackrel{d}{=}$ means "by definition."

generate the sequences

$$y_{i+1} = y_i + h\left[f(t_i, y_i) + \cdots + \frac{h^{p-1}}{p!} f^{(p-1)}(t_i, y_i)\right],$$

$$t_{i+1} = t_0 + (i+1)h, \; i = 0, 1, \ldots, n-1,$$

where $t_0 = a$, $y_0 = y(a)$; y_{i+1} is an approximation to $y(t_{i+1})$.

Note that Euler's method is the Taylor series method of order $p = 1$.

EXAMPLE 1 Approximate the solution to $y' = y$, $y(0) = 1.00$ at $t = 1.00$ using the Taylor series method of order $p = 2$ with $h = 0.25$.

SOLUTION The recursion relation for y_i is

$$y_{i+1} = y_i + h\left[f(t_i, y_i) + \frac{h}{2} f^{(1)}(t_i, y_i)\right].$$

Since

$$f(t, y) = y,$$

$$f^{(1)}(t, y) = f_t(t, y) + f_y(t, y)f(t, y) = 0 + 1 \cdot y = y,$$

the algorithm becomes

$$y_{i+1} = y_i + h\left[y_i + \frac{h}{2} y_i\right] = y_i\left(1 + h + \frac{h^2}{2}\right),$$

$$t_{i+1} = (i+1)\,h, \qquad t_0 = 0, \qquad y_0 = 1.00.$$

Now using the fact that $h = 0.25$, we have

$$1 + h + \frac{h^2}{2} = 1 + 0.25 + \frac{0.25^2}{2} = 1.28125,$$

and therefore the sequences generated are

$$y_1 = y_0\left(1 + h + \frac{h^2}{2}\right) = 1.00 \cdot 1.28125 = 1.28125, \qquad t_1 = 0.25;$$

$$y_2 = 1.28125 y_1 = 1.28125 \cdot 1.28125 = 1.64160, \qquad t_2 = 0.50;$$

$$y_3 = 1.28125 y_2 = 2.10330, \qquad t_3 = 0.75;$$

$$y_4 = 1.28125 y_3 = 2.69486, \qquad t_4 = 1.00.$$

Table 3.1 Approximations to e with $p = 2$, $h = 0.25$

n	$h = 1/n$	y_n	|Error|
8	1.25×10^{-1}	2.71184	6.4×10^{-3}
16	6.25×10^{-2}	2.71659	1.7×10^{-3}
32	3.125×10^{-2}	2.71785	4.3×10^{-4}
64	1.5625×10^{-2}	2.71817	1.1×10^{-4}
128	7.8125×10^{-3}	2.71825	2.7×10^{-5}

The actual solution is $y(t) = e^t$ and $y(1) = 2.71828$, which gives an absolute error of 2.3×10^{-2}. If one is willing to do more work, Table 3.1 can be constructed. A little calculation shows that the absolute error at $t = 1.00$ is proportional to h^2; in fact, for $n \geq 16$ we have

$$|\text{Error}|_{t=1.00} \cong 0.5h^2,$$

as the theory predicted.

As a matter of interest, the same rather large value $h = 0.25$ could be used and the order increased to attain better accuracy instead of keeping the order fixed and decreasing h. For example, for $p = 3, 4$ the formulas become

$$p = 3: \quad y_{i+1} = y_i\left(1 + h + \frac{h^2}{2} + \frac{h^3}{6}\right),$$

$$p = 4: \quad y_{i+1} = y_i\left(1 + h + \frac{h^2}{2} + \frac{h^3}{6} + \frac{h^4}{24}\right),$$

and we get the results shown in Table 3.2. Therefore the same order of accuracy is obtained for $p = 4$ and $h = 0.25$ as for $p = 2$ and $h = 7.8125 \times 10^{-3}$. Of course, the higher total derivatives $f^{(n)}$ are easy to compute for $f(t, y) = y$. ■

EXAMPLE 2 Approximate the solution to $y' = 2ty^2$, $y(0) = 0.5$ at $t = 1.0$ using the Taylor series algorithm of order $p = 2$ with $h = 0.1$.

Table 3.2 Approximations to e with $p = 3, 4$, $h = 0.25$

Order p	y_4	|Error|
3	2.71683	1.5×10^{-3}
4	2.71821	7.2×10^{-5}

SOLUTION Here $f(t, y) = 2ty^2$, so

$$f_t = 2y^2, \qquad f_y = 4ty.$$

Therefore

$$f^{(1)} = f_t + f_y f = 2y^2 + 8t^2y^3$$

and the algorithm for the sequence y_i is

$$y_{i+1} = y_i + h\left[2t_iy_i^2 + \frac{h}{2}(2y_i^2 + 8t_i^2y_i^3)\right],$$

$$t_{i+1} = (i+1)h, \qquad t_0 = 0, \qquad y_0 = 0.5.$$

The results are presented in Table 3.3. The actual solution (obtained by separation of variables) is

$$y(t) = \frac{1}{2 - t^2},$$

so $y(1) = 1$ and the absolute error in the approximate solution at $t = 1$ is 1.9×10^{-2}. Using 100 steps ($h = 0.01$) instead of 10 steps ($h = 0.1$) would have produced an error of 2.6×10^{-4}. Again it looks as if the error is proportional to h^2. ■

In the above example, if we had wanted to use the Taylor series method of order $p = 3$, we would have needed to compute $f^{(2)}$. Since

$$f^{(1)} = 2y^2 + 8t^2y^3,$$

Table 3.3 **Approximate solution to $y' = 2ty^2$, $y(0) = 0.5$ with $h = 0.1$**

i	t_i	y_i	i	t_i	y_i
0	0.0	0.50000	6	0.6	0.60828
1	0.1	0.50250	7	0.7	0.65962
2	0.2	0.51013	8	0.8	0.73051
3	0.3	0.52335	9	0.9	0.83120
4	0.4	0.54304	10	1.0	0.98108
5	0.5	0.57060			

it follows that

$$f_t^{(1)} = 16ty^3, \qquad f_y^{(1)} = 4y + 24t^2y^2,$$

and so

$$f^{(2)} = f_t^{(1)} + f_y^{(1)}f = 16ty^3 + (4y + 24t^2y^2)(2ty^2).$$

The algorithm for the sequence y_i is then

$$y_{i+1} = y_i + h\left[2t_iy_i^2 + \frac{h}{2}(2y_i^2 + 8t_i^2y_i^3) + \frac{h^2}{6}(24t_iy_i^3 + 48t_i^3y_i^4)\right],$$

$$t_{i+1} = (i + 1)h, \qquad t_0 = 0, \qquad y_0 = 0.5.$$

With $h = 0.1$ an error of order $10^{-3} = 0.1^3$ for the approximation of $y(1)$ would be expected.

Taylor series methods can be quite effective if the total derivatives $f^{(n)}$ of $f(t, y)$ are not too difficult to evaluate. In Example 1 above, they were easy to evaluate and in Example 2, not much more difficult. If, on the other hand, the function is something like

$$f(t, y) = \cos t \sin y,$$

then

$$f^{(1)} = -\sin t \sin y + (\cos t \cos y) \cos t \sin y,$$

$$f^{(2)} = -\cos t \sin y - 2 \cos t \sin t \cos y \sin y + (-\sin t \cos y + \cos^2 t \cos^2 y - \cos^2 t \sin^2 y)\cos t \sin y,$$

etc., and we see that the total derivatives of f can become quite messy.

EXERCISES 3.3

In Exercises 1–10 use the Taylor series method of order $p = 2$ with $h = 0.1$ to compute an approximate solution at $t = 0.5$ to each initial value problem. In each case find the exact solution and compare it with your numerical result.

1. $y' = 2y - 1, \quad y(0) = 1$

2. $y' = 1 + y^2, \quad y(0) = 0$

3. $y' = -y, \quad y(0) = 1$

4. $y' = y - t, \quad y(0) = 2$

5. $y' = -ty^2, \quad y(0) = 2$

6. $y' = t/y, \quad y(0) = 1$

7. $y' - y = \sin 2t, \quad y(0) = 0$

8. $y' = t, \quad y(0) = 0$

9. $y' = y(1 - y), \quad y(0) = 2$

10. $y' = y^2 - 3y + 2, \quad y(0) = 2$

11. For the linear equation $y' = P_1(t)y + Q_1(t)$ show that the derivatives needed in the Taylor series method can be obtained from the recursion

$$y^{(r)} = P_r(t)y + Q_r(t),$$

where

$$P_r(t) = P'_{r-1}(t) + P_1(t)P_{r-1}(t),$$
$$Q_r(t) = Q'_{r-1}(t) + Q_1(t)P_{r-1}(t), \quad r = 2,3, \ldots .$$

12. Apply the method of Exercise 11 to develop a third order Taylor series formula to approximate the solution of the initial value problem

$$y' = ty + 1, \qquad y(0) = 0.$$

Implement your formula by using a step size of $h = 0.25$ to approximate $y(1)$.

13. Approximate Dawson's integral $e^{-t^2}\int_0^t e^{-s^2}\,ds$ on the interval $[0, 0.5]$ using Taylor series methods of order $p = 1, 2, 3$ with $h = 0.1$. Dawson's integral is the solution to the initial value problem

$$y' = 1 - 2ty, \qquad y(0) = 0,$$

and its values at $t = 0.1, 0.2, \ldots, 0.5$ correct to five decimals are [5]:

t	$y(t)$
0.1	0.09934
0.2	0.19475
0.3	0.28263
0.4	0.35994
0.5	0.42444

14. Consider the initial value problem

$$y' = -2y, \qquad y(0) = 1.$$

a) Solve the problem analytically to find $y(1)$.

b) Write a computer program that implements the Taylor series algorithm of order $p = 2$ for this problem. Let your program begin with a fixed step size $h = 1/N$ and repeatedly halve it until a given accuracy, say $10^{-\alpha}$ is achieved; i.e., continue to halve until

$$|y(1) - y_N| < 10^{-\alpha}.$$

Let $\alpha = 3$ and start with $N = 2$.

c) Repeat part (b) using Taylor algorithms of order $p = 3$ and 4 and note the differences in the final value as p increases for fixed N. The example points out that Taylor algorithms can be quite effective when the derivatives are easy to evaluate.

3.4 RUNGE-KUTTA METHODS

We now introduce another class of methods, the Runge–Kutta methods, for solving the initial value problem (3.2.1, 3.2.2).[3] These methods are designed to approximate Taylor series methods but have the advantage of not requiring explicit evaluations of the derivatives of $f(t, y)$. The basic idea is to use a linear combination of values of $f(t, y)$ to approximate $y(t)$. This linear combination is matched up as well as possible with the Taylor series for $y(t)$ to obtain methods of the highest possible order p. An example with one evaluation of f is Euler's method; an example with two evaluations of f is the improved Euler method.

We proceed now to derive a family of Runge–Kutta formulas using two evaluations of $f(t, y)$ per step. It will turn out that the improved Euler method is a special case of these formulas. The techniques employed in the derivation extend easily to the development of all Runge–Kutta type formulas. To begin with, we are given the values t_i and y_i and the object is to choose the points $\hat{t}_i$ and $\hat{y}_i$ and the constants a_1 and a_2 so as to match

$$y_{i+1} = y_i + h[a_1 f(t_i, y_i) + a_2 f(\hat{t}_i, \hat{y}_i)] \tag{3.4.1}$$

with the Taylor expansion

$$\begin{aligned} y(t_{i+1}) &= y_i + h\left[y'(t_i) + y''(t_i)\frac{h}{2} + y'''(t_i)\frac{h^2}{6} + \cdots \right] \\ &= y_i + h\left[f(t_i, y_i) + f^{(1)}(t_i, y_i)\frac{h}{2} + f^{(2)}(t_i, y_i)\frac{h^2}{6} + \cdots \right] \\ &= y_i + h\left[f + (f_t + f_y f)\frac{h}{2} + (f^2 f_{yy} + 2ff_{ty} + f_{tt} + f_t f_y + ff_y^2)\frac{h^2}{6} + \cdots \right]. \end{aligned}$$

In the last expression and in what follows, all arguments of f and its derivatives will be suppressed when they are evaluated at the point (t_i, y_i). It will be convenient to

3. Named after the German mathematicians Carl Runge (1856–1927) and Wilhelm Kutta (1867–1944).

express $\hat{t}_i$ and $\hat{y}_i$ as

$$\hat{t}_i = t_i + b_1 h, \qquad \hat{y}_i = y_i + b_2 hf,$$

and so the object is to match

$$\theta = a_1 f + a_2 f(t_i + b_1 h, y_i + b_2 hf) \tag{3.4.2}$$

with

$$f + \frac{h}{2}(f_t + f_y f) + \frac{h^2}{6}(f^2 f_{yy} + 2ff_{ty} + f_{tt} + f_t f_y + ff_y^2) + \cdots . \tag{3.4.3}$$

Expanding (3.4.2) in a Taylor series about (t_i, y_i), we have

$$\begin{aligned}\theta &= (a_1 + a_2)f + a_2 h(b_1 f_t + b_2 ff_y) \\ &\quad + \frac{a_2 h^2}{2}(b_2^2 f^2 f_{yy} + 2b_1 b_2 ff_{ty} + b_1^2 f_{tt}) + O(h^3).\end{aligned} \tag{3.4.4}$$

Equating coefficients of like powers of h in (3.4.4) and in the previous Taylor expansion (3.4.3) gives

$$\begin{aligned}h^0&: \quad a_1 + a_2 = 1; \\ h^1&: \quad a_2 b_1 = \tfrac{1}{2}, \quad a_2 b_2 = \tfrac{1}{2}; \\ h^2&: \quad a_2 b_2^2 = \tfrac{1}{3}, \quad a_2 b_1 b_2 = \tfrac{1}{3}, \quad a_2 b_1^2 = \tfrac{1}{3}; \\ &\quad ? = \frac{f_t f_y}{6}, \quad ? = \frac{ff_y^2}{6},\end{aligned}$$

where the question marks indicate that there are no terms in (3.4.4) that match up with the given terms in the Taylor expansion. Therefore a match up can only be obtained through first powers of h. This means that the two-point evaluation will agree *exactly* with the result obtained from the Taylor series method of order $p = 2$. The relations for h^0 and h^1 give three equations in the four unknowns a_1, a_2, b_1, b_2; choose $a_2 = c$, an arbitrary parameter. Then the system of equations

$$a_1 + a_2 = 1, \qquad a_2 b_1 = \tfrac{1}{2}, \qquad a_2 b_2 = \tfrac{1}{2}$$

can be solved exactly to give

$$a_2 = c, \quad a_1 = 1 - c, \quad b_1 = b_2 = \frac{1}{2c}, \quad c \neq 0.$$

Hence

$$\theta = (1 - c)f(t_i, y_i) + cf\left(t_1 + \frac{h}{2c}, y + \frac{h}{2c} f(t_i, y_i)\right)$$

gives a method of order $p = 2$ if $c \neq 0$ and f is sufficiently smooth. We state this in the following algorithm.

Runge–Kutta algorithm of order $p = 2$

To obtain an approximate solution of order $p = 2$ to the initial value problem (3.2.1, 3.2.2) on $[a, b]$, let $h = (b - a)/n$ and generate the sequences

$$y_{i+1} = y_i + h\left[(1 - c)f(t_i, y_i) + cf\left(t_i + \frac{h}{2c}, y_i + \frac{h}{2c} f(t_i, y_i)\right)\right],$$

$$t_{i+1} = a + (i + 1)h, \quad i = 0, 1, \ldots, n - 1,$$

where $c \neq 0$, $t_0 = a$, $y_0 = y(a)$; y_{i+1} is an approximation to $y(t_i)$.

If $c = \frac{1}{2}$, we recognize this as the improved Euler method discussed in Chapter 1. The choice $c = 1$ gives another method called the *Euler–Cauchy method.*

EXAMPLE 1 Approximate the solution to $y' = y^2 + 1$, $y(0) = 0.0$ at $t = 0.1, 0.2, \ldots, 1.0$ using the Euler–Cauchy method with $h = 0.1$.

SOLUTION The recursion relation for y_i is

$$y_{i+1} = y_i + hf\left(t_i + \frac{h}{2}, y_i + \frac{h}{2} f(t_i, y_i)\right) \tag{3.4.5}$$

$$= y_i + h\left\{1 + \left[y_i + \frac{h}{2}(1 + y_i^2)\right]^2\right\}, \qquad y_0 = 0.0, \tag{3.4.6}$$

and the resulting approximations are given in Table 3.4. How does this compare with the actual solution, $y(t) = \tan t$? ■

Table 3.4 **Approximate solution to $y' = y^2 + 1$, $y(0) = 0.0$ with $h = 0.1$**

t_i	y_i	t_i	y_i	t_i	y_i
0.0	0.00000	0.4	0.42223	0.8	1.02534
0.1	0.10025	0.5	0.54539	0.9	1.25256
0.2	0.20252	0.6	0.68263	1.0	1.54327
0.3	0.30900	0.7	0.83977		

Higher order Runge–Kutta formulas can be derived in a similar manner by considering linear combinations of 3, 4, or more evaluations of f and trying to match them up with the appropriate Taylor series expansion of $y(t)$. Notice that one evaluation of f produced a method of order $p = 1$ and two evaluations produced a method of order $p = 2$. In fact, k evaluations per step produce methods of order $p = k$ for $k = 1, 2, 3$, and 4 but not for $k = 5$. For this reason fourth order formulas are quite common, and since the local error is $O(h^5)$, they are quite accurate. As in the second order case, where the parameter c was arbitrary, there is a family of fourth order formulas that depend on several arbitrary parameters. One choice of these parameters leads to the so-called *classical* formulas and a Runge–Kutta algorithm of order $p = 4$.

Runge–Kutta algorithm of order $p = 4$

To obtain an approximate solution of order $p = 4$ to the initial value problem (3.2.1, 3.2.2) on $[a, b]$, let $h = (b - a)/n$ and generate the sequences

$$
\begin{aligned}
y_{i+1} &= y_i + \frac{h}{6}(k_1 + 2k_2 + 2k_3 + k_4), \\
t_{i+1} &= t_0 + (i + 1)h, \qquad i = 0, 1, \ldots, n - 1,
\end{aligned}
$$

where

$$
\begin{aligned}
k_1 &= f(t_i, y_i), \\
k_2 &= f\left(t_i + \frac{h}{2}, y_i + \frac{h}{2}k_1\right), \\
k_3 &= f\left(t_i + \frac{h}{2}, y_i + \frac{h}{2}k_2\right), \\
k_4 &= f(t_i + h, y_i + hk_3),
\end{aligned}
$$

and $t_0 = a$, $y_0 = y(a)$. At $t = t_i$, y_i is an approximation to $y(t_i)$.

EXAMPLE 2 Approximate the solution to

$$y' = 2ty^2, \qquad y(0) = 0.5,$$

using the classical fourth order Runge–Kutta formula with $h = 0.1$ and $n = 10$. Compare the output with Table 3.3.

SOLUTION For the reader's benefit, the first two steps are calculated below. The output from the algorithm is given step by step in Table 3.5.

Table 3.5 Approximate solution to $y' = 2ty^2$, $y(0) = 0.5$ with $h = 0.1$

t	y	k_1	k_2	k_3	k_4	$\frac{h}{6}(k_1 + 2k_2 + 2k_3 + k_4)$
0.0000	0.5000	0.0000	0.0250	0.0251	0.0505	0.0025
0.1000	0.5025	0.0505	0.0765	0.0769	0.1041	0.0077
0.2000	0.5102	0.1041	0.1328	0.1336	0.1645	0.0134
0.3000	0.5236	0.1645	0.1980	0.1992	0.2363	0.0199
0.4000	0.5435	0.2363	0.2775	0.2796	0.3265	0.0280
0.5000	0.5714	0.3265	0.3800	0.3835	0.4462	0.0383
0.6000	0.6098	0.4462	0.5194	0.5254	0.6141	0.0525
0.7000	0.6623	0.6140	0.7203	0.7314	0.8653	0.0730
0.8000	0.7353	0.8651	1.0304	1.0524	1.2717	0.1050
0.9000	0.8403	1.2711	1.5523	1.6010	2.0017	0.1597
1.0000	0.9999					

Step 1.

$$k_1 = f(0, 0.5) = 2(0)(0.5)^2 = 0.0000,$$

$$k_2 = f\left(0 + \frac{0.1}{2}, 0.5 + \frac{0.1}{2}k_1\right) = 2(0.05)(0.5)^2 = 0.0250,$$

$$k_3 = f\left(0 + \frac{0.1}{2}, 0.5 + \frac{0.1}{2}k_2\right) = 2(0.05)(0.5013)^2 = 0.0251,$$

$$k_4 = f(0.1, 0.5 + 0.1k_3) = 2(0.1)(0.5025)^2 = 0.0505,$$

$$y_1 = 0.5 + \frac{0.1}{6}(k_1 + 2k_2 + 2k_3 + k_4) = 0.5025,\ t_1 = 0.1,$$

Step 2.

$$k_1 = f(0.1, 0.5025) = 2(0.1)(0.5025)^2 = 0.0505,$$

$$k_2 = f\left(0.1 + \frac{0.1}{2}, 0.5025 + \frac{0.1}{2}k_1\right) = 2(0.15)(0.5050)^2 = 0.0765,$$

$$k_3 = f\left(0.1 + \frac{0.1}{2}, 0.5025 + \frac{0.1}{2}k_2\right) = 2(0.15)(0.5063)^2 = 0.0769,$$

$$k_4 = f(0.2, 0.5025 + 0.1k_3) = 2(0.2)(0.5102)^2 = 0.1041,$$

$$y_2 = 0.5025 + \frac{0.1}{6}(k_1 + 2k_2 + 2k_3 + k_4) = 0.5102,\ t_2 = 0.2, \text{ etc.}$$

As one would expect, the result is considerably more accurate than that obtained by using the Taylor algorithm of order $p = 2$. In fact, the absolute error here at $t = 1$ is 5.5×10^{-6} (the correct answer to seven significant digits is 0.9999945). ■

As we pointed out in the beginning of the chapter, the methods developed here extend readily to systems of first order differential equations. If we want to approximate the solution to the initial value problem

$$\mathbf{y}' = \mathbf{f}(t, \mathbf{y}), \tag{3.4.7}$$

$$\mathbf{y}(a) = \mathbf{y}_0 \tag{3.4.8}$$

where $\mathbf{y}$, $\mathbf{f}$, and $\mathbf{y}_0$ are N dimensional vectors, the Runge–Kutta algorithm of order $p = 4$ is easily extended.

Runge–Kutta algorithm of order $p = 4$ for systems of first order equations

To obtain an approximate solution of order $p = 4$ to the initial value problem (3.4.7, 3.4.8) on $[a, b]$, let $h = (b - a)/n$ and generate the sequences

$$\mathbf{y}_{i+1} = \mathbf{y}_i + \frac{h}{6}(\mathbf{k}_1 + 2\mathbf{k}_2 + 2\mathbf{k}_3 + \mathbf{k}_4),$$

$$t_{i+1} = t_0 + (i + 1)h, \qquad i = 0, 1, 2, \ldots, n - 1,$$

where

$$\begin{aligned} \mathbf{k}_1 &= \mathbf{f}(t_i, \mathbf{y}_i), \\ \mathbf{k}_2 &= \mathbf{f}\left(t_i + \frac{h}{2}, \mathbf{y}_i + \frac{h}{2}\mathbf{k}_1\right), \\ \mathbf{k}_3 &= \mathbf{f}\left(t_i + \frac{h}{2}, \mathbf{y}_i + \frac{h}{2}\mathbf{k}_2\right), \\ \mathbf{k}_4 &= \mathbf{f}(t_i + h, \mathbf{y}_i + h\mathbf{k}_3), \end{aligned} \tag{3.4.9}$$

and $t_0 = a$, $\mathbf{y}_0 = \mathbf{y}(a)$. At $t = t_i$, $\mathbf{y}_i$ is an approximation to $\mathbf{y}(t_i)$.

EXAMPLE 3 Write out the necessary formulas to extend the Runge–Kutta algorithm of order $p = 4$ to approximate the solution to the initial value problem

$$y_1' = y_2, \qquad y_2' = y_1 + t,$$

$$y_1(0) = 1, \qquad y_2(0) = 1.$$

SOLUTION First of all, using the notation of (3.4.7, 3.4.8) we can write

$$\mathbf{y} = \begin{bmatrix} y_1 \\ y_2 \end{bmatrix}, \qquad \mathbf{f} = \begin{bmatrix} f_1 \\ f_2 \end{bmatrix} = \begin{bmatrix} y_2 \\ y_1 + t \end{bmatrix}, \qquad \mathbf{y}_0 = \begin{bmatrix} 1 \\ 1 \end{bmatrix}.$$

So

$$\mathbf{k}_1 = \mathbf{f}(t, \mathbf{y}) = \begin{bmatrix} y_2 \\ y_1 + t \end{bmatrix},$$

which implies

$$\mathbf{y} + \frac{h}{2}\mathbf{k}_1 = \begin{bmatrix} y_1 \\ y_2 \end{bmatrix} + \frac{h}{2}\begin{bmatrix} y_2 \\ y_1 + t \end{bmatrix} = \begin{bmatrix} y_1 + \frac{h}{2} y_2 \\ y_2 + \frac{h}{2}(y_1 + t) \end{bmatrix},$$

and therefore

$$\mathbf{k}_2 = \mathbf{f}\left(t + \frac{h}{2}, \mathbf{y} + \frac{h}{2}\mathbf{k}_1\right) = \begin{bmatrix} y_2 + \frac{h}{2}(y_1 + t) \\ y_1 + \frac{h}{2} y_2 + t + \frac{h}{2} \end{bmatrix},$$

etc. At the ith step, the values of t, y_1, and y_2 are assumed to be evaluated at $(t_i, \mathbf{y}_i)$. ■

The methods given also apply to higher order differential equations in which the highest order derivative can be solved for explicitly since equations of this type can always be written in the form (3.4.7, 3.4.8).

EXAMPLE 4 Write the initial value problem

$$y'' + ty' + y = \sin t, \qquad y(0) = 1, \qquad y'(0) = 2$$

in the form (3.4.7, 3.4.8).

SOLUTION Let

$$y_1 = y, \qquad y_2 = y';$$

then

$$y_1' = y' = y_2,$$
$$y_2' = y'' = -y_1 - ty_2 + \sin t$$

with $y_1(0) = 1$ and $y_2(0) = 2$.

Using the notation of (3.4.7, 3.4.8), we get

$$\mathbf{y} = \begin{bmatrix} y_1 \\ y_2 \end{bmatrix}, \qquad \mathbf{f} = \begin{bmatrix} f_1 \\ f_2 \end{bmatrix} = \begin{bmatrix} y_2 \\ -y - ty_2 + \sin t \end{bmatrix}, \qquad \mathbf{y}_0 = \begin{bmatrix} 1 \\ 2 \end{bmatrix}. \qquad \blacksquare$$

The Fortran subroutine RK4, which implements the 4th order Runge–Kutta algorithm, is given in Fig. 3.1. It is followed by a sample program in Example 5, which illustrates its use. For convenience we have included the name of the subroutine F (defining the differential equations to be solved) in the argument list of RK4. This allows the user to choose the name for the subroutine. This could be useful, for example, if two different differential equations were to be solved in the same program. To illustrate this capability, we have used the name DERIV for the differential equation subroutine F in Example 5. A programming point worth noting here is that whenever the name of a subroutine appears in the argument list of another subroutine in a Fortran program, that name must appear in an EXTERNAL statement in the calling program.

EXAMPLE 5 Using the subroutine RK4, set up a program to approximate the solution to the initial value problem

$$y_1' = y_2, \qquad y_2' = y_1, \qquad y_1(0) = 1.0, \qquad y_2(0) = 1.0.$$

Use a step size of $h = 0.1$ and print out the solution at $t = 0.0$, 0.5, and 1.0 (see Fig. 3.2). ■

An issue of practical importance arises here. How does one intelligently pick a step size h for any of the numerical methods, and for a given h what can be said about the accuracy of the computed result? It can be shown that for any of the methods given, the error at any fixed point has the qualitative behavior illustrated in Fig. 3.3.

```
      SUBROUTINE RK4(F,T,Y,H,N)
C
C SUBROUTINE RK4 IMPLEMENTS THE CLASSICAL 4TH ORDER
C RUNGE-KUTTA FORMULAS OVER ONE STEP OF LENGTH H.
C
C INPUT:
C
C     F - NAME OF A SUBROUTINE SUBPROGRAM DEFINING THE
C         DIFFERENTIAL EQUATIONS. F MUST HAVE THE FORM
```

Figure 3.1 A listing of the subroutine RK4.

```
C
C             SUBROUTINE F(T,Y,YP,N)
C             REAL Y(N),YP(N),T,Y
C             INTEGER N
C             YP(1)= . . .
C                 .
C                 .
C                 .
C             YP(N)= . . .
C             RETURN
C             END
C       T - VALUE OF THE INDEPENDENT VARIABLE
C       Y - N-DIMENSIONAL SOLUTION VECTOR EVALUATED AT T
C       H - STEP SIZE TO BE USED (H.NE.0)
C       N - NUMBER OF COMPONENTS IN SOLUTION VECTOR Y
C
C OUTPUT:
C
C       T - NEW VALUE OF INDEPENDENT VARIABLE (INPUT VALUE
C           PLUS H)
C       Y - SOLUTION VECTOR EVALUATED AT T
C
C NOTE: Y AND T MUST APPEAR AS VARIABLES IN THE CALLING
C PROGRAM, AND F MUST APPEAR IN AN EXTERNAL STATEMENT IN
C THE CALLING PROGRAM.
C
      REAL T,Y(N),H,K1(10),K2(10),K3(10), K4(10), YT(10)
      INTEGER N,I
C
      CALL F(T,Y,K1,N)
      DO 10 I=1,N
          YT(I)=Y(I)+0.5*H*K1(I)
   10 CONTINUE
      CALL F (T+0.5*H,YT,K2,N)
      DO 20 I=1,N
          YT(I)=Y(I)+0.5*H*K2(I)
   20 CONTINUE
      CALL F(T+0.5*H,YT,K3,N)
      DO 30 I=1,N
          YT(I)=Y(I)+H*K3(I)
   30 CONTINUE
      CALL F(T+H,YT,K4,N)
      DO 40 I=1,N
          Y(I)=Y(I)+H*(K1(I)+2.0*(K2(I)+K3(I))+K4(I))/6.0
   40 CONTINUE
      T=T+H
      RETURN
      END
```

Figure 3.1 (*continued*)

```
C SAMPLE PROGRAM TO ILLUSTRATE SUBROUTINE RK4
C
      REAL Y(2),T,H
      INTEGER NEQN,NP,NPRINT,I
      EXTERNAL DERIV
C
      Y(1)=1.0
      Y(2)=1.0
      T=0.0
      WRITE(*,10)T,Y(1),Y(2)
      H=0.1
      NEQN=2
      NP=0
      NPRINT=5
      DO 5 I=1,10
         NP=NP+1
         CALL RK4(DERIV,T,Y,H,NEQN)
         IF(NP.LT.NPRINT)GO TO 5
         WRITE(*,10)T,Y(1),Y(2)
         NP=0
   5  CONTINUE
      STOP
  10  FORMAT(1X,F10.2,2E15.4)
      END
C
      SUBROUTINE DERIV(T,Y,YP,N)
      REAL T,Y(N),YP(N)
      INTEGER N
      YP(1)=Y(2)
      YP(2)=Y(1)
      RETURN
      END
```

T	Y(1)	Y(2)
0.	0.1000E+01	0.1000E+01
0.50	0.1649E+01	0.1649E+01
1.00	0.2718E+01	0.2718E+01

Figure 3.2 **Sample program and output using RK4 to solve the problem of Example 5.**

Values of $h < h^*$ tend to produce poor results due to round-off error (error induced because the arithmetic is not exact), while values of $h > h^*$ produce poor results due to discretization error (error induced because the algorithm is not exact).

In practice, the following simple approach is useful: Choose a reasonable step size h, for instance 0.1, and perform the integration. Reduce h by a factor of 2 and

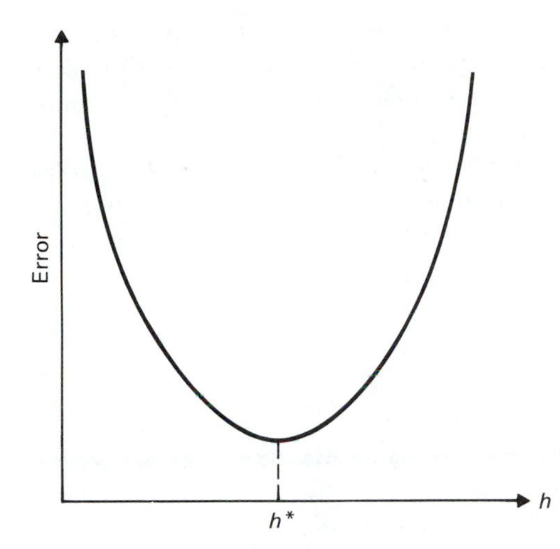

Figure 3.3 Qualitative behavior of absolute error at a fixed point as a function of the step size h.

integrate again. Compare the two solutions at all common values of t for which the solution is desired. If they agree to the required accuracy, accept the second result. If not, halve again, integrate, and compare. Continue until two consecutive results agree to the required accuracy.

Of course, if the choice $h = 0.1$ produces a solution that agrees with the solution for $h = 0.05$ to more accuracy than you need, a larger h value should be considered to start with. In general, this procedure should keep the step size h as large as possible, so as to avoid round-off error problems. It should be pointed out, however, that this process gives an indication of the accuracy of the computed results, *but not a proof*.

Another point of some practical importance should be made. What if a formula of order $p = 2$ is used to solve an initial value problem whose solution has two continuous derivatives, but not three? Examination of the local truncation error will show that the formula is then of order $p = 1$ and therefore the rate of convergence is of order h and not h^2. The point here is that numerical procedures can be used on problems whose solutions are not smooth, but the rate of convergence may be reduced.

A common example of problems whose solutions will not be smooth are those where the coefficients have a jump discontinuity at some point. In numerically solving such a problem one should *not* integrate across the point. For example, suppose we are solving the problem

$$y'' + y = f(t), \quad 0 \leq t \leq 2, \quad y(0) = 0, \quad y'(0) = 1,$$
$$f(t) = \begin{cases} 1 \text{ for } 0 \leq t < 1, \\ 0 \text{ for } 1 \leq t \leq 2. \end{cases}$$

A good procedure would be to integrate from $t = 0$ to $t = 1$ and then from $t = 1$ to $t = 2$, and not to integrate across the discontinuity in $f(t)$ at $t = 1$. On each subinterval ($0 \leq t \leq 1$ and $1 \leq t \leq 2$), the differential equation has smooth solutions and convergence rates will be as advertised.

In Chapter 8 we will discuss numerical methods for solving initial value problems in more detail. There we will analyze the types of errors that are present in numerical solutions and will discuss techniques for estimating and controlling some of them.

EXERCISES 3.4

1. Derive the expansion (3.4.4) in the text. (*Hint:* Proceed by a succession of one-variable expansions, e.g.,

$$f(t + \Delta, y + \delta) = f(t, y + \delta) + f_t(t, y + \delta)\,\Delta + \cdots$$
$$= f(t, y) + f_y(t, y)\,\delta + f_t(t, y)\,\Delta + f_{ty}(t, y)\,\Delta\delta + \cdots.)$$

Write out the system of equations (3.4.9) as in Example 3 for each of the initial value problems in Exercises 2–5.

2. $y_1' = y_2, \quad y_2' = y_1, \quad y_1(0) = 1, \quad y_2(0) = 0$

3. $y_1' = -4y_1 - y_2, \quad y_2' = y_1 - 2y_2, \quad y_1(0) = 0, \quad y_2(0) = -1$

4. $y'' + ty' + y = \sin t, \quad y(0) = 1, \quad y'(0) = 2$

5. $y'' + y' = \ln(t), \quad y(1) = 0, \quad y'(1) = -1$

Approximate the solution to the initial value problems in Exercises 6–10 at $t = 0.3$ by using the Euler–Cauchy method with step size $h = 0.1$. Do your calculations by hand and round all results to four decimal places. Compare your numerical results with the true solution at $t = 0.1$, 0.2, and 0.3.

6. $y' = y, \quad y(0) = 1$

7. $y' = t - y, \quad y(0) = 0$

8. $y' = 1 + y^2, \quad y(0) = 0$

9. $y' = -y + t + 4, \quad y(0) = 2$

10. $y' = \dfrac{4t}{y} - ty, \quad y(0) = 3$

Approximate the solution to the initial value problems in Exercises 11–15 at $t = 0.3$ by using the classical Runge–Kutta method with step size $h = 0.1$. Do your calculations by hand and round all results to four decimal places. Organize your calculations as in Example 2. Compare your numerical results with the true solution at $t = 0.1$, 0.2, and 0.3 and with the corresponding results in Exercises 6–10.

11. $y' = y, \quad y(0) = 1$
12. $y' = t - y, \quad y(0) = 0$
13. $y' = 1 + y^2, \quad y(0) = 0$
14. $y' = -y + t + 4, \quad y(0) = 2$
15. $y' = \dfrac{4t}{y} - ty, \quad y(0) = 3$

Use the subroutine RK4 to obtain an approximate solution to the initial value problems in Exercises 16–31. Use $h = 0.1$ and tabulate each solution at the t-values indicated.

16. $y' = -4y + 1, \quad y(0) = 2; \quad t = 0.0, 0.5, 1.0$
17. $y' = y - t, \quad y(0) = 2; \quad t = 0.0, 0.5, 1.0$
18. $y' = 1 + y^2, \quad y(0) = 0; \quad t = 0.0, 0.5, 1.0$
19. $y' = 1 - 2ty, \quad y(0) = 0; \quad t = 0.0, 0.5, 1.0$
20. $y' = ty^{1/3}, \quad y(0) = 1; \quad t = 0.0, 0.5, 1.0$
21. $y' = -ty^2, \quad y(0) = 2; \quad t = 0.0, 0.5, 1.0$
22. $y'' - 3y' + 2y = e^{-t}, \quad y(1) = 0, \quad y'(1) = 0; \quad t = 1.0, 1.5, 2.0$
23. $y'' + 4y' + 8y = \sin t, \quad y(0) = 1, \quad y'(0) = 0; \quad t = 0.0, 0.2, 0.4, \ldots, 1.0$
24. $y'' + y = \sin t, \quad y(0) = 0, \quad y'(0) = 2; \quad t = 0.0, 1.0, 2.0, 3.0$
25. $y'' + y' = \ln t, \quad y(1) = 0, \quad y'(1) = -1; \quad t = 1.0, 2.0, \ldots, 10.0$
26. $y'' + 6y' + 2y = t^2 + 1, \quad y(0) = 1, \quad y'(0) = 1; \quad t = 0.0, 0.5, 1.0$
27. $x' = -4x - y, \quad x(0) = 0, \quad y' = x - 2y, \quad y(0) = -1; \quad t = 0.0, 0.1, \ldots, 1.0$
28. $x' = 2x + 6y, \quad x(0) = 1, \quad y' = -2x - 5y, \quad y(0) = 1; \quad t = 0.0, 0.3, 0.6, \ldots, 5.0$
29. $x' = -3x - 4y, \quad x(0) = 0, \quad y' = 3x + 2y, \quad y(0) = 1; \quad t = 0.0, 0.2, 0.4, \ldots, 1.0$
30. $x' = y, \quad x(0) = 0, y' = 3x + 2y, \quad y(0) = 1; \quad t = 0.0, 0.5, 1.0$
31. $x' = 3x - 2y, \quad x(0) = 2, \quad y' = -2x + 6y, \quad y(0) = 0; \quad t = 0.0, 0.5, 1.0$

SUMMARY

A general algorithm for generating approximate solutions to a first order initial value problem can be written

$$y_{i+1} = y_i + h\theta(t_i, y_i),$$

where y_i denotes the approximate solution at $t = t_i$. The local error, $h\tau_i$, at t_i is defined by the equation,

$$y(t_{i+1}) = y(t_i) + h\theta(t_i, y(t_i)) + h\tau_i,$$

where $y(t_i)$ denotes the actual value of the solution at t_i. If $\tau_i = O(h^p)$, the algorithm is said to be of order p. Although the local error is $O(h^{p+1})$, because of the accumulation of local errors, the actual error of the approximated solution at a fixed point t is $O(h^p)$. For instance, the Euler algorithm is of order $p = 1$, and the improved Euler algorithm is of order $p = 2$.

In this chapter, a method is first developed which can produce an algorithm of any desired order. It is based on the fact that, given the initial value problem

$$\frac{dy}{dt} = f(t, y), \qquad y(t_0) = y_0,$$

then if $f(t, y)$ is sufficiently smooth one can compute by implicit differentiation the values of $y'(t_0)$, $y''(t_0)$, etc. Consequently, one can form the Taylor polynomial of order p which approximates the solution near $t = t_0$. This leads to the Taylor series algorithm of order p.

The second class of numerical methods developed in this chapter are the Runge–Kutta methods, which are some of the most widely used approximation schemes. They have the advantage of not requiring explicit evaluations of higher derivatives and use instead linear combinations of values of $f(t, y)$ to approximate the solution. The Runge–Kutta method of order $p = 2$ is derived, and the algorithm for the Runge–Kutta method of order $p = 4$ is given. This method is very popular because it is easily programmed and gives high accuracy for small step sizes h.

MISCELLANEOUS EXERCISES

3.1. Consider the problem

$$y' = 2|t|y, \qquad y(-1) = 1/e$$

on the interval $[-1,1]$. Find the analytical solution $y(t)$ and show that $y'(t)$ is continuous on $[-1, 1]$ but $y''(t)$ is not. Study the behavior of the error of Euler's method at some fixed points in $[-1, 1]$ as $h \to 0$.

3.2. When $f(t, y)$ depends only on t, show that the classical fourth order Runge–Kutta formula reduces to Simpson's rule

$$\int_a^{a+h} f(t)\, dt = \frac{h}{6}\left[f(a) + 4f\left(a + \frac{h}{2}\right) + f(a + h)\right].$$

What is the order of Simpson's rule; i.e., what is the highest degree polynomial $P(t)$ which the rule integrates exactly? To what quadrature rule does the Runge–Kutta method of order 2 correspond when $c = \frac{1}{2}$? When $c = 1$?

3.3. The differential equation $y' = y - 0.5y^2$ represents a simple population growth model. Let $y(0) = 1$.

a) Write a program using RK4 to find an approximate solution on $[0, 2]$. Obtain four-figure accuracy.

b) Solve the problem analytically and compare with your results in (a).

3.4. Consider a pendulum consisting of a mass m at the end of a weightless rod of length l. Assume that there is no air resistance and no friction at the pivot point. Physical arguments similar to those in Chapter 2 can be used to show that the equation of motion is

$$\frac{d^2\theta}{dt^2} + \frac{g}{l}\sin\theta = 0, \tag{3.5.1}$$

where g is the acceleration of gravity and $\theta = \theta(t)$ is the angular displacement (in radians) of the pendulum from the vertical equilibrium position at time t. (Notice that the mass m does not occur in the equation.) Henceforth take $g/l = 4.0$ sec^{-2}.

a) Obtain numerical solutions to (3.5.1) for $0 \leq t \leq 10$ by using RK4 with $h = 0.1$ and the initial conditions $\theta(0) = \theta_0$ and $\theta'(0) = 0$, where $\theta_0 = \pi/20, \pi/10, \pi/5, \pi/2$. You will discover (as suggested by physics) that the motion is periodic. Find an approximate value for the period and construct phase plane plots in the (θ, θ') plane when $\theta_0 = \pi/10$ and $\theta_0 = \pi/2$.

b) If $\theta = \theta(t)$ is small, $(g/l)\sin\theta$ can be replaced by $g\theta/l$ in (3.5.1). For the cases $\theta_0 = \pi/20$ and $\theta_0 = \pi/10$ solve this problem analytically and compare your results with the numerical results you obtained in (a).

c) If viscous damping due to air resistance or friction at the pivot is added, Eq. (3.5.1) becomes

$$\frac{d^2\theta}{dt^2} + c\frac{d\theta}{dt} + \frac{g}{l}\sin\theta = 0.$$

Obtain numerical solutions with the same initial conditions as in (a), where $\theta_0 = \pi/10, \pi/5$, and $\pi/2$, and $c = 1.0$ sec^{-1}. Construct a phase plane plot in the case $\theta_0 = \pi/2$.

3.5. A very simple system of differential equations, which has been used to model the growth of two populations where one is a predator and the other its prey, is

$$\frac{dR}{dt} = R(a - bP), \qquad \frac{dP}{dt} = P(-c + dR),$$

where a, b, c, and d are all positive constants. Here $P = P(t)$ is the number of predators (measured in thousands) at time t (measured in years); $R = R(t)$ is the prey population similarly measured. For $a = 5.0$, $b = 1.0$, $c = 4.0$, $d = 0.8$ use RK4 with $h = 0.1$ for $0.0 \leq t \leq 2.0$ to approximate $R(t)$ and $P(t)$ numerically when

a) $R(0) = 10, \quad P(0) = 1.$

b) $R(0) = 5, \quad P(0) = 6.$

c) $R(0) = 1, \quad P(0) = 10.$

In each case construct a phase plane plot in the RP-plane. (The biological assumptions leading to such a model will be discussed in Chapter 7.)

3.6. Approximate the solution to the second order differential equation

$$y'' + t^2y' + (\sin t)y = 3e^t,$$

$$y(0) = y'(0) = 1,$$

using RK4 with $h = 0.1$. Tabulate your results for $t = 0.0, 0.2, \ldots, 1.0$.

3.7. An interesting fact about Runge–Kutta methods is that the error depends on the form of the equation as well as on the solution itself. Verify that $y(t) = (t - 1)^2$ is a solution to both initial value problems

$$y' = 2(t - 1), \qquad y(0) = 1$$

and

$$y' = \frac{2y}{t - 1}, \qquad y(0) = 1.$$

Show that the Euler–Cauchy method is exact for the first equation but not for the second.

3.8. Consider the initial value problem for the damped pendulum

$$\frac{d^2\theta}{dt^2} + \frac{d\theta}{dt} + 4 \sin \theta = 0,$$

$$\theta(0) = \frac{\pi}{10}, \qquad \theta'(0) = 0.$$

Write a program using the subroutine RK4 with step size $h = 0.05$ to approximate the first time $\hat{t}$ when the pendulum reaches the rest position $\theta(\hat{t}) = 0$.

3.9. Let $y(T)$ and $Y(h)$ denote exact and approximate solutions at $t = T$ to an initial value problem, $y' = f(t, y)$, $y(t_0) = y_0$. Suppose that if h is sufficiently small, the error can be estimated by

$$y(T) \cong Y(h) + kh^m,$$

where k is a constant and m depends on the numerical method used to compute the approximate solution. Replace h by $2h$ in this estimate and derive the extrapolation formula,

$$y(T) \cong Y(h) + \frac{1}{(2^m - 1)} [Y(h) - Y(2h)].$$

3.10. Compare exact and approximate solutions at $t = 1$ to the initial value problem,

$$\frac{dy}{dt} = \frac{e^t}{y}, \qquad y(0) = 1,$$

with $h = \frac{1}{8}$ and $h = \frac{1}{16}$ using the Euler, improved Euler, and classical Runge–Kutta methods. Then compute

$$y(1) \cong Y\left(\frac{1}{16}\right) + \frac{1}{(2^m - 1)} \left[Y\left(\frac{1}{16}\right) - Y\left(\frac{1}{8}\right) \right]$$

with $Y(\frac{1}{16})$ and $Y(\frac{1}{8})$ the approximate values generated by the different methods. Take $m = 1$ for the Euler method, $m = 2$ for the improved Euler method, and $m = 4$ for the classical Runge–Kutta method. Use as many significant figures as possible. Discuss your results.

3.11. Linear resonance of an undamped mass–spring system can be suppressed by a suitable periodic instantaneous reduction of the velocity. As was shown in Section 2.11, all solutions to the forced-oscillator equation,

$$\frac{d^2y}{dt^2} + 4\pi^2 y = 4\pi \cos(2\pi t), \tag{3.5.2}$$

are unbounded. The purpose of this exercise is to show that the growth of $y(t)$, the displacement of the mass from equilibrium, can be controlled and to study the effect of this control on numerical approximations to $y(t)$. The controlled displacement will be defined on intervals $j \leq t \leq j + 1$, $j = 0, 1, 2, \ldots$ by $y(t) = y^j(t)$ where the $y^j(t)$ are solutions to (3.5.2) that satisfy the constraints,

$$y^0(0) = 0, \qquad \frac{dy^0}{dt}(0) = 0,$$

$$y^1(1) = y^0(1), \qquad \frac{dy^1}{dt}(1) = \frac{dy^0}{dt}(1) - 2\pi,$$

$$y^j(j) = y^{j-1}(j), \qquad \frac{dy^j}{dt}(j) = \frac{dy^{j-1}}{dt}(j) - 2\pi.$$

a) Find the exact solutions, $\{y^j(t)\}$, $j = 0, 1, 2, \ldots$

b) Use RK4 with $h = \frac{1}{20}$ to compute an approximate solution on $0 \leq t \leq 4$. This can be done by integrating from $t = 0$ to $t = 1$, then from $t = 1$ to $t = 2$, etc. Plot the computed values of $y(t)$ versus t and $y'(t)$ versus t on $0 \leq t \leq 4$. The exact solution is periodic. What can be said about the approximate solution?

c) Plot the difference between the exact and computed values of $y(t)$ versus t and $y'(t)$ versus t on $0 \leq t \leq 4$. Discuss the differences between the graphs on $0 \leq t \leq 1$ and $3 \leq t \leq 4$.

3.12. All the solutions to

$$\frac{d^2y}{dt^2} + \tan y = 0$$

are periodic if $-\pi/2 < y(t) < \pi/2$. Use RK4 with $h = \frac{1}{20}$ and compute three approximate solutions $y^1(t)$, $y^2(t)$, and $y^3(t)$ on $0 \leq t \leq 10$ such that

$$y^1(0) = 0, \qquad \frac{dy^1}{dt}(0) = 0.1,$$

$$y^2(0) = 0, \qquad \frac{dy^2}{dt}(0) = 1,$$

$$y^3(0) = 0, \qquad \frac{dy^3}{dt}(0) = 2.$$

Plot the three integral curves in the y, dy/dt plane on the same piece of paper. Does the period of a solution depend on its amplitude?

REFERENCES

A classical introductory text on numerical analysis, which has an excellent discussion of numerical methods for ordinary differential equations in Chapter 14, is

1. P. Henrici, *Elements of Numerical Analysis,* Wiley, New York, 1968.

Another good discussion is Chapter 4 of

2. R.C. Buck and E.F. Buck, *Introduction to Differential Equations,* Houghton Mifflin, Boston, 1976.

More advanced discussions can be found in

3. L.F. Shampine and R.C. Allen, *Numerical Computing: An Introduction,* W.B. Saunders, Philadelphia, 1973.
4. G.E. Forsythe, M.A. Malcolm, and C.B. Moler, *Computer Methods for Mathematical Computations,* Prentice-Hall, Englewood Cliffs, N.J., 1977.

A useful reference is

5. M. Abramowitz and I. Stegun, eds., *Handbook of Mathematical Tables,* No. 55 in NBS Applied Math. Series, U.S. Government Printing Office, Washington, D.C., 1966.

4

THE LAPLACE TRANSFORM

4.1 INTRODUCTION

In this chapter we introduce another technique, which comes under the general classification of *transform methods,* for solving the initial value problem for constant coefficient linear differential equations. Transform methods are used extensively in the analysis of linear systems; the common features of these methods are:

1. The solution of the system $y(t)$, which exists in the t-domain, is transformed into a function $Y(s)$ of another independent variable s; call this the s-domain. The initial conditions are usually incorporated in the transformation.
2. The system of differential or integral equations, which $y(t)$ satisfies in the t-domain, is in turn transformed into a system of algebraic equations in the s-domain. If one can solve these equations for $Y(s)$, one obtains a relation of the form $Y(s) = \Phi(s)$.
3. If one can find a function $\phi(t)$ whose transform is $\Phi(s)$, then one can apply the inverse transformation that takes us from the s-domain to the t-domain and assert that $y(t) = \phi(t)$ is the required solution.

Different transform methods are used, depending on the nature of the t-domain (e.g., the whole line or a half-line) and on the class of functions admissible as solutions (e.g., continuous, differentiable, or analytic). The two most commonly used transforms are the Fourier transform and the Laplace transform, and we will study the latter. It was first introduced in 1782 by the great French mathematician

Pierre Laplace (1749–1827), but its application and the techniques associated with it were not developed until about a hundred years later.

The Laplace transform, used as a technique for solving constant coefficient linear ordinary differential equations, is particularly effective for problems where the forcing function is discontinuous or has corners. For systems of linear differential equations its use obviates some of the algebraic difficulties associated with such systems. But in addition to this, the Laplace transform is used extensively to study input–output relations in systems analysis, to analyze feedback control systems, and to solve certain classes of partial differential equations of mathematical physics. It is one of the important tools of applied mathematics.

4.2 THE LAPLACE TRANSFORM OF e^{at}

Given a function $f(t)$, defined for $0 \leq t < \infty$, define the *Laplace transform* of $f(t)$ by

$$L\{f\} = \int_0^\infty e^{-st} f(t)\, dt = F(s), \tag{4.2.1}$$

where s is a complex variable, whenever the improper integral exists. Recall that the existence of the integral means that the limit

$$\lim_{R\to\infty} \int_0^R e^{-st} f(t)\, dt$$

exists. The fact that the variable s is complex is not significant in many applications, but the reader should be aware that the Laplace transform takes functions $f(t)$ defined on the half line $0 \leq t < \infty$ into functions $F(s)$ defined on the complex plane.

The Laplace transform of $f(t) = e^{at}$ will now be worked out in detail, first of all because of its importance and frequent occurrence, and second, because it easily allows us to point out the interpretation of the variable s. Proceeding formally from the definition (4.2.1) above, we get

$$L\{e^{at}\} = \int_0^\infty e^{-st} e^{at}\, dt = \int_0^\infty e^{-(s-a)t}\, dt.$$

For the transform to exist, the improper integral must exist, so we must determine whether

$$\lim_{R\to\infty} \int_0^R e^{-(s-a)t}\, dt = \frac{1}{s-a} - \lim_{R\to\infty} \frac{e^{-(s-a)R}}{s-a} \tag{4.2.2}$$

exists. Since a is real and s is complex, we let $s = x + iy$ and therefore $s - a = (x - a) + iy$ and

$$e^{-(s-a)R} = e^{-(x-a)R}e^{-iyR} = e^{-(x-a)R}[\cos(yR) - i\sin(yR)].$$

Also note that the modulus of $e^{-(s-a)R}$ is

$$|e^{-(s-a)R}| = e^{-(x-a)R}|\cos(yR) - i\sin(yR)| = e^{-(x-a)R}$$

since $|u + iv| = \sqrt{u^2 + v^2}$. One can now immediately see that the limit in (4.2.2)

1. does not exist if $x - a < 0$, since $\lim_{R\to\infty} e^{-(x-a)R} = \infty$;
2. does not exist if $x - a = 0$, and hence $s - a = iy$, since

$$\lim_{R\to\infty} \frac{\cos(yR) - i\sin(yR)}{iy}$$

   does not exist for any value of y;
3. does exist if $x - a > 0$, since $\lim_{R\to\infty} e^{-(x-a)R} = 0$.

We conclude that if $x = \text{Re}(s) > a$, then

$$\frac{1}{s - a} - \lim_{R\to\infty} \frac{e^{-(s-a)R}}{s - a} = \frac{1}{s - a},$$

and our first Laplace transform is

$$L\{e^{at}\} = F(s) = \frac{1}{s - a}, \qquad \text{Re}(s) > a. \tag{4.2.3}$$

Therefore L takes the real function e^{at} into the complex rational function $1/(s - a)$.

The reader should note the following: If it had been merely assumed that s was real, we would have easily derived the formula

$$L\{e^{at}\} = F(s) = \frac{1}{s - a}, \qquad s > a,$$

since the limit in (4.2.2) exists only if $s - a > 0$. A general rule of thumb is that in computing transforms of elementary functions, one can proceed as if s were real, and in any inequalities defining an s-region replace s by $\text{Re}(s)$ (see Fig. 4.1).

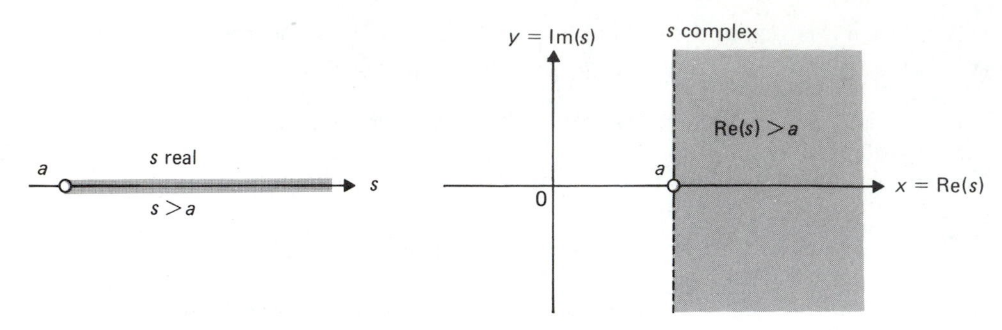

Figure 4.1

EXERCISES 4.2

Find the Laplace transform of each of the functions in Exercises 1–8. Determine and sketch the region in the complex plane in which it is valid.

1. e^{2t} **2.** $e^{-0.5t}$ **3.** e^{-7t} **4.** $4e^{t}$

5. e^{10t} **6.** e^{2-t} **7.** e^{1+3t} **8.** e^{2t-4}

9. By following the analysis given to find $L\{e^{at}\}$ for a real, prove that for k real we have

$$L\{e^{ikt}\} = \frac{1}{s - ik}, \qquad \mathrm{Re}(s) > 0.$$

10. Prove the formula

$$L\{e^{at}\} = \frac{1}{s - a}, \qquad \mathrm{Re}(s) > \mathrm{Re}(a),$$

where a is a complex number.

11. What is $L\{1\}$ and for what region in the complex plane is it valid? What is $L\{6\}$?

12. Using properties of the integral and the formula for $L\{e^{at}\}$, establish the formula

$$L\{Ae^{\alpha t} + Be^{\beta t}\} = \frac{A}{s - \alpha} + \frac{B}{s - \beta},$$

valid for $\mathrm{Re}(s) > \max(\alpha, \beta)$, where α, β are real and A, B are arbitrary constants.

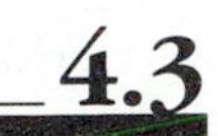

4.3 AN APPLICATION TO FIRST ORDER EQUATIONS

At this point, rather than developing a collection of formulas for the Laplace transform of various functions, we will state two of its important properties and then proceed directly to use it to solve a first order differential equation.

Property of linearity. If $f(t) = Ag(t) + Bh(t)$, where A and B are constants and $g(t)$ and $h(t)$ have Laplace transforms $G(s)$ and $H(s)$, respectively, valid for $\text{Re}(s) > a$, then

$$L\{f(t)\} = F(s) = AL\{g(t)\} + BL\{h(t)\} = AG(s) + BH(s), \qquad \text{Re}(s) > a. \tag{4.3.1}$$

The reader can see that this follows directly from the definition and the linearity of the integral.

Since the Laplace transform is to be used to study ordinary differential equations, we ask the question: Given $L\{f(t)\} = F(s)$ defined for $\text{Re}(s) > a$, can it be used to obtain $L\{f'(t)\}$, the Laplace transform of its derivative? If it is assumed that $f'(t)$ is continuous or piecewise continuous for $0 \leq t < \infty$, then the answer is yes and it depends on the following integration by parts:

$$\int_0^R e^{-st}f'(t)\,dt = e^{-st}f(t)\big|_0^R + s\int_0^R e^{-st}f(t)\,dt$$

$$= e^{-sR}f(R) - f(0) + s\int_0^R e^{-st}f(t)\,dt.$$

The limit of the last integral as $R \to \infty$ is $sF(s)$, valid for $\text{Re}(s) > a$. If it is further assumed that $f(t)$ satisfies $\lim_{R\to\infty} e^{-sR}f(R) = 0$ for $\text{Re}(s) > a$, we obtain the second important property.

The differentiation formula.

$$L\{f'(t)\} = sL\{f(t)\} - f(0) = sF(s) - f(0). \tag{4.3.2}$$

Note that if $F(s)$ exists for $\text{Re}(s) > a$, then so does $L\{f'(t)\}$, and the above result was obtained under the assumptions that $f'(t)$ was continuous or piecewise continuous and $\lim_{t\to\infty} e^{-st} f(t) = 0$ for $\text{Re}(s) > a$. The class of functions that satisfy the last relation are called *functions of exponential order,* and they represent a large class of functions whose Laplace transforms exist. This will be discussed in the next section.

With our meager collection of tools, consisting of the formula (4.2.3) for $L\{e^{at}\}$, the property of linearity (4.3.1), and the differentiation formula (4.3.2), we will show how to use the Laplace transform to solve the initial value problem

$$y' + ky = be^{at}, \qquad y(0) = A, \qquad a \neq -k. \tag{4.3.3}$$

Although the problem is relatively simple, the reader should *carefully* follow each step in the solution process, since the problem will be a prototype for all that follows. Let $y(t)$ be the solution of (4.3.3) and $Y(s) = L\{y(t)\}$ be its Laplace transform. Apply the transform to both sides of the equation, use the property of linearity and the formula for $L\{e^{at}\}$ to obtain

$$L\{y' + ky\} = L\{y'\} + kL\{y\} = L\{be^{at}\} = \frac{b}{s-a}.$$

Now applying the differentiation formula to the first term on the left, we get

$$sL\{y\} - y(0) + kL\{y\} = \frac{b}{s-a},$$

or, since $L\{y\} = Y(s)$ and $y(0) = A$,

$$sY(s) - A + kY(s) = \frac{b}{s-a}.$$

Solving the last expression for $Y(s)$ gives

$$Y(s) = \frac{A}{s+k} + \frac{b}{(s+k)(s-a)}. \tag{4.3.4}$$

Thus, the first step is done, namely, by using the properties of the transform and algebraic operations an expression for $Y(s)$, the transform of the solution $y(t)$ has been obtained.

It is important to note in (4.3.4) that the initial value $y(0) = A$ is incorporated into the expression for $Y(s)$. The fact that the initial data are a part of the transform is an attractive feature of the method since one does not have to perform additional operations at the end to find values of arbitrary constants. Of course the initial data enter through the differentiation formula.

As it stands, the expression for $Y(s)$ is not in a form we recognize because of the second term. But if this term is decomposed by using partial fractions, then

$$Y(s) = \frac{A}{s+k} + \frac{b}{k+a}\left[\frac{1}{s-a} - \frac{1}{s+k}\right]$$

or

$$Y(s) = \left(A - \frac{b}{k+a}\right)\frac{1}{s+k} + \frac{b}{k+a}\frac{1}{s-a}. \tag{4.3.5}$$

Now, using the formula for $L\{e^{at}\}$ and linearity, we see that

$$y(t) = \left(A - \frac{b}{k+a}\right)e^{-kt} + \frac{b}{k+a}e^{at} \tag{4.3.6}$$

is a function whose Laplace transform is $Y(s)$ and that it is the solution of the initial value problem.

The astute reader may well ask how we got from the expression (4.3.5) to the solution (4.3.6) since the Laplace transform only tells us how to get from $y(t)$ to $Y(s) = L\{y(t)\}$. The answer is that associated with the operator L is an operator L^{-1}, called the *inverse Laplace transform,* and it has the following properties:

Uniqueness property. If $f(t)$ is continuous and has a transform $F(s) = L\{f(t)\}$, then

$$L^{-1}\{F(s)\} = f(t).$$

Linearity property. If $g(t)$ and $h(t)$ are continuous and have Laplace transforms $G(s)$ and $H(s)$, respectively, and A and B are constants, then

$$L^{-1}\{AG(s) + BH(s)\} = Ag(t) + Bh(t).$$

One sees now that applying L^{-1} to both sides of (4.3.5) gives us the solution $y(t)$.

We finally remark that, given a certain general class of functions $F(s)$ in the complex plane, there is an integral that permits us to calculate $f(t) = L^{-1}\{F(s)\}$. It is a rather complicated line integral in the complex plane, and if it had to be used each time to get back to the t-domain, then much of the attractiveness of the Laplace transform method would be lost, especially for problems that we can solve using the methods already developed in this book. The uniqueness property above is what saves matters, for it means that for continuous functions $f(t)$ one can go back and forth via a table once $F(s)$ has been calculated. Our table now has only one entry

$f(t)$	$F(s) = L\{f(t)\}$
e^{at}	$\dfrac{1}{s-a}$

or equivalently

$F(s)$	$f(t) = L^{-1}\{F(s)\}$
$\dfrac{1}{s-a}$	e^{at}

In the next section the table will be expanded to include most functions that will arise in the analysis of linear constant coefficient differential equations.

EXERCISES 4.3

Use the Laplace transform to solve the initial value problems in Exercises 1–10.

1. $y' + 3y = 4e^{-2t}, \quad y(0) = 1$

2. $y' - y = e^{-t}, \quad y(0) = -2$

3. $y' = 2e^{3t}, \quad y(0) = 1$

4. $y' + 3y = 6, \quad y(0) = 3$

5. $y' - \frac{3}{2}y = e^{t/2}, \quad y(0) = 0$

6. $y' = e^{2-t}, \quad y(0) = -4$

7. $y' + y = -e^{t}, \quad y(0) = 2$

8. $y' + y = 6e^{1+t}, \quad y(0) = 1$

9. $y' + 4y = 8, \quad y(0) = 0$

10. $y' - 8y = 10, \quad y(0) = 2$

11. a) Find the solution of the initial value problem

$$y' + y = 2e^{-t}, \quad y(0) = 0$$

by using the variation of parameters method or otherwise.

b) Use the Laplace transform in (a) and show that

$$Y(s) = \frac{2}{(s+1)^2}.$$

From this infer that $L\{te^{-t}\} = 1/(s+1)^2$.

c) Generalize the above to find the solution of

$$y' - ky = be^{kt}, \quad y(0) = 0$$

and from this infer that $L\{te^{kt}\} = 1/(s-k)^2$.

Use the result of Exercise 11(c) to solve the initial value problems in Exercises 12–17.

12. $y' + y = 4e^{-t}, \quad y(0) = 1$

13. $y' - 2y = e^{2t}, \quad y(0) = 4$

14. $y' - 3y = 6e^{3t}, \quad y(0) = 0$

15. $y' + 4y = 4e^{-4t}, \quad y(0) = 4$

16. $y' - 4y = 4e^{4t}, \quad y(0) = 4$

17. $y' + 2y = 6e^{3-2t}, \quad y(0) = 0$

Use the Laplace transform and the property of linearity to solve the initial value problems in Exercises 18–23.

18. $y' + 2y = 6e^{t} - 3e^{4t}, \quad y(0) = 1$

19. $y' - y = 1 + 4e^{3t}, \quad y(0) = 5$

20. $y' + y = \cosh t, \quad y(0) = 1$ (see Exercise 11 above).

21. $y' + 4y = 2e^{t} - 4e^{-t}, \quad y(0) = 0$

22. $y' - 2y = 4 + e^{t}, \quad y(0) = 1$

23. $y' + 2y = e^{-t} + 2e^{-2t}, \quad y(0) = 3$ (see Exercise 11 above)

24. Follow the proof of the differentiation formula and integrate by parts twice to prove that

$$L\{f''(t)\} = s^2F(s) - sf(0) - f'(0),$$

where $L\{f(t)\} = F(s)$.

4.4 FURTHER PROPERTIES AND TRANSFORM FORMULAS

In the previous section, when the differentiation formula was derived, we mentioned the class of functions of exponential order. We now define it.

Definition

The class E_α of functions of exponential order α is the set of all continuous functions on $0 \leq t < \infty$ having the property that, if $f(t)$ is in E_α, there exists a constant α and a positive constant M such that

$$|f(t)| \leq Me^{\alpha t}, \qquad 0 \leq t < \infty.$$

In a later section the requirement that members of E_α be continuous will be dropped, but for now such generality is not needed.

In a nutshell, the functions of exponential order α are functions that grow no faster than the exponential $e^{\alpha t}$. For instance,

1. $f(t) = 3e^{2t}$ belongs to E_2 since $|f(t)| \leq 3e^{2t}$, so $M = 3$ and $\alpha = 2$.
2. $f(t) = t^n$ belongs to E_α for any $\alpha > 0$ since

$$\frac{\alpha^n t^n}{n!} \leq 1 + \alpha t + \cdots + \frac{\alpha^n t^n}{n!} + \cdots = e^{\alpha t},$$

   which implies that $|t^n| \leq (n!/\alpha^n)e^{\alpha t}$ and so $M = n!/\alpha^n$.
3. $f(t) = 4e^{-t} \cos 3t$ belongs to E_{-1} since

$$|4e^{-t} \cos 3t| \leq 4e^{-t}.$$

4. $f(t) = e^{t^2}$ does not belong to E_α for any α, since $e^{t^2} \leq Me^{\alpha t}$ implies that $e^{t^2 - \alpha t} \leq M$. But if $\alpha > 0$, then for $t > \alpha$ the exponent $t^2 - \alpha t$ is positive and grows without bound, so the exponential cannot be bounded. If $\alpha \leq 0$, then $t^2 - \alpha t$ is positive for $t > 0$ and, again, the exponential is unbounded.

A little thought shows that the set of all possible solutions of homogeneous constant coefficient linear differential equations is of exponential order. These solutions are linear combinations of functions like e^{at}, $e^{at} \cos bt$, $e^{at} \sin bt$, and integral powers of t times them.

The important property of functions of exponential order is that they are Laplace transformable.

If $f(t)$ belongs to E_α, then $L\{f(t)\} = F(s)$ exists for $\mathrm{Re}(s) > \alpha$.

The proof of this statement follows from these simple estimates:

$$\left| \int_0^\infty f(t)e^{-st}\, dt \right| \leq \int_0^\infty |f(t)|\,|e^{-st}|\, dt$$

$$\leq M \int_0^\infty e^{\alpha t} e^{-\operatorname{Re}(s)t}\, dt = M \int_0^\infty e^{-[\operatorname{Re}(s)-\alpha]t}\, dt.$$

The last improper integral converges if $\operatorname{Re}(s) - \alpha > 0$, which implies that

$$L\{f\} = \int_0^\infty f(t)e^{-st}\, dt$$

converges also. Furthermore, since the improper integral equals $M/[\operatorname{Re}(s) - \alpha]$, this also shows the following:

If $f(t)$ belongs to E_α, then $F(s)$ approaches zero as $\operatorname{Re}(s)$ approaches infinity.

Therefore functions like 1, s^n, $n > 0$, $s^2/(s^2 - 3)$, or e^{2s} cannot be Laplace transforms of functions of exponential order.

We now proceed to compute the Laplace transforms of a few elementary functions, then give a few other properties of the transform which allow the table of transforms to be extended.

EXAMPLE 1 Show that $L\{\sin bt\} = \dfrac{b}{s^2 + b^2}, \qquad \operatorname{Re}(s) > 0.$

SOLUTION One can find this directly by computing

$$L\{\sin bt\} = \int_0^\infty e^{-st} \sin bt\, dt$$

through integration by parts. However, this integral and the transform of $\cos bt$ can be found by using complex exponentials as well. We remind the reader from our discussion in Chapter 2 that the differentiation (and hence the integration) formula for the exponential e^{zt} is the same regardless of whether z is real or complex. Therefore from the previous formula for $L\{e^{at}\}$ it follows that

$$L\{e^{ibt}\} = \frac{1}{s - ib}, \qquad \operatorname{Re}(s) > 0,$$

and furthermore

$$\frac{1}{s - ib} = \frac{1}{s - ib}\,\frac{s + ib}{s + ib} = \frac{s}{s^2 + b^2} + i\,\frac{b}{s^2 + b^2}.$$

Now use linearity and the fact that

$$L\{e^{ibt}\} = L\{\cos bt + i \sin bt\} = L\{\cos bt\} + i\, L\{\sin bt\}.$$

For s real one can equate real and imaginary parts of the two expressions; the formulas are valid for complex s as well. ■

EXAMPLE 2 Show that $L\{\cos bt\} = \dfrac{s}{s^2 + b^2}, \qquad \mathrm{Re}(s) > 0.$

SOLUTION This follows from the previous example. ■

EXAMPLE 3 Show that $L\{\cosh bt\} = \dfrac{s}{s^2 - b^2}, \qquad \mathrm{Re}(s) > |b|.$

SOLUTION Recall that $\cosh bt = \frac{1}{2}(e^{bt} + e^{-bt})$; then, by the linearity property,

$$L\{\cosh bt\} = \frac{1}{2}\,[L\{e^{bt}\} + L\{e^{-bt}\}] = \frac{1}{2}\left[\frac{1}{s - b} + \frac{1}{s + b}\right] = \frac{s}{s^2 - b^2}.$$ ■

EXAMPLE 4 Show that $L\{\sinh bt\} = \dfrac{b}{s^2 - b^2}, \qquad \mathrm{Re}(s) > |b|.$

SOLUTION This follows from the identity $\sinh bt = \frac{1}{2}(e^{bt} - e^{-bt})$. ■

EXAMPLE 5 Show that $L\{t\} = \dfrac{1}{s^2}, \qquad \mathrm{Re}(s) > 0.$

SOLUTION A direct way to prove this is to use the definition of the transform and integrate by parts:

$$L\{t\} = \int_0^\infty e^{-st}t\,dt = -\frac{1}{s}\,te^{-st}\Big|_0^\infty + \frac{1}{s}\int_0^\infty e^{-st}\,dt.$$

Since t belongs to E^α for any $\alpha > 0$, then the first term equals zero when $\mathrm{Re}(s) > 0$. The second term equals

$$\frac{1}{s}\,L\{1\} = \frac{1}{s}\,L\{e^{0t}\} = \frac{1}{s^2}.$$ ■

A more general result, left to the reader to prove, is the formula

$$L\{t^n\} = \frac{n!}{s^{n+1}}, \qquad \mathrm{Re}(s) > 0,$$

where n is a nonnegative integer. It is interesting to note that all of the formulas above essentially depend on the transform of e^{at}.

We next state a property that allows the table of transforms to be extended considerably. It is called the *shift property*.

If $L\{f(t)\} = F(s)$, then $L\{e^{\alpha t}f(t)\} = F(s - \alpha)$.

Therefore the computation of the Laplace transform of $e^{\alpha t}f(t)$ is simply accomplished by shifting the argument of $F(s)$ by α units. If $F(s)$ is valid for $\text{Re}(s) > a$, then $F(s - \alpha)$ is valid for $\text{Re}(s) > a + \alpha$. A proof of the shift property is immediate since

$$L\{e^{\alpha t}f(t)\} = \int_0^\infty e^{-st}e^{\alpha t}f(t)\,dt = \int_0^\infty e^{-(s-\alpha)t}f(t)\,dt,$$

and the last expression is merely the Laplace transform of $f(t)$ evaluated at $s - \alpha$.

EXAMPLE 6 Show that $L\{e^{2t}\cos 3t\} = \dfrac{s-2}{(s-2)^2+9}, \qquad \text{Re}(s) > 2.$

SOLUTION Since $L\{\cos 3t\} = s/(s^2 + 3^2)$, from Example 2 above, the shift from s to $(s - 2)$ gives the transform. ■

EXAMPLE 7 Show that $L\{e^{\alpha t}t\} = \dfrac{1}{(s-\alpha)^2}, \qquad \text{Re}(s) > \alpha.$

SOLUTION From Example 5, $L\{t\} = 1/s^2$, and the required transform is obtained by the shift from s to $(s - \alpha)$. ■

EXAMPLE 8 Find $L^{-1}\left\{\dfrac{3s-2}{s^2+6s+25}\right\}$.

SOLUTION Completing the square in the denominator and applying a little algebra gives

$$\frac{3s-2}{s^2+6s+25} = \frac{3s-2}{(s+3)^2+16} = \frac{3(s+3)}{(s+3)^2+16} - \frac{11}{(s+3)^2+16}$$
$$= 3\,\frac{s+3}{(s+3)^2+16} - \frac{11}{4}\,\frac{4}{(s+3)^2+16}.$$

Therefore

$$L^{-1}\left\{\frac{3s-2}{s^2+6s+2s}\right\} = 3e^{-3t}\cos 4t - \frac{11}{4}e^{-3t}\sin 4t. \quad ■$$

The reader should follow the algebraic steps carefully in the last example since such manipulations will be performed frequently in the coming sections.

EXAMPLE 9 Find $L^{-1}\left\{\dfrac{2s+3}{s^2-2s+3}\right\}$.

SOLUTION Proceeding as in the previous example, we have

$$\begin{aligned}\frac{2s+3}{s^2-2s+3} &= \frac{2(s-1)}{(s-1)^2+2} + \frac{5}{(s-1)^2+2}\\ &= 2\frac{s-1}{(s-1)^2+(\sqrt{2})^2} + \frac{5}{\sqrt{2}}\frac{\sqrt{2}}{(s-1)^2+(\sqrt{2})^2}.\end{aligned}$$

Therefore

$$L^{-1}\left\{\frac{2s+3}{s^2-2s+3}\right\} = 2e^t\cos\sqrt{2}t + \frac{5}{\sqrt{2}}e^t\sin\sqrt{2}t.$$ ■

Two other properties of the Laplace transform, which are occasionally useful, are:

1. If $f(t)$ belongs to E_α and its Laplace transform is $F(s)$, then

$$L\left\{\int_0^t f(r)\,dr\right\} = \frac{1}{s}F(s), \qquad \operatorname{Re}(s) > \alpha.$$

2. If $f(t)$ belongs to E_α, then its Laplace transform $F(s)$ has derivatives of all orders given by the formula

$$\frac{d^n}{ds^n}F(s) = L\{(-t)^n f(t)\}.$$

These two properties provide alternative ways of computing other transforms from the given ones; examples of this are left for the exercises.

Finally, we state a more general form of the differentiation property of Laplace transforms. Its proof is a repeated application of integration by parts combined with the argument given in the previous derivation.

The differentiation formulas. If $f(t), f'(t), \ldots, f^{(n)}(t)$ belong to E_α and $L\{f(t)\} = F(s)$, then, for $\operatorname{Re}(s) > \alpha$, the following is valid:

$$\begin{aligned}L\{f'(t)\} &= sF(s) - f(0),\\ L\{f''(t)\} &= s^2F(s) - sf(0) - f'(0),\\ L\{f^{(n)}(t)\} &= s^nF(s) - s^{n-1}f(0) - \cdots - sf^{(n-2)}(0) - f^{(n-1)}(0).\end{aligned}$$

One sees, therefore, that the transform of the nth derivative depends only on the transform of the function and n initial data values.

EXAMPLE 10 If $y(t)$ is the solution of a third order differential equation with initial conditions

$$y(0) = 1, \qquad y'(0) = -3, \qquad y''(0) = 2,$$

and $Y(s)$ is its Laplace transform, then

$$\begin{aligned} L\{y'(t)\} &= sY(s) - 1, \\ L\{y''(t)\} &= s^2Y(s) - s + 3, \\ L\{y'''(t)\} &= s^3Y(s) - s^2 + 3s - 2. \end{aligned}$$

■

We include in this section a short table of Laplace transforms (Table 4.1). The entries 15–19 in the table have not been previously discussed but will be derived later. They are added at this point for the reader's convenience.

EXERCISES 4.4

Use Table 4.1 to find the Laplace transforms of the functions in Exercises 1–18. Give the region of validity.

1. t^3 **2.** $\sin 3t$ **3.** $4e^{2t} \sin t$
4. $-2t \sin t$ **5.** $e^{-3t} \cosh t$ **6.** $e^{2t}t \cos t$
7. $4t^2 \sin 3t$ **8.** $3e^{-t} + 4 \sin 2t$ **9.** $9 \sinh t \sin 3t$
10. $5 \cos 2t$ **11.** $e^{-t} \cos 4t$ **12.** $t \sin 2t$
13. $3 \sinh 7t$ **14.** $e^{-t}t^2$ **15.** $\int_0^t r \sin r \, dr$
16. $5 + 4t$ **17.** $e^{2t} \cos (t + \pi/4)$ **18.** $t^4 \cosh 2t$

Use Table 4.1 to find the inverse Laplace transform of the functions in Exercises 19–36.

19. $\dfrac{4}{s^3}$ **20.** $\dfrac{3}{s + 1}$ **21.** $\dfrac{5}{s - 2}$

22. $\dfrac{3}{s^2 + 3}$ **23.** $\dfrac{4s}{s^2 + 3}$ **24.** $\dfrac{8}{s^2 - 16}$

25. $\dfrac{1}{s^2 - 2s + 5}$

Table 4.1 Short table of Laplace transforms.

	$f(t)$	$F(s)$	Region of validity
1.	1	$\frac{1}{s}$	$\mathrm{Re}(s) > 0$
2.	$t^n,\ n = 0, 1, 2, \cdots$	$\frac{n!}{s^{n+1}}$	$\mathrm{Re}(s) > 0$
3.	e^{at}	$\frac{1}{s-a}$	$\mathrm{Re}(s) > a$
4.	$\sin bt$	$\frac{b}{s^2+b^2}$	$\mathrm{Re}(s) > 0$
5.	$\cos bt$	$\frac{s}{s^2+b^2}$	$\mathrm{Re}(s) > 0$
6.	$\sinh bt$	$\frac{b}{s^2-b^2}$	$\mathrm{Re}(s) > \lvert b\rvert$
7.	$\cosh bt$	$\frac{s}{s^2-b^2}$	$\mathrm{Re}(s) > \lvert b\rvert$
8.	$e^{at}\sin bt$	$\frac{b}{(s-a)^2+b^2}$	$\mathrm{Re}(s) > a$
9.	$e^{at}\cos bt$	$\frac{s-a}{(s-a)^2+b^2}$	$\mathrm{Re}(s) > a$
10.	$t \sin bt$	$\frac{2bs}{(s^2+b^2)^2}$	$\mathrm{Re}(s) > 0$
11.	$t \cos bt$	$\frac{s^2-b^2}{(s^2+b^2)^2}$	$\mathrm{Re}(s) > 0$
12.	$e^{at}f(t)$	$F(s-a)$	$\mathrm{Re}(s) > \alpha + a$ if $F(s)$ valid for $\mathrm{Re}(s) > \alpha$
13.	$\int_0^t f(r)\,dr$	$\frac{1}{s}F(s)$	$\mathrm{Re}(s) > \max(a, 0)$ if $F(s)$ valid for $\mathrm{Re}(s) > a$
14.	$(-t)^n f(t)$	$\frac{d^n}{ds^n}F(s)$	valid for same region as $F(s)$
15.	$\frac{1}{b}\sin bt - t\cos bt$	$\frac{2b^2}{(s^2+b^2)^2}$	$\mathrm{Re}(s) > 0$
16.	$f(t-c)u(t-c)$	$e^{-sc}F(s)$	valid for same region as $F(s)$
17.	$f(t+T) = f(t)$	$\dfrac{\int_0^T e^{-sr}f(r)\,dr}{1-e^{-sT}}$	$\mathrm{Re}(s) > 0$
18.	$\delta(t-t_0)$	e^{-st_0}	
19.	$\delta(t-t_0)f(t)$	$e^{-st_0}f(t_0)$	
20.	$f^{(n)}(t)$ $n = 1, 2, 3, \ldots$	$s^nF(s) - s^{n-1}f(0) - s^{n-2}f^{(1)}(0)$ $- \cdots - f^{(n-1)}(0)$	

26. $\dfrac{s+4}{s^2+8s+25}$

27. $\dfrac{6}{s^2+2s+4}$

28. $\dfrac{4s-9}{s^2+9}$

29. $\dfrac{3s+1}{s^2-6s+18}$

30. $\dfrac{6s}{(s^2+4)^2}$

31. $\dfrac{-9}{(s-2)^2}$

32. $\dfrac{12}{(s+\pi)^4}$

33. $\dfrac{6}{s-3}+\dfrac{s+4}{s^2+25}$

34. $\dfrac{4}{(s+1)^2}-\dfrac{2s+3}{s^2+s+5/2}$

35. $\dfrac{2}{(s^2+1)^2}$

36. $\dfrac{-(s^2+9)}{(s^2-9)^2}$

37. Establish entry (4) of Table 4.1 by direct integration by parts.

38. Establish entry (2) of Table 4.1 by using entries (14) and (1) or induction.

39. Establish entry (11) of Table 4.1 by using entries (14) and (5).

40. Using entries (13) and (10) of Table 4.1 and integration by parts, derive the formula

$$L^{-1}\left\{\frac{2b^2}{(s^2+b^2)^2}\right\}=\frac{1}{b}\sin bt-t\cos bt.$$

41. Establish entry (13) of Table 4.1.

$$\left(\textit{Hint: } \frac{d}{dt}\int_0^t f(r)\,dr=f(t).\right)$$

Given that $y(t)$ is the solution of a second order differential equation and satisfies the following initial conditions, find $L\{y'(t)\}$ and $L\{y''(t)\}$ in terms of $Y(s)=L\{y(t)\}$ in Exercises 42–47.

42. $y(0)=0,\quad y'(0)=1$

43. $y(0)=3,\quad y'(0)=-2$

44. $y(0)=1,\quad y'(0)=0$

45. $y(0)=-1,\quad y'(0)=1$

46. $y(0)=0,\quad y'(0)=0$

47. $y(0)=-2,\quad y'(0)=-1$

In Exercises 48–53 find $L\{y^{(n)}(t)\}$, $n=1,2,3,4$, in terms of $Y(s)=L\{y(t)\}$ if $y(t)$ satisfies the given initial conditions.

48. $y(0)=0,\ y'(0)=-1,\ y''(0)=0,\ y^{(3)}(0)=1$

49. $y(0)=y'(0)=1,\ y''(0)=y^{(3)}(0)=-1$

50. $y(0)=1,\ y'(0)=y''(0)=0,\ y^{(3)}(0)=2$

51. $y(0)=y'(0)=2,\ y''(0)=0,\ y^{(3)}(0)=1$

52. $y(0)=1,\ y'(0)=-1,\ y''(0)=y^{(3)}(0)=2$

53. $y(0)=1,\ y'(0)=2,\ y''(0)=3,\ y^{(3)}(0)=4$

54. The following formula for the integration of the transform,

$$L\left\{\frac{f(t)}{t}\right\}=\int_s^\infty F(r)\,dr,$$

where $F(s) = L\{f(t)\}$ is similar to entry (13) of Table 4.1. The formula is valid if $f(t)$ is of class E_α and $\lim_{t\to 0+} [f(t)/t]$ exists. Use the formula to show that

a) $L\left\{\dfrac{\sin kt}{t}\right\} = \text{arc cot}\,\dfrac{s}{k}$

b) $L\left\{\displaystyle\int_0^t \frac{\sin \tau}{\tau}\,d\tau\right\} = \dfrac{1}{s}\,\text{arc cot}\, s$

c) $L\left\{\dfrac{1 - e^{-t}}{t}\right\} = \ln\left(1 + \dfrac{1}{s}\right)$

Verify that the following functions in Exercises 55–60 are of class E_α and find a suitable value of α and positive constant M:

55. $t^2 \sin t$

56. 2^{3t-1}

57. $6(t^2 + 1)^{200}$

58. $3 \cosh 2t$

59. $\exp\,[t^{2/3}]$

60. $\tanh t$

4.5 SOLVING CONSTANT COEFFICIENT LINEAR EQUATIONS

In this section we introduce the reader to the use of the Laplace transform in solving constant coefficient linear systems. One of the inherent difficulties of the Laplace transform method is the need to compute lengthy and elaborate partial fraction decompositions. In the following examples we have tried to keep this to a minimum. The idea here is to clearly demonstrate the method free from a lot of computational obfuscation.

EXAMPLE 1 Solve the initial value problem

$$y'' + 9y = \cos 3t, \qquad y(0) = 1, \qquad y'(0) = -1.$$

SOLUTION If $Y(s) = L\{y(t)\}$, then from the initial data and the differentiation formulas it follows that

$$L\{y''(t)\} = s^2Y(s) - s + 1.$$

Applying the Laplace transform to both sides of the differential equation and using entry (5) from Table 4.1 gives

$$s^2Y(s) - s + 1 + 9Y(s) = L\{\cos 3t\} = \frac{s}{s^2 + 9}.$$

Now solve for $Y(s)$ to obtain

$$Y(s) = \frac{s-1}{s^2+9} + \frac{s}{(s^2+9)^2}.$$

The denominator is $s^2 + 9 = s^2 + 3^2$, and an examination of the last term tells us that the key entries in Table 4.1 are (4), (5), and (10). A little rewriting gives

$$Y(s) = \frac{s}{s^2+9} - \frac{1}{3}\frac{3}{s^2+9} + \frac{1}{6}\frac{6s}{(s^2+9)^2},$$

and now the inverse transform can be applied (by using Table 4.1 directly!) to obtain

$$y(t) = \cos 3t - \tfrac{1}{3}\sin 3t + \tfrac{1}{6}t \sin 3t,$$

the desired solution. ■

EXAMPLE 2 Solve the initial value problem

$$y'' + 3y' + 2y = 6, \qquad y(0) = 0, \qquad y'(0) = 2.$$

SOLUTION Proceeding as above, first use the initial data to compute

$$L\{y''(t)\} = s^2Y(s) - 2, \qquad L\{y'(t)\} = sY(s);$$

applying the Laplace transform to the equation gives

$$s^2Y(s) - 2 + 3sY(s) + 2Y(s) = L\{6\} = \frac{6}{s}.$$

Now solve for $Y(s)$ to obtain

$$Y(s) = \frac{2}{s^2+3s+2} + \frac{6}{s(s^2+3s+2)}$$

or

$$Y(s) = \frac{2}{(s+1)(s+2)} + \frac{6}{s(s+1)(s+2)} = \frac{2s+6}{s(s+1)(s+2)}.$$

From the theory of partial fractions we know that

$$\frac{2s+6}{s(s+1)(s+2)} = \frac{A}{s} + \frac{B}{s+1} + \frac{C}{s+2};$$

combining terms on the right side and equating the numerators gives the relation

$$2s + 6 = A(s+1)(s+2) + Bs(s+2) + Cs(s+1).$$

Equating like powers of s leads to the equations

$$A + B + C = 0, \qquad 3A + 2B + C = 2, \qquad 2A = 6.$$

These have the solution $A = 3$, $B = -4$, $C = 1$, which could have also been obtained by letting $s = 0$, $s = -1$, and $s = -2$, successively, in the previous equation. Therefore

$$Y(s) = \frac{3}{s} - \frac{4}{s+1} + \frac{1}{s+2},$$

and now Table 4.1 gives

$$y(t) = 3 - 4e^{-t} + e^{-2t},$$

the required solution. ■

At this point the reader may well ask: What is the advantage of using Laplace transforms to solve problems such as Example 1 and Example 2? Example 1 is a case of resonance, so we know that the solution is of the form

$$y(t) = A\cos 3t + B\sin 3t + Ct\cos 3t + Dt\sin 3t,$$

and substitution and comparison of coefficients would give the answer. From the standpoint of algebra it looks like the transform method has a slight edge, especially since the initial conditions are built in.

In Example 2 this advantage is certainly not evident since we had to factor the characteristic polynomial $s^2 + 3s + 2 = (s+1)(s+2)$ anyway, and this would indicate that the trial solution is of the form

$$y(t) = Ae^{-t} + Be^{-2t} + C.$$

In view of the time spent on the partial fraction decomposition, direct substitution would probably win the day. In the next section some formulas will be introduced, which reduce the computational time but, at best, the matter will be a toss-up.

In the authors' opinion, the real advantage of the Laplace transform method occurs when one is faced with a constant coefficient differential equation in which the forcing term (the right-hand side) is a discontinuous function or one with corners. For instance, the function

$$f(t) = \begin{cases} 0 \text{ for } t < 1, \\ I \text{ for } t \geq 1, \end{cases}$$

would ideally model, say, a current or force that is zero for $t < 1$, and then is switched on to full power I at $t = 1$. Finding the solution of the initial value problem

$$y'' + ay' + by = f(t), \qquad y(0) = A, \qquad y'(0) = B$$

cannot be done by comparison of coefficients, since the right-hand side is not a polynomial, or an exponential, or a sine, or a cosine function. Variation of parameters will work if used with care, but the Laplace transform method is the sure winner. This class of problems will be considered in Sections 4.8 and 4.9.

Another class of problems where the Laplace transform has a clear advantage is the solving of systems of linear differential equations. The reason for this is that some theoretical algebraic considerations are bypassed and, again, that the initial conditions are contained in the transformation. This is demonstrated in the next example.

EXAMPLE 3 Find the solution $x(t)$, $y(t)$ of the initial value problem

$$\begin{aligned} \dot{x} &= 2x + y, & x(0) &= 1, \\ \dot{y} &= 3x + 4y, & y(0) &= 0. \end{aligned}$$

SOLUTION This is a constant coefficient system of two linear differential equations, and so we must apply the Laplace transform to each equation. Letting $X(s) = L\{x(t)\}$ and $Y(s) = L\{y(t)\}$, we get from the differentiation formula and the initial data that

$$\begin{aligned} L\{\dot{x}(t)\} &= sX(s) - 1, \\ L\{\dot{y}(t)\} &= sY(s). \end{aligned}$$

Now apply the transform to each differential equation to get

$$\begin{aligned} sX(s) - 1 &= 2X(s) + Y(s), \\ sY(s) &= 3X(s) + 4Y(s), \end{aligned}$$

or, equivalently,

$$\begin{aligned} (s - 2)X(s) - Y(s) &= 1, \\ -3X(s) + (s - 4)Y(s) &= 0. \end{aligned}$$

This is a system of linear algebraic equations, which must be solved for $X(s)$ and $Y(s)$; using Cramer's rule we obtain

$$\begin{aligned} X(s) &= \frac{\begin{vmatrix} 1 & -1 \\ 0 & s - 4 \end{vmatrix}}{\begin{vmatrix} s - 2 & -1 \\ -3 & s - 4 \end{vmatrix}} \\ &= \frac{s - 4}{s^2 - 6s + 5} = \frac{s - 4}{(s - 1)(s - 5)}, \end{aligned}$$

$$Y(s) = \frac{\begin{vmatrix} s-2 & 1 \\ -3 & 0 \end{vmatrix}}{\begin{vmatrix} s-2 & -1 \\ -3 & s-4 \end{vmatrix}}$$

$$= \frac{3}{(s-1)(s-5)}.$$

Finally, a partial fraction decomposition gives

$$X(s) = \frac{s-4}{(s-1)(s-5)}$$

$$= \frac{3/4}{s-1} + \frac{1/4}{s-5},$$

$$Y(s) = \frac{3}{(s-1)(s-5)}$$

$$= \frac{-3/4}{s-1} + \frac{3/4}{s-5},$$

and therefore

$$x(t) = \frac{3}{4}e^{t} + \frac{1}{4}e^{5t}, \qquad y(t) = -\frac{3}{4}e^{t} + \frac{3}{4}e^{5t}. \quad \blacksquare$$

After studying linear systems in Chapter 5, the reader will see how efficient the Laplace transform is as a solution technique and how many of the algebraic difficulties are circumvented.

As a simple application of linear systems of differential equations consider the following.

A mixing problem. Two 100-gallon tanks are connected in the manner depicted in Fig. 4.2, where the arrows indicate the flow of liquid from and to the outside and between the two tanks. Fresh water is flowing into tank A from the

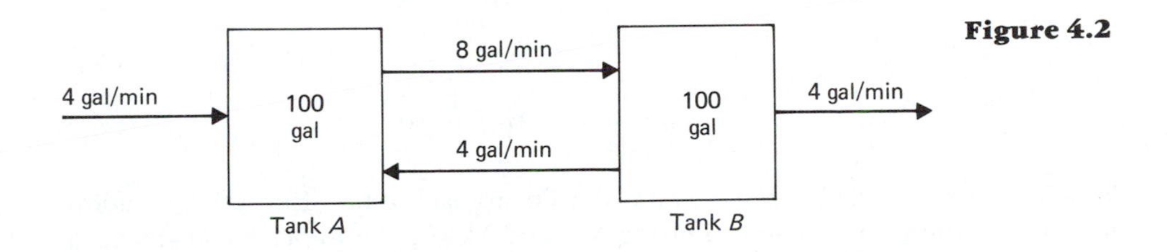

Figure 4.2

outside, and at $t = 0$ tank A contains 100 gallons of fresh water, whereas tank B contains 100 gallons of brine consisting of 10 pounds of dissolved salt (that is, 0.1 pound of salt per gallon). Find the amounts of salt $x(t)$, $y(t)$ (in pounds) in tanks A and B, respectively, as functions of time assuming instantaneous stirring. The solution will lead to a simple linear system of differential equations.

Since $x(t)$ is the number of pounds of salt in tank A at time t, the rate of change of $x(t)$ is

$$\dot{x}(t) = (\text{Rate in}) - (\text{Rate out}),$$

in units of lb/min. To determine how many pounds of salt per minute are going into A, we notice that the only salt contribution to A is from tank B since the 4 gal/min coming from the outside is fresh water. At time t, tank B contains $y(t)$ pounds of salt dissolved in 100 gallons of water, and this flows into tank A at the rate of 4 gal/min. Hence for tank A we can write:

$$(\text{Rate in}) = \frac{y(t)}{100}\,(\text{lb/gal}) \cdot 4\,(\text{gal/min}) = \frac{4}{100}\,y(t)\,(\text{lb/min}),$$

whereas

$$(\text{Rate out}) = \frac{x(t)}{100}\,(\text{lb/gal}) \cdot 8\,(\text{gal/min}) = \frac{8}{100}\,x(t)\,(\text{lb/min}).$$

Also, at $t = 0$ tank A contains no salt, so $x(0) = 0$. This gives the first differential equation and initial condition

$$\dot{x} = \frac{4}{100}y - \frac{8}{100}x, \qquad x(0) = 0$$

or

$$\dot{x} = -\frac{2}{25}x + \frac{1}{25}y, \qquad x(0) = 0. \tag{4.5.1}$$

The same analysis of tank B leads to the second differential equation and initial condition

$$\dot{y} = \frac{8}{100}x - \frac{8}{100}y, \qquad y(0) = 10$$

or

$$\dot{y} = \frac{2}{25}x - \frac{2}{25}y, \qquad y(0) = 10. \tag{4.5.2}$$

Equations (4.5.1) and (4.5.2) are first order linear systems of differential equations with initial conditions at $t = 0$. Letting $X(s)$ and $Y(s)$ be the Laplace transforms of

$x(t)$ and $y(t)$, respectively, and applying the differentiation formula, we get

$$sX(s) = -\frac{2}{25}X(s) + \frac{1}{25}Y(s),$$

$$sY(s) - 10 = \frac{2}{25}X(s) - \frac{2}{25}Y(s),$$

or

$$(s + \frac{2}{25})X(s) - \frac{1}{25}Y(s) = 0,$$

$$-\frac{2}{25}X(s) + (s + \frac{2}{25})Y(s) = 10,$$

which must be solved for $X(s)$ and $Y(s)$. From Cramer's rule we obtain

$$X(s) = \frac{10}{25}\frac{1}{(s + 2/25)^2 - (\sqrt{2}/25)^2} = \frac{10}{\sqrt{2}}\frac{\sqrt{2}/25}{(s + 2/25)^2 - (\sqrt{2}/25)^2},$$

$$Y(s) = 10\frac{s + 2/25}{(s + 2/25)^2 - (\sqrt{2}/25)^2},$$

and now the shift formula and entries (6) and (7) of Table 4.1 give the solution

$$x(t) = \frac{10}{\sqrt{2}}e^{-2t/25}\sinh\frac{\sqrt{2}}{25}t, \qquad y(t) = 10e^{-2t/25}\cosh\frac{\sqrt{2}}{25}t$$

for the amounts of salt in pounds at time t in tanks A and B, respectively. By converting the hyperbolic sine and cosine into their exponential form, the reader can further analyze the system and easily see that both $x(t)$ and $y(t)$ approach zero as t approaches infinity. After a long time both tanks will essentially contain fresh water.

EXERCISES 4.5

Use the Laplace transform to solve the initial value problems in Exercises 1–10.

1. $y'' + 9y = 0,\ \ y(0) = 1,\ \ y'(0) = 2$

2. $y'' - y' - 6y = 0,\ \ y(0) = 0,\ \ y'(0) = 3$

3. $y'' + y = \sin t,\ \ y(0) = 2,\ \ y'(0) = -1$

4. $y'' + 5y' + 6y = 4e^t,\ \ y(0) = y'(0) = 0$

5. $y'' - 4y = 8,\ \ y(0) = 2,\ \ y'(0) = 0$

6. $y'' + 2y' + 5y = 0,\ \ y(0) = 6,\ \ y'(0) = -3$

7. $y'' + 7y' + 12y = 12,\ \ y(0) = y'(0) = 0$

8. $y'' + 4y = t,\ \ y(0) = 4,\ \ y'(0) = 0$

9. $y'' + y' - 2y = 2e^{4t},\ \ y(0) = 0,\ \ y'(0) = 3$

10. $y'' - 2y' = 0,\ \ y(0) = y'(0) = 1$

In Exercises 11–18 use the Laplace transform to find the solutions $x(t)$, $y(t)$ of the initial value problems.

11. $\dot{x} = x + y, \quad x(0) = 3, \quad \dot{y} = 9x + y, \quad y(0) = -3$

12. $\dot{x} = 4x + 5y, \quad x(0) = -1, \quad \dot{y} = -4x - 4y, \quad y(0) = 2$

13. $\dot{x} = x + 3y, \quad x(0) = 1, \quad \dot{y} = 3x + y, \quad y(0) = 0$

14. $\dot{x} = 2x + 4y, \quad x(0) = 1, \quad \dot{y} = -4x + 2y, \quad y(0) = -1$

15. $\dot{x} = 3x - 3y, \quad x(0) = 4, \quad \dot{y} = 6x - 3y, \quad y(0) = 3$

16. $\dot{x} = 2x - y, \quad x(0) = 1, \quad \dot{y} = x + 4y, \quad y(0) = 0$

17. $\dot{x} = x + 3y, \quad x(0) = 2, \quad \dot{y} = 5x + 3y, \quad y(0) = 1$

18. $\dot{x} = 2x + 3y, \quad x(0) = -1, \quad \dot{y} = 2x + y, \quad y(0) = 1$

19. Given the inhomogeneous system of linear differential equations

$$\dot{x} = ax + by + f(t), \qquad x(0) = 0,$$
$$\dot{y} = cx + dy + g(t), \qquad y(0) = 0,$$

show that the Laplace transforms of the solutions $x(t)$, $y(t)$ are

$$X(s) = \frac{(s - d)F(s)}{P(s)} + \frac{bG(s)}{P(s)},$$

$$Y(s) = \frac{(s - a)G(s)}{P(s)} + \frac{cF(s)}{P(s)},$$

where $F(s)$ and $G(s)$ are the Laplace transforms of $f(t)$ and $g(t)$, respectively, and

$$P(s) = s^2 - (a + d)s + (ad - bc).$$

(The above initial value problem would represent a system at rest being excited by driving forces $f(t)$ and $g(t)$.)

Use the results of Exercise 19 or any other method to find the solutions $x(t)$, $y(t)$ of the initial value problems in Exercises 20–23.

20. $\dot{x} = x + y, \quad x(0) = 0, \quad \dot{y} = 9x + y + e^t, \quad y(0) = 0$

21. $\dot{x} = -2x + y + 1, \quad x(0) = 0, \quad \dot{y} = x - 2y, \quad y(0) = 0$

22. $\dot{x} = 4x + 5y + e^{-4t}, \quad x(0) = 0, \quad \dot{y} = -4x - 4y + e^{4t}, \quad y(0) = 0$

23. $\dot{x} = x - y + 2e^t, \quad x(0) = 0, \quad \dot{y} = -x + y + e^t, \quad y(0) = 0$

Exercises 24–27 discuss a mixing problem with flow configuration given by Fig. 4.3 and with the following initial data: Tank A contains 100 gal of brine consisting of 10 lb of dissolved salt. Fresh water is flowing in at the rate of 6 gal/min. Tank B contains 100 gal of fresh water.

24. Find the amounts of salt $x(t)$, $y(t)$ (in pounds) in tanks A and B. Show that both approach zero as t approaches infinity.

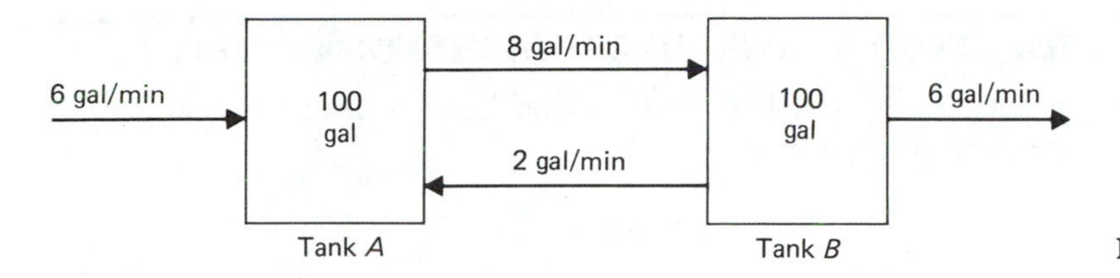

Figure 4.3

25. Instead of salt, let the dissolved quantity be 10 lb of a toxic substance. Then one would be interested in finding the time T at which the system has been flushed sufficiently, so that the outflow from tank B is at a "safe" level. Using the result from Exercise 24, show that the maximum amount of toxic substance in tank B is approximately 3.85 lb (a concentration of 0.0385 lb/gal) and that this maximum occurs after approximately 13.7 minutes. Also show that for $T \cong 138$ min the concentration in tank B is less than 0.0004 lb/gal.

26. Suppose that to reduce the toxicity in tank B faster, we pump fresh water directly into it from the outside at a rate of 4 gal/min. To maintain the volume at 100 gal we must therefore pump water out at the rate of 10 gal/min instead of 6 gal/min. Find the amount of toxic substance $y(t)$ in tank B. Show that for $T \cong 98$ min the concentration in tank B is less than 0.0004 lb/gal.

27. Show that the strategy in Exercise 26 is equivalent to draining tank A to the outside at a rate of 4 gal/min and pumping fresh water into it at a rate of 10 gal/min instead of 6 gal/min. Do this by finding $L\{y(t)\}$.

28. A closed two-tank system has the configuration depicted in Fig. 4.4. The initial data for each tank are: Tank A contains 8 lb of chemical X dissolved in 100 gal of fresh water. Tank B contains 4 lb of chemical X dissolved in 100 gal of fresh water.

a) Find the amounts $x(t)$, $y(t)$ of chemical X in tanks A and B.

b) Show that $x(t)$ is a decreasing function and $y(t)$ is an increasing function of time, and both approach 6 lb as t approaches infinity.

c) If we require only that the amount of $y(T)$ of chemical X equal $(6 - \epsilon)$ lb, where $0 < \epsilon < 2$, show that $T = (25/2) \ln (2/\epsilon)$. Graph T as a function of ϵ.

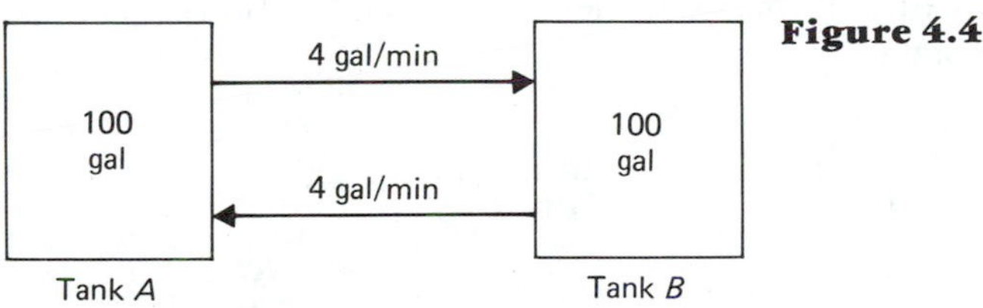

Figure 4.4

4.6 PARTIAL FRACTIONS AND THE HEAVISIDE FORMULAS

As the reader has seen in the last few sections, the Laplace transforms we have obtained are expressions of the form

$$F(s) = P(s)/Q(s),$$

where P and Q are polynomials in s, the degree of P is less than the degree of Q, and P and Q have no common factors. This is a rational function of s, and the object of a partial fraction decomposition is to reduce it to a sum of simpler rational functions whose inverse Laplace transforms can be easily looked up in Table 4.1.

Since $Q(s)$ is a polynomial, we know that over the real number field its only possible factors are

1. $(s - c)$, where c is a real simple root, and hence $(s - c)^2$ is not a factor of $Q(s)$;
2. $(s - c)^k$, $k = 1, 2, \ldots, r$, so c is a real root of multiplicity r, and hence $(s - c)^{r+1}$ is not a factor of $Q(s)$;
3. $(s - a)^2 + b^2$, $b \neq 0$, which is a simple irreducible quadratic factor, and hence $[(s - a)^2 + b^2]^2$ is not a factor of $Q(s)$; and
4. $[(s - a)^2 + b^2]^k$, $k = 1, 2, \ldots, q$, which is an irreducible quadratic factor of order q, and hence $[(s - a)^2 + b^2]^{q+1}$ is not a factor of $Q(s)$.

By irreducible we mean "having only complex roots," and the reader can see clearly that $(s - a)^2 + b^2$ cannot have real roots if a and b are real, $b \neq 0$. By completing the square, we can write any quadratic polynomial with complex roots in this form.

Case 1. If $Q(s)$ has simple roots $c_1, c_2, \ldots, c_n$, then the partial fraction expansion of $P(s)/Q(s)$ will be of the form

$$\frac{P(s)}{Q(s)} = \frac{A_1}{s - c_1} + \frac{A_2}{s - c_2} + \cdots + \frac{A_n}{s - c_n} + N(s),$$

where $N(s)$ is a collection of terms (if any) not involving $s - c_i$, $i = 1, 2, \ldots, n$.

Case 2. If $Q(s)$ has a root c of multiplicity r, then the partial fraction expansion of $P(s)/Q(s)$ will be of the form

$$\frac{P(s)}{Q(s)} = \frac{A_1}{s - c} + \frac{A_2}{(s - c)^2} + \cdots + \frac{A_r}{(s - c)^r} + N(s),$$

where, again, $N(s)$ is a collection of terms (if any) not involving $s - c$.

Case 3. If $Q(s)$ has a simple irreducible quadratic factor $(s - a)^2 + b^2$, then in the partial fraction expansion there will be a term of the form

$$\frac{Bs + C}{(s - a)^2 + b^2},$$

where B and C are constants to be determined.

Having written down the decomposition of $P(s)/Q(s)$ based on the three cases given above, it remains to evaluate the constants A_i, B, and C. The reader may have developed a favorite method while doing partial fraction expansions for evaluating integrals; the examples below will be done by combining terms and comparing coefficients of like powers of s. At the end of the section we will mention Heaviside's formulas for Case 1 and Case 2, which are a very efficient way of evaluating the constants.

EXAMPLE 1 Find A_1 and A_2 in the decomposition

$$\begin{aligned}\frac{P(s)}{Q(s)} &= \frac{3s + 2}{s^2 + 2s - 3} \\ &= \frac{A_1}{s - 1} + \frac{A_2}{s + 3}.\end{aligned}$$

SOLUTION

$$\begin{aligned}\frac{A_1}{s - 1} + \frac{A_2}{s + 3} &= \frac{A_1(s + 3) + A_2(s - 1)}{(s - 1)(s + 3)} \\ &= \frac{(A_1 + A_2)s + (3A_1 - A_2)}{s^2 + 2s - 3} \\ &= \frac{3s + 2}{s^2 + 2s - 3}.\end{aligned}$$

Consequently a comparison of coefficients gives

$$A_1 + A_2 = 3, \qquad 3A_1 - A_2 = 2,$$

with solution $A_1 = \frac{5}{4}$, $A_2 = \frac{7}{4}$. ■

EXAMPLE 2 Find A_1, B, and C in the decomposition

$$\begin{aligned}\frac{P(s)}{Q(s)} &= \frac{s^2 + 1}{(s - 1)(s^2 + 4)} \\ &= \frac{A_1}{s - 1} + \frac{Bs + C}{s^2 + 4}.\end{aligned}$$

SOLUTION

$$\frac{A_1}{s-1} + \frac{Bs + C}{s^2 + 4} = \frac{A_1(s^2 + 4) + Bs(s-1) + C(s-1)}{(s-1)(s^2+4)}$$

$$= \frac{(A_1 + B)s^2 + (-B + C)s + (4A_1 - C)}{(s-1)(s^2+4)} = \frac{s^2+1}{(s-1)(s^2+4)}.$$

Comparing like powers of s implies

$$A_1 + B = 1, \qquad -B + C = 0, \qquad 4A_1 - C = 1$$

with solution $A_1 = \frac{2}{5}$, $B = C = \frac{3}{5}$. ■

EXAMPLE 3 Find

$$L^{-1}\left\{\frac{1}{s^4 - 1}\right\}.$$

SOLUTION Since $s^4 - 1 = (s-1)(s+1)(s^2+1)$, we must first find the constants A_1, A_2, B, and C in the decomposition

$$\frac{1}{s^4 - 1} = \frac{A_1}{s-1} + \frac{A_2}{s+1} + \frac{Bs + C}{s^2 + 1}.$$

The right-hand side can be combined to give

$$\frac{[A_1(s+1)(s^2+1) + A_2(s-1)(s^2+1) + Bs(s^2-1) + C(s^2-1)]}{(s^4-1)} = \frac{1}{s^4-1}.$$

Combining like powers of s and comparing them with the right-hand side gives the equations

$$A_1 + A_2 + B = 0, \qquad A_1 - A_2 + C = 0,$$

$$A_1 + A_2 - B = 0, \qquad A_1 - A_2 - C = 1.$$

The solutions are $A_1 = \frac{1}{4}$, $A_2 = -\frac{1}{4}$, $B = 0$, $C = -\frac{1}{2}$, and therefore

$$L^{-1}\left\{\frac{1}{s^4-1}\right\} = L^{-1}\left\{\frac{1}{4}\frac{1}{s-1} - \frac{1}{4}\frac{1}{s+1} - \frac{1}{2}\frac{1}{s^2+1}\right\}$$

$$= \frac{1}{4}e^t - \frac{1}{4}e^{-t} - \frac{1}{2}\sin t.$$ ■

EXAMPLE 4 Find A_1, A_2, and A_3 in the decomposition

$$\frac{P(s)}{Q(s)} = \frac{s^2 + 2}{(s-4)^3} = \frac{A_1}{s-4} + \frac{A_2}{(s-4)^2} + \frac{A_3}{(s-4)^3}.$$

SOLUTION

$$\frac{A_1}{s-4} + \frac{A_2}{(s-4)^2} + \frac{A_3}{(s-4)^3} = \frac{A_1(s-4)^2 + A_2(s-4) + A_3}{(s-4)^3}$$

$$= \frac{A_1 s^2 + (-8A_1 + A_2)s + (16A_1 - 4A_2 + A_3)}{(s-4)^3}$$

$$= \frac{s^2+2}{(s-4)^3}.$$

Consequently

$$A_1 = 1, \qquad -8A_1 + A_2 = 0, \qquad 16A_1 - 4A_2 + A_3 = 2,$$

with solution $A_1 = 1$, $A_2 = 8$, and $A_3 = 18$. ■

In the calculation of the inverse Laplace transform of a term of the form

$$\frac{Bs + C}{(s-a)^2 + b^2},$$

after B and C have been determined, one needs to modify the numerator so as to be able to write the above as a linear combination of

$$\frac{s-a}{(s-a)^2+b^2} = L^{-1}\{e^{at}\cos bt\} \qquad \text{and} \qquad \frac{b}{(s-a)^2+b^2} = L^{-1}\{e^{at}\sin bt\}.$$

The next two examples demonstrate the procedure.

EXAMPLE 5 Find

$$L^{-1}\left\{\frac{3s+5}{s^2+2s+6}\right\}.$$

SOLUTION

$$\frac{3s+5}{s^2+2s+6} = \frac{3s+5}{(s+1)^2+5} = \frac{3(s+1)+2}{(s+1)^2+(\sqrt{5})^2}$$

$$= 3\,\frac{s+1}{(s+1)^2+(\sqrt{5})^2} + \frac{2}{\sqrt{5}}\,\frac{\sqrt{5}}{(s+1)^2+(\sqrt{5})^2}.$$

Therefore

$$L^{-1}\left\{\frac{3s+5}{s^2+2s+6}\right\} = 3e^{-t}\cos\sqrt{5}t + \frac{2}{\sqrt{5}}e^{-t}\sin\sqrt{5}t.$$ ■

EXAMPLE 6 Find

$$L^{-1}\left\{\frac{3s^2 + 4}{s^2(s^2 - 3s + 3)}\right\}.$$

SOLUTION Since $s^2 - 3s + 3 = (s - \frac{3}{2})^2 + \frac{3}{4}$, the quadratic factor is irreducible so the first task is to find the constants A_1, A_2, B, and C in the decomposition. Thus

$$\frac{3s^2 + 4}{s^2(s^3 - 3s + 3)} = \frac{A_1}{s} + \frac{A_2}{s^2} + \frac{Bs + C}{s^2 - 3s + 3}.$$

Combining terms in the right-hand side and equating numerators gives

$$A_1 s(s^2 - 3s + 3) + A_2(s^2 - 3s + 3) + (Bs + C)s^2 = 3s^2 + 4.$$

Comparing like powers of s leads to the equations

$$A_1 + B = 0, \qquad -3A_1 + A_2 + C = 3, \qquad 3A_1 - 3A_2 = 0, \qquad 3A_2 = 4$$

with solutions $A_1 = A_2 = \frac{4}{3}$, $B = -\frac{4}{3}$ and $C = \frac{17}{3}$. We now have to find

$$L^{-1}\left\{\frac{4}{3}\frac{1}{s} + \frac{4}{3}\frac{1}{s^2} + \frac{1}{3}\frac{-4s + 17}{\left(s - \frac{3}{2}\right)^2 + \left(\frac{\sqrt{3}}{2}\right)^2}\right\},$$

and it remains to modify the numerator of the last term. One easily calculates

$$-4s + 17 = -4\left(s - \frac{3}{2}\right) + 11 = -4\left(s - \frac{3}{2}\right) + \frac{22}{\sqrt{3}}\frac{\sqrt{3}}{2},$$

and therefore

$$L^{-1}\left\{\frac{3s^2 + 4}{s^2(s^2 - 3s + 3)}\right\} = L^{-1}\left\{\frac{4}{3}\frac{1}{s} + \frac{4}{3}\frac{1}{s^2} - \frac{4}{3}\frac{s - \frac{3}{2}}{\left(s - \frac{3}{2}\right)^2 + \left(\frac{\sqrt{3}}{2}\right)^2}\right.$$

$$\left. + \frac{22}{3\sqrt{3}}\frac{\sqrt{3}/2}{\left(s - \frac{3}{2}\right)^2 + \left(\frac{\sqrt{3}}{2}\right)^2}\right\}$$

$$= \frac{4}{3} + \frac{4}{3}t - \frac{4}{3}e^{3t/2}\cos\frac{\sqrt{3}}{2}t + \frac{22}{3\sqrt{3}}e^{3t/2}\sin\frac{\sqrt{3}}{2}t. \quad \blacksquare$$

Case 4 (a few remarks). If $Q(s)$ has an irreducible quadratic factor $(s - a)^2 + b^2$ of order $q \geq 1$, it can be shown that this corresponds to transforming a differential equation of order $2q$ or more. In practice, $q = 1$ (Case 3) and $q = 2$ are the only values that frequently occur.

Corresponding to a factor $[(s - a)^2 + b^2]^2$ in $Q(s)$, the partial fraction expansion of $P(s)/Q(s)$ contains a term of the form

$$\frac{As + B}{(s - a)^2 + b^2} + \frac{Cs + D}{[(s - a)^2 + b^2]^2},$$

and the values of A, B, C, and D have to be found by "brute force." The first term above presents no problem since its inverse transform will be a sum of $e^{at} \cos bt$ and $e^{at} \sin bt$. If the second term is written as

$$\frac{C}{2b} \frac{2b(s - a)}{[(s - a)^2 + b^2]^2} + \frac{D + Ca}{[(s - a)^2 + b^2]^2},$$

then the shift formula and entry (10) of Table 4.1 show that the first part has an inverse transform

$$\frac{C}{2b} e^{at} t \sin bt.$$

For the second part we need the identity

$$\frac{1}{[(s - a)^2 + b^2]^2} = \frac{1}{2b^3} \frac{b}{(s - a)^2 + b^2} - \frac{1}{2b^2} \frac{(s - a)^2 - b^2}{[(s - a)^2 + b^2]^2},$$

and now the shift formula and entries (8) and (11) give the inverse transform

$$(D + Ca) \left[\frac{1}{2b^3} e^{at} \sin bt - \frac{1}{2b^2} e^{at} t \cos bt \right].$$

To help the reader avoid these calculations in the future, the following entry has been added to Table 4.1 as entry (15):

	$f(t)$	$F(s)$	Region of validity
15.	$\frac{1}{b} \sin bt - t \cos bt$	$\frac{2b^2}{(s^2 + b^2)^2}$	$\mathrm{Re}(s) > 0$

EXAMPLE 7 Find

$$L^{-1} \left\{ \frac{3s + 2}{[(s + 1)^2 + 1]^2} \right\}.$$

SOLUTION A little arithmetical juggling gives

$$\frac{3s + 2}{[(s + 1)^2 + 1]^2} = \frac{3}{2} \frac{2(s + 1)}{[(s + 1)^2 + 1]^2} - \frac{1}{2} \frac{2}{[(s + 1)^2 + 1]^2},$$

and the shift formula and entries (10) and (15) give

$$L^{-1}\left\{\frac{3s+2}{[(s+1)^2+1]^2}\right\} = \frac{3}{2}e^{-t}t\sin t - \frac{1}{2}e^{-t}(\sin t - t\cos t). \qquad \blacksquare$$

At the beginning of the section we mentioned Heaviside's formulas. Oliver Heaviside (1850–1925) was a reclusive English engineer who developed heuristically many of the techniques described in this chapter and called it "the operational calculus." At the time his work was derided by many of the scientists of the day, but later researchers justified the correctness of his calculations and pointed out their relation to the Laplace transform introduced a hundred years earlier.

HEAVISIDE FORMULA FOR CASE 1

If $Q(s)$ has simple roots $c_1, c_2, \ldots, c_n$, then

$$\frac{P(s)}{Q(s)} = \frac{A_1}{s-c_1} + \frac{A_2}{s-c_2} + \cdots + \frac{A_n}{s-c_n} + N(s),$$

where

$$A_i = \frac{P(c_i)}{Q'(c_i)}, \quad \text{or equivalently} \quad A_i = \frac{P(c_i)}{Q_i(c_i)},$$

with $Q_i(s) = Q(s)/(s-c_i)$, $i = 1, 2, \ldots, n$. As before $N(s)$ represents *the terms (if any) not involving* $s - c_i$, $i = 1, 2, \ldots, n$.

This is an easy result to verify (which the reader may wish to do) and can be summarized simply: To find the A_i in the simple root case either

1. differentiate $Q(s)$ and evaluate $P(s)/Q'(s)$ at $s = c_i$, or
2. cancel the factor $s - c_i$ from the denominator of $P(s)/Q(s)$ and evaluate what remains at $s = c_i$.

EXAMPLE 8 Find A_1 and A_2 in the decomposition

$$\frac{P(s)}{Q(s)} = \frac{s-1}{s^2+s-6} = \frac{A_1}{s+3} + \frac{A_2}{s-2}.$$

SOLUTION $P(s) = s - 1$, $Q'(s) = 2s + 1$, so

$$A_1 = \frac{P(-3)}{Q'(-3)} = \frac{-4}{-6+1} = \frac{4}{5}, \qquad A_2 = \frac{P(2)}{Q'(2)} = \frac{1}{5}. \qquad \blacksquare$$

EXAMPLE 9 Find A_1 in the decomposition

$$\frac{P(s)}{Q(s)} = \frac{s^2+3}{(s-2)(s^2+9)} = \frac{A_1}{s-2} + \frac{Bs+C}{s^2+9}.$$

SOLUTION Here it is easier to cancel $s - 2$ so $Q_1(s) = s^2 + 9$ and $P(s) = s^2 + 3$, and so

$$A_1 = \frac{P(2)}{Q_1(2)} = \frac{7}{13}. \qquad \blacksquare$$

EXAMPLE 10 Find A_1, A_2, A_3, and A_4 in the decomposition

$$\frac{P(s)}{Q(s)} = \frac{1}{s^4 - 10s^2 + 9} = \frac{A_1}{s+1} + \frac{A_2}{s-1} + \frac{A_3}{s+3} + \frac{A_4}{s-3}.$$

SOLUTION $P(s) = 1$, $Q'(s) = 4s^3 - 20s$, and so

$$A_1 = \frac{1}{Q'(-1)} = \frac{1}{16}, \qquad A_2 = \frac{1}{Q'(1)} = -\frac{1}{16}$$

$$A_3 = \frac{1}{Q'(-3)} = -\frac{1}{48}, \qquad A_4 = \frac{1}{Q'(3)} = \frac{1}{48}.$$

This example really shows the efficiency of the formula. $\blacksquare$

HEAVISIDE FORMULA FOR CASE 2

If $Q(s)$ has a root c of multiplicity r then

$$\frac{P(s)}{Q(s)} = \frac{A_1}{s-c} + \frac{A_2}{(s-c)^2} + \cdots + \frac{A_r}{(s-c)^r} + N(s),$$

where

$$A_r = \phi(c), \qquad A_{r-1} = \frac{\phi'(c)}{1!}, \qquad A_{r-2} = \frac{\phi''(c)}{2!}, \qquad \ldots, \qquad A_1 = \frac{\phi^{(r-1)}(c)}{(r-1)!},$$

with

$$\phi(s) = (s-c)^r \frac{P(s)}{Q(s)}.$$

As before $N(s)$ represents the terms (if any) not involving $s - c$. The reader may wish to verify the formula by first multiplying the expression given for $P(s)/Q(s)$ by $(s - c)^r$, then evaluating the remaining expression and its higher derivatives at $s = c$.

EXAMPLE 11 Find A_1, A_2, and A_3 in the decomposition

$$\frac{P(s)}{Q(s)} = \frac{4s^2 - s + 2}{(s-1)^3} = \frac{A_1}{s-1} + \frac{A_2}{(s-1)^2} + \frac{A_3}{(s-1)^3}.$$

SOLUTION Set

$$\phi(s) = (s-1)^3 \frac{P(s)}{Q(s)} = 4s^2 - s + 2,$$

and then

$$A_3 = \phi(1) = 5, \qquad A_2 = \phi'(1) = 7, \qquad A_1 = \frac{\phi''(1)}{2!} = 4.$$ ■

EXAMPLE 12 Find

$$L^{-1}\left\{\frac{s^2 - 15}{s(s+4)^2}\right\}.$$

SOLUTION Write

$$\frac{s^2 - 15}{s(s+4)^2} = \frac{A_0}{s} + \frac{A_1}{s+4} + \frac{A_2}{(s+4)^2},$$

and then multiply by s and evaluate the remainder at $s = 0$ to get

$$A_0 = \left.\frac{s^2 - 15}{(s+4)^2}\right|_{s=0} = \frac{-15}{16}.$$

Furthermore, to find A_1 and A_2 we cancel $(s+4)^2$ to get

$$\phi(s) = \frac{s^2 - 15}{s} = s - \frac{15}{s},$$

and therefore

$$A_2 = \phi(-4) = -\frac{1}{4}, \quad \text{and} \quad A_1 = \phi'(-4) = \frac{31}{16}.$$

Hence

$$L^{-1}\left\{\frac{s^2 - 15}{s(s+4)^2}\right\} = L^{-1}\left\{-\frac{15}{16}\frac{1}{s} + \frac{31}{16}\frac{1}{s+4} - \frac{1}{4}\frac{1}{(s+4)^2}\right\}$$

$$= -\frac{15}{16} + \frac{31}{16}e^{-4t} - \frac{1}{4}te^{-4t}.$$ ■

There are Heaviside formulas for Cases 3 and 4, but they are far more complicated than the ones given. In applications where one has only one or two terms involving irreducible quadratics, one can use the above Heaviside formulas to calculate the coefficients corresponding to any simple or multiple roots. This will reduce the algebraic complications, and one can now either combine terms and compare like powers of s, or just substitute various selected values for s to obtain much simpler linear equations for the remaining coefficients.

EXERCISES 4.6

Using the Heaviside formulas and/or albegraic manipulation, find the unknown constants in the partial fraction expansions of Exercises 1–8.

1. $\dfrac{3s-1}{(s-2)(s+1)} = \dfrac{A_1}{s-2} + \dfrac{A_2}{s+1}$

2. $\dfrac{4s^2+2}{s(s-1)(s-3)} = \dfrac{A_1}{s} + \dfrac{A_2}{s-1} + \dfrac{A_3}{s-3}$

3. $\dfrac{2s+5}{(s+2)^2} = \dfrac{A_1}{s+2} + \dfrac{A_2}{(s+2)^2}$

4. $\dfrac{s^2-3s+1}{s^3} = \dfrac{A_1}{s} + \dfrac{A_2}{s^2} + \dfrac{A_3}{s^3}$

5. $\dfrac{4s^2+9}{(s-1)(s^2+2s+10)} = \dfrac{A_1}{s-1} + A_2\dfrac{s+1}{(s+1)^2+3^2} + A_3\dfrac{3}{(s+1)^2+3^2}$

(*Hint:* First write the expansion in the form

$$\frac{A_1}{s-1} + \frac{Bs+C}{(s+1)^2+3^2}$$

and find A_1, B, and C.)

6. $\dfrac{s+4}{s^2(s^2+4)} = \dfrac{A_1}{s} + \dfrac{A_2}{s^2} + A_3\dfrac{s}{s^2+4} + A_4\dfrac{2}{s^2+4}$

7. $\dfrac{6s}{(s-2)(s^2-2s+3)} = \dfrac{A_1}{s-2} + A_2\dfrac{s-1}{(s-1)^2+(\sqrt{2})^2} + A_3\dfrac{\sqrt{2}}{(s-1)^2+(\sqrt{2})^2}$

8. $\dfrac{s^3+2s^2-5s+14}{(s^2+9)^2} = A_1\dfrac{s}{s^2+9} + A_2\dfrac{3}{s^2+9} + A_3\dfrac{6s}{(s^2+9)^2} + A_4\dfrac{18}{(s^2+9)^2}$

(*Hint:* First write the expansion in the form

$$\frac{As+B}{s^2+9} + \frac{Cs+D}{(s^2+9)^2}$$

and clear out the denominators to get

$$(As+B)(s^2+9) + Cs + D = s^3 + 2s^2 - 5s + 14.$$

Now compare like powers of s to find A, B, C, and D.)

9. Write down the inverse Laplace transforms of whichever expansions you computed in Exercises 1–8.

In Exercises 10–19 find the inverse Laplace transforms.

10. $L^{-1}\left\{\dfrac{2s+1}{s^2+5s+6}\right\}$

11. $L^{-1}\left\{\dfrac{s+10}{s^3-2s^2-15s}\right\}$

12. $L^{-1}\left\{\dfrac{s-5}{s^2+2s+26}\right\}$

13. $L^{-1}\left\{\dfrac{s^2-5}{s^3+s^2+9s+9}\right\}$

14. $L^{-1}\left\{\dfrac{4s^2+2}{s^4-4s^2}\right\}$

15. $L^{-1}\left\{\dfrac{21}{(s-2)(s^2+4s+9)}\right\}$

16. $L^{-1}\left\{\dfrac{3s-1}{(s^2-2s+5)^2}\right\}$

17. $L^{-1}\left\{\dfrac{3s^2+8s+3}{(s^2+1)(s^2+9)}\right\}$

18. $L^{-1}\left\{\dfrac{1}{s^2+4}+\dfrac{5}{(s^2+4)^2}\right\}$

19. $L^{-1}\left\{\dfrac{3s+51}{(s+17)^5}\right\}$ (Look before you leap!)

20. Using the factorization in the complex numbers

$$(s-a)^2+b^2=[s-(a-bi)][s-(a+bi)],$$

one can apply Heaviside formula 1 to treat the single irreducible quadratic in Case 3. After the partial fraction expansion is accomplished, one uses the fact that

$$L^{-1}\left\{\frac{1}{s-(a\pm bi)}\right\}=e^{(a\pm bi)t}$$
$$=e^{at}(\cos bt\pm i\sin bt),$$

valid for $\operatorname{Re}(s)>a$. Use these facts to derive the formulas:

a) $L^{-1}\left\{\dfrac{b}{s^2+b^2}\right\}=\sin bt$

b) $L^{-1}\left\{\dfrac{s}{s^2+b^2}\right\}=\cos bt$

c) $L^{-1}\left\{\dfrac{b}{(s-a)^2+b^2}\right\}=e^{at}\sin bt$

d) $L^{-1}\left\{\dfrac{s-a}{(s-a)^2+b^2}\right\}=e^{at}\cos bt$

Using the Laplace transform solve the inhomogeneous initial value problems of Exercises 21–27.

21. $y''+3y'+2y=4e^{2t},\quad y(0)=2,\quad y'(0)=0$

22. $y''+5y'+6y=3e^{-2t},\quad y(0)=0,\quad y'(0)=1$

23. $y''+9y=2\cos 3t,\quad y(0)=1,\quad y'(0)=1$

24. $y''+25y=6e^{-t},\quad y(0)=1,\quad y'(0)=0$

25. $y''+2y'+5y=4t,\quad y(0)=0,\quad y'(0)=0$

26. $y''-4y'+13y=5\sin t,\quad y(0)=1,\quad y'(0)=4$

27. $y''+4y'+4y=-e^{-2t},\quad y(0)=2,\quad y'(0)=0$

Use the Laplace transform to solve the inhomogeneous systems of differential equations of Exercises 28–33.

28. $\dot{x} = x + 3y + t, \quad x(0) = 0, \quad \dot{y} = 3x + y + e^{-3t}, \quad y(0) = 0$

29. $\dot{x} = 2x - 2y + e^{-t}, \quad x(0) = 1, \quad \dot{y} = 4x - 2y - 2t, \quad y(0) = 1$

30. $\dot{x} = x - 2y - 4, \quad x(0) = 0, \quad \dot{y} = 2x + y + \sin t, \quad y(0) = 0$

31. $\dot{x} = 3x - 4y, \quad x(0) = 1, \quad \dot{y} = 4x - 5y + 2e^{-t}, \quad y(0) = 0$

32. $\dot{x} = x + 3y + 1, \quad x(0) = 0, \quad \dot{y} = 5x + 3y + 2\sin t, \quad y(0) = 1$

33. $\dot{x} = 4x + 5y + 3\cos 2t, \quad x(0) = 0, \quad \dot{y} = -4x - 4y, \quad y(0) = 1$

4.7 THE CONVOLUTION INTEGRAL: WEIGHTING FUNCTIONS

A natural question to ask is the following one: If $f(t)$ has a Laplace transform $F(s)$ and $g(t)$ has a Laplace transform $G(s)$, what function $h(t)$ has the Laplace transform $H(s) = F(s)G(s)$? A moment's reflection clearly shows the answer is not $h(t) = f(t)g(t)$. For instance, if $f(t) = g(t) = 1$, then $f(t)g(t) = 1$, but

$$H(s) = \frac{1}{s}\frac{1}{s} = \frac{1}{s^2},$$

whose inverse Laplace transform is $t \neq f(t)g(t)$.

The answer to the question is that $h(t)$ is an expression involving the integral of $f(t)$ and $g(t)$, called the *convolution integral*, or *convolution of f and g*:

$$f*g = \int_0^t f(t - \tau)g(\tau)\, d\tau.$$

Note that in the standard notation on the left the dependence on the variable t is understood and is omitted. To emphasize the dependence, it is sometimes useful to write $(f*g)(t)$.

EXAMPLE 1 Compute $t*\cos t$.

SOLUTION

$$t*\cos t = \int_0^t (t - \tau)\cos\tau\, d\tau = t\int_0^t \cos\tau\, d\tau - \int_0^t \tau\cos\tau\, d\tau$$

$$= t\sin t - \cos t - t\sin t + 1 = 1 - \cos t.$$

■

The simple change of variables $\tau = t - s$ in the convolution integral gives

$$f*g = -\int_t^0 f(s)g(t-s)\,ds = \int_0^t g(t-s)f(s)\,ds = g*f,$$

so we see that the order of the functions is immaterial.

The convolution integral, or, equivalently, the operation $(*)$, can be thought of as a generalized product between functions, and simple calculations show that $f*g$ satisfies

$$\begin{aligned} f*g &= g*f \quad \text{(commutativity)},\\ f*(g+h) &= f*g + f*h \quad \text{(distributivity)},\\ f*(g*h) &= (f*g)*h \quad \text{(associativity)}. \end{aligned}$$

However, the important property is the one mentioned at the beginning of this section, the convolution property.

Convolution property.
If $L\{f(t)\} = F(s)$ and $L\{g(t)\} = G(s)$ exist for $\text{Re}(s) > a$, then

$$L\{f*g\} = F(s)G(s), \qquad \text{Re}(s) > a,$$

or, equivalently,

$$f*g = L^{-1}\{F(s)G(s)\}.$$

The proof of this very important property is left until the end of this section, since we prefer first to give some applications of the property and a physical motivation of the convolution integral itself.

A very useful application of the convolution property is in the calculation of inverse Laplace transforms $L^{-1}\{H(s)\}$, where $H(s)$ is recognized as the product of two simple functions $F(s)$ and $G(s)$. An integration will have to be performed at the end, but that may be easier than working out a complicated partial fraction expansion.

EXAMPLE 2 Compute

$$L^{-1}\left\{\frac{1}{s^3(s^2+1)}\right\}.$$

SOLUTION This example, which can be easily done by using Heaviside's formulas, is mainly intended to demonstrate the convolution integral approach. One can write

$$\frac{1}{s^3(s^2+1)} = \left(\frac{1}{s^3}\right)\left(\frac{1}{s^2+1}\right),$$

and so

$$F(s) = \frac{1}{s^3} = L\left\{\frac{t^2}{2}\right\}$$

and

$$G(s) = \frac{1}{s^2+1} = L\{\sin t\}.$$

Therefore

$$\begin{aligned} L^{-1}\left\{\frac{1}{s^3(s^2+1)}\right\} &= \frac{t^2}{2} * \sin t = \frac{1}{2}\int_0^t (t-\tau)^2 \sin\tau\, d\tau \\ &= \frac{1}{2}t^2\int_0^t \sin\tau\, d\tau - t\int_0^t \tau\sin\tau\, d\tau + \frac{1}{2}\int_0^t \tau^2\sin\tau\, d\tau \\ &= -1 + \frac{t^2}{2} + \cos t. \end{aligned}$$

■

EXAMPLE 3 Compute

$$L^{-1}\left\{\frac{2b^2}{[s^2+b^2]^2}\right\}.$$

SOLUTION This example, which was discussed in the previous section, is ideally set up for using the convolution property. Just write

$$\frac{2b^2}{[s^2+b^2]^2} = 2\left(\frac{b}{s^2+b^2}\right)\left(\frac{b}{s^2+b^2}\right),$$

and since each of the terms in parentheses is the transform of $\sin bt$, we get

$$\begin{aligned} L^{-1}\left\{\frac{2b^2}{[s^2+b^2]^2}\right\} &= 2[\sin bt * \sin bt] \\ &= 2\int_0^t \sin b(t-\tau)\sin b\tau\, d\tau \\ &= 2\sin bt\int_0^t \cos b\tau\sin b\tau\, d\tau - 2\cos bt\int_0^t \sin^2 b\tau\, d\tau \\ &= \frac{1}{b}\sin^3 bt - t\cos bt + \frac{1}{2b}\sin 2bt\cos bt. \end{aligned}$$

Since $\sin 2bt = 2\sin bt\cos bt$, the answer previously given will be obtained after a little trigonometrical manipulation. ■

The convolution integral approach is especially efficient when the denominator of $F(s)$ is a product of irreducible quadratic factors.

The real importance of the convolution integral is that it is one of the building blocks in the analysis of linear systems. This is best illustrated by an example. Consider the general initial value problem

$$y'' + ay' + by = f(t), \qquad y(0) = 0, \qquad y'(0) = 0,$$

which can be regarded as a mechanical or electrical system at rest excited by an input or driving term $f(t)$. First apply the Laplace transform to obtain

$$s^2Y(s) + asY(s) + bY(s) = L\{f(t)\} = F(s)$$

and hence

$$Y(s) = \frac{F(s)}{s^2 + as + b} = P(s)F(s).$$

The function $P(s) = (s^2 + as + b)^{-1}$ is called the *transfer function* of the system described by the differential equation above.

In the s-domain the system could be described by the diagram in Fig. 4.5, where the box represents the action of the system at rest on the input, namely, multiplication by the transfer function. Now apply the inverse Laplace transform to get

$$y(t) = L^{-1}\{Y(s)\} = L^{-1}\{P(s)(F(s)\},$$

and by the convolution property

$$y(t) = p(t)*f(t),$$

where $p(t) = L^{-1}\{P(s)\}$ is called the *weighting function* of the system, and $y(t)$ is called the *zero-state output.*

In the t-domain the system could then be described by the diagram in Fig. 4.6, where the box represents the action of the system at rest on the input, namely, convolution with the weighting function. In the last section of this chapter this analysis will be extended further, but for now we wish to study the effect of the weighting function on the input by examining the convolution integral a little closer.

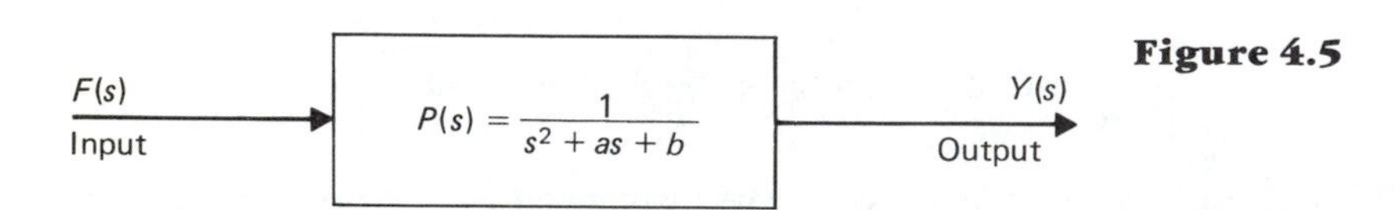

Figure 4.5

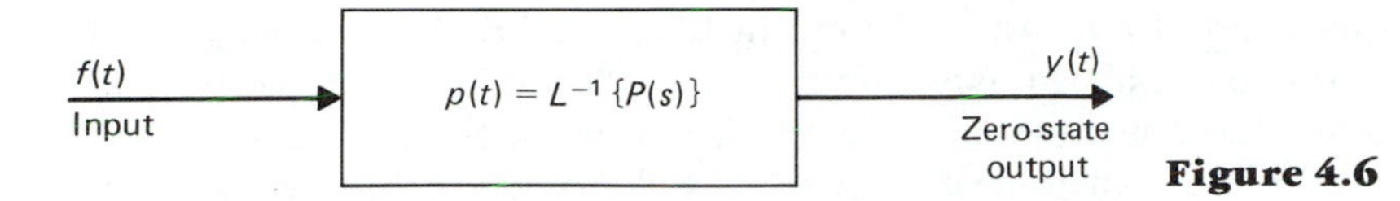

Figure 4.6

If the equation $y(t) = p(t)*f(t)$, which describes the zero-state output, is written in the integral form, then

$$y(t) = \int_0^t p(\tau)f(t - \tau)\, d\tau$$

since $p*f = f*p$. Since $0 \leq \tau \leq t$, the integral can be interpreted by saying that the input τ units in the past, namely $f(t - \tau)$, is weighted by the value of the weighting function evaluated at time τ. The idea of regarding the zero-state output as a weighted integral in turn allows us to discuss the *memory* of the system.

Suppose, for example, that the input $f(t)$ is a bounded function and that the weighting function $p(t)$ is a function that goes to zero as τ approaches ∞ and is negligible for $\tau \geq T$. Its graph might look like that in Fig. 4.7. Then $y(t)$ for $t > T$ can be expressed as

$$y(t) = \int_0^T p(\tau)f(t - \tau)\, d\tau + \int_T^t p(\tau)f(t - \tau)\, d\tau,$$

and by the assumptions on $p(\tau)$ the second integral can be considered negligible. A way of saying this is that values of the input for more than T units in the past have very little effect on the output; i.e., for any t, the contributions of $f(t - \tau)$ to the output are negligible for $\tau \geq T$.

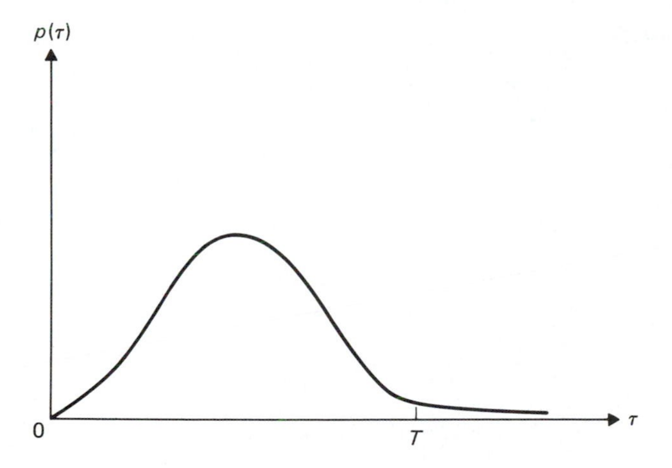

Figure 4.7

On the other hand, if $p(\tau)$ is a function which is, say, strictly increasing, then the second integral in the above expression for $y(t)$ will not be negligible for any $T > 0$. In this case, the system is said to have *infinite memory,* and for any t and any τ, where $0 \leq \tau \leq t$, the contribution of $f(t - \tau)$ will strongly influence the zero-state output. For practical "real-world" systems, one would not expect this kind of behavior.

We conclude this section with a proof of the convolution property that depends on the following very useful formula for the change of order of integration in a double integral:

$$\int_a^b \left[\int_a^t F(t, \tau)\, d\tau \right] dt = \int_a^b \left[\int_\tau^b F(t, \tau)\, dt \right] d\tau.$$

The justification for this formula follows easily from a diagram of the region of integration shown in Fig. 4.8. In the first integral the region of integration is described by $a \leq \tau \leq t$ and $a \leq t \leq b$, whereas in the second integral it is described by $\tau \leq t \leq b$ and $a \leq \tau \leq b$. These clearly describe the same region.

We can now proceed directly to prove the convolution property. By the definition of the convolution integral and of the Laplace transform, we have

$$\begin{aligned} L\{f*g\} &= \int_0^\infty e^{-st} \left[\int_0^t f(t - \tau) g(\tau)\, d\tau \right] dt \\ &= \int_0^\infty \left[\int_0^t e^{-st} f(t - \tau) g(\tau)\, d\tau \right] dt. \end{aligned}$$

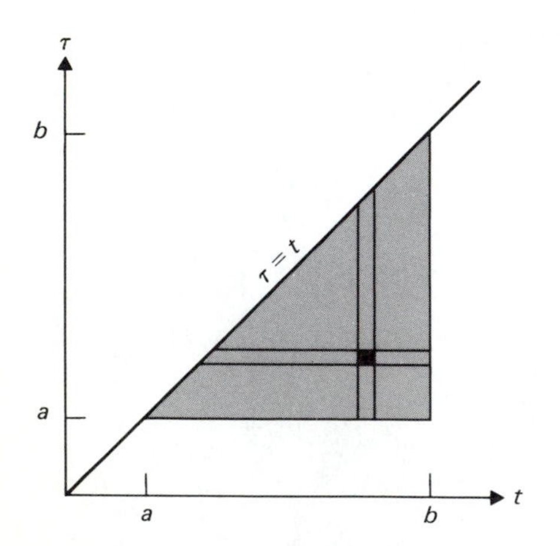

Figure 4.8

Apply the formula for the interchange of the order of integration with $a = 0, b = \infty$ to obtain

$$L\{f*g\} = \int_0^\infty \left[\int_\tau^\infty e^{-st}f(t-\tau)g(\tau)\,dt\right]d\tau.$$

In the bracketed integral, make the change of variables $t = \tau + r$ to obtain finally

$$\begin{aligned} L\{f*g\} &= \int_0^\infty \left[\int_0^\infty e^{-s(\tau+r)}f(r)g(\tau)\,dr\right]d\tau \\ &= \int_0^\infty e^{-s\tau}g(\tau)\left[\int_0^\infty e^{-sr}f(r)\,dr\right]d\tau \\ &= \left[\int_0^\infty e^{-sr}f(r)\,dr\right]\left[\int_0^\infty e^{-s\tau}g(\tau)\,d\tau\right] \\ &= L\{f\}L\{g\} = F(s)G(s), \end{aligned}$$

which is the desired result.

EXERCISES 4.7

In Exercises 1–10 compute the convolution integrals $f*g$ directly, then verify your result by computing $L^{-1}\{F(s)G(s)\}$.

1. $t*t$ **2.** $t*e^{-t}$ **3.** e^t*e^{-t} **4.** $e^{-t}*\cos t$ **5.** $1*\sin 2t$
6. $t^2*\cos t$ **7.** $e^{-3t}*e^t$ **8.** $t*t^2$ **9.** $\sin 3t*\sin t$ **10.** $e^t*\sin 2t$

Use the convolution property to compute the inverse Laplace transforms in Exercises 11–18.

11. $L^{-1}\left\{\dfrac{1}{s^2}\cdot\dfrac{1}{s+4}\right\}$ **12.** $L^{-1}\left\{\dfrac{1}{s^2}\cdot\dfrac{1}{s^2+9}\right\}$ **13.** $L^{-1}\left\{\dfrac{1}{s^2+1}\cdot\dfrac{1}{s^2+9}\right\}$

14. $L^{-1}\left\{\dfrac{s^2}{(s^2+4)^2}\right\}$ **15.** $L^{-1}\left\{\dfrac{s^2}{s^4-1}\right\}$ **16.** $L^{-1}\left\{\dfrac{s}{(s^2+16)^2}\right\}$

17. $L^{-1}\left\{\dfrac{s}{s^2-1}\cdot\dfrac{1}{s^2-1}\right\}$ **18.** $L^{-1}\left\{\dfrac{1}{s^4(s^2+1)}\right\}$

In Exercises 19–26 find the transfer function and weighting function of the systems described by the differential equations.

19. $y'' - 3y' + 10y = f(t)$ **20.** $y'' + 16y = f(t)$ **21.** $y'' - 9y = f(t)$
22. $y'' + 4y' + 10y = f(t)$ **23.** $y'' + 4y' + 4y = f(t)$ **24.** $y'' + 7y' + 12y = f(t)$
25. $y'' + 3y' + 3y = f(t)$ **26.** $y'' + 6y = f(t)$

27. Assume that the weighting function of a system is

$$p(t) = \begin{cases} t & \text{if } 0 \le t \le \frac{1}{2}, \\ 1 - t & \text{if } \frac{1}{2} \le t \le 1, \\ 0 & \text{if } t > 1, \end{cases}$$

and the input is $f(t) = e^t$.

a) Show that the zero-state output for $t > 1$ is $Kf(t-1) = Ke^{t-1}$, where K is some constant, and hence the system has memory $T = 1$.

b) Using directly the definition of the Laplace transform, find the transfer function $P(s)$.

28. a) Using the formula for the change of order of integration in a double integral, show that

$$\int_0^t \int_0^r f(\tau)\, d\tau\, dr = \int_0^t (t - \tau) f(\tau)\, d\tau.$$

b) Using the convolution property and the above result, verify the following formulas:

$$L^{-1}\left\{\frac{1}{s} F(s)\right\} = \int_0^t f(\tau)\, d\tau,$$

$$L^{-1}\left\{\frac{1}{s^2} F(s)\right\} = \int_0^t \int_0^r f(\tau)\, d\tau\, dr.$$

29. A Volterra integral equation of convolution or renewal type is of the form

$$y(t) = f(t) + \int_0^t k(t - \tau) y(\tau)\, d\tau,$$

where $f(t)$ and $k(t)$ (called the *kernel of the equation*) are given functions, and $y(t)$ is the solution sought. It is ideally suited for analysis using the Laplace transform. Assuming that $f(t)$ and $k(t)$ have Laplace transforms $F(s)$ and $K(s)$, respectively, show that the Laplace transform $Y(s)$ satisfies the relation

$$Y(s) = \frac{F(s)}{1 - K(s)}.$$

Use the result of Exercise 29 to find the solution $y(t)$ of the integral equations in Exercises 30–33.

30. $y(t) = 4t + \int_0^t \sin(t - \tau)y(\tau)\,d\tau$

31. $y(t) = \cos t - \int_0^t e^{t-\tau}y(\tau)\,d\tau$

32. $y(t) = a \sin t + c\int_0^t \sin(t - \tau)y(\tau)\,d\tau$; consider the three cases $c < 1$, $c = 1$, and $c > 1$.

33. $y(t) = \sin t - \int_0^t (t - \tau)y(\tau)\,d\tau$

4.8 THE UNIT STEP FUNCTION: TRANSFORMS OF NONSMOOTH FUNCTIONS

It was stated in Section 4.5 that the Laplace transform is a very effective tool for solving differential equations where the forcing term is a nonsmooth function. Such a function could have one or more jump discontinuities (i.e., the graph of the function could have breaks in it) or it could be a continuous function whose first derivative is discontinuous. In the latter case the graph of the function would have "corners" in it—a simple example of such a function would be $f(t) = |t|$. We see that $f'(t) = -1$ if $t < 0$ and $f'(t) = 1$ if $t > 0$, whereas $f(t)$ is continuous for all t.

One of the basic tools for describing such nonsmooth functions and for computing their Laplace transforms is the *unit step,* or *Heaviside, function.* It is described as follows:

$$u(t) = \begin{cases} 0 & \text{if } t < 0, \\ 1 & \text{if } t \geq 0. \end{cases}$$

The reader should note that in some texts $u(0)$ may be defined to be 0 instead of 1, or the equality sign at $t = 0$ may be dropped completely, so that $u(0)$ is undefined. In later applications the Laplace transform of the product of $u(t)$ with a given function will be computed, so the value of $u(0)$ will not affect the result. It follows that for $c > 0$

$$u(t - c) = \begin{cases} 0 & \text{if } t < c, \\ 1 & \text{if } t \geq c, \end{cases}$$

so that $u(t - c)$ has a simple jump discontinuity of value 1 at $t = c$ (see Fig. 4.9).

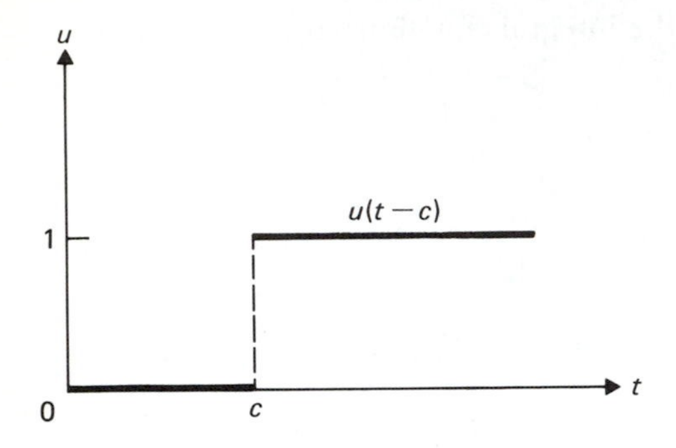

Figure 4.9

To use the unit step function as a tool, first observe the following important property: If $f(t)$ is a function defined for $t \geq 0$ and if $c > 0$, then

$$f(t-c)u(t-c) = \begin{cases} 0 & \text{if } t < c, \\ f(t-c) & \text{if } t \geq c, \end{cases}$$

so the effect of multiplying $f(t-c)$ by $u(t-c)$ is to shift the graph of $f(t)$ by c units to the right (see Fig. 4.10). This basic property will be used to construct some non-smooth functions in the following examples.

EXAMPLE 1 A pulse of magnitude 4 and duration from $t = 1$ to $t = 2$ could be described by the function

$$f(t) = \begin{cases} 0 & \text{if } t < 1, \\ 4 & \text{if } 1 \leq t < 2, \\ 0 & \text{if } t \geq 2, \end{cases}$$

as shown in Fig. 4.11.

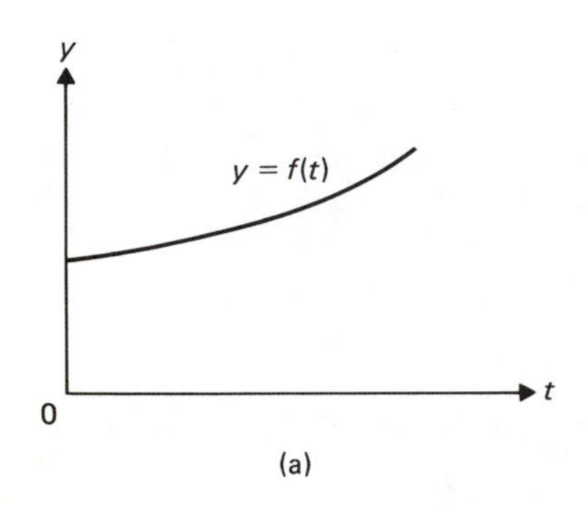

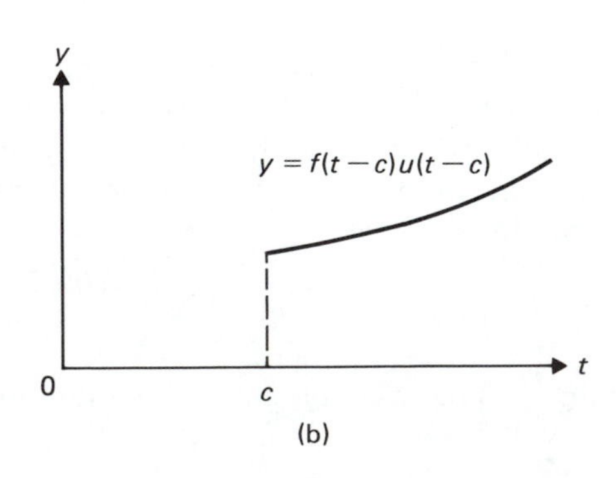

Figure 4.10

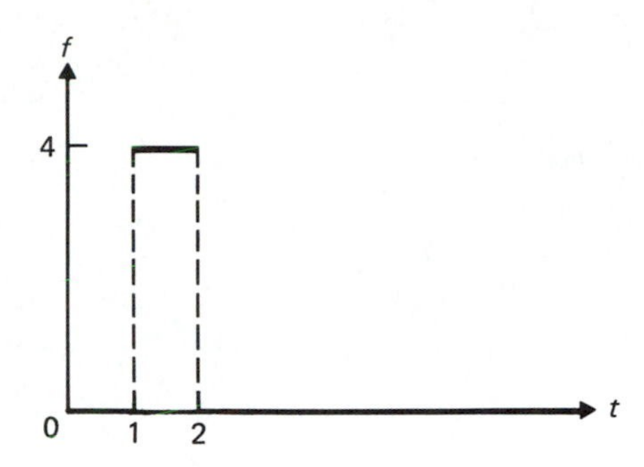

Figure 4.11

From the property above with $f(t) = 4$ it follows that the function $4u(t-1)$ describes a function that is zero for $t < 1$ and equals 4 for $t \geq 1$ (see Fig. 4.12).

To "lop off" the contribution for $t \geq 2$, a function must be subtracted that is zero for $t < 2$ and equals 4 for $t \geq 2$. But that function, by the same reasoning, is $4u(t-2)$; hence

$$f(t) = 4u(t-1) - 4u(t-2).$$ ■

EXAMPLE 2 The preceding example leads to another useful fact, which the reader can easily verify: If $a < b$, then

$$g(t)[u(t-a) - u(t-b)] = \begin{cases} 0 & \text{if } t < a, \\ g(t) & \text{if } a \leq t < b, \\ 0 & \text{if } t \geq b. \end{cases}$$

This fact is used to construct the triangular pulse (Fig. 4.13) given by

$$f(t) = \begin{cases} t & \text{if } 0 \leq t \leq 1, \\ 2-t & \text{if } 1 < t \leq 2, \\ 0 & \text{if } t > 2. \end{cases}$$

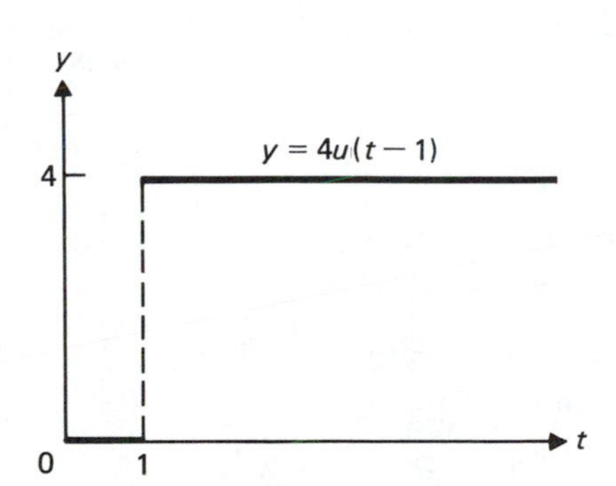

Figure 4.12

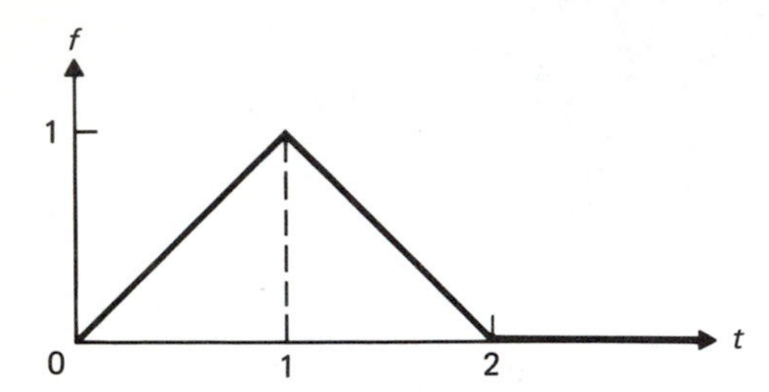

Figure 4.13

This is a continuous function but its derivative is discontinuous at $t = 1$ and $t = 2$, where the graph has corners.

Using the fact above we see that

$$t[u(t) - u(t-1)] = \begin{cases} t & \text{if} \quad 0 \le t < 1, \\ 0 & \text{if} \quad t \ge 1, \end{cases}$$

and

$$(2-t)[u(t-1) - u(t-2)] = \begin{cases} 0 & \text{if} \quad t < 1, \\ 2-t & \text{if} \quad 1 \le t < 2, \\ 0 & \text{if} \quad t \ge 2. \end{cases}$$

Adding the two parts together, we conclude that

$$\begin{aligned} f(t) &= t[u(t) - u(t-1)] + (2-t)[u(t-1) - u(t-2)] \\ &= t + (2-2t)u(t-1) - (2-t)u(t-2), \quad t \ge 0, \end{aligned}$$

is the required pulse. ■

EXAMPLE 3 Find the representation in terms of unit step functions of the function

$$f(t) = \begin{cases} 2 & \text{if} \quad 0 \le t < 2, \\ -2 & \text{if} \quad 2 \le t < 4, \\ e^{-t} & \text{if} \quad t \ge 4. \end{cases}$$

SOLUTION The function is composed of three parts

$$2[u(t) - u(t-2)] = \begin{cases} 2 & \text{if} \quad 0 \le t < 2, \\ 0 & \text{if} \quad t \ge 2; \end{cases}$$

$$(-2)[u(t-2) - u(t-4)] = \begin{cases} 0 & \text{if} \quad t < 2, \\ -2 & \text{if} \quad 2 \le t < 4, \\ 0 & \text{if} \quad t \ge 4; \end{cases}$$

and

$$e^{-t}[u(t-4)] = \begin{cases} 0 & \text{if } 0 \leq t < 4, \\ e^{-t} & \text{if } t \geq 4. \end{cases}$$

Adding the three parts together gives

$$\begin{aligned} f(t) &= 2[u(t) - u(t-2)] + (-2)[u(t-2) - u(t-4)] + e^{-t}[u(t-4)] \\ &= 2 - 4u(t-2) + (2 + e^{-t})u(t-4). \end{aligned}$$ ■

EXAMPLE 4 Using the facts above and proceeding interval by interval, it is easy to construct the representation of the following square wave $S(t)$ with period $2a$ shown in Fig. 4.14.

Its representation is given by the series

$$\begin{aligned} S(t) = {} & (1)[u(t) - u(t-a)] + (-1)[u(t-a) - u(t-2a)] \\ & + (1)[u(t-2a) - u(t-3a)] \\ & + (-1)[u(t-3a) - u(t-4a)] \\ & + (1)[u(t-4a) - u(t-5a)] \\ & + \cdots, \quad t \geq 0. \end{aligned}$$

Combining the terms gives

$$\begin{aligned} S(t) &= 1 - 2u(t-a) + 2u(t-2a) - 2u(t-3a) + 2u(t-4a) + \cdots \\ &= 1 + 2\sum_{n=1}^{\infty} (-1)^n u(t-na), \quad t \geq 0. \end{aligned}$$

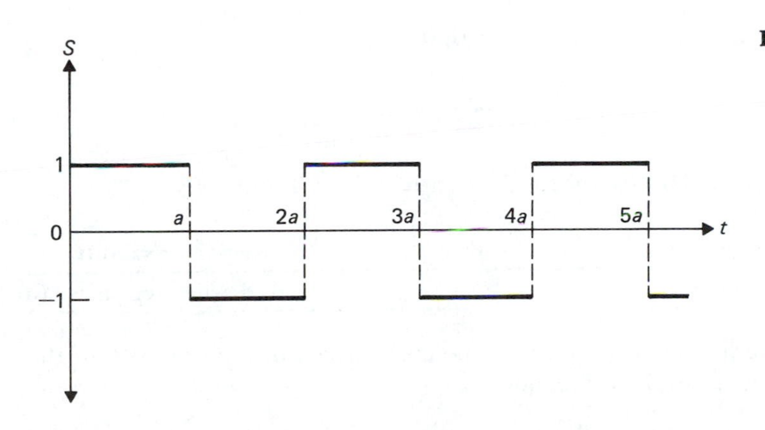

Figure 4.14

A simpler representation is obtained by noting the periodicity of $f(t)$ and just describing $S(t)$ over the first period:

$$S(t) = 1 - 2u(t - a) + u(t - 2a), \quad 0 \le t \le 2a,$$
$$S(t + 2a) = S(t), \quad t \ge 0. \qquad \blacksquare$$

The above examples should indicate to the reader the usefulness of the unit step function in representing nonsmooth functions.

We now come to the business of computing the Laplace transform of nonsmooth functions, such as those in the above examples. The first task is to compute the Laplace transform of $u(t - c)$, but this follows directly from the definition. Since $u(t - c)$ is zero for $t < c$ and equals 1 for $t \ge c$, then

$$\begin{aligned} L\{u(t - c)\} &= \int_0^\infty e^{-st}u(t - c)\,dt \\ &= \int_c^\infty e^{-st}\,dt = \frac{e^{-sc}}{s}, \end{aligned}$$

valid for $\mathrm{Re}(s) > 0$. Notice that the Laplace transform of $u(t - c)$ is simply the Laplace transform of $f(t) = 1$ multiplied by e^{-sc}.

We find that it is an equally simple matter to compute the Laplace transform of $f(t - c)u(t - c)$. Suppose $f(t)$ has a Laplace transform $F(s)$ valid for $\mathrm{Re}(s) > a$. Then,

$$\begin{aligned} L\{f(t - c)u(t - c)\} &= \int_0^\infty e^{-st}f(t - c)u(t - c)\,dt \\ &= \int_c^\infty e^{-st}f(t - c)\,dt. \end{aligned}$$

Making the change of variables $r = t - c$ gives

$$\begin{aligned} L\{f(t - c)u(t - c)\} &= \int_0^\infty e^{-s(r+c)}f(r)\,dr \\ &= e^{-sc}\int_0^\infty e^{-sr}f(r)\,dr = e^{-sc}L\{f\} = e^{-sc}F(s), \end{aligned}$$

also valid for $\mathrm{Re}(s) > a$. This useful result is entry (16) of Table 4.1:

	$f(t)$	$F(s)$	Region of validity
16.	$f(t - c)u(t - c)$	$e^{-sc}F(s)$	valid for same region as $F(s)$

The reader should notice that this entry is the counterpart in the s-domain of the shift property (entry 12) in the t-domain.

We are now ready to consider some differential equations with nonsmooth forcing terms.

EXAMPLE 5 Find the solution of the initial value problem

$$y'' + y = f(t), \qquad y(0) = 1, \qquad y'(0) = 0,$$

where $f(t)$ is the rectangular pulse given in Example 1 above.

SOLUTION We substitute the expression derived for $f(t)$ to obtain

$$y'' + y = 4u(t-1) - 4u(t-2), \qquad y(0) = 1, \qquad y'(0) = 0.$$

If $Y(s)$ is the Laplace transform of $y(t)$, use the differentiation formulas and entry (16) above with $f(t) = 1$ to get

$$s^2Y(s) - s + Y(s) = 4\frac{e^{-s}}{s} - 4\frac{e^{-2s}}{s}.$$

Solving for $Y(s)$ and computing the required partial fraction expansions gives

$$Y(s) = \frac{s}{s^2+1} + 4e^{-s}\left[\frac{1}{s} - \frac{s}{s^2+1}\right] - 4e^{-2s}\left[\frac{1}{s} - \frac{s}{s^2+1}\right].$$

From Table 4.1,

$$L^{-1}\left\{\frac{1}{s} - \frac{s}{s^2+1}\right\} = 1 - \cos t,$$

and from entry (16) it follows again that the solution is

$$y(t) = \cos t + 4[1 - \cos(t-1)]u(t-1) - 4[1 - \cos(t-2)]u(t-2).$$

The reader can verify from the above expression that the solution can also be written as

$$y(t) = \begin{cases} \cos t & \text{if } 0 \le t \le 1, \\ \cos t + 4 - 4\cos(t-1) & \text{if } 1 \le t \le 2, \\ \cos t - 4\cos(t-1) + 4\cos(t-2) & \text{if } t \ge 2. \end{cases}$$ ■

EXAMPLE 6 The next example is a slight variant of the previous example. Find the zero-state output for the equation

$$y'' + y = \phi(t),$$

where $\phi(t)$ is the function (see Fig. 4.15)

$$\phi(t) = t[1 - u(t-1)].$$

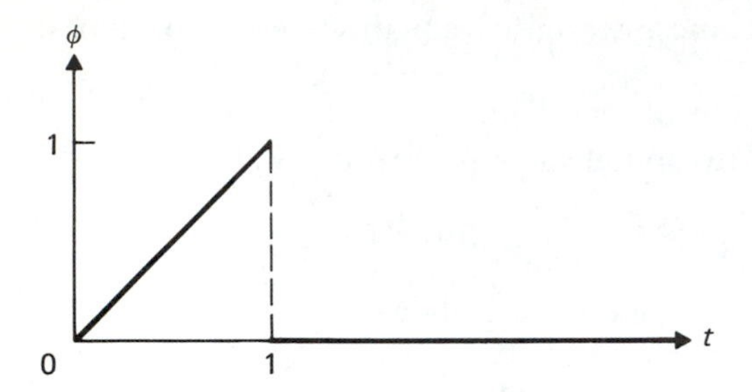

Figure 4.15

SOLUTION As it stands, $\phi(t)$ is not in a form suitable for application of entry (16), but it can be rewritten as

$$\phi(t) = t - tu(t-1) = t - [(t-1)+1]u(t-1).$$

The second term in the last expression is in the form $f(t-1)u(t-1)$, where $f(t) = t+1$, and now the Laplace transform can be applied. Since zero-state output implies that the initial conditions are $y(0) = y'(0) = 0$, applying the Laplace transform gives

$$s^2Y(s) + Y(s) = L\{\phi(t)\} = \frac{1}{s^2} - e^{-s}\left[\frac{1}{s^2} + \frac{1}{s}\right].$$

It is left to the reader to show that

$$y(t) = t - \sin t - [t - \sin(t-1) - \cos(t-1)]u(t-1)$$

is the zero-state output corresponding to the input $\phi(t)$. The solution $y(t)$ can also be written as

$$y(t) = \begin{cases} t - \sin t & \text{if} \quad 0 \le t \le 1, \\ -\sin t + \sin(t-1) + \cos(t-1) & \text{if} \quad t > 1, \end{cases}$$

and the reader should note that both $y(t)$ and $y'(t)$ are continuous at the discontinuity $t = 1$ of $\phi(t)$. By evaluating both expressions above and their derivatives at $t = 1$ one sees that they agree at $t = 1$, and

$$y(1) = 1 - \sin 1, \qquad y'(1) = 1 - \cos 1.$$

However, a further differentiation shows that the second derivative of the solution is not continuous at $t = 1$. A graph of the solution in a neighborhood of $t = 1$ is shown in Fig. 4.16. ■

The above example indicates that discontinuous forcing terms $\phi(t)$ will not result in discontinuous solutions, but will produce discontinuities in the higher derivatives of the solution. This can be easily seen if one writes the zero-state out-

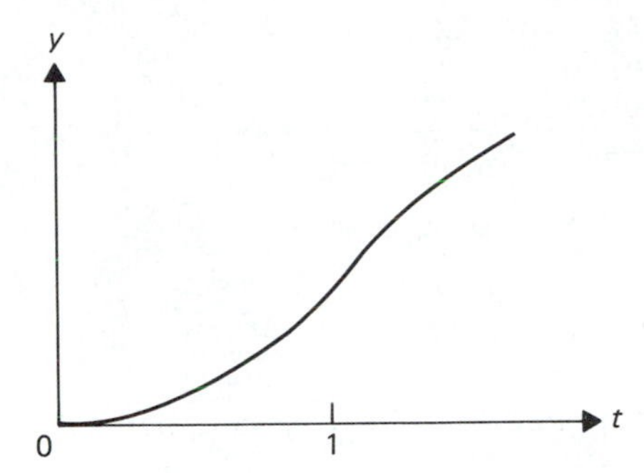

Figure 4.16 Graph of $y(t)$.

put (which is a particular solution) in the convolution form

$$y(t) = \int_0^t p(t - \tau)\phi(\tau)\, d\tau.$$

The weighting function $p(t)$ is a solution of the homogeneous equation, so it is a smooth function, and so if $\phi(t)$ is a piecewise continuous function then $y(t)$ will be continuous.

Furthermore by Leibniz's formula for differentiating an integral we have

$$y'(t) = p(0)\phi(t) + \int_0^t p'(t - \tau)\phi(\tau)\, d\tau.$$

If $p(0) = 0$, as is the case when $p(t) = L^{-1}\{(s^2 + as + b)^{-1}\}$, then by the same argument given above we see that $y'(t)$ will also be continuous whenever $\phi(t)$ is a piecewise continuous function.

The last example really shows the effectiveness of the Laplace transform.

EXAMPLE 7 Find the zero-state output for

$$y'' + y = S(t),$$

where $S(t)$ is the square wave of period $2a$ given in Example 4 above.

SOLUTION If one uses its series representation

$$S(t) = 1 + 2 \sum_{n=1}^{\infty} (-1)^n u(t - na),$$

then $L\{S(t)\}$ can be computed by transforming term by term and by using the linearity of the transform. Therefore

$$L\{S(t)\} = \frac{1}{s} + 2 \sum_{n-1}^{\infty} (-1)^n \frac{e^{-sna}}{s};$$

now applying the transform to the differential equation gives

$$s^2Y(s) + Y(s) = \frac{1}{s} + 2\sum_{n=1}^{\infty}(-1)^n\frac{e^{-sna}}{s}$$

or

$$Y(s) = \frac{1}{s(s^2+1)} + 2\sum_{n=1}^{\infty}(-1)^n\frac{e^{-sna}}{s(s^2+1)}.$$

But the inverse Laplace transform can also be applied term by term, and hence

$$y(t) = (1 - \cos t) + 2\sum_{n=1}^{\infty}(-1)^n[1 - \cos(t - na)]u(t - na)$$

is the desired output. Notice that for any given value of t the sum will be finite since $(t - na) < 0$ for all n sufficiently large, and the corresponding value of $u(t - na)$ will be zero. ■

In the computation of the Laplace transform of the periodic square wave its given series representation was used. More often one is given $f(t)$ explicitly defined for $0 \leq t \leq T$, and the periodicity condition $f(t + T) = f(t)$. A general formula will now be developed for the transform of a periodic function and added to the table of Laplace transforms.

If $f(t + T) = f(t)$, where T is the period, then it follows easily that $f(t + nT) = f(t)$ for any positive (or negative) integer n. Now write the integral for $L\{f\}$ as a sum of integrals over intervals of length T as follows:

$$L\{f\} = \int_0^{\infty} e^{-st}f(t)\,dt = \sum_{n=0}^{\infty}\int_{nT}^{(n+1)T} e^{-st}f(t)\,dt = \sum_{n=0}^{\infty} e^{-snT}\int_0^T e^{-sr}f(r)\,dr$$

since $f(t)$ is periodic with period T, or

$$L\{f\} = \left[\int_0^T e^{-sr}f(r)\,dr\right]\sum_{n=0}^{\infty}(e^{-sT})^n.$$

Now make use of the fact that

$$\sum_{n=0}^{\infty} z^n = 1 + z + z^2 + \cdots = \frac{1}{1-z} \qquad \text{if } \operatorname{Re}(z) < 1,$$

and let $z = e^{-sT}$. Then $|z| < 1$ if $\operatorname{Re}(s) > 0$, and the formula gives

$$\sum_{n=0}^{\infty}(e^{-sT})^n = \frac{1}{1 - e^{-sT}}.$$

Thus another useful entry can be added to Table 4.1:

	$f(t)$	$F(s)$	Region of validity
17.	$f(t+T)=f(t)$	$\dfrac{\int_0^T e^{-sr}f(r)\,dr}{1-e^{-sT}}$	$\mathrm{Re}(s)>0$

If entry (17) is used instead of the series to compute the transform of the square wave $S(t)$, one obtains

$$L\{S(t)\} = \frac{\int_0^{2a} e^{-sr}S(r)\,dr}{1-e^{-2sa}} = \frac{\int_0^{a} e^{-sr}(1)\,dr + \int_a^{2a} e^{-sr}\,(-1)\,dr}{1-e^{-2sa}}$$

$$= \frac{1}{s}\frac{(1-e^{-sa})^2}{1-e^{-2sa}} = \frac{1}{s}\left[\frac{1-e^{-sa}}{1+e^{-sa}}\right] = \frac{1}{s}\tanh\frac{sa}{2}\,.$$

The Laplace transform of the zero-state solution $y'' + y = S(t)$ is then

$$Y(s) = \left(\frac{1}{s^2+1}\right)\left(\frac{1}{s}\tanh\frac{sa}{2}\right) = L\{\sin t\}L\{S(t)\},$$

and now the convolution property gives another representation of the solution

$$y(t) = (\sin t)*(S(t)) = \int_0^t \sin(t-\tau)S(\tau)\,d\tau.$$

The series representation obtained above for $y(t)$ is therefore an explicit representation of the convolution integral. One easily sees from the representation that the solution is a continuous and differentiable function.

This section is concluded with a comment that modifies our previous discussion of E_α, the functions of exponential order, and of the Laplace transform. As has been clearly demonstrated in this section, one can drop the requirement of continuity for functions whose Laplace transform exists. Instead, it can be replaced with the weaker requirement that it be sectionally continuous, defined as follows.

Definition

A function is *sectionally continuous* on $0 \le t < \infty$ if it is continuous on every finite interval with exception of at most a finite number of points, where it has a simple jump discontinuity.

A little contemplation will convince the reader that every sectionally continuous function can be represented by a sum of continuous functions, each multiplied by the difference of two unit step functions.

Given the above, we next modify the definition of functions of exponential order.

Definition

A function $f(t)$, $0 \leq t < \infty$, is said to be of *exponential order for t sufficiently large* if there exist positive constants M and T and a constant α such that

$$|f(t)| \leq Me^{\alpha t}, \quad t \geq T.$$

Finally, putting the two definitions together provides us with a more general class of functions whose Laplace transforms exist.

If $f(t)$, $0 \leq t < \infty$, is sectionally continuous and is of exponential order for t sufficiently large, then its Laplace transform exists.

The proof of this statement follows from the estimate:

$$\begin{aligned}\left|\int_0^\infty f(t)e^{-st}\,dt\right| &\leq \int_0^\infty |f(t)|\,|e^{-st}|\,dt \\ &\leq \int_0^T |f(t)|\,|e^{-st}|\,dt + \int_T^\infty |f(t)|\,|e^{-st}|\,dt.\end{aligned}$$

In the last expression, the first integral is bounded since $f(t)$ is sectionally continuous, whereas for T sufficiently large, the second integral converges by an argument exactly like the one given previously in Section 4.4.

EXERCISES 4.8

For each of the functions in Exercises 1–8, sketch its graph, find an expression for it in terms of the unit step function $u(t)$, and find its Laplace transform. All functions are defined only for $t \geq 0$.

1. $f(t) = \begin{cases} 1 & \text{if} \quad t < 2, \\ -1 & \text{if} \quad 2 \leq t < 4, \\ 0 & \text{if} \quad t \geq 4 \end{cases}$

2. $f(t) = \begin{cases} 0 & \text{if} \quad t < 1, \\ 2 & \text{if} \quad 1 \leq t < 2, \\ -2 & \text{if} \quad 2 \leq t < 3, \\ 0 & \text{if} \quad t \geq 3 \end{cases}$

3. $f(t) = \begin{cases} t & \text{if } t < 1, \\ 2 - t & \text{if } 1 \le t < 3, \\ t - 4 & \text{if } 3 \le t < 4, \\ 0 & \text{if } t \ge 4 \end{cases}$

4. $f_n(t) = \begin{cases} nt & \text{if } t < 1/n, \\ 1 & \text{if } t \ge 1/n \end{cases}$
Discuss what happens when $n \to \infty$.

5. $f(t) = \begin{cases} 0 & \text{if } t < 2\pi, \\ \sin t & \text{if } t \ge 2\pi \end{cases}$

6. $f(t) = \begin{cases} 4e^{1-t} & \text{if } t < 1, \\ 4 & \text{if } t \ge 1 \end{cases}$

7. $f(t) = \begin{cases} |\sin t| & \text{if } t < 2\pi, \\ 0 & \text{if } t \ge 2\pi \end{cases}$

8. $f(t) = \begin{cases} t - 1 & \text{if } t < 2, \\ t - 3 & \text{if } 2 \le t < 4, \\ 0 & \text{if } t \ge 4 \end{cases}$

Sketch the graphs of the functions in Exercises 9–16 and find their Laplace transforms.

9. $u(t - 1) + 4u(t - 2)$
10. $2tu(t - \frac{1}{2})$
11. $e^{t-2}u(t - 3)$
12. $u(t - a) \sin t$; discuss the cases $a = \pi$, $a = 2\pi$.
13. $t + (t - 1)u(t - 1)$
14. $t^2u(t - 1)$
15. $(t - 2)u(t - 1) - (t - 1)u(t - 2)$
16. $t^2[2 - u(t - 2)]$

Find the solution of the initial value problems in Exercises 17–24.

17. $y'' + 3y' + 2y = f(t)$, $y(0) = 0$, $y'(0) = 0$, where $f(t)$ is given in Exercise 1 above. Also, describe the solution without using unit step functions.

18. $y'' + 9y = f(t)$, $y(0) = 1$, $y'(0) = 0$, where $f(t)$ is given in Exercise 2 above. Also, describe the solution without using unit step functions.

19. $y'' + y = f(t)$, $y(0) = 0$, $y'(0) = 0$, where $f(t)$ is given in Exercise 5 above. Are $y(t)$, $y'(t)$, and $y''(t)$ continuous at $t = 2\pi$?

20. $y'' + 4y' + 3y = f(t)$, $y(0) = 2$, $y'(0) = 0$, where $f(t)$ is given in Exercise 6 above.

21. $y'' + y = f_n(t)$, $y(0) = 0$, $y'(0) = 0$, where $f_n(t)$ is given in Exercise 4 above. Does the solution approach the solution of $y'' + y = 1$, $y(0) = 0$, $y'(0) = 0$, as $n \to \infty$?

22. $y'' + 2y' + y = f(t)$, $y(0) = 1$, $y'(0) = 0$ where $f(t)$ is given in Exercise 8 above.

23. $y'' - 4y = u(t - 2) - u(t - 4)$, $y(0) = 0$, $y'(0) = 1$

24. $y'' + \omega^2 y = f(t)$, $y(0) = 0$, $y'(0) = 0$ where $f(t)$ is given in Exercise 7 above. Discuss the cases $\omega^2 \ne 1$, $\omega^2 = 1$.

Sketch the graphs of the periodic functions in Exercises 25–28 and find their Laplace transforms. Do this in two ways: by expressing the function as a series of terms involving unit step functions then calculating the transform term by term, and by using entry (17).

25. $f(t) = \begin{cases} 2 & \text{if } 0 \le t < 1, \\ 0 & \text{if } 1 \le t < 2, \end{cases} \quad f(t+2) = f(t)$

26. $f(t) = 2 - t$ if $0 \le t < 4$, $\quad f(t+4) = f(t)$

27. $f(t) = \begin{cases} 0 & \text{if } 0 \le t < \frac{1}{2}, \\ 1 & \text{if } \frac{1}{2} \le t < \frac{3}{2}, \\ 0 & \text{if } \frac{3}{2} \le t < 2, \end{cases} \quad f(t+2) = f(t)$

28. $f(t) = |\cos t|$ if $t \ge 0$

Find the solutions of the initial value problems in Exercises 29–32.

29. $y'' - 9y = \frac{9}{2} f(t)$, $\quad y(0) = 0$, $\quad y'(0) = 2$, where $f(t)$ is given in Exercise 25 above.

30. $y'' + 4y = f(t)$, $\quad y(0) = 0$, $\quad y'(0) = 0$, where $f(t)$ is given in Exercise 26 above.

31. $y'' + y = |\sin t|$, $\quad y(0) = 4$, $\quad y'(0) = 0$

32. $y'' + y = f(t)$, $\quad y(0) = 0$, $\quad y'(0) = 0$, where $f(t)$ is the "spiky sine,"

$$f(t) = \begin{cases} \dfrac{2}{\pi} t & \text{if } 0 \le t \le \dfrac{\pi}{2}, \\ 2 - \dfrac{2}{\pi} t & \text{if } \dfrac{\pi}{2} \le t \le \dfrac{3\pi}{2}, \\ \dfrac{2}{\pi} t - 4 & \text{if } \dfrac{3\pi}{2} \le t \le 2\pi, \end{cases}$$

and $f(t + 2\pi) = f(t)$.

4.9 THE UNIT IMPULSE FUNCTION: TRANSFER FUNCTIONS

Given a mechanical or electrical system, we wish to model a force or voltage of large magnitude which occurs over a very short period of time. A function that describes this behavior is

$$\delta_\epsilon(t - t_0) = \begin{cases} 0 & \text{if } t < t_0 - \epsilon, \\ \dfrac{1}{2\epsilon} & \text{if } t_0 - \epsilon \le t < t_0 + \epsilon, \\ 0 & \text{if } t \ge t_0 + \epsilon, \end{cases}$$

where $\epsilon > 0$ is a small positive number and $t_0 > 0$. Using the unit step function one can also write

$$\delta_\epsilon(t - t_0) = \frac{1}{2\epsilon}[u(t - t_0 + \epsilon) - u(t - t_0 - \epsilon)],$$

and the graph of $\delta_\epsilon(t - t_0)$ is shown in Fig. 4.17. Clearly, as ϵ approaches zero, the rectangular pulse $\delta_\epsilon(t - t_0)$ gets taller and thinner but the limit does not exist.

In spite of the limit $\lim_{\epsilon\to 0} \delta_\epsilon(t - t_0)$ not existing in the usual sense, one can derive some interesting properties of the function. First of all, since $\delta_\epsilon(t - t_0) = 0$ for all t outside the interval $t_0 - \epsilon \le t \le t_0 + \epsilon$, it follows that

$$\int_{-\infty}^{\infty} \delta_\epsilon(t - t_0)\, dt = \int_{t_0-\epsilon}^{t_0+\epsilon} \frac{1}{2\epsilon}\, dt = 1. \tag{4.9.1}$$

If $\delta_\epsilon(t - t_0)$ is thought of as a force, then (4.9.1) says the *total impulse* is unity.

Second, given any continuous function $f(t)$ defined on $-\infty < t < \infty$, we have

$$\int_{-\infty}^{\infty} \delta_\epsilon(t - t_0)f(t)\, dt = \int_{t_0-\epsilon}^{t_0+\epsilon} \frac{1}{2\epsilon} f(t)\, dt = \frac{1}{2\epsilon}\int_{t_0-\epsilon}^{t_0+\epsilon} f(t)\, dt.$$

Recall the mean value theorem for integrals that states $\int_a^b f(t)\, dt = (b - a)f(\tau)$ for some τ in $a \le t \le b$, and consequently

$$\int_{-\infty}^{\infty} \delta_\epsilon(t - t_0)f(t)\, dt = \frac{1}{2\epsilon}[2\epsilon f(\tau_\epsilon)] = f(\tau_\epsilon) \tag{4.9.2}$$

for some τ_ϵ in $t_0 - \epsilon \le t \le t_0 + \epsilon$.

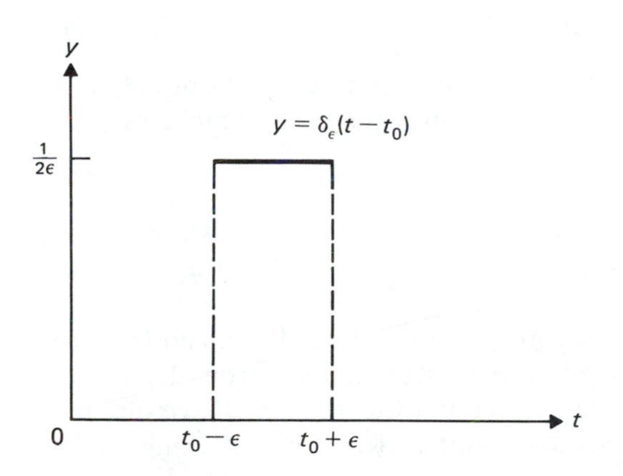

Figure 4.17

Third, the representation of $\delta_\epsilon(t - t_0)$ in terms of unit step functions implies immediately that for $t_0 > 0$

$$L\{\delta_\epsilon(t - t_0)\} = \frac{1}{2\epsilon}\left[\frac{e^{-s(t_0-\epsilon)}}{s} - \frac{e^{-s(t_0+\epsilon)}}{s}\right] = e^{-st_0}\left[\frac{\sinh s\epsilon}{s\epsilon}\right]. \qquad \textbf{(4.9.3)}$$

For the last property, $\epsilon > 0$ must be sufficiently small so that $t_0 - \epsilon > 0$, in order for the Laplace transform to be defined.

Now suppose that ϵ is allowed to approach zero in each of the properties (4.9.1), (4.9.2), and (4.9.3). For the first two properties some not-so-trivial mathematical justification of the limiting process is required but it can be accomplished. If $\delta(t - t_0)$ denotes the result of letting ϵ go to zero, then (4.9.1) becomes

$$\int_{-\infty}^{\infty} \delta(t - t_0)\, dt = 1. \qquad \textbf{(4.9.4)}$$

For the property (4.9.2), τ_ϵ must remain in the interval $t_0 - \epsilon \le t \le t_0 + \epsilon$ as ϵ goes to zero. Consequently $\lim_{\epsilon\to 0} \tau_\epsilon = t_0$ and

$$\int_{-\infty}^{\infty} \delta(t - t_0) f(t)\, dt = f(t_0) \qquad \textbf{(4.9.5)}$$

for any continuous function defined for $-\infty < t < \infty$. Actually, $f(t)$ has to be continuous only in some neighborhood of $t = t_0$, and it can then be defined to be zero outside this interval. Property (4.9.5) is sometimes called the *sifting property*.

In the last property, (4.9.3), which defines the Laplace transform of $\delta_\epsilon(t - t_0)$, observe by l'Hôpital's rule that

$$\lim_{\epsilon\to 0} \frac{\sinh s\epsilon}{s\epsilon} = \lim_{\epsilon\to 0} \frac{\cosh s\epsilon}{1} = 1.$$

Therefore

$$L\{\delta(t - t_0)\} = e^{-st_0} \qquad \textbf{(4.9.6)}$$

is valid for $t_0 > 0$ since the transform of $\delta_\epsilon(t - t_0)$ can be calculated only when $t_0 - \epsilon > 0$ or $t_0 > \epsilon > 0$. But nothing prevents us from letting $t_0 \to 0$, and this gives

$$L\{\delta(t)\} = 1. \qquad \textbf{(4.9.7)}$$

This clearly points out that $\delta(t)$ is not at all a "nice" function since, if it were, its Laplace transform would be a function that goes to zero as Re(s) approaches infinity.

The "function" $\delta(t)$ defined by the relations (4.9.4), (4.9.5), (4.9.6.), and (4.9.7) is called the *unit impulse function* or *Dirac delta function,* named after the Nobel Laureate physicist Paul Dirac (born 1902). It is not a function in any traditional sense but is an example of what modern-day mathematicians call a "generalized function." The theory of such functions was only recently made rigorous, but physicists, engineers, and applied mathematicians have been using them heuristically for years.

The users of such functions have developed such definitions as: $\delta(t)$ is a function that is zero for all $t \neq 0$ and satisfies

$$\int_{-\infty}^{\infty} \delta(t)\, dt = 1$$

and

$$\int_{-\infty}^{\infty} \delta(t - t_0) f(t)\, dt = f(t_0).$$

One can see why this would trouble a mathematician, but a physicist would be content to think of $\delta(t)$ as a unit impulse or a point mass.

EXAMPLE 1 Find the zero-state output for the equation

$$y'' - y = \delta(t - 1).$$

SOLUTION The right-hand side can be regarded as a unit impulse administered at $t = 1$. If $Y(s)$ is the Laplace transform of $y(t)$, then using formula (4.9.6) we have

$$s^2 Y(s) - Y(s) = e^{-s}.$$

Therefore

$$Y(s) = e^{-s}\left[\frac{1}{s^2 - 1}\right],$$

and hence

$$y(t) = [\sinh(t - 1)]u(t - 1),$$

or

$$y(t) = \begin{cases} 0 & \text{if } 0 \le t \le 1, \\ \sinh(t - 1) & \text{if } t \ge 1. \end{cases}$$

We see that $y(t)$ is continuous at $t = 1$ but $y'(t)$ is not, and that $y(t)$ is a displacement of the solution $y(t) = \sinh t$ (which is the weighting function) one unit to the right. ∎

Another definition of the unit impulse function comes from the representation of $\delta_\epsilon(t - t_0)$ in terms of unit step functions. Recall that

$$\delta_\epsilon(t - t_0) = \frac{1}{2\epsilon}[u(t - t_0 + \epsilon) - u(t - t_0 - \epsilon)];$$

letting $t - t_0 = \tau + \epsilon$, we obtain

$$\delta_\epsilon(\tau + \epsilon) = \frac{u(\tau + 2\epsilon) - u(\tau)}{2\epsilon}.$$

The right-hand side is the difference quotient for the unit step function, so letting ϵ approach zero leads to the definition

$$\delta(t) \text{ is the derivative of } u(t).$$

This of course makes no sense at $t = 0$, where $u(t)$ is not even continuous, much less differentiable. But engineers and control systems analysts frequently use this definition; mathematically it turns out that $\delta(t)$ is a "generalized derivative" of $u(t)$.

The unit impulse function is a useful tool in the analysis of linear systems. Recall that in our discussion in Section 4.7, where we studied the zero-state initial value problem

$$y'' + ay' + by = f(t), \qquad y(0) = 0, \qquad y'(0) = 0,$$

we obtained the relation

$$Y(s) = \frac{F(s)}{s^2 + as + b} = P(s)F(s).$$

The function $P(s) = (s^2 + as + b)^{-1}$ was called the *transfer function.* But if $f(t) = \delta(t)$, then $F(s) = 1$, so if the *solution* of the differential equation with $f(t) = \delta(t)$ is defined as the *impulse response,* this leads to the statement

Definition

The transfer function $P(s)$ is the Laplace transform of the impulse response.

Another definition of the impulse response would be the zero-state output corresponding to the input $\delta(t)$.

Recall also that the weighting function $p(t)$ was defined by the relation $p(t) = L^{-1}\{P(s)\}$ and the zero-state output was

$$y(t) = L^{-1}\{Y(s)\} = L^{-1}\{P(s)F(s)\}.$$

But if $F(s) = 1$, then $y(t) = p(t)$, and this leads to the statement

Definition

The weighting function $p(t)$ is the zero-state output corresponding to the unit impulse function input.

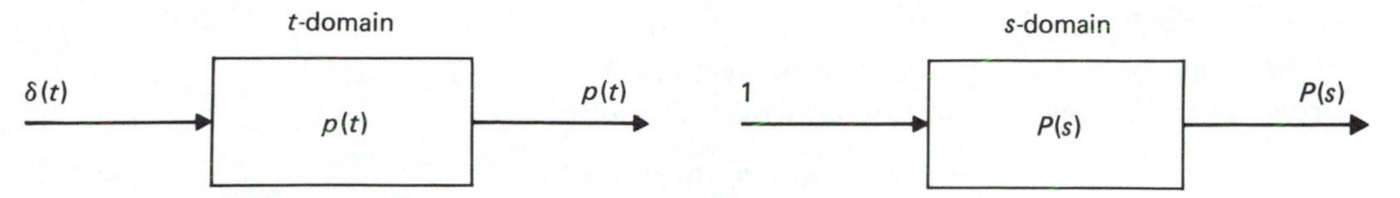

Figure 4.18

Using the terminology developed in the last paragraph, we can say that the weighting function is simply the impulse response. In the t-domain and s-domain the system diagrams would look like those in Fig. 4.18. In the first diagram, the "black box" performs a convolution; in the second it multiplies.

EXERCISES 4.9

An initial value problem of the form

$$y'' + ay' + by = \delta(t - t_0), \qquad t_0 > 0, \qquad y(0) = A, \qquad y'(0) = B,$$

can be interpreted as a mechanical or electrical system subjected to a unit impulse input at $t = t_0$. Using the formula (4.9.6), we can easily solve it by Laplace transform methods, and the solution will be a displacement of the solution with zero input. Solve the initial value problems of Exercises 1–8.

1. $y'' + 4y = \delta(t - \pi), \quad y(0) = 1, \quad y'(0) = 0$
2. $y'' + 6y' + 9y = 3 + \delta(t - 1), \quad y(0) = 0, \quad y'(0) = 0$
3. $y'' + 2y' + 5y = 8e^{-t} + \delta(t - 2), \quad y(0) = 2, \quad y'(0) = 0$
4. $y'' - 16y = 4\delta(t - 1) + 8t, \quad y(0) = 0, \quad y'(0) = 0$
5. $y'' + 9y = \delta(t - \pi/3), \quad y(0) = 1, \quad y'(0) = 0$
6. $y'' + 2y + y = \delta(t - 1), \quad y(0) = 0, \quad y'(0) = 1$
7. $y'' + 4y' + \frac{25}{4}y = 6e^{t} + 3\delta(t - 1), \quad y(0) = 0, \quad y'(0) = 0$
8. $y'' + \frac{1}{16}y = 2\delta(t - 4\pi), \quad y(0) = 1, \quad y'(0) = 1$
9. Using the property (4.9.5) of the unit impulse function and the definition of the Laplace transform, show that

$$L\{\delta(t - t_0)f(t)\} = e^{-st_0}f(t_0),$$

where $t_0 > 0$ and $f(t)$ is defined to be zero for $t < 0$.

Use the result of Exercise 9 to solve the initial value problems of Exercises 10–15.

10. $y'' + 3y' + 3y = \delta(t - 1)t^2, \quad y(0) = 0, \quad y'(0) = 0$

11. $y'' + 16y = \delta(t - \pi/8) \sin 4t, \quad y(0) = 2, \quad y'(0) = 0$

12. $y'' + y = \delta(t - \alpha) \cos t, \quad \alpha > 0, \quad y(0) = 0, \quad y'(0) = 1$
As α approaches $\pi/2$, does the solution approach the solution of

$$y'' + y = 0, y(0) = 0, y'(0) = 1?$$

13. $y'' - 4y = \delta(t - 1)e^{2t}, \quad y(0) = 0, \quad y'(0) = 0$

14. $y'' + 5y' + 6y = 3\delta(t - 1)t + u(t - 1), \quad y(0) = 0, \quad y'(0) = 0$

15. $y'' - \delta(t - 1)y = 0, \quad y(0) = 0, \quad y'(0) = 0$
Discuss whether the solution you obtain makes sense.

SUMMARY

Transform methods work as follows: the unknown solution $y(t)$ of a differential equation is transformed into a function $Y(s)$, where s is complex variable, which satisfies a system of algebraic equations. If one can solve these equations for $Y(s)$ and one can reverse the procedure, i.e., invert the transformation, then one obtains $y(t)$.

One extensively used transform method employs the Laplace transform, and it plays an important role in the analysis of circuit theory problems. This chapter develops the methodology of the Laplace transform and shows how it is used to solve second order linear constant coefficient differential equations, as well as systems of first order linear differential equations. For the latter the use of the Laplace transform is an extremely efficient technique.

The Laplace transform is also useful for solving inhomogeneous constant coefficient linear differential equations where the right-hand side or forcing term is not a smooth function. For instance, it could be a step function, a square or sawtooth wave, or an impulse function representing a force of large magnitude occurring over a very short time period. The tools needed to analyze this class of problems are the unit step function or Heaviside function, the unit impulse function or Dirac delta function, and their Laplace transforms. These are developed in this chapter and used to solve problems of the type described above.

MISCELLANEOUS EXERCISES

4.1. Noting that $d^{n+1}(t^n)/dt^{n+1} = 0$, use the differentiation formula directly to show that $L\{t^n\} = n!/s^{n+1}$.

4.2. The transform formula

$$L\{t^n y(t)\} = (-1)^n \frac{d^n}{ds^n} Y(s),$$

where $L\{y(t)\} = Y(s)$, allows one to "solve" the special class of linear differential equations whose coefficients are polynomials in t. Applying the transform, one obtains a linear differential equation in $Y(s)$ whose coefficients are polynomials in s (if it is easier to solve than the original, one is in luck!).

a) Use the above formula to show that

$$L\{t^2 y'(t)\} = sY''(s) + 2Y'(s)$$

and

$$L\{ty''(t)\} = -s^2 Y'(s) - 2sY(s) + y(0).$$

b) Use the Laplace transform to solve

$$ty'' - ty' - y = 0, \qquad y(0) = 0, \qquad y'(0) = 4.$$

c) Use the Laplace transform to show that the solution of

$$ty'' + y' + ty = 0, \qquad y(0) = 1$$

is

$$y(t) = J_0(t) = \sum_{n=0}^{\infty} \frac{(-1)^n}{(2^n n!)^2} t^{2n},$$

the Bessel function of order zero of the first kind. Do this by showing first that

$$L\{y(t)\} = Y(s) = \frac{c}{\sqrt{s^2+1}} = \frac{c}{s}\left(1 + \frac{1}{s^2}\right)^{-1/2},$$

where c is an arbitrary constant. Expand the second expression by using the binomial series, then invert and use the initial condition.

4.3. A transform that is used to analyze discrete or sampled data systems is the z-transform. It is defined as follows: If T is a fixed positive number and $f(t)$ a function defined for $t \geq 0$, then its z-transform $F(z)$ is given by

$$F(z) = \sum_{n=0}^{\infty} f(nT) z^{-n},$$

where z is a complex number. One may regard the fixed number T as the sampling interval. Hence, functions in the t-domain, $0 \leq t < \infty$, are transformed into functions in the z-domain which is the complex plane.

a) If $f(t) = 1$, $t \geq 0$, show that $F(z) = z/(z - 1)$, valid for $|z| > 1$. (*Hint:* Recall that

$$1 + u + u^2 + \cdots + u^n + \cdots = \frac{1}{1 - u},$$

valid for $|u| < 1$.)

b) If $f(t) = t$, $t \geq 0$, show that $F(z) = zT/(z - 1)^2$. (*Hint:* Convergent power series can be differentiated term by term.)

c) If $f(t) = e^{2t}$, $t \geq 0$, show that $F(z) = z/(z - e^{2T})$, valid for $|z| > e^{2T}$.

4.4. When the z-transform is used, translation through integer multiples of the sampling interval T corresponds to differentiation. If $f(t) = 0$ for $t < 0$ and $Z[f(t)] = F(z)$ is the z-transform of $f(t)$, establish the formulas

$$Z[f(t - kT)] = z^{-k}F(z)$$

and

$$Z[f(t + kT)] = z^kF(z) - [z^kf(0) + z^{k-1}f(T) + \cdots + zf((k - 1)T)],$$

where k is a positive integer.

4.5. Use the z-transform to verify that the given function $Y(z)$ is the z-transform of the solution $y(t)$ of the following finite difference equations where $T = 1$.

a) $y(t + 1) + 2y(t) = 0, \quad y(0) = 4; \quad Y(z) = \dfrac{4z}{z + 2} = 4 - \dfrac{8}{z + 2}$

b) $y(t + 1) + 2y(t) = 1, \quad y(0) = 0;$

$$Y(z) = \frac{z}{(z + 2)(z - 1)} = \frac{2}{3}\frac{1}{z + 2} + \frac{1}{3}\frac{1}{z - 1}$$

c) $y(t + 2) + 3y(t + 1) - 10y(t) = 0, \quad y(0) = 0, y(1) = 1;$

$$Y(z) = \frac{z}{z^2 + 3z - 10} = \frac{2}{7}\frac{1}{z - 2} + \frac{5}{7}\frac{1}{z + 5}$$

Remark: The last two problems show that the z-transform corresponds to the Laplace transform when solving finite difference equations.

4.6. Using the z-transform show that:

a) If $f(t) = a[u(t) - u(t - T)]$, then its z-transform is

$$F(z) = a\sum_{n=0}^{\infty} [u(nT) - u(nT - T)]z^{-n}$$

$$= a.$$

b) If $f(t) = a^{(t-T)/T}u(t - T)$, then its z-transform is

$$F(z) = \sum_{n=0}^{\infty} a^{n-1}u(nT - T)z^{-n}$$
$$= z^{-1}\sum_{n=0}^{\infty}\left(\frac{a}{z}\right)^n = \frac{1}{z-a},$$

valid for $|z| > |a|$.

4.7. Use the results above to find the solutions of the following difference equations by employing the inverse z-transform.

a) In Problem 4.5(a) show that the solution at the sample values $t = 0, 1, 2, \ldots$ is given by

$$y(t) = \begin{cases} 4 & \text{if } t = 0, \\ -8(-2)^{t-1} = 4(-2)^t & \text{if } t = 1, 2, \ldots \end{cases}$$

b) In Problem 4.5(b) show that

$$y(t) = \begin{cases} 0 & \text{if } t = 0, \\ \frac{2}{3}(2)^{t-1} + \frac{1}{3} & \text{if } t = 1, 2, \ldots \end{cases}$$

c) In Problem 4.5(c) show that

$$y(t) = \begin{cases} 0 & \text{if } t = 0, \\ \frac{2}{7}(2)^{t-1} + \frac{5}{7}(-5)^{t-1} & \text{if } t = 1, 2, \ldots \end{cases}$$

REFERENCES

A standard text on the Laplace transform with an extensive discussion of its application to partial differential equations is

1. R.V. Churchill, *Operational Mathematics,* 2nd ed., McGraw-Hill, New York, 1958.

A book that discusses a variety of transforms as well as the Laplace transform is

2. W. Kaplan, *Operational Methods for Linear Systems,* Addison-Wesley, Reading, Mass., 1962.

For a discussion of systems theory including transform methods see

3. J.J. DiStefano III, A.R. Stubberud, and I.J. Williams, *Feedback and Control Systems,* Schaum, New York, 1967.
4. C.A. Desoer and L.A. Zadeh, *Linear Systems Theory,* McGraw-Hill, New York, 1963.
5. E.V. Bohn, *The Transform Analysis of Linear Systems,* Addison-Wesley, Reading, Mass., 1963.
6. S.C. Gupta, *Transform and State Variable Methods in Linear Systems,* John Wiley and Sons, New York, 1966.

5

LINEAR SYSTEMS OF DIFFERENTIAL EQUATIONS

5.1 INTRODUCTION

A physical system with several interdependent quantities can often be modeled by a system of simultaneous ordinary differential equations. If the independent variable is denoted by t and the dependent variables by $y_1, y_2, \ldots, y_n$, a typical system is

$$\begin{aligned}
\frac{dy_1}{dt} &= g_1(t, y_1, y_2, \ldots, y_n), \\
\frac{dy_2}{dt} &= g_2(t, y_1, y_2, \ldots, y_n), \\
&\vdots \\
\frac{dy_n}{dt} &= g_n(t, y_1, y_2, \ldots, y_n).
\end{aligned} \tag{5.1.1}$$

The system of differential equations is said to be *linear* if each of the functions $g_j(t, y_1, y_2, \ldots, y_n)$ is a linear function of the dependent variables, hence,

$$\begin{aligned}
\frac{dy_1}{dt} &= a_{11}(t)y_1 + a_{12}(t)y_2 + \cdots + a_{1n}(t)y_n + f_1(t), \\
\frac{dy_2}{dt} &= a_{21}(t)y_1 + a_{22}(t)y_2 + \cdots + a_{2n}(t)y_n + f_2(t), \\
&\vdots \\
\frac{dy_n}{dt} &= a_{n1}(t)y_1 + a_{n2}(t)y_2 + \cdots + a_{nn}(t)y_n + f_n(t).
\end{aligned} \tag{5.1.2}$$

The coefficients $a_{ij}(t)$ and $f_j(t)$, $i, j = 1, 2, \ldots, n$, are assumed to be continuous functions of t on some nonempty interval. If all the $f_j(t)$ are identically zero, the system is called *homogeneous;* otherwise, the system is called *nonhomogeneous.* For convenience and simplicity we shall consider only linear systems with constant coefficients in this chapter.

Numerous physical problems can be modeled by linear systems with constant coefficients. Three such problems are considered in the following examples.

EXAMPLE 1 Consider the consecutive unimolecular reactions

$$A \xrightarrow{k_1} B \xrightarrow{k_2} C,$$

where A, B, and C are compounds and k_1 and k_2 are reaction rates. Let $y_1 = [A]$, $y_2 = [B]$, and $y_3 = [C]$ denote the concentrations of the respective compounds in grams per cubic centimeter. If the rate of each reaction is proportional to the concentration of reactant present, the reaction can be modeled by the linear system of differential equations

$$\frac{dy_1}{dt} = -k_1 y_1, \tag{5.1.3}$$

$$\frac{dy_2}{dt} = -k_2 y_2 + k_1 y_1, \tag{5.1.4}$$

$$\frac{dy_3}{dt} = k_2 y_2. \tag{5.1.5}$$

By adding the above equations, we get

$$\frac{d}{dt}(y_1 + y_2 + y_3) = 0,$$

and, after integrating, the mass conservation law

$$y_1 + y_2 + y_3 = \text{const.}$$

In this example the equations are fairly easy to solve. Suppose the concentrations of the reactants are specified at time $t = 0$, say,

$$y_1(0) = a, \qquad y_2(0) = b, \qquad y_3(0) = c.$$

Then from Eq. (5.1.3) and the initial condition $y_1(0) = a$, we obtain, using the methods of Chapter 1,

$$y_1(t) = ae^{-k_1 t}.$$

Substituting this value for $y_1(t)$ into Eq. (5.1.4) gives

$$\frac{dy_2}{dt} = -k_2 y_2(t) + k_1 a e^{-k_1 t}, \qquad y_2(0) = b.$$

If $k_1 \neq k_2$, this initial value problem has the solution

$$y_2(t) = \frac{k_1 a}{k_2 - k_1} e^{-k_1 t} + \left(b - \frac{k_1 a}{k_2 - k_1}\right) e^{-k_2 t}. \tag{5.1.6}$$

Finally, from the mass conservation law we get

$$\begin{aligned} y_3(t) &= a + b + c - y_1(t) - y_2(t) \\ &= a + b + c - \frac{ak_2}{k_2 - k_1} e^{-k_1 t} - \left(b - \frac{k_1 a}{k_2 - k_1}\right) e^{-k_2 t}. \end{aligned}$$

Of course, $y_3(t)$ could have also been obtained by solving Eq. (5.1.5), using (5.1.6) and the initial condition $y_3(0) = c$. ■

EXAMPLE 2 Consider the simple *LRC* circuit shown in Fig. 5.1. If the current in the circuit is denoted by I and the charge on the capacitor by Q, an application of Kirchhoff's voltage law gives

$$L\frac{dI}{dt} + RI + \frac{Q}{C} = 0,$$

where

$$\frac{dQ}{dt} = I.$$

This can be written as the linear system

$$\begin{aligned} \frac{dQ}{dt} &= I, \\ \frac{dI}{dt} &= -\frac{1}{LC}Q - \frac{R}{L}I. \end{aligned}$$

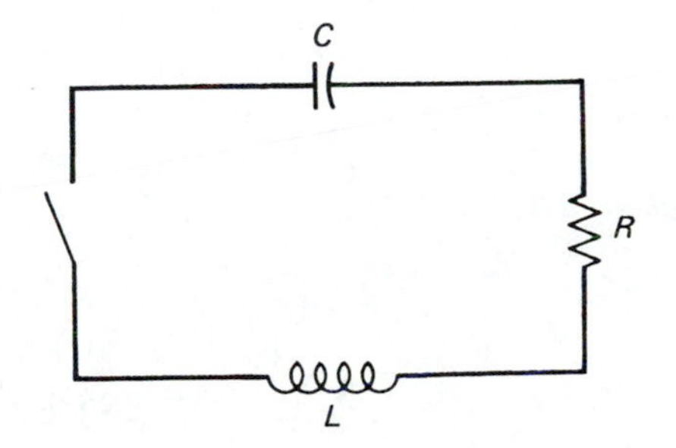

Figure 5.1

Suppose that the switch is closed at time $t = 0$ and that $I(0) = 0$, $Q(0) = Q_0$. If $R^2 < 4L/C$, the solution to this initial value problem is given by

$$Q = Q_0 e^{-Rt/2L}\left[\cos \omega t + \frac{R}{2L\omega}\sin \omega t\right],$$

$$I = \frac{-Q_0}{\omega LC} e^{-Rt/2L} \sin \omega t,$$

where

$$\omega = \left(\frac{4L}{C} - R^2\right)^{1/2} \bigg/ 2L.$$

We shall not go into the details here of how to construct the solution since it was shown earlier that the system is equivalent to the second order equation

$$L\frac{d^2Q}{dt} + R\frac{dQ}{dt} + \frac{Q}{C} = 0,$$

which was studied in Chapter 2. The point to be stressed here is that the circuit problem is quite naturally formulated as a linear system of first order differential equations. ■

EXAMPLE 3 Consider a mechanical system consisting of two equal masses and three identical springs. The masses are on a smooth level surface and are connected to each other and to two walls as shown in Fig. 5.2. It is assumed that the motion is along a straight line and that the only forces on the masses are spring forces. Let y_1 and y_2 denote the displacements of the masses from their equilibrium positions. By Hooke's law, the spring forces are proportional to the displacements from the equilibrium positions, and so the forces acting on the masses are those shown in the figure. By Newton's second law, the time rate of change of the momentum of each mass is equal to the sum of the forces acting on it. This leads to the pair of differential equations

$$m\frac{d^2y_1}{dt^2} = -ky_1 + k(y_2 - y_1),$$

$$m\frac{d^2y_2}{dt^2} = -ky_2 - k(y_2 - y_1).$$

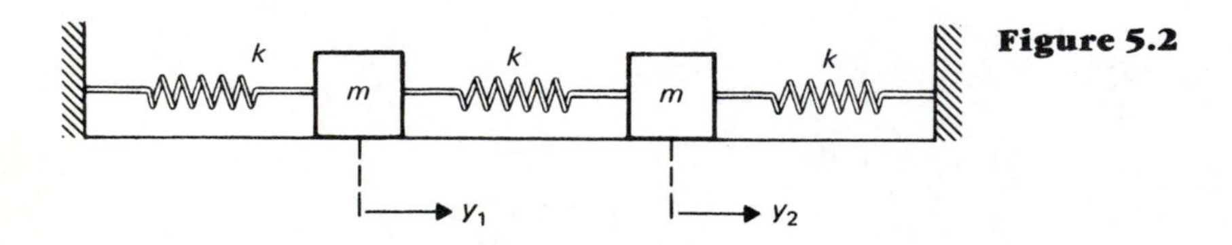

Figure 5.2

This system of two coupled second order differential equations can be analyzed as formulated, but we shall rewrite it to illustrate a method of representing higher order equations as a system of first order equations. To do this, introduce the new dependent variables, $y_3 = dy_1/dt$, $y_4 = dy_2/dt$, to get

$$\begin{aligned}
\frac{dy_1}{dt} &= y_3, \\
\frac{dy_2}{dt} &= y_4, \\
\frac{dy_3}{dt} &= -2\frac{k}{m}y_1 + \frac{k}{m}y_2, \\
\frac{dy_4}{dt} &= \frac{k}{m}y_1 - 2\frac{k}{m}y_2,
\end{aligned} \tag{5.1.7}$$

where $d^2y_1/dt^2 = dy_3/dt$ and $d^2y_2/dt^2 = dy_4/dt$.

From symmetry one can expect two simple motions called *normal modes of vibrations*. The first motion has the two masses moving together as a rigid body, the second—as mirror images. Let us examine them in turn. To induce the masses to move together, they are displaced in the same direction by equal amounts and then released with zero initial velocity. Hence,

$$y_1(0) = y_2(0) = a, \qquad y_3(0) = y_4(0) = 0.$$

The solution to the system (5.1.7) subject to these initial conditions is

$$\begin{aligned}
y_1(t) &= y_2(t) = a \cos \omega t, \\
y_3(t) &= y_4(t) = -a\omega \sin \omega t,
\end{aligned}$$

where the frequency $\omega = (k/m)^{1/2}$ is a *natural frequency* of the system. To induce the masses to move as mirror images, they are displaced in opposite directions by the same amount and then released with zero initial velocity. Hence,

$$y_1(0) = -y_2(0) = b, \qquad y_3(0) = y_4(0) = 0,$$

and the corresponding solution is

$$\begin{aligned}
y_1(t) &= -y_2(t) = b \cos \sigma t, \\
y_3(t) &= -y_4(t) = -b\sigma \sin \sigma t,
\end{aligned}$$

where σ, the second natural frequency, will be found by substitution in (5.1.7). The first two equations are automatically satisfied, and from the second two we have

$$\begin{aligned}
-b\sigma^2 \cos \sigma t &= -2\frac{k}{m} b \cos \sigma t - \frac{k}{m} b \cos \sigma t, \\
b\sigma^2 \cos \sigma t &= \frac{k}{m} b \cos \sigma t + 2\frac{k}{m} b \cos \sigma t.
\end{aligned}$$

Either of these equations can be used to find $\sigma = (3k/m)^{1/2} = 3^{1/2}\omega$. Therefore the frequency of the second normal mode is higher than that of the first normal mode. This could have been predicted since in the second mode the middle spring is active. ■

EXERCISES 5.1

1. Solve the problem of Example 1 when $k_1 = k_2$.
2. Verify by substitution that

$$y_1(t) = a \cos \omega t + a' \sin \omega t + b \cos \sigma t + b' \sin \sigma t,$$

$$y_2(t) = a \cos \omega t + a' \sin \omega t - b \cos \sigma t - b' \sin \sigma t,$$

is a solution to the mass–spring system of Example 3, where a, a', b, b' are arbitrary constants, $\omega = (k/m)^{1/2}$, $\sigma = (3k/m)^{1/2}$.

3. Use Laplace transforms to solve (5.1.7) for the two normal mode solutions.

5.2 VECTORS AND MATRICES

This section contains a résumé of some topics of linear algebra needed for the study of linear systems of differential equations. Some results will be proved, others stated without proof; the omitted proofs are assigned as exercises or as reading projects.

We start with n-vectors, which are ordered n-tuples

$$\mathbf{y} = \begin{bmatrix} y_1 \\ y_2 \\ \vdots \\ y_n \end{bmatrix} = \text{col}(y_1, y_2, \ldots, y_n)$$

of real or complex numbers. The number y_i is called the ith component of $\mathbf{y}$. The algebra of n-vectors is the same as the algebra of the two- or three-dimensional vectors used in mechanics and calculus courses. For example, two vectors are equal if and only if all their components are pairwise equal. Vector addition and subtraction and the product of a vector and a number (also called a *scalar*) are defined by

$$\mathbf{y} \pm \mathbf{x} = \begin{bmatrix} y_1 \pm x_1 \\ y_2 \pm x_2 \\ \vdots \\ y_n \pm x_n \end{bmatrix}, \qquad c\mathbf{y} = \begin{bmatrix} cy_1 \\ cy_2 \\ \vdots \\ cy_n \end{bmatrix}.$$

The *scalar* or *dot product* of two vectors $\mathbf{x}$ and $\mathbf{y}$ is

$$\mathbf{x} \cdot \mathbf{y} = x_1y_1 + x_2y_2 + \cdots + x_ny_n.$$

Vectors $\mathbf{y}^1, \mathbf{y}^2, \ldots, \mathbf{y}^k$ are *linearly dependent* if there are numbers $c_1, c_2, \ldots, c_k$, *not all zero,* such that

$$c_1\mathbf{y}^1 + c_2\mathbf{y}^2 + \cdots + c_k\mathbf{y}^k = \mathbf{0},$$

where $\mathbf{0}$ denotes the n-vector whose components are all 0. If the vectors $\mathbf{y}^1, \mathbf{y}^2, \ldots, \mathbf{y}^k$ are not linearly dependent, they are *linearly independent;* hence,

$$c_1\mathbf{y}^1 + c_2\mathbf{y}^2 + \cdots + c_k\mathbf{y}^k = \mathbf{0}$$

is satisfied only if $c_1 = c_2 = \cdots = c_k = 0$.

EXAMPLE 1 Show that the vectors $\mathbf{y}^1, \mathbf{y}^2, \mathbf{y}^3$ are linearly independent while the vectors $\mathbf{y}^1, \mathbf{y}^2, \mathbf{y}^3, \mathbf{y}^4$ are linearly dependent, where

$$\mathbf{y}^1 = \begin{bmatrix} 1 \\ 1 \\ 1 \\ 1 \end{bmatrix}, \quad \mathbf{y}^2 = \begin{bmatrix} 1 \\ 0 \\ 1 \\ 0 \end{bmatrix}, \quad \mathbf{y}^3 = \begin{bmatrix} 1 \\ 1 \\ 0 \\ 1 \end{bmatrix}, \quad \mathbf{y}^4 = \begin{bmatrix} 0 \\ 2 \\ -1 \\ 2 \end{bmatrix}.$$

SOLUTION First note that

$$c_1\mathbf{y}^1 + c_2\mathbf{y}^2 + c_3\mathbf{y}^3 = c_1 \begin{bmatrix} 1 \\ 1 \\ 1 \\ 1 \end{bmatrix} + c_2 \begin{bmatrix} 1 \\ 0 \\ 1 \\ 0 \end{bmatrix} + c_3 \begin{bmatrix} 1 \\ 1 \\ 0 \\ 1 \end{bmatrix}$$

$$= \begin{bmatrix} c_1 + c_2 + c_3 \\ c_1 + c_3 \\ c_1 + c_2 \\ c_1 + c_3 \end{bmatrix} = \begin{bmatrix} 0 \\ 0 \\ 0 \\ 0 \end{bmatrix}$$

implies that

$$\begin{aligned} c_1 + c_2 + c_3 &= 0, \\ c_1 + c_3 &= 0, \\ c_1 + c_2 &= 0. \end{aligned}$$

Clearly, $c_1 = c_2 = c_3 = 0$ and therefore $\mathbf{y}^1, \mathbf{y}^2$, and $\mathbf{y}^3$ are linearly independent. On the other hand, it is easy to check that $\mathbf{y}^4 = \mathbf{y}^1 - 2\mathbf{y}^2 + \mathbf{y}^3$. Equivalently, if $c_1 = 1$,

$c_2 = -2$, $c_3 = 1$, $c_4 = -1$, then

$$\mathbf{y}^1 - 2\mathbf{y}^2 + \mathbf{y}^3 - \mathbf{y}^4 = \begin{bmatrix} 1 \\ 1 \\ 1 \\ 1 \end{bmatrix} - 2\begin{bmatrix} 1 \\ 0 \\ 1 \\ 0 \end{bmatrix} + \begin{bmatrix} 1 \\ 1 \\ 0 \\ 1 \end{bmatrix} - \begin{bmatrix} 0 \\ 2 \\ -1 \\ 2 \end{bmatrix}$$

$$= \begin{bmatrix} 1-2+1-0 \\ 1-0+1-2 \\ 1-2+0+1 \\ 1-0+1-2 \end{bmatrix} = \begin{bmatrix} 0 \\ 0 \\ 0 \\ 0 \end{bmatrix}.$$

Therefore $\mathbf{y}^1, \mathbf{y}^2, \mathbf{y}^3, \mathbf{y}^4$ are linearly dependent. ■

The solutions to systems of linear differential equations can be represented as *vector-valued functions.* For example,

$$\mathbf{y}(t) = \begin{bmatrix} y_1(t) \\ y_2(t) \\ y_3(t) \\ y_4(t) \end{bmatrix} = \begin{bmatrix} a\cos t \\ a\cos t \\ -a\sin t \\ -a\sin t \end{bmatrix}$$

is a normal mode solution to the mass–spring system of Example 3 if $k/m = 1$. Vector-valued functions are integrated and differentiated by components. For example,

$$\int_0^t \begin{bmatrix} a\cos s \\ a\cos s \\ -a\sin s \\ -a\sin s \end{bmatrix} ds = \begin{bmatrix} a\sin t \\ a\sin t \\ a(\cos t - 1) \\ a(\cos t - 1) \end{bmatrix}$$

and

$$\frac{d}{dt}\begin{bmatrix} a\sin t \\ a\sin t \\ a(\cos t - 1) \\ a(\cos t - 1) \end{bmatrix} = \begin{bmatrix} a\cos t \\ a\cos t \\ -a\sin t \\ -a\sin t \end{bmatrix}.$$

Vector-valued functions satisfy the same algebraic rules as constant vectors, but the definitions of linear dependence and independence must be modified.

The functions $\mathbf{y}^1(t), \mathbf{y}^2(t), \ldots, \mathbf{y}^k(t)$ are said to be *linearly dependent on the interval* $t_1 \le t \le t_2$ if there exist *constants* $c_1, c_2, \ldots, c_k$, not all zero, such that

$$c_1\mathbf{y}^1(t) + c_2\mathbf{y}^2(t) + \cdots + c_k\mathbf{y}^k(t) \equiv \mathbf{0}, \quad t_1 \le t \le t_2.$$

Otherwise they are said to be *linearly independent on the interval.*

Therefore if the relation

$$c_1\mathbf{y}^1(t) + c_2\mathbf{y}^2(t) + \cdots + c_k\mathbf{y}^k(t) \equiv 0, \quad t_1 \le t \le t_2,$$

implies that the constants $c_1, c_2, \ldots, c_n$ are all zero, then $\mathbf{y}^1(t), \mathbf{y}^2(t), \ldots, \mathbf{y}^k(t)$ are linearly independent on $t_1 \le t \le t_2$.

EXAMPLE 2 Show that

$$\mathbf{y}^1(t) = \begin{bmatrix} t \\ t \end{bmatrix}, \qquad \mathbf{y}^2(t) = \begin{bmatrix} t^2 \\ t^2 \end{bmatrix}$$

are linearly independent vector-valued functions on any nonempty interval, but for each fixed t they are dependent.

SOLUTION If c_1 and c_2 are constants satisfying

$$c_1 \begin{bmatrix} t \\ t \end{bmatrix} + c_2 \begin{bmatrix} t^2 \\ t^2 \end{bmatrix} = \begin{bmatrix} c_1t + c_2t^2 \\ c_1t + c_2t^2 \end{bmatrix} \equiv \begin{bmatrix} 0 \\ 0 \end{bmatrix}$$

for $t_1 \le t \le t_2$, with $t_1 < t_2$, it follows that $c_1 = c_2 = 0$. This is a consequence of the fact that a polynomial of degree two can vanish at most twice on any interval. On the other hand, for any *fixed* t,

$$t \begin{bmatrix} t \\ t \end{bmatrix} - \begin{bmatrix} t^2 \\ t^2 \end{bmatrix} = \begin{bmatrix} 0 \\ 0 \end{bmatrix},$$

but the coefficients $t, -1$ are not both constants as required. ■

An $n \times n$ matrix is an array of n^2 numbers a_{ij}, where $i, j = 1, 2, \ldots, n$. It will be denoted by a capital letter and displayed as follows:

$$A = \begin{bmatrix} a_{11} & a_{12} & \ldots & a_{1n} \\ a_{21} & a_{22} & \ldots & a_{2n} \\ \vdots & \vdots & & \vdots \\ a_{n1} & a_{n2} & \ldots & a_{nn} \end{bmatrix}.$$

It is also written as $A = (a_{ij})$. The entry a_{ij} can be real or complex, with the first subscript referring to its row, the second to its column. Since the number of columns equals the number of rows, an $n \times n$ matrix is called a *square* matrix.

Equality of matrices means pairwise equality of entries. Addition, subtraction, and scalar multiplication are defined by

$$A \pm B = \begin{bmatrix} a_{11} \pm b_{11} & a_{12} \pm b_{12} & \ldots & a_{1n} \pm b_{1n} \\ a_{21} \pm b_{21} & a_{22} \pm b_{22} & \ldots & a_{2n} \pm b_{2n} \\ \vdots & \vdots & & \vdots \\ a_{n1} \pm b_{n1} & a_{n2} \pm b_{n2} & \ldots & a_{nn} \pm b_{nn} \end{bmatrix}$$

and

$$cA = \begin{bmatrix} ca_{11} & ca_{12} & \dots & ca_{1n} \\ ca_{21} & ca_{22} & \dots & ca_{2n} \\ \vdots & \vdots & & \vdots \\ ca_{n1} & ca_{n2} & \dots & ca_{nn} \end{bmatrix},$$

or, more concisely, $(a_{ij}) \pm (b_{ij}) = (a_{ij} \pm b_{ij})$ and $c(a_{ij}) = (ca_{ij})$.

The product of two $n \times n$ matrices $A = (a_{ij})$ and $B = (b_{ij})$ is the $n \times n$ matrix AB defined by

$$AB = \left(\sum_{k=1}^{n} a_{ik} b_{kj} \right).$$

Unlike real or complex numbers, AB is not in general equal to BA. The rule for multiplying matrices can be easily remembered by noting that the (i, j)th entry of AB is the scalar, or dot, product of the ith row of A and the jth column of B. The vectors formed from the rows of A are denoted $\mathbf{A}_j$, those formed from its columns are denoted $\mathbf{A}^j$. Therefore if c_{ij} is the (i, j)th entry of $C = AB$, then $c_{ij} = \mathbf{A}_i \cdot \mathbf{B}^j$.

EXAMPLE 3 Let

$$A = \begin{bmatrix} 2 & 3 & 0 \\ 0 & -2 & 0 \\ 1 & 0 & 4 \end{bmatrix}, \quad B = \begin{bmatrix} -1 & 2 & 1 \\ 1 & -2 & -1 \\ 1 & 0 & 1 \end{bmatrix},$$

and form $3A - 4B$, AB, and BA.

SOLUTION Since $3A - 4B = (3a_{ij} - 4b_{ij})$, we can write

$$3A - 4B = \begin{bmatrix} 6+4 & 9-8 & 0-4 \\ 0-4 & -6+8 & 0+4 \\ 3-4 & 0+0 & 12-4 \end{bmatrix} = \begin{bmatrix} 10 & 1 & -4 \\ -4 & 2 & 4 \\ -1 & 0 & 8 \end{bmatrix}.$$

From the rule for matrix multiplication it follows that

$$AB = \begin{bmatrix} 1 & -2 & -1 \\ -2 & 4 & 2 \\ 3 & 2 & 5 \end{bmatrix}, \quad BA = \begin{bmatrix} -1 & -7 & 4 \\ 1 & 7 & -4 \\ 3 & 3 & 4 \end{bmatrix}.$$

A typical step in the computation of AB is the scalar, or dot, product of the third row of A by the first column of B to obtain the (3, 1)th entry of AB:

$$\sum_{k=1}^{3} a_{3k}b_{k1} = \begin{bmatrix} 1 \\ 0 \\ 4 \end{bmatrix} \cdot \begin{bmatrix} -1 \\ 1 \\ 1 \end{bmatrix} = -1 + 0 + 4 = 3.$$

Note that $AB \neq BA$. ■

The $n \times n$ matrix whose (i, j)th entry is 1 if $i = j$ and 0 if $i \neq j$ is called the *identity* matrix and is denoted by I. It is simply the square matrix with ones on the main diagonal and zeros elsewhere. It has the easily verified property of

$$IA = AI = A$$

for any $n \times n$ matrix A. For example,

$$\begin{bmatrix} 1 & 0 \\ 0 & 1 \end{bmatrix}\begin{bmatrix} 2 & 3 \\ 3 & 1 \end{bmatrix} = \begin{bmatrix} 2 & 3 \\ 3 & 1 \end{bmatrix}\begin{bmatrix} 1 & 0 \\ 0 & 1 \end{bmatrix} = \begin{bmatrix} 2 & 3 \\ 3 & 1 \end{bmatrix}.$$

The following algebraic rules for vectors and matrices are used extensively:

$$\mathbf{y} + \mathbf{z} = \mathbf{z} + \mathbf{y}, \qquad A + B = B + A;$$
$$(\mathbf{y} + \mathbf{w}) + \mathbf{z} = \mathbf{y} + (\mathbf{w} + \mathbf{z}), \qquad (A + B) + C = A + (B + C);$$
$$c(\mathbf{y} + \mathbf{z}) = c\mathbf{y} + c\mathbf{z}, \qquad c(A + B) = cA + cB, \quad c \text{ a scalar};$$
$$A(BC) = (AB)C;$$
$$A(B + C) = AB + AC.$$

The multiplication of a vector $\mathbf{y}$ by a matrix A is a vector $A\mathbf{y}$ whose ith component is the scalar, or dot, product of the ith row of A with the vector $\mathbf{y}$. Thus $(A\mathbf{y})_i = \mathbf{A}_i \cdot \mathbf{y} = \sum_{k=1}^{n} a_{ik}y_k$.

EXAMPLE 4 Let

$$A = \begin{bmatrix} 2 & 1 & 4 \\ 4 & 0 & -2 \\ 5 & 0 & 1 \end{bmatrix}, \qquad \mathbf{y} = \begin{bmatrix} 0 \\ 1 \\ -3 \end{bmatrix}, \qquad I = \begin{bmatrix} 1 & 0 & 0 \\ 0 & 1 & 0 \\ 0 & 0 & 1 \end{bmatrix}.$$

Then

$$A\mathbf{y} = \begin{bmatrix} 2 & 1 & 4 \\ 4 & 0 & -2 \\ 5 & 0 & 1 \end{bmatrix}\begin{bmatrix} 0 \\ 1 \\ -3 \end{bmatrix} = \begin{bmatrix} 0 + 1 - 12 \\ 0 + 0 + 6 \\ 0 + 0 - 3 \end{bmatrix} = \begin{bmatrix} -11 \\ 6 \\ -3 \end{bmatrix},$$

$$I\mathbf{y} = \begin{bmatrix} 1 & 0 & 0 \\ 0 & 1 & 0 \\ 0 & 0 & 1 \end{bmatrix}\begin{bmatrix} 0 \\ 1 \\ -3 \end{bmatrix} = \begin{bmatrix} 0 \\ 1 \\ -3 \end{bmatrix}.$$

The last computation illustrates a simple but useful fact, namely that $\mathbf{y} = I\mathbf{y}$ for all n-vectors $\mathbf{y}$. ■

The vectors formed from the columns of a matrix A can be either linearly dependent or linearly independent. To examine this, write

$$c_1\mathbf{A}^1 + c_2\mathbf{A}^2 + \cdots + c_n\mathbf{A}^n = \mathbf{0}$$

or, more concisely, $A\mathbf{c} = \mathbf{0}$. Clearly, $A\mathbf{c} = \mathbf{0}$ for some $\mathbf{c} \neq \mathbf{0}$ implies linear dependence of the columns of A, while $A\mathbf{c} = \mathbf{0}$ only if $\mathbf{c} = \mathbf{0}$ implies linear independence.

EXAMPLE 5 Use the matrices of Example 3 and show that the column vectors of A are linearly independent and that the column vectors of B are linearly dependent.

SOLUTION To show that the column vectors of A are linearly independent one writes

$$c_1\mathbf{A}^1 + c_2\mathbf{A}^2 + c_3\mathbf{A}^3 = c_1\begin{bmatrix} 2 \\ 0 \\ 1 \end{bmatrix} + c_2\begin{bmatrix} 3 \\ -2 \\ 0 \end{bmatrix} + c_3\begin{bmatrix} 0 \\ 0 \\ 4 \end{bmatrix} = \begin{bmatrix} 0 \\ 0 \\ 0 \end{bmatrix}$$

or

$$\begin{aligned} 2c_1 + 3c_2 + 0c_3 &= 0, \\ 0c_1 - 2c_2 + 0c_3 &= 0, \\ c_1 + 0c_2 + 4c_3 &= 0. \end{aligned}$$

Clearly, $c_1 = c_2 = c_3 = 0$. On the other hand, the column vectors of B are dependent since

$$\mathbf{B}^1 + \mathbf{B}^2 - \mathbf{B}^3 = \begin{bmatrix} -1 \\ 1 \\ 1 \end{bmatrix} + \begin{bmatrix} 2 \\ -2 \\ 0 \end{bmatrix} - \begin{bmatrix} 1 \\ -1 \\ 1 \end{bmatrix} = \begin{bmatrix} 0 \\ 0 \\ 0 \end{bmatrix}.$$ ■

It is a theorem of linear algebra that the n column vectors of an $n \times n$ matrix A are linearly independent if and only if the determinant, $\det(A)$, is nonzero. This theorem gives a test for checking the independence of n vectors of dimension n: form an $n \times n$ matrix whose columns are the vectors and compute its determinant. Unfortunately, the determinant of a square matrix of arbitrary dimension is both difficult to define and to evaluate. For simplicity, only determinants of 2×2 and 3×3 matrices will be considered. They are given for the 2×2 case by

$$\det(A) = \begin{vmatrix} a_{11} & a_{12} \\ a_{21} & a_{22} \end{vmatrix} = a_{11}a_{22} - a_{21}a_{12}$$

and for the 3×3 case by

$$\det(B) = \begin{vmatrix} b_{11} & b_{12} & b_{13} \\ b_{21} & b_{22} & b_{23} \\ b_{31} & b_{32} & b_{33} \end{vmatrix} = b_{11}\begin{vmatrix} b_{22} & b_{23} \\ b_{32} & b_{33} \end{vmatrix} - b_{12}\begin{vmatrix} b_{21} & b_{23} \\ b_{31} & b_{33} \end{vmatrix} + b_{13}\begin{vmatrix} b_{21} & b_{22} \\ b_{31} & b_{32} \end{vmatrix}$$

$$= b_{11}(b_{22}b_{33} - b_{32}b_{23}) - b_{12}(b_{21}b_{33} - b_{31}b_{23}) + b_{13}(b_{21}b_{32} - b_{31}b_{22}).$$

EXAMPLE 6 Evaluate the determinants of the matrices A and B of Example 3.

SOLUTION

$$\det(A) = \begin{vmatrix} 2 & 3 & 0 \\ 0 & -2 & 0 \\ 1 & 0 & 4 \end{vmatrix} = 2\begin{vmatrix} -2 & 0 \\ 0 & 4 \end{vmatrix} - 3\begin{vmatrix} 0 & 0 \\ 1 & 4 \end{vmatrix} + 0\begin{vmatrix} 0 & -2 \\ 1 & 0 \end{vmatrix}$$

$$= 2(-8 - 0) - 3(0 - 0) + 0(0 + 2) = -16,$$

$$\det(B) = \begin{vmatrix} -1 & 2 & 1 \\ 1 & -2 & -1 \\ 1 & 0 & 1 \end{vmatrix}$$

$$= -1\begin{vmatrix} -2 & -1 \\ 0 & 1 \end{vmatrix} - 2\begin{vmatrix} 1 & -1 \\ 1 & 1 \end{vmatrix} + 1\begin{vmatrix} 1 & -2 \\ 1 & 0 \end{vmatrix}$$

$$= -1(-2 - 0) - 2(1 + 1) + 1(0 + 2) = 0.$$

■

EXAMPLE 7 Use the determinant test to determine the dependence or independence of the following vectors:

$$\mathbf{x} = \begin{bmatrix} 2 \\ 2 \\ 0 \end{bmatrix}, \quad \mathbf{y} = \begin{bmatrix} 0 \\ -1 \\ 2 \end{bmatrix}, \quad \mathbf{z} = \begin{bmatrix} 3 \\ 0 \\ 2 \end{bmatrix}.$$

SOLUTION Form a matrix C with columns $\mathbf{C}^1 = \mathbf{x}$, $\mathbf{C}^2 = \mathbf{y}$, $\mathbf{C}^3 = \mathbf{z}$ and evaluate

$$\det(C) = \begin{vmatrix} 2 & 0 & 3 \\ 2 & -1 & 0 \\ 0 & 2 & 2 \end{vmatrix}$$

$$= 2\begin{vmatrix} -1 & 0 \\ 2 & 2 \end{vmatrix} - 0\begin{vmatrix} 2 & 0 \\ 0 & 2 \end{vmatrix} + 3\begin{vmatrix} 2 & -1 \\ 0 & 2 \end{vmatrix}$$

$$= (2)(-2) - 0 + (3)(4) = 8.$$

Since $\det(C) \neq 0$, the vectors are linearly independent. ■

The relationship between the linear independence of the columns of a square matrix and the nonvanishing of its determinant plays a key role in the theory of linear algebraic equations. The algebraic system of equations

$$\begin{aligned} a_{11}x_1 + a_{12}x_2 + a_{13}x_3 &= b_1, \\ a_{21}x_1 + a_{22}x_2 + a_{23}x_3 &= b_2, \\ a_{31}x_1 + a_{32}x_2 + a_{33}x_3 &= b_3 \end{aligned} \tag{5.2.1}$$

can be written in two ways as

$$A\mathbf{x} = \mathbf{b} \qquad \text{and} \qquad x_1\mathbf{A}^1 + x_2\mathbf{A}^2 + x_3\mathbf{A}^3 = \mathbf{b},$$

where the coefficient matrix is $A = (a_{ij})$. By Cramer's rule, if $\det(A) \neq 0$, there is a unique solution vector $\mathbf{x}$ whose components are

$$x_1 = \frac{\begin{vmatrix} b_1 & a_{12} & a_{13} \\ b_2 & a_{22} & a_{23} \\ b_3 & a_{32} & a_{33} \end{vmatrix}}{\det(A)}, \qquad x_2 = \frac{\begin{vmatrix} a_{11} & b_1 & a_{13} \\ a_{21} & b_2 & a_{23} \\ a_{31} & b_3 & a_{33} \end{vmatrix}}{\det(A)}, \tag{5.2.2}$$

$$x_3 = \frac{\begin{vmatrix} a_{11} & a_{12} & b_1 \\ a_{21} & a_{22} & b_2 \\ a_{31} & a_{32} & b_3 \end{vmatrix}}{\det(A)}.$$

It follows from (5.2.2) that $\mathbf{x}$ can be written as

$$\mathbf{x} = b_1\mathbf{C}^1 + b_2\mathbf{C}^2 + b_3\mathbf{C}^3 \quad \text{or} \quad \mathbf{x} = C\mathbf{b}, \tag{5.2.3}$$

where C is a 3×3 matrix. For example, the components of the first column of C are the factors of b_1 obtained from computing each of the determinants in (5.2.2):

$$\mathbf{C}^1 = \frac{1}{\det(A)} \begin{bmatrix} a_{22}a_{33} - a_{32}a_{23} \\ -(a_{21}a_{33} - a_{31}a_{23}) \\ a_{21}a_{32} - a_{31}a_{22} \end{bmatrix}.$$

This is illustrated in the following example.

EXAMPLE 8 Write the solution to the following system as $\mathbf{x} = C\mathbf{b}$.

$$\begin{aligned} -x_1 + 5x_2 - 3x_3 &= b_1, \\ x_1 - 3x_2 + 2x_3 &= b_2, \\ -x_2 + x_3 &= b_3, \end{aligned} \qquad \text{or} \qquad \begin{bmatrix} -1 & 5 & -3 \\ 1 & -3 & 2 \\ 0 & -1 & 1 \end{bmatrix} \begin{bmatrix} x_1 \\ x_2 \\ x_3 \end{bmatrix} = \begin{bmatrix} b_1 \\ b_2 \\ b_3 \end{bmatrix}.$$

SOLUTION This could be solved by systematic elimination, but we notice that when the three equations are added, the coefficients of x_1 and x_3 sum to 0. Hence,

$$x_2 = b_1 + b_2 + b_3.$$

From the third equation, we get

$$x_3 = x_2 + b_3 = b_1 + b_2 + 2b_3,$$

and from the second equation

$$x_1 = 3x_2 - 2x_3 + b_2 = b_1 + 2b_2 - b_3.$$

Thus

$$\mathbf{x} = \begin{bmatrix} x_1 \\ x_2 \\ x_3 \end{bmatrix} = \begin{bmatrix} 1 & 2 & -1 \\ 1 & 1 & 1 \\ 1 & 1 & 2 \end{bmatrix} \begin{bmatrix} b_1 \\ b_2 \\ b_3 \end{bmatrix} = C\mathbf{b}.$$

Another route to the same solution is via Cramer's rule:

$$x_1 = \frac{\begin{vmatrix} b_1 & 5 & -3 \\ b_2 & -3 & 2 \\ b_3 & -1 & 1 \end{vmatrix}}{\det(A)}, \qquad x_2 = \frac{\begin{vmatrix} -1 & b_1 & -3 \\ 1 & b_2 & 2 \\ 0 & b_3 & 1 \end{vmatrix}}{\det(A)},$$

$$x_3 = \frac{\begin{vmatrix} -1 & 5 & b_1 \\ 1 & -3 & b_2 \\ 0 & -1 & b_3 \end{vmatrix}}{\det(A)},$$

where the ith column of C has as its components the factors of b_i in each of the above expressions. We leave these computations to the reader. ■

We now generalize the previous discussion and introduce the notion of the inverse of a square matrix. Two n-dimensional linear systems of equations

$$A\mathbf{x} = \mathbf{b} \quad \text{and} \quad C\mathbf{b} = \mathbf{x}$$

with $\det(A) \neq 0$, $\det(C) \neq 0$ are dual in the sense that either $\mathbf{b}$ or $\mathbf{x}$ can be considered as being given and then either $\mathbf{x}$ or $\mathbf{b}$ is the corresponding solution. By substitution we have (recall that $\mathbf{b} = I\mathbf{b}$ and $\mathbf{x} = I\mathbf{x}$) that

$$A\mathbf{x} = AC\mathbf{b} = \mathbf{b} \quad \text{or} \quad (AC - I)\mathbf{b} = \mathbf{0},$$
$$C\mathbf{b} = CA\mathbf{x} = \mathbf{x} \quad \text{or} \quad (CA - I)\mathbf{x} = \mathbf{0}.$$

Because of the arbitrariness of **b** (or **x**), these equations imply that

$$AC = I \qquad \text{and} \qquad CA = I.$$

If these equations are satisfied, then C is called the *inverse* of A and is denoted by A^{-1}. The last pair of equations implies that

$$AA^{-1} = A^{-1}A = I, \tag{5.2.4}$$

and it is not hard to show that a matrix A^{-1} satisfying (5.2.4) is unique.

In general, an algebraic formula for the inverse of a matrix is rather complicated. For $n = 2$, the inverse is especially easy to determine:

$$\text{If } A = \begin{bmatrix} a & b \\ c & d \end{bmatrix}, \quad \text{then } A^{-1} = \frac{1}{ad - bc}\begin{bmatrix} d & -b \\ -c & a \end{bmatrix}.$$

The verification of this formula is left to the reader.

The next topics in our summary of linear algebra are eigenvalues and eigenvectors, strange terms coming from the German words *Eigenwert* and *Eigenvektor* meaning *characteristic value* and *characteristic vector,* respectively. They occur for instance in the study of normal modes of vibrations.

If A is a square matrix and $\mathbf{x}$ is a nonzero vector, such that $A\mathbf{x} = \lambda\mathbf{x}$ for some real or complex number λ, then $\mathbf{x}$ is called an *eigenvector* of A corresponding to the *eigenvalue* λ. The defining equation can also be written as

$$(A - \lambda I)\mathbf{x} = \mathbf{0} \tag{5.2.5}$$

with the aid of the identity $\mathbf{x} = I\mathbf{x}$. Equation (5.2.5) will have a nonzero solution if and only if the column vectors of $A - \lambda I$ are linearly dependent. But, as we have seen, this is equivalent to the requirement that

$$\det(A - \lambda I) = 0. \tag{5.2.6}$$

If A is an $n \times n$ matrix, then Eq. (5.2.6) is a polynomial of degree n in λ called the *characteristic equation* of A. It must be satisfied by any eigenvalue of A, and since a polynomial of degree n has at most n distinct roots, an $n \times n$ matrix has at most n distinct eigenvalues. The process of finding eigenvalues and eigenvectors is illustrated in the following example.

EXAMPLE 9 Find the eigenvalues and eigenvectors of

$$A = \begin{bmatrix} 0 & 1 & 0 \\ 0 & 0 & 1 \\ -2 & 1 & 2 \end{bmatrix}.$$

SOLUTION The characteristic equation is

$$0 = \det(A - \lambda I) = \det\left[\begin{bmatrix} 0 & 1 & 0 \\ 0 & 0 & 1 \\ -2 & 1 & 2 \end{bmatrix} - \begin{bmatrix} \lambda & 0 & 0 \\ 0 & \lambda & 0 \\ 0 & 0 & \lambda \end{bmatrix}\right]$$

$$= \begin{vmatrix} -\lambda & 1 & 0 \\ 0 & -\lambda & 1 \\ -2 & 1 & 2-\lambda \end{vmatrix} = -\lambda^3 + 2\lambda^2 + \lambda - 2.$$

Since

$$-\lambda^3 + 2\lambda^2 + \lambda - 2 = -(\lambda - 1)(\lambda + 1)(\lambda - 2),$$

the roots $\lambda = 1$, $\lambda = -1$, and $\lambda = 2$ are the eigenvalues of A. To find the corresponding eigenvectors, we select nontrivial solutions to the homogeneous equations $(A - \lambda I)\mathbf{x} = 0$, where λ is an eigenvalue. The equations are

$$(A - I)\mathbf{x} = \mathbf{0}, \qquad (A + I)\mathbf{x} = \mathbf{0}, \qquad (A - 2I)\mathbf{x} = \mathbf{0},$$

or

$$\begin{array}{ccc} (\lambda = 1) & (\lambda = -1) & (\lambda = 2) \\ -x_1 + x_2 = 0, & x_1 + x_2 = 0, & -2x_1 + x_2 = 0, \\ -x_2 + x_3 = 0, & x_2 + x_3 = 0, & -2x_2 + x_3 = 0, \\ -2x_1 + x_2 + x_3 = 0; & -2x_1 + x_2 + 3x_3 = 0; & -2x_1 + x_2 = 0. \end{array}$$

One set of nontrivial solutions can be constructed by taking $x_1 = 1$ and solving for x_2 and x_3. The respective eigenvectors are

$$\begin{array}{ccc} (\lambda = 1) & (\lambda = -1) & (\lambda = 2) \\ \mathbf{x}^1 = \begin{bmatrix} 1 \\ 1 \\ 1 \end{bmatrix}, & \mathbf{x}^2 = \begin{bmatrix} 1 \\ -1 \\ 1 \end{bmatrix}, & \mathbf{x}^3 = \begin{bmatrix} 1 \\ 2 \\ 4 \end{bmatrix}. \end{array}$$

Any other set of nontrivial solutions can be obtained by another choice of x_1 and, clearly, this is equivalent to multiplying the above solutions by a constant. This is a general result: an eigenvector remains an eigenvector upon multiplication by any nonzero scalar. ■

If an $n \times n$ matrix has n distinct eigenvalues, it can be shown that the corresponding n eigenvectors are linearly independent. But if there are multiple eigenvalues (corresponding to multiple roots of the characteristic equation), there may or may not be n linearly independent eigenvectors. This matter will not be discussed further here, but it will come up in subsequent sections.

Our final topic is matrix valued functions. Their algebra is the same as that of constant matrices and, just as with vector valued functions, they are differentiated and integrated entry by entry. For example, if $\Theta(t)$ is the matrix

$$\Theta(t) = \begin{bmatrix} \cos t & \sin t \\ -\sin t & \cos t \end{bmatrix},$$

then

$$\frac{d}{dt}\Theta(t) = \begin{bmatrix} -\sin t & \cos t \\ -\cos t & -\sin t \end{bmatrix}.$$

Note that the matrix $\Theta(t)$ is a solution to a *matrix differential equation,*

$$\frac{d\Theta}{dt} = J\Theta, \qquad J = \begin{bmatrix} 0 & 1 \\ -1 & 0 \end{bmatrix},$$

since

$$J\Theta = \begin{bmatrix} 0 & 1 \\ -1 & 0 \end{bmatrix} \begin{bmatrix} \cos t & \sin t \\ -\sin t & \cos t \end{bmatrix} = \begin{bmatrix} -\sin t & \cos t \\ -\cos t & -\sin t \end{bmatrix} = \frac{d\Theta}{dt}.$$

Matrix differential equations will be encountered in Section 5.3.

The theorem that the n columns of an $n \times n$ matrix A are linearly independent if and only if the determinant $\det(A)$ is nonzero is not valid for general matrix valued functions. However, if the columns of the matrix are solutions to a homogeneous linear system of differential equations, then the theorem is valid (see [1, p. 69]) and the determinant test for linear dependence or linear independence can be employed.

This completes our résumé of linear algebra. We have presented a few concepts that are needed in an introductory study of linear systems of ordinary differential equations. Many important topics were not mentioned; they can be found in more advanced differential equation texts such as [1].

EXERCISES 5.2

For each of the pairs of vectors **x**, **y** in Exercises 1–6 find $\mathbf{x} + \mathbf{y}$, $\mathbf{x} - \mathbf{y}$, $\mathbf{x} \cdot \mathbf{y}$, and determine whether they are linearly independent or linearly dependent.

1. $\mathbf{x} = \begin{bmatrix} 1 \\ 0 \end{bmatrix}, \quad \mathbf{y} = \begin{bmatrix} 2 \\ 0 \end{bmatrix}$

2. $\mathbf{x} = \begin{bmatrix} 1 \\ 2 \end{bmatrix}, \quad \mathbf{y} = \begin{bmatrix} 3 \\ 4 \end{bmatrix}$

3. $\mathbf{x} = \begin{bmatrix} 1 \\ 2 \end{bmatrix}$, $\mathbf{y} = \begin{bmatrix} 2 \\ 4 \end{bmatrix}$

4. $\mathbf{x} = \begin{bmatrix} 1 \\ 2 \\ 3 \end{bmatrix}$, $\mathbf{y} = \begin{bmatrix} 4 \\ 5 \\ 6 \end{bmatrix}$

5. $\mathbf{x} = \begin{bmatrix} 1 \\ 0 \\ 1 \end{bmatrix}$, $\mathbf{y} = \begin{bmatrix} 2 \\ 3 \\ -1 \end{bmatrix}$

6. $\mathbf{x} = \begin{bmatrix} 1 \\ -1 \\ 2 \end{bmatrix}$, $\mathbf{y} = \begin{bmatrix} 1 \\ -1 \\ -1 \end{bmatrix}$

Determine which of the sets of vectors in Exercises 7–12 are linearly independent and which are linearly dependent.

7. $\begin{bmatrix} 1 \\ 0 \\ 0 \end{bmatrix}, \begin{bmatrix} 0 \\ 1 \\ 0 \end{bmatrix}, \begin{bmatrix} 0 \\ 0 \\ 1 \end{bmatrix}$

8. $\begin{bmatrix} 3 \\ 2 \\ 1 \end{bmatrix}, \begin{bmatrix} 2 \\ -1 \\ 2 \end{bmatrix}, \begin{bmatrix} -1 \\ 2 \\ 1 \end{bmatrix}$

9. $\begin{bmatrix} 2 \\ -2 \\ 1 \end{bmatrix}, \begin{bmatrix} 1 \\ 3 \\ 0 \end{bmatrix}, \begin{bmatrix} 0 \\ 0 \\ 1 \end{bmatrix}$

10. $\begin{bmatrix} 1 \\ 2 \end{bmatrix}, \begin{bmatrix} 3 \\ 4 \end{bmatrix}, \begin{bmatrix} 5 \\ 6 \end{bmatrix}$

11. $\begin{bmatrix} 1 \\ 0 \\ 0 \end{bmatrix}, \begin{bmatrix} -2 \\ 1 \\ 0 \end{bmatrix}, \begin{bmatrix} 3 \\ -1 \\ 1 \end{bmatrix}, \begin{bmatrix} 7 \\ -2 \\ 1 \end{bmatrix}$

12. $\begin{bmatrix} 1 \\ 3 \\ -2 \end{bmatrix}, \begin{bmatrix} 0 \\ 0 \\ 1 \end{bmatrix}, \begin{bmatrix} 1 \\ 3 \\ -1 \end{bmatrix}$

For each of the vector functions $\mathbf{x}(t)$ in Exercises 13–18 find $d\mathbf{x}(t)/dt$ and $\int_0^t \mathbf{x}(s)\, ds$.

13. $\mathbf{x}(t) = \begin{bmatrix} t \\ t^2 \end{bmatrix}$

14. $\mathbf{x}(t) = \begin{bmatrix} t \\ \sin t \\ e^t \end{bmatrix}$

15. $\mathbf{x}(t) = \begin{bmatrix} te^t \\ \sin 2t \\ 2t + 1 \end{bmatrix}$

16. $\mathbf{x}(t) = \begin{bmatrix} e^t \\ 2e^t \\ -e^t \end{bmatrix}$

17. $\mathbf{x}(t) = \begin{bmatrix} \sin t + \cos t \\ \sin t - \cos t \end{bmatrix}$

18. $\mathbf{x}(t) = \begin{bmatrix} 1 \\ e^{2t} \end{bmatrix}$

Determine which of the sets of vector functions in Exercises 19–22 are linearly independent and which are linearly dependent for $0 \leq t \leq 1$.

19. $\begin{bmatrix} 1 \\ t \end{bmatrix}, \begin{bmatrix} t \\ t^2 \end{bmatrix}$

20. $\begin{bmatrix} e^t \\ -e^t \end{bmatrix}, \begin{bmatrix} e^{2t} \\ 3e^{2t} \end{bmatrix}$

21. $\begin{bmatrix} 2e^t \\ -e^t \\ 3e^t \end{bmatrix}, \begin{bmatrix} e^{-t} \\ 4e^{-t} \\ -e^{-t} \end{bmatrix}, \begin{bmatrix} 3e^{-2t} \\ e^{-2t} \\ -e^{-2t} \end{bmatrix}$

22. $\begin{bmatrix} e^t \\ 1 \\ e^t \end{bmatrix}, \begin{bmatrix} e^{-t} \\ 1 \\ -e^{-t} \end{bmatrix}, \begin{bmatrix} \cosh t \\ 1 \\ \sinh t \end{bmatrix}$

23. Find the Laplace transform $\mathbf{X}(s)$ for each of the vector functions $\mathbf{x}(t)$ in Exercises 13–18.

24. Let

$$A = \begin{bmatrix} 1 & 2 & 1 \\ 2 & 1 & 0 \\ 0 & 1 & 1 \end{bmatrix}, \qquad B = \begin{bmatrix} 2 & 1 & 2 \\ 1 & 0 & 1 \\ 0 & 1 & 0 \end{bmatrix}.$$

Find

a) $A + B$ **b)** $A - B$ **c)** $A - 2B$ **d)** $2A - B$
e) AB **f)** BA

25. If

$$A = \begin{bmatrix} 2 & 1 & 1 \\ 3 & 1 & 0 \\ 0 & 2 & 1 \end{bmatrix}, \qquad B = \begin{bmatrix} 1 & 1 & 1 \\ 2 & 3 & 1 \\ 1 & 0 & 1 \end{bmatrix},$$

find

a) $3A + B$ **b)** $2A - 3B$ **c)** AB **d)** BA

26. Let

$$A = \begin{bmatrix} 1 & 2 & 3 \\ 1 & 2 & 1 \\ 0 & 1 & 0 \end{bmatrix}, \qquad B = \begin{bmatrix} 1 & 1 & 1 \\ 6 & 1 & 4 \\ 1 & 4 & 2 \end{bmatrix}, \qquad C = \begin{bmatrix} 0 & 1 & 3 \\ 0 & 4 & 1 \\ 2 & 0 & 4 \end{bmatrix}.$$

Verify that

a) $A + (B + C) = (A + B) + C$ **b)** $A(B + C) = AB + AC$
c) $(A + B)C = AC + BC$ **d)** $A(BC) = (AB)C$
e) $2(A + B) = 2A + 2B$

27. The transpose of a matrix $A = (a_{ij})$ denoted by A^T is defined by $A^T = (a_{ji})$. It is the matrix obtained from A by interchanging the rows and columns of A. If

$$A = \begin{bmatrix} 1 & 2 \\ 3 & 4 \end{bmatrix}, \qquad B = \begin{bmatrix} 5 & 6 \\ 7 & 8 \end{bmatrix},$$

find

a) A^T **b)** B^T **c)** $(A + B)^T$ **d)** $A^T + B^T$
e) $(AB)^T$ **f)** B^TA^T

28. Verify that three columns of the identity matrix

$$I = \begin{bmatrix} 1 & 0 & 0 \\ 0 & 1 & 0 \\ 0 & 0 & 1 \end{bmatrix}$$

form a linearly independent set. Verify that $IA = AI = A$ if

$$A = \begin{bmatrix} 1 & 2 & 3 \\ 4 & 5 & 6 \\ 7 & 8 & 9 \end{bmatrix}.$$

Find the determinant of each of the matrices in Exercises 29–34. In which of the matrices do the columns form a linearly independent set of vectors?

29. $\begin{bmatrix} 1 & 2 \\ 3 & 4 \end{bmatrix}$

30. $\begin{bmatrix} 2 & 1 \\ 4 & 2 \end{bmatrix}$

31. $\begin{bmatrix} 1 & 2 & 3 \\ 2 & 4 & 5 \\ 3 & 5 & 6 \end{bmatrix}$

32. $\begin{bmatrix} 1 & 2 & 1 \\ -2 & 1 & 8 \\ 1 & -2 & -7 \end{bmatrix}$

33. $\begin{bmatrix} 1 & 0 & 0 \\ 0 & 1 & 0 \\ 0 & 0 & 1 \end{bmatrix}$

34. $\begin{bmatrix} 1 & 2 & 1 \\ 3 & 1 & 4 \\ 1 & 0 & 1 \end{bmatrix}$

35. Complete the details of Example 8 by finding C via Cramer's rule.

Solve each of the systems of algebraic equations $A\mathbf{x} = \mathbf{b}$ in Exercises 36–40 and find the matrix C so that the solution $\mathbf{x}$ can be written in the form $\mathbf{x} = C\mathbf{b}$.

36. $\begin{bmatrix} 1 & 1 \\ 2 & 1 \end{bmatrix}\begin{bmatrix} x_1 \\ x_2 \end{bmatrix} = \begin{bmatrix} 1 \\ 0 \end{bmatrix}$

37. $\begin{bmatrix} 1 & -3 \\ 2 & 5 \end{bmatrix}\begin{bmatrix} x_1 \\ x_2 \end{bmatrix} = \begin{bmatrix} -5 \\ 12 \end{bmatrix}$

38. $\begin{bmatrix} 7 & 4 \\ 5 & -3 \end{bmatrix}\begin{bmatrix} x_1 \\ x_2 \end{bmatrix} = \begin{bmatrix} 26 \\ 1 \end{bmatrix}$

39. $\begin{bmatrix} 2 & 1 & 3 \\ 3 & 4 & 1 \\ 1 & 0 & -2 \end{bmatrix}\begin{bmatrix} x_1 \\ x_2 \\ x_3 \end{bmatrix} = \begin{bmatrix} 6 \\ 8 \\ -1 \end{bmatrix}$

40. $\begin{bmatrix} 1 & 1 & -1 \\ 3 & -1 & 0 \\ 9 & 1 & 1 \end{bmatrix}\begin{bmatrix} x_1 \\ x_2 \\ x_3 \end{bmatrix} = \begin{bmatrix} 0 \\ 3 \\ 10 \end{bmatrix}$

41. Verify that $\begin{bmatrix} a & b \\ c & d \end{bmatrix}^{-1} = \dfrac{1}{ad - bc}\begin{bmatrix} d & -b \\ -c & a \end{bmatrix}$.

Find the inverse of each of the 2×2 matrices in Exercises 42–47. Verify that $AA^{-1} = A^{-1}A = I$.

42. $\begin{bmatrix} 3 & 5 \\ 1 & 2 \end{bmatrix}$

43. $\begin{bmatrix} 2 & 5 \\ 1 & 3 \end{bmatrix}$

44. $\begin{bmatrix} 3 & 2 \\ 1 & 2 \end{bmatrix}$

45. $\begin{bmatrix} 25 & 36 \\ 9 & 13 \end{bmatrix}$

46. $\begin{bmatrix} 1 & 2 \\ 3 & 4 \end{bmatrix}$

47. $\begin{bmatrix} 3 & -5 \\ -1 & 2 \end{bmatrix}$

48. Verify that the inverse of

$$\begin{bmatrix} 1 & 2 & 3 \\ 2 & 3 & 4 \\ 3 & 4 & 6 \end{bmatrix} \quad \text{is} \quad \begin{bmatrix} -2 & 0 & 1 \\ 0 & 3 & -2 \\ 1 & -2 & 1 \end{bmatrix}.$$

Find the characteristic polynomial, the eigenvalues, and eigenvectors of the matrices in Exercises 49–54.

49. $\begin{bmatrix} 2 & -1 \\ -1 & 2 \end{bmatrix}$

50. $\begin{bmatrix} 1 & -2 \\ -3 & 2 \end{bmatrix}$

51. $\begin{bmatrix} 1 & 1 \\ 2 & 2 \end{bmatrix}$

52. $\begin{bmatrix} 3 & -1 & -1 \\ -12 & 0 & 5 \\ 4 & -2 & -1 \end{bmatrix}$

53. $\begin{bmatrix} 2 & 2 & 1 \\ 1 & 3 & 1 \\ 1 & 2 & 2 \end{bmatrix}$

54. $\begin{bmatrix} 1 & 1 & -2 \\ -1 & 2 & 1 \\ 0 & 1 & -1 \end{bmatrix}$

For each of the matrix valued functions $A(t)$ in Exercises 55–60 find $dA(t)/dt$.

55. $A(t) = \begin{bmatrix} e^t & e^{-t} \\ 2e^t & -e^t \end{bmatrix}$

56. $A(t) = \begin{bmatrix} 1 & e^{2t} \\ 2 & -2e^{2t} \end{bmatrix}$

57. $A(t) = \begin{bmatrix} e^{2t}\cos t & 2e^{2t}\sin t \\ 3e^{2t}\cos t & e^{2t}\sin t \end{bmatrix}$

58. $A(t) = \begin{bmatrix} e^t & e^{2t} & e^{5t} \\ e^t & -2e^{2t} & e^{5t} \\ -e^t & -3e^{2t} & 3e^{5t} \end{bmatrix}$

59. $A(t) = \begin{bmatrix} 0 & 2\cos t & 5\sin t \\ 2e^t & 2\cos t & 5\sin t \\ e^t & 3\cos t & -5\sin t \end{bmatrix}$

60. $A(t) = \begin{bmatrix} 0 & e^{-t} & 3te^{-t} \\ 2e^t & -4e^{-t} & (3+6t)e^{-t} \\ e^t & -2e^{-t} & (3+3t)e^{-t} \end{bmatrix}$

In each of the problems in Exercises 61–63 verify that the given matrix satisfies the given matrix differential equation.

61. $\dfrac{d\Theta}{dt} = \begin{bmatrix} 2 & 1 \\ 3 & 4 \end{bmatrix}\Theta, \quad \Theta(t) = \begin{bmatrix} e^t & e^{5t} \\ -e^t & 3e^{5t} \end{bmatrix}$

62. $\dfrac{d\Theta}{dt} = \begin{bmatrix} 1 & 1 \\ -2 & 3 \end{bmatrix}\Theta, \quad \Theta(t) = e^{2t}\begin{bmatrix} \cos t & \sin t \\ \cos t - \sin t & \cos t + \sin t \end{bmatrix}$

63. $\dfrac{d\Theta}{dt} = \begin{bmatrix} 2 & -1 & 2 \\ 1 & 0 & 2 \\ -2 & 1 & -1 \end{bmatrix}\Theta, \quad \Theta(t) = \begin{bmatrix} 0 & \cos t + \sin t & 2\sin t \\ 2e^t & \cos t + \sin t & 2\sin t \\ e^t & -\sin t & \cos t - \sin t \end{bmatrix}$

64. Not every square matrix has an inverse. Show that this is the case for

$$A = \begin{bmatrix} 1 & 2 \\ 1 & 2 \end{bmatrix}.$$

65. Find the Laplace transformation of the matrices in Exercises 55–60.

5.3 HOMOGENEOUS LINEAR SYSTEMS WITH CONSTANT COEFFICIENTS

We begin with two-dimensional constant coefficient homogeneous linear systems, so that the ideas and techniques will not be obscured by algebraic complexities. However, our choice of topics and notation allows their extension to higher dimensional systems. The differential equations

$$\frac{dy_1}{dt} = a_{11}y_1 + a_{12}y_2, \qquad \frac{dy_2}{dt} = a_{21}y_1 + a_{22}y_2 \tag{5.3.1}$$

in vector–matrix notation are

$$\frac{d\mathbf{y}}{dt} = A\mathbf{y},$$

where

$$A = \begin{bmatrix} a_{11} & a_{12} \\ a_{21} & a_{22} \end{bmatrix}, \qquad \mathbf{y} = \begin{bmatrix} y_1 \\ y_2 \end{bmatrix}.$$

In finding a solution to Eqs. (5.3.1) we are guided by our experience with both first and second order linear differential equations; we set

$$\mathbf{y} = \begin{bmatrix} y_1 \\ y_2 \end{bmatrix} = \begin{bmatrix} \alpha e^{\lambda t} \\ \beta e^{\lambda t} \end{bmatrix} = \begin{bmatrix} \alpha \\ \beta \end{bmatrix} e^{\lambda t} \tag{5.3.2}$$

and try to determine α, β, and λ, so that we have a solution. Before studying the general problem, we shall discuss a specific example in some detail.

EXAMPLE 1 Consider the system

$$\frac{dy_1}{dt} = y_1 + y_2, \qquad \frac{dy_2}{dt} = 4y_1 - 2y_2 \tag{5.3.3}$$

or

$$\frac{d}{dt}\begin{bmatrix} y_1 \\ y_2 \end{bmatrix} = \begin{bmatrix} 1 & 1 \\ 4 & -2 \end{bmatrix}\begin{bmatrix} y_1 \\ y_2 \end{bmatrix}.$$

Substituting (5.3.2) into (5.3.3) gives

$$\alpha\lambda e^{\lambda t} = \alpha e^{\lambda t} + \beta e^{\lambda t}, \qquad \beta\lambda e^{\lambda t} = 4\alpha e^{\lambda t} - 2\beta e^{\lambda t},$$

which, upon dividing by $e^{\lambda t}$, simplifies to

$$(1 - \lambda)\alpha + \beta = 0, \qquad 4\alpha - (2 + \lambda)\beta = 0. \tag{5.3.4}$$

The Eqs. (5.3.4) can be thought of as the equations of two straight lines in the $\alpha\beta$ plane intersecting at the origin (see Fig. 5.3). This intersection corresponds to the trivial solution $y_1 = 0$, $y_2 = 0$. In order to construct a nontrivial solution, λ is chosen so that the lines coincide. Since both lines contain the origin, they will be coincident if and only if their slopes are equal. This is equivalent to the algebraic condition

$$\begin{vmatrix} 1 - \lambda & 1 \\ 4 & -2 - \lambda \end{vmatrix} = 0, \tag{5.3.5}$$

or

$$\lambda^2 + \lambda - 6 = (\lambda + 3)(\lambda - 2) = 0.$$

An important observation is that the last equation is just the characteristic equation of the coefficient matrix

$$A = \begin{bmatrix} 1 & 1 \\ 4 & -2 \end{bmatrix}.$$

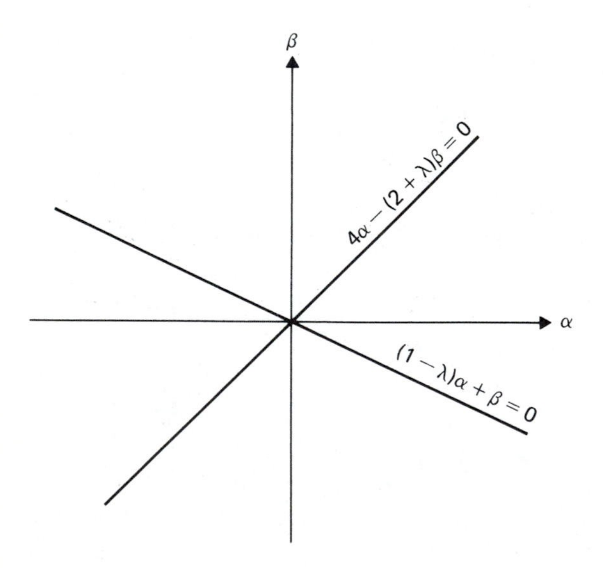

Figure 5.3 Graphs of equations (5.3.4).

Hence we can take $\lambda = -3$ or $\lambda = 2$. If $\lambda = -3$, α and β must satisfy $4\alpha + \beta = 0$; if $\lambda = 2$, α and β must satisfy $-\alpha + \beta = 0$ (see Fig. 5.4). Any point on either line (other than the origin) will determine a nontrivial solution to the system of differential equations (5.3.3).

First consider the case where $\lambda = -3$. Geometrically, it is clear that all points on the line $4\alpha + \beta = 0$ are scalar multiples of any nontrivial point, for example, $\alpha = 1, \beta = -4$. Therefore we obtain a one-parameter family of solutions

$$\mathbf{y}^1(t) = c_1 \begin{bmatrix} 1 \\ -4 \end{bmatrix} e^{-3t}, \tag{5.3.6}$$

where c_1 is an arbitrary constant. In contrast to the scalar equation $dy/dt = ry$ with solution $y = ce^{rt}$, the coefficient of the exponential function is a vector whose direction is determined by the coefficient matrix of the system of differential equations. Repeating the argument for $\lambda = 2$ gives a second one-parameter family of solutions

$$\mathbf{y}^2(t) = c_2 \begin{bmatrix} 1 \\ 1 \end{bmatrix} e^{2t}. \tag{5.3.7}$$

The two values of λ, -3 and 2, are eigenvalues of the coefficient matrix

$$A = \begin{bmatrix} 1 & 1 \\ 4 & -2 \end{bmatrix},$$

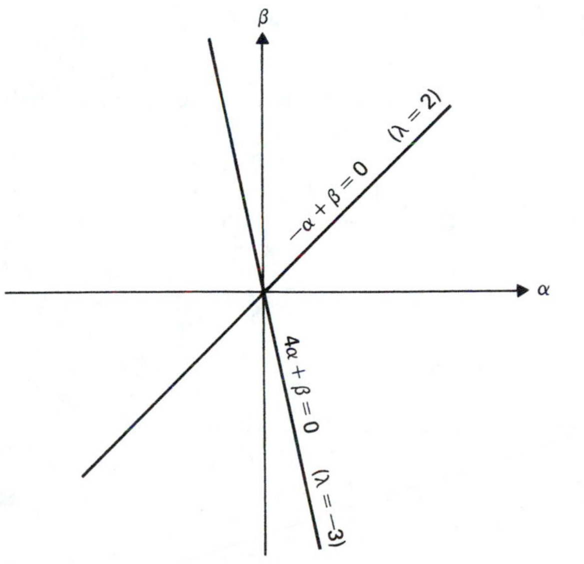

Figure 5.4

and the corresponding vectors

$$\begin{bmatrix} 1 \\ -4 \end{bmatrix} \quad \text{and} \quad \begin{bmatrix} 1 \\ 1 \end{bmatrix}$$

are eigenvectors of A. ■

EXAMPLE 2 Show that the two solutions obtained (with $c_1 = c_2 = 1$) in Example 1 are linearly independent.

SOLUTION The solution vectors $\mathbf{y}^1(t)$ and $\mathbf{y}^2(t)$ are linearly independent, as can be shown by the determinant test (see p. 350):

$$\begin{vmatrix} y_1^1(t) & y_1^2(t) \\ y_2^1(t) & y_2^2(t) \end{vmatrix} = \begin{vmatrix} e^{-3t} & e^{2t} \\ -4e^{-3t} & e^{2t} \end{vmatrix} = 5e^{-t} \neq 0. \tag{5.3.8}$$

A linear combination of these solutions,

$$\mathbf{y}(t) = \mathbf{y}^1(t) + \mathbf{y}^2(t) = c_1 \begin{bmatrix} 1 \\ -4 \end{bmatrix} e^{-3t} + c_2 \begin{bmatrix} 1 \\ 1 \end{bmatrix} e^{2t}, \tag{5.3.9}$$

is easily shown to be a solution to the differential equation (5.3.3). Furthermore, it is a *general solution,* since, as we shall show, the constants c_1 and c_2 can be chosen to match arbitrary initial data. ■

EXAMPLE 3 Find the solution to (5.3.3) satisfying

$$\mathbf{y}(0) = \begin{bmatrix} 6 \\ -4 \end{bmatrix}.$$

SOLUTION Substituting $t = 0$ into (5.3.9) implies that c_1 and c_2 must satisfy the vector equation

$$c_1 \begin{bmatrix} 1 \\ -4 \end{bmatrix} + c_2 \begin{bmatrix} 1 \\ 1 \end{bmatrix} = \begin{bmatrix} 6 \\ -4 \end{bmatrix}$$

or

$$c_1 + c_2 = 6,$$
$$-4c_1 + c_2 = -4.$$

Solving this system gives $c_1 = 2$, $c_2 = 4$, and the required solution is

$$\mathbf{y}(t) = 2 \begin{bmatrix} 1 \\ -4 \end{bmatrix} e^{-3t} + 4 \begin{bmatrix} 1 \\ 1 \end{bmatrix} e^{2t}.$$

■

EXAMPLE 4 Find the solution to (5.3.3) satisfying the general initial conditions

$$\mathbf{y}(t_0) = \mathbf{y}_0 = \begin{bmatrix} y_1^0 \\ y_2^0 \end{bmatrix}.$$

SOLUTION By using (5.3.9), we obtain the equations

$$\mathbf{y}(t_0) = \begin{bmatrix} y_1^0 \\ y_2^0 \end{bmatrix} = c_1 \begin{bmatrix} 1 \\ -4 \end{bmatrix} e^{-3t_0} + c_2 \begin{bmatrix} 1 \\ 1 \end{bmatrix} e^{2t_0}$$

or

$$y_1^0 = c_1 e^{-3t_0} + c_2 e^{2t_0},$$
$$y_2^0 = -4c_1 e^{-3t_0} + c_2 e^{2t_0}.$$

Using Cramer's rule, we write

$$c_1 = \frac{\begin{vmatrix} y_1^0 & e^{2t_0} \\ y_2^0 & e^{2t_0} \end{vmatrix}}{\begin{vmatrix} e^{-3t_0} & e^{2t_0} \\ -4e^{-3t_0} & e^{2t_0} \end{vmatrix}} = \frac{1}{5}(y_1^0 - y_2^0)e^{3t_0},$$

$$c_2 = \frac{\begin{vmatrix} e^{-3t_0} & y_1^0 \\ -4e^{-3t_0} & y_2^0 \end{vmatrix}}{\begin{vmatrix} e^{-3t_0} & e^{2t_0} \\ -4e^{-3t_0} & e^{2t_0} \end{vmatrix}} = \frac{1}{5}(y_2^0 + 4y_1^0)e^{-2t_0};$$

thus the constants c_1 and c_2 are uniquely determined. ■

Let us now return to the general system

$$\frac{d}{dt}\begin{bmatrix} y_1 \\ y_2 \end{bmatrix} = \begin{bmatrix} a_{11} & a_{12} \\ a_{21} & a_{22} \end{bmatrix} \begin{bmatrix} y_1 \\ y_2 \end{bmatrix} \quad \text{or} \quad \frac{d\mathbf{y}}{dt} = A\mathbf{y}. \tag{5.3.10}$$

Set

$$\mathbf{y} = \begin{bmatrix} y_1 \\ y_2 \end{bmatrix} = \begin{bmatrix} b_1 \\ b_2 \end{bmatrix} e^{\lambda t} = \mathbf{b}e^{\lambda t} \tag{5.3.11}$$

and substitute into the differential equation to obtain

$$\lambda \mathbf{b} e^{\lambda t} = A\mathbf{b}e^{\lambda t}.$$

Cancelling the scalar quantity $e^{\lambda t}$ and writing $\mathbf{b} = I\mathbf{b}$ gives $\lambda I\mathbf{b} = A\mathbf{b}$ or, equivalently,

$$\begin{aligned}(a_{11} - \lambda)b_1 + a_{12}b_2 &= 0,\\ a_{21}b_1 + (a_{22} - \lambda)b_2 &= 0,\end{aligned} \tag{5.3.12}$$

or $(A - \lambda I)\mathbf{b} = \mathbf{0}$. Hence, λ is an eigenvalue and $\mathbf{b}$ is an eigenvector of the matrix A. Equations (5.3.12) have solutions b_1 and b_2 (not both zero) if and only if λ satisfies the *characteristic equation*

$$\det(A - \lambda I) = \begin{vmatrix} a_{11} - \lambda & a_{12} \\ a_{21} & a_{22} - \lambda \end{vmatrix} = 0. \tag{5.3.13}$$

This quadratic equation can have distinct real, distinct complex, or equal real roots. Just as in the study of second order linear differential equations with constant coefficients, we must examine each of the three types of eigenvalues. Each will provide a different type of solutions to the system of differential equations (5.3.10).

TYPE 1—DISTINCT REAL EIGENVALUES

To construct the solutions (5.3.11), we must find values of b_1 and b_2 for each of λ_1 and λ_2. From (5.3.12) it is clear that the following algebraic systems must be solved:

$$(A - \lambda_1 I)\mathbf{b} = 0 \quad \text{and} \quad (A - \lambda_2 I)\mathbf{b} = 0.$$

Denoting these solutions by $\mathbf{b}^1$ and $\mathbf{b}^2$, respectively, we find the two solutions of (5.3.10):

$$\mathbf{y}^1(t) = \mathbf{b}^1 e^{\lambda_1 t} = \begin{bmatrix} b_1^1 \\ b_2^1 \end{bmatrix} e^{\lambda_1 t}$$

and

$$\mathbf{y}^2(t) = \mathbf{b}^2 e^{\lambda_2 t} = \begin{bmatrix} b_1^2 \\ b_2^2 \end{bmatrix} e^{\lambda_2 t}.$$

As mentioned previously, the corresponding eigenvectors are linearly independent since the eigenvalues are distinct. By the determinant test,

$$\begin{vmatrix} y_1^1(t) & y_1^2(t) \\ y_2^1(t) & y_2^2(t) \end{vmatrix} = \begin{vmatrix} b_1^1 e^{\lambda_1 t} & b_1^2 e^{\lambda_2 t} \\ b_2^1 e^{\lambda_1 t} & b_2^2 e^{\lambda_2 t} \end{vmatrix} = e^{\lambda_1 t} e^{\lambda_2 t} \begin{vmatrix} b_1^1 & b_1^2 \\ b_2^1 & b_2^2 \end{vmatrix} \neq 0$$

(the last determinant is nonzero since $\mathbf{b}^1$ and $\mathbf{b}^2$ are linearly independent), and the solutions $\mathbf{y}^1(t)$, $\mathbf{y}^2(t)$ are linearly independent. Hence, the coefficients c_1 and c_2 in the linear combination

$$\mathbf{y}(t) = c_1\mathbf{y}^1(t) + c_2\mathbf{y}^2(t) \tag{5.3.14}$$

can be chosen to satisfy arbitrary initial data. We conclude that (5.3.14) is a *general solution* to the homogeneous differential equation (5.3.10).

All of this can be summarized in the following algorithm.

Algorithm for solving $dy/dt = Ay$ (distinct real eigenvalue case)

1. Determine the eigenvalues λ_1 and λ_2 that are roots of the characteristic equation

$$\det(A - \lambda I) = 0.$$

2. Find the corresponding eigenvectors $\mathbf{b}^1$ and $\mathbf{b}^2$ that are solutions of

$$(A - \lambda_j I)\mathbf{b} = 0, \quad j = 1, 2.$$

3. Two linearly independent solutions are given by

$$\mathbf{y}^1(t) = \mathbf{b}^1 e^{\lambda_1 t}, \qquad \mathbf{y}^2(t) = \mathbf{b}^2 e^{\lambda_2 t}$$

and the general solution is

$$\mathbf{y}(t) = c_1\mathbf{y}^1(t) + c_2\mathbf{y}^2(t).$$

4. If initial data $\mathbf{y}(t_0) = \mathbf{y}^0$ are given, find c_1, c_2 by solving

$$\mathbf{y}^0 = c_1\mathbf{y}^1(t_0) + c_2\mathbf{y}^2(t_0).$$

EXAMPLE 5 Find the general solution of $\dfrac{d\mathbf{y}}{dt} = \begin{bmatrix} 2 & 6 \\ -2 & -5 \end{bmatrix}\mathbf{y}$.

SOLUTION The eigenvalues are roots of the quadratic equation

$$\begin{vmatrix} 2-\lambda & 6 \\ -2 & -5-\lambda \end{vmatrix} = (2-\lambda)(-5-\lambda) + 12 = \lambda^2 + 3\lambda + 2 = 0,$$

hence $\lambda_1 = -1, \lambda_2 = -2$. To obtain the solution corresponding to $\lambda_1 = -1$, solve

$$[2-(-1)]b_1 + 6b_2 = 0, \qquad -2b_1 + [-5-(-1)]b_2 = 0$$

to get $b_1 = -2$, $b_2 = 1$, and therefore

$$\mathbf{b}^1 = \begin{bmatrix} -2 \\ 1 \end{bmatrix}.$$

The corresponding solution to the differential equation is

$$\mathbf{y}^1(t) = \begin{bmatrix} -2 \\ 1 \end{bmatrix} e^{-t}.$$

To get the solution corresponding to $\lambda_2 = -2$, solve

$$[2 - (-2)]b_1 + 6b_2 = 0, \qquad -2b_1 + [-5 - (-2)]b_2 = 0$$

to obtain

$$\mathbf{b}^2 = \begin{bmatrix} -3 \\ 2 \end{bmatrix},$$

which gives

$$\mathbf{y}^2(t) = \begin{bmatrix} -3 \\ 2 \end{bmatrix} e^{-2t}.$$

The general solution is

$$\begin{aligned} \mathbf{y}(t) &= c_1\mathbf{y}^1(t) + c_2\mathbf{y}^2(t) \\ &= c_1 \begin{bmatrix} -2 \\ 1 \end{bmatrix} e^{-t} + c_2 \begin{bmatrix} -3 \\ 2 \end{bmatrix} e^{-2t}. \end{aligned}$$

■

EXAMPLE 6 Find the solution of the initial value problem

$$\frac{d\mathbf{y}}{dt} = \begin{bmatrix} -5 & 4 \\ 1 & -2 \end{bmatrix} \mathbf{y}, \qquad \mathbf{y}(0) = \begin{bmatrix} 3 \\ -2 \end{bmatrix}.$$

SOLUTION The eigenvalues are the roots of

$$\begin{vmatrix} -5 - \lambda & 4 \\ 1 & -2 - \lambda \end{vmatrix} = \lambda^2 + 7\lambda + 6 = 0,$$

so $\lambda_1 = -1$, $\lambda_2 = -6$. A vector $\mathbf{b}^1$, found by solving the system of equations

$$[-5 - (-1)]b_1 + 4b_2 = 0, \qquad b_1 + [-2 - (-1)]b_2 = 0,$$

is

$$\mathbf{b}^1 = \begin{bmatrix} 1 \\ 1 \end{bmatrix}.$$

Similarly,

$$\mathbf{b}^2 = \begin{bmatrix} 4 \\ -1 \end{bmatrix}$$

satisfies

$$[-5 - (-6)]b_1 + 4b_2 = 0, \qquad b_1 + [-2 - (-6)]b_2 = 0.$$

The corresponding solutions are

$$\mathbf{y}^1(t) = \begin{bmatrix} 1 \\ 1 \end{bmatrix} e^{-t}, \qquad \mathbf{y}^2(t) = \begin{bmatrix} 4 \\ -1 \end{bmatrix} e^{-6t},$$

and the general solution is

$$\mathbf{y}(t) = c_1\mathbf{y}^1(t) + c_2\mathbf{y}^2(t) = c_1\begin{bmatrix}1\\1\end{bmatrix}e^{-t} + c_2\begin{bmatrix}4\\-1\end{bmatrix}e^{-6t}.$$

To obtain the solution satisfying the initial conditions, c_1 and c_2 must satisfy

$$\mathbf{y}(0) = \begin{bmatrix}3\\-2\end{bmatrix} = c_1\begin{bmatrix}1\\1\end{bmatrix} + c_2\begin{bmatrix}4\\-1\end{bmatrix}$$

or

$$3 = c_1 + 4c_2, \qquad -2 = c_1 - c_2.$$

Solving these gives $c_1 = -1$, $c_2 = 1$, and the desired solution is

$$\begin{aligned}\mathbf{y}(t) &= (-1)\begin{bmatrix}1\\1\end{bmatrix}e^{-t} + (1)\begin{bmatrix}4\\-1\end{bmatrix}e^{-6t}\\ &= \begin{bmatrix}-1\\-1\end{bmatrix}e^{-t} + \begin{bmatrix}4\\-1\end{bmatrix}e^{-6t}.\end{aligned}$$

■

TYPE 2—DISTINCT COMPLEX CONJUGATE EIGENVALUES

In this case both λ and its complex conjugate λ^* are eigenvalues of the coefficient matrix A. If b_1 and b_2 are solutions of (5.3.12) corresponding to λ, then b_1^* and b_2^* are also solutions of (5.3.12) corresponding to λ^*. To see this, write

$$(a_{11} - \lambda)b_1 + a_{12}b_2 = 0, \qquad a_{21}b_1 + (a_{22} - \lambda)b_2 = 0,$$

and take complex conjugates of both equations to get

$$(a_{11} - \lambda)^*b_1^* + a_{12}^*b_2^* = 0^*, \qquad a_{21}^*b_1^* + (a_{22} - \lambda)^*b_2^* = 0^*.$$

Since a_{11}, a_{12}, a_{21}, and a_{22} are real, this becomes

$$(a_{11} - \lambda^*)b_1^* + a_{12}b_2^* = 0, \qquad a_{21}b_1^* + (a_{22} - \lambda^*)b_2^* = 0,$$

thus b_1^* and b_2^* are solutions corresponding to λ^*. The general solution in this case can now be written in its complex form as

$$\begin{aligned}y_1(t) &= c_1b_1e^{\lambda t} + c_2b_1^*e^{\lambda^* t},\\ y_2(t) &= c_1b_2e^{\lambda t} + c_2b_2^*e^{\lambda^* t},\end{aligned}$$

or

$$\mathbf{y}(t) = c_1\begin{bmatrix}b_1\\b_2\end{bmatrix}e^{\lambda t} + c_2\begin{bmatrix}b_1^*\\b_2^*\end{bmatrix}e^{\lambda^* t}. \tag{5.3.15}$$

Since the complex eigenvalues are distinct, the eigenvectors are linearly independent.

The real form of the general solution can be obtained from the complex form by some clever choices of the constants c_1 and c_2 in (5.3.15). The procedure will be illustrated by the following example.

EXAMPLE 7 Find the general real solution of

$$\frac{d\mathbf{y}}{dt} = \begin{bmatrix} 1 & 5 \\ -1 & -3 \end{bmatrix} \mathbf{y}. \tag{5.3.16}$$

SOLUTION The characteristic equation

$$\begin{vmatrix} 1-\lambda & 5 \\ -1 & -3-\lambda \end{vmatrix} = \lambda^2 + 2\lambda + 2 = 0$$

has the roots $\lambda = -1 + i$ and $\lambda^* = -1 - i$. Equations (5.3.12) become

$$(2 - i)b_1 + 5b_2 = 0, \qquad -b_1 - (2 + i)b_2 = 0,$$

and the solution $b_1 = 5$, $b_2 = -2 + i$ is chosen. Therefore $b_1^* = 5$, $b_2^* = -2 - i$, and the general solution has the form

$$\begin{aligned} y_1(t) &= 5c_1e^{(-1+i)t} + 5c_2e^{(-1-i)t}, \\ y_2(t) &= (-2 + i)c_1e^{(-1+i)t} + (-2 - i)c_2e^{(-1-i)t}. \end{aligned}$$

To find the real form of the general solution, a procedure is followed similar to that used in Chapter 2 to find the general real solution from the general complex solution for second order linear equations. Consider the two cases: (1) $c_1 = c_2 = 1/2$ and (2) $c_1 = -c_2 = 1/2i$ and make use of the Euler formula

$$e^{i\theta} = \cos\theta + i\sin\theta, \tag{5.3.17}$$

which implies that

$$\cos\theta = \frac{e^{i\theta} + e^{-i\theta}}{2}, \qquad \sin\theta = \frac{e^{i\theta} - e^{-i\theta}}{2i}.$$

Case 1. $c_1 = c_2 = 1/2$.

$$\begin{aligned} y_1(t) &= \frac{5}{2}e^{-t}e^{it} + \frac{5}{2}e^{-t}e^{-it} = 5e^{-t}\left(\frac{e^{it} + e^{-it}}{2}\right) = 5e^{-t}\cos t = \mathrm{Re}[5e^{(-1+i)t}], \\ y_2(t) &= \frac{1}{2}(-2 + i)e^{-t}e^{it} + \frac{1}{2}(-2 - i)e^{-t}e^{-it} \\ &= e^{-t}\left[(-2)\left(\frac{e^{it} + e^{-it}}{2}\right) + i\left(\frac{e^{it} - e^{-it}}{2}\right)\right] \\ &= e^{-t}\left[(-2)\left(\frac{e^{it} + e^{-it}}{2}\right) - \left(\frac{e^{it} - e^{-it}}{2i}\right)\right] \\ &= e^{-t}(-2\cos t - \sin t) = \mathrm{Re}[(-2 + i)e^{(-1+i)t}]. \end{aligned}$$

The choice for c_1 and c_2 in Case 1 is motivated by the fact that if a is a complex number $a = \alpha + i\beta$, then $a^* = \alpha - i\beta$ and

$$\frac{1}{2}a + \frac{1}{2}a^* = \alpha = \text{Re}(a).$$

Case 2. $c_1 = -c_2 = 1/2i$.

$$\begin{aligned} y_1(t) &= \frac{5}{2i} e^{-t}e^{-it} - \frac{5}{2i} e^{-t}e^{-it} = 5e^{-t}\left(\frac{e^{it} - e^{-it}}{2i}\right) \\ &= 5e^{-t} \sin t = \text{Im}[5e^{(-1+i)t}], \\ y_2(t) &= \frac{1}{2i}(-2 + i)e^{-t}e^{it} - \frac{1}{2i}(-2 - i)e^{-t}e^{-it} \\ &= e^{-t}\left[(-2)\left(\frac{e^{it} - e^{-it}}{2i}\right) + \left(\frac{e^{it} + e^{-it}}{2}\right)\right]. \\ &= e^{-t}(-2 \sin t + \cos t) = \text{Im}[(-2 + i)e^{(-1+i)t}]. \end{aligned}$$

As in Case 1, the choice of c_1 and c_2 is based on the fact that for $a = \alpha + i\beta$,

$$\frac{1}{2i}a - \frac{1}{2i}a^* = \beta = \text{Im}(a).$$

Clearly, the vector functions

$$\mathbf{y}^1(t) = \begin{bmatrix} 5 \cos t \\ -2 \cos t - \sin t \end{bmatrix} e^{-t}, \qquad \mathbf{y}^2(t) = \begin{bmatrix} 5 \sin t \\ \cos t - 2 \sin t \end{bmatrix} e^{-t}$$

are solutions to (5.3.16) and they are real. The general real solution is then given by

$$\mathbf{y}(t) = c_1 \begin{bmatrix} 5 \cos t \\ -2 \cos t - \sin t \end{bmatrix} e^{-t} + c_2 \begin{bmatrix} 5 \sin t \\ \cos t - 2 \sin t \end{bmatrix} e^{-t},$$

and the ability to solve an arbitrary initial value problem depends on the fact that for any value of t_0

$$\begin{aligned} &\begin{vmatrix} 5e^{-t_0} \cos t_0 & 5e^{-t_0} \sin t_0 \\ -e^{-t_0}(2 \cos t_0 + \sin t_0) & e^{-t_0}(\cos t_0 - 2 \sin t_0) \end{vmatrix} \\ &\quad = 5e^{-2t_0}[\cos t_0(\cos t_0 - 2 \sin t_0) + (2 \cos t_0 + \sin t_0)\sin t_0] \\ &\quad = 5e^{-2t_0} \neq 0. \end{aligned}$$

■

The technique outlined in Example 7 can be applied in general to (5.3.15). Let $\lambda = p + iq$, $b_1 = u_1 + iv_1$, $b_2 = u_2 + iv_2$. Then we can consider two cases.

Case 1. $c_1 = c = 1/2$.

$$y_1^1(t) = \operatorname{Re}(b_1 e^{\lambda t}) = e^{pt}(u_1 \cos qt - v_1 \sin qt),$$
$$y_2^1(t) = \operatorname{Re}(b_2 e^{\lambda t}) = e^{pt}(u_2 \cos qt - v_2 \sin qt).$$

Case 2. $c_1 = -c_2 = 1/2i$.

$$y_1^2(t) = \operatorname{Im}(b_1 e^{\lambda t}) = e^{pt}(u_1 \sin qt + v_1 \cos qt),$$
$$y_2^2(t) = \operatorname{Im}(b_2 e^{\lambda t}) = e^{pt}(u_2 \sin qt + v_2 \cos qt).$$

The general solution can then be written in the form

$$\mathbf{y}(t) = e^{pt}\left[c_1\begin{bmatrix} u_1 \cos qt - v_1 \sin qt \\ u_2 \cos qt - v_2 \sin qt \end{bmatrix} + c_2\begin{bmatrix} u_1 \sin qt + v_1 \cos qt \\ u_2 \sin qt + v_2 \cos qt \end{bmatrix}\right],$$

where c_1 and c_2 are arbitrary constants. That this is the general solution follows again from the fact that the determinant

$$e^{2pt}\begin{vmatrix} u_1 \cos qt - v_1 \sin qt & u_1 \sin qt + v_1 \cos qt \\ u_2 \cos qt - v_2 \sin qt & u_2 \sin qt + v_2 \cos qt \end{vmatrix} = e^{2pt}(u_1 v_2 - v_1 u_2) \neq 0,$$

since the vectors

$$\begin{bmatrix} b_1 \\ b_2 \end{bmatrix}, \begin{bmatrix} b_1^* \\ b_2^* \end{bmatrix}$$

are linearly independent (see Exercises 24 and 25, p. 381).

All of this can be summarized in the following algorithm.

Algorithm for solving $d\mathbf{y}/dt = A\mathbf{y}$ (complex eigenvalue case)

1. Determine the eigenvalues λ and λ^* that are roots of the equation

$$\det(A - \lambda I) = 0.$$

2. Find the corresponding eigenvectors

$$\mathbf{b} = \begin{bmatrix} b_1 \\ b_2 \end{bmatrix} \quad \text{and} \quad \mathbf{b}^* = \begin{bmatrix} b_1^* \\ b_2^* \end{bmatrix}$$

that are solutions of

$$(A - \lambda I)b = 0, \qquad (A - \lambda^* I)b^* = 0.$$

3. Two linearly independent real solutions are given by

$$\mathbf{y}^1(t) = \begin{bmatrix} u_1 \cos qt - v_1 \sin qt \\ u_2 \cos qt - v_2 \sin qt \end{bmatrix} e^{pt},$$

$$\mathbf{y}^2(t) = \begin{bmatrix} u_1 \sin qt + v_1 \cos qt \\ u_2 \sin qt + v_2 \cos qt \end{bmatrix} e^{pt},$$

where $\lambda = p + iq$, $b_1 = u_1 + iv_1$, $b_2 = u_2 + iv_2$, and the general solution is given by

$$\mathbf{y}(t) = c_1\mathbf{y}^1(t) + c_2\mathbf{y}^2(t).$$

4. If initial data $\mathbf{y}(t_0) = \mathbf{y}^0$ are given, find c_1, c_2 from

$$\begin{bmatrix} y_1^0 \\ y_2^0 \end{bmatrix} = c_1\mathbf{y}^1(t_0) + c_2\mathbf{y}^2(t_0).$$

EXAMPLE 8 Solve the initial value problem

$$\frac{d\mathbf{y}}{dt} = \begin{bmatrix} 3 & -3 \\ 3 & -1 \end{bmatrix} \mathbf{y},\ \mathbf{y}(0) = \begin{bmatrix} 3 \\ 0 \end{bmatrix}.$$

SOLUTION The characteristic equation

$$\begin{vmatrix} 3 - \lambda & -3 \\ 3 & -1 - \lambda \end{vmatrix} = \lambda^2 - 2\lambda + 6 = 0$$

has roots $\lambda = 1 + \sqrt{5}i$ and $\lambda^* = 1 - \sqrt{5}i$, and therefore $p = 1$, $q = \sqrt{5}$. Equations (5.3.12) to determine the eigenvector $\mathbf{b}$ are

$$(2 - \sqrt{5}i)b_1 - 3b_2 = 0, \qquad 3b_1 + (-2 - \sqrt{5}i)b_2 = 0,$$

and the solutions $b_1 = 3$, $b_2 = 2 - \sqrt{5}i$ are chosen. Therefore $u_1 = 3$, $v_1 = 0$, $u_2 = 2$, $v_2 = -\sqrt{5}$, and two linearly independent solutions are

$$\mathbf{y}^1(t) = \begin{bmatrix} 3 \cos \sqrt{5}t \\ 2 \cos \sqrt{5}t + \sqrt{5} \sin \sqrt{5}t \end{bmatrix} e^t,$$

$$\mathbf{y}^2(t) = \begin{bmatrix} 3 \sin \sqrt{5}t \\ 2 \sin \sqrt{5}t - \sqrt{5} \cos \sqrt{5}t \end{bmatrix} e^t.$$

The general solution is $\mathbf{y}(t) = c_1\mathbf{y}^1(t) + c_2\mathbf{y}^2(t)$. Using the initial data we obtain the equation

$$\mathbf{y}(0) = c_1\mathbf{y}^1(0) + c_2\mathbf{y}^2(0) = c_1\begin{bmatrix} 3 \\ 2 \end{bmatrix} + c_2\begin{bmatrix} 0 \\ -\sqrt{5} \end{bmatrix} = \begin{bmatrix} 3 \\ 0 \end{bmatrix}.$$

The solution is $c_1 = 1$, $c_2 = 2\sqrt{5}/5$, and

$$\mathbf{y}(t) = \begin{bmatrix} 3 \cos \sqrt{5}t + \frac{6}{5}\sqrt{5} \sin \sqrt{5}t \\ \frac{9}{5}\sqrt{5} \sin \sqrt{5}t \end{bmatrix} e^t$$

is the solution of the initial value problem. ■

TYPE 3—EQUAL REAL EIGENVALUES

Here we consider two situations. The first is with equations (5.3.10) uncoupled, i.e., when $a_{11} = a_{22} = p$ and $a_{12} = a_{21} = 0$. Then the equations become

$$\frac{d}{dt}\begin{bmatrix} y_1 \\ y_2 \end{bmatrix} = \begin{bmatrix} p & 0 \\ 0 & p \end{bmatrix}\begin{bmatrix} y_1 \\ y_2 \end{bmatrix}, \tag{5.3.18}$$

and they can be solved separately to obtain

$$\mathbf{y}(t) = \begin{bmatrix} c_1 e^{pt} \\ c_2 e^{pt} \end{bmatrix} = c_1 \begin{bmatrix} 1 \\ 0 \end{bmatrix} e^{pt} + c_2 \begin{bmatrix} 0 \\ 1 \end{bmatrix} e^{pt}.$$

This is clearly the general solution.

If equations (5.3.10) are coupled, there is more work to do. Denoting the double root by λ, one can obtain an associated eigenvector $\mathbf{b}$ from (5.3.12) and have the one solution

$$\mathbf{y}^1(t) = \mathbf{b}e^{\lambda t}. \tag{5.3.19}$$

To obtain a second solution, one might proceed as in the case of second order linear equations with constant coefficients and try

$$\mathbf{y}^2 = \alpha t e^{\lambda t} \tag{5.3.20}$$

for some constant vector α. Substituting into (5.3.10) gives

$$\alpha e^{\lambda t} + \lambda \alpha t e^{\lambda t} = A \alpha t e^{\lambda t}. \tag{5.3.21}$$

In order for this equation to be satisfied for all t, it is necessary for the coefficients of $te^{\lambda t}$ and $e^{\lambda t}$ to be the same on both sides of the equation. So

$$\alpha = 0, \qquad \lambda\alpha = A\alpha;$$

hence $\alpha = 0$, and there is no nonzero solution of the form (5.3.20) to (5.3.10). Since Eq. (5.3.21) contains terms both in $te^{\lambda t}$ and $e^{\lambda t}$, it might be more reasonable to try a solution of the form

$$\mathbf{y}^2(t) = \zeta e^{\lambda t} + \sigma t e^{\lambda t}, \tag{5.3.22}$$

where ζ and σ are constant vectors. Substituting this expression for $\mathbf{y}$ into (5.3.10), one obtains

$$\lambda\zeta e^{\lambda t} + \sigma e^{\lambda t} + \lambda\sigma t e^{\lambda t} = A\zeta e^{\lambda t} + A\sigma t e^{\lambda t}.$$

Equating coefficients of $te^{\lambda t}$ and $e^{\lambda t}$ gives the equations

$$\lambda\zeta + \sigma = A\zeta \quad \text{or} \quad (A - \lambda I)\zeta = \sigma \tag{5.3.23}$$

and

$$\lambda\sigma = A\sigma \quad \text{or} \quad (A - \lambda I)\sigma = \mathbf{0}. \tag{5.3.24}$$

To solve Eqs. (5.3.23, 5.3.24), first note that the second equation says that σ is an eigenvector of A corresponding to the eigenvalue λ, so we might as well choose $\sigma = \mathbf{b}$, the eigenvector that we have already found. The vector ζ can now be found by solving Eq. (5.3.23) with $\sigma = \mathbf{b}$. However, if $\mathbf{b}$ has not already been found, it is easier to proceed as follows:

Find ζ, so that (5.3.23) determines σ with $\sigma_1^2 + \sigma_2^2 > 0$ to insure $\sigma \neq \mathbf{0}$. This σ will automatically satisfy (5.3.24) (see Exercise 38, p. 382).

Therefore a second solution is of the form

$$\mathbf{y}^2(t) = \zeta e^{\lambda t} + \sigma t e^{\lambda t},$$

and a general solution is (if we let $\mathbf{b} = \sigma$)

$$\mathbf{y}(t) = c_1 \sigma e^{\lambda t} + c_2(\zeta + \sigma t)e^{\lambda t}. \tag{5.3.25}$$

To show that (5.3.25) is indeed the general solution to (5.3.10), consider the arbitrary initial data

$$\mathbf{y}(t_0) = \mathbf{y}^0 \tag{5.3.26}$$

and set

$$y_1^0 = \sigma_1 e^{\lambda t_0} c_1 + (\zeta_1 + \sigma_1 t_0)e^{\lambda t_0} c_2,$$
$$y_2^0 = \sigma_2 e^{\lambda t_0} c_1 + (\zeta_2 + \sigma_2 t_0)e^{\lambda t_0} c_2.$$

The equations have a unique solution c_1, c_2 if and only if the determinant

$$\begin{vmatrix} \sigma_1 e^{\lambda t_0} & (\zeta_1 + \sigma_1 t_0)e^{\lambda t_0} \\ \sigma_2 e^{\lambda t_0} & (\zeta_2 + \sigma_2 t_0)e^{\lambda t_0} \end{vmatrix} = e^{2\lambda t_0}(\sigma_1 \zeta_2 - \sigma_2 \zeta_1) \neq 0.$$

If the determinant were to vanish, then $\sigma_1 \zeta_2 - \sigma_2 \zeta_1 = 0$ or, equivalently,

$$\begin{vmatrix} \sigma_1 & \zeta_1 \\ \sigma_2 & \zeta_2 \end{vmatrix} = 0,$$

which implies that vectors σ and ζ are linearly dependent. This means there exists a constant $c \neq 0$ such that $\zeta = c\sigma$. Substituting this into Eq. (5.3.23) gives

$$c(A - \lambda I)\sigma = \sigma,$$

which, when compared with Eq. (5.3.24), implies $\sigma = 0$. But this is impossible since σ is an eigenvector with $\sigma_1^2 + \sigma_2^2 > 0$. All of this can be summarized in the following algorithm.

Algorithm for solving $dy/dt = Ay$ (real equal eigenvalues case)

1. Determine the eigenvalue λ that is a root of the equation

$$\det (A - \lambda I) = 0.$$

2. Determine vectors $\zeta \neq \mathbf{0}$ and $\sigma \neq \mathbf{0}$ such that
$$(A - \lambda I)\zeta = \sigma, \qquad (A - \lambda I)\sigma = \mathbf{0}.$$
3. Two linearly independent solutions are given by
$$\mathbf{y}^1(t) = \sigma e^{\lambda t}, \qquad \mathbf{y}^2(t) = (\zeta + \sigma t)e^{\lambda t},$$
and the general solution is
$$\mathbf{y}(t) = c_1\mathbf{y}^1(t) + c_2\mathbf{y}^2(t).$$
4. If initial data $\mathbf{y}(t_0) = \mathbf{y}^0$ are given, find c_1, c_2 from
$$\mathbf{y}^0 = c_1\mathbf{y}^1(t_0) + c_2\mathbf{y}^2(t_0).$$

EXAMPLE 9 Obtain the general solution of
$$\mathbf{y}' = \begin{bmatrix} 3 & -4 \\ 1 & -1 \end{bmatrix} \mathbf{y}.$$

SOLUTION From (5.3.13), we get
$$\begin{vmatrix} 3 - \lambda & -4 \\ 1 & -1 - \lambda \end{vmatrix} = \lambda^2 - 2\lambda + 1 = 0,$$
hence $\lambda = 1$ is a double root. From (5.3.23) it follows that
$$\begin{bmatrix} 3 - 1 & -4 \\ 1 & -1 - 1 \end{bmatrix} \begin{bmatrix} \zeta_1 \\ \zeta_2 \end{bmatrix} = \begin{bmatrix} \sigma_1 \\ \sigma_2 \end{bmatrix}$$
or
$$\begin{bmatrix} 2 & -4 \\ 1 & -2 \end{bmatrix} \begin{bmatrix} \zeta_1 \\ \zeta_2 \end{bmatrix} = \begin{bmatrix} \sigma_1 \\ \sigma_2 \end{bmatrix}.$$
Now the task is to choose ζ_1 and ζ_2, not both zero, that will give values σ_1 and σ_2 that are not both zero. The choice $\zeta_1 = 1$, $\zeta_2 = 0$ gives $\sigma_1 = 2$, $\sigma_2 = 1$. As a check, note from (5.3.24) that
$$\begin{bmatrix} 2 & -4 \\ 1 & -2 \end{bmatrix} \begin{bmatrix} 2 \\ 1 \end{bmatrix} = \begin{bmatrix} 0 \\ 0 \end{bmatrix}.$$
Therefore $\sigma = \begin{bmatrix} 2 \\ 1 \end{bmatrix}$ is an eigenvector of A corresponding to $\lambda = 1$, and so
$$\mathbf{y}^1(t) = \begin{bmatrix} 2 \\ 1 \end{bmatrix} e^t.$$
A second solution is given by
$$\mathbf{y}^2(t) = \left[\begin{bmatrix} 1 \\ 0 \end{bmatrix} + \begin{bmatrix} 2 \\ 1 \end{bmatrix} t\right] e^t$$

and the general solution is

$$\mathbf{y}(t) = c_1 \begin{bmatrix} 2 \\ 1 \end{bmatrix} e^t + c_2 \left[\begin{bmatrix} 1 \\ 0 \end{bmatrix} + \begin{bmatrix} 2 \\ 1 \end{bmatrix} t \right] e^t. \tag{5.3.27}$$ ∎

EXAMPLE 10 Find the solution in Example 9 that satisfies

$$\mathbf{y}(0) = \begin{bmatrix} 1 \\ 1 \end{bmatrix}.$$

SOLUTION Setting t=0 in the previous solution gives

$$\begin{bmatrix} 1 \\ 1 \end{bmatrix} = c_1 \begin{bmatrix} 2 \\ 1 \end{bmatrix} + c_2 \begin{bmatrix} 1 \\ 0 \end{bmatrix}$$

or

$$1 = 2c_1 + c_2, \qquad 1 = c_1.$$

So $c_1 = 1$, $c_2 = -1$, and the solution is

$$\mathbf{y}(t) = \left[\begin{bmatrix} 2 \\ 1 \end{bmatrix} - \begin{bmatrix} 1 \\ 0 \end{bmatrix} - \begin{bmatrix} 2 \\ 1 \end{bmatrix} t \right] e^t = \left[\begin{bmatrix} 1 \\ 1 \end{bmatrix} - \begin{bmatrix} 2 \\ 1 \end{bmatrix} t \right] e^t.$$ ∎

Although we will not do much with systems of linear differential equations of dimension greater than two, it is useful to work a few examples. In general, for systems of size $n \geq 3$ the algebraic difficulties become formidable, particularly for the repeated eigenvalue case, and quite often the best approach is the Laplace transform or a numerical approximation.

EXAMPLE 11 Find the general solution of

$$\mathbf{y}' = \begin{bmatrix} 1 & -1 & -1 \\ 0 & 1 & 3 \\ 0 & 3 & 1 \end{bmatrix} \mathbf{y}.$$

SOLUTION First determine the eigenvalues of the coefficient matrix by solving

$$\begin{vmatrix} 1-\lambda & -1 & -1 \\ 0 & 1-\lambda & 3 \\ 0 & 3 & 1-\lambda \end{vmatrix} = -\lambda^3 + 3\lambda^2 + 6\lambda - 8$$

$$= -(\lambda - 1)(\lambda - 4)(\lambda + 2) = 0;$$

so the eigenvalues are $\lambda = 1, 4, -2$. Since they are distinct, the corresponding eigenvectors will be linearly independent. For $\lambda = 1$, an eigenvector $\mathbf{b}^1$ is obtained as a solution to

$$(A - 1I)\mathbf{b} = 0 \qquad \text{or} \qquad \begin{bmatrix} 0 & -1 & -1 \\ 0 & 0 & 3 \\ 0 & 3 & 0 \end{bmatrix} \begin{bmatrix} b_1 \\ b_2 \\ b_3 \end{bmatrix} = \begin{bmatrix} 0 \\ 0 \\ 0 \end{bmatrix},$$

and a solution is

$$\mathbf{b}^1 = \begin{bmatrix} 1 \\ 0 \\ 0 \end{bmatrix}.$$

For $\lambda = 4$,

$$(A - 4I)\mathbf{b} = 0 \quad \text{or} \quad \begin{bmatrix} -3 & -1 & -1 \\ 0 & -3 & 3 \\ 0 & 3 & -3 \end{bmatrix} \begin{bmatrix} b_1 \\ b_2 \\ b_3 \end{bmatrix} = \begin{bmatrix} 0 \\ 0 \\ 0 \end{bmatrix},$$

and a choice for $\mathbf{b}^2$ is

$$\mathbf{b}^2 = \begin{bmatrix} 2 \\ -3 \\ -3 \end{bmatrix}.$$

Finally, for $\lambda = -2$, solve

$$[A - (-2)I]\mathbf{b} = 0 \quad \text{or} \quad \begin{bmatrix} 3 & -1 & -1 \\ 0 & 3 & 3 \\ 0 & 3 & 3 \end{bmatrix} \begin{bmatrix} b_1 \\ b_2 \\ b_3 \end{bmatrix} = \begin{bmatrix} 0 \\ 0 \\ 0 \end{bmatrix}$$

to get

$$\mathbf{b}^3 = \begin{bmatrix} 0 \\ -1 \\ 1 \end{bmatrix}.$$

Three linearly independent solutions are

$$\mathbf{y}^1(t) = \begin{bmatrix} 1 \\ 0 \\ 0 \end{bmatrix} e^t, \qquad \mathbf{y}^2(t) = \begin{bmatrix} 2 \\ -3 \\ -3 \end{bmatrix} e^{4t}, \qquad \mathbf{y}^3(t) = \begin{bmatrix} 0 \\ -1 \\ 1 \end{bmatrix} e^{-2t},$$

and the general solution is

$$\mathbf{y}(t) = c_1 \begin{bmatrix} 1 \\ 0 \\ 0 \end{bmatrix} e^t + c_2 \begin{bmatrix} 2 \\ -3 \\ -3 \end{bmatrix} e^{4t} + c_3 \begin{bmatrix} 0 \\ -1 \\ 1 \end{bmatrix} e^{-2t}.$$

■

EXAMPLE 12 Obtain the general solution of

$$\mathbf{y}' = \begin{bmatrix} 1 & -1 & -1 \\ 1 & 1 & 0 \\ 3 & 0 & 1 \end{bmatrix} \mathbf{y}.$$

SOLUTION The eigenvalues are found as solutions of

$$\begin{vmatrix} 1-\lambda & -1 & -1 \\ 1 & 1-\lambda & 0 \\ 3 & 0 & 1-\lambda \end{vmatrix} = (1-\lambda)^3 + 3(1-\lambda) + (1-\lambda)$$

$$= (1-\lambda)(\lambda^2 - 2\lambda + 5) = 0;$$

hence $\lambda = 1, 1 \pm 2i$. For $\lambda = 1$, solve

$$(A - 1I)\mathbf{b} = 0 \quad \text{or} \quad \begin{bmatrix} 0 & -1 & -1 \\ 1 & 0 & 0 \\ 3 & 0 & 0 \end{bmatrix} \begin{bmatrix} b_1 \\ b_2 \\ b_3 \end{bmatrix} = \begin{bmatrix} 0 \\ 0 \\ 0 \end{bmatrix}$$

to get

$$\mathbf{b}^1 = \begin{bmatrix} 0 \\ 1 \\ -1 \end{bmatrix}.$$

For $\lambda = 1 + 2i$,

$$[A - (1+2i)I]\mathbf{b} = 0 \quad \text{or} \quad \begin{bmatrix} -2i & -1 & -1 \\ 1 & -2i & 0 \\ 3 & 0 & -2i \end{bmatrix} \begin{bmatrix} b_1 \\ b_2 \\ b_3 \end{bmatrix} = \begin{bmatrix} 0 \\ 0 \\ 0 \end{bmatrix},$$

and the solution is

$$\mathbf{b}^2 = \begin{bmatrix} 2i \\ 1 \\ 3 \end{bmatrix},$$

and for $\lambda = 1 - 2i$

$$\mathbf{b}^3 = \mathbf{b}^{2*} = \begin{bmatrix} -2i \\ 1 \\ 3 \end{bmatrix}.$$

Three linearly independent solutions are:

$$\mathbf{y}^1(t) = \begin{bmatrix} 0 \\ 1 \\ -1 \end{bmatrix} e^t, \quad \mathbf{y}^2(t) = \begin{bmatrix} 2i \\ 1 \\ 3 \end{bmatrix} e^{(1+2i)t} \quad \mathbf{y}^3(t) = \begin{bmatrix} -2i \\ 1 \\ 3 \end{bmatrix} e^{(1-2i)t},$$

and the general complex solution is

$$\mathbf{y}(t) = c_1 \begin{bmatrix} 0 \\ 1 \\ -1 \end{bmatrix} e^t + c_2 \begin{bmatrix} 2i \\ 1 \\ 3 \end{bmatrix} e^{(1+2i)t} + c_3 \begin{bmatrix} -2i \\ 1 \\ 3 \end{bmatrix} e^{(1-2i)t}.$$

To get the general real solution, proceed in a manner similar to that in Example 7 (p. 370). The choices $c_2 = c_3 = 1/2$ and $c_2 = -c_3 = 1/2i$ give

$$\frac{1}{2}\mathbf{y}^2(t) + \frac{1}{2}\mathbf{y}^3(t) = \begin{bmatrix} -2\sin 2t \\ \cos 2t \\ 3\cos 2t \end{bmatrix} e^t$$

and

$$\frac{1}{2i}\mathbf{y}^2(t) - \frac{1}{2i}\mathbf{y}^3(t) = \begin{bmatrix} 2\cos 2t \\ \sin 2t \\ 3\sin 2t \end{bmatrix} e^t.$$

Therefore the general real solution is

$$\mathbf{y}(t) = c_1 \begin{bmatrix} 0 \\ 1 \\ -1 \end{bmatrix} e^t + c_2 \begin{bmatrix} -2\sin 2t \\ \cos 2t \\ 3\cos 2t \end{bmatrix} e^t + c_3 \begin{bmatrix} 2\cos 2t \\ \sin 2t \\ 3\sin 2t \end{bmatrix} e^t. \qquad \blacksquare$$

EXERCISES 5.3

Find the general solution of the following Type 1 systems.

1. $\mathbf{y}' = \begin{bmatrix} 3 & -2 \\ 2 & -2 \end{bmatrix}\mathbf{y}$ **2.** $\mathbf{y}' = \begin{bmatrix} 0 & 1 \\ 8 & -2 \end{bmatrix}\mathbf{y}$ **3.** $\mathbf{y}' = \begin{bmatrix} -3 & 1 \\ 1 & -2 \end{bmatrix}\mathbf{y}$

4. $\mathbf{y}' = \begin{bmatrix} -5 & 4 \\ 1 & -2 \end{bmatrix}\mathbf{y}$ **5.** $\mathbf{y}' = \begin{bmatrix} -3 & 2 \\ 2 & -3 \end{bmatrix}\mathbf{y}$ **6.** $\mathbf{y}' = \begin{bmatrix} 1 & 0 \\ 1 & 2 \end{bmatrix}\mathbf{y}$

In each of the following, solve the given Type 1 initial value problem.

7. $\mathbf{y}' = \begin{bmatrix} -3 & 4 \\ -2 & 3 \end{bmatrix}\mathbf{y}, \quad \mathbf{y}(0) = \begin{bmatrix} 1 \\ 0 \end{bmatrix}$ **8.** $\mathbf{y}' = \begin{bmatrix} 5 & -2 \\ 6 & -2 \end{bmatrix}\mathbf{y}, \quad \mathbf{y}(2) = \begin{bmatrix} 1 \\ 0 \end{bmatrix}$

9. $\mathbf{y}' = \begin{bmatrix} -1 & 6 \\ 1 & -2 \end{bmatrix}\mathbf{y}, \quad \mathbf{y}(0) = \begin{bmatrix} 1 \\ 1 \end{bmatrix}$ **10.** $\mathbf{y}' = \begin{bmatrix} 7 & 6 \\ 2 & 6 \end{bmatrix}\mathbf{y}, \quad \mathbf{y}(0) = \begin{bmatrix} 0 \\ 1 \end{bmatrix}$

11. $\mathbf{y}' = \begin{bmatrix} 2 & -1 \\ 3 & -2 \end{bmatrix} \mathbf{y}, \quad \mathbf{y}(0) = \begin{bmatrix} 1 \\ 4 \end{bmatrix}$

In each of the following, find the general real solution of the given Type 2 system of differential equations.

12. $\mathbf{y}' = \begin{bmatrix} 3 & -5 \\ 2 & 1 \end{bmatrix} \mathbf{y}$

13. $\mathbf{y}' = \begin{bmatrix} -1 & -4 \\ 1 & -1 \end{bmatrix} \mathbf{y}$

14. $\mathbf{y}' = \begin{bmatrix} 2 & -5 \\ 2 & -4 \end{bmatrix} \mathbf{y}$

15. $\mathbf{y}' = \begin{bmatrix} 4 & -2 \\ 5 & 2 \end{bmatrix} \mathbf{y}$

16. $\mathbf{y}' = \begin{bmatrix} 8 & 16 \\ -5 & -8 \end{bmatrix} \mathbf{y}$

17. $\mathbf{y}' = \begin{bmatrix} 3 & 5 \\ -1 & -1 \end{bmatrix} \mathbf{y}$

In each of the following, solve the given Type 2 initial value problem.

18. $\mathbf{y}' = \begin{bmatrix} 4 & -4 \\ 5 & -4 \end{bmatrix} \mathbf{y}, \quad \mathbf{y}(\pi) = \begin{bmatrix} 0 \\ 1 \end{bmatrix}$

19. $\mathbf{y}' = \begin{bmatrix} 4 & 2 \\ -13 & -6 \end{bmatrix} \mathbf{y}, \quad \mathbf{y}(0) = \begin{bmatrix} 1 \\ 0 \end{bmatrix}$

20. $\mathbf{y}' = \begin{bmatrix} 3 & 2 \\ -5 & 1 \end{bmatrix} \mathbf{y}, \quad \mathbf{y}(\tfrac{3}{2}\pi) = \begin{bmatrix} 1 \\ 0 \end{bmatrix}$

21. $\mathbf{y}' = \begin{bmatrix} 2 & -4 \\ 1 & 2 \end{bmatrix} \mathbf{y}, \quad \mathbf{y}(\pi) = \begin{bmatrix} 1 \\ 1 \end{bmatrix}$

22. $\mathbf{y}' = \begin{bmatrix} 2 & -1 \\ 3 & 4 \end{bmatrix} \mathbf{y}, \quad \mathbf{y}(0) = \begin{bmatrix} 1 \\ 1 \end{bmatrix}$

23. $\mathbf{y}' = \begin{bmatrix} 0 & 1 \\ -1 & 0 \end{bmatrix} \mathbf{y}, \quad \mathbf{y}(\pi) = \begin{bmatrix} 1 \\ 1 \end{bmatrix}$

24. Show that if λ is a complex number and $A\mathbf{b} = \lambda\mathbf{b}$, $A\mathbf{b}^* = \lambda^*\mathbf{b}^*$, then $\mathbf{b}$ and $\mathbf{b}^*$ are linearly independent.

25. If $b_1 = u_1 + iv_1$, $b_2 = u_2 + iv_2$, show that

$$\begin{vmatrix} b_1 & b_1^* \\ b_2 & b_2^* \end{vmatrix} = -2i \begin{vmatrix} u_1 & v_1 \\ u_2 & v_2 \end{vmatrix}.$$

In each of the following, find the general solution of the given Type 3 differential equations.

26. $\mathbf{y}' = \begin{bmatrix} -3 & 2 \\ -2 & 1 \end{bmatrix} \mathbf{y}$

27. $\mathbf{y}' = \begin{bmatrix} 2 & 1 \\ -1 & 4 \end{bmatrix} \mathbf{y}$

28. $\mathbf{y}' = \begin{bmatrix} 4 & 8 \\ -2 & 12 \end{bmatrix} \mathbf{y}$

29. $\mathbf{y}' = \begin{bmatrix} 3 & -4 \\ 1 & -1 \end{bmatrix} \mathbf{y}$

30. $\mathbf{y}' = \begin{bmatrix} 4 & 4 \\ -9 & -8 \end{bmatrix} \mathbf{y}$

31. $\mathbf{y}' = \begin{bmatrix} 2 & -1 \\ 4 & 6 \end{bmatrix} \mathbf{y}$

In each of the following, solve the given Type 3 initial value problem.

32. $\mathbf{y}' = \begin{bmatrix} 0 & 1 \\ -9 & 6 \end{bmatrix} \mathbf{y}, \quad \mathbf{y}(0) = \begin{bmatrix} 1 \\ 2 \end{bmatrix}$

33. $\mathbf{y}' = \begin{bmatrix} 1 & -4 \\ 4 & -7 \end{bmatrix} \mathbf{y}, \quad \mathbf{y}(0) = \begin{bmatrix} 3 \\ 1 \end{bmatrix}$

34. $\mathbf{y}' = \begin{bmatrix} 5 & 0 \\ 1 & 5 \end{bmatrix} \mathbf{y}, \quad \mathbf{y}(0) = \begin{bmatrix} 1 \\ 1 \end{bmatrix}$

35. $\mathbf{y}' = \begin{bmatrix} -4 & -1 \\ 1 & -2 \end{bmatrix} \mathbf{y}, \quad \mathbf{y}(0) = \begin{bmatrix} 1 \\ 0 \end{bmatrix}$

36. $\mathbf{y}' = \begin{bmatrix} 1 & -2 \\ 2 & -3 \end{bmatrix} \mathbf{y}, \quad \mathbf{y}(0) = \begin{bmatrix} 0 \\ 1 \end{bmatrix}$

37. $\mathbf{y}' = \begin{bmatrix} 5 & -7 \\ 7 & -9 \end{bmatrix} \mathbf{y}, \quad \mathbf{y}(1) = \begin{bmatrix} 0 \\ 1 \end{bmatrix}$

38. Show that if λ is a double eigenvalue of the 2×2 matrix A and if $(A - \lambda I)\boldsymbol{\zeta} = \boldsymbol{\sigma}$, where $\boldsymbol{\sigma} \neq \mathbf{0}$, then $(A - \lambda I)\boldsymbol{\sigma} = \mathbf{0}$. (*Hint:* Show that $(A - \lambda I)^2$ is the zero matrix, where

$$\lambda = (a_{11} + a_{22})/2.)$$

39. Use the Laplace transform to solve the following.

a) The system of Example 11 with initial data

$$\mathbf{y}(0) = \begin{bmatrix} 2 \\ 0 \\ -6 \end{bmatrix};$$

b) The system of Example 12 with initial data

$$\mathbf{y}(0) = \begin{bmatrix} 2 \\ 0 \\ 0 \end{bmatrix}.$$

Use the eigenvalue–eigenvector method or the Laplace transform to find the solution of the following initial value problems.

40. $\mathbf{y}' = \begin{bmatrix} 1 & 0 & 1 \\ 0 & 1 & 2 \\ 1 & 2 & 5 \end{bmatrix} \mathbf{y}, \quad \mathbf{y}(0) = \begin{bmatrix} 1 \\ 0 \\ 1 \end{bmatrix}$

41. $\mathbf{y}' = \begin{bmatrix} 2 & 0 & 9 \\ 0 & 3 & 0 \\ 1 & 0 & 2 \end{bmatrix} \mathbf{y}, \quad \mathbf{y}(0) = \begin{bmatrix} 0 \\ 1 \\ 0 \end{bmatrix}$

42. $\mathbf{y}' = \begin{bmatrix} -2 & 0 & 0 \\ 0 & 3 & -1 \\ 0 & 1 & 3 \end{bmatrix} \mathbf{y}, \quad \mathbf{y}(0) = \begin{bmatrix} 2 \\ 0 \\ 0 \end{bmatrix}$

43. $\mathbf{y}' = \begin{bmatrix} 0 & 1 & 0 \\ 0 & -2 & -5 \\ 0 & 1 & 2 \end{bmatrix} \mathbf{y}, \quad \mathbf{y}(0) = \begin{bmatrix} 0 \\ 5 \\ -2 \end{bmatrix}$

44. $\mathbf{y}' = \begin{bmatrix} 2 & 1 & 1 \\ 1 & 1 & 0 \\ 1 & 0 & 1 \end{bmatrix} \mathbf{y}, \quad \mathbf{y}(0) = \begin{bmatrix} 1 \\ 0 \\ -1 \end{bmatrix}$

5.4 THE FUNDAMENTAL MATRIX

In the previous section we developed techniques for solving the first order linear system of differential equations

$$\mathbf{y}' = A\mathbf{y}, \qquad \mathbf{y}(t_0) = \mathbf{y}^0. \tag{5.4.1}$$

The solution could be obtained by analyzing the eigenvalue–eigenvector structure of the matrix A or directly by Laplace transform methods. But suppose that several different initial conditions $\mathbf{y}(t_0) = \mathbf{y}^1, \mathbf{y}(t_0) = \mathbf{y}^2, \ldots$, etc., are given. Does one have to solve each problem separately or can a general solution be developed in which the various initial conditions can be substituted to obtain the answer directly?

Fortunately, such a general solution can be found, and to accomplish this the concept of the *fundamental* or *transition state matrix* of the system (5.4.1) must be developed. With the fundamental matrix in hand, any initial value problem associated with $\mathbf{y}' = A\mathbf{y}$ can be solved as well as the inhomogeneous problem

$$\mathbf{y}' = A\mathbf{y} + \mathbf{f}(t), \qquad \mathbf{y}(t_0) = \mathbf{y}^0.$$

This last topic will be discussed in the next section.

To motivate the discussion, we go back to the one-dimensional version of the system (5.4.1), namely

$$y' = ay, \qquad y(t_0) = y_0, \tag{5.4.2}$$

where $y = y(t)$ is a scalar function and a is a given constant. For the function $\phi(t) = e^{at}$, the following conclusions are true:

1. $\phi' = a\phi$ since $\phi'(t) = ae^{at} = a\phi(t)$;
2. $\phi(0) = 1$;
3. The solution of $y' = ay$, $y(0) = y_0$, is given by

$$y(t) = \phi(t)y_0 = e^{at}\, y_0;$$

4. The solution of $y' = ay$, $y(t_0) = y_0$ is given by

$$y(t) = \phi(t - t_0)y_0 = e^{a(t-t_0)}y_0.$$

The reader can easily verify points (3) and (4) by differentiation and direct substitution. In view of this, $\phi(t)$ could be called the fundamental solution of (5.4.2). We also note at this point that $\phi(t)$ has the series representation

$$\phi(t) = e^{at} = 1 + at + \frac{1}{2!}a^2t^2 + \cdots + \frac{1}{m!}a^mt^m + \cdots,$$

which converges for any a and for $-\infty < t < \infty$.

Now, given the first order system (5.4.1), consider the matrix function defined formally by

$$\Phi(t) = e^{tA} = I + tA + \frac{1}{2!}t^2A^2 + \cdots + \frac{1}{m!}t^mA^m + \cdots, \tag{5.4.3}$$

where A is the $n \times n$ coefficient matrix of (5.4.1) and its powers are defined recursively by

$$A^2 = A \cdot A, \quad A^3 = A \cdot A^2, \quad \ldots, \quad A^m = A \cdot A^{m-1}.$$

Recall that

$$I = \begin{bmatrix} 1 & & 0 \\ & 1 & \\ 0 & & 1 \end{bmatrix}$$

is the $n \times n$ identity matrix with ones on the main diagonal and zeros elsewhere. The proof that the power series defined above converges to a matrix function for any constant $n \times n$ matrix A and $-\infty < t < \infty$ is beyond the scope of this book. It is hoped that the following examples will convince the reader that it does.

EXAMPLE 1 Let $A = \begin{bmatrix} 1 & 0 \\ 1 & 2 \end{bmatrix}$; then

$$A^2 = A \cdot A = \begin{bmatrix} 1 & 0 \\ 1 & 2 \end{bmatrix}\begin{bmatrix} 1 & 0 \\ 1 & 2 \end{bmatrix} = \begin{bmatrix} 1 & 0 \\ 3 & 4 \end{bmatrix} = \begin{bmatrix} 1 & 0 \\ 2^2 - 1 & 2^2 \end{bmatrix}$$

$$A^3 = A \cdot A^2 = \begin{bmatrix} 1 & 0 \\ 1 & 2 \end{bmatrix}\begin{bmatrix} 1 & 0 \\ 3 & 4 \end{bmatrix} = \begin{bmatrix} 1 & 0 \\ 7 & 8 \end{bmatrix} = \begin{bmatrix} 1 & 0 \\ 2^3 - 1 & 2^3 \end{bmatrix}$$

$$\vdots \qquad\qquad\qquad\qquad \vdots$$

$$A^m = A \cdot A^{m-1} = \begin{bmatrix} 1 & 0 \\ 1 & 2 \end{bmatrix}\begin{bmatrix} 1 & 0 \\ 2^{m-1} - 1 & 2^{m-1} \end{bmatrix} = \begin{bmatrix} 1 & 0 \\ 2^m - 1 & 2^m \end{bmatrix}.$$

Therefore

$$e^{tA} = \begin{bmatrix} 1 & 0 \\ 0 & 1 \end{bmatrix} + t\begin{bmatrix} 1 & 0 \\ 1 & 2 \end{bmatrix} + \frac{t^2}{2!}\begin{bmatrix} 1 & 0 \\ 2^2 - 1 & 2^2 \end{bmatrix} + \frac{t^3}{3!}\begin{bmatrix} 1 & 0 \\ 2^3 - 1 & 2^3 \end{bmatrix}$$

$$+ \cdots + \frac{t^m}{m!}\begin{bmatrix} 1 & 0 \\ 2^m - 1 & 2^m \end{bmatrix} + \cdots$$

$$= \begin{bmatrix} \displaystyle\sum_{m=0}^{\infty} \frac{t^m}{m!} & 0 \\ \displaystyle\sum_{m=0}^{\infty} \frac{(2t)^m}{m!} - \sum_{m=0}^{\infty} \frac{t^m}{m!} & \displaystyle\sum_{m=0}^{\infty} \frac{(2t)^m}{m!} \end{bmatrix} = \begin{bmatrix} e^t & 0 \\ e^{2t} - e^t & e^{2t} \end{bmatrix}. \qquad \blacksquare$$

EXAMPLE 2 An even simpler example is the case when A is a diagonal matrix

$$A = \begin{bmatrix} \lambda_1 & & & 0 \\ & \lambda_2 & & \\ & & \ddots & \\ 0 & & & \lambda_n \end{bmatrix}.$$

Then

$$A^m = \begin{bmatrix} \lambda_1^m & & & 0 \\ & \lambda_2^m & & \\ & & \ddots & \\ 0 & & & \lambda_n^m \end{bmatrix}$$

and

$$e^{tA} = \begin{bmatrix} \sum_{m=0}^{\infty} \frac{(\lambda_1 t)^m}{m!} & & & 0 \\ & \sum_{m=0}^{\infty} \frac{(\lambda_2 t)^m}{m!} & & \\ & & \ddots & \\ 0 & & & \sum_{m=0}^{\infty} \frac{(\lambda_n t)^m}{m!} \end{bmatrix} = \begin{bmatrix} e^{\lambda_1 t} & & & 0 \\ & e^{\lambda_2 t} & & \\ & & \ddots & \\ 0 & & & e^{\lambda_n t} \end{bmatrix}. \qquad \blacksquare$$

If a power series converges, then it can be differentiated termwise, and this is also true for the power series (5.4.3) that defines e^{tA}. Hence

$$\begin{aligned} \frac{d}{dt}\Phi(t) &= \frac{d}{dt}e^{tA} = \frac{d}{dt}\left[I + tA + \frac{t^2}{2!}A^2 + \cdots + \frac{t^m}{m!}A^m + \cdots\right] \\ &= 0 + A + tA^2 + \frac{t^2}{2!}A^3 + \cdots + \frac{t^{m-1}}{(m-1)!}A^m + \cdots \\ &= A\left(I + tA + \frac{t^2}{2!}A^2 + \cdots + \frac{t^{m-1}}{(m-1)!}A^{m-1} + \cdots\right) \\ &= Ae^{tA} = A\Phi(t), \end{aligned}$$

or

$$\Phi'(t) = A\Phi(t),$$

which is the matrix analogue of conclusion (1) above (p. 383). For instance, in Examples 1 and 2 above,

$$\frac{d}{dt}\begin{bmatrix} e^t & 0 \\ e^{2t} - e^t & e^{2t} \end{bmatrix} = \begin{bmatrix} e^t & 0 \\ 2e^{2t} - e^t & 2e^{2t} \end{bmatrix} = \begin{bmatrix} 1 & 0 \\ 1 & 2 \end{bmatrix}\begin{bmatrix} e^t & 0 \\ e^{2t} - e^t & e^{2t} \end{bmatrix}$$

and

$$\frac{d}{dt}\begin{bmatrix} e^{\lambda_1 t} & & & 0 \\ & e^{\lambda_2 t} & & \\ & & \ddots & \\ 0 & & & e^{\lambda_n t} \end{bmatrix} = \begin{bmatrix} \lambda_1 e^{\lambda_1 t} & & & 0 \\ & \lambda_2 e^{\lambda_2 t} & & \\ & & \ddots & \\ 0 & & & \lambda_n e^{\lambda_n t} \end{bmatrix}$$

$$= \begin{bmatrix} \lambda_1 & & & 0 \\ & \lambda_2 & & \\ & & \ddots & \\ 0 & & & \lambda_n \end{bmatrix} \begin{bmatrix} e^{\lambda_1 t} & & & 0 \\ & e^{\lambda_2 t} & & \\ & & \ddots & \\ 0 & & & e^{\lambda_n t} \end{bmatrix}.$$

Furthermore

$$\Phi(0) = e^{0A} = I + 0A + \frac{0^2}{2!}A^2 + \cdots + \frac{0^m}{m!}A^m + \cdots = I,$$

the identity matrix, which is the matrix analogue of conclusion (2) above.

Finally, given the vector $\mathbf{y}^0$, consider the vector obtained by multiplying it by $\Phi(t)$:

$$\mathbf{y}(t) = \Phi(t)\mathbf{y}^0 = e^{tA}\mathbf{y}^0 = I\mathbf{y}^0 + tA\mathbf{y}^0 + \frac{t^2}{2!}A^2\mathbf{y}^0 + \cdots + \frac{t^m}{m!}A^m\mathbf{y}^0 + \cdots.$$

Since $I\mathbf{y}^0 = \mathbf{y}^0$, we have

$$\mathbf{y}(0) = \Phi(0)\mathbf{y}^0 = I\mathbf{y}^0 = \mathbf{y}^0,$$

and now differentiating $\mathbf{y}(t)$ gives

$$\begin{aligned} \mathbf{y}'(t) &= \frac{d}{dt}\left(I\mathbf{y}^0 + tA\mathbf{y}^0 + \frac{t^2}{2!}A^2\mathbf{y}^0 + \cdots + \frac{t^m}{m!}A^m\mathbf{y}^0 + \cdots\right) \\ &= A\left(I\mathbf{y}^0 + tA\mathbf{y}^0 + \frac{t^2}{2!}A^2\mathbf{y}^0 + \cdots + \frac{t^{m-1}}{(m-1)!}A^{m-1}\mathbf{y}^0 + \cdots\right) \\ &= A\Phi(t)\mathbf{y}^0 = A\mathbf{y}(t). \end{aligned}$$

The last two relations imply that

$$\mathbf{y}(t) = \Phi(t)\mathbf{y}^0$$

is the solution to the initial value problem

$$\mathbf{y}' = A\mathbf{y}, \qquad \mathbf{y}(0) = \mathbf{y}^0.$$

This is the matrix version of conclusion (3). It is left to the reader to verify that $\mathbf{y}(t) = \Phi(t - t_0)\mathbf{y}^0$ is the solution to the initial value problem

$$\mathbf{y}' = A\mathbf{y}, \qquad \mathbf{y}(t_0) = \mathbf{y}^0.$$

We conclude that $\Phi(t) = e^{tA}$ satisfies all four of the relations satisfied by $\phi(t) = e^{at}$ in the scalar case. For instance, in Example 1 above

$$\Phi(t) = \begin{bmatrix} e^t & 0 \\ e^{2t} - e^t & e^{2t} \end{bmatrix} \quad \text{and} \quad \Phi(0) = \begin{bmatrix} 1 & 0 \\ 0 & 1 \end{bmatrix} = I.$$

Let

$$\mathbf{y}(t) = \begin{bmatrix} y_1(t) \\ y_2(t) \end{bmatrix} = \begin{bmatrix} e^t & 0 \\ e^{2t} - e^t & e^{2t} \end{bmatrix} \begin{bmatrix} y_1^0 \\ y_2^0 \end{bmatrix} = \Phi(t)\mathbf{y}^0;$$

then

$$\mathbf{y}(0) = \begin{bmatrix} y_1(0) \\ y_2(0) \end{bmatrix} = \begin{bmatrix} 1 & 0 \\ 0 & 1 \end{bmatrix} \begin{bmatrix} y_1^0 \\ y_2^0 \end{bmatrix} = \begin{bmatrix} y_1^0 \\ y_2^0 \end{bmatrix} = \mathbf{y}^0$$

and

$$\begin{aligned} \mathbf{y}'(t) &= \begin{bmatrix} y_1'(t) \\ y_2'(t) \end{bmatrix} = \begin{bmatrix} e^t & 0 \\ 2e^{2t} - e^t & 2e^{2t} \end{bmatrix} \begin{bmatrix} y_1^0 \\ y_2^0 \end{bmatrix} \\ &= \begin{bmatrix} 1 & 0 \\ 1 & 2 \end{bmatrix} \begin{bmatrix} e^t & 0 \\ e^{2t} - e^t & e^{2t} \end{bmatrix} \begin{bmatrix} y_1^0 \\ y_2^0 \end{bmatrix} \\ &= \begin{bmatrix} 1 & 0 \\ 1 & 2 \end{bmatrix} \begin{bmatrix} y_1(t) \\ y_2(t) \end{bmatrix} = \begin{bmatrix} 1 & 0 \\ 1 & 2 \end{bmatrix} \mathbf{y}(t). \end{aligned}$$

Hence, if the above relations are expressed by components, then

$$\left.\begin{aligned} y_1(t) &= y_1^0 e^t \\ y_2(t) &= y_1^0(e^{2t} - e^t) + y_2^0 e^{2t} \end{aligned}\right\} = \Phi(t) \begin{bmatrix} y_1^0 \\ y_2^0 \end{bmatrix}$$

is the solution to the initial value problem

$$\left.\begin{aligned} y_1' &= y_1 \\ y_2' &= y_1 + 2y_2 \end{aligned}\right\} = \begin{bmatrix} 1 & 0 \\ 1 & 2 \end{bmatrix} \begin{bmatrix} y_1 \\ y_2 \end{bmatrix}$$

$$y_1(0) = y_1^0, \qquad y_2(0) = y_2^0.$$

It is left to the reader to verify for Example 2 that

$$\mathbf{y}(t) = \begin{bmatrix} e^{\lambda_1 t} & & 0 \\ & e^{\lambda_2 t} & \\ & & \ddots & \\ 0 & & & e^{\lambda_n t} \end{bmatrix} \begin{bmatrix} y_1^0 \\ y_2^0 \\ \vdots \\ y_n^0 \end{bmatrix}$$

is the solution of the system

$$\begin{bmatrix} y_1' \\ y_2' \\ \vdots \\ y_n' \end{bmatrix} = \begin{bmatrix} \lambda_1 y_1 \\ \lambda_2 y_2 \\ \vdots \\ \lambda_n y_n \end{bmatrix} = \begin{bmatrix} \lambda_1 & & & 0 \\ & \lambda_2 & & \\ & & \ddots & \\ 0 & & & \lambda_n \end{bmatrix} \begin{bmatrix} y_1 \\ y_2 \\ \vdots \\ y_n \end{bmatrix}$$

with initial conditions

$$y_1(0) = y_1^0, \qquad y_2(0) = y_2^0, \qquad \ldots, \qquad y_n(0) = y_n^0.$$

Our conclusions are summarized in the following statement:

For the *fundamental matrix*

$$\Phi(t) = e^{tA} = \sum_{m=0}^{\infty} \frac{t^m A^m}{m!}$$

the following is true:

1. $\Phi' = A\Phi$;
2. $\Phi(0) = I$; and
3. the solution of $\mathbf{y}' = A\mathbf{y}$, $\mathbf{y}(t_0) = \mathbf{y}^0$ is given by $\mathbf{y}(t) = \Phi(t - t_0)\mathbf{y}^0$.

There is one remaining problem, and that is to find the fundamental matrix $\Phi(t)$. The power series representation is clearly not useful since it involves computing successive powers of A. For all but the simplest matrices, writing down the first m terms of the power series will not give an inkling as to the nature of $\Phi(t)$.

Hence another way to compute $\Phi(t)$ is needed, and the key to the method lies in the following observation.

The jth column $\Phi^j(t)$ of $\Phi(t)$ is the solution of the initial value problem

$$\mathbf{y}' = A\mathbf{y}, \qquad \mathbf{y}(0) = \mathbf{e}^j,$$

where $\mathbf{e}^j$ is the vector with one in the jth place and zeros elsewhere, i.e., the jth unit coordinate vector.

The proof of this assertion follows easily by noting first of all that $\Phi(0) = I$ implies that the jth column, $\Phi^j(t)$, of $\Phi(t)$ satisfies $\Phi^j(0) = \mathbf{e}^j$. Secondly, from the property of the fundamental matrix it follows that solutions to the above problem can be written in the form $\mathbf{y} = \Phi(t)\mathbf{e}^j$. But this is just $\Phi^j(t)$ since if $\Phi(t) = (\phi_{ij}(t))$, then

$$y_k(t) = \sum_{i=1}^{n} \phi_{ki}(t)e_i^j = \phi_{kj}(t)$$

because $e_i^j = 0$ unless $i = j$ when $e_j^j = 1$. Put more simply, the kth entry in the vector $\mathbf{y}(t)$ is the dot product of the kth row of $\Phi(t)$ with $\mathbf{e}^j$, and this will give the kth entry in the jth column of $\Phi(t)$.

In Example 1

$$\begin{aligned} y_1' &= y_1 \\ y_2' &= y_1 + 2y_2, \end{aligned} \qquad \Phi(t) = \begin{bmatrix} e^t & 0 \\ e^{2t} - e^t & e^{2t} \end{bmatrix},$$

and so the columns of $\Phi(t)$ are

$$\Phi^1(t) = \begin{bmatrix} e^t \\ e^{2t} - e^t \end{bmatrix}, \qquad \Phi^2(t) = \begin{bmatrix} 0 \\ e^{2t} \end{bmatrix}.$$

It is seen, for instance, that the entries of $\Phi^1(t)$, $y_1(t) = e^t$, $y_2(t) = e^{2t} - e^t$, satisfy

$$\begin{aligned} y_1'(t) &= e^t = y_1(t), \\ y_2'(t) &= 2e^{2t} - e^t = e^t + 2(e^{2t} - e^t) = y_1(t) + 2y_2(t), \end{aligned}$$

and, furthermore,

$$\begin{bmatrix} y_1(0) \\ y_2(0) \end{bmatrix} = \begin{bmatrix} 1 \\ 0 \end{bmatrix} = \Phi^1(0) = \mathbf{e}^1.$$

The reader can see directly that $y_1(t) = 0$, $y_2(t) = e^{2t}$ also satisfy the system of differential equations and

$$\begin{bmatrix} y_1(0) \\ y_2(0) \end{bmatrix} = \begin{bmatrix} 0 \\ 1 \end{bmatrix} = \Phi^2(0) = \mathbf{e}^2.$$

In Example 2, the system $y_k' = \lambda_k y_k$, $k = 1, 2, \ldots, n$, was given and the jth column of $\Phi(t)$ was

$$\Phi^j(t) = \text{col}(0, \ldots, 0, \underset{\substack{j\text{th} \\ \text{place}}}{e^{\lambda_j t}}, 0, \ldots, 0),$$

which satisfies $\Phi^j(0) = \mathbf{e}^j$. It corresponds to the solution $y_j(t) = e^{\lambda_j t}$, $y_k(t) = 0$, $k \neq j$, of the system.

Our findings are summarized in the following algorithm.

Algorithm for finding $\Phi(t) = e^{tA}$, where A is an $n \times n$ matrix
Given the linear system $\mathbf{y}' = A\mathbf{y}$, where A is an $n \times n$ constant matrix, $\Phi^j(t)$, the jth column of the fundamental matrix $\Phi(t)$, is obtained by solving the initial value problem

$$\mathbf{y}' = A\mathbf{y}, \qquad \mathbf{y}(0) = \mathbf{e}^j,$$

where $\mathbf{e}^j$ is the jth unit coordinate vector in R^n, that is, $y_j(0) = 1$, $y_k(0) = 0$, $k \neq j$.

EXAMPLE 3 Find the fundamental solution matrix for

$$A = \begin{bmatrix} -1 & 6 \\ 1 & -2 \end{bmatrix}$$

and then solve the initial value problem

$$\mathbf{y}' = A\mathbf{y}, \qquad \mathbf{y}(1) = \begin{bmatrix} 2 \\ -3 \end{bmatrix}.$$

SOLUTION Since

$$\det\begin{bmatrix} -1-\lambda & 6 \\ 1 & -2-\lambda \end{bmatrix} = \lambda^2 + 3\lambda - 4 = (\lambda + 4)(\lambda - 1),$$

the eigenvalues of A are $\lambda = -4$ and $\lambda = 1$. The eigenvector corresponding to $\lambda = -4$ is $\begin{bmatrix} -2 \\ 1 \end{bmatrix}$, so

$$\mathbf{y}^1(t) = e^{-4t}\begin{bmatrix} -2 \\ 1 \end{bmatrix} = \begin{bmatrix} -2e^{-4t} \\ e^{-4t} \end{bmatrix}$$

is a solution. The eigenvector corresponding to $\lambda = 1$ is $\begin{bmatrix} 3 \\ 1 \end{bmatrix}$, so

$$\mathbf{y}^2(t) = e^t\begin{bmatrix} 3 \\ 1 \end{bmatrix} = \begin{bmatrix} 3e^t \\ e^t \end{bmatrix}$$

is another linearly independent solution, and

$$\begin{aligned} \mathbf{y}(t) = \begin{bmatrix} y_1(t) \\ y_2(t) \end{bmatrix} &= c_1\begin{bmatrix} -2e^{-4t} \\ e^{-4t} \end{bmatrix} + c_2\begin{bmatrix} 3e^t \\ e^t \end{bmatrix} \\ &= \begin{bmatrix} -2c_1e^{-4t} + 3c_2e^t \\ c_1e^{-4t} + c_2e^t \end{bmatrix} \end{aligned}$$

is the general solution of the system

$$y_1' = -y_1 + 6y_2$$
$$y_2' = y_1 - 2y_2.$$

To obtain $\Phi(t)$, first find its first column $\Phi^1(t)$ by letting

$$\mathbf{y}(0) = \begin{bmatrix} y_1(0) \\ y_2(0) \end{bmatrix} = \begin{bmatrix} -2c_1 + 3c_2 \\ c_1 + c_2 \end{bmatrix} = \begin{bmatrix} 1 \\ 0 \end{bmatrix} = \mathbf{e}^1.$$

Solving for c_1 and c_2 gives $c_1 = -\frac{1}{5}$, $c_2 = \frac{1}{5}$, so

$$\Phi^1(t) = \begin{bmatrix} \frac{2}{5}e^{-4t} + \frac{3}{5}e^t \\ -\frac{1}{5}e^{-4t} + \frac{1}{5}e^t \end{bmatrix}.$$

Now, to get the second column $\Phi^2(t)$, solve

$$\mathbf{y}(0) = \begin{bmatrix} y_1(0) \\ y_2(0) \end{bmatrix} = \begin{bmatrix} -2c_1 + 3c_2 \\ c_1 + c_2 \end{bmatrix} = \begin{bmatrix} 0 \\ 1 \end{bmatrix} = \mathbf{e}^2$$

to get

$$\Phi^2(t) = \begin{bmatrix} -\frac{6}{5}e^{-4t} + \frac{6}{5}e^t \\ \frac{3}{5}e^{-4t} + \frac{2}{5}e^t \end{bmatrix}.$$

Therefore

$$\Phi(t) = e^{tA} = \begin{bmatrix} \frac{2}{5}e^{-4t} + \frac{3}{5}e^t & -\frac{6}{5}e^{-4t} + \frac{6}{5}e^t \\ -\frac{1}{5}e^{-4t} + \frac{1}{5}e^t & \frac{3}{5}e^{-4t} + \frac{2}{5}e^t \end{bmatrix}.$$

The solution of the system with initial conditions $y_1(1) = 2$, $y_2(1) = -3$ is

$$\begin{bmatrix} y_1(t) \\ y_2(t) \end{bmatrix} = \Phi(t-1)\begin{bmatrix} 2 \\ -3 \end{bmatrix}$$
$$= \begin{bmatrix} \frac{2}{5}e^{-4(t-1)} + \frac{3}{5}e^{t-1} & -\frac{6}{5}e^{-4(t-1)} + \frac{6}{5}e^{t-1} \\ -\frac{1}{5}e^{-4(t-1)} + \frac{1}{5}e^{t-1} & \frac{3}{5}e^{-4(t-1)} + \frac{2}{5}e^{t-1} \end{bmatrix}\begin{bmatrix} 2 \\ -3 \end{bmatrix}$$
$$= \frac{1}{5}\begin{bmatrix} 22e^{-4(t-1)} - 12e^{t-1} \\ -11e^{-4(t-1)} - 4e^{t-1} \end{bmatrix}.$$

■

EXAMPLE 4 Find the fundamental matrix for

$$A = \begin{bmatrix} 2 & 1 \\ -1 & 4 \end{bmatrix}$$

by using the Laplace transform.

SOLUTION We first solve the initial value problem

$$y_1'(t) = 2y_1(t) + y_2(t), \quad y_1(0) = 1,$$
$$y_2'(t) = -y_1(t) + 4y_2(t), \quad y_2(0) = 0.$$

Applying the Laplace transform gives

$$sY_1(s) - 1 = 2Y_1(s) + Y_2(s),$$
$$sY_2(s) - 0 = -Y_1(s) + 4Y_2(s).$$

Solving the above system for $Y_1(s)$ and $Y_2(s)$, one obtains

$$Y_1(s) = \frac{s-4}{(s-3)^2} = \frac{1}{s-3} - \frac{1}{(s-3)^2},$$
$$Y_2(s) = \frac{-1}{(s-3)^2}.$$

Applying the inverse transform, one finds that

$$\mathbf{\Phi}^1(t) = \begin{bmatrix} y_1(t) \\ y_2(t) \end{bmatrix} = \begin{bmatrix} e^{3t} - te^{3t} \\ -te^{3t} \end{bmatrix}.$$

If the order of the initial values 1 and 0 is reversed in the system of equations for $Y_1(s)$ and $Y_2(s)$, then

$$Y_1(s) = \frac{1}{(s-3)^2},$$
$$Y_2(s) = \frac{s-2}{(s-3)^2} = \frac{1}{s-3} + \frac{1}{(s-3)^2}.$$

This implies that

$$\mathbf{\Phi}^2(t) = \begin{bmatrix} y_1(t) \\ y_2(t) \end{bmatrix} = \begin{bmatrix} te^{3t} \\ e^{3t} + te^{3t} \end{bmatrix},$$

and the fundamental matrix of the system is therefore

$$\mathbf{\Phi}(t) = \begin{bmatrix} e^{3t} - te^{3t} & te^{3t} \\ -te^{3t} & e^{3t} + te^{3t} \end{bmatrix}.$$

■

EXERCISES 5.4

1. Verify that the function $\mathbf{y}(t) = \mathbf{\Phi}(t - t_0)\mathbf{y}^0$ is the solution to the initial value problem

$$\mathbf{y}' = A\mathbf{y}, \qquad \mathbf{y}(t_0) = \mathbf{y}^0.$$

In each of the following, find the fundamental matrix $\mathbf{\Phi}(t)$ by first finding a general solution to the system of differential equations and then finding solutions $\mathbf{\Phi}^1(t)$ and $\mathbf{\Phi}^2(t)$ that satisfy

$$\mathbf{\Phi}^1(0) = \begin{bmatrix} 1 \\ 0 \end{bmatrix}, \qquad \mathbf{\Phi}^2(0) = \begin{bmatrix} 0 \\ 1 \end{bmatrix}.$$

2. $\mathbf{y}' = \begin{bmatrix} 4 & -3 \\ 8 & -6 \end{bmatrix} \mathbf{y}$ **3.** $\mathbf{y}' = \begin{bmatrix} 0 & 1 \\ 8 & -2 \end{bmatrix} \mathbf{y}$ **4.** $\mathbf{y}' = \begin{bmatrix} 2 & 1 \\ 0 & 2 \end{bmatrix} \mathbf{y}$

5. $\mathbf{y}' = \begin{bmatrix} 5 & 3 \\ -3 & -1 \end{bmatrix} \mathbf{y}$ **6.** $\mathbf{y}' = \begin{bmatrix} -1 & -5 \\ 1 & 1 \end{bmatrix} \mathbf{y}$ **7.** $\mathbf{y}' = \begin{bmatrix} 1 & -1 \\ 5 & -1 \end{bmatrix} \mathbf{y}$

8. $\mathbf{y}' = \begin{bmatrix} 0 & 3 \\ 1 & -2 \end{bmatrix} \mathbf{y}$ **9.** $\mathbf{y}' = \begin{bmatrix} 1 & -1 \\ 4 & 1 \end{bmatrix} \mathbf{y}$ **10.** $\mathbf{y}' = \begin{bmatrix} -6 & -2 \\ 17 & 0 \end{bmatrix} \mathbf{y}$

11. $\mathbf{y}' = \begin{bmatrix} 1 & -2 \\ 2 & -3 \end{bmatrix} \mathbf{y}$

Use the Laplace transform to find the fundamental matrix $\mathbf{\Phi}(t)$ in each of the following.

12. $\mathbf{y}' = \begin{bmatrix} 2 & 1 \\ 3 & 4 \end{bmatrix} \mathbf{y}$ **13.** $\mathbf{y}' = \begin{bmatrix} 3 & 1 \\ -2 & 1 \end{bmatrix} \mathbf{y}$ **14.** $\mathbf{y}' = \begin{bmatrix} 1 & -1 \\ -3 & 3 \end{bmatrix} \mathbf{y}$

15. $\mathbf{y}' = \begin{bmatrix} 5 & -1 \\ 1 & 3 \end{bmatrix} \mathbf{y}$ **16.** $\mathbf{y}' = \begin{bmatrix} 2 & -53 \\ 1 & -2 \end{bmatrix} \mathbf{y}$ **17.** $\mathbf{y}' = \begin{bmatrix} 3 & 1 \\ 0 & 3 \end{bmatrix} \mathbf{y}$

18. $\mathbf{y}' = \begin{bmatrix} 2 & 3 \\ 7 & -2 \end{bmatrix} \mathbf{y}$ **19.** $\mathbf{y}' = \begin{bmatrix} 1 & 1 \\ -3 & 1 \end{bmatrix} \mathbf{y}$

First find the fundamental matrix, then use it to solve the following initial value problems.

20. $\mathbf{y}' = \begin{bmatrix} -1 & -5 \\ 1 & 3 \end{bmatrix} \mathbf{y}, \quad \mathbf{y}(0) = \begin{bmatrix} 5 \\ 2 \end{bmatrix}, \quad \mathbf{y}(\pi) = \begin{bmatrix} 3 \\ 1 \end{bmatrix}$

21. $\mathbf{y}' = \begin{bmatrix} 4 & -1 \\ 1 & -2 \end{bmatrix} \mathbf{y}, \quad \mathbf{y}(0) = \begin{bmatrix} 2 \\ 1 \end{bmatrix}, \quad \mathbf{y}(1) = \begin{bmatrix} 1 \\ 2 \end{bmatrix}$

22. $\mathbf{y}' = \begin{bmatrix} 1 & 3 \\ 12 & 1 \end{bmatrix} \mathbf{y}, \quad \mathbf{y}(0) = \begin{bmatrix} 3 \\ -2 \end{bmatrix}, \quad \mathbf{y}(2) = \begin{bmatrix} 1 \\ 0 \end{bmatrix}$

23. $\mathbf{y}' = \begin{bmatrix} 3 & -2 \\ 5 & -1 \end{bmatrix} \mathbf{y}, \quad \mathbf{y}(0) = \begin{bmatrix} 1 \\ -1 \end{bmatrix}, \quad \mathbf{y}(1) = \begin{bmatrix} 1 \\ 1 \end{bmatrix}$

24. $\mathbf{y}' = \begin{bmatrix} 1 & -5 \\ 2 & -1 \end{bmatrix} \mathbf{y}, \quad \mathbf{y}(0) = \begin{bmatrix} -1 \\ 1 \end{bmatrix}, \quad \mathbf{y}\left(\frac{\pi}{6}\right) = \begin{bmatrix} 2 \\ 1 \end{bmatrix}$

25. $\mathbf{y}' = \begin{bmatrix} -1 & 3 \\ -1 & -5 \end{bmatrix} \mathbf{y}, \quad \mathbf{y}(0) = \begin{bmatrix} -2 \\ 2 \end{bmatrix}, \quad \mathbf{y}(\ln 2) = \begin{bmatrix} 1 \\ 3 \end{bmatrix}$

26. $\mathbf{y}' = \begin{bmatrix} 5 & 0 & -6 \\ 2 & -1 & -2 \\ 4 & -2 & -4 \end{bmatrix} \mathbf{y}, \quad \mathbf{y}(0) = \begin{bmatrix} 1 \\ 0 \\ 1 \end{bmatrix}, \quad \mathbf{y}(-1) = \begin{bmatrix} -1 \\ 1 \\ 0 \end{bmatrix}$

27. $\mathbf{y}' = \begin{bmatrix} 1 & 0 & 0 \\ 2 & 3 & -5 \\ 6 & 2 & -3 \end{bmatrix} \mathbf{y}, \quad \mathbf{y}(0) = \begin{bmatrix} 3 \\ 1 \\ 1 \end{bmatrix}, \quad , \mathbf{y}(\pi) = \begin{bmatrix} 2 \\ -1 \\ 1 \end{bmatrix}$

28. Find conditions on the quantities a, b, c, and d that guarantee that all solutions of

$$\mathbf{y}' = \begin{bmatrix} a & b \\ c & d \end{bmatrix} \mathbf{y}, \qquad \mathbf{y}(0) = \mathbf{y}_0$$

satisfy

a) $|\mathbf{y}(t)|$ is bounded for $t \geq 0$,

b) $\lim_{t \to \infty} |\mathbf{y}(t)| = 0$.

Note: $|\mathbf{y}(t)| = [y_1(t)^2 + y_2(t)^2]^{1/2}$, the usual Euclidean length of a vector.

29. Given the matrix

$$A = \begin{bmatrix} 0 & a \\ -a & 0 \end{bmatrix},$$

show that

$$\Theta(t) = (\cos at)I + (1/a \sin at)A$$

satisfies properties **(1)**, **(2)**, and **(3)** of a fundamental matrix $\Phi(t)$. Then show that $\Theta(t) = \Phi(t)$.

30. Given the matrix

$$A = \begin{bmatrix} 0 & a \\ a & 0 \end{bmatrix},$$

show that

$$\Theta(t) = (\cosh at)I + (1/a \sinh at)A$$

satisfies properties **(1)**, **(2)**, and **(3)** of a fundamental matrix $\Phi(t)$. Then show that $\Theta(t) = \Phi(t)$.

31. Given the initial value problem

$$\mathbf{y}' = A\mathbf{y} + \mathbf{k}, \qquad \mathbf{y}(0) = \mathbf{y}_0,$$

where $\mathbf{k}$ is a constant vector, suppose that $\mathbf{b}$ is a vector satisfying $A\mathbf{b} = -\mathbf{k}$.

Show that the solution of the initial value problem is

$$\mathbf{y}(t) = \Phi(t)\,(\mathbf{y}_0 - \mathbf{b}) + \mathbf{b},$$

where $\Phi(t)$ is the fundamental matrix associated with A.

Use the technique described in Exercise 31 to solve the following initial value problems.

32. $\mathbf{y}' = \begin{bmatrix} -4 & 3 \\ 2 & -3 \end{bmatrix} \mathbf{y} + \begin{bmatrix} 2 \\ -2 \end{bmatrix}, \quad \mathbf{y}(0) = \begin{bmatrix} 1 \\ 0 \end{bmatrix}$

33. $\mathbf{y}' = \begin{bmatrix} 2 & 4 \\ 3 & -2 \end{bmatrix} \mathbf{y} + \begin{bmatrix} 4 \\ 1 \end{bmatrix}, \quad \mathbf{y}(0) = \begin{bmatrix} 1 \\ 2 \end{bmatrix}$

34. $\mathbf{y}' = \begin{bmatrix} -1 & -1 \\ 1 & -3 \end{bmatrix} \mathbf{y} + \begin{bmatrix} 3 \\ 2 \end{bmatrix}, \quad \mathbf{y}(0) = \begin{bmatrix} 0 \\ 1 \end{bmatrix}$

35. $\mathbf{y}' = \begin{bmatrix} 2 & 1 \\ -2 & 4 \end{bmatrix} \mathbf{y} + \begin{bmatrix} 5 \\ 0 \end{bmatrix}, \quad \mathbf{y}(0) = \begin{bmatrix} -3 \\ 1 \end{bmatrix}$

36. A closed two-tank system has the configuration depicted in Fig. 5.5. Tank A has x_0 lb of chemical X dissolved in 200 gal of fresh water. Tank B has y_0 lb of chemical X dissolved in 200 gal of fresh water.

- **a)** Write down the system of differential equations satisfied by $x(t)$ and $y(t)$, the amounts of chemical X at time t in tanks A and B, respectively.
- **b)** Find the fundamental matrix of the system.
- **c)** Show that the amount of chemical X in either tank approaches $\frac{1}{2}(x_0 + y_0)$ as t approaches infinity.

37. A two-tank system has the configuration shown in Fig. 5.6. Tank A initially contains 200 gal of fresh water, and tank B has y_0 lb of chemical X dissolved in 200 gal of fresh water. Fresh water is flowing into tank A from the outside.

- **a)** Write down the system of differential equations and initial conditions satisfied by $x(t)$ and $y(t)$, the amounts of chemical X at time t in tanks A and B, respectively.
- **b)** Find the fundamental matrix of the system.
- **c)** Show that $x(t)$ and $y(t)$ approach zero as t approaches infinity.

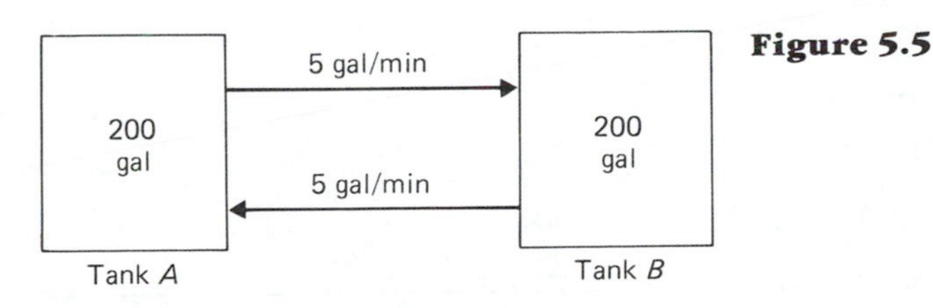

Figure 5.5

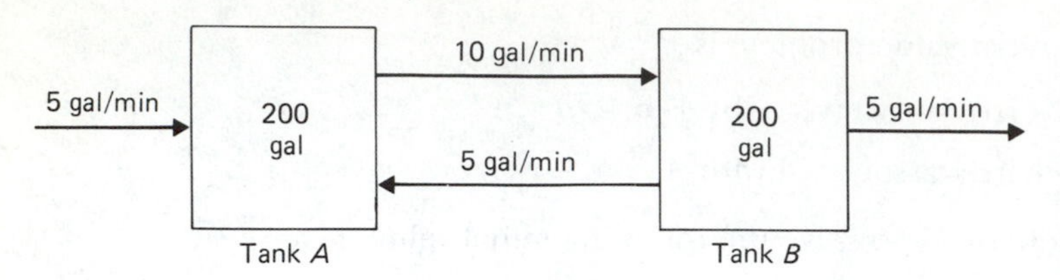

Figure 5.6

38. In Exercise 37, suppose instead that the water flowing into tank A contains $\frac{1}{10}$ lb/gal of chemical X. Find $x(t)$ and $y(t)$, the amounts of chemical X at time t in tanks A and B, respectively, and discuss their behavior as t approaches infinity. (*Hint:* You will need the results of Exercise 31.)

39. For the electrical network shown in Fig. 5.7 we have the relation $dQ_2/dt = I_2$, and Kirchhoff's laws give the following equations:

$$I = I_1 + I_2, \qquad L\frac{dI}{dt} + R_1 I_1 = E,$$

$$R_2 I_2 + \frac{Q_2}{C} - R_1 I_1 = 0.$$

a) Use the equations above to show that Q_2 and I satisfy the system

$$\frac{d}{dt}\begin{bmatrix} I_2 \\ Q \end{bmatrix} = A\begin{bmatrix} I_2 \\ Q \end{bmatrix} + \begin{bmatrix} E/L \\ 0 \end{bmatrix},$$

where

$$A = \frac{1}{CL(R_1 + R_2)}\begin{bmatrix} -CR_1R_2 & -R_1 \\ CLR_1 & -L \end{bmatrix}.$$

b) If $R_1 = 100$ ohms, $R_2 = 400$ ohms, $C = \frac{3}{2} \times 10^{-3}$ farads, $E = 40$ volts, and $L = 20$ henrys, find the general solution of the system. What are the steady-state values of Q_2 and I? (*Hint*: You will need the results of Exercise 31.)

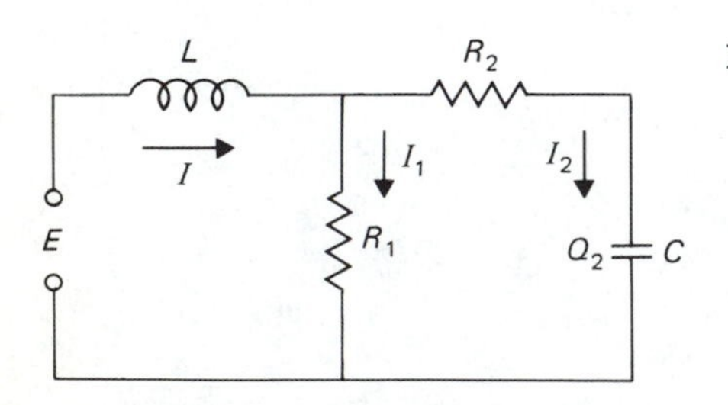

Figure 5.7

5.5 THE INHOMOGENEOUS LINEAR SYSTEM—VARIATION OF PARAMETERS

The next problem that we consider is finding solutions of the inhomogeneous linear system:

$$\begin{aligned}
y_1' &= a_{11}y_1 + a_{12}y_2 + \cdots + a_{1n}y_n + f_1(t), && y_1(t_0) = y_1^0,\\
y_2' &= a_{21}y_1 + a_{22}y_2 + \cdots + a_{2n}y_n + f_2(t), && y_2(t_0) = y_2^0,\\
&\vdots && \vdots\\
y_n' &= a_{n1}y_1 + a_{n2}y_2 + \cdots + a_{nn}y_n + f_n(t), && y_n(t_0) = y_n^0,
\end{aligned}$$

where the functions $f_1(t), f_2(t), \ldots, f_n(t)$ are continuous in some neighborhood of $t = t_0$. If $\mathbf{f}(t)$ is the vector function

$$\mathbf{f}(t) = \text{col}(f_1(t), f_2(t), \ldots, f_n(t))$$

and $\mathbf{y}^0 = \text{col}(y_1^0, y_2^0, \ldots, y_n^0)$, then the above system can be simply written as

$$\mathbf{y}' = A\mathbf{y} + \mathbf{f}(t), \qquad \mathbf{y}(t_0) = \mathbf{y}^0, \tag{5.5.1}$$

where $A = (a_{ij})$ is the $n \times n$ constant coefficient matrix.

As in the case of first order and second order equations the task is to find any particular solution $\mathbf{y}_p(t)$ of the system of differential equations, i.e., one must find a function $\mathbf{y}_p(t)$ that satisfies

$$\mathbf{y}_p'(t) = A\mathbf{y}_p(t) + \mathbf{f}(t).$$

Then to solve the initial value problem the particular solution is added to the solution $\mathbf{y}_h(t)$ of the homogeneous system

$$\mathbf{y}' = A\mathbf{y}, \qquad \mathbf{y}(t_0) = \mathbf{y}^0 - \mathbf{y}_p(t_0).$$

From the previous discussion we have that

$$\mathbf{y}_h(t) = \Phi(t - t_0)[\mathbf{y}^0 - \mathbf{y}_p(t_0)],$$

and so the solution of the initial value problem (5.5.1) is

$$\mathbf{y}(t) = \mathbf{y}_h(t) + \mathbf{y}_p(t) = \Phi(t - t_0)(\mathbf{y}^0 - \mathbf{y}_p(t_0)) + \mathbf{y}_p(t).$$

To verify this, first note that

$$\begin{aligned}
\mathbf{y}(t_0) &= \Phi(0)[\mathbf{y}^0 - \mathbf{y}_p(t_0)] + \mathbf{y}_p(t_0)\\
&= I(\mathbf{y}^0 - \mathbf{y}_p(t_0)) + \mathbf{y}_p(t_0) = \mathbf{y}^0
\end{aligned}$$

and

$$\begin{aligned}\mathbf{y}'(t) &= \Phi'(t - t_0)[\mathbf{y}^0 - \mathbf{y}_p(t_0)] + \mathbf{y}_p'(t)\\ &= A\Phi(t - t_0)[\mathbf{y}^0 - \mathbf{y}_p(t_0)] + A\mathbf{y}_p(t) + \mathbf{f}(t)\\ &= A\{\Phi(t - t_0)[\mathbf{y}^0 - \mathbf{y}_p(t_0)] + \mathbf{y}_p(t)\} + \mathbf{f}(t)\\ &= A\mathbf{y}(t) + \mathbf{f}(t),\end{aligned}$$

since $\Phi'(t) = A\Phi(t)$. Thus $\mathbf{y}(t)$ is the desired solution.

How do we find a particular solution $\mathbf{y}_p(t)$? As in the case of first and second order equations, there is a formula called the variation of parameters formula that is sure-fire. One can also use the Laplace transform and convert the problem into an algebraic one in which the initial conditions can be incorporated and thus solve the entire problem in one (sometimes long!) step. Of course this method can be used only when $\mathbf{f}(t)$ is Laplace transformable.

Another method, usually not mentioned in most textbooks when discussing systems, is the method of comparison of coefficients, which might be better called the method of "judicious guessing." The reader, who has used the technique to find solutions of inhomogeneous second order equations, will easily see the extension to inhomogeneous systems in the following examples.

EXAMPLE 1 Find a particular solution of

$$\begin{aligned}y_1' &= 3y_1 + 2y_2 - 3e^t,\\ y_2' &= y_1 - 4y_2.\end{aligned}$$

SOLUTION In this case,

$$\mathbf{f}(t) = \begin{bmatrix} f_1(t) \\ f_2(t) \end{bmatrix} = \begin{bmatrix} -3e^t \\ 0 \end{bmatrix},$$

and so a trial solution would be

$$\mathbf{y}_p(t) = \begin{bmatrix} y_1(t) \\ y_2(t) \end{bmatrix} = \begin{bmatrix} ae^t \\ be^t \end{bmatrix},$$

corresponding to the fact that for a scalar differential equation one would try a multiple of e^t. Substitution into the system of equations gives

$$\begin{aligned}ae^t &= 3ae^t + 2be^t - 3e^t,\\ be^t &= ae^t - 4be^t;\end{aligned}$$

cancelling the factor of e^t leads to the system of algebraic equations

$$\begin{aligned}-2a - 2b &= -3,\\ a - 5b &= 0.\end{aligned}$$

The solution is $a = \frac{5}{4}$, $b = \frac{1}{4}$, and therefore

$$\mathbf{y}_p(t) = \begin{bmatrix} \frac{5}{4} e^t \\ \frac{1}{4} e^t \end{bmatrix}.$$

Note that since the inhomogeneous term $-3e^t$ appears only in the first equation, one might have used a trial solution

$$\mathbf{y}_p(t) = \begin{bmatrix} ae^t \\ 0 \end{bmatrix}.$$

As the answer shows, this would not have worked. ■

EXAMPLE 2 Find a particular solution of

$$y_1' = 3y_1 + 2y_2 + 7t,$$
$$y_2' = y_1 - 4y_2 - 3.$$

SOLUTION The inhomogeneous terms are polynomials of degree ≤ 1; this suggests

$$\mathbf{y}_p(t) = \begin{bmatrix} y_1(t) \\ y_2(t) \end{bmatrix} = \begin{bmatrix} at + b \\ ct + d \end{bmatrix}.$$

Substitution into the system of equations gives

$$a = 3at + 3b + 2ct + 2d + 7t,$$
$$c = at + b - 4ct - 4d - 3.$$

If the coefficients of t and the constant terms are equated in each equation, four linear equations in the unknowns a, b, c, and d are obtained. These are

$$\begin{aligned} a - 3b \qquad\quad - 2d &= 0, \\ 3a \qquad + 2c \qquad\quad &= -7, \\ b - c - 4d &= 3, \\ a \qquad - 4c \qquad\quad &= 0, \end{aligned}$$

and the solution is

$$a = -2, \qquad b = -\frac{3}{14}, \qquad c = -\frac{1}{2}, \qquad d = -\frac{19}{28}.$$

Hence

$$\mathbf{y}_p(t) = \begin{bmatrix} -2t - \frac{3}{14} \\ -\frac{1}{2}t - \frac{19}{28} \end{bmatrix}.$$ ■

Example 2 points out one of the drawbacks of the method, namely, that for even the simplest inhomogeneous terms the algebraic calculations become formidable. For instance, in the examples above if

$$\mathbf{f}(t) = \begin{bmatrix} f_1(t) \\ f_2(t) \end{bmatrix} = \begin{bmatrix} 2t \\ \sin t \end{bmatrix},$$

the trial solution would have the form

$$\mathbf{y}_p(t) = \begin{bmatrix} y_1(t) \\ y_2(t) \end{bmatrix} = \begin{bmatrix} at + b + c \cos t + d \sin t \\ et + f + g \cos t + h \sin t \end{bmatrix}.$$

After substitution into the differential equation and equating constant terms and the coefficients of t, $\sin t$, and $\cos t$, respectively, one would have eight linear equations in the eight unknowns a, b, c, d, e, f, g, and h.

Also note that if $f(t)$ contains a term $e^{\lambda t}$ and λ is an eigenvalue of A, then this is the case of resonance and the trial solutions will contain terms of the form $(at + b)e^{\lambda t}$ in each component. This again will increase the level of algebraic difficulties, so we must conclude that the method of "judicious guessing" should be used in only the simplest cases. Therefore a general technique must be developed; such a technique will depend on knowing $e^{tA} = \mathbf{\Phi}(t)$, the fundamental matrix of the homogeneous system.

First observe that e^{tA} satisfies the following property.

For any two real numbers t and s,

$$e^{tA}e^{sA} = e^{sA}e^{tA} = e^{(t+s)A}. \qquad \textbf{(5.5.2)}$$

The above relation can be proved by substituting the series representations for e^{tA} and e^{sA} and comparing like powers of A. It is not a completely trivial assertion since, in general, if A and B are $n \times n$ matrices,

$$e^Ae^B \neq e^Be^A \neq e^{A+B},$$

in contrast to the scalar case. From relation (5.5.2) it follows by letting $s = -t$ that

$$e^{tA}e^{-tA} = e^{-tA}e^{tA} = e^{(t-t)A} = e^{0A} = I,$$

or equivalently

$$\mathbf{\Phi}(t)\,\mathbf{\Phi}(-t) = \mathbf{\Phi}(-t)\,\mathbf{\Phi}(t) = I.$$

In the terminology of linear algebra, these last relations state that $e^{-tA} = \mathbf{\Phi}(-t)$ is the inverse matrix of $e^{tA} = \mathbf{\Phi}(t)$.

EXAMPLE 3 For the given A, find $\Phi(t)$ and verify that $\Phi(t)\,\Phi(-t) = I$.

a) If

$$A = \begin{bmatrix} 0 & 1 \\ -1 & 0 \end{bmatrix},$$

then

$$e^{tA} = \Phi(t) = \begin{bmatrix} \cos t & \sin t \\ -\sin t & \cos t \end{bmatrix},$$

and therefore

$$\Phi(-t) = \begin{bmatrix} \cos(-t) & \sin(-t) \\ -\sin(-t) & \cos(-t) \end{bmatrix} = \begin{bmatrix} \cos t & -\sin t \\ \sin t & \cos t \end{bmatrix}.$$

Thus

$$\begin{aligned} \Phi(t)\,\Phi(-t) &= \begin{bmatrix} \cos^2 t + \sin^2 t & -\cos t \sin t + \sin t \cos t \\ -\sin t \cos t + \cos t \sin t & \sin^2 t + \cos^2 t \end{bmatrix} \\ &= \begin{bmatrix} 1 & 0 \\ 0 & 1 \end{bmatrix} = I. \end{aligned}$$

b) If

$$A = \begin{bmatrix} 2 & 0 & 0 \\ 0 & 1 & 0 \\ 0 & 0 & -3 \end{bmatrix},$$

then

$$e^{tA} = \Phi(t) = \begin{bmatrix} e^{2t} & 0 & 0 \\ 0 & e^{t} & 0 \\ 0 & 0 & e^{-3t} \end{bmatrix}$$

and therefore

$$\Phi(-t) = \begin{bmatrix} e^{-2t} & 0 & 0 \\ 0 & e^{-t} & 0 \\ 0 & 0 & e^{3t} \end{bmatrix}.$$

Then

$$\Phi(t)\,\Phi(-t) = \begin{bmatrix} e^{2t}e^{-2t} & 0 & 0 \\ 0 & e^{t}e^{-t} & 0 \\ 0 & 0 & e^{-3t}e^{3t} \end{bmatrix} = \begin{bmatrix} 1 & 0 & 0 \\ 0 & 1 & 0 \\ 0 & 0 & 1 \end{bmatrix} = I. \qquad \blacksquare$$

The fact that the inverse of the matrix $\Phi(t)$ can be obtained by merely changing the sign of t is an extremely convenient property, as anyone who has computed the inverse of a square matrix will know!

We are now ready to develop the *variation of parameters* formula for finding a particular solution of the initial value problem

$$\mathbf{y}' = A\mathbf{y} + \mathbf{f}(t), \qquad \mathbf{y}(t_0) = \mathbf{y}^0. \tag{5.5.3}$$

From the previous discussion we know that any solution of the homogeneous equation $\mathbf{y}' = A\mathbf{y}$ can be written as $\mathbf{y}(t) = e^{tA}\mathbf{c} = \Phi(t)\mathbf{c}$, where $\mathbf{c}$ is a constant vector. The *basic assumption* of the method of variation of parameters is that a particular solution of the inhomogeneous equation can be written as

$$\mathbf{y}_p(t) = e^{tA}\mathbf{c}(t), \tag{5.5.4}$$

where $\mathbf{c}(t)$ is a *time-dependent vector.*

To see where the assumption leads, substitute the above expression for $\mathbf{y}_p(t)$ into (5.5.3) to obtain

$$\mathbf{y}_p'(t) = Ae^{tA}\mathbf{c}(t) + e^{tA}\mathbf{c}'(t) = A\mathbf{y}_p(t) + \mathbf{f}(t) = Ae^{tA}\mathbf{c}(t) + \mathbf{f}(t).$$

Cancelling the term $Ae^{tA}\mathbf{c}(t)$ and multiplying both sides of the resulting equation by e^{-tA} gives

$$e^{-tA}[e^{tA}\mathbf{c}'(t)] = e^{-tA}e^{tA}\mathbf{c}'(t) = I\mathbf{c}'(t) = \mathbf{c}'(t) = e^{-tA}\mathbf{f}(t).$$

This is the simplest of all first order differential equations and it can be integrated to find $\mathbf{c}(t)$:

$$\mathbf{c}(t) = \int_{t_0}^{t} e^{-sA}\mathbf{f}(s)\, ds.$$

Note that the lower limit of integration is chosen conveniently so that $\mathbf{c}(t_0) = \mathbf{0}$, although this is not necessary; any antiderivative of $\mathbf{c}'(t)$ would suffice.

Finally, if the expression for $\mathbf{c}(t)$ is substituted into (5.5.4), we have

$$\mathbf{y}_p(t) = e^{tA}\int_{t_0}^{t} e^{-sA}\mathbf{f}(s)\, ds. \tag{5.5.5}$$

It can be verified directly that $\mathbf{y}_p(t)$ is the solution of the initial value problem

$$\mathbf{y}'(t) = A\mathbf{y} + \mathbf{f}(t), \qquad \mathbf{y}(t_0) = \mathbf{0}.$$

It is obvious that $\mathbf{y}_p(t_0) = \mathbf{0}$. Differentiating (5.5.5) gives

$$\mathbf{y}_p'(t) = Ae^{tA}\int_{t_0}^{t} e^{-sA}\mathbf{f}(s)\, ds + e^{tA}e^{-tA}\mathbf{f}(t) = A\mathbf{y}_p(t) + \mathbf{f}(t),$$

and hence the assertion is verified and $\mathbf{y}_p(t)$ is a particular solution.

The solution of the original initial value problem (5.5.3) will be the sum of a particular solution and a solution of the homogeneous equation. Since $\mathbf{y}_p(t_0) = \mathbf{0}$ and a solution of $\mathbf{y}' = A\mathbf{y}$, $\mathbf{y}(t_0) = \mathbf{y}^0$ is $\mathbf{y}(t) = e^{(t-t_0)A}\,\mathbf{y}^0 = \mathbf{\Phi}(t - t_0)\mathbf{y}^0$, one can now state the following algorithm.

Variation of parameters algorithm

The solution of the inhomogeneous initial value problem

$$\mathbf{y}' = A\mathbf{y} + \mathbf{f}(t), \qquad \mathbf{y}(t_0) = \mathbf{y}^0$$

can be found as follows:

1. Find the fundamental matrix $\mathbf{\Phi}(t) = e^{tA}$ of the homogeneous system $\mathbf{y}' = A\mathbf{y}$ by either the eigenvalue–eigenvector method or by using the Laplace transform.
2. Compute, if possible, the integral

$$\int_{t_0}^{t} e^{-sA}\mathbf{f}(s)\,ds = \int_{t_0}^{t} \mathbf{\Phi}(-s)\,\mathbf{f}(s)\,ds.$$

3. A particular solution $\mathbf{y}_p(t)$ satisfying $\mathbf{y}_p(t_0) = 0$ is given by the expression

$$\mathbf{y}_p(t) = e^{tA}\int_{t_0}^{t} e^{-sA}\,\mathbf{f}(s)\,ds = \mathbf{\Phi}(t)\int_{t_0}^{t} \mathbf{\Phi}(-s)\,\mathbf{f}(s)\,ds.$$

4. The solution of the initial value problem is

$$\mathbf{y}(t) = e^{(t-t_0)A}\,\mathbf{y}^0 + \mathbf{y}_p(t) = \mathbf{\Phi}(t - t_0)\mathbf{y}^0 + \mathbf{y}_p(t),$$

where the first term is the solution of the homogeneous equation satisfying the given initial conditions.

From our previous discussion it follows that equivalent expressions for the solution are

$$\mathbf{y}(t) = e^{(t-t_0)A}\mathbf{y}^0 + \int_{t_0}^{t} e^{(t-s)A}\mathbf{f}(s)\,ds$$

or

$$\mathbf{y}(t) = \mathbf{\Phi}(t - t_0)\mathbf{y}^0 + \int_{t_0}^{t} \mathbf{\Phi}(t - s)\,\mathbf{f}(s)\,ds.$$

The last expression contains a convolution integral whose significance will be pointed out shortly.

EXAMPLE 4 Solve the initial value problem

$$y_1' = 2y_1 + t, \qquad y_1(0) = 2,$$
$$y_2' = -3y_2 + \sin t, \qquad y_2(0) = -1.$$

SOLUTION The example can be solved directly, but its simplicity makes it easier to show how to use the variation of parameters formula. Here $t_0 = 0$,

$$A = \begin{bmatrix} 2 & 0 \\ 0 & -3 \end{bmatrix}, \qquad \mathbf{f}(t) = \begin{bmatrix} t \\ \sin t \end{bmatrix}, \qquad \mathbf{y}^0 = \begin{bmatrix} 2 \\ -1 \end{bmatrix},$$

and therefore

$$e^{tA} = \begin{bmatrix} e^{2t} & 0 \\ 0 & e^{-3t} \end{bmatrix} \quad \text{and} \quad e^{-sA} = \begin{bmatrix} e^{-2s} & 0 \\ 0 & e^{3s} \end{bmatrix}.$$

Hence

$$\int_0^t e^{-sA}\mathbf{f}(s)\,ds = \int_0^t \begin{bmatrix} e^{-2s} & 0 \\ 0 & e^{3s} \end{bmatrix} \begin{bmatrix} s \\ \sin s \end{bmatrix} ds$$

$$= \int_0^t \begin{bmatrix} se^{-2s} \\ e^{3s}\sin s \end{bmatrix} ds = \begin{bmatrix} -\frac{1}{2}te^{-2t} - \frac{1}{4}e^{-2t} + \frac{1}{4} \\ \frac{3}{10}e^{3t}\sin t - \frac{1}{10}e^{3t}\cos t + \frac{1}{10} \end{bmatrix},$$

where the integration of a vector is accomplished by integrating each component. A particular solution is therefore

$$\mathbf{y}_p(t) = e^{tA}\int_0^t e^{-sA}\mathbf{f}(s) = \begin{bmatrix} y_{1p}(t) \\ y_{2p}(t) \end{bmatrix}$$

$$= \begin{bmatrix} e^{2t} & 0 \\ 0 & e^{-3t} \end{bmatrix} \begin{bmatrix} -\frac{1}{2}te^{-2t} - \frac{1}{4}\,3_{10}\,e^{-2t} + \frac{1}{4} \\ \frac{3}{10}e^{3t}\,\frac{3}{10}\sin t - \frac{1}{10}e^{3t}\cos t + \frac{1}{10} \end{bmatrix}$$

$$= \begin{bmatrix} -\frac{1}{2}t - \frac{1}{4} + \frac{1}{4}e^{2t} \\ \frac{3}{10}\sin t - \frac{1}{10}\cos t + \frac{1}{10}e^{-3t} \end{bmatrix}.$$

Since

$$e^{tA}\mathbf{y}^0 = \begin{bmatrix} e^{2t} & 0 \\ 0 & e^{-3t} \end{bmatrix} \begin{bmatrix} 2 \\ -1 \end{bmatrix} = \begin{bmatrix} 2e^{2t} \\ -e^{-3t} \end{bmatrix},$$

this vector is added to the previous one to get the required solution

$$\mathbf{y}(t) = \begin{bmatrix} y_1(t) \\ y_2(t) \end{bmatrix} = \begin{bmatrix} \frac{9}{4}e^{2t} - \frac{1}{2}t - \frac{1}{4} \\ -\frac{9}{10}e^{-3t} + \frac{3}{10}\sin t - \frac{1}{10}\cos t \end{bmatrix}. \qquad \blacksquare$$

EXAMPLE 5 Find a particular solution of

$$y_1' = 2y_1 - y_2 + 1,$$
$$y_2' = 4y_1 - 2y_2 + t^{-1}.$$

SOLUTION This example is an excellent one for the method of variation of parameters since, because of the presence of the term t^{-1}, neither judicious guessing nor the use of the Laplace transform (to be discussed shortly) will work. The first task is to find e^{tA}. Since

$$A = \begin{bmatrix} 2 & -1 \\ 4 & -2 \end{bmatrix}$$

has a double eigenvalue $\lambda = 0$, a solution of the homogeneous problem is of the form

$$y(t) = \begin{bmatrix} y_1(t) \\ y_2(t) \end{bmatrix} = \begin{bmatrix} a \\ b \end{bmatrix} + t\begin{bmatrix} c \\ d \end{bmatrix} = \begin{bmatrix} a + ct \\ b + dt \end{bmatrix}.$$

The first column $\mathbf{\Phi}^1(t)$ of e^{tA} is found by solving

$$y_1' = 2y_1 - y_2, \qquad y_1(0) = 1,$$
$$y_2' = 4y_1 - 2y_2, \qquad y_2(0) = 0.$$

The initial condition implies that $a = 1$, $b = 0$, and substituting $y_1(t) = 1 + ct$, $y_2(t) = dt$ into the differential equations gives $d = 2c$ and $c = 2$. Therefore

$$\mathbf{\Phi}^1(t) = \begin{bmatrix} 1 + 2t \\ 4t \end{bmatrix}.$$

Now, changing the initial conditions to

$$y_1(0) = 0, \qquad y_2(0) = 1,$$

implies that $a = 0$, $b = 1$, and substitution gives $d = 2c$ and $c = -1$. Therefore the second column of e^{tA} is

$$\mathbf{\Phi}^2(t) = \begin{bmatrix} -t \\ 1 - 2t \end{bmatrix}$$

and

$$e^{tA} = \begin{bmatrix} 1 + 2t & -t \\ 4t & 1 - 2t \end{bmatrix}, \qquad e^{-sA} = \begin{bmatrix} 1 - 2s & s \\ -4s & 1 + 2s \end{bmatrix}.$$

Since $\mathbf{f}(t) = \begin{bmatrix} 1 \\ t^{-1} \end{bmatrix}$, any value for t_0 may be selected except $t_0 = 0$ because of the presence of t^{-1}; for convenience we choose $t_0 = 1$. Hence

$$\int_1^t e^{-sA}\mathbf{f}(s)\, ds = \int_1^t \begin{bmatrix} 1 - 2s & s \\ -4s & 1 + 2s \end{bmatrix} \begin{bmatrix} 1 \\ s^{-1} \end{bmatrix} ds$$

$$= \int_1^t \begin{bmatrix} 2 - 2s \\ -4s + s^{-1} + 2 \end{bmatrix} ds = \begin{bmatrix} 2t - t^2 - 1 \\ -2t^2 + \ln t + 2t \end{bmatrix},$$

and a particular solution is

$$\mathbf{y}_p(t) = e^{tA} \int_1^t e^{-sA}\mathbf{f}(s)\, ds = \begin{bmatrix} y_{1p}(t) \\ y_{2p}(t) \end{bmatrix}$$

$$= \begin{bmatrix} 1 + 2t & -t \\ 4t & 1 - 2t \end{bmatrix} \begin{bmatrix} 2t - t^2 - 1 \\ -2t^2 + \ln t + 2t \end{bmatrix}$$

$$= \begin{bmatrix} -1 + t^2 - t \ln t \\ -2t + 2t^2 + (1 - 2t) \ln t \end{bmatrix}.$$

A sigh of relief is not uncommon after one has solved a problem by the method of variation of parameters! ■

In the last example it was mentioned that one could also use the Laplace transform to solve an inhomogeneous system when the inhomogeneous term $\mathbf{f}(t)$ is Laplace transformable. It is an extremely effective method and avoids in part the lengthy integrations of the variation of parameters method. The price to be paid is the sometimes tedious algebraic manipulations and partial fractions expansions. We illustrate the method with an example.

EXAMPLE 6 Using Laplace transforms solve the initial value problem

$$y_1' = -4y_1 - y_2 + e^{-t}, \qquad y_1(0) = 1$$

$$y_2' = y_1 - 2y_2 + \sin 2t, \qquad y_2(0) = 2.$$

SOLUTION Let $Y_1(s)$ and $Y_2(s)$ be the Laplace transforms of $y_1(t)$ and $y_2(t)$. Apply the transform to each equation to obtain, after using the differentiation formula,

$$sY_1(s) - 1 = -4Y_1(s) - Y_2(s) + \frac{1}{s + 1},$$

$$sY_2(s) - 2 = Y_1(s) - 2Y_2(s) + \frac{2}{s^2 + 4},$$

or, equivalently,

$$(s+4)Y_1(s)+Y_2(s)=1+\frac{1}{s+1}$$

$$-Y_1(s)+(s+2)Y_2(s)=2+\frac{2}{s^2+4}.$$

Solving this system of equations for $Y_1(s)$ and $Y_2(s)$ gives

$$Y_1(s)=\frac{s}{(s+3)^2}+\frac{s+2}{(s+3)^2(s+1)}-\frac{2}{(s+3)^2(s^2+4)},$$

$$Y_2(s)=\frac{2s+9}{(s+3)^2}+\frac{1}{(s+3)^2(s+1)}+\frac{2s+8}{(s+3)^2(s^2+4)}.$$

Each of the above terms must be expanded in a partial fraction expansion by using Heaviside's formulas or otherwise. After a reasonable amount of time and scratch paper, one obtains

$$Y_1(s)=\frac{459}{676}\cdot\frac{1}{s+3}-\frac{69}{26}\cdot\frac{1}{(s+3)^2}+\frac{1}{4}\cdot\frac{1}{s+1}+\frac{12}{169}\cdot\frac{s}{s^2+4}-\frac{5}{169}\cdot\frac{2}{s^2+4};$$

$$Y_2(s)=\frac{1335}{676}\cdot\frac{1}{s+3}+\frac{69}{26}\cdot\frac{1}{(s+3)^2}+\frac{1}{4}\cdot\frac{1}{s+1}-\frac{38}{169}\cdot\frac{s}{s^2+4}+\frac{44}{169}\cdot\frac{2}{s^2+4}.$$

Applying the inverse transform gives the desired answer

$$y_1(t)=\frac{459}{676}e^{-3t}-\frac{69}{26}te^{-3t}+\frac{1}{4}e^{-t}+\frac{12}{169}\cos 2t-\frac{5}{169}\sin 2t,$$

$$y_2(t)=\frac{1335}{676}e^{-3t}+\frac{69}{26}te^{-3t}+\frac{1}{4}e^{-t}-\frac{38}{169}\cos 2t+\frac{44}{169}\sin 2t.$$ ■

The partial fractions expansions and arithmetic in the last example were lengthy, but using the variation of parameters formula one would get the following expression for a particular solution:

$$\begin{bmatrix} e^{-3t}-te^{-3t} & -te^{-3t} \\ te^{-3t} & e^{-3t}+te^{-3t}\end{bmatrix}\int_0^t \begin{bmatrix} e^{2s}+se^{2s}+se^{3s}\sin 2s \\ -se^{2s}+(e^{3s}-se^{3s})\sin 2s\end{bmatrix} ds.$$

Computationally, the choice of method is a toss-up with the transform method possibly having a slight edge. Note the special advantage in using the Laplace transform of not having to find the fundamental matrix e^{tA}.

In the discussion preceding the examples we gave the following alternative expression for a particular solution when $t_0 = 0$:

$$\mathbf{y}_p(t) = \int_0^t \Phi(t - s)\, \mathbf{f}(s)\, ds,$$

where $\Phi(t) = e^{tA}$ is the fundamental matrix associated with the matrix A. Observe that $\mathbf{y}_p(t)$ satisfies the initial condition $\mathbf{y}_p(0) = \mathbf{0}$ and can be written simply as the convolution integral

$$\mathbf{y}_p(t) = \Phi * \mathbf{f}.$$

To conclude this section we will derive this expression by using transform techniques.

First of all, recall that the fundamental matrix $\Phi(t)$ satisfies the matrix differential equation and initial condition

$$\Phi' = A\Phi, \qquad \Phi(0) = I.$$

If $\hat{\Phi}(s)$ is the Laplace transform of $\Phi(t)$, the differentiation formula of the Laplace transform can be applied to obtain

$$s\hat{\Phi}(s) - I = A\hat{\Phi}(s). \tag{5.5.6}$$

Note that the Laplace transform of the matrix $\Phi(t)$ will just be the matrix whose components are the transforms of the individual entries of $\Phi(t)$. For example: If

$$A = \begin{bmatrix} -3 & 0 \\ 0 & 2 \end{bmatrix},$$

then

$$\Phi(t) = \begin{bmatrix} e^{-3t} & 0 \\ 0 & e^{2t} \end{bmatrix},$$

and hence

$$\hat{\Phi}(s) = \begin{bmatrix} \dfrac{1}{s+3} & 0 \\ 0 & \dfrac{1}{s-2} \end{bmatrix}$$

and

$$s\hat{\Phi}(s) - I = \begin{bmatrix} \dfrac{s}{s+3} & 0 \\ 0 & \dfrac{s}{s-2} \end{bmatrix} - \begin{bmatrix} 1 & 0 \\ 0 & 1 \end{bmatrix}$$

$$= \begin{bmatrix} \dfrac{-3}{s+3} & 0 \\ 0 & \dfrac{2}{s-2} \end{bmatrix}$$

$$= \begin{bmatrix} -3 & 0 \\ 0 & 2 \end{bmatrix} \begin{bmatrix} \dfrac{1}{s+3} & 0 \\ 0 & \dfrac{1}{s-2} \end{bmatrix} = A\hat{\Phi}(s).$$

Returning to Eq. (5.5.6), we see that it can be rewritten as $(sI - A)\hat{\Phi}(s) = I$, which implies that

$$\hat{\Phi}(s) = (sI - A)^{-1}. \tag{5.5.7}$$

This says that $\hat{\Phi}(s)$, the Laplace transform of $\Phi(t)$, is the inverse matrix of the matrix $sI - A$. In the above example, for instance,

$$sI - A = s\begin{bmatrix} 1 & 0 \\ 0 & 1 \end{bmatrix} - \begin{bmatrix} -3 & 0 \\ 0 & 2 \end{bmatrix} = \begin{bmatrix} s+3 & 0 \\ 0 & s-2 \end{bmatrix}$$

and

$$(sI - A)^{-1} = \begin{bmatrix} \dfrac{1}{s+3} & 0 \\ 0 & \dfrac{1}{s-2} \end{bmatrix} = \hat{\Phi}(s).$$

The relation (5.5.7) is what is needed to find an expression for $\mathbf{y}_p(t)$, the particular solution.

Recall that $\mathbf{y}_p(t)$ satisfies the initial value problem

$$\mathbf{y}_p' = A\mathbf{y}_p + \mathbf{f}(t), \qquad \mathbf{y}_p(0) = \mathbf{0},$$

so if $\mathbf{f}(t)$ is Laplace transformable with Laplace transform $\mathbf{F}(s)$, then

$$s\mathbf{Y}_p(s) - \mathbf{0} = A\mathbf{Y}_p(s) + \mathbf{F}(s),$$

where $\mathbf{Y}_p(s)$ is the transform of $\mathbf{y}_p(t)$. The last relation implies that

$$(sI - A)\mathbf{Y}_p(s) = \mathbf{F}(s)$$

or

$$\mathbf{Y}_p(s) = (sI - A)^{-1}\mathbf{F}(s).$$

But, by using the relation (5.5.7), we see that this is equivalent to the expression

$$\mathbf{Y}_p(s) = \hat{\mathbf{\Phi}}(s)\,\mathbf{F}(s).$$

The convolution property states that the inverse transform of the product of two transforms is the convolution integral of their inverse transforms; hence

$$\mathbf{y}_p(t) = \mathbf{\Phi} * \mathbf{f} = \int_0^t \mathbf{\Phi}(t - s)\,\mathbf{f}(s)\,ds.$$

This is the result we wanted.

In the language of systems analysis given in Chapter 4, the last two relations state respectively that for the system at rest

$$\mathbf{y}' = A\mathbf{y} + \mathbf{f}(t), \qquad \mathbf{y}(0) = \mathbf{0},$$

the following is true.

In the s-domain, $\hat{\mathbf{\Phi}}(s) = (sI - A)^{-1}$ is the transfer function and $\mathbf{Y}_p(s)$ is the output corresponding to input $\mathbf{F}(s)$.

In the t- domain, $\mathbf{\Phi}(t) = e^{tA} = L^{-1}\{\hat{\mathbf{\Phi}}(s)\}$ is the weighting function and $\mathbf{y}_p(t)$ is the zero state output corresponding to the input $\mathbf{f}(t)$.

This can be represented diagrammatically as shown in Fig. 5.8, where the first box represents multiplication of the input by the transfer function and the second box represents the convolution of the input with the weighting function.

Note that the above analysis has given us another way to compute the fundamental matrix $\mathbf{\Phi}(t)$. Namely, compute

$$(sI - A)^{-1} = L\{\mathbf{\Phi}(t)\} = \hat{\mathbf{\Phi}}(s)$$

and then compute

$$\mathbf{\Phi}(t) = L^{-1}\{\hat{\mathbf{\Phi}}(s)\}.$$

Aside from the algebraic problems in computing the inverse transform, the drawback is in the computing of the matrix inverse $(sI - A)^{-1}$. For dimensions $n = 2$ or 3 the method can be very useful.

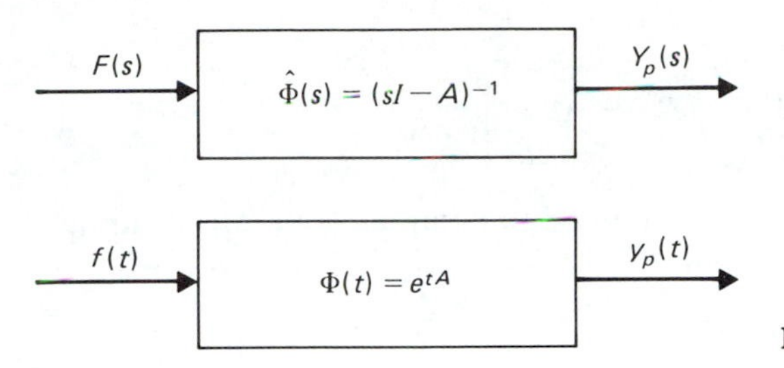

Figure 5.8

EXAMPLE 7 In Example 5,

$$A = \begin{bmatrix} 2 & -1 \\ 4 & -2 \end{bmatrix};$$

hence

$$(sI - A) = \begin{bmatrix} s-2 & 1 \\ -4 & s+2 \end{bmatrix},$$

and

$$\hat{\mathbf{\Phi}}(s) = (sI - A)^{-1} = \frac{1}{s^2}\begin{bmatrix} s+2 & -1 \\ 4 & s-2 \end{bmatrix} = \begin{bmatrix} \dfrac{1}{s} + \dfrac{2}{s^2} & -\dfrac{1}{s^2} \\ \dfrac{4}{s^2} & \dfrac{1}{s} - \dfrac{2}{s^2} \end{bmatrix}.$$

Consequently,

$$\mathbf{\Phi}(t) = L^{-1}\{\hat{\mathbf{\Phi}}(s)\} = \begin{bmatrix} 1+2t & -t \\ 4t & 1-2t \end{bmatrix}.$$ ■

EXAMPLE 8 If $A = \begin{bmatrix} 3 & -4 \\ 1 & -1 \end{bmatrix}$, then $(sI - A) = \begin{bmatrix} s-3 & 4 \\ -1 & s+1 \end{bmatrix}$ and

$$(sI - A)^{-1} = \frac{1}{(s-1)^2}\begin{bmatrix} s+1 & -4 \\ 1 & s-3 \end{bmatrix}$$

$$= \begin{bmatrix} \dfrac{1}{s-1} + \dfrac{2}{(s-1)^2} & \dfrac{-4}{(s-1)^2} \\ \dfrac{1}{(s-1)^2} & \dfrac{1}{s-1} - \dfrac{2}{(s-1)^2} \end{bmatrix}.$$

Hence

$$\Phi(t) = \begin{bmatrix} e^t + 2te^t & -4te^t \\ te^t & e^t - 2te^t \end{bmatrix}. \qquad \blacksquare$$

As the examples show, the method avoids the need to compute solutions of initial value problems to find each column of $\Phi(t)$.

EXERCISES 5.5

Solve the following initial value problems by using the method of comparison of coefficients. The fundamental matrices have been found in Exercises 5.4.

1. $\mathbf{y}' = \begin{bmatrix} 0 & 1 \\ 8 & -2 \end{bmatrix} \mathbf{y} + \begin{bmatrix} 3 \\ 1 \end{bmatrix}, \quad \mathbf{y}(0) = \begin{bmatrix} 1 \\ 10 \end{bmatrix}$

2. $\mathbf{y}' = \begin{bmatrix} -1 & -5 \\ 1 & 1 \end{bmatrix} \mathbf{y} + \begin{bmatrix} -2t \\ 1 \end{bmatrix}, \quad \mathbf{y}(0) = \begin{bmatrix} 1 \\ -1 \end{bmatrix}$

3. $\mathbf{y}' = \begin{bmatrix} 2 & 1 \\ 0 & 2 \end{bmatrix} \mathbf{y} + \begin{bmatrix} 2 \\ 3e^{-t} \end{bmatrix}, \quad \mathbf{y}(0) = \begin{bmatrix} \frac{1}{2} \\ 0 \end{bmatrix}$

4. $\mathbf{y}' = \begin{bmatrix} 2 & 3 \\ 7 & -2 \end{bmatrix} \mathbf{y} + \begin{bmatrix} 0 \\ 3 \sin t \end{bmatrix}, \quad \mathbf{y}(0) = \begin{bmatrix} 0 \\ 0 \end{bmatrix}$

(*Hint:* If you are familiar with Gaussian elimination it is an excellent way to solve the resulting algebraic system of four equations in four unknowns.)

5. $\mathbf{y}' = \begin{bmatrix} 1 & -1 \\ -3 & 3 \end{bmatrix} \mathbf{y} + \begin{bmatrix} e^{2t} \\ 0 \end{bmatrix}, \quad \mathbf{y}(0) = \begin{bmatrix} 0 \\ 1 \end{bmatrix}$

6. $\mathbf{y}' = \begin{bmatrix} -6 & -2 \\ 17 & 0 \end{bmatrix} \mathbf{y} + \begin{bmatrix} 0 \\ 75 \sin t \end{bmatrix}, \quad \mathbf{y}(0) = \begin{bmatrix} 0 \\ 0 \end{bmatrix}$

Solve the following initial value problems by using the variation of parameters method. The fundamental matrices have been found in Exercises 5.4.

7. $\mathbf{y}' = \begin{bmatrix} 2 & 1 \\ 3 & 4 \end{bmatrix} \mathbf{y} + \begin{bmatrix} 2e^{2t} \\ -2e^{2t} \end{bmatrix}, \quad \mathbf{y}(0) = \begin{bmatrix} 3 \\ -4 \end{bmatrix}$

8. $\mathbf{y}' = \begin{bmatrix} 1 & 3 \\ 12 & 1 \end{bmatrix} \mathbf{y} + \begin{bmatrix} 3e^{7t} \\ 4 \end{bmatrix}, \quad y(0) = \begin{bmatrix} 1 \\ 2 \end{bmatrix}$

9. $\mathbf{y}' = \begin{bmatrix} -1 & -5 \\ 1 & 3 \end{bmatrix} \mathbf{y} + \begin{bmatrix} 3 \\ 5 \cos t \end{bmatrix}, \quad \mathbf{y}(\pi) = \begin{bmatrix} 0 \\ 0 \end{bmatrix}$

10. $\mathbf{y}' = \begin{bmatrix} 3 & 1 \\ -2 & 1 \end{bmatrix} \mathbf{y} + \begin{bmatrix} e^{2t} \\ -e^{2t} \end{bmatrix}, \quad \mathbf{y}(0) = \begin{bmatrix} 2 \\ -1 \end{bmatrix}$

11. $\mathbf{y}' = \begin{bmatrix} 5 & -1 \\ 1 & 3 \end{bmatrix} \mathbf{y} + \begin{bmatrix} 2e^{-t} \\ e^{t} \end{bmatrix}, \quad \mathbf{y}(2) = \begin{bmatrix} -3 \\ 2 \end{bmatrix}$

12. $\mathbf{y}' = \begin{bmatrix} -4 & -1 \\ 1 & -2 \end{bmatrix} \mathbf{y} + \begin{bmatrix} e^{2t} \\ 4e^{3t} \end{bmatrix}, \quad \mathbf{y}(-2) = \begin{bmatrix} 3 \\ 4 \end{bmatrix}$

Find the transfer function (the Laplace transform $\hat{\mathbf{\Phi}}(s)$ of the fundamental matrix $\mathbf{\Phi}(t)$) by computing $(sI - A)^{-1}$ directly for the following matrices A. Then find $\mathbf{\Phi}(t)$ by using the inverse Laplace transform.

13. $A = \begin{bmatrix} 3 & 0 \\ 0 & 1 \end{bmatrix}$

14. $A = \begin{bmatrix} 2 & 1 \\ 4 & 2 \end{bmatrix}$

15. $A = \begin{bmatrix} 1 & 5 \\ -2 & -1 \end{bmatrix}$

16. $A = \begin{bmatrix} 1 & 2 \\ -4 & -3 \end{bmatrix}$

17. $A = \begin{bmatrix} 1 & 0 & 0 \\ 0 & 2 & 1 \\ 0 & 0 & 2 \end{bmatrix}$

18. $A = \begin{bmatrix} 3 & 1 & 0 \\ 0 & 3 & 1 \\ 0 & 0 & 3 \end{bmatrix}$

19. The configuration in Fig. 5.5 of Exercise 36, Section 5.4, is changed so that tank A has water containing $\frac{1}{10}$ lb/gal of chemical X flowing into the tank at a rate of 2 gal/min. Tank B has liquid flowing out at a rate of 2 gal/min. If $x_0 = 20$ lb, $y_0 = 5$ lb, find $x(t)$ and $y(t)$, the amounts of chemical X at time t in tanks A and B, respectively. What is the steady-state solution?

20. The configuration in Fig. 5.6 of Exercise 37, Section 5.4, is changed so that the water flowing into tank A contains $\frac{1}{10}e^{-t}$ lb/gal of chemical X. Find $x(t)$ and $y(t)$, the amounts of chemical X at time t in tanks A and B, respectively.

21. The network in Fig. 5.7 of Exercise 39, Section 5.4, is changed so that the voltage is 40 sin t volts. If the remaining values of R_1, R_2, C, and L are as given in part (b) of Exercise 39, Section 5.4, find the general solution and the steady-state solution.

SUMMARY

The simplest example of a linear system of ordinary differential equations is

$$\frac{dx}{dt} = ax + by + f(t), \qquad \frac{dy}{dt} = cx + dy + g(t),$$

whose solution is a pair of functions $(x(t), y(t))$. By using the notation of vectors and matrices one can write the system as

$$\frac{d\mathbf{x}}{dt} = A\mathbf{x} + \mathbf{f}(t),$$

where

$$\mathbf{x} = \begin{bmatrix} x \\ y \end{bmatrix}, \quad A = \begin{bmatrix} a & b \\ c & d \end{bmatrix}, \quad \mathbf{f}(t) = \begin{bmatrix} f(t) \\ g(t) \end{bmatrix}.$$

Mass–spring systems involving two or more masses, chemical reactions involving two or more compounds, coupled *LRC* circuits, and multiple-tank mixing problems can be analyzed using linear systems of differential equations.

The chapter begins with a brief introduction to vectors and matrices, then studies the nature of the space of solutions of the linear homogeneous system ($\mathbf{f}(t) \equiv \mathbf{0}$). As in the case of the second order linear differential equation, one must develop the concept of a linearly independent set of solutions that will be used to describe the general solution of the system. Analogously this reduces to an algebraic problem associated with the analysis of a polynomial associated with the matrix of coefficients A, and the various cases must be examined.

For the inhomogeneous system, where $\mathbf{f}(t)$ is present, a general expression for the solution is developed from the general solution of the homogeneous system. As in the second order case, this is called the variation of parameters algorithm, and it depends on knowing the fundamental matrix e^{tA} of the homogeneous system $d\mathbf{x}/dt = A\mathbf{x}$. This is the matrix analogue of e^{at}, the fundamental solution of the scalar linear differential equation $dx/dt = ax$. Properties of e^{tA} and techniques for finding it are given.

MISCELLANEOUS EXERCISES

Mass–spring systems without damping, such as those described in Example 3 of Section 5.1, can be expressed by systems of second order differential equations

$$\ddot{\mathbf{y}} + A\mathbf{y} = \mathbf{f}(t),$$

where $\mathbf{y} = \operatorname{col}(y_1, \ldots, y_n)$ describes the configuration of the system, $\mathbf{f}(t) = \operatorname{col}(f_1(t), \ldots, f_n(t))$ is the vector of forcing functions, A is an $n \times n$ matrix, and n is the number of degrees of freedom.

For the homogeneous equation $\mathbf{f}(t) = \mathbf{0}$, setting $\mathbf{y}(t) = e^{i\lambda t}\,\mathbf{b}$ leads to the algebraic system

$$(-\lambda^2 I + A)\mathbf{b} = 0,$$

which λ and $\mathbf{b}$ must satisfy. Hence $\mathbf{b}$ is a real eigenvector of A corresponding to the real eigenvalue λ^2, and the condition $\det(-\lambda^2 I + A) = 0$ must be satisfied to have nontrivial solutions. Each value λ^2 that satisfies the last equation will determine two real solutions: $(\cos \lambda t)\mathbf{b}$ and $(\sin \lambda t)\mathbf{b}$, and the general solution will be a linear combination of all such solutions.

5.1.

a) Letting $\omega^2 = k/m$, show that for the system described in Example 3 of Section 5.1

$$A = \begin{bmatrix} 2\omega^2 & -\omega^2 \\ -\omega^2 & 2\omega^2 \end{bmatrix}.$$

b) Show that the eigenvalues of A are $\lambda_1 = \omega^2$ and $\lambda_2 = 3\omega^2$ with respective eigenvalues $\mathbf{b} = \text{col}(1,1)$ and $\mathbf{b} = \text{col}(1,-1)$. Write down the general solution of the system.

c) Solve the initial value problems:
i) $\mathbf{y}(0) = \text{col}(a, a)$, $\dot{\mathbf{y}}(0) = \text{col}(0, 0)$
ii) $\mathbf{y}(0) = \text{col}(a, -a)$, $\dot{\mathbf{y}}(0) = \text{col}(0, 0)$

5.2. Modify Example 3 of Section 5.1 so that the spring constants are k_1, k_2, k_3, and write down the corresponding second order system. Let $k_1 = k_3$, $\omega^2 = k_1/m$, $\eta^2 = k_3/m$, $\sigma^2 = \omega^2 + 2\eta^2$; find the equations for the eigenvalues and eigenvectors of the second order system, and write down the general solution.

The Laplace transform can be used to find a particular solution of the nonhomogeneous second order system

$$\ddot{\mathbf{y}} + A\mathbf{y} = \mathbf{f}(t).$$

If $\mathbf{Y}(s)$ and $\mathbf{F}(s)$ are the transforms of $\mathbf{y}(t)$ and $\mathbf{f}(t)$, respectively, and the initial conditions are chosen as $\mathbf{y}(0) = \dot{\mathbf{y}}(0) = \mathbf{0}$, then the application of the transform gives

$$s^2\mathbf{Y}(s) + A\mathbf{Y}(s) = \mathbf{F}(s).$$

Therefore

$$\mathbf{Y}(s) = (s^2 I + A)^{-1}\mathbf{F}(s) = \hat{U}(s)\mathbf{F}(s),$$

and if $U(t) = L^{-1}\{\hat{U}(s)\}$, then by the convolution formula a particular solution is

$$\mathbf{y}_p(t) = \int_0^t U(t - \tau)f(\tau)\, d\tau.$$

5.3. For the second part of Problem 5.2, find the matrix $U(t)$ and use it to find a particular solution when $\mathbf{f}(t) = \text{col}(\sin \omega t, \sin \sigma t)$.

The differential equation describing a forced mass–spring system

$$m_1\ddot{y} + k_1 y = F \sin \omega t$$

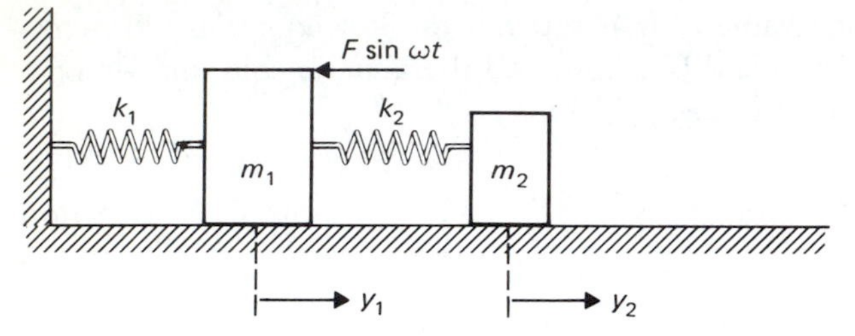

Figure 5.9

is in resonance if $\omega^2 = k_1/m_1$. The resonance frequency can be shifted by adding a small mass–spring system as a dynamic damper. This system is shown in Fig. 5.9, and the differential equations describing it are

$$m_1\ddot{y}_1 + k_1 y_1 - k_2(y_2 - y_1) = F \sin \omega t,$$

$$m_2\ddot{y}_2 + k_2(y_2 - y_1) = 0.$$

5.4.

a) Letting $\mathbf{y} = \text{col}(y_1, y_2)$ and $\alpha^2 = k_1/m_1$, $\beta^2 = k_2/m_2$, $\gamma^2 = k_2/m_1$, write the system in the form $\ddot{\mathbf{y}} + A\mathbf{y} = \mathbf{f}(t)$.

b) A particular solution can be found by letting $\mathbf{y}_p(t) = \mathbf{g} \sin \omega t$, substituting it into the system, and finding $\mathbf{g}$. Show that if β^2 is chosen to be equal to ω^2, the first component of the particular solution is zero.

c) The result in part (b) shows that the resonant oscillation is transferred to the dynamic damper by choosing $\alpha^2 = \beta^2$, that is, $k_1/m_1 = k_2/m_2$. This is the defining relation for a *tuned dynamic damper.* Find the eigenvalues λ^2 of A in this case.

d) Let $\omega^2 = \alpha^2 = \beta^2 = 2$, $\gamma^2 = 1$, and $F/m_1 = 4$. First find a particular solution $\mathbf{y}_p(t) = \mathbf{g} \sin \sqrt{2}t$, then find the general solution. Use it to solve the initial value problem $\mathbf{y}(0) = \dot{\mathbf{y}}(0) = \mathbf{0}$.

In the circuit of Fig. 5.10, the currents I_1 and I_2 are coupled by the mutual inductance of the two coils. The system of differential equations describing the system are derived by applying Kirchhoff's voltage law to the two loops:

$$L_1 \frac{dI_1}{dt} - M \frac{dI_2}{dt} + R_1 I_1 = E(t),$$

$$-M \frac{dI_1}{dt} + L_2 \frac{dI_2}{dt} + R_2 I_2 = 0.$$

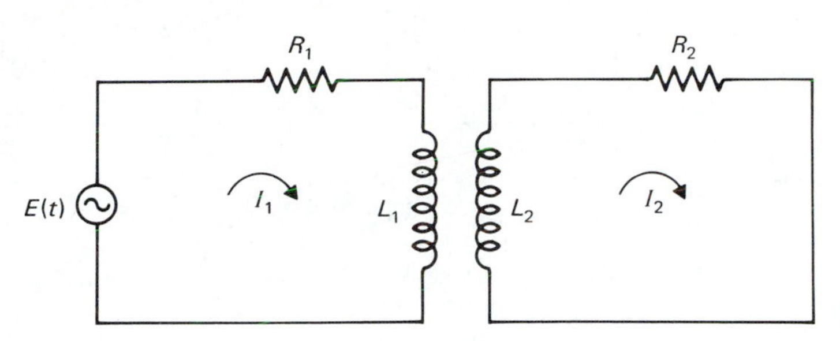

Figure 5.10

5.5.

a) Assume that $M^2 \neq L_1L_2$, and by solving the above system for dI_1/dt and dI_2/dt write the system in the normal form

$$\frac{d\mathbf{I}}{dt} = A\mathbf{I} + \mathbf{F}(t),$$

where $\mathbf{I} = \text{col}(I_1, I_2)$. Find the characteristic equation of the matrix A.

b) Set $R_1 = R_2 = 15$ ohms, $L_1 = L_2 = 4$ henrys, $M = 1$ henry, and $E(t) = 4 \sin 2t$ volts. Given the initial conditions $I_1(0) = I_2(0) = 0$ amperes, solve the initial value problem by using Laplace transforms.

c) An ideal transformer is described by the equation $M^2 = L_1L_2$. Let $R_1 = R_2 = 1$ ohm, $L_1 = L_2 = M = 1$ henry, and $E(t) = 4 \sin t$ volts, and find the general solution of the system.

5.6. Time-varying resistors are employed, for instance, in circuits modeling carbon microphones. In the circuit described in Exercise 5.5, let $R_1 = 15(1 + p \sin \omega t)$ ohms, $|p| < 1$, $R_2 = 15$ ohms, $L_1 = L_2 = 4$ henrys, $M = 1$ henry, $E(t) = 150$ volts. Assume that the initial conditions are $I_1(0) = I_2(0) = 0$ amperes.

a) Write a computer program (or adapt the model program in Chapter 3) using RK4 to compute an approximate solution on $0 \leq t \leq T$. Note that the system must be written in normal form.

b) Let $p = \omega = 0$ and generate an approximate solution on $0 \leq t \leq 4$ using a step size $h = 0.1$. Plot $I_1(t)$ and $I_2(t)$ versus t and $I_1(t)$ versus $I_2(t)$ on separate graphs.

c) Let $p = 0.5$, $\omega = 2$ and generate an approximate solution on $0 \leq t \leq 10$ using a step size $h = 0.1$. Plot $I_1(t)$ and $I_2(t)$ versus t and $I_1(t)$ versus $I_2(t)$ on separate graphs. The varying resistance *modulates* the constant ($p = 0$) currents.

5.7. Two tanks with inlet and outlet pipes are connected together as shown in Fig. 5.11. Each tank contains V gallons of brine and the flow rates have been selected to keep the volumes constant. At the start of the process, the brine in tank A con-

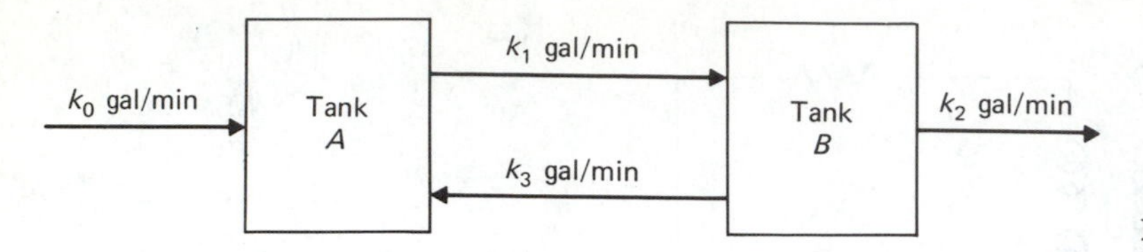

Figure 5.11

tains Q_1 pounds of salt and that in tank B contains Q_2 pounds of salt. The concentration of the incoming mixture is $C_0(t)$ pounds of salt per gallon. Let C_1 and C_2 denote the concentrations (in pounds per gallon) of salt in tanks A and B, respectively. Instantaneous mixing is assumed.

a) Show that the k_i's must satisfy the relations $k_0 = k_2$, $k_1 = k_0 + k_3$, $k_1 = k_2 + k_3$, and that C_1 and C_2 satisfy the initial value problem:

$$\frac{d}{dt}(VC_1) = -k_1C_1 + (k_1 - k_2)C_2 + k_2C_0(t), \qquad C_1(0) = Q_1/V,$$

$$\frac{d}{dt}(VC_2) = k_1C_1 - k_1C_2, \qquad C_2(0) = Q_2/V.$$

(The reader may wish to review the previous discussion of tank problems in Chapters 1 and 4.) Subsequent eigenvalue computations can be simplified by introducing a new independent variable $\tau = t/V$. This leads to the system:

$$\frac{dC_1}{d\tau} = -k_1C_1 + (k_1 - k_2)C_2 + k_2C_0(V\tau), \qquad C_1(0) = Q_1/V,$$

$$\frac{dC_2}{d\tau} = k_1C_1 - k_1C_2, \qquad C_2(0) = Q_2/V.$$

Let $V = 300$, $k_1 = 9$, $k_2 = 4$.

b) Find the fundamental matrix for the corresponding homogeneous system.

c) Write down the variation of parameters formula for a general $C_0(t)$ and general initial conditions.

d) Let $C_0(t) = 3e^{-t/100}$, $Q_1 = 300$ lb, $Q_2 = 1200$ lb, and solve the initial value problem using the Laplace transform, using the method of judicious guessing, and evaluating the integral in (c).

5.8. Three tanks are connected in a closed system as shown in Fig. 5.12. Each tank contains V gallons of brine, and the flow rates have been selected to keep the volumes constant. Let C_i denote the concentration (in pounds per gallon) of salt in the ith tank, $i = 1, 2, 3$.

a) Derive the system of differential equations describing the system.

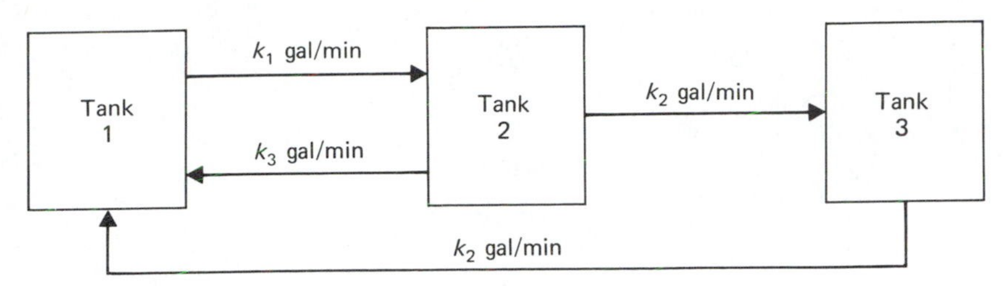

Figure 5.12

b) Show that $C_1 = C_2 = C_3 = a > 0$ is a solution for any positive number a, and therefore $\lambda = 0$ is an eigenvalue of the coefficient matrix.

c) Let $k_2 = 1$, $k_3 = b \geq 0$, and $k_1 = 1 + b$. Derive formulas for the nonzero eigenvalues and characterize them (real distinct, real equal, complex) as functions of the parameter b.

d) Show that $C_1(t) + C_2(t) + C_3(t) =$ constant for all t, and use this constraint to derive a two-dimensional system for $C_2(t)$ and $C_3(t)$.

e) Find the general solutions with

$$C_1(0) + C_2(0) + C_3(0) = 6$$

when

i) $k_1 = 2$, $k_2 = k_3 = 1$

ii) $k_1 = 3/2$, $k_2 = 1$, $k_3 = 1/2$

iii) $k_1 = 1 + \sqrt{3}/2$, $k_2 = 1$, $k_3 = \sqrt{3}/2$.

f) If one of the pumps that runs at a constant rate is replaced by a variable speed pump, then it is not possible to solve the systems of differential equations exactly. Set $C_1(0) = 5$, $C_2(0) = C_3(0) = 0$, $k_1 = 2$ gal/min, $k_3 = 1$ gal/min, and $k_2 = 1 + 0.5 \cos 2t$ gal/min. Construct an approximate solution using RK4 with $h = 0.1$. Plot C_1 versus t. A computer-generated plot is given in Fig. 5.13. Compare your result to the solution corresponding to $k_2 = 1$ gal/min with the other data unchanged. Does a pulsating pump produce better mixing?

5.9. The biological processes of living organisms can often be modeled by splitting them into one or more distinct stages, or *compartments*. Each compartment is characterized by its constituent material or by its spatial assignment. It is assumed that the material in each compartment is homogeneous and that each compartment has a steady-state flux of material into and out of it.

A compartment system is usually studied experimentally by injecting a pulse or tracer of labeled material into one or more of the compartments. This introduces a transient into the system, which can be used to study its steady-state properties. To accomplish this, the interactions between individual compartments must be determined experimentally.

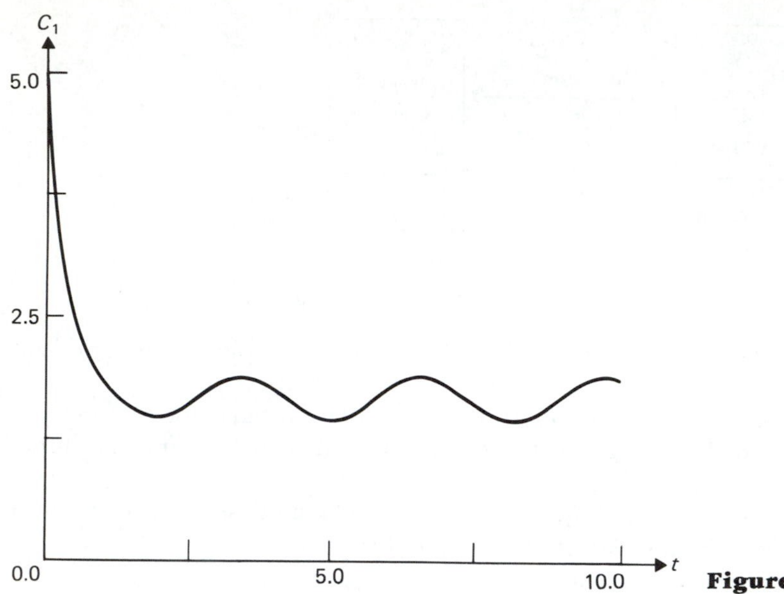

Figure 5.13 A plot of $C_1(t)$.

As an example of this approach, suppose one wanted to analyze the elimination or clearance of creatinine in a dog. Creatinine is a nitrogenous waste product and is a fairly constant component of urine. The compartment system of Fig. 5.14 models this process and the experiment designed to study it: Here l_{21} is the fraction per unit time of creatinine in the plasma compartment transferred to the tissue compartment; l_{12} is the fraction per unit time of creatinine in the tissue compartment transferred to the plasma compartment; l_{01} is the fraction per unit time of creatinine eliminated from the system, and ρ is a bolus input (e.g., a simple injection into a vein) of creatinine introduced at time $t = 0$. It is assumed that the level of creatinine in the plasma at $t = 0$ is null.

If x_1 is the concentration of creatinine in the plasma and x_2 is its concentration in the tissue, the compartment model is described by the following system:

$$\frac{dx_1}{dt} = -(l_{01} + l_{21})x_1 + l_{12}x_2, \qquad x_1(0) = \rho,$$

$$\frac{dx_2}{dt} = l_{21}x_1 - l_{12}x_2, \qquad x_2(0) = 0.$$

To determine the rates l_{ij}, an experiment was performed in which 0.6 g/l of creatinine were injected into the veins of ten dogs, and blood samples were drawn and analyzed at specified times. The constants were found to be

$$l_{01} = 0.0395 \text{ min}^{-1}, \qquad l_{12} = 0.0479 \text{ min}^{-1}, \qquad l_{21} = 0.0391 \text{ min}^{-1}.$$

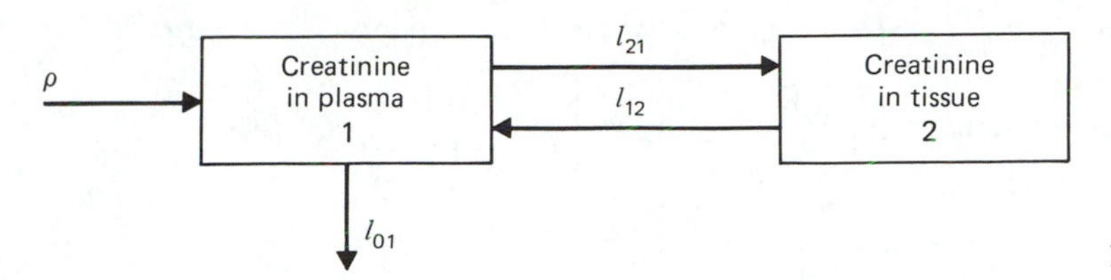

Figure 5.14

a) Find the fundamental solution matrix of the system.
b) Evaluate the solution at enough points to enable you to sketch a graph of the concentration of creatinine in the plasma and in the tissue for 10 minutes. (For a discussion of this experiment and compartment models see S. I. Rubinow, *Introduction to Mathematical Biology,* Wiley, New York, 1975, Chap. 3.)

5.10. Published data on the disappearance of connecting peptide, denoted by *C*-peptide, in the plasma of humans indicate that its decay can be modeled by a two-compartment system consisting of an intravascular and an extravascular pool, as shown in Fig. 5.15.

a) Denoting the concentration of intravascular *C*-peptide by x_1 and that of extravascular *C*-peptide by x_2, write down the system of differential equations describing the model.
b) Find the fundamental matrix of the system.
c) Assuming $x_1(0) = 4.6$ pmol/ml, $x_2(0) = 0$, evaluate the solution at enough points to plot $x_1(t)$ for $0 \leq t \leq 60$ min.

(See R. P. Eaton, R. C. Allen, D. S. Schade, K. M. Erickson, and J. Standefer, *Prehepatic Insulin Production in Man: Kinetic Analysis Using Peripheral Connecting Peptide Behavior,* Jour. of Endocrinology and Metabolism, 51 (1980), pp. 520–528.)

5.11. Show that $\mathbf{y}(t) = \Sigma_{n=0}^{\infty} t^{2n+1}\mathbf{a}^{(2n)}$ is a solution to

$$\frac{d\mathbf{y}}{dt} = \frac{1}{t}\begin{bmatrix} t^2 & 1 \\ 1 & -t^2 \end{bmatrix}\mathbf{y}, \quad (t > 0),$$

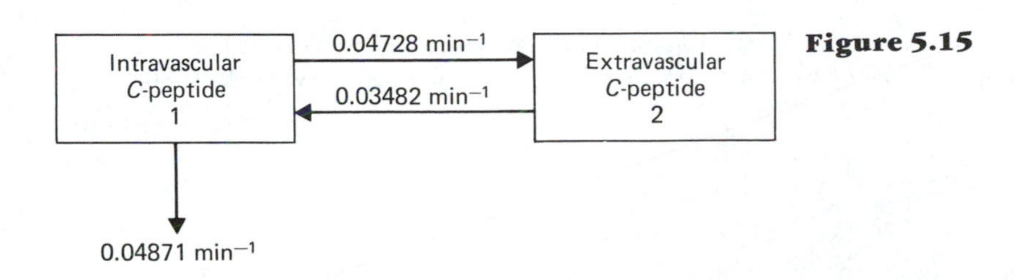

Figure 5.15

where $\mathbf{a}^{(0)} = A_0\mathbf{a}^{(0)}$, $[(2n+1)I - A_0]\mathbf{a}^{(2n)} = A_2\mathbf{a}^{(2n-2)}$, $n = 1, 2, 3, \ldots$, and

$$A_0 = \begin{bmatrix} 0 & 1 \\ 1 & 0 \end{bmatrix}, \qquad A_2 = \begin{bmatrix} 1 & 0 \\ 0 & -1 \end{bmatrix}.$$

Then show that

$$\mathbf{a}^{(2n)} = \frac{1}{4n(n+1)} \begin{bmatrix} (2n+1) & 1 \\ 1 & -(2n+1) \end{bmatrix} \mathbf{a}^{(2n-2)}, \qquad n = 1, 2, 3, \ldots.$$

5.12. Write a computer program to compute the vector

$$\mathbf{y}^{(M)}(t) = \sum_{n=0}^{M} t^{2n+1}\mathbf{a}^{(2n)},$$

where the $\mathbf{a}^{2n}$ are defined in Exercise 5.11. Take $\mathbf{a}^{(0)} = \text{col}(1, 1)$, $t = 2$, and find an M, so that convergence to four significant figures *seems* to be obtained.

5.13. Integrate with RK4:

$$\frac{d\mathbf{y}}{dt} = \frac{1}{t}\begin{bmatrix} t^2 & 1 \\ 1 & -t^2 \end{bmatrix}\mathbf{y}$$

on $h \leq t \leq 2$ with

a) $\mathbf{y}(0) = \begin{bmatrix} h \\ h \end{bmatrix}$, $h = 0.1$, and $h = 0.2$;

b) $\mathbf{y}(0) = \begin{bmatrix} h \\ h \end{bmatrix} + \frac{1}{4}\begin{bmatrix} h^3 \\ -h^3 \end{bmatrix}$, $h = 0.1$, and $h = 0.2$.

c) Compare your answers with the result of Exercise 5.12. Which initial condition more closely matches the series solution? Why?

REFERENCES

1. E. Coddington and N. Levinson, *Theory of Ordinary Differential Equations,* McGraw-Hill, New York, 1975.
2. B. Noble, J.W. Daniel, *Applied Linear Algebra,* Prentice-Hall, Englewood Cliffs, N.J., 1988.
3. P.C. Shields, *Elementary Linear Algebra,* Worth Publishers, New York, 1980.
4. G. Strang, *Linear Algebra and Its Applications,* Academic Press, New York, 1980.

6

NONCONSTANT COEFFICIENT SECOND ORDER LINEAR EQUATIONS AND SERIES SOLUTIONS

6.1 INTRODUCTION

In Chapter 2, the general theory for the initial value problem

$$a(t)y'' + b(t)y' + c(t)y = 0,$$

$$y(t_0) = r, \qquad y'(t_0) = s,$$

was discussed. There it was assumed that $a(t)$, $b(t)$, and $c(t)$ were continuous functions on some open interval containing t_0 and that $a(t)$ did not vanish at t_0. The theory stated that the solution $y(t)$ could be expressed uniquely as

$$y(t) = c_1y_1(t) + c_2y_2(t),$$

where $y_1(t)$ and $y_2(t)$ were a linearly independent pair of solutions of the differential equation, and the constants c_1 and c_2 depended on the initial values r and s. Furthermore, the existence of such a pair of linearly independent solutions $y_1(t)$ and $y_2(t)$ was also guaranteed by the theory.

Much of the remainder of Chapter 2 was devoted to a discussion of the constant coefficient case

$$ay'' + by' + cy = 0.$$

In this case finding a pair of linearly independent solutions depended solely on solving the characteristic equation

$$a\lambda^2 + b\lambda + c = 0.$$

The type of solutions depended on the nature of the roots of the characteristic equation (real and unequal, real and equal, or complex) but the solutions could be *explicitly found.*

The reader may well ask whether this simplicity of the solution procedure carries over in some analogous fashion to the nonconstant coefficient case. The answer is a definite NO! With the exception of the special case of the Euler differential equation,

$$t^2y'' + bty' + cy = 0$$

(see Example 2 and Exercise 14 of Section 2.6), there are no general techniques to reduce the solving of nonconstant coefficient linear differential equations to an algebraic process.

However, if we are willing to extend our notion of the solution of a differential equation to allow for the solution to be expressed as an infinite series, then there is a solution procedure that is applicable to a large class of nonconstant coefficient linear differential equations. The infinite series will be a power series in the independent variable, possibly multiplied by a known function, and the coefficients of the power series can be determined recursively. Given the generality of the problem, one could hardly wish for a more satisfactory outcome.

For instance, a solution of

$$y'' + \frac{1}{t}y' + y = 0$$

is given by the power series

$$y(t) = 1 - \frac{t^2}{2^2} + \frac{t^4}{2^4(2!)^2} - \frac{t^6}{2^6(3!)^2} + \cdots = \sum_{n=0}^{\infty} (-1)^n \frac{t^{2n}}{2^{2n}(n!)^2},$$

as we shall see later in this chapter. The function given by the power series, denoted by $J_0(t)$, is an oscillatory function with an infinite number of zeros on the positive axis, and there are tables and numerical procedures to estimate it for any value of its argument. This is not such a marked contrast from the familiar function $\cos t$, which is a solution of $y'' + y = 0$ and which can also be represented by a power series.

The function $J_0(t)$, called the *Bessel function of the first kind of order zero,* occurs in a number of problems in heat conduction in solids and vibration of membranes. It is an example of what are called *special functions,* many of which arise as series solutions of second order linear differential equations. A few of these special functions will be discussed at the end of this chapter—they are useful tools for the kitbag of any applied scientist.

6.2 SERIES SOLUTIONS—PART 1

In the remainder of this chapter we will denote the independent variable by x rather than t. Therefore the second order linear equation will be written as

$$a(x)y'' + b(x)y' + c(x)y = 0,$$

with solutions $y = y(x)$. The use of x rather than t is somewhat traditional, since many of the problems associated with series solutions or special functions arise from physical problems where x is a spatial rather than a temporal variable.

To motivate the use of infinite series, we start with a simple example that can be solved explicitly.

EXAMPLE 1 Consider the first order differential equation

$$y' = 2xy,$$

whose general solution is $y(x) = Ae^{x^2}$, where A is an arbitrary constant. Assume that the solution can be represented by a convergent power series; hence

$$y(x) = \sum_{n=0}^{\infty} a_n x^n$$

and the task is to find the a_n.

SOLUTION Since a convergent power series can be differentiated term by term and also multiplied by $2x$, the series can be substituted into the differential equation to obtain

$$y'(x) = \sum_{n=0}^{\infty} na_n x^{n-1} = 2xy(x) = \sum_{n=0}^{\infty} 2a_n x^{n+1}.$$

Expanding both sides gives

$$[0 \cdot a_0 + 1 \cdot a_1 + 2a_2x + 3a_3x^2 + 4a_4x^3 + 5a_5x^4 + 6a_6x^5 + \cdots]$$
$$= [2a_0x + 2a_1x^2 + 2a_2x^3 + 2a_3x^4 + 2a_4x^5 + 2a_5x^6 + \cdots],$$

and by comparing the coefficients of like powers of x we have

$$0 \cdot a_0 = 0, \quad 1 \cdot a_1 = 0, \quad 2a_2 = 2a_0, \quad 3a_3 = 2a_1,$$
$$4a_4 = 2a_2, \quad 5a_5 = 2a_3, \quad 6a_6 = 2a_4, \quad \ldots .$$

This implies that a_0 is arbitrary and

$$a_1 = 0, \quad a_2 = a_0, \quad a_3 = \frac{2}{3}a_1 = 0, \quad a_4 = \frac{a_2}{2} = \frac{a_0}{2},$$
$$a_5 = \frac{2}{5}a_3 = 0, \quad a_6 = \frac{a_4}{3} = \frac{a_0}{3 \cdot 2}, \quad \ldots ;$$

hence

$$y(x) = a_0 \left[1 + x^2 + \frac{x^4}{2!} + \frac{x^6}{3!} + \cdots \right],$$

which is the first four terms of the Taylor series for $a_0 e^{x^2}$.

The reader can easily see that the general recursion relation for the a_n is $na_n = 2a_{n-2}$, $n = 2, 3, \ldots$. Since a_0 is arbitrary and a_1 is zero, this implies that

$$a_{2n+1} = 0, \qquad n = 0, 1, 2, \ldots,$$

$$a_{2n} = \frac{2a_{2n-2}}{2n} = \frac{1}{n} \cdot \frac{2a_{2n-4}}{2n-2} = \frac{1}{n(n-1)} \cdot \frac{2a_{2n-6}}{2n-4} = \cdots = \frac{1}{n!} a_0.$$

Therefore

$$y(x) = \sum_{n=0}^{\infty} \frac{a_0}{n!} x^{2n} = a_0 \sum_{n=0}^{\infty} \frac{x^{2n}}{n!} = a_0 e^{x^2},$$

and the power series solution is exactly the Taylor series of the general solution. ∎

We now return to the second order equation

$$y'' + p(x)y' + q(x)y = 0 \tag{6.2.1}$$

and consider first the simplest case, in which the coefficient functions $p(x)$ and $q(x)$ are smooth, well-behaved functions in some neighborhood of a given point, which is assumed to be $x = 0$ for simplicity of notation. Consequently, we will assume that $p(x)$ and $q(x)$ have infinite power series representations

$$p(x) = \sum_{n=0}^{\infty} p_n x^n, \qquad q(x) = \sum_{n=0}^{\infty} q_n x^n,$$

where these series converge in some neighborhood of $x = 0$. This leads to the following definition.

Definition

If the coefficients $p(x)$ and $q(x)$ of equation (6.2.1) can be represented by power series that are convergent in some neighborhood of $x = 0$, then $x = 0$ is said to be an *ordinary point* of the differential equation.

With the above assumptions it is natural to suppose that the solution $y(x)$ will also have a power series representation convergent in some neighborhood of $x = 0$, and therefore

$$y(x) = \sum_{n=0}^{\infty} a_n x^n. \tag{6.2.2}$$

Since a power series can be differentiated term by term, the solution procedure is to substitute the series for $y(x)$ into (6.2.1), with $p(x)$ and $q(x)$ replaced by their power series representation, and compare coefficients of like powers of x. This will lead to a recursion relation by which the a_n's can be determined step by step, and possibly a general expression can be derived that will give all the a_n's. (In the case where the ordinary point is $x_0 \neq 0$, one would use instead $y(x) = \sum_{n=0}^{\infty} a_n(x - x_0)^n$, and the algebraic procedure is the same.) We summarize the above discussion.

Solution procedure for ordinary point case

1. Replace the coefficients $p(x)$ and $q(x)$ in the differential equation (6.2.1) by their power series representations if necessary.
2. Differentiate the series (6.2.2) successively term by term and substitute the series for $y(x)$, $y'(x)$ and $y''(x)$ into the differential equation.
3. Combine terms and compare coefficients of like powers of x. This will lead to a recursion relation by which the coefficients a_n can be determined step by step. Possibly a general recursion relation can be obtained by which all the a_n's can be determined.

This procedure is best illustrated by an example.

EXAMPLE 2 Find the series solution of *Airy's equation*

$$y'' - xy = 0. \tag{6.2.3}$$

SOLUTION Since $p(x) = 0$ and $q(x) = -x$ the point $x = 0$ is an ordinary point. From (6.2.2) we obtain the relations

$$y'(x) = \sum_{n=1}^{\infty} na_n x^{n-1}, \qquad y''(x) = \sum_{n=2}^{\infty} n(n-1)a_n x^{n-2}$$

through term-by-term differentiation. After multiplying the series for $y(x)$ by x, substitution into (6.2.3) gives

$$\sum_{n=2}^{\infty} n(n-1)a_n x^{n-2} - \sum_{n=0}^{\infty} a_n x^{n+1} = 0. \tag{6.2.4}$$

If just the first few terms of the solution are desired, one can expand the above sums to get

$$(2a_2 + 6a_3x + 12a_4x^2 + 20a_5x^3 + 30a_6x^4 + 42a_7x^5 + \cdots) \\ - (a_0x + a_1x^2 + a_2x^3 + a_3x^4 + a_4x^5 + \cdots) = 0.$$

Now, combining the coefficients of like powers of x and setting them equal to zero gives the relations

$$2a_2 = 0, \quad 6a_3 - a_0 = 0, \quad 12a_4 - a_1 = 0, \quad 20a_5 - a_2 = 0,$$
$$30a_6 - a_3 = 0, \quad 42a_7 - a_4 = 0, \quad \ldots,$$

and consequently

$$a_2 = 0, \quad a_3 = \frac{a_0}{6}, \quad a_4 = \frac{a_1}{12}, \quad a_5 = \frac{a_2}{20} = 0,$$
$$a_6 = \frac{a_3}{30} = \frac{a_0}{180}, \quad a_7 = \frac{a_4}{42} = \frac{a_1}{504}, \ldots .$$

Therefore

$$\begin{aligned} y(x) &= a_0 + a_1 x + 0 + \frac{a_0}{6}x^3 + \frac{a_1}{12}x^4 + 0 + \frac{a_0}{180}x^6 + \frac{a_1}{504}x^7 + \cdots \\ &= a_0\left[1 + \frac{1}{6}x^3 + \frac{1}{180}x^6 + \cdots\right] + a_1\left[x + \frac{1}{12}x^4 + \frac{1}{504}x^7 + \cdots\right], \end{aligned}$$

where a_0 and a_1 are arbitrary constants. ■

Examining the last expression we see that it can be written in the form

$$y(x) = a_0 y_1(x) + a_1 y_2(x),$$

where $y_1(x)$ will have a power series representation in $3n$th powers of x, where $n = 0, 1, 2, \ldots$, and $y_2(x)$ in $(3n + 1)$st powers of x, where $n = 0, 1, 2, \ldots$. Clearly, the two expressions are linearly independent and they represent two linearly independent solutions of Airy's equation. The standard notation for them is

$$\mathrm{Ai}(x) = 1 + \frac{1}{6}x^3 + \frac{1}{180}x^6 + \cdots,$$
$$\mathrm{Bi}(x) = x + \frac{1}{12}x^4 + \frac{1}{504}x^7 + \cdots;$$

Ai(x) and Bi(x) are called the Airy functions, and the reader is referred to [10], where they are extensively tabulated and many of their properties are given. The two arbitrary constants a_0 and a_1 can be determined only if initial or boundary conditions are given.

If a general expression for the series representations of Ai(x) and Bi(x) is desired, we must return to relation (6.2.4) and try to derive a general recursion relation from it. The procedure used is to shift indices in the two series until they can be combined. Note that the first series leads off with x^0, whereas the second leads off with x, so write

$$2a_2 + \sum_{n=3}^{\infty} n(n-1)a_n x^{n-2} - \sum_{n=0}^{\infty} a_n x^{n+1} = 0.$$

Now both series lead off with x, but the first starts with $n = 3$, whereas the second starts with $n = 0$, so reindex the second as

$$2a_2 + \sum_{n=3}^{\infty} n(n-1)a_n x^{n-2} - \sum_{n=3}^{\infty} a_{n-3} x^{n-2} = 0.$$

Both sums can now be combined to obtain

$$2a_2 + \sum_{n=3}^{\infty} [n(n-1)a_n - a_{n-3}]x^{n-2} = 0.$$

From the above follow the relations

$$a_2 = 0, \qquad a_n = \frac{a_{n-3}}{n(n-1)}, \quad n = 3, 4, 5, \ldots, \tag{6.2.5}$$

which clearly imply that

$a_2, a_5, a_8, \ldots, a_{3n+2}, \ldots$ are all zero,

$a_3, a_6, a_9, \ldots, a_{3n}, \ldots$ all depend on a_0,

$a_4, a_7, a_{10}, \ldots, a_{3n+1}, \ldots$ all depend on a_1.

The relations (6.2.5) are the desired general recursion relations.

The matter of reindexing the series, so as to be able to combine terms, is somewhat arbitrary. For instance, in the expression

$$2a_2 + \sum_{n=3}^{\infty} n(n-1)a_n x^{n-2} - \sum_{n=0}^{\infty} a_n x^{n+1} = 0$$

one could reindex the first series instead, so that it starts with $n = 0$. This would give

$$2a_2 + \sum_{n=0}^{\infty} (n+3)(n+2)a_{n+3} x^{n+1} - \sum_{n=0}^{\infty} a_n x^{n+1} = 0,$$

and the recursion relation would be

$$a_2 = 0, \qquad a_{n+3} = \frac{a_n}{(n+3)(n+2)}, \quad n = 0, 1, 2, \ldots.$$

It is easily seen that this would give exactly the same result.

The reader should now study the analysis below to see how to use (6.2.5) to get the general series expressions for Ai(x) and Bi(x). First,

$$a_n = \frac{a_{n-3}}{n(n-1)}$$

implies that

$$a_{3n} = \frac{a_{3n-3}}{3n(3n-1)} = \frac{a_{3(n-1)}}{3^2 n(n-\frac{1}{3})},$$

and now apply the recursion relation to a_{3n-3}:

$$a_{3n-3} = \frac{a_{3n-6}}{(3n-3)(3n-4)}$$

or

$$a_{3(n-1)} = \frac{a_{3(n-2)}}{3(n-1)(3n-4)} = \frac{a_{3(n-2)}}{3^2(n-1)(n-\frac{4}{3})}.$$

Substitute the last expression for $a_{3(n-1)}$ in the expression for a_{3n} to get

$$a_{3n} = \frac{a_{3(n-2)}}{3^4 n(n-1)(n-\frac{1}{3})(n-\frac{4}{3})}.$$

Proceeding backwards in this fashion, one finally obtains

$$a_{3n} = \frac{a_0}{3^{2n}\, n!(n-\frac{1}{3})(n-\frac{4}{3})\cdots(\frac{2}{3})}, \qquad n = 1, 2, \ldots,$$

and since a_0 is arbitrary, the choice of $a_0 = 1$ gives

$$\text{Ai}(x) = 1 + \sum_{n=1}^{\infty} \frac{1}{3^{2n}\, n!(n-\frac{1}{3})(n-\frac{4}{3})\cdots(\frac{2}{3})}\, x^{3n}.$$

The reader may wish to work out the second case starting with a_{3n+1} and letting $a_1 = 1$ to get

$$\text{Bi}(x) = x + \sum_{n=1}^{\infty} \frac{1}{3^{2n}\, n!(n+\frac{1}{3})(n-\frac{2}{3})\cdots(\frac{4}{3})}\, x^{3n+1},$$

and to show, by using the ratio test, that both series converge for $-\infty < x < \infty$.

A brief word should be said about convergence of the power series obtained by the method just described. Suppose one is given the linear differential equation

$$y'' + p(x)y' + q(x)y = f(x),$$

where $p(x)$, $q(x)$, and $f(x)$ have power series representations convergent for $|x - x_0| < r$, $r > 0$, so that x_0 is an *ordinary point*. Then the power series for the solution, obtained formally by substituting the power series

$$y(x) = \sum_{n=0}^{\infty} a_n(x - x_0)^n$$

and recursively solving for the coefficients, will also converge for $|x - x_0| < r$. This useful result, which is proved in more advanced texts, means that there is no need to test for convergence of the power series obtained for the solution. It will have the same radius of convergence as the smallest radius of convergence of the power series for the coefficients.

EXERCISES 6.2

1. Use the power series method to find the general solutions of
   a) $y'' + y = 0$;
   b) $y'' - 4y = 0$.

   Verify that you obtain the series for $\sin x$ and $\cos x$ in (a), and for e^{2x} and e^{-2x} in (b).

In Exercises 2–5, find the recurrence relation for the coefficients and the first six nonzero terms of the series solutions with the ordinary point, $x_0 = 0$.

2. $y'' - xy' - 2y = 0$
3. $y'' + x^2y = 0$
4. $y'' + y' + xy = 0, \quad y(0) = 1, \quad y'(0) = 0$
5. $y'' - (1 + x^2)y = 0, \quad y(0) = -2, \quad y'(0) = 2$

By expressing the coefficients in a power series, substituting $y(x) = \sum_{n=0}^{\infty} a_n x^n$, and equating like powers of x, find the first five nonzero terms of the series solutions of Exercises 6–8.

6. $y'' + 2y' + (\sin x)y = 0$
7. $y'' + e^x y = 0, \quad y(0) = 1, \quad y'(0) = -1$
8. $y'' + (\alpha + \beta \cos 2x)y = 0$ (Mathieu's equation: α, β are parameters.)

In Exercises 9–13, the ordinary point $x_0 \neq 0$ so the series solution will be of the form $y(x) = \sum_{n=0}^{\infty} a_n(x - x_0)^n$. Find its first five nonzero terms.

9. $y'' + (x - 1)y' + 2y = 0$ with the ordinary point $x_0 = 1$.
10. $y'' + (x^2 - 1)y = 0$ with the ordinary point $x_0 = 1$.
    (*Hint:* $x^2 - 1 = 2(x - 1) + (x - 1)^2$.)
11. $y'' + xy' + (\ln x)y = 0$ with the ordinary point $x_0 = 1$.
    (*Hint:* $x = 1 + (x - 1)$.)

12. $y'' + (\ln x)y = 0$ with the ordinary point $x_0 = 1$.

13. $y'' + (\sin x)y = 0$ with the ordinary point $x_0 = \pi/2$.

14. Series solutions can be used occasionally to find approximations to solutions of nonlinear equations. Find the first three nonzero terms of the series solutions of the following:

a) $y' = y^2 + (1 + x^2)$, $y(0) = 0$. (*Hint:* The initial conditions imply that $y(x) = a_1 x + a_2 x^2 + \cdots$, and a useful fact is

$$\left(\sum_{k=1}^{n} c_k\right)^2 = \sum_{k=1}^{n} c_k^2 + 2 \sum_{j \neq k} c_j c_k.)$$

b) $y' = 1/y + x^2$, $y(0) = 1$. (*Hint:* $y(x) = 1 + a_1 x + a_2 x^2 + \cdots$; recall that

$$(1 + u)^{-1} = 1 - u + u^2 - \cdots + (-1)^n u^n + \cdots, \quad |u| < 1.)$$

Method of Taylor Series. Given the initial value problem

$$y'' + p(x)y' + q(x)y = 0, \qquad y(0) = a, \qquad y'(0) = b,$$

where $p(x)$ and $q(x)$ are smooth functions near $x = 0$, one can find $y''(0)$ by direct substitution:

$$\begin{aligned} y''(0) &= -p(0)y'(0) - q(0)y(0) \\ &= -bp(0) - aq(0). \end{aligned}$$

Differentiating the equation and then substituting again will give $y'''(0)$. For example,

$$y'''(0) + p'(0)y'(0) + p(0)y''(0) + q'(0)y(0) + q(0)y'(0) = 0$$

implies that

$$y'''(0) = -bp'(0) - p(0)y''(0) - aq'(0) - bq(0).$$

Proceeding in this manner, one can use the derivatives obtained to develop the Taylor series of the solution:

$$y(x) = y(0) + y'(0)x + \frac{y''(0)}{2!}x^2 + \cdots + \frac{y^{(n)}(0)}{n!}x^n + \cdots.$$

In Exercises 15–20, use this method to find the first four nonzero terms of the series solutions.

15. $y'' + x^2 y = 0$, $y(0) = 0$, $y'(0) = -1$

16. Exercise 4 above

17. $y'' + (e^{2x})y' + y = 0$, $y(0) = 0$, $y'(0) = 1$

18. Exercise 5 above

19. $y'' + xy = e^{2x}, \quad y(0) = 0, \quad y'(0) = 1$

20. $y'' - xy' + y = 4 \sin x, \quad y(0) = 1, \quad y'(0) = -1$

21. Obtain the general series expression for the Airy function $\mathrm{Bi}(x)$.

6.3 SERIES SOLUTIONS—PART 2

In the previous section we considered series solutions for

$$y'' + p(x)y' + q(x)y = 0,$$

where it was assumed that $p(x)$ and $q(x)$ were smooth functions and had convergent power series expansions in some neighborhood of a given point x_0. In this case the desired series solutions were obtained by substitution and comparison of coefficients of like powers of $x - x_0$. Does this procedure work when $p(x)$ or $q(x)$ are singular at $x = x_0$? The answer is NO, except when the singularity is of a special kind, but fortunately this class of equations contains many of the equations that arise in applied mathematics and mathematical physics.

Suppose as before that the given point in question is $x_0 = 0$, and suppose further that the differential equation can be written in the form

$$y'' + \frac{p(x)}{x} y' + \frac{q(x)}{x^2} y = 0, \tag{6.3.1}$$

where $p(x)$ and $q(x)$ are smooth functions in a neighborhood of $x = 0$; hence

$$p(x) = \sum_{n=0}^{\infty} p_n x^n, \qquad q(x) = \sum_{n=0}^{\infty} q_n x^n.$$

In this case $x = 0$ is said to be a *regular singular point,* and it is this type of singularity for which the method of power series, suitably modified, also works.

Definition

If the functions $p(x)$ and $q(x)$ of equation (6.3.1) can be represented by power series convergent in some neighborhood of $x = 0$, then $x = 0$ is said to be a *regular singular point* of the differential equation.

The following are examples of differential equations with a regular singular point at $x = 0$. Note that one must sometimes modify the coefficients to get the equation into the form of (6.3.1).

1. $x^2y'' + y = 0$ or $y'' + \dfrac{1}{x^2} y = 0$ with $p(x) = 0$, $q(x) = 1$;

2. $4x^2y'' - 3xy' + 2y = 0$ or $y'' + \dfrac{-3/4}{x}y' + \dfrac{1/2}{x^2}y = 0$ with $p(x) = -\dfrac{3}{4}$, $q(x) = \dfrac{1}{2}$;
3. $xy'' + (x + 2)y' + e^xy = 0$ or $y'' + \dfrac{x + 2}{x}y' + \dfrac{xe^x}{x^2}y = 0$ with $p(x) = x + 2$, $q(x) = xe^x$;
4. $y'' + 2y' + \dfrac{4}{x}y = 0$ or $y'' + \dfrac{2x}{x}y' + \dfrac{4x}{x^2}y = 0$ with $p(x) = 2x$, $q(x) = 4x$.

THE METHOD OF FROBENIUS

The *method of Frobenius* provides a technique for finding a series solution, but the series will be of the form

$$y(x) = x^r \sum_{n=0}^{\infty} a_n x^n = \sum_{n=0}^{\infty} a_n x^{n+r}, \tag{6.3.2}$$

where r is some number *(not necessarily an integer!)* to be determined from the differential equation. This is the modification incurred by the fact that $x = 0$ is a regular singular point. Term-by-term differentiation of the expression for $y(x)$ gives

$$y'(x) = \sum_{n=0}^{\infty} (n + r)a_n x^{n+r-1}$$

and

$$y''(x) = \sum_{n=0}^{\infty} (n + r)(n + r - 1)a_n x^{n+r-2}.$$

Now write the differential equation (6.3.1) in the form

$$x^2y'' + xp(x)y' + q(x)y = 0$$

and substitute the expansions for $p(x)$, $q(x)$, $y(x)$, $y'(x)$, and $y''(x)$ to obtain

$$\sum_{n=0}^{\infty} (n + r)(n + r - 1)a_n x^{n+r} + \left(\sum_{n=0}^{\infty} p_n x^n\right) \sum_{n=0}^{\infty} (n + r)a_n x^{n+r} + \left(\sum_{n=0}^{\infty} q_n x^n\right) \sum_{n=0}^{\infty} a_n x^{n+r} = 0.$$

The term x^r can be factored out of each term, and combining the coefficients of like powers of x we obtain the expression

$$\begin{aligned} &x^r[r(r-1)a_0 + p_0ra_0 + q_0a_0] \\ &\quad + x^r[(1+r)ra_1 + p_0(1+r)a_1 + p_1ra_0 + q_0a_1 + q_1a_0]x \\ &\quad + x^r[\quad]x^2 + \cdots + x^r[\quad]x^n + \cdots = 0, \end{aligned}$$

where we have explicitly written down the coefficients of $x^0 = 1$ and x. Equating to zero the first bracketed term and assuming $a_0 \neq 0$ gives

$$r(r-1) + p_0r + q_0 = r^2 + (p_0 - 1)r + q_0 = 0. \qquad \textbf{(6.3.3)}$$

This quadratic polynomial in r is called the *indicial equation,* and its two roots, r_1 and r_2, will be the admissible values of the exponent r in the expression (6.3.2) for the solution. Note that p_0 and q_0 are merely the respective values of $p(0)$ and $q(0)$ and can be easily obtained from the differential equation.

The theory of the regular singular point case, which the reader may wish to examine in more detail in some of the references, leads to the following solution procedure.

Solution procedure for the regular singular point case

1. Find $p_0 = p(0)$; and $q_0 = q(0)$, then find the roots r_1 and r_2 of the indicial equation (6.3.3).
2. Choose as r_1 that root of the indicial equation for which the real part of $r_1 - r_2$ is nonnegative; if r_1 and r_2 are real and unequal then r_1 would be the largest root, for instance. Then for $r = r_1$, let the solution be represented by the series

$$y_1(x) = x^{r_1} \sum_{n=0}^{\infty} a_n x^n.$$

3. Replace $p(x)$ and $q(x)$ by their power series representations, substitute $y_1(x)$ into the differential equation and proceed as in the ordinary point case. One can always obtain recursively all the coefficients a_n and the series obtained will converge in some deleted neighborhood of $x = 0$. By a deleted neighborhood we mean the set of all x satisfying $0 < |x| < \alpha$ for some $\alpha > 0$. In some cases the series may also converge at $x = 0$.
4. If r_2 is the second root and $r_1 - r_2$ is not zero or a positive integer, then one can proceed as in (1). The series

$$y_2(x) = x^{r_2} \sum_{n=0}^{\infty} b_n x^n$$

with the b_n's determined recursively converges in some deleted neighborhood of $x = 0$ and is a solution. The two solutions, $y_1(x)$ and $y_2(x)$, are linearly independent.

5. If $r_1 - r_2$ equals zero or a positive integer, then there are two possibilities: Either a solution of the form $y_2(x)$ above exists or the second solution is of the form

$$y_2(x) = y_1(x)\,\beta \ln x + x^{r_2} \sum_{n=0}^{\infty} b_n x^n,$$

with β a constant dependent on $y_1(x)$ and $p(x)$.

In the case where the regular singular point $x_0 \neq 0$ one would use $p_0 = p(x_0)$ and $q_0 = q(x_0)$ in the indicial equation. The solution corresponding to the root r_1 would then be of the form

$$y_1(x) = (x - x_0)^{r_1} \sum_{n=0}^{\infty} a_n\,(x - x_0)$$

and the solution procedure above applies. (See Exercise 8 of Section 6.3.)

The possibility (5) above is the so-called *logarithmic case,* and if $r_1 - r_2$ equals zero, it always occurs since the indicial equation has only one root. If $r_1 - r_2$ is a positive integer, there is a technique using $p(x)$ and $y_1(x)$ to determine whether the logarithmic case occurs; it uses the reduction of order formula to be discussed shortly, and the interested reader is referred to [2] or [3] for details.

The problem with the case where the real part of $r_1 - r_2$ is a positive integer is that if one substitutes the series $y_2(x) = x^{r_2} \Sigma_{n=0}^{\infty} b_n x^n$ into the differential equation, then one cannot find a recursion relation for all the b_n. Specifically, for some positive integer m one can no longer solve for b_m in terms of the previous b_j and the process comes to a halt. One must resort to other techniques to find a second linearly independent solution.

A simple example of the logarithmic case, which the reader has seen before, is the Euler differential equation

$$x^2y'' + axy' + by = 0, \quad \text{where } a, b = \text{const.}$$

Clearly, $x = 0$ is a regular singular point and $p(x) = a$, $q(x) = b$, so the indicial equation is

$$r^2 + (a - 1)r + b = 0.$$

If the roots of the indicial equation are r_1 and r_2 and both are real and equal, then two linearly independent solutions are

$$y_1(x) = x^{r_1} \quad \text{and} \quad y_2(x) = x^{r_1} \ln x = y_1(x) \ln x.$$

This is an example of case (5) above with $\beta = 1$ and $\Sigma_{n=0}^{\infty} b_n x^n = 0$.

The Euler differential equation can be used to illustrate two observations about regular singular points. The first is that solutions need not be undefined at a regular singular point. For instance, a linearly independent pair of solutions of

$$y'' - \frac{2}{x}y' + \frac{2}{x^2}y = 0$$

are $y_1(x) = x$ and $y_2(x) = x^2$ and both are smooth, well-defined functions at the regular singular point $x_0 = 0$. The example also illustrates the second observation, namely that the solutions at a regular singular point need not be full infinite series, but may merely be polynomials. Some examples of equations with a regular singular point and their corresponding indicial equations are given below.

EXAMPLE 1 Find the indicial equation, its roots, and the general form of the solution for

a) $y'' - [1/(2x)]y' + [(1 + x^2)/(2x^2)]y = 0;$
b) $y'' + (2/x)y' + xy = 0;$
c) $y'' + (3/x)y' + [(1 + x)/x^2]y = 0.$

SOLUTION

a) Because $p(x) = -1/2$ and $q(x) = 1/2 + (1/2)x^2$; we have $p_0 = -1/2$, $q_0 = 1/2$. The indicial equation is

$$r^2 + \left(-\frac{1}{2} - 1\right)r + \frac{1}{2} = (r - 1)\left(r - \frac{1}{2}\right) = 0$$

with roots $r_1 = 1$ and $r_2 = 1/2$, and $r_1 - r_2 = 1/2$, which is not zero or a positive integer. We conclude that two linearly independent series solutions,

$$y_1(x) = x\sum_{n=0}^{\infty} a_n x^n \quad \text{and} \quad y_2(x) = x^{1/2}\sum_{n=0}^{\infty} b_n x^n,$$

can be found.

b) If the last term is written as $x = x^3/x^2$, it is seen that $p(x) = 2$, $q(x) = x^3$, hence $p_0 = 2$, $q_0 = 0$. The indicial equation is

$$r^2 + (2 - 1)r = r(r + 1) = 0$$

with roots $r_1 = 0$ and $r_2 = -1$, so $r_1 - r_2 = 1$ a positive integer. Corresponding to the root $r_1 = 0$ there is a solution $y_1(x) = \Sigma_{n=0}^{\infty} a_n x^n$, and the second solution is either a logarithmic case or has the form $y_2(x) = x^{-1}\,\Sigma_{n=0}^{\infty} b_n x^n$, corresponding to $r_2 = -1$, which turns out to be the case.

c) We have $p(x) = 3$, $q(x) = 1 + x$, and $p_0 = 3$, $q_0 = 1$. The indicial equation is

$$r^2 + (3 - 1)r + 1 = (r + 1)^2 = 0$$

with roots $r_1 = r_2 = -1$ and $r_1 - r_2 = 0$. This is a logarithmic case, and the solutions are of the form

$$y_1(x) = x^{-1} \sum_{n=0}^{\infty} a_n x^n$$

corresponding to $r_1 = -1$, and

$$y_2(x) = y_1(x)\, \beta \ln x + x^{-1} \sum_{n=0}^{\infty} b_n x^n. \qquad \blacksquare$$

Once the value of r_1 is determined from the indicial equation, finding the series solution for $y_1(x)$ proceeds as in the case of an ordinary point. The series is substituted and like powers of x are compared after shifting indices if necessary.

EXAMPLE 2 Find $y_1(x)$ for the differential equation of Example 1(c) above:

$$x^2y'' + 3xy' + (1 + x)y = 0.$$

SOLUTION Set $y_1(x) = \sum_{n=0}^{\infty} a_n x^{n-1}$, then differentiate the series twice, term by term, and substitute it into the differential equation to get

$$x^2 \sum_{n=0}^{\infty} (n-1)(n-2)a_n x^{n-3} + 3x \sum_{n=0}^{\infty} (n-1)a_n x^{n-2} + (1 + x) \sum_{n=0}^{\infty} a_n x^{n-1} = 0.$$

Multiply each series by its coefficient and combine terms to obtain

$$\sum_{n=0}^{\infty} [(n-1)(n-2) + 3(n-1) + 1]a_n x^{n-1} + \sum_{n=0}^{\infty} a_n x^n$$

$$= \sum_{n=0}^{\infty} n^2 a_n x^{n-1} + \sum_{n=0}^{\infty} a_n x^n = 0.$$

When $n = 0$ in the first series, the coefficient of x^{-1} is zero, so we can shift its index by one to get

$$\sum_{n=0}^{\infty} (n+1)^2 a_{n+1} x^n + \sum_{n=0}^{\infty} a_n x^n = \sum_{n=0}^{\infty} [(n+1)^2 a_{n+1} + a_n]x^n = 0.$$

Hence $a_{n+1} = -a_n/(n + 1)^2$, $n = 0, 1, 2, \ldots$, which implies

$$a_1 = -a_0, \qquad a_2 = \frac{-a_1}{2^2} = \frac{a_0}{2^2}, \qquad a_3 = \frac{-a_2}{3^2} = \frac{-a_0}{3^2 \cdot 2^2}, \text{ etc.,}$$

and in general,

$$a_n = \frac{(-1)^n a_0}{n^2 \cdot \cdot \cdot 3^2 \cdot 2^2} = \frac{(-1)^n a_0}{(n!)^2}.$$

Since a_0 is arbitrary, it may be set equal to unity, and the first solution is

$$y_1(x) = x^{-1} \sum_{n=0}^{\infty} \frac{(-1)^n}{(n!)^2} x^n.$$

Usually writing down the first few terms of the series expansion, after combining those series that lead off with the same power of x, will give a clue as to how to shift the indices. Careful bookkeeping does the rest! ■

The following exercises deal, for the most part, with the nonlogarithmic case. To obtain the second solution in the logarithmic case, a technique called *the method of reduction of order* is often used. That is the subject of the next section.

EXERCISES 6.3

The equations of Exercises 1–7 have a regular singular point at $x = 0$. Find the indicial equation, determine its roots, and state what general form the series solutions will have.

1. $2x^2y'' + (3x - 2x^2)y' - (x + 1)y = 0$

2. $x^2y'' + xy' + (x^2 - \frac{1}{9})y = 0$

3. $4xy'' + 2y' - y = 0$

4. $xy'' + y' + x^2y = 0$

5. $x^2y'' + 5xy' + 4(\cos x)y = 0$

6. $x^2y'' + (2x + x^2)y' - 2y = 0$

7. $16x^2y'' + 24xy' - 3y = 0$

8. If a differential equation has a regular singular point at $x = a$, then it can be written in the form

$$y'' + \frac{p(x)}{x - a} y' + \frac{q(x)}{(x - a)^2} y = 0,$$

where $p(x)$ and $q(x)$ are smooth functions in a neighborhood of $x = a$. By assuming a solution of the form

$$y(x) = (x - a)^r \sum_{n=0}^{\infty} a_n (x - a)^n,$$

show that r must be a root of the indicial equation

$$r^2 + (p_0 - 1)r + q_0 = 0,$$

where $p_0 = p(a)$, $q_0 = q(a)$.

In Exercises 9–12 find the regular singular points, the corresponding indicial equation, and its roots, and state what general form the series solution will have.

9. $(x + 1)y'' + \frac{3}{2}y' + y = 0$

10. $(x - 1)y'' + xy' + y = 0$

11. $2(x - 2)^2y'' - \dfrac{6(x - 2)}{x}y' + 3y = 0$

12. $(1 - x^2)y'' - 2xy' + 6y = 0$

The equations in Exercises 13–17 have a regular singular point at $x = 0$ and are the case $r_1 - r_2 \neq 0$ or a positive integer. Find the recursion relation and the first four nonzero terms of the series expansion for each of the two linearly independent solutions. Find a general expression for each solution if possible.

13. $2xy'' + y' - y = 0$

14. $2x^2y'' - xy' + (1 - x^2)y = 0$

15. $4xy'' + 2y' + y = 0$

16. $3x^2y'' + 4xy' - 2y = 0$

17. $2x^2y'' - 5(\sin x)\, y' + 3y = 0$; find only the first three nonzero terms of each solution.

The equations in Exercises 18–22 have a regular singular point at $x = 0$ and $r_1 - r_2$ is a positive integer (the nonlogarithmic case). Both series solutions can often be found by substituting $y(x) = x^{r_2} \sum_{n=0}^{\infty} a_n x^n$, where r_2 is the *smaller* root of the indicial equation. Find the recursion relation and the first four nonzero terms of the series expansion for each of the two linearly independent solutions. Find a general expression for each solution if possible.

18. $xy'' + (3 + x^3)y' + 3x^2y = 0$

19. $xy'' + 2y' + x^2y = 0$

20. $xy'' + (4 + x)y' + 2y = 0$

21. $x(1 - x)y'' - (4 + x)y' + 4y = 0$

22. $x(1 - x)y'' - (4 + x)y' + 4y = 0$, solve for the regular singular point $x = 1$.

6.4 THE METHOD OF REDUCTION OF ORDER AND THE LOGARITHMIC CASE OF A REGULAR SINGULAR POINT

Suppose we are given the second order linear equation

$$y'' + p(x)y' + q(x)y = 0, \tag{6.4.1}$$

where $p(x)$ and $q(x)$ are continuous for all x in some neighborhood of $x = x_0$ but may be discontinuous or singular at x_0 itself. For instance, this will be the case if x_0 is a regular singular point. Suppose further the happy circumstance that a nontrivial solution $y_1(x)$ of (6.4.1) is known. This solution could be the result of a fortuitous guess or of diligent labor, as in the case of a regular singular point, where $y_1(x)$ is the series solution corresponding to r_1, the largest root of the indicial equation.

The method of reduction of order deals with the following question. Given a nontrivial solution $y_1(x)$ of (6.4.1), can one find a second linearly independent solution $y_2(x)$? The answer is YES, and the method is especially useful when $y_1(x)$ is given by an infinite series and one wishes to obtain the first few terms of the series representation of $y_2(x)$. For the logarithmic case of the regular singular point it is an especially effective tool.

Suppose $y_1(x)$ is a solution of (6.4.1). The method of reduction of order, like the method of variation of parameters, assumes that the second solution can be expressed as $y_2(x) = y_1(x)\,u(x)$, where $u(x)$ is to be determined. Therefore

$$y_2 = y_1 u, \qquad y_2' = y_1 u' + y_1' u, \qquad y_2'' = y_1 u'' + 2y_1' u' + y_1'' u,$$

and since $y_2(x)$ is assumed to be a solution, this implies that

$$y_2'' + p(x)y_2' + q(x)y_2 = [y_1'' + p(x)y_1' + q(x)y_1]u + y_1 u'' + (2y_1' + p(x)y_1)u' = 0.$$

Since $y_1(x)$ is a solution, the coefficient of u is zero, and so $u = u(x)$ must satisfy the relation

$$y_1 u'' + [2y_1' + p(x)y_1]u' = 0$$

or

$$u'' + \left[2\frac{y_1'}{y_1} + p(x)\right]u' = 0.$$

The last relation is a first order linear homogeneous equation in $w = u'$ and can be solved by using the methods of Chapter 1. The solution is

$$u'(x) = \exp\left[-2\int^x \frac{y_1'(s)}{y_1(s)}\,ds - \int^x p(s)\,ds\right] = \frac{\exp\left[-\int^x p(s)\,ds\right]}{y_1(x)^2}.$$

(We have ignored the immaterial constant of integration.) Integrating once more gives $u(x)$ and therefore $y_2(x)$. Following is a summary of the previous analysis.

Method of reduction of order. Given a solution $y_1(x)$ of equation (6.4.1), to find a second linearly independent solution $y_2(x)$, let $y_2(x) = y_1(x)u(x)$. Substitute and solve the differential equation obtained for $u(x)$ to get

$$y_2(x) = y_1(x)\int^x \frac{\exp\left[-\int^r p(s)\,ds\right]}{y_1(r)^2}\,dr, \tag{6.4.2}$$

which is the *reduction of order* formula.

To show that $y_1(x)$ and $y_2(x)$ are a linearly independent pair of solutions, we compute their Wronskian. The formulas above for u' and y_2' imply that

$$y_2'(x) = y_1'(x) \int^x \frac{\exp\left[-\int^r p(s)\,ds\right]}{y_1(r)^2}\,dr + \frac{\exp\left[-\int^x p(s)\,ds\right]}{y_1(x)},$$

and therefore

$$\begin{aligned} y_1(x)y_2'(x) &= y_1'(x)y_1(x) \int^x \frac{\exp\left[-\int^r x\,p(s)\,ds\right]}{y_1(r)^2}\,dr + \exp\left[-\int^x p(s)\,ds\right] \\ &= y_1'(x)y_2(x) + \exp\left[-\int^x p(s)\,ds\right]. \end{aligned}$$

Consequently

$$W(y_1, y_2)(x) = y_1(x)y_2'(x) - y_1'(x)y_2(x) = \exp\left[-\int^x p(s)\,ds\right] \neq 0.$$

Therefore $y_1(x)$ and $y_2(x)$ are a linearly independent pair. Note that if $p(x) = 0$, so that the differential equation is

$$y'' + q(x)y = 0,$$

then the reduction of order formula is simply

$$y_2(x) = y_1(x) \int^x \frac{1}{y_1(r)^2}\,dr.$$

EXAMPLE 1 Use the reduction of order formula to construct a second solution to each of the following differential equations.

a) The constant coefficient equation

$$y'' + 2by' + b^2y = 0;$$

b) the Euler equation

$$x^2y'' - 7xy' + 16y = 0, \qquad x \neq 0;$$

c) the nonconstant coefficient equation

$$y'' - \frac{2}{x^2 + 1}y = 0.$$

SOLUTION

a) This equation has a solution $y_1(x) = e^{-bx}$ and is the case where the characteristic polynomial has a double root $\lambda = b$. Since $p(x) = 2b$, the reduction of order formula gives

$$y_2(x) = e^{-bx} \int^x \frac{\exp\left[-\int^r 2b\, ds\right]}{(e^{-br})^2}\, dr = e^{-bx} \int^x \frac{e^{-2br}}{e^{-2br}}\, dr = xe^{-bx}.$$

b) A trial solution of the type $y_1(x) = x^k$ yields $k = 4$. Rewriting the equation as

$$y'' - \frac{7}{x} y' + \frac{16}{x^2} y = 0, \quad x \neq 0,$$

we see that $p(x) = -7/x$ and

$$\exp\left[-\int^r \left(-\frac{7}{s}\right) ds\right] = \exp\,[7 \ln r] = r^7.$$

Hence

$$y_2(x) = x^4 \int^x \frac{r^7}{(r^4)^2}\, dr = x^4 \ln x.$$

c) Careful examination reveals that $y_1(x) = x^2 + 1$ is a solution. Since $p(x) = 0$, the second linearly independent solution is

$$y_2(x) = (x^2 + 1) \int^x \frac{1}{(r^2 + 1)^2}\, dr = (x^2 + 1) \left[\frac{x}{2(x^2 + 1)} + \frac{1}{2} \tan^{-1} x\right]$$

$$= \frac{x}{2} + \frac{1}{2} (x^2 + 1) \tan^{-1} x.$$ ■

It is often the case that the integral in the reduction of order formula cannot be evaluated. Nevertheless the expression for the second solution may still be useful, for instance to determine the asymptotic behavior of the general solution.

Returning to the topic of series solutions of linear differential equations, it was mentioned above that the first solution $y_1(x)$ could always be expressed as an infinite series. The use of the reduction of order formula to find the second linearly independent solution $y_2(x)$ will then involve the squaring and inverting of an infinite series, followed by a term-by-term integration. There is a systematic way to accomplish this, and the procedure is a direct generalization of arithmetical operations on polynomials.

The Cauchy product formula for series. Given the two power series

$$\sum_{n=0}^{\infty} a_n(x - x_0)^n, \qquad \sum_{n=0}^{\infty} b_n(x - x_0)^n,$$

their product is given by the power series

$$\begin{aligned}\left[\sum_{n=0}^{\infty} a_n(x - x_0)^n\right]&\left[\sum_{n=0}^{\infty} b_n(x - x_0)^n\right] \\ &= \sum_{n=0}^{\infty}\left(\sum_{j=0}^{n} a_j b_{n-j}\right)(x - x_0)^n \\ &= a_0b_0 + (a_0b_1 + a_1b_0)(x - x_0) + (a_0b_2 + a_1b_1 + a_2b_0)(x - x_0)^2 \\ &\quad + \cdots + (a_0b_n + a_1b_{n-1} + \cdots + a_nb_0)(x - x_0)^n + \cdots.\end{aligned}$$

The formula is merely a systematic way of combining in the product all those terms that have the same power of $x - x_0$. If the two series are the same, this gives the formula for squaring a series:

The Cauchy product formula for squaring a series

$$\begin{aligned}\left[\sum_{n=0}^{\infty} a_n(x - x_0)^n\right]^2 &= a_0^2 + 2a_0a_1(x - x_0) + (a_1^2 + 2a_0a_2)(x - x_0)^2 \\ &\quad + (2a_0a_3 + 2a_1a_2)(x - x_0)^3 \\ &\quad + (a_2^2 + 2a_0a_4 + 2a_1a_3)(x - x_0)^4 + \cdots \\ &\quad + (a_n^2 + 2a_0a_{2n} + \cdots + 2a_{n-1}a_{n+1})(x - x_0)^{2n} \\ &\quad + (2a_0a_{2n+1} + \cdots + 2a_na_{n+1})(x - x_0)^{2n+1} + \cdots.\end{aligned}$$

EXAMPLE 2 Find the first four nonzero terms of the series for $\cos^2 x$.

SOLUTION Since

$$\cos x = \sum_{n=0}^{\infty} (-1)^n \frac{x^{2n}}{(2n)!},$$

then $a_1 = a_3 = a_5 = \cdots = 0$ and

$$a_0 = 1, \qquad a_2 = -\frac{1}{2!} = -\frac{1}{2}, \qquad a_4 = \frac{1}{4!} = \frac{1}{24}$$

$$a_0 = -\frac{1}{6!} = -\frac{1}{720}, \ldots.$$

Therefore the series for $\cos^2 x$ will have only even terms and the Cauchy product formula can be used to compute its first four nonzero terms:

$$\cos^2 x = (1)^2 + \left[(0)^2 + 2(1)\left(-\frac{1}{2}\right)\right]x^2 + \left[\left(-\frac{1}{2}\right)^2 + 2(1)\left(\frac{1}{24}\right) + 2(0)(0)\right]x^4$$
$$+ \left[(0)^2 + 2(1)\left(-\frac{1}{720}\right) + 2(0)(0) + 2\left(-\frac{1}{2}\right)\left(\frac{1}{24}\right)\right]x^6 + \cdots$$
$$= 1 - x^2 + \frac{1}{3}x^4 - \frac{2}{45}x^6 + \cdots . \qquad \blacksquare$$

In the reduction of order formula there appears an expression of the form $1/y_1(x)^2$, so if $y_1(x)$ is represented by a power series, it can be squared by using the above formula. But then it must be inverted or, equivalently, the series representation of an expression of the form

$$\frac{1}{\sum_{n=0}^{\infty} b_n(x - x_0)^n}$$

must be computed. To accomplish this, let the last expression equal the series $\sum_{n=0}^{\infty} q_n(x - x_0)^n$, where the q_n are to be determined. Therefore we have the relation

$$\left[\sum_{n=0}^{\infty} b_n(x - x_0)^n\right]\left[\sum_{n=0}^{\infty} q_n(x - x_0)^n\right] = 1,$$

and now use the Cauchy product formula on the left side and then equate the coefficients of like powers of $x - x_0$ to obtain

$$b_0q_0 = 1, \quad b_0q_1 + b_1q_0 = 0, \quad b_0q_2 + b_1q_1 + b_2q_0 = 0, \ldots,$$
$$b_0q_n + b_1q_{n-1} + \cdots + b_nq_0 = 0, \ldots .$$

Now solve for q_0 in the first equation, substitute its value in the second equation, and solve for q_1, use q_0 and q_1 in the third equation to find q_2, etc. The above formulas are a systematic recursive approach to what amounts to "long division."

It is assumed in the above division process that $b_0 \neq 0$. If $b_0, b_1, \ldots, b_{k-1}$ are all zero, but $b_k \neq 0$, then the above quotient may be written as

$$\frac{1}{\sum_{n=0}^{\infty} b_n(x - x_0)^n} = \frac{1}{(x - x_0)^k}\,\frac{1}{\sum_{n=0}^{\infty} b_{k+n}(x - x_0)^n},$$

and the division algorithm applied to the second term.

EXAMPLE 3 Find the first few terms of $1/g(x)^2$, where

$$g(x) = 1 + \sum_{n=1}^{\infty} \frac{1}{n+2} x^n.$$

SOLUTION We have

$$a_0 = 1, \quad a_1 = \frac{1}{3}, \quad a_2 = \frac{1}{4}, \quad a_3 = \frac{1}{5}, \ldots, \quad a_n = \frac{1}{n+2},$$

and by the Cauchy product formula

$$\begin{aligned} g(x)^2 &= (1 + \tfrac{1}{3}x + \tfrac{1}{4}x^2 + \cdots)(1 + \tfrac{1}{3}x + \tfrac{1}{4}x^2 + \cdots \\ &= 1^2 + \left[2 \cdot 1 \cdot \frac{1}{3}\right]x + \left[\left(\frac{1}{3}\right)^2 + 2 \cdot 1 \cdot \frac{1}{4}\right]x^2 \\ &\quad + \left[2 \cdot 1 \cdot \frac{1}{5} + 2 \cdot \frac{1}{3} \cdot \frac{1}{4}\right]x^3 + \cdots \\ &= 1 + \frac{2}{3}x + \frac{11}{18}x^2 + \frac{17}{30}x^3 + \cdots. \end{aligned}$$

To find the first few terms of $1/g(x)^2 = q_0 + q_1x + q_2x^2 + \cdots$, one must solve the equation

$$(1 + \tfrac{2}{3}x + \tfrac{11}{18}x^2 + \cdots)(q_0 + q_1x + q_2x^2 + \cdots) = 1.$$

Therefore

$$1 \cdot q_0 = 1, \quad 1 \cdot q_1 + \frac{2}{3}q_0 = 0, \quad 1 \cdot q_2 + \frac{2}{3}q_1 + \frac{11}{18}q_0 = 0,$$

$$1 \cdot q_3 + \frac{2}{3}q_2 + \frac{11}{18}q_1 + \frac{17}{30}q_0 = 0, \ldots,$$

and we successively solve these to obtain

$$q_0 = 1, \quad q_1 = -\frac{2}{3}, \quad q_2 = -\frac{1}{6}, \quad q_3 = -\frac{13}{270}, \quad \ldots.$$

Therefore

$$\frac{1}{g(x)^2} = 1 - \frac{2}{3}x - \frac{1}{6}x^2 - \frac{13}{270}x^3 + \cdots. \qquad \blacksquare$$

After the above detour into the multiplication and inversion of power series, it is time to return to our principal topic—series solutions of linear differential equations around a regular singular point. The method of reduction of order will be used to obtain the first few terms of the series representation of the second solution $y_2(x)$, when the roots r_1 and r_2 of the indicial equation satisfy $r_1 - r_2 = 0$, the logarithmic case. This will be accomplished via some examples.

EXAMPLE 4 Find the first few terms of the logarithmic solution to

$$y'' + \frac{1}{x}y' + y = 0.$$

SOLUTION In Section 6.1 it was stated that a solution is

$$y_1(x) = J_0(x) = \sum_{n=0}^{\infty} \frac{(-1)^n}{2^{2n}(n!)^2} x^{2n} = 1 - \frac{1}{4}x^2 + \frac{1}{64}x^4 - \frac{1}{2304}x^6 + \cdots.$$

Since $p(x) = 1/x$, the reduction of order formula gives

$$y_2(x) = J_0(x) \int^x \frac{\exp\left[-\int^r \frac{1}{s}\,ds\right]}{J_0(r)^2}\,dr = J_0(x) \int^x \frac{1}{rJ_0(r)^2}\,dr.$$

By rewriting the equation in the form

$$y'' + \frac{1}{x}y' + \frac{x^2}{x^2}y = 0,$$

we see that $x = 0$ is a regular singular point, and since $p_0 = 1$, $q_0 = 0$, the indicial equation is $r^2 + (1 - 1)r + 0 = r^2 = 0$ with roots $r_1 = r_2 = 0$. This is the logarithmic case.

To find the first few terms of $y_2(x)$, first calculate $J_0(x)^2$. Since

$$a_0 = 1, \quad a_1 = 0, \quad a_2 = -\frac{1}{4}, \quad a_3 = 0, \quad a_4 = \frac{1}{64}, \quad \ldots,$$

we have

$$\begin{aligned} J_0(x)^2 &= 1^2 + (2 \cdot 1 \cdot 0)x + \left[0^2 + 2 \cdot 1 \cdot \left(-\frac{1}{4}\right)\right]x^2 \\ &\quad + \left[2 \cdot 1 \cdot 0 + 2 \cdot 0 \cdot \left(-\frac{1}{4}\right)\right]x^3 \\ &\quad + \left[\left(-\frac{1}{4}\right)^2 + 2 \cdot 1 \cdot \frac{1}{64} + 2 \cdot 0 \cdot 0\right]x^4 + \cdots \\ &= 1 - \frac{1}{2}x^2 + \frac{3}{32}x^4 + \cdots. \end{aligned}$$

To find $1/J_0(x)^2$, the following relations have to be solved:

$$1 \cdot q_0 = 1, \quad 1 \cdot q_1 + 0 \cdot q_0 = 0, \quad 1 \cdot q_2 + 0 \cdot q_1 + \left(-\frac{1}{2}\right)q_0 = 0,$$

$$1 \cdot q_3 + 0 \cdot q_2 + \left(-\frac{1}{2}\right)q_1 + 0 \cdot q_0 = 0,$$

$$1 \cdot q_4 + 0 \cdot q_3 + \left(-\frac{1}{2}\right)q_2 + 0 \cdot q_1 + \frac{3}{32}q_0 = 0, \ldots$$

to obtain

$$q_0 = 1, \quad q_1 = 0, \quad q_2 = \frac{1}{2}, \quad q_3 = 0, \quad q_4 = \frac{5}{32}, \ldots .$$

Therefore

$$\begin{aligned} y_2(x) &= J_0(x) \int^x \frac{1}{r}\left(1 + \frac{1}{2}r^2 + \frac{5}{32}r^4 + \cdots\right) dr \\ &= J_0(x)\left[\ln x + \frac{1}{4}x^2 + \frac{5}{128}x^4 + \cdots\right]. \end{aligned}$$

■

EXAMPLE 5 Find the first few terms of the logarithmic solution to

$$y'' + \frac{3}{x}y' + \frac{1 + x}{x^2}y = 0.$$

SOLUTION In the final example of the previous section, a solution was found to be

$$y_1(x) = x^{-1}\left[1 - x + \frac{x^2}{4} - \frac{x^3}{36} + \cdots\right].$$

The point $x = 0$ is a regular singular point and the roots of the indicial equation are $r_1 = r_2 = -1$, so again this is the logarithmic case. Since $p(x) = 3/x$, then

$$\exp\left[-\int^r \frac{3}{s}\, ds\right] = \frac{1}{r^3},$$

and the second solution is

$$y_2(x) = y_1(x) \int^x \frac{1}{r^3 y_1(r)^2} dr = y_1(x) \int^x \frac{1}{r} \frac{1}{\left[1 - r + \dfrac{r^2}{4} - \dfrac{r^3}{36} + \cdots\right]^2} dr.$$

Letting $g(r)$ be the expression in the brackets, we use the Cauchy product formula to get

$$g(r)^2 = 1 - 2r + \tfrac{3}{2}r^2 - \tfrac{5}{9}r^3 + \cdots.$$

To find $1/g(r)^2$ we must solve the relations

$$1 \cdot q_0 = 1, \quad 1 \cdot q_1 + (-2)q_0 = 0,$$

$$1 \cdot q_2 + (-2)q_1 + \left(\frac{3}{2}\right)q_0 = 0,$$

$$1 \cdot q_3 + (-2)q_2 + \left(\frac{3}{2}\right)q_1 + \left(-\frac{5}{9}\right)q_0 = 0,$$

to obtain

$$\frac{1}{g(r)^2} = 1 + 2r + \frac{5}{2}r^2 + \frac{23}{9}r^3 + \cdots.$$

Therefore

$$y_2(x) = y_1(x) \int^x \frac{1}{r}\left[1 + 2r + \frac{5}{2}r^2 + \frac{23}{9}r^3 + \cdots\right] dr$$

$$= y_1(x) \ln x + y_1(x)\left[2x + \frac{5}{4}x^2 + \frac{23}{27}x^3 + \cdots\right].$$

A further multiplication of series would give

$$y_2(x) = y_1(x) \ln x + [2 - \frac{3}{4}x + \frac{11}{108}x^2 + \cdots]. \qquad \blacksquare$$

EXAMPLE 6 Find a solution to $y'' + (2/x)y' + xy = 0$ that is singular at $x = 0$.

SOLUTION The indicial equation is $r^2 + r = 0$ with roots $r_1 = 0$, $r_2 = -1$. Since $r_1 - r_2 = 1$ (a positive integer), this may or may not be a logarithmic case. The solution $y_1(x)$ corresponding to $r_1 = 0$ has the series representation (see Exercise 19 of the previous section)

$$y_1(x) = \sum_{n=0}^{\infty} a_n x^n.$$

Since $p(x) = 2/x$ then

$$\exp\left[\int^r \frac{2}{s}\, ds\right] = \frac{1}{r^2},$$

and by the reduction of order formula gives

$$y_2(x) = y_1(x) \int^x \frac{1}{r^2 y_1(r)^2}\, dr = y_1(x) \int^x \frac{1}{r^2\left(\sum_{n=0}^{\infty} a_n r^n\right)^2}\, dr$$

$$= y_1(x) \int^x \frac{1}{r^2}\,[b_0 + b_1 r + b_2 r^2 + \cdots]\, dr$$

$$= y_1(x)[-b_0 x^{-1} + b_1 \ln x + b_2 x + \cdots],$$

where

$$\frac{1}{\left(\sum_{n=0}^{\infty} a_n x^n\right)^2} = \sum_{n=0}^{\infty} b_n x^n.$$

There will be a logarithmic term if $b_1 \neq 0$; otherwise the solution will be of the form

$$y_2(x) = x^{-1} \sum_{n=0}^{\infty} c_n x^n$$

corresponding to the root $r_2 = -1$. The reader may wish to find the first few a_n's and consequently show that $b_1 = 0$. Therefore the singular solution is of the form above, and not the logarithmic case. ■

EXERCISES 6.4

Given the linear differential equations of Exercises 1–7 and one solution $y_1(x)$, use the reduction of order formula to find a second linearly independent solution $y_2(x)$.

1. $y'' - 4y' + 13y = 0, \quad y_1(x) = e^{2x} \cos 3x$

2. $y'' - 9y = 0, \quad y_1(x) = e^{3x}$

3. $y'' - xy' + y = 0, \quad y_1(x) = x$

4. $x^2y'' + 4xy' - 10y = 0, \quad y_1(x) = x^2$

5. $y'' - \frac{6}{x^2}y = 0, \quad y_1(x) = \frac{1}{x^2}$

6. $(1 - x^2)y'' - 2xy' + 2y = 0, \quad y_1(x) = x$

7. $y'' - 2(1 + \tan^2 x)y = 0, \quad y_1(x) = \tan x$

Given the following series

$$P(x) = 1 + \sum_{n=1}^{\infty} \frac{1}{n^2} x^n,$$

$$Q(x) = \sum_{n=0}^{\infty} \frac{(-1)^n}{n!} x^n,$$

compute the first four nonzero terms of the expressions in Exercises 8–13.

8. $P(x)^2$ **9.** $\dfrac{1}{P(x)^2}$ **10.** $\dfrac{1}{Q(x)^2}$ **11.** $P(x)\, Q(x)$

12. $\dfrac{Q(x)}{P(x)}$ **13.** $P(x)\, Q^2(x)$

Given the functions $p(x)$ and the series $y_1(x)$, compute the first three nonzero terms of the expression

$$\int^x \frac{\exp\left[-\int^r p(s)\, ds\right]}{y_1(r)^2}\, dr$$

in Exercises 14–18.

14. $p(x) = \dfrac{3}{x}$,

$$y_1(x) = 1 - \frac{x^2}{2} + \frac{x^4}{24} - \frac{x^6}{720} + \cdots$$

15. $p(x) = -\dfrac{2}{x}$,

$$y_1(x) = 1 + \tfrac{1}{2} x + \tfrac{1}{4} x^2 + \tfrac{1}{16} x^3 + \cdots$$

16. $p(x) = \dfrac{1}{2x}$, $y_1(x) = 1 + x + x^2 + x^3 + \cdots$

17. $p(x) = \dfrac{1}{x}$, $y_1(x) = x - \dfrac{x^3}{3} + \dfrac{x^5}{9} - \cdots$

18. $p(x) = -\dfrac{1}{3x}$, $y_1(x) = 1 - \dfrac{x^2}{2!} + \dfrac{x^4}{4!} + \cdots$

The equations of Exercises 19–24 have a regular singular point at $x = 0$ and are examples of the logarithmic case. Find the recursion relation and the first four nonzero terms of the series expansion of the solution corresponding to the root r_1. If possible, find a general expression for this solution. Then use the method of reduction of order to find the first few terms of the second linearly independent solution.

19. $xy'' + y' - xy = 0$
20. $x^2y'' + xy' + xy = 0$
21. $x^2y'' + xy' + (x - 1)y = 0$
22. $xy'' + y = 0$
23. $x^2y'' - 3xy' + (4x + 4)y = 0$
24. $x^2y'' + 5xy' + (3 - x^2)y = 0$

6.5 SOME SPECIAL FUNCTIONS AND TOOLS OF THE TRADE

THE GAMMA FUNCTION

This function arises so frequently in series representations of solutions that it is worth discussing briefly. The *Gamma function,* $\Gamma(x)$, is defined by the definite integral

$$\Gamma(x) = \int_0^\infty e^{-t}t^{x-1}\,dt,$$

which converges for all *positive* x. One sees immediately that $\Gamma(1) = 1$, and a simple integration by parts shows that

$$\Gamma(x + 1) = \int_0^\infty e^{-t}t^x\,dt = -e^{-t}t^x\Big|_0^\infty + x\int_0^\infty e^{-t}t^{x-1}\,dt = 0 + x\Gamma(x).$$

The *fundamental identity* $\Gamma(x + 1) = x\,\Gamma(x)$ implies, in particular, that for $x = n$, a positive integer, the following holds:

$$\begin{aligned}\Gamma(n + 1) &= n\Gamma(n) = n(n - 1)\Gamma(n - 1)\\ &= \cdots = n(n - 1)\cdots 2\cdot 1 = n!\end{aligned}$$

since $\Gamma(1) = 1$. Therefore the Gamma function extends the definition of the factorial $x!$, previously defined only for $x = 1, 2, \ldots$, to a function $\Gamma(x)$ that is defined for all values of $x > 0$ and that agrees with $n!$ when $x = n + 1$. Its importance will become clearer in the next section when we discuss the Bessel function.

The fundamental identity also implies that one needs to know only the values of $\Gamma(x + 1)$ for $0 \le x \le 1$ to be able to compute $\Gamma(x)$ for any $x > 0$. For if $x > 2$, one can always find a positive integer n and real number α, $0 \le \alpha \le 1$, so that $x = n + \alpha + 1$. Then by successively using the fundamental identity we obtain

$$\begin{aligned}\Gamma(x) &= \Gamma(n + \alpha + 1) = (n + \alpha)\Gamma(n + \alpha)\\ &= (n + \alpha)(n + \alpha - 1)\Gamma(n + \alpha - 1) = \cdots\\ &= (n + \alpha)(n + \alpha - 1)\cdots(1 + \alpha)\Gamma(1 + \alpha);\end{aligned} \tag{6.5.1}$$

hence computing $\Gamma(x)$ becomes a matter of knowing $\Gamma(1 + \alpha)$ from a table, then performing a series of multiplications. Note that $\Gamma(x) = \Gamma(x + 1)/x$ and $\Gamma(1) = 1$ implies that $\Gamma(x) \to \infty$ as $x \to 0^+$.

EXAMPLE 1 Find $\Gamma(3.77)$.

SOLUTION

$$\begin{aligned}\Gamma(3.77) &= 2.77\,\Gamma(2.77) = (2.77)(1.77)\,\Gamma(1.77)\\ &\cong (2.77)(1.77)(0.92376) = 4.52910,\end{aligned}$$

where the approximate value of $\Gamma(1.77)$ was obtained from a table. Extensive tables of $\Gamma(x)$ can be found, for instance, in [11]. ■

Observe that by using the fundamental identity in reverse, $\Gamma(x)$ can be defined for $x < 0$, where x is not a negative integer. For instance, to find $\Gamma(-\frac{4}{3})$, we write

$$\left(-\frac{4}{3}\right)\Gamma\left(-\frac{4}{3}\right) = \Gamma\left(-\frac{4}{3} + 1\right) = \Gamma\left(-\frac{1}{3}\right),$$

$$\left(-\frac{1}{3}\right)\Gamma\left(-\frac{1}{3}\right) = \Gamma\left(-\frac{1}{3} + 1\right) = \Gamma\left(\frac{2}{3}\right),$$

and

$$\left(\frac{2}{3}\right)\Gamma\left(\frac{2}{3}\right) = \Gamma\left(\frac{5}{3}\right).$$

All of this implies that

$$\Gamma\left(-\frac{4}{3}\right) = \left(-\frac{3}{4}\right)\left(-3\right)\left(\frac{3}{2}\right)\Gamma\left(\frac{5}{3}\right) = \frac{27}{8}\Gamma\left(\frac{5}{3}\right),$$

and the last value can be obtained from tables.

Finally we note the useful and important fact that $\Gamma(\frac{1}{2}) = \sqrt{\pi}$. This result is often proved in advanced calculus texts.

THE BESSEL FUNCTION

A differential equation that arises frequently in boundary value problems and problems in mathematical physics is the *Bessel equation:*

$$y'' + \frac{1}{x}y' + \frac{x^2 - \alpha^2}{x^2}y = 0, \qquad \textbf{(6.5.2)}$$

where α is a given parameter. For this discussion, α will be assumed to be a real number. The solutions of (6.5.2) occur in many types of potential problems involving cylindrical boundaries, as well as in such areas as elasticity theory, fluid mechanics, and electromagnetic field theory.

The Bessel equation is so important that literally hundreds of volumes of tables of its solutions, their derivatives, and their zeros for both integer and fractional values of α have been published. For instance, in the late 1940s the Harvard Computation Laboratory published 12 volumes, each of about 650 pages, giving values of solutions to (6.5.2) for $\alpha = 0, 1, 2, \ldots, 135$. This was done by means of computers that by today's standards would be called prehistoric; today many of those tables have been replaced by stored computational packages and subroutines.

One sees immediately that $x = 0$ is a regular singular point of the Bessel equation and that $p(x) = 1$, $q(x) = -\alpha^2 + x^2$. Hence $p_0 = 1$, $q_0 = -\alpha^2$, and its indicial equation is

$$r^2 + (1 - 1)r - \alpha^2 = r^2 - \alpha^2 = 0.$$

Therefore its roots are $r_1 = \alpha$ and $r_2 = -\alpha$ and $r_1 - r_2 = 2\alpha$, which means there are two distinct cases:

1. $\alpha \neq 0$ and not a positive integer. Then it can be shown that there are two linearly independent solutions of the form

$$J_\alpha(x) = x^\alpha \sum_{n=0}^{\infty} a_n x^n, \qquad J_{-\alpha}(x) = x^{-\alpha} \sum_{n=0}^{\infty} b_n x^n$$

   called *Bessel functions of fractional or nonintegral order α and $-\alpha$.*

2. $\alpha = 0$ or a positive integer, say $\alpha = m$. In this case one solution is of the form

$$J_m(x) = x^m \sum_{n=0}^{\infty} a_n x^n;$$

   it is called the *Bessel function of order m.* The second linearly independent solution is of the logarithmic type and is denoted by $Y_m(x)$, the *Bessel function of the second kind of order m.*

In the earlier sections of this chapter we introduced $J_0(x)$. In what follows we will develop the series representation of $J_\alpha(x)$, $\text{Re}(\alpha) \geq 0$, and list some of its important properties.

Write the Bessel equations as

$$x^2 y'' + xy' + (-\alpha^2 + x^2)y = 0,$$

let $y(x) = x^\alpha \sum_{n=0}^\infty a_n x^n = \sum_{n=0}^\infty a_n x^{n+\alpha}$, and substitute the series to obtain

$$x^2 \sum_{n=0}^\infty [(n+\alpha)(n+\alpha-1)]a_n x^{n+\alpha-2}$$

$$+ x \sum_{n=0}^\infty (n+\alpha)a_n x^{n+\alpha-1} - \alpha^2 \sum_{n=0}^\infty a_n x^{n+\alpha} + x^2 \sum_{n=0}^\infty a_n x^{n+\alpha} = 0.$$

The first three series can be combined to get

$$\sum_{n=0}^\infty [(n+\alpha)^2 - \alpha^2]a_n x^{n+\alpha} + \sum_{n=0}^\infty a_n x^{n+\alpha+2} = 0.$$

Note that in the first series the term corresponding to $n = 0$ is zero. Therefore write down the term for $n = 1$ and reindex to obtain

$$[(1+\alpha)^2 - \alpha^2]a_1 x^{1+\alpha} + \sum_{n=0}^\infty \{[(n+2+\alpha)^2 - \alpha^2]a_{n+2} + a_n\}x^{n+\alpha+2} = 0.$$

Since $\alpha \geq 0$, this implies that $a_1 = 0$ and that

$$a_{n+2} = -\frac{a_n}{(n+2+\alpha)^2 - \alpha^2} = -\frac{a_n}{(n+2)(n+2+2\alpha)}, \quad n = 0, 1, \ldots.$$

Hence $a_1 = a_3 = \cdots = a_{2n+1} = 0$ and

$$a_2 = -\frac{a_0}{2(2+2\alpha)} = -\frac{a_0}{2^2(1+\alpha)},$$

$$a_4 = -\frac{a_2}{4(4+2\alpha)} = \frac{a_0}{2^4 \cdot 2(2+\alpha)(1+\alpha)},$$

$$a_6 = -\frac{a_4}{6(6+2\alpha)} = -\frac{a_0}{2^6 \cdot 3 \cdot 2(3+\alpha)(2+\alpha)(1+\alpha)}, \ldots,$$

$$a_{2n} = (-1)^n \frac{a_0}{2^{2n} n!(n+\alpha)(n-1+\alpha)\cdots(1+\alpha)}, \quad n = 1, 2, \ldots.$$

We conclude that

$$J_\alpha(x) = a_0 x^\alpha \sum_{n=0}^\infty \frac{(-1)^n}{2^{2n} n!(n+\alpha)(n-1+\alpha)\cdots(1+\alpha)} x^{2n}.$$

We note that the series can be written in powers of $x/2$, and if we examine the denominator of the general term and compare it to the expression (6.5.1) for the

Gamma function, we see that considerable simplification is obtained by letting $a_0 = [2^\alpha \Gamma(\alpha + 1)]^{-1}$. We obtain the compact form

$$J_\alpha(x) = \left(\frac{x}{2}\right)^\alpha \sum_{n=0}^{\infty} \frac{(-1)^n}{n!\Gamma(n + \alpha + 1)} \left(\frac{x}{2}\right)^{2n}.$$

In comparing this expression with the previous one we can see the notational convenience of using the Gamma function. We now examine the various possibilities for different values of α.

Case 1. Since $\alpha = m$, where $m = 0$ or m is a positive integer, we have

$$\Gamma(n + m + 1) = (n + m)!,$$

and we obtain the *Bessel function of order m:*

$$J_m(x) = \left(\frac{x}{2}\right)^m \sum_{n=0}^{\infty} \frac{(-1)^n}{n!(n + m)!} \left(\frac{x}{2}\right)^n.$$

From the series it follows

$$J_0(0) = 1, \quad J_m(0) = 0, \quad m = 1, 2, \ldots .$$

A more extensive analysis shows that $J_m(x)$ is an oscillatory function that approaches zero as $x \to \infty$, very similar to a damped cosine or sine function. For instance,

$$J_0(x_j) = 0 \quad \text{for} \quad x_j \cong 2.405, 5.520, 8.654, 11.97, \ldots$$

$$J_1(x_j) = 0 \quad \text{for} \quad x_j \cong 3.832, 7.016, 10.17, 13.32, \ldots$$

$$J_2(x_j) = 0 \quad \text{for} \quad x_j \cong 5.136, 8.417, 11.62, 14.80, \ldots .$$

A graph of $J_m(x)$ for $m = 0, 1, 2$ is shown in Fig. 6.1; these three Bessel functions occur often in applications. Tables of the Bessel functions and their zeros can be found, for instance, in [11].

As mentioned above, when $\alpha = m$, $m = 0$ or a positive integer, the second linearly independent solution of the Bessel equation is $Y_m(x)$, the *Bessel function of the second kind of order m,* and it is of logarithmic type. Its series representation is very complicated, but it is also an oscillatory function which, however, becomes unbounded as $x \to 0$. A graph of $Y_0(x)$ and $Y_1(x)$ is shown in Fig. 6.2.

There are many interesting properties and identities for the Bessel functions that we do not have space to discuss; the reader could browse through [5] or [8]

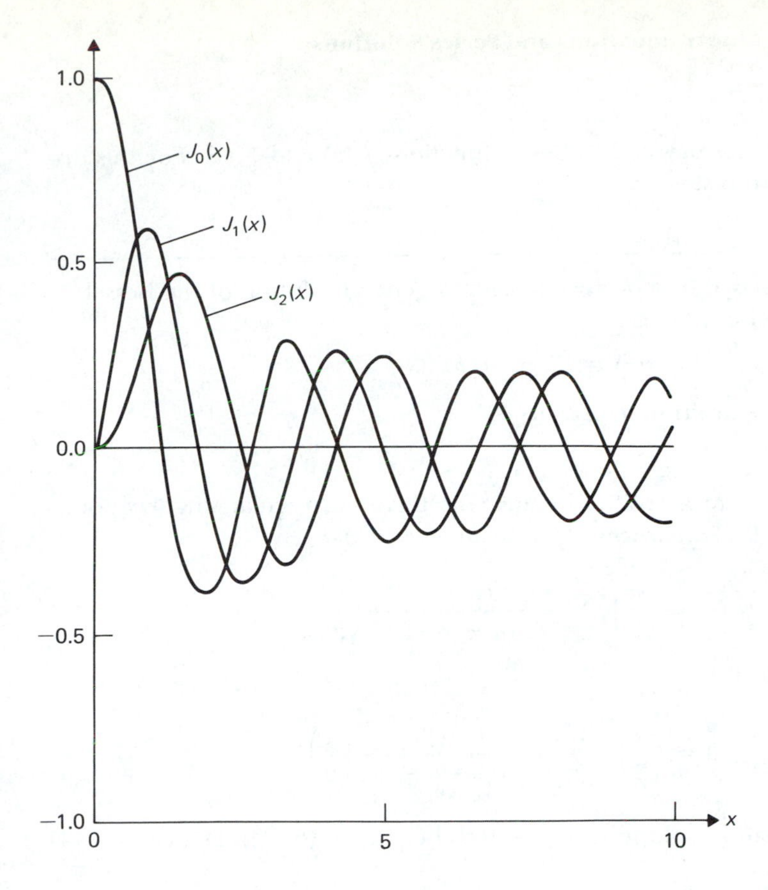

Figure 6.1 Graphs of $J_0(x)$, $J_1(x)$, and $J_2(x)$.

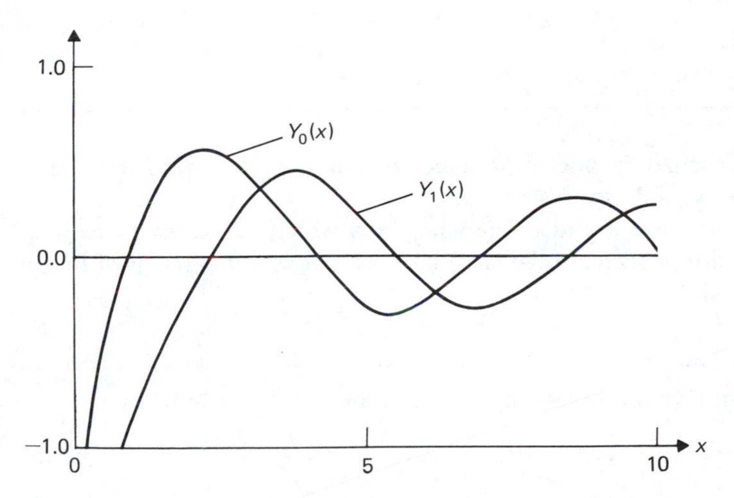

Figure 6.2 Graphs of $Y_0(x)$ and $Y_1(x)$.

to see the vast literature devoted to Bessel functions $J_m(x)$ and $Y_m(x)$. For this discussion we only need to state that:

When $\alpha = m$, $m = 0$ or a positive integer, a general solution of the Bessel equation (6.5.2) is

$$y(x) = aJ_m(x) + bY_m(x),$$

where a and b are arbitrary constants.

Case 2. When $\alpha \neq 0$ or a positive integer, the theory of regular singular points tells us that the two *Bessel functions of fractional order* α *and* $-\alpha$,

$$J_\alpha(x) = \left(\frac{x}{2}\right)^{\alpha} \sum_{n=0}^{\infty} \frac{(-1)^n}{n!\Gamma(n + \alpha + 1)} \left(\frac{x}{2}\right)^{2n}$$

and

$$J_{-\alpha}(x) = \left(\frac{x}{2}\right)^{-\alpha} \sum_{n=0}^{\infty} \frac{(-1)^n}{n!\Gamma(n - \alpha + 1)} \left(\frac{x}{2}\right)^{2n},$$

are a fundamental pair of solutions of the Bessel equation (6.5.2). Hence:

When $\alpha \neq 0$ or a positive integer, the general solution of the Bessel equation (6.5.2) is

$$y(x) = aJ_\alpha(x) + bJ_{-\alpha}(x),$$

where a and b are arbitrary constants.

The function $J_\alpha(x)$ is oscillatory and approaches zero as $x \to \infty$, and for large x, $J_\alpha(x)$ looks very much like a damped sine wave.

Finally, a frequently useful and time-saving fact, which allows us to express many solutions of equations with regular singular points at $x = 0$ in terms of Bessel functions, is the following.

Given an equation with a regular singular point at $x = 0$ in the form

$$x^2y'' + (1 - 2s)xy' + [(s^2 - r^2\alpha^2) + a^2r^2x^{2r}]y = 0, \qquad \textbf{(6.5.3)}$$

where s, r, a, and α are given constants, every solution can be written in the form

$$y(x) = c_1 x^s J_\alpha(ax^r) + c_2 x^s J_{-\alpha}(ax^r),$$

where c_1 and c_2 are arbitrary constants. If $\alpha = 0$ or a positive integer, then we replace $J_{-\alpha}(ax^r)$ with $Y_\alpha(ax^r)$.

This result is obtained by using a change of variable, and the verification is left to the reader in Exercise 9 (p. 465).

EXAMPLE 2

a) In the previous section on regular singular points, the series expression for one solution of

$$x^2 y'' + 3xy' + (1 + x)y = 0$$

and the first few terms of the second solution (of logarithmic type) were developed. By comparison with the differential equation (6.5.3), we find:

$$1 - 2s = 3, \quad s^2 - r^2\alpha^2 = 1, \quad a^2 r^2 = 1, \quad 2r = 1,$$

which, when solved, gives $r = \frac{1}{2}$, $a = 2$, $s = -1$, $\alpha = 0$. Therefore a fundamental pair of solutions is

$$y_1(x) = x^{-1} J_0(2x^{1/2}), \quad y_2(x) = x^{-1} Y_0(2x^{1/2}).$$

b) Let $s = \frac{1}{2}$, $r = a = 1$ in (6.5.3) to obtain

$$x^2 y'' + [(\tfrac{1}{4} - \alpha^2) + x^2]y = 0$$

and the above fact tells us that a fundamental pair of solutions are $x^{1/2} J_\alpha(x)$ and either $x^{1/2} J_{-\alpha}(x)$ or $x^{1/2} Y_\alpha(x)$. However, if $\alpha = \frac{1}{2}$, the above equation, after canceling x^2, becomes $y'' + y = 0$, whose fundamental pair of solutions are $\sin x$ and $\cos x$. This implies that $J_{1/2}(x)$ and $J_{-1/2}(x)$ can be expressed in terms of $x^{-1/2} \sin x$ and $x^{-1/2} \cos x$. In fact it can be shown that (see Exercise 7, p. 464)

$$J_{1/2}(x) = \sqrt{\frac{2}{\pi x}} \sin x, \quad J_{-1/2}(x) = \sqrt{\frac{2}{\pi x}} \cos x,$$

so $J_{1/2}(x)$ and $J_{-1/2}(x)$ can be computed without the use of special tables. ■

The above examples show the advantage of the general form (6.5.3), namely, that the solutions of a large class of equations with a regular singular point at $x = 0$ can be expressed in terms of well-tabulated functions. Hence, the computation of complicated series solutions is avoided, which is a real benefit!

THE AIRY FUNCTIONS

The *Airy functions* $\mathrm{Ai}(x)$, $\mathrm{Bi}(x)$ are the fundamental pair of solutions of *Airy's equation*

$$y'' - xy = 0, \tag{6.5.4}$$

as was shown before. Similarly, $\mathrm{Ai}(-x)$, $\mathrm{Bi}(-x)$ denote the fundamental pair of solutions of

$$y'' + xy = 0, \tag{6.5.5}$$

and both pairs of functions are tabulated, for example, in [10]. By writing (6.5.5) in the form

$$x^2y'' + x^3y = 0$$

we see that it is in the general form (6.5.3) with $1 - 2s = 0$, $s^2 - r^2\alpha^2 = 0$, $a^2r^2 = 1$, and $2r = 3$. The solution of these equations is $r = \frac{3}{2}$, $s = \frac{1}{2}$, $a = \frac{2}{3}$, and $\alpha = \frac{1}{3}$; therefore $\mathrm{Ai}(-x)$ and $\mathrm{Bi}(-x)$ can be written as linear combinations of

$$\sqrt{x}\,J_{1/3}\left(\frac{2}{3}\,x^{3/2}\right) \quad \text{and} \quad \sqrt{x}\,J_{-1/3}\left(\frac{2}{3}\,x^{3/2}\right).$$

In fact, for $x > 0$,

$$\mathrm{Ai}(-x) = \frac{1}{3}\sqrt{x}\left[J_{1/3}\left(\frac{2}{3}\,x^{3/2}\right) + J_{-1/3}\left(\frac{2}{3}\,x^{3/2}\right)\right],$$

$$\mathrm{Bi}(-x) = -\sqrt{\frac{x}{3}}\left[J_{1/3}\left(\frac{2}{3}\,x^{3/2}\right) - J_{-1/3}\left(\frac{2}{3}\,x^{3/2}\right)\right],$$

and again we see how ubiquitous are the Bessel functions!

THE LEGENDRE POLYNOMIALS

The differential equation

$$(1 - x^2)y'' - 2xy' + \lambda y = 0, \tag{6.5.6}$$

with λ a given parameter, is called *Legendre's equation;* it occurs frequently in the analysis of potential problems on spherical domains. One sees that $x = 0$ is an ordinary point since

$$p(x) = \frac{-2x}{1 - x^2}, \qquad q(x) = \frac{\lambda}{1 - x^2}$$

clearly are smooth functions for $|x| < 1$. From the equation in the form

$$y'' + \frac{1}{x-1}\frac{2x}{1+x}y' + \frac{1}{(x-1)^2}\frac{1-x}{1+x}\lambda y = 0, \tag{6.5.7}$$

it follows that $x = 1$ is a regular singular point (and so is $x = -1$ by a similar rewriting).

For the regular singular point at $x = 1$ it is seen from (6.5.7) that

$$p(x) = \frac{2x}{1+x} \quad \text{and} \quad q(x) = \left(\frac{1-x}{1+x}\right)\lambda.$$

Therefore $p(1) = 1$, $q(1) = 0$, and the indicial equation is

$$r^2 + (1 - 1)r + 0 = r^2 = 0$$

with roots $r_1 = r_2 = 0$. This is the logarithmic case, and two linearly independent solutions valid in a deleted neighborhood of $x = 1$ are of the form

$$y_1(x) = \sum_{n=0}^{\infty} a_n(x-1)^n, \qquad y_2(x) = y_1(x)\beta \ln|x-1| + \sum_{n=0}^{\infty} b_n(x-1)^n.$$

A similar analysis can be made for $x = -1$, which is also a logarithmic case. The interesting case occurs when we examine the ordinary point $x = 0$.

Since $x = 0$ is an ordinary point, both linearly independent solutions have a series representation $y(x) = \Sigma_{n=0}^{\infty} a_n x^n$. Differentiating the series term by term and substituting it in (6.5.6) gives

$$\sum_{n=0}^{\infty} n(n-1)a_n x^{n-2} + \sum_{n=0}^{\infty} [-n(n-1) - 2n + \lambda]a_n x^n = 0.$$

Now shift the index by two in the first sum and combine terms to obtain

$$\sum_{n=0}^{\infty} \{(n+2)(n+1)a_{n+2} + [-n(n+1) + \lambda]a_n\}x^n = 0,$$

which leads to the recurrence relation

$$a_{n+2} = \frac{n(n+1) - \lambda}{(n+2)(n+1)} a_n, \quad n = 0, 1, 2, \ldots. \tag{6.5.8}$$

It is seen that $a_2, a_4, \ldots, a_{2n} \ldots$ depend on a_0, while $a_3, a_5, \ldots, a_{2n+1}, \ldots$ depend on a_1, with a_0, a_1 arbitrary. This leads to two series, one in even powers of x and the other in odd powers of x, which are a linearly independent pair of solutions of Legendre's equation in a neighborhood of $x = 0$.

However, the interesting case in many applications is when the parameter $\lambda = m(m + 1)$ for some nonnegative integer m, in which case the relation (6.5.8) tells us that:

1. If m is even, then $a_2, a_4, \ldots, a_m$ will not be zero, but $a_{m+2}, a_{m+4}, \ldots$ will all be zero. Therefore one solution will be an even polynomial $P_m(x)$ of degree m and the other solution will be a power series in odd powers of x.
2. If m is odd, then $a_3, a_5, \ldots, a_m$ will not be zero, but $a_{m+2}, a_{m+4}, \ldots$ will all be zero. Therefore one solution will be an odd polynomial $P_m(x)$ of degree m and the other solution will be a power series in even powers of x.

The polynomial solutions $P_m(x)$ described above are called the *Legendre polynomials* corresponding to $\lambda = m(m + 1)$. They are extremely important because they are the *only* solutions of Legendre's equation that are defined for $x = 0$ and are bounded at $x = \pm 1$ (the power series solutions diverge at $x = \pm 1$ like $\ln(1 \mp x)$).

The recursion relation (6.5.8) may be used to construct some Legendre polynomials, for instance,

1. If $\lambda = 20 = 4 \cdot 5$, then $m = 4$ and

$$a_{0+2} = a_2 = \frac{(0)(1) - 20}{(0 + 2)(0 + 1)} a_0 = -10a_0,$$

$$a_{2+2} = a_4 = \frac{(2)(3) - 20}{(2 + 2)(2 + 1)} a_2 = \frac{-14}{12}(-10a_0) = \frac{35}{3} a_0,$$

hence $P_4(x) = a_0(1 - 10x^2 + \frac{35}{3}x^4)$. It is conventional to choose a_0 so $P_4(1) = 1$; therefore

$$P_4(x) = \frac{3}{8}(1 - 10x^2 + \frac{35}{3}x^4).$$

2. If $\lambda = 30 = 5 \cdot 6$, then $m = 5$ and

$$a_{1+2} = a_3 = \frac{(1)(2) - 30}{(1 + 2)(1 + 1)} a_1 = -\frac{14}{3} a_1,$$

$$a_{3+2} = a_5 = \frac{(3)(4) - 30}{(3 + 2)(3 + 1)} a_3 = \frac{-18}{20}\frac{-14}{3} a_1 = \frac{21}{5} a_1,$$

and choosing a_1 so that $P_5(1) = 1$, we have

$$P_5(x) = \frac{15}{8}(x - \frac{14}{3}x^3 + \frac{21}{5}x^5).$$

The reader may check that

$$P_0(x) = 1, \quad P_1(x) = x, \quad P_2(x) = \frac{1}{2}(3x^2 - 1), \quad P_3(x) = \frac{3}{2}\left(\frac{5}{3}x^3 - x\right),$$

and for a further check, may wish to use the *Rodrigues' formula,*

$$P_m(x) = \frac{1}{2^m m!} \frac{d^m}{dx^m} (x^2 - 1)^m.$$

This formula generates all the Legendre polynomials by successive differentiation.

Finally, we remark that the Legendre polynomials are an example of a *family of orthogonal polynomials on* $-1 \leq x \leq 1$. By this we mean that

$$\int_{-1}^{1} P_n(x) P_m(x)\, dx = 0 \quad \text{if} \quad n \neq m.$$

This fact is extremely useful in boundary value problems and in approximation theory.

EXERCISES 6.5

1. Given
   a) $\Gamma(1.185) = 0.92229$; evaluate $\Gamma(5.185)$ and $\Gamma(-3.815)$.
   b) $\Gamma(1.910) = 0.96523$; evaluate $\Gamma(4.1910)$ and $\Gamma(-2.090)$.
2. Stirling's formula for an asymptotic approximation of $n! = \Gamma(n + 1)$ is

   $$\Gamma(n + 1) \approx n^n e^{-n} \sqrt{2\pi n},$$

   meaning that

   $$\lim_{n\to\infty} \frac{\Gamma(n + 1)}{n^n e^{-n} \sqrt{2\pi n}} = 1.$$

   This means the approximation is good as n gets large, in the sense of a small relative error *not* a small absolute error. Use Stirling's formula to find an approximation of $100! \cong 9.3326 \times 10^{157}$ and compute the relative and absolute errors. (*Hint:* For computing, use the logarithmic form

   $$\ln \Gamma(n + 1) \approx \left(n + \frac{1}{2}\right) \ln n - n + \frac{1}{2} \ln 2\pi.)$$

3. The Maclaurin series for e^{-x} is

   $$e^{-x} = \sum_{n=0}^{\infty} \frac{(-1)^n}{n!} x^n;$$

   and since it is an alternating series, the error in stopping at N terms is less than the magnitude of the $(N + 1)$st term.
   a) Use Stirling's formula to approximate the error in estimating e^{-10} with the first 20, 25, and 30 terms of the series.

b) How many terms would be needed to compute e^{-20} with an error of less than 10^{-10}?

One can infer from the above that for computational purposes, power series can be very inefficient!

4. By differentiating the series for the Bessel function term by term, show that

$$xJ'_p(x) = pJ_p(x) - xJ_{p+1}(x),$$
$$xJ'_p(x) = -pJ_p(x) + xJ_{p-1}(x).$$

5. The two relations in Exercise 4 imply the important recursion relation

$$J_{p+1}(x) = \frac{2p}{x}J_p(x) - J_{p-1}(x),$$

which allows one to compute higher order Bessel functions in terms of lower order ones. Use it and the expressions given on p. 459 for $J_{1/2}(x)$ and $J_{-1/2}(x)$ to obtain the values of

a) $J_{3/2}(1.76)$,
b) $J_{-3/2}(0.587)$,
c) $J_{5/2}(6.78)$.

6. Use the substitution $y = u/\sqrt{x}$ directly in Bessel's equation to obtain the differential equation

$$u'' + \left(1 - \frac{\alpha^2 - \frac{1}{4}}{x^2}\right)u = 0$$

with solutions $x^{1/2}J_\alpha(x)$.

7. Using the fact that $\Gamma(\frac{1}{2}) = \sqrt{\pi}$, show directly from the series representation for $J_{1/2}(x)$ and $J_{-1/2}(x)$ that

$$\sin x = \sqrt{\frac{\pi x}{2}}J_{1/2}(x), \qquad \cos x = \sqrt{\frac{\pi x}{2}}J_{-1/2}(x).$$

(You will need the identity 6.5.1.)

8. The integral representation of $J_n(x)$,

$$J_n(x) = \frac{1}{\pi}\int_0^\pi \cos(n\theta - x\sin\theta)\,d\theta, \qquad n = 0, 1, 2, \ldots,$$

was obtained by F. W. Bessel (1784–1846) in his study of astronomical orbits.

a) Use it to show that $J_{-n}(x) = (-1)^n J_n(x)$ and $J'_0(x) = -J_1(x)$.
b) Use it and a quadrature formula (e.g., Simpson's rule) to show that 2.405 is an approximate zero of $J_0(x)$.
c) Similarly, show that 3.833 is an approximate zero of $J_1(x)$.

9. Given the Bessel equation

$$z^2w'' + zw' + (z^2 - \alpha^2)w = 0,$$

let $z = ax^r$, $y = x^s w$ and obtain the differential equation (6.5.3).

Express the solutions of the differential equations in Exercises 10–14 in terms of Bessel functions by using (6.5.3).

10. $x^2y'' + 5xy' + (3 + 4x^2)y = 0$

11. $y'' + 4xy = 0$

12. $x^2y'' + 3xy' + (1 + x)y = 0$

13. $2x^2y'' - xy' + (1 + x^2)y = 0$

14. $xy'' + y = 0$

15. Using Rodrigues' formula directly,

a) find $P_2(x)$, $P_3(x)$, and $P_4(x)$;

b) show that $\int_{-1}^{1} x^r P_m(x)\,dx = 0$, $r = 0, 1, \ldots, m - 1$. (*Hint:* Use successive integration by parts.)

16. Show that the Legendre equation can be written in the form

$$\frac{d}{dx}[(1 - x^2)y'] = -\lambda y$$

and use this to show the orthogonality relation

$$\int_{-1}^{1} P_n(x)P_m(x)\,dx = 0, \quad n \neq m.$$

(*Hint:* Let $\lambda = m(m + 1)$ and $y(x) = P_m(x)$ in the differential equation, then multiply both sides by $P_n(x)$ and integrate on $-1 \leq x \leq 1$ by parts. Repeat the process with $\lambda = n(n + 1)$, etc.)

17. The function $y(\theta) = P_n(\cos\theta)$, $0 < \theta < \pi$, where $P_n(x)$ is a Legendre polynomial, arises in potential theory for a spherical body. Show that $y(\theta)$ satisfies

$$y'' + (\cot\theta)y' + n(n + 1)y = 0.$$

18. The Chebyshev equation is

$$(1 - x^2)y'' - xy' + \lambda^2 y = 0.$$

a) Show that $x = \pm 1$ are regular singular points, and determine the nature of the two linearly independent solutions valid near them.

b) Show that $x = 0$ is an ordinary point and that if $\lambda^2 = m^2$, m an integer, one of the linearly independent solutions valid near $x = 0$ is a polynomial $T_m(x)$. Find $T_m(x)$, $m = 0, 1, 2, 3, 4$.

19. The Hermite equation is

$$y'' - 2xy' + 2\lambda y = 0.$$

Show that $x = 0$ is an ordinary point and that if $\lambda = m$, m a nonnegative integer, then one of the linearly independent solutions is a polynomial $H_m(x)$. Find $H_m(x)$, $m = 0, 1, 2, 3, 4$.

SUMMARY

The problem of finding two linearly independent solutions or a general solution of the second order linear differential equation

$$a(t)\frac{d^2y}{dt^2} + b(t)\frac{dy}{dt} + c(t)y = 0$$

when the coefficients $a(t)$, $b(t)$, and $c(t)$ are constants is a simple algebraic task. However, when they are not, there is no general technique for finding solutions. But for a large class of equations, a method wherein the solution is represented by a power series can be employed.

The first case discussed is where $a(t)$, $b(t)$, and $c(t)$ are smooth, well-behaved functions in the neighborhood of some point t_0, and $a(t_0) \neq 0$. Then any solution $y(t)$ can be represented by a power series $y(t) = \Sigma_0^\infty a_n(t - t_0)^n$ that is substituted into the differential equation, and the coefficients a_n are found term by term via a recursion relation.

The second case discussed is where $a(t) = (t - t_0)^2$, $b(t) = (t - t_0)p(t)$, and $c(t) = q(t)$ where $p(t)$ and $q(t)$ are smooth well-behaved functions in a neighborhood of $t = t_0$. In this case a solution can be represented by an infinite series

$$y(t) = (t - t_0)^r \sum_0^\infty a_n (t - t_0)^n$$

where the value of r is determined by a quadratic equation. This series can be substituted into the differential equation and the coefficients a_n determined recursively. A second linearly independent solution can be found, but its series form depends on the nature of the roots of the quadratic equation.

Many of the special functions of mathematical physics, such as the Bessel functions or the Legendre polynomials, arise as series solutions of second order differential equations of the type described above. The chapter concludes with a brief description of some of these functions.

MISCELLANEOUS EXERCISES

6.1. A solution $y(x)$ of an ordinary differential equation is said to be *oscillatory* if it vanishes infinitely often on a half line $x_0 \leq x < \infty$. The following comparison theorem is useful in determining whether solutions are oscillatory.

Given the two second order linear equations

$$y'' + p(x)y = 0 \tag{1}$$

$$y'' + q(x)y = 0 \tag{2}$$

with $q(x) \geq p(x)$ for $x \geq x_0$. If the solutions of (1) are oscillatory, then so are those of (2), and the zeros of the solutions of (2) are closer together than those of the solutions of (1).

Use the theorem to study the Airy equation

$$y'' + xy = 0, \quad x \geq 1,$$

and by comparison with equations of the form $y'' + k^2y = 0$ show that its solutions are oscillatory. Furthermore show that:

a) The zeros of its solutions are less than π units apart.

b) The distance between successive zeros approaches zero as x approaches ∞.

c) There are either 1 or 2 zeros in the interval $1 \leq x \leq 4$. (A check with tables shows that $\mathrm{Ai}(-x)$ vanishes once and $\mathrm{Bi}(-x)$ vanishes twice in the interval.)

6.2. By transforming the Bessel equation to the form

$$y'' + \left[1 - \frac{\alpha^2 - \frac{1}{4}}{x^2}\right] y = 0$$

(whose solutions are $x^{1/2}J_\alpha(x)$ and $x^{1/2}J_{-\alpha}(x)$ or $x^{1/2}Y_\alpha(x)$), use the comparison theorem above to show that:

a) The Bessel functions are oscillatory.

b) The zeros of $J_\alpha(x)$ are separated by more than π units if $\alpha^2 < \frac{1}{4}$.

6.3. Show that $x = 0$ is a regular singular point of the differential equation

$$x^2y'' + (\sin x)y' - (\cos x)y = 0,$$

and that the roots of the indicial equation are $r_1 = 1$ and $r_2 = -1$. By expanding $\sin x$ and $\cos x$ in their Taylor series at $x = 0$ and by retaining enough terms, determine the first three nonzero terms of the series solution corresponding to r_1.

6.4. In the previous problem use the method of reduction of order to determine whether the second solution corresponding to $r_2 = -1$ is of logarithmic type or not.

6.5. Find the series solution of Laguerre's equation

$$xy'' + (1 - x)y' + \lambda y = 0$$

near the regular singular point $x_0 = 0$. Show that the solution reduces to a polynomial $L_m(x)$, called a Laguerre polynomial, if $\lambda = m$, a positive integer. Find $L_1(x)$, $L_2(x)$, and $L_3(x)$.

6.6. The Fourier–Legendre series expansion of a function $f(x)$, $-1 < x < 1$, is given by

$$f(x) = \sum_{n=0}^{\infty} a_n P_n(x), \qquad a_n = \frac{2n+1}{2} \int_{-1}^{1} f(x)\, P_n(x)\, dx.$$

It will converge to $f(x)$, for instance, for any function $f(x)$ that is continuous and has a piecewise continuous first derivative.

a) Using the orthogonality relation (see Exercise 16 of Section 6.5) and the fact that $\int_{-1}^{1} P_n(x)^2\, dx = 2/(2n+1)$, obtain the above formula for a_n.

b) Show that if $f(x)$ is an even (odd) function, only even (odd) indexed terms will appear in the series.

c) Use the formula above to compute the first three nonzero terms of the Fourier–Legendre series for $f(x) = |x|$, $-1 < x < 1$. Graph your result.

d) Do the same as in (c) for $f(x) = e^x$, $-1 < x < 1$.

6.7. Find the solution of the boundary value problem

$$(1-x)^2 y'' - 2xy' + 12y = 0, \qquad y(0) = 0, \qquad y(\tfrac{1}{2}) = 4.$$

6.8. Find the solution of the boundary value problem

$$y'' + \frac{1}{x} y' + \frac{x^2 - (1/4)}{x^2} y = 0, \qquad y(0) \text{ bounded}, \qquad y\left(\frac{\pi}{2}\right) = 1.$$

6.9. Using (6.5.3), determine for what values of α the boundary value problem

$$x^2 y'' - 3xy' + \left(\frac{15}{4} + x^2\right) y = 0, \qquad y\left(\frac{\pi}{2}\right) = 0, \qquad y(\alpha\pi) = 0,$$

will have nontrivial solutions or only the zero solution.

6.10. Derive the formula for the Laplace transform of t^α, where α is not necessarily an integer:

$$L\{t^\alpha\} = \frac{\Gamma(\alpha+1)}{s^{\alpha+1}}; \quad \alpha > -1, \quad \text{Re}(s) > 0.$$

REFERENCES

A majority of books on ordinary differential equations discuss series solutions and special functions. For a more detailed discussion than is given here we could suggest

1. W.E. Boyce and R.C. DiPrima, *Elementary Differential Equations,* 4th ed., Wiley, New York, 1986.
2. E.A. Coddington, *An Introduction to Ordinary Differential Equations*, Prentice-Hall, Englewood Cliffs, N.J., 1961.

3. E.D. Rainville, *Intermediate Differential Equations,* 2nd ed., Macmillan, New York, 1964.
4. G.F. Simmons, *Differential Equations with Applications and Historical Notes,* McGraw-Hill, New York, 1972.

To get an idea of the depth of analysis devoted to special functions at the turn of this century, we recommend browsing through these two classics:

5. G.N. Watson, *A Treatise on the Theory of Bessel Functions,* 2nd ed., Cambridge University Press, Cambridge, 1944.
6. E.T. Whittaker and G.N. Watson, *A Course of Modern Analysis,* 4th ed., Cambridge University Press, Cambridge, 1965.

For a more accessible discussion of special functions we suggest

7. H. Hochstadt, *Special Functions of Mathematical Physics,* Holt, New York, 1961.
8. L.C. Andrews, *Special Functions for Engineers and Applied Mathematicians,* Macmillan, London, 1985.
9. H. Hochstadt, *The Functions of Mathematical Physics,* Wiley Interscience, New York, 1971.
10. F.W.J. Olver, *Introduction to Asymptotics and Special Functions,* Academic Press, New York, 1974.

For a book of tables and formulas of special functions the standby is

11. M. Abramowitz and I.A. Stegun, *Handbook of Mathematical Functions,* Applied Mathematics Series 55, National Bureau of Standards, Washington, D.C., 1964. (Also available from Dover Publications.)

7

NONLINEAR DIFFERENTIAL EQUATIONS

The study of nonlinear phenomena is an essential part of the education of an engineer, physical scientist, or mathematical biologist. Self-sustained oscillations, amplitude-dependent periodicities, and growth-limited populations are all nonlinear phenomena that can be modeled by nonlinear differential equations. This does not mean that the study of linear differential equations is unimportant. They successfully model low-current *LCR* circuits, small-amplitude mass–spring systems, and radioactive decay. They are also employed to approximate nonlinear differential equations. But much of nature is inherently nonlinear and must be modeled and studied accordingly. We shall start the study of nonlinear differential equations in this chapter. It is a fascinating and difficult subject but, fortunately, many of the essential ideas are accessible to undergraduates.

Most nonlinear differential equations cannot be solved explicitly; i.e., their solutions cannot be written in terms of a finite combination of elementary functions, such as polynomials or exponentials. Because of this, they are often studied by geometric methods as well as by numerical and analytical approximate methods. Geometric methods are employed to investigate qualitative properties, such as the behavior of solutions near equilibrium states (constant solutions), to characterize bounded and unbounded solution regimes, and to establish the existence of periodic solutions. Numerical methods are employed to construct approximate solutions to initial value problems and as an exploratory tool to help understand the qualitative properties of the solutions.

7.1 AN INTRODUCTORY EXAMPLE—THE PENDULUM

Our study begins with a description of the slightly damped or undamped motion of a simple pendulum. This motion is described by a nonlinear ordinary differential equation, and in this section we will describe (somewhat heuristically) two methods used to study nonlinear equations. In later sections the discussion of these methods will be greatly expanded.

A simple pendulum is a bob of mass m suspended from a fixed point 0 by a light rod of length a. If it is allowed to swing only in a vertical plane, its position is described by the angle x, the inclination of the rod to the downward vertical (Fig. 7.1). The angular momentum of the bob, $y = ma\,dx/dt$, and the angle x completely describe the *state* of the system at any time.

By Newton's second law, the time rate of change of the angular momentum is equal to the force acting perpendicular to the rod. If F represents the force due to friction and air resistance, then the total force is $-mg \sin x - F$. As in the case of the mass–spring systems discussed in Chapter 2, suppose we assume that the force F is proportional to the angular velocity and hence to the angular momentum. If the constant of proportionality is c, $c > 0$, then the first order nonlinear system of differential equations governing the motion of the simple pendulum is

$$\frac{dx}{dt} = \frac{1}{ma}y, \qquad \frac{dy}{dt} = -mg \sin x - cy. \tag{7.1.1}$$

The system is nonlinear because of the presence of the term $\sin x$, a nonlinear function of the dependent variable x.

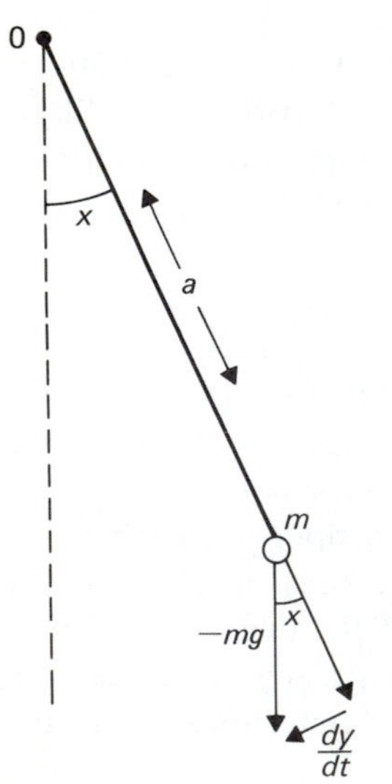

Figure 7.1 The simple pendulum.

An equivalent second order nonlinear differential equation can be obtained from (7.1.1) by differentiating both sides of the first equation and substituting the second expression for dy/dt with y replaced by $ma\, dx/dt$. This gives

$$\frac{d^2x}{dt^2} = \frac{1}{ma}\frac{dy}{dt} = \frac{1}{ma}\left(-mg \sin x - cma\frac{dx}{dt}\right)$$

or

$$\frac{d^2x}{dt^2} + c\frac{dx}{dt} + \frac{g}{a}\sin x = 0, \tag{7.1.2}$$

which is called the *damped nonlinear simple pendulum equation.* We assume that the damping is light, which requires that the constant c satisfy the relation $0 < c^2 < 4g/a$.

An *equilibrium state* of the nonlinear system (7.1.1) is a constant solution, and immediate inspection shows that $x(t) = 0$, $y(t) = 0$ is an equilibrium state. It corresponds to the pendulum at rest with the bob directly below the suspension point. If it is assumed furthermore that the motion of the pendulum is restricted to small oscillations about the equilibrium state, then it seems reasonable to use the approximation $\sin x \cong x$ (valid for $|x| < 0.1$). This results in the approximate system

$$\frac{dx}{dt} = \frac{1}{ma}y, \qquad \frac{dy}{dt} = -mgx - cy, \tag{7.1.3}$$

that can be completely solved for given values of the constants m, g, a, and c by using the methods of Chapter 5.

The system (7.1.3) is the *linearized* system corresponding to the system (7.1.1) at the equilibrium point $(x, y) = (0, 0)$. The question is: Do the solutions of the linearized system (7.1.3) approximate the solutions of the nonlinear system (7.1.1), and equally important, are their limiting qualitative behaviors the same? The answer is YES, and the particular conditions for which this is true will be discussed in the next section. Here we will give only some computer generated plots of $x(t)$ versus $y(t)$ for particular values of the parameters (see Fig. 7.2).

The variables x and y are called the *phase* of the motion and the xy-plane is called the *phase plane.* The curves plotted in Fig. 7.2 are called *phase paths,* or *orbits,* and the arrows indicate the evolution of the system, i.e., the direction of increasing time. The plots support the reasonable physical conclusion that for small initial disturbances from rest, the pendulum will have a damped oscillatory motion that approaches the rest position. Thus we can say $(0, 0)$ is a *stable equilibrium state* (the precise notion of stability will be defined later).

It should be mentioned that there is a second equilibrium state $x(t) = \pi$, $y(t) = 0$. This can be thought of as the case where the pendulum is (very) carefully balanced on its end with the bob directly above the suspension point. Physical intuition tells us that $(\pi, 0)$ should be an *unstable equilibrium state* since a small distur-

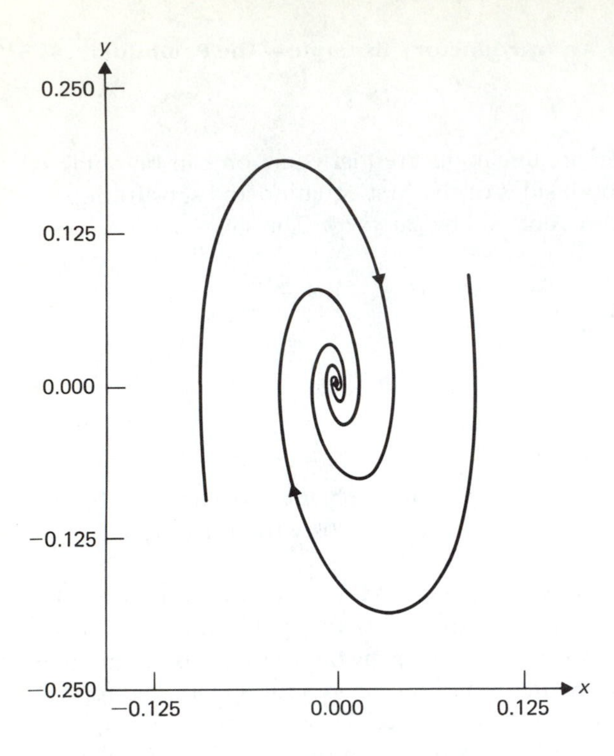

Figure 7.2(a) The nonlinear system ($c = 0.5$, $m = 2$, $a = 1$, $g = 32$).

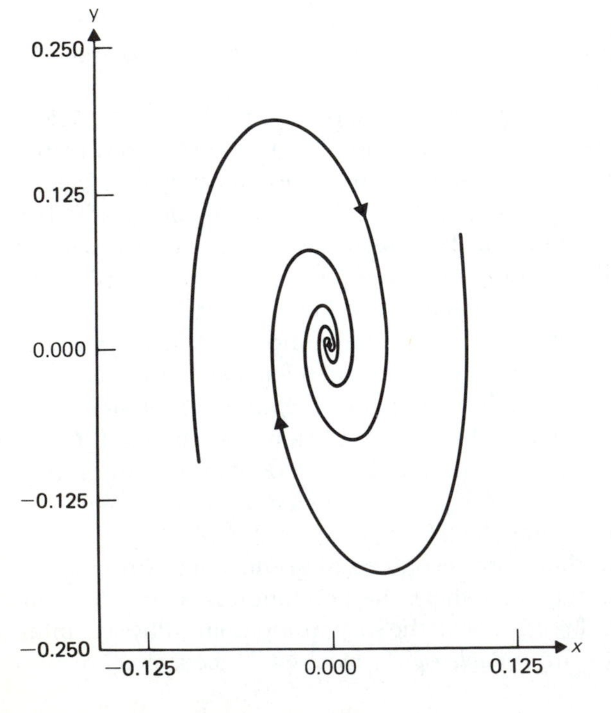

Figure 7.2(b) The linearized system ($c = 0.5$, $m = 2$, $a = 1$, $g = 32$).

bance results in a motion that does not return to the equilibrium state. Finally, because of the periodicity of sin x we should expect that the equilibrium states $(2n\pi, 0)$, $n = \pm 1, \pm 2, \ldots$, will be copies of $(0, 0)$.

Let us now return to the system (7.1.1) with $c = 0$ (i.e., the case of no damping or friction force) to demonstrate another quite distinct method for studying certain nonlinear systems. The system is therefore

$$\frac{dx}{dt} = \frac{1}{ma} y, \qquad \frac{dy}{dt} = -mg \sin x, \tag{7.1.4}$$

which is equivalent to the simple undamped pendulum equation

$$\frac{d^2x}{dt^2} + \frac{g}{a} \sin x = 0. \tag{7.1.5}$$

Since the system is autonomous (does not depend explicitly on time), we can associate with it the first order differential equation

$$\frac{dy}{dx} = \frac{dy/dt}{dx/dt} = \frac{-mg \sin x}{y/ma}.$$

This equation is separable and has an implicit solution

$$E(x, y) = \frac{1}{2m} y^2 + mga(1 - \cos x) = \text{const.} \tag{7.1.6}$$

By graphing the curves described by (7.1.6) for various choices of the constant, a phase plane portrait of the system (7.1.4) can be obtained; it is shown in Fig. 7.3. A simple technique for obtaining the graphs shown will be given in a later section.

What is the physical significance of the curves shown in Fig. 7.3? By rewriting (7.1.6) as

$$E\left(x, \frac{dx}{dt}\right) = \frac{1}{2} m \left(a \frac{dx}{dt}\right)^2 + mga(1 - \cos x) = \text{const},$$

we can see that the first term represents the *kinetic energy* of the pendulum, and the second term represents the *potential energy*. Hence E is simply the *total energy* of the pendulum, and the last expression, and consequently (7.1.6), simply expresses the fact that for the simple undamped pendulum, the total energy is always constant. It follows that the curves sketched in the phase plane portrait shown in Fig. 7.3 are simply curves of constant energy for various choices of the total energy value.

A system whose total energy is constant is called a *conservative system*, and the method described above to construct the phase plane portrait is called the *energy method*. It is typically used to study systems of the form

$$\dot{x} = y, \qquad \dot{y} = -g(x),$$

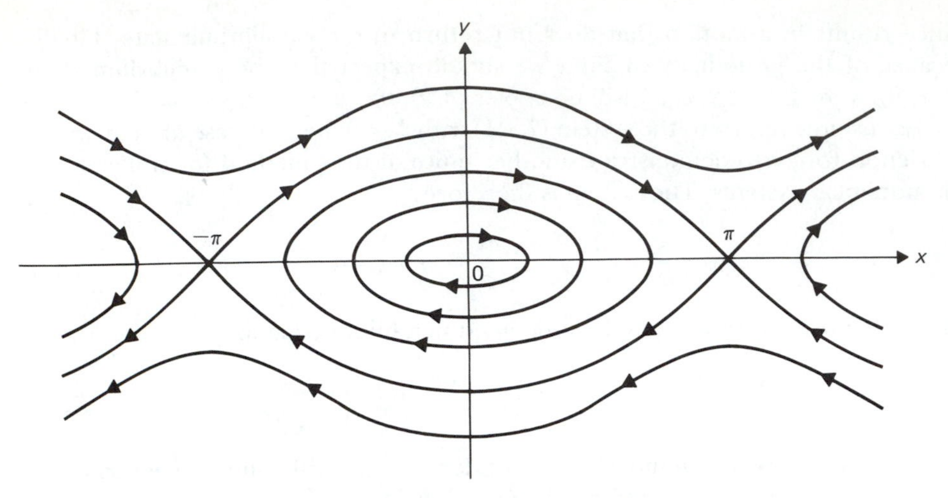

Figure 7.3 Phase plane portrait of the simple pendulum.

corresponding to second order undamped equations of the form

$$\frac{d^2x}{dt^2} + g(x) = 0.$$

Letting $g(x) = (g/a) \sin x$, we see that the simple undamped pendulum equation is of this type.

We have given a brief overview of two methods frequently used to study nonlinear systems of differential equations. The first consists of finding the equilibrium points of the system and studying the linearized system near each equilibrium point. The second is used for conservative systems and consists of finding the curves of constant energy. An amplified discussion of each method will be given in the following sections.

EXERCISES 7.1

1. For periodic motion, the average kinetic and potential energies are defined as

$$\overline{\text{KE}} = \frac{1}{T} \int_0^T \frac{1}{2} ma^2 \left(\frac{dx}{dt}\right)^2 dt,$$

$$\overline{\text{PE}} = \frac{1}{T} \int_0^T mga(1 - \cos x)\, dt,$$

where T is the period. Show that if (7.1.5) is replaced by

$$\frac{d^2x}{dt^2} + \omega^2 x = 0,$$

where

$$\omega^2 = \frac{g}{a},$$

then $T = 2\pi/\omega$ and $\overline{KE} = \overline{PE}$. (Note: $\cos x \cong 1 - x^2/2$ if x is small, where x is in radians.)

2. The lightly damped approximate pendulum equation is given by

$$\frac{d^2x}{dt^2} + c\frac{dx}{dt} + \omega^2 x = 0, \quad c^2 < 4\omega^2, \quad 0 < c.$$

Its characteristic equation has roots

$$\lambda = -\frac{c}{2} + i\sigma, \qquad \lambda^* = -\frac{c}{2} - i\sigma,$$

where

$$\sigma = \left(\omega^2 - \frac{c^2}{4}\right)^{1/2}.$$

Let $T^* = 2\pi/\sigma$ denote the pseudoperiod. Replace T by T^* in the above formulas for $\overline{KE}$ and $\overline{PE}$ and compute $\Delta = \overline{KE} - \overline{PE}$. Take $\omega^2 = 1$ and plot Δ as a function of c.

3. Some of the orbits shown in Fig. 7.3 are closed curves that represent periodic motions of the simple undamped pendulum. If α is the value of x, where a closed orbit crosses the positive x-axis (hence α is the maximum angular displacement), then it can be shown that the period T of the periodic motion is given by the expression

$$T = 4\left(\frac{a}{2g}\right)^{1/2} \int_0^{\alpha} (\cos x - \cos \alpha)^{-1/2}\, dx.$$

Show that an equivalent expression is

$$T = 2\left(\frac{a}{g}\right)^{1/2} \int_0^{\alpha} \left[\sin^2\left(\frac{\alpha}{2}\right) - \sin^2\left(\frac{x}{2}\right)\right]^{-1/2} dx,$$

and use this expression to show that as α approaches π, T approaches ∞.

4. In the second formula given in Exercise 3 let $\sin(x/2) = k \sin \phi$, $k = \sin(\alpha/2)$ and show that

$$T = 4\left(\frac{a}{g}\right)^{1/2} \int_0^{\pi/2} (1 - k^2 \sin^2 \phi)^{-1/2}\, d\phi.$$

The integral

$$K(k) = \int_0^{\pi/2} (1 - k^2 \sin^2 \phi)^{-1/2} \, d\phi$$

is called an *elliptic integral of the first kind.* It has been extensively studied and its values for selected k's can be found in most handbooks [14].

5. Using values from a handbook, plot $K(k) - \pi/2$ and $\alpha - \sin \alpha$ versus α for various values of α on the same graph. Comment on the relationship between the small-angle approximation $\sin \alpha \cong \alpha$ and $T \cong 2\pi(a/g)^{1/2}$.

6. The Schuler pendulum has period 24 hours and is of importance in the analysis of inertial guidance systems. Using $g = 9.80$ m/sec^2, plot the length (in meters) of a Schuler pendulum as a function of α, the maximum displacement.

7. Let $\eta = x/2$, $g/a = 1$ and transform the pendulum equation (7.1.5) into

$$\frac{d^2\eta}{dt^2} + \sin \eta \cos \eta = 0.$$

Select a value of the energy so that

$$\left(\frac{d\eta}{dt}\right)^2 = \cos^2 \eta$$

and derive formulas for η and $d\eta/dt$ as functions of t. Sketch the corresponding orbits in the phase plane.

8. If an external force is imposed on the pendulum, the differential equation governing the motion is

$$\frac{d^2x}{dt^2} + c\frac{dx}{dt} + \omega^2 \sin x = F(t).$$

Take $c = 0.1$, $\omega^2 = 4.0$, $F(t) = 2 \cos t$, $x(0) = 0.1$, $\dot{x}(0) = 0$, and compute a numerical solution with RK4 on $0 \leq t \leq 10$ with $h = 0.1$. Plot $x(t)$ versus t and $\dot{x}(t)$ versus $x(t)$ on separate sheets of graph paper.

9. Show that the time rate of change of the total energy $E(x, \dot{x})$ for the damped nonlinear pendulum equation

$$\frac{d^2x}{dt^2} + c\frac{dx}{dt} + \omega^2 \sin x = 0$$

is given by

$$\frac{dE}{dt} = -c\left(\frac{dx}{dt}\right)^2 = -2c \text{ (Kinetic energy)},$$

where

$$E(x, \dot{x}) = \frac{1}{2}\dot{x}^2 + \omega^2(1 - \cos x).$$

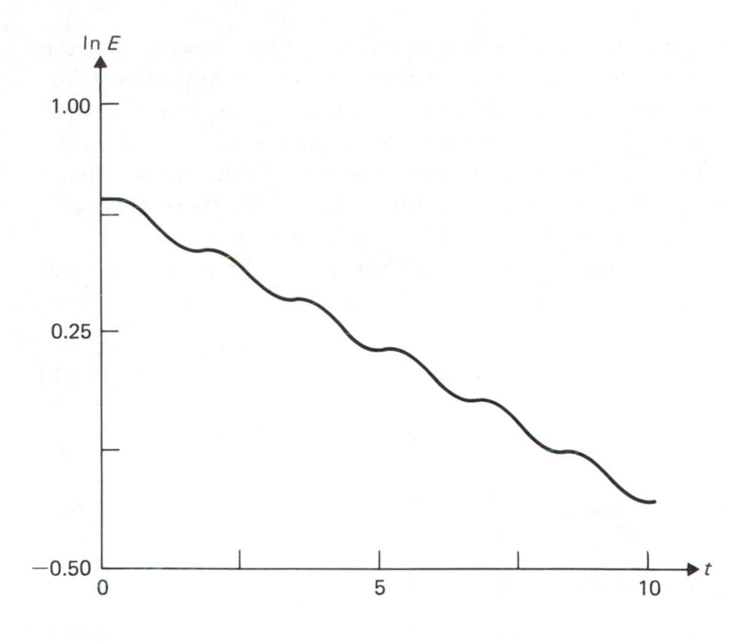

Figure 7.4 Plot of ln E versus t for Exercise 9 ($x_0 = 1.0$, $\dot{x}_0 = 0.5$, $c = 0.1$, $\omega^2 = 4.0$).

If the pendulum is oscillating, the averaged total energy $\overline{E}$ is almost equal to one-half the averaged kinetic energy when c and x are small. Hence

$$\frac{d\overline{E}}{dt} \cong -c\overline{E} \qquad \text{and} \qquad \ln \overline{E} \cong -ct + \text{const.}$$

A plot of the logarithm of the *instantaneous energy* versus time should be a curve that oscillates around the plot of the logarithm of the *averaged energy* versus time. (The concept of a time-dependent averaged variable is developed in the method of averaging in [9].) Investigate this by integrating the damped nonlinear pendulum equation numerically with RK4 on $0 \leq t \leq 10$, $c = 0.1$, $\omega^2 = 4$, and $h = 0.1$. Take $x = x_0$, $\dot{x} = 0.5$ at $t = 0$, $x_0 = 0.0$, 0.1, 0.5, 1.0. Compute E and its logarithm at each step and plot ln E versus t for each choice of x_0. A sample plot is given in Fig. 7.4.

7.2 PHASE PATHS, EQUILIBRIUM STATES, AND ALMOST LINEAR SYSTEMS

In this section we begin the study of two-dimensional autonomous systems,

$$\frac{dx}{dt} = P(x, y), \qquad \frac{dy}{dt} = Q(x, y), \tag{7.2.1}$$

where $P(x, y)$, $Q(x, y)$ are continuously differentiable functions defined in a region of the xy-plane. According to the existence theorem found in the Appendix 1, for any point (x_0, y_0) in the region of differentiability and for any t_0, there is a unique solution $x = x(t)$, $y = y(t)$ of (7.2.1) such that $x(t_0) = x_0$, $y(t_0) = y_0$. A curve in the *phase plane* (the xy-plane) that is described parametrically by a solution is called a *trajectory:* on a trajectory, position is given as a function of time. More generally, the graph of a solution in the phase plane is a *phase path,* or *orbit.*

The following example should help distinguish between the concepts of a trajectory and a phase path. The equation

$$\frac{d^2x}{dt^2} + x = 0 \tag{7.2.2}$$

is equivalent to the system

$$\frac{dx}{dt} = y, \qquad \frac{dy}{dt} = -x, \tag{7.2.3}$$

as is easily verified. A general solution of (7.2.2) is

$$x(t) = C\sin(t - \theta), \tag{7.2.4}$$

where C and θ are arbitrary constants and t is the independent variable. Therefore a general solution of the system (7.2.3) is

$$x(t) = C\sin(t - \theta), \qquad y(t) = C\cos(t - \theta), \qquad -\infty < t < \infty. \tag{7.2.5}$$

Clearly, the graph of the solution (7.2.5) is the circle

$$x^2 + y^2 = C^2.$$

The circle is a phase path, while the graph of (7.2.5) is a trajectory. This may seem pedantic, but if you were tracking a satellite, its *trajectory* (position as a function of time) is far more important than its path. Other examples are easy to imagine.

For many problems, the constant solutions, if any, are significant. Constant solutions are called *critical points,* or *equilibrium points* of the system (7.2.1). Since the derivatives must vanish if the solution is constant, it follows that $(\bar{x}, \bar{y})$ are the coordinates of an equilibrium point if and only if

$$P(\bar{x}, \bar{y}) = 0 \qquad \text{and} \qquad Q(\bar{x}, \bar{y}) = 0. \tag{7.2.6}$$

EXAMPLE 1 Find the equilibrium points of the system

$$\frac{dx}{dt} = x(2 + x + y), \qquad \frac{dy}{dt} = y(2 - x + y).$$

SOLUTION The equilibrium points are the solutions to the nonlinear algebraic equations

$$x(2 + x + y) = 0, \qquad y(2 - x + y) = 0.$$

There are four possibilities:

a) $\bar{x} = 0, \quad \bar{y} = 0$
b) $\bar{x} = 0, \quad 2 - \bar{x} + \bar{y} = 0$
c) $2 + \bar{x} + \bar{y} = 0, \quad \bar{y} = 0$
d) $2 + \bar{x} + \bar{y} = 0, \quad 2 - \bar{x} + \bar{y} = 0$

Solving in each of the cases gives

a) $(0, 0)$, b) $(0, -2)$, c) $(-2, 0)$, d) $(0, -2)$;

hence there are three distinct equilibrium points. ■

The solutions near an equilibrium point $(\bar{x}, \bar{y})$ can be studied by creating a linear differential equation that approximates the original differential equation near the equilibrium point. To do this, let $x = \bar{x} + u$, $y = \bar{y} + v$ and substitute them into the system (7.2.1) to obtain

$$\frac{dx}{dt} = 0 + \frac{du}{dt} = P(\bar{x} + u, \bar{y} + v),$$

$$\frac{dy}{dt} = 0 + \frac{dv}{dt} = Q(\bar{x} + u, \bar{y} + v).$$

It was assumed that $P(x, y)$ and $Q(x, y)$ are continuously differentiable and, consequently, the Taylor expansion gives

$$P(\bar{x} + u, \bar{y} + v) = P(\bar{x}, \bar{y}) + P_x(\bar{x}, \bar{y})u + P_y(\bar{x}, \bar{y})v + R(u, v),$$

$$Q(\bar{x} + u, \bar{y} + v) = Q(\bar{x}, \bar{y}) + Q_x(\bar{x}, \bar{y})u + Q_y(\bar{x}, \bar{y})v + S(u, v),$$

where the subscripts denote partial derivatives, and $R(u, v)$ and $S(u, v)$ are the remainder terms. Since $(\bar{x}, \bar{y})$ is an equilibrium point, $P(\bar{x}, \bar{y}) = 0$, $Q(\bar{x}, \bar{y}) = 0$, and therefore u and v satisfy the differential equations

$$\frac{du}{dt} = au + bv + R(u, v), \qquad \frac{dv}{dt} = cu + dv + S(u, v), \tag{7.2.7}$$

where the constants a, b, c, d are defined by

$$a = P_x(\bar{x}, \bar{y}), \quad b = P_y(\bar{x}, \bar{y}), \quad c = Q_x(\bar{x}, \bar{y}), \quad d = Q_y(\bar{x}, \bar{y}). \tag{7.2.8}$$

If $P(x, y)$ and $Q(x, y)$ have continuous second derivatives, then the remainders $R(u, v)$ and $S(u, v)$ are small near $u = 0$, $v = 0$. In fact, Taylor's theorem states that $R(u, v)$ and $S(u, v)$ vanish so rapidly that

$$\lim_{(u,\, v)\to(0,\, 0)} \frac{R(u, v)}{(u^2 + v^2)^{1/2}} = 0, \qquad \lim_{(u,\, v)\to(0,\, 0)} \frac{S(u, v)}{(u^2 + v^2)^{1/2}} = 0. \tag{7.2.9}$$

If, in addition, the condition

$$ad - bc \neq 0 \tag{7.2.10}$$

is satisfied, then it can be shown that the equilibrium point $(\bar{x}, \bar{y})$ is *isolated.* By this it is meant that there is some neighborhood of $(\bar{x}, \bar{y})$ containing no other equilibria.

Definition

A differential equation of the form (7.2.7), for which the conditions (7.2.9) and (7.2.10) are satisfied, is said to be *almost linear.*

The solutions to an almost linear differential equation are approximated in a neighborhood of $u = 0$, $v = 0$ by the solutions of the corresponding linear system

$$\frac{du}{dt} = au + bv, \qquad \frac{dv}{dt} = cu + dv. \tag{7.2.11}$$

Therefore a description of the solutions of the linear system (7.2.11) near (0, 0) will help analyze the behavior of solutions of the original nonlinear system (7.2.1) near the equilibrium point $(\bar{x}, \bar{y})$.

A precise formulation of the sense in which the solutions to an almost linear system are approximated by the solutions of the corresponding linear system can be given by using techniques developed in more advanced texts [3, 4, 5]. We shall illustrate it with a carefully chosen example of an almost linear system that can be solved explicitly—a rare occurrence!

EXAMPLE 2 Show that the system

$$\frac{dx}{dt} = -y(1 - x^2 - y^2), \qquad \frac{dy}{dt} = x(1 - x^2 - y^2)$$

has an isolated equilibrium point $\bar{x} = 0$, $\bar{y} = 0$, as well as a circle of equilibrium points, $\bar{x}^2 + \bar{y}^2 = 1$.

SOLUTION The trivial change of variables, $x = 0 + u$, $y = 0 + v$ gives the almost linear system

$$\frac{du}{dt} = -v + v(u^2 + v^2), \qquad \frac{dv}{dt} = +u - u(u^2 + v^2),$$

and therefore $R(u, v) = v(u^2 + v^2)$ and $S(u, v) = -u(u^2 + v^2)$. The reader can easily check that both conditions (7.2.9) and (7.2.10) are satisfied. Now multiply the first equation by u, the second by v, and then add them to obtain

$$u\frac{du}{dt} + v\frac{dv}{dt} = -uv + vu + uv(u^2 + v^2) - vu(u^2 + v^2) = 0.$$

The last relation is equivalent to

$$\frac{1}{2}\frac{d}{dt}(u^2 + v^2) = 0,$$

which implies that $u^2 + v^2 = c^2$, a constant; i.e., the orbits of the almost linear system are circles. The orbits of the corresponding linear system

$$\frac{du}{dt} = -v, \qquad \frac{dv}{dt} = u$$

are also circles. It can be easily shown (see Exercise 17, p. 487) that

$$u_1(t) = \rho\cos(\omega t + \phi_0), \qquad v_1(t) = \rho\sin(\omega t + \phi_0),$$

where $\omega = 1 - \rho^2$, $0 \leq \rho < 1$, is the solution to the almost linear system within the unit circle. Furthermore,

$$u_2(t) = \rho\cos(t + \phi_0), \qquad v_2(t) = \rho\sin(t + \phi_0)$$

is the corresponding solution to the linear system, and both solutions satisfy the same initial conditions

$$u(0) = u_0 = \rho\cos\phi_0, \qquad v(0) = v_0 = \rho\sin\phi_0.$$

Both solutions lie on the circle $u^2 + v^2 = \rho^2$, but the dependence of ω on ρ means that they will drift apart as t increases. However, since

$$|u_1(t) - u_2(t)| \leq 2\rho, \qquad |v_1(t) - v_2(t)| \leq 2\rho,$$

their difference is bounded by 2ρ, the diameter of the circle. Furthermore, if ρ is small, ω will be almost unity, so the solutions will drift apart very slowly. We conclude that the closer we are to the equilibrium point (0, 0), the better the solutions of the linear system approximate the solutions of the almost linear system. ■

EXAMPLE 3 Find the equilibrium points and construct approximate solutions by solving the corresponding linear differential equations for the system

$$\frac{dx}{dt} = x, \qquad \frac{dy}{dt} = y(1 - x + y).$$

SOLUTION Solving the equations

$$P(x, y) = x = 0, \qquad Q(x, y) = y(1 - x + y) = 0,$$

gives the two equilibrium points (0, 0) and (0, −1). Furthermore

$$P_x(\bar{x}, \bar{y}) = 1, \qquad P_y(\bar{x}, \bar{y}) = 0,$$
$$Q_x(\bar{x}, \bar{y}) = -\bar{y}, \qquad Q_y(\bar{x}, \bar{y}) = 1 - \bar{x} + 2\bar{y}.$$

We first study $(\bar{x}, \bar{y}) = (0, 0)$ and let $x = 0 + u$, $y = 0 + v$, which gives

$$a = P_x(0, 0) = 1, \qquad b = P_y(0, 0) = 0,$$
$$c = Q_x(0, 0) = 0, \qquad d = Q_y(0, 0) = 1.$$

Therefore

$$\frac{du}{dt} = u, \qquad \frac{dv}{dt} = v + (v^2 - uv)$$

and

$$\frac{du}{dt} = u, \qquad \frac{dv}{dt} = v,$$

are the almost linear and linear differential equations associated with the equilibrium point (0, 0). To see that (0, 0) is an isolated equilibrium point, the conditions (7.2.9) and (7.2.10) will be verified. Clearly, $ad - bc = 1 \neq 0$. To show that the remainder terms $R(u, v) = 0$, $S(u, v) = v^2 - uv$ satisfy (7.2.9), polar coordinates

$$u = \rho \cos \phi, \qquad v = \rho \sin \phi$$

are introduced. Then

$$\frac{S(u, v)}{(u^2 + v^2)^{1/2}} = \frac{S(\rho \cos \phi, \rho \sin \phi)}{\rho} = \frac{\rho^2 \sin^2 \phi - \rho^2 \cos \phi \sin \phi}{\rho}$$
$$= \rho(\sin^2 \phi - \cos \phi \sin \phi) \to 0 \quad \text{as} \quad \rho \to 0,$$

so both conditions are satisfied. The theory asserts that the solution to the linear system

$$u(t) = u_0 e^t, \qquad v(t) = v_0 e^t$$

is an approximate solution to the almost linear system for a limited time interval if $u_0^2 + v_0^2$ is small.

The second equilibrium point is $(\bar{x}, \bar{y}) = (0, -1)$, and since

$$a = P_x(0, -1) = 1, \qquad b = P_y(0, -1) = 0,$$
$$Q_x(0, -1) = 1, \qquad Q_y(0, -1) = -1,$$

substituting $x = 0 + u$, $y = -1 + v$ gives

$$\frac{du}{dt} = u, \qquad \frac{dv}{dt} = u - v + (v^2 - uv),$$

and

$$\frac{du}{dt} = u, \qquad \frac{dv}{dt} = u - v,$$

the almost linear and linear systems of differential equations associated with the equilibrium point $(0, -1)$. Observe that since $P(x, y)$ and $Q(x, y)$ in this problem are polynomials of degree not more than 2, then $R(u, v)$ and $S(u, v)$ do not depend on the coordinates of the equilibrium point being studied. The theory asserts that the solution to the linear system

$$u = u_0 e^t, \qquad v = \left(v_0 - \frac{u_0}{2}\right) e^{-t} + \frac{u_0}{2} e^t$$

is an approximate solution to the almost linear system for a limited time interval if $u_0^2 + v_0^2$ is small. ■

The *local nature* of the approximation of solutions near $(u, v) = (0, 0)$ of the almost linear system by the solutions of the linear system must be emphasized. Furthermore, it is not necessarily the case that the phase paths of the two systems are similar. For instance, the phase paths of the first could be a family of spirals focused at $(0, 0)$, while for the second system they could be a family of circles centered at $(0, 0)$. Hence local approximation would be good for short time intervals but not long ones. We shall also study the possible phase plane portraits of both linear and almost linear two-dimensional autonomous systems.

EXERCISES 7.2

Find the equilibrium points of the following systems.

1. $\dfrac{dx}{dt} = x + 2y + 1, \quad \dfrac{dy}{dt} = 3x + y - 2$

2. $\dfrac{dx}{dt} = 2x + 5y - 11, \quad \dfrac{dy}{dt} = x + y + 1$

3. $\dfrac{dx}{dt} = xy + x - 4y + 6, \quad \dfrac{dy}{dt} = x - y + 2$

4. $\dfrac{dx}{dt} = -x + y + 3, \quad \dfrac{dy}{dt} = 2xy + 8x - 12$

5. $\dfrac{dx}{dt} = 4x^2 + y^2 - 17, \quad \dfrac{dy}{dt} = x^2 + y^2 - 5$

6. $\dfrac{dx}{dt} = x^2 + y^2 - 2x - 2y, \quad \dfrac{dy}{dt} = x + y - 2$

Find the equilibrium points and construct approximate solutions near them by solving the corresponding system of linear differential equations.

7. $\dfrac{dx}{dt} = 2x - y - 1, \quad \dfrac{dy}{dt} = 3x + 6y - 24$

8. $\dfrac{dx}{dt} = 3x + 4y + 1, \quad \dfrac{dy}{dt} = 2x + y - 1$

9. $\dfrac{dx}{dt} = y, \quad \dfrac{dy}{dt} = 1 - e^x$

10. $\dfrac{dx}{dt} = 1 - y^3, \quad \dfrac{dy}{dt} = x$

11. $\dfrac{dx}{dt} = y + x^2, \quad \dfrac{dy}{dt} = x^2 - x - 2$

12. $\dfrac{dx}{dt} = 4 - y^2, \quad \dfrac{dy}{dt} = 2x$

Verify that the following systems are almost linear with respect to origin. (*Hint:* Use polar coordinates.)

13. $\dfrac{dx}{dt} = -4x + 5y + x \sin y, \quad \dfrac{dy}{dt} = x + 4y + y \sin x$

14. $\dfrac{dx}{dt} = 3x + 2y + x^2 + 2xy^4, \quad \dfrac{dy}{dt} = -2x + y + y^2$

Consider the initial value problem

$$\frac{dx}{dt} = P(x, y), \qquad \frac{dy}{dt} = Q(x, y),$$

$$x(0) = \bar{x} + u_0, \qquad y(0) = \bar{y} + v_0,$$

where u_0 and v_0 are measurement errors and $(\bar{x}, \bar{y})$ is an equilibrium point of the system. Let

$$x(t) = \bar{x} + u(t), \qquad y(t) = \bar{y} + v(t),$$

then construct approximations for $u(t)$ and $v(t)$ by solving a suitable linear system of differential equations. Find conditions on the eigenvalues of the linear system so that the error is bounded for all positive t. Apply your analysis to the following initial value problems.

15. $\dfrac{dx}{dt} = -x - 3y + x \sin y, \quad x(0) = 0 + 10^{-4}$

$\dfrac{dy}{dt} = 3x - y + 1 - \cos y, \quad y(0) = 0 - 10^{-4}$

16. $\dfrac{dx}{dt} = x + 5y + xy, \quad x(0) = 0 + 10^{-10}$

$\dfrac{dy}{dt} = 5x + y + x^2 + y^2, \quad y(0) = 0$

17. Show that $x(t) = \rho \cos(\omega t + \phi)$, $y(t) = \rho \sin(\omega t + \phi)$, $\omega = 1 - \rho^2$, with ρ, ϕ arbitrary constants, is a solution to the nonlinear system

$$\frac{dx}{dt} = -y(1 - x^2 - y^2), \qquad \frac{dy}{dt} = x(1 - x^2 - y^2),$$

$$x(0) = x_0, \qquad y(0) = y_0, \qquad \rho^2 = x_0^2 + y_0^2.$$

18. Show that if $y_0 > x_0 \geq 0$, then the difference between the solutions of the two systems

$$\frac{dx}{dt} = x, \qquad \frac{dy}{dt} = y$$

and

$$\frac{dx}{dt} = x, \qquad \frac{dy}{dt} = y(1 - x + y),$$

where $x(0) = x_0$, $y(0) = y_0$, diverges as t increases. (*Hint:* Let $y(t) = w(t)e^t$, $x(t) = x_0e^t$ denote the solution to the nonlinear system. Derive the differential equation satisfied by w and show that $w(t) > y_0$ for t large.)

7.3 THE PHASE PLANE AND STABILITY

The *phase plane* of solutions to a two-dimensional autonomous system

$$\frac{dx}{dt} = P(x, y), \qquad \frac{dy}{dt} = Q(x, y) \tag{7.3.1}$$

can be considered to be generated by particles flowing in the plane with the velocity at the point (x, y) given by

$$\mathbf{V} = \frac{dx}{dt}\mathbf{e}_1 + \frac{dy}{dt}\mathbf{e}_2 = P(x, y)\,\mathbf{e}_1 + Q(x, y)\,\mathbf{e}_2. \tag{7.3.2}$$

The curves generated by the particles are the graphs of a one-parameter family of solutions $x = x(t - s)$, $y = y(t - s)$, where s is arbitrary. This *time shift invariance* of the solutions is a consequence of the system being autonomous; i.e., the independent variable t does not appear explicitly. To prove this invariance, observe that if

$$\frac{dx}{dt}(t) = P(x(t), y(t)), \qquad \frac{dy}{dt}(t) = Q(x(t), y(t)),$$

then

$$\frac{d}{dt}x(t - s) = \frac{dx(t - s)}{d(t - s)} = P(x(t - s), y(t - s)),$$

$$\frac{d}{dt}y(t - s) = \frac{dx(t - s)}{d(t - s)} = Q(x(t - s), y(t - s)).$$

The phase plane of system (7.3.1) will also include equilibrium states, if any. Some will be stagnation points around which the flow swirls, while others will be sources from which all nearby paths exit, or sinks into which they enter. There are also equilibrium states from which exactly two paths exit and into which two paths enter. Our understanding of the geometric structure of the phase paths is facilitated by constructing phase plane portraits consisting of sketches of selected phase paths and equilibrium states. It is convenient to extend the definition of a path, so that equilibrium states may be included. A path is said to *enter* an equilibrium point $(\bar{x}, \bar{y})$ if

$$\lim_{t \to -\infty} x(t) = \bar{x}, \qquad \lim_{t \to -\infty} y(t) = \bar{y}, \qquad \text{and} \lim_{t \to -\infty} \frac{y(t) - \bar{y}}{x(t) - \bar{x}}$$

exists or diverges to $\pm\infty$. A path is said to *leave* an equilibrium state $(\bar{x}, \bar{y})$ if

$$\lim_{t \to -\infty} x(t) = \bar{x}, \qquad \lim_{t \to -\infty} y(t) = \bar{y}, \qquad \text{and} \lim_{t \to -\infty} \frac{y(t) - \bar{y}}{x(t) - \bar{x}}$$

exists or diverges to $\pm\infty$. Look ahead to Fig. 7.8 to see an example where two paths enter and two paths leave the (unstable) equilibrium state (0, 0).

EXAMPLE 1 Discuss the equilibrium states of the system

$$\frac{dx}{dt} = x, \qquad \frac{dy}{dt} = y.$$

SOLUTION The only equilibrium state is $\bar{x} = 0$, $\bar{y} = 0$, and there is a family of paths covering the entire xy-plane (with exception of the origin) that leave the equilibrium state in every possible direction. These paths are the graphs of the family of solutions parametrized by α:

$$x(t) = (\cos \alpha)e^t, \qquad y(t) = (\sin \alpha)e^t, \quad 0 \leq \alpha < 2\pi.$$

Clearly, $\lim_{t \to -\infty} x(t) = \lim_{t \to -\infty} y(t) = 0$ and

$$\lim_{t \to -\infty} \frac{(\sin \alpha)e^t - 0}{(\cos \alpha)e^t - 0} = \tan \alpha,$$

so the paths can leave in every possible direction. Figure 7.5 shows the phase plane of the system. ■

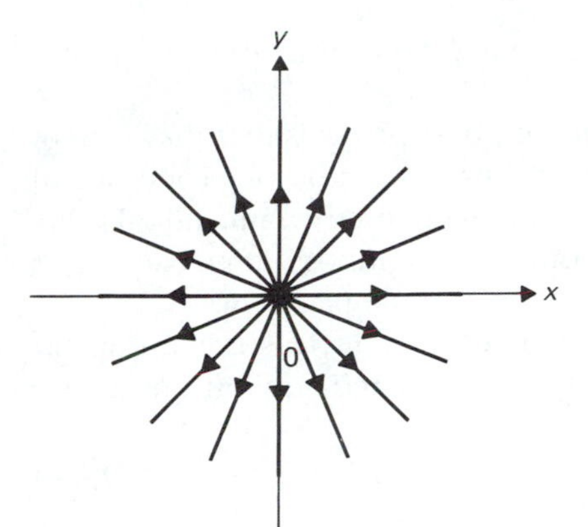

Figure 7.5 A phase plane portrait for Example 1.

Associated with the two-dimensional autonomous system (7.3.1) is the first order differential equation

$$\frac{dy}{dx} = \frac{Q(x, y)}{P(x, y)} \qquad \text{or} \qquad Q(x, y)\,dx - P(x, y)\,dy = 0. \tag{7.3.3}$$

The graphs of solutions to (7.3.3) are called *integral curves* and an implicit solution $F(x,y) = \text{const}$ is called an *integral.*

Unlike phase paths, integral curves are not directed since t, the independent variable in (7.3.1), has been eliminated in (7.3.3). The integral curves contain the phase paths, but the converse is not true. This is illustrated by the system in Example 1 whose associated first order equation

$$\frac{dy}{dx} = \frac{y}{x} \qquad \text{or} \qquad y\,dx - x\,dy = 0$$

has an implicit solution, $\arctan(y/x) = \text{const}$ or $y = mx$. Each choice of m determines a straight line through the origin that contains two phase paths from the family previously given. These are

$$x(t) = (\cos\alpha)e^t, \qquad y(t) = (\sin\alpha)e^t$$

and

$$x(t) = [\cos(\alpha + \pi)]e^t = -(\cos\alpha)e^t,$$
$$y(t) = [\sin(\alpha + \pi)]e^t = -(\sin\alpha)e^t,$$

where α, $0 \le \alpha < \pi$, is chosen to satisfy $m = (\sin\alpha)/(\cos\alpha)$ with the choice of $\alpha = \pi/2$ corresponding to the y-axis.

A property of integral curves, and therefore of phase paths, is that they cannot intersect except at equilibrium states. Otherwise there would be two solutions of the first order equation (7.3.3), $y = y_1(x)$, $y = y_2(x)$, taking on the same initial value $y_0 = y_1(x_0)$, $y_0 = y_2(x_0)$, where (x_0, y_0) is the hypothesized point of intersection. This would violate the uniqueness of solutions of the initial value problem.

The easiest class of two-dimensional autonomous systems to study are linear constant coefficient systems that have the origin as an isolated equilibrium state

$$\frac{dx}{dt} = ax + by, \qquad \frac{dy}{dt} = cx + dy, \qquad ad - bc \neq 0. \tag{7.3.4}$$

The eigenvalue–eigenvector method of Chapter 5 provides us with a systematic approach to the construction of phase plane portraits for these systems. Let $x = \alpha e^{\lambda t}$, $y = \beta e^{\lambda t}$. Then, after substituting and dividing by $e^{\lambda t}$, we have

$$\lambda\alpha = a\alpha + b\beta, \qquad \lambda\beta = c\alpha + d\beta.$$

The eigenvalues λ_1 and λ_2 are roots of the characteristic equation

$$0 = \begin{vmatrix} a - \lambda & b \\ c & d - \lambda \end{vmatrix} = \lambda^2 - (a + d)\lambda + (ad - bc).$$

Note that the assumption $ad - bc \neq 0$ implies that no eigenvalue can be zero.

The equilibrium state $\bar{x} = 0$, $\bar{y} = 0$ is classified by the type of eigenvalue. If the eigenvalues are real, there are five possibilities:

Real 1: $\lambda_1 < \lambda_2 < 0$, Real 2: $0 < \lambda_1 < \lambda_2$,

$$\text{Real 3:}\quad \lambda_2 < 0 < \lambda_1, \qquad \text{Real 4:}\quad \lambda_1 = \lambda_2 < 0,$$
$$\text{Real 5:}\quad 0 < \lambda_1 = \lambda_2.$$

If the eigenvalues are complex, $\lambda = p + iq$, $\lambda^* = p - iq$, there are three possibilities:

$$\text{Complex 1:}\quad p = \text{Re}(\lambda) < 0,$$
$$\text{Complex 2:}\quad p = \text{Re}(\lambda) > 0,$$
$$\text{Complex 3:}\quad p = \text{Re}(\lambda) = 0.$$

Each case will be examined in some detail and a typical phase plane portrait will be sketched.

Real 1: $\lambda_1 < \lambda_2 < 0$. By using the theory developed in Chapter 5 it can be established that there are two linearly independent eigenvectors and that a general solution to (7.3.4) can be written in the form

$$x(t) = c_1\alpha_1 e^{\lambda_1 t} + c_2\alpha_2 e^{\lambda_2 t},$$
$$y(t) = c_1\beta_1 e^{\lambda_1 t} + c_2\beta_2 e^{\lambda_2 t}$$

or

$$\begin{bmatrix} x(t) \\ y(t) \end{bmatrix} = c_1\begin{bmatrix} \alpha_1 \\ \beta_1 \end{bmatrix} e^{\lambda_1 t} + c_2\begin{bmatrix} \alpha_2 \\ \beta_2 \end{bmatrix} e^{\lambda_2 t}.$$

The vectors in parentheses are the eigenvectors associated with λ_1 and λ_2, respectively. Points on the half-lines determined by the eigenvectors flow toward the origin as t increases since the exponentials $e^{\lambda_1 t}$ and $e^{\lambda_2 t}$ tend to zero. Points in the sectors between these half-lines flow toward the origin asymptotically to the half-line associated with λ_2 since $\lambda_1 < \lambda_2 < 0$ implies that $e^{\lambda_1 t}$ damps out more rapidly than $e^{\lambda_2 t}$. A more formal justification is given by the computation

$$\frac{y(t)}{x(t)} = \frac{e^{\lambda_2 t}(c_1\beta_1 e^{(\lambda_1 - \lambda_2)t(\lambda_1-\lambda_2)t} + c_2\beta_2)}{e^{\lambda_2 t}(c_1\alpha_1 e^{(\lambda_1-\lambda_2)t} + c_2\alpha_2)} \to \frac{\beta_2}{\alpha_2}$$

as $t \to \infty$ since $\lambda_1 - \lambda_2 < 0$.

EXAMPLE 2 Find the general solution and sketch a phase plane portrait for the system

$$\frac{dx}{dt} = -6x + 3y, \qquad \frac{dy}{dt} = -2x - y.$$

SOLUTION The characteristic equation

$$0 = \begin{vmatrix} -6-\lambda & 3 \\ -2 & -1-\lambda \end{vmatrix} = \lambda^2 + 7\lambda + 12 = (\lambda + 3)(\lambda + 4)$$

has roots $\lambda_1 = -4$, $\lambda_2 = -3$. The eigenvectors are found by solving the algebraic system

$$(-6-\lambda)\alpha + 3\beta = 0, \qquad -2\alpha + (-1-\lambda)\beta = 0$$

with $\lambda = \lambda_1 = -4$ and $\lambda = \lambda_2 = -3$. This leads to the equations

$$-2\alpha_1 + 3\beta_1 = 0 \qquad \text{and} \qquad -3\alpha_2 + 3\beta_2 = 0.$$

Nontrivial solutions are $\alpha_1 = 3$, $\beta_1 = 2$, and $\alpha_2 = 1$, $\beta_2 = 1$; hence a general solution is

$$\begin{bmatrix} x(t) \\ y(t) \end{bmatrix} = c_1 \begin{bmatrix} 3 \\ 2 \end{bmatrix} e^{-4t} + c_2 \begin{bmatrix} 1 \\ 1 \end{bmatrix} e^{-3t}.$$

If c_1 and c_2 are not both zero, the first term disappears more rapidly than the second, so as the phase path approaches the origin it becomes asymptotic to the line $y = x$ determined by the second eigenvector. A phase plane portrait is sketched in Fig. 7.6. ■

If all the phase paths enter the equilibrium state as in the case above, the equilibrium state is called a *stable node*. If all the phase paths leave the equilibrium state, the equilibrium state is called an *unstable node*. The latter is illustrated by the next type.

Real 2: $0 < \lambda_2 < \lambda_1$. Since the eigenvalues λ_1 and λ_2 are both positive, the phase paths enter the origin as $t \to -\infty$. The preceding phase plane portrait described for the case $\lambda_1 < \lambda_2 < 0$ is unchanged except that the arrows showing the direction of increasing time point outward instead of inward.

EXAMPLE 3 By replacing t by $-t$ in Example 2 we obtain the system

$$\frac{dx}{dt} = 6x - 3y, \qquad \frac{dy}{dt} = 2x + y.$$

Find a general solution and sketch a phase plane portrait.

SOLUTION The eigenvalues are $\lambda_1 = 4$, $\lambda_2 = 3$, and a general solution is

$$\begin{bmatrix} x(t) \\ y(t) \end{bmatrix} = c_1 \begin{bmatrix} 3 \\ 2 \end{bmatrix} e^{4t} + c_2 \begin{bmatrix} 1 \\ 1 \end{bmatrix} e^{3t}.$$

A phase plane portrait is sketched in Fig. 7.7. ■

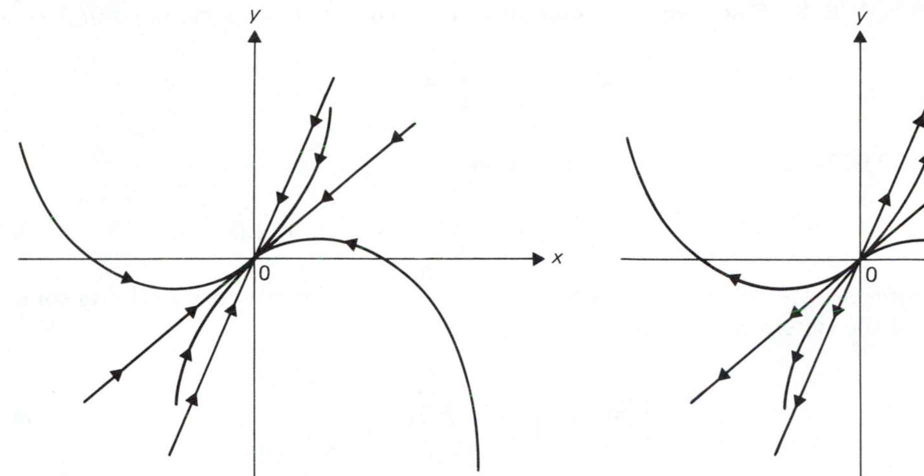

Figure 7.6 A phase plane portrait with $\lambda_1 < \lambda_2 < 0$.

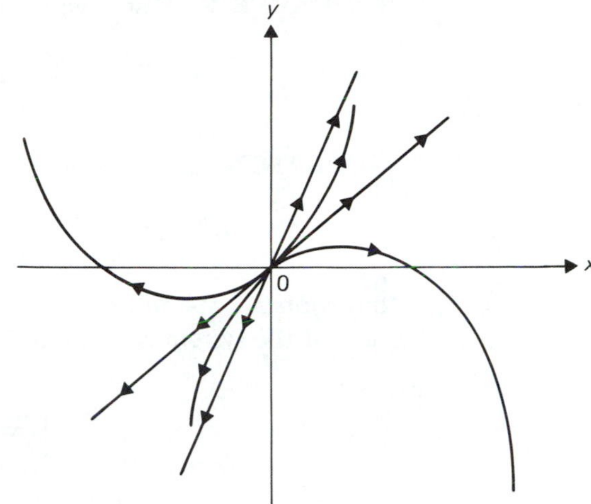

Figure 7.7 A phase plane portrait with $0 < \lambda_1 < \lambda_2$.

Real 3: $\lambda_2 < 0 < \lambda_1$. If the eigenvalues are real and of opposite sign, the equilibrium state is *unstable* and is called a *saddle point*. The points on the half-lines corresponding to the negative eigenvalue flow toward the origin, and those on the half-lines corresponding to the positive eigenvalue flow away from the origin as time increases. A general solution with c_1, $c_2 \neq 0$ is of the form

$$\begin{bmatrix} x(t) \\ y(t) \end{bmatrix} = c_1 \begin{bmatrix} \alpha_1 \\ \beta_1 \end{bmatrix} e^{\lambda_1 t} + c_2 \begin{bmatrix} \alpha_2 \\ \beta_2 \end{bmatrix} e^{\lambda_2 t},$$

and therefore

$$\frac{y(t)}{x(t)} = \frac{e^{\lambda_1 t}(c_1\beta_1 + c_2\beta_2 e^{(\lambda_2-\lambda_1)t})}{e^{\lambda_1 t}(c_1\alpha_1 + c_2\alpha_2 e^{(\lambda_2-\lambda_1)t})} \to \frac{\beta_1}{\alpha_1}$$

as $t \to \infty$ since $\lambda_2 - \lambda_1 < 0$, and

$$\frac{y(t)}{x(t)} = \frac{e^{\lambda_2 t}(c_1\beta_1 e^{(\lambda_1-\lambda_2)t} + c_2\beta_2)}{e^{\lambda_2 t}(c_1\alpha_1 e^{(\lambda_1-\lambda_2)t} + c_2\alpha_2)} \to \frac{\beta_2}{\alpha_2}$$

as $t \to -\infty$ since $\lambda_1 - \lambda_2 > 0$. Therefore points on the sectors between the half-lines flow away from the origin asymptotically toward the half-line associated with the positive eigenvalue as $t \to +\infty$ and asymptotically toward the half-line associated with the negative eigenvalue as $t \to -\infty$.

EXAMPLE 4 Find a general solution and sketch the phase plane portrait for

$$\frac{dx}{dt} = x + 3y, \quad \frac{dy}{dt} = 3x + y.$$

SOLUTION The characteristic equation

$$0 = \begin{vmatrix} 1 - \lambda & 3 \\ 3 & 1 - \lambda \end{vmatrix} = \lambda^2 - 2\lambda - 8 = (\lambda + 2)(\lambda - 4)$$

has roots $\lambda_1 = 4$ and $\lambda_2 = -2$, and by computing the eigenvectors a general solution of the system is found to be

$$\begin{bmatrix} x(t) \\ y(t) \end{bmatrix} = c_1 \begin{bmatrix} 1 \\ 1 \end{bmatrix} e^{4t} + c_2 \begin{bmatrix} 1 \\ -1 \end{bmatrix} e^{-2t}.$$ ■

A phase plane portrait is sketched in Fig. 7.8. This sketch should be compared with the sketch of paths near the unstable equilibrium states $(\pi, 0)$ or $(-\pi, 0)$ of the pendulum equation in Fig. 7.3.

Real 4: $\lambda_1 = \lambda_2 < 0$. If the eigenvalues are real, equal, and negative, the equilibrium state is a *stable node.* If λ is the common value, then there are two subcases that require separate discussion:

1. There are two linearly independent eigenvectors.
2. There is one linearly independent eigenvector.

We consider first subcase (1). Let

$$\begin{bmatrix} \alpha_1 \\ \beta_1 \end{bmatrix} \quad \text{and} \quad \begin{bmatrix} \alpha_2 \\ \beta_2 \end{bmatrix}$$

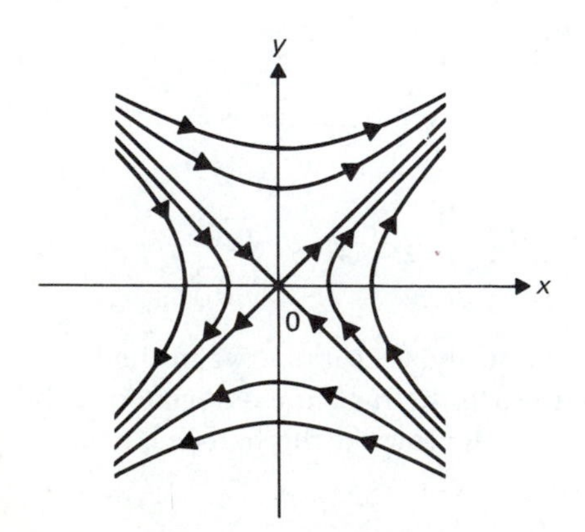

Figure 7.8 A phase plane portrait with $\lambda_2 < 0 < \lambda_1$.

be the two linearly independent eigenvectors. The algebraic system satisfied by the components of the eigenvectors,

$$\begin{cases} (a-\lambda)\alpha_1 + b\beta_1 = 0 \\ c\alpha_1 + (d-\lambda)\beta_1 = 0 \end{cases} \quad \text{and} \quad \begin{cases} (a-\lambda)\alpha_2 + b\beta_2 = 0 \\ c\alpha_2 + (d-\lambda)\beta_2 = 0, \end{cases}$$

can be regrouped (λ is the same throughout) to give

$$\begin{cases} (a-\lambda)\alpha_1 + b\beta_1 = 0 \\ (a-\lambda)\alpha_2 + b\beta_2 = 0 \end{cases} \quad \text{and} \quad \begin{cases} c\alpha_1 + (d-\lambda)\beta_1 = 0 \\ c\alpha_2 + (d-\lambda)\beta_2 = 0. \end{cases}$$

The linear independence of the eigenvectors implies that

$$\begin{vmatrix} \alpha_1 & \alpha_2 \\ \beta_1 & \beta_2 \end{vmatrix} = \alpha_1\beta_2 - \alpha_2\beta_1 \neq 0,$$

and therefore each of the last pairs of equations can have only trivial solutions, hence $a - \lambda = 0$, $b = 0$, $c = 0$, $d - \lambda = 0$. This means that when there is a double eigenvalue and two linearly independent eigenvectors, the system must be of the form

$$\frac{dx}{dt} = \lambda x, \qquad \frac{dy}{dt} = \lambda y.$$

A general solution is

$$\begin{bmatrix} x(t) \\ y(t) \end{bmatrix} = c_1\begin{bmatrix} 1 \\ 0 \end{bmatrix} e^{\lambda t} + c_2\begin{bmatrix} 0 \\ 1 \end{bmatrix} e^{\lambda t} = \begin{bmatrix} c_1 \\ c_2 \end{bmatrix} e^{\lambda t},$$

and since $\lambda < 0$, all points flow into the origin on straight lines determined by the choice of c_1 and c_2 (see Fig. 7.9).

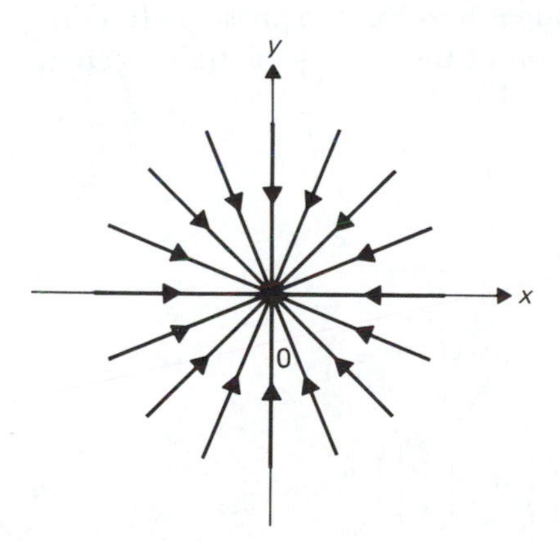

Figure 7.9 **A phase plane portrait for subcase 1, $\lambda_1 = \lambda_2 < 0$.**

If there is only one linearly independent eigenvector, a general solution can be written in the form

$$\begin{bmatrix} x(t) \\ y(t) \end{bmatrix} = c_1 \begin{bmatrix} \alpha \\ \beta \end{bmatrix} e^{\lambda t} + c_2 \left(\begin{bmatrix} \eta \\ \sigma \end{bmatrix} + t \begin{bmatrix} \alpha \\ \beta \end{bmatrix} \right) e^{\lambda t}.$$

The points on the two half-lines determined by the one eigenvector flow toward the origin as t increases. All other points flow toward the origin asymptotically to these half-lines since the $te^{\lambda t}$ term in the general solution dominates:

$$\frac{y(t)}{x(t)} = \frac{e^{\lambda t}(c_1\beta + c_2\sigma + c_2 t\beta)}{e^{\lambda t}(c_1\alpha + c_2\eta + c_2 t\alpha)} = \frac{c_2\beta + (c_1\beta + c_2\sigma)/t}{c_2\alpha + (c_1\alpha + c_2\eta)/t} \rightarrow \frac{\beta}{\alpha}$$

as $t \rightarrow \infty$.

EXAMPLE 5 Find a general solution and sketch the phase plane portrait for

$$\frac{dx}{dt} = -4x + y, \qquad \frac{dy}{dt} = -x - 2y.$$

SOLUTION The characteristic equation $\lambda^2 + 6\lambda + 9 = (\lambda + 3)^2 = 0$ has a double root $\lambda = -3$. Following the procedure discussed in Chapter 5, we set $\lambda = -3$ in the expressions

$$\begin{aligned} (-4 - \lambda)\alpha + \beta &= 0, & (-4 - \lambda)\eta + \sigma &= \alpha, \\ -\alpha + (-2 - \lambda)\beta &= 0, & -\eta + (-2 - \lambda)\sigma &= \beta, \end{aligned}$$

and obtain $\alpha = 1$, $\beta = 1$, $\eta = 0$, $\sigma = 1$. A general solution is

$$\begin{bmatrix} x(t) \\ y(t) \end{bmatrix} = c_1 \begin{bmatrix} 1 \\ 1 \end{bmatrix} e^{-3t} + c_2 \left(\begin{bmatrix} 0 \\ 1 \end{bmatrix} + t \begin{bmatrix} 1 \\ 1 \end{bmatrix} \right) e^{-3t}.$$

A phase plane portrait is sketched in Fig. 7.10. ■

Real 5: $0 < \lambda_1 = \lambda_2$. If the common eigenvalue is positive, the phase paths enter the origin as $t \rightarrow -\infty$. Once again, the direction of the arrows of the preceding phase plane portraits is reversed (see Figs. 7.5 and 7.11).

EXAMPLE 6 Solve

$$\frac{dx}{dt} = 4x - y, \qquad \frac{dy}{dt} = x + 2y.$$

SOLUTION The characteristic equation $\lambda^2 - 6\lambda + 9 = (\lambda - 3)^2 = 0$ has a double root $\lambda = 3$, and a general solution is

$$\begin{bmatrix} x(t) \\ y(t) \end{bmatrix} = c_1 \begin{bmatrix} 1 \\ 1 \end{bmatrix} e^{3t} + c_2 \left(\begin{bmatrix} 2 \\ 1 \end{bmatrix} + t \begin{bmatrix} 1 \\ 1 \end{bmatrix} \right) e^{3t}.$$ ■

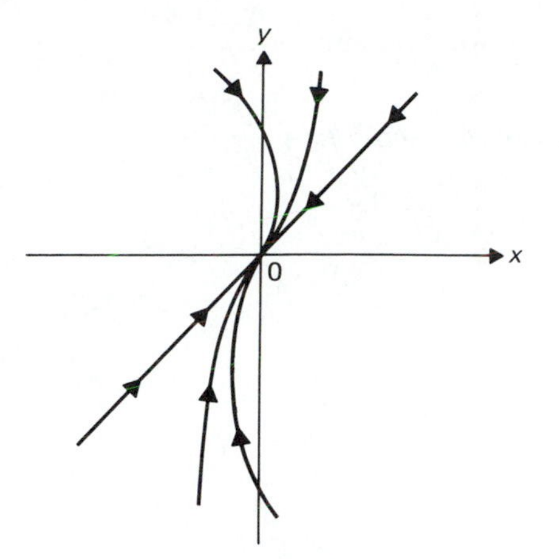

Figure 7.10 A phase plane portrait for subcase 2, $\lambda_1 = \lambda_2 < 0$.

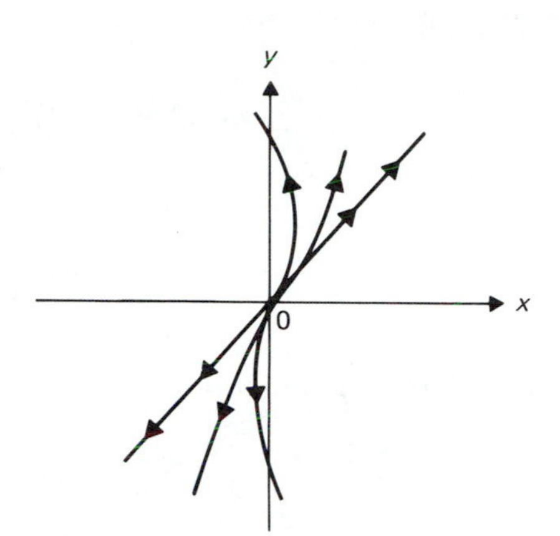

Figure 7.11 A phase plane portrait for subcase 2, $0 < \lambda_1 < \lambda_2$.

Let us now consider the remaining cases for which the eigenvalues are complex. To avoid being lost in a jungle of algebra, only systems of the form (called the *real canonical form*)

$$\frac{dx}{dt} = px - qy, \qquad \frac{dy}{dt} = qx + py$$

will be considered. It is known that any two-dimensional linear system with complex eigenvalues can be transformed into such a system by a change of coordinates. The characteristic equation of the above system is

$$\begin{vmatrix} p - \lambda & -q \\ q & p - \lambda \end{vmatrix} = (p - \lambda)^2 + q^2 = 0,$$

and the eigenvalues are $\lambda = p + iq$, $\lambda^* = p - iq$. Since there are no real eigenvectors, another approach to obtain the geometry of the phase plane is needed.

One approach that works is to write the system using polar coordinates. Let $x = r \cos \theta$, $y = r \sin \theta$; differentiating and substituting into the above system gives

$$\frac{dx}{dt} = \frac{dr}{dt} \cos \theta - r \sin \theta \frac{d\theta}{dt} = pr \cos \theta - qr \sin \theta,$$

$$\frac{dy}{dt} = \frac{dr}{dt} \sin \theta + r \cos \theta \frac{d\theta}{dt} = qr \cos \theta + pr \sin \theta.$$

To solve for dr/dt, multiply the first equation by $\cos \theta$, the second by $\sin \theta$ and add to get

$$\frac{dr}{dt}(\cos^2 \theta + \sin^2 \theta) = pr(\cos^2 \theta + \sin^2 \theta) \qquad \text{or} \qquad \frac{dr}{dt} = pr.$$

Proceeding in a similar manner to solve for $d\theta/dt$, one obtains

$$r\frac{d\theta}{dt}(\sin^2\theta + \cos^2\theta) = qr(\sin^2\theta + \cos^2\theta)$$

or, if $r \neq 0$,

$$\frac{d\theta}{dt} = q.$$

Therefore a general solution is

$$r = r_0e^{pt}, \qquad \theta = qt + \theta_0,$$

or

$$x(t) = r_0e^{pt}\cos(qt + \theta_0), \qquad y(t) = r_0e^{pt}\sin(qt + \theta_0),$$

where r_0 and θ_0 can be determined if initial conditions are given. The phase paths are spirals if $p \neq 0$; they are circles if $p = 0$. The direction of the flow is determined by q and is clockwise if q is negative, counterclockwise if q is positive. The complex eigenvalue cases can now be summarized.

Complex 1: $p = \text{Re}(\lambda) < 0$. The origin is a *stable spiral point.* The phase paths spiral in toward the origin with the orientation determined by the sign of q.

Complex 2: $p = \text{Re}(\lambda) > 0$. The origin is an *unstable spiral point.* The phase paths spiral out from the origin with the orientation determined by the sign of q.

Complex 3: $p = \text{Re}(\lambda) = 0$. The origin is a *center.* The phase paths are circles with their orientation determined by the sign of q. They are the graphs of periodic solutions to the differential equation.

EXAMPLE 7 For each of the following differential equations sketch a phase plane portrait.

$$\textit{Complex 1:}\quad \frac{dx}{dt} = -3x + y, \qquad \frac{dy}{dt} = -x - 3y;$$

$$\textit{Complex 2:}\quad \frac{dx}{dt} = 3x + 2y, \qquad \frac{dy}{dt} = -5x + y;$$

$$\textit{Complex 3:}\quad \frac{dx}{dt} = x - 2y, \qquad \frac{dy}{dt} = 2x - y.$$

SOLUTION Since $p = -3$, in the equations of complex 1, the origin is a stable spiral point, and since $q = -1$, the flow is clockwise. A general solution is

$$x(t) = r_0 e^{-3t}\cos(-t + \theta_0), \qquad y(t) = r_0 e^{-3t}\sin(-t + \theta_0),$$

and a phase plane portrait is sketched in Fig. 7.12(a).

The system of complex 2 is not in the real canonical form that was discussed

Figure 7.12 Phase plane portraits with complex λ.

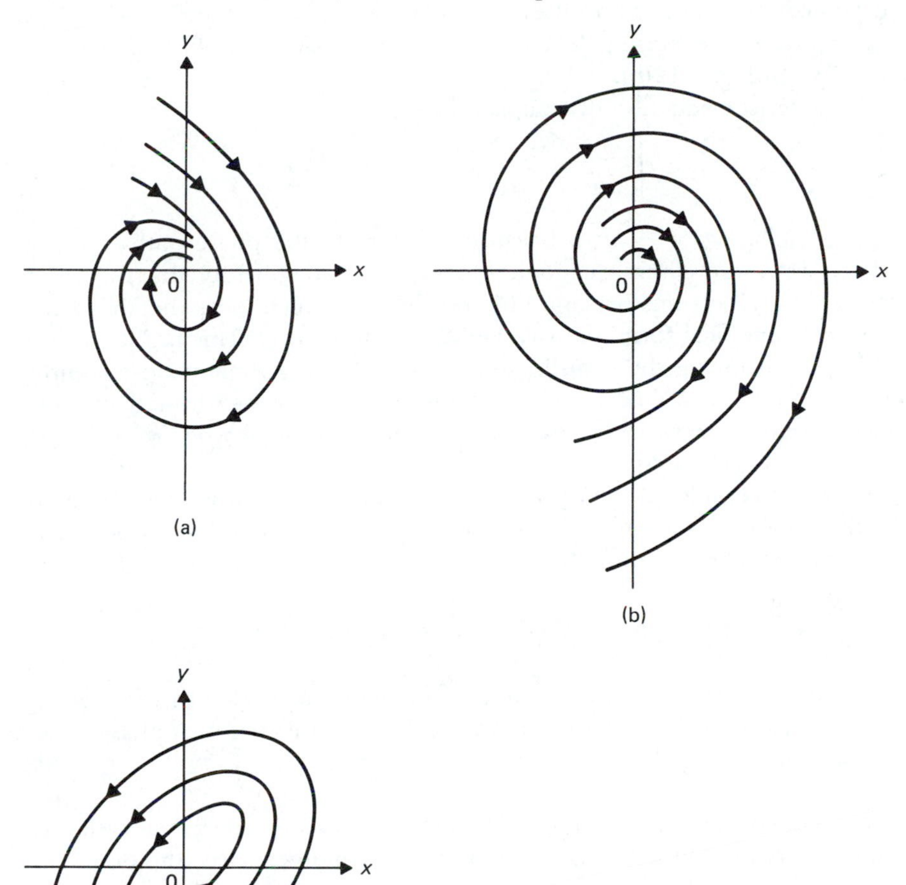

above, but once its eigenvalues are computed, the qualitative behavior of the phase paths is known. The characteristic equation

$$0 = \begin{vmatrix} 3-\lambda & 2 \\ -5 & 1-\lambda \end{vmatrix} = \lambda^2 - 4\lambda + 13$$

has roots $\lambda = 2 + 3i$, $\lambda^* = 2 - 3i$. Since $\text{Re}(\lambda) = 2 > 0$, the phase paths spiral outward. Because the system is not in canonical form, the direction of the flow cannot be determined by the sign of q since it is not known. However, the direction is easily found by determining whether y is decreasing or increasing as a phase path crosses the x-axis. In the second differential equation, if $x > 0$ and $y = 0$, then $dy/dt < 0$; hence y is decreasing. The flow must be clockwise. A phase plane portrait is sketched in Fig. 7.12(b).

The characteristic equation of complex 3,

$$0 = \begin{vmatrix} 1-\lambda & -2 \\ 2 & -1-\lambda \end{vmatrix} = \lambda^2 + 3,$$

has roots $\lambda = \sqrt{3}i$, $\lambda^* = -\sqrt{3}i$. Because $\text{Re}(\lambda) = 0$, the phase paths are not spirals. If the system were in real canonical form, we could conclude that the phase paths were circles. The transformation of coordinates, which takes the above system to the real canonical form and vice versa, is a linear one. This has the effect of stretching or shrinking the coordinate axes and rotating them. Consequently, the circular phase paths will be transformed into elliptical paths whose major and minor axes will not necessarily be the xy-axes. Since $dy/dt > 0$ when $x > 0$, $y = 0$, the flow is counterclockwise.

In the case of complex 3 the equations of the family of ellipses can be determined by finding the equations of the integral curves. The implicit solution of the associated differential equation

$$\frac{dy}{dx} = \frac{2x - y}{x - 2y} \qquad \text{or} \qquad 2x\,dx + 2y\,dy - (y\,dx + x\,dy) = 0$$

is $x^2 - xy + y^2 = \text{const}$. By a rotation of axes it can be shown that this represents a family of ellipses centered at (0, 0): These are the integral curves. A phase plane portrait is sketched in Fig. 7.12(c). ■

Up to now the terms *stable* and *unstable* have been used in the classification of equilibrium points without defining them precisely. We now proceed to do so.

Definition

An equilibrium point $(\bar{x}, \bar{y})$ of the system (7.3.1) is said to be *stable* if for each number $\epsilon > 0$ there is a number $\delta > 0$ such that if $x(t)$, $y(t)$ is any solution to (7.3.1) with

$$|x(t_0) - \bar{x}| + |y(t_0) - \bar{y}| < \delta,$$

then the solution exists for all $t \geq t_0$ and

$$|x(t) - \overline{x}| + |y(t) - \overline{y}| < \epsilon \quad \text{for all } t \geq t_0.$$

An equilibrium point $(\overline{x}, \overline{y})$ is said to be *unstable* if it is not stable.

Definition

An equilibrium point $(\overline{x}, \overline{y})$ of the system (7.3.1) is said to be *asymptotically stable* if it is stable and if there is a number $\eta > 0$ such that if

$$|x(t_0) - \overline{x}| + |y(t_0) - \overline{y}| < \eta,$$

then

$$\lim_{t \to \infty} \{|x(t) - \overline{x}| + |y(t) - \overline{y}|\} = 0.$$

These concepts are illustrated in Fig. 7.13. If the above definitions are applied to the linear system we have just analyzed, the results for the equilibrium point (0, 0) can be summarized in Table 7.1.

In this section we found that it is easy to study the qualitative properties of linear two-dimensional systems having the origin as an isolated equilibrium state. Their general solutions were constructed by the eigenvalue–eigenvector method and then their phase plane portraits were sketched. With this as a background, we are ready to analyze almost linear systems, the topic of the next section.

Figure 7.13

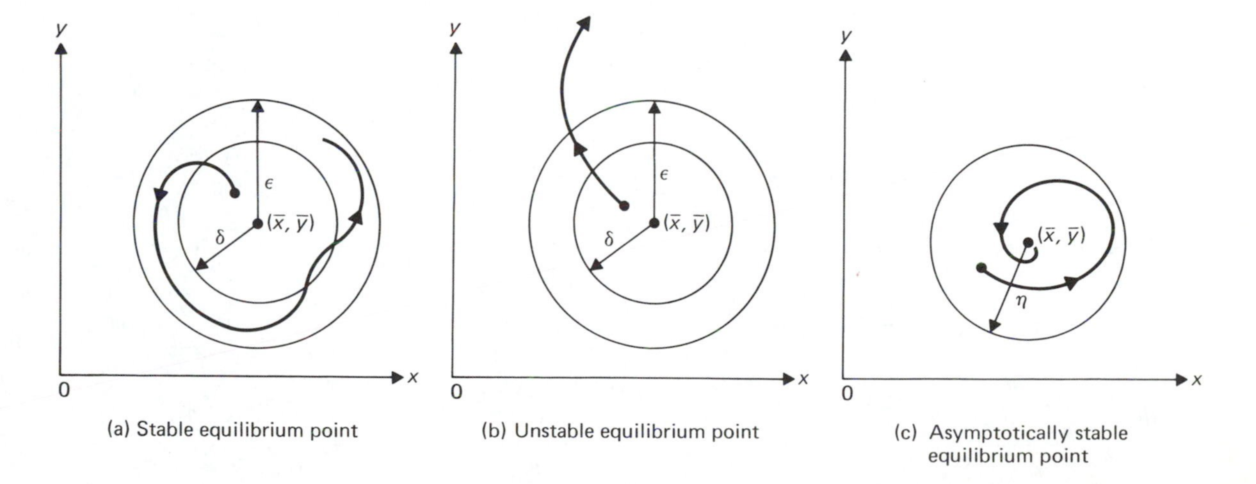

(a) Stable equilibrium point

(b) Unstable equilibrium point

(c) Asymptotically stable equilibrium point

Table 7.1 Classification of equilibrium states

Case		Type	Stability
Real(1)	$\lambda_1 < \lambda_2 < 0$	node	asymptotically stable
Real(2)	$0 < \lambda_2 < \lambda_1$	node	unstable
Real(3)	$\lambda_2 < 0 < \lambda_1$	saddle point	unstable
Real(4)	$\lambda_1 = \lambda_2 < 0$	node	asymptotically stable
Real(5)	$0 < \lambda_1 = \lambda_2$	node	unstable
Complex(1)	$\mathrm{Re}(\lambda) < 0$	spiral	asymptotically stable
Complex(2)	$\mathrm{Re}(\lambda) > 0$	spiral	unstable
Complex(3)	$\mathrm{Re}(\lambda) = 0$	center	stable

EXERCISES 7.3

Use Table 7.1 to describe the equilibrium state (0, 0). Find a general solution and sketch a phase plane portrait.

1. $\dfrac{dx}{dt} = x + 12y, \quad \dfrac{dy}{dt} = 3x + y$

2. $\dfrac{dx}{dt} = 5x - 2y, \quad \dfrac{dy}{dt} = 4x - y$

3. $\dfrac{dx}{dt} = -4x + 3y, \quad \dfrac{dy}{dt} = -2x + y$

4. $\dfrac{dx}{dt} = 2x + 5y, \quad \dfrac{dy}{dt} = x - 2y$

5. $\dfrac{dx}{dt} = x - 4y, \quad \dfrac{dy}{dt} = x + y$

6. $\dfrac{dx}{dt} = 3x + 2y, \quad \dfrac{dy}{dt} = -5x + y$

7. $\dfrac{dx}{dt} = -2x + y, \quad \dfrac{dy}{dt} = -4x + 3y$

8. $\dfrac{dx}{dt} = 3x - 4y, \quad \dfrac{dy}{dt} = 2x - 3y$

9. $\dfrac{dx}{dt} = -3x + 4y, \quad \dfrac{dy}{dt} = -x + y$

10. $\dfrac{dx}{dt} = -y, \quad \dfrac{dy}{dt} = x - 2y$

11. $\frac{dx}{dt} = x - y, \quad \frac{dy}{dt} = x + 3y$

12. $\frac{dx}{dt} = 2x - 5y, \quad \frac{dy}{dt} = x - 2y$

13. $\frac{dx}{dt} = -8y, \quad \frac{dy}{dt} = 2x$

14. $\frac{dx}{dt} = -2x + y, \quad \frac{dy}{dt} = -3x - 6y$

15. $\frac{dx}{dt} = -2x, \quad \frac{dy}{dt} = -2y$

16. $\frac{dx}{dt} = 3x, \quad \frac{dy}{dt} = -y$

17. $\frac{dx}{dt} = 3x + 8y, \quad \frac{dy}{dt} = -x - 3y$

18. $\frac{dx}{dt} = 5x + 2y, \quad \frac{dy}{dt} = 4x + 3y$

19. $\frac{dx}{dt} = -3x - y, \quad \frac{dy}{dt} = 5x + y$

20. $\frac{dx}{dt} = -2x + y, \quad \frac{dy}{dt} = -x - y$

7.4 ALMOST LINEAR SYSTEMS

The very important mathematical result for almost linear systems

$$\frac{dx}{dt} = ax + by + R(x, y), \qquad \frac{dy}{dt} = cx + dy + S(x, y),$$

is that the stability of the origin is determined (with one exception) by the associated linear system of differential equations. If the origin is an asymptotically stable equilibrium point of the associated linear system, then it is an asymptotically stable equilibrium point of the almost linear system. If the origin is an unstable equilibrium point of the linear system, then it is an unstable equilibrium point of the almost linear system. The exceptional case is when the origin is a center of the linear system for it may or may not be a stable equilibrium point for the almost linear system. One can imagine the family of ellipses centered at the origin being "smeared" in the almost linear system and possibly becoming a family of spirals that could be unstable or asymptotically stable. On the other hand, the center structure of the linear system could be preserved in the almost linear system, as Example 2 in Section 7.2 demonstrates.

Equally important is that the phase portrait near the origin of the almost linear system is locally approximated (in the sense described previously) by the phase portrait of the linear system in the unstable and asymptotically stable cases. These statements allow us to refer to the origin as a stable or unstable node, or spiral, or a saddle point of the almost linear system when it is of that type for the linear system. The justification for the above statements can be found in more advanced texts such as [4].

EXAMPLE 1 Analyze the damped pendulum equation

$$\ddot{x} + c\dot{x} + \sin x = 0, \quad 0 < c < 2 .$$

SOLUTION It is equivalent to the system

$$\frac{dx}{dt} = y, \qquad \frac{dy}{dt} = -\sin x - cy,$$

and its equilibrium states are either asymptotically stable or unstable. To see this, first linearize with respect to the equilibrium states $x = 0$, $y = 0$, and then $x = \pi$, $y = 0$, and verify in each case that it is an almost linear system. Letting $x = 0 + u$, $y = 0 + v$, we get

$$\frac{du}{dt} = v, \qquad \frac{dv}{dt} = -u - cv + (u - \sin u),$$

and therefore $R(u, v) = 0$ and $S(u, v) = u - \sin u$. Since

$$\frac{|u - \sin u|}{\sqrt{u^2 + v^2}} \le \frac{|u - \sin u|}{|u|} = \left|1 - \frac{\sin u}{u}\right| \to 0$$

as $u \to 0$, the system is almost linear. The same system would have been directly obtained by using the Taylor expansion for $\sin u$ near $u = 0$ as follows:

$$\begin{aligned}\frac{dv}{dt} &= -cv - \sin u = -cv - \left(u - \frac{u^3}{3!} + \frac{u^5}{5!} \cdots\right) \\ &= -u - cv + S(u, v).\end{aligned}$$

The characteristic equation of the associated linear system is

$$\begin{vmatrix} 0 - \lambda & 1 \\ -1 & -c - \lambda \end{vmatrix} = \lambda^2 + c\lambda + 1 = 0.$$

If $c^2 < 4$, the eigenvalues are

$$\lambda = -\frac{c}{2} + i\,\frac{(4 - c^2)^{1/2}}{2}, \qquad \lambda^* = -\frac{c}{2} - i\,\frac{(4 - c^2)^{1/2}}{2}$$

Since $c > 0$, the origin is an asymptotically stable spiral point of the linear system. Therefore the origin is an asymptotically stable spiral point of the damped pendulum. To analyze the equilibrium point $(\pi, 0)$, set $x = \pi + u$, $y = 0 + v$ and obtain

$$\frac{du}{dt} = v, \qquad \frac{dv}{dt} = -cv - \sin(u + \pi) = u - cv + (\sin u - u).$$

The characteristic equation of the associated linear system is

$$\begin{vmatrix} 0 - \lambda & 1 \\ 1 & -c - \lambda \end{vmatrix} = \lambda^2 + c\lambda - 1 = 0.$$

The eigenvalues are

$$\lambda_2 = -\frac{c}{2} - \frac{(c^2 + 4)^{1/2}}{2} < 0 < -\frac{c}{2} + \frac{(c^2 + 4)^{1/2}}{2} = \lambda_1,$$

and $(u, v) = (0, 0)$ is a saddle point of the linear system. By the theory, the point $(x, y) = (\pi, 0)$ is therefore an unstable equilibrium point (a saddle) of the damped pendulum. ■

Given a two-dimensional autonomous nonlinear system that cannot be solved explicitly, how does one analyze it? We suggest the following systematic approach.

1. Find and classify the equilibrium states.
2. Solve the linear systems associated with the equilibrium states to obtain approximate solutions to the nonlinear system and to determine stability.
3. Sketch a local phase plane portrait; i.e., sketch phase paths near the equilibrium states.
4. Seek implicit solutions to the corresponding first order differential equation to obtain integral curves (usually difficult).
5. (Optional) Use a computer graphics package to sketch a phase portrait. A rough sketch can often be made by determining regions where dx/dt and dy/dt are univalent.

EXAMPLE 2 Analyze

$$\frac{dx}{dt} = y - x^3, \qquad \frac{dy}{dt} = 1 - xy.$$

SOLUTION To find the equilibrium points solve the equations

$$P(x, y) = y - x^3 = 0, \qquad Q(x, y) = 1 - xy = 0.$$

Eliminating y from both equations, one obtains $0 = 1 - x^4$ with solutions $\bar{x} = 1$ and $\bar{x} = -1$. The corresponding values of y are $\bar{y} = 1$ and $\bar{y} = -1$, so the two equilibrium points are $(1, 1)$ and $(-1, -1)$. Furthermore,

$$P_x(\bar{x}, \bar{y}) = -3\bar{x}^2, \qquad P_y(\bar{x}, \bar{y}) = 1,$$

$$Q_x(\bar{x}, \bar{y}) = -\bar{y}, \qquad Q_y(\bar{x}, \bar{y}) = -\bar{x}.$$

To analyze the point (1, 1), let $(\bar{x}, \bar{y}) = (1, 1)$ and obtain the associated linear system

$$\frac{du}{dt} = P_x(1, 1)u + P_y(1, 1)v = -3u + v,$$

$$\frac{dv}{dt} = Q_x(1, 1)u + Q_y(1, 1)v = -u - v.$$

The characteristic equation

$$0 = \begin{vmatrix} -3 - \lambda & 1 \\ -1 & -1 - \lambda \end{vmatrix} = \lambda^2 + 4\lambda + 4 = (\lambda + 2)^2$$

has a double root, $\lambda = -2$, and therefore (0, 0) is an asymptotically stable node of the linear system. Consequently, the equilibrium state (1, 1) is an asymptotically stable node. Since $x = 1 + u$, $y = 1 + v$, an approximate solution is of the form

$$\begin{bmatrix} x(t) \\ y(t) \end{bmatrix} = \begin{bmatrix} 1 \\ 1 \end{bmatrix} + c_1 \begin{bmatrix} \alpha \\ \beta \end{bmatrix} e^{-2t} + c_2 \left(\begin{bmatrix} \eta \\ \sigma \end{bmatrix} + t \begin{bmatrix} \alpha \\ \beta \end{bmatrix} \right) e^{-2t},$$

where α, β, η, and σ satisfy the system of equations

$$[-3 - (-2)]\alpha + \beta = 0, \qquad [-3 - (-2)]\eta + \sigma = \alpha,$$
$$-\alpha + [-1 - (-2)]\beta = 0, \qquad -\eta + [-1 - (-2)]\sigma = \beta.$$

A solution is $\alpha = \beta = \sigma = 1$ and $\eta = 0$, and therefore an approximate solution to the nonlinear system near (1, 1) is

$$\begin{bmatrix} x(t) \\ y(t) \end{bmatrix} = \begin{bmatrix} 1 \\ 1 \end{bmatrix} + c_1 \begin{bmatrix} 1 \\ 1 \end{bmatrix} e^{-2t} + c_2 \left(\begin{bmatrix} 0 \\ 1 \end{bmatrix} + t \begin{bmatrix} 1 \\ 1 \end{bmatrix} \right) e^{-2t}.$$

To analyze the point $(-1, -1)$, let $\bar{x} = -1$, $\bar{y} = -1$, and proceed as above to obtain the associated linear system

$$\frac{du}{dt} = -3u + v, \qquad \frac{dv}{dt} = u + v.$$

The characteristic equation

$$0 = \begin{vmatrix} -3 - \lambda & 1 \\ 1 & 1 - \lambda \end{vmatrix} = \lambda^2 + 2\lambda - 4$$

has roots $\lambda_1 = -1 + \sqrt{5}$, $\lambda_2 = -1 - \sqrt{5}$, and therefore the origin is a saddle point of the linear system. Consequently, the equilibrium state $(-1, -1)$ is a saddle point and, therefore, is an unstable state. It is left for the reader to show that an approximate solution to the nonlinear system near $(-1, -1)$ is

$$\begin{bmatrix} x(t) \\ y(t) \end{bmatrix} = \begin{bmatrix} -1 \\ -1 \end{bmatrix} + c_1 \begin{bmatrix} 1 \\ 2 + \sqrt{5} \end{bmatrix} e^{(-1+\sqrt{5})t} + c_2 \begin{bmatrix} 1 \\ 2 - \sqrt{5} \end{bmatrix} e^{(-1-\sqrt{5})t}.$$

A local phase plane portrait is sketched in Fig. 7.14. The curves $y = x^3$ and $xy = 1$ on which $dx/dt = 0$ and $dy/dt = 0$, respectively, are included. In this example it is possible to follow the paths leaving the saddle $(-1, -1)$ that flow up and to the right. They must cross the $dy/dt = 0$ curve, turn down, cross the $dx/dt = 0$ curve, and flow into the attracting node $(1, 1)$. This is confirmed in the computer generated phase plane portrait shown in Fig. 7.15. ■

The reader may have noticed that in the discussion we dealt only with the associated linear system and did not verify the conditions

$$\lim_{(u,\, v)\to(0,\, 0)} \frac{R(u, v)}{\sqrt{u^2 + v^2}} = \lim_{(u,\, v)\to(0,\, 0)} \frac{S(u, v)}{\sqrt{u^2 + v^2}} = 0$$

on the remainder terms for the almost linear system. These conditions are automatically satisfied whenever the right-hand side of the original differential equation is a continuously twice differentiable function in a neighborhood of the equilibrium point. Certainly this is the case if it consists of a linear part plus polynomial terms in x and y of degree 2 or higher.

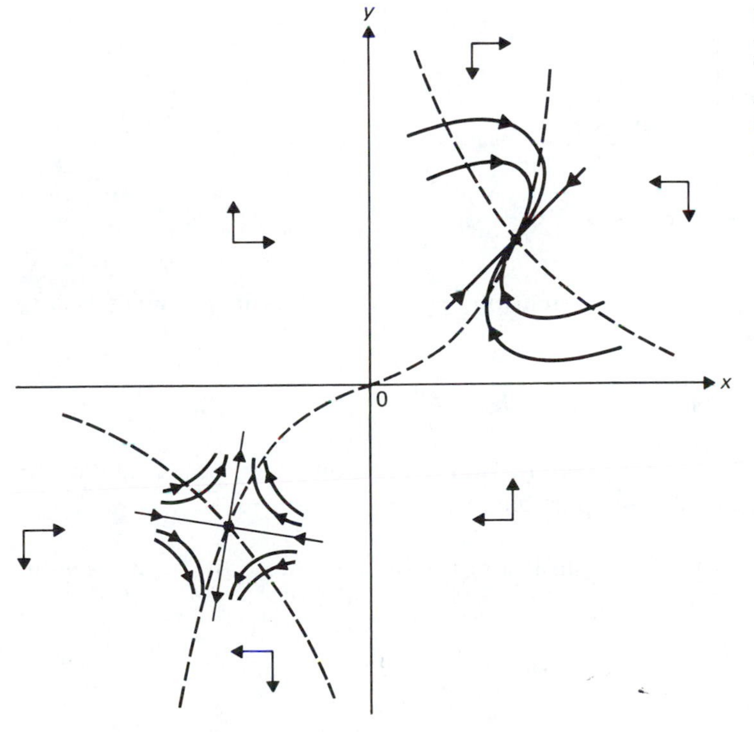

Figure 7.14 Phase plane portrait of $\dot{x} = y - x^3$, $\dot{y} = 1 - xy$. The arrow pairs indicate whether x and y are increasing or decreasing in each region determined by the curves $y = x^3$ and $xy = 1$.

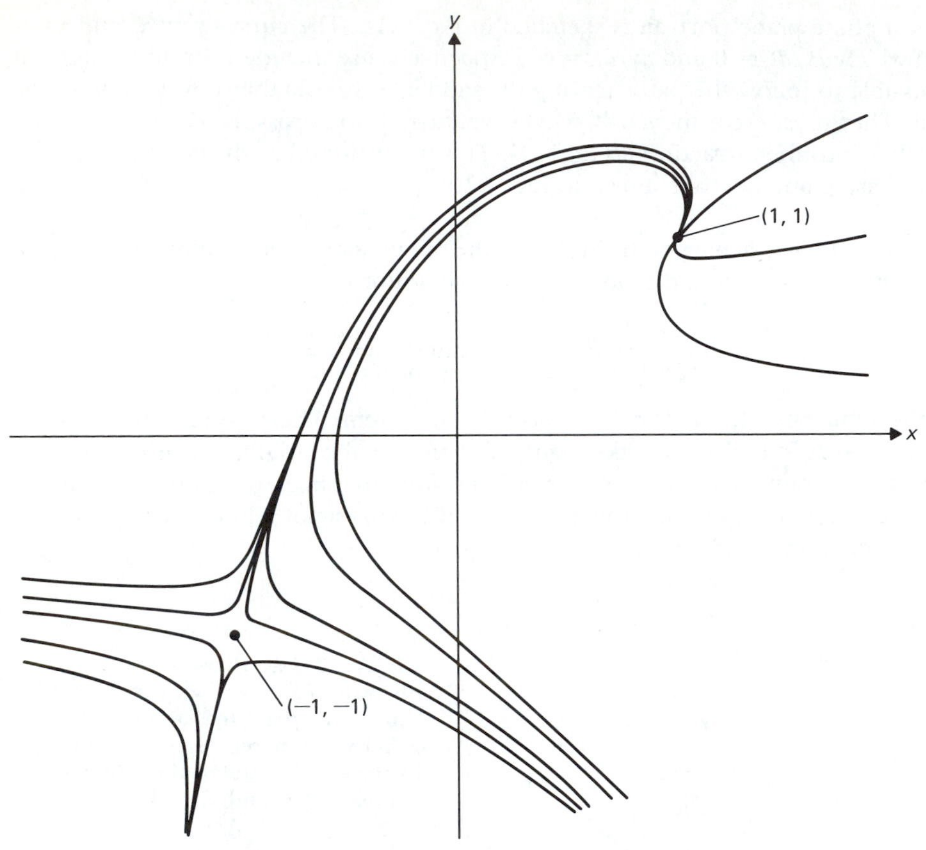

Figure 7.15 Plot generated using RKF45 of $\dot{x} = y - x^3$, $\dot{y} = 1 - xy$.

EXAMPLE 3 The following system of equations occurs in the analysis of a forced nonlinear oscillator:

$$\frac{dx}{dt} = x(4 - x^2 - y^2) - h, \qquad \frac{dy}{dt} = y(4 - x^2 - y^2),$$

where h is the amplitude of a periodic forcing function. Describe its equilibrium points and sketch a phase plane portrait.

SOLUTION If $h = 0$, the origin is an unstable node since the corresponding linear system is

$$\frac{du}{dt} = 4u, \qquad \frac{dv}{dt} = 4v.$$

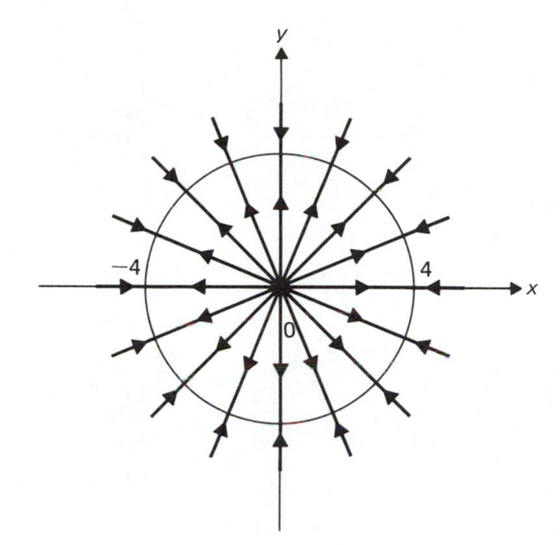

Figure 7.16 Global phase plane portrait of $\dot{r} = r(4 - r^2)$, $\dot{\theta} = 0$.

When $h = 0$, the circle $x^2 + y^2 = 4$ is a set of stable equilibrium points; this is easily shown if polar coordinates are employed. Set $x = r\cos\theta$, $y = r\sin\theta$, then differentiate and substitute to obtain

$$\frac{dx}{dt} = \frac{dr}{dt}\cos\theta - r\frac{d\theta}{dt}\sin\theta = (r\cos\theta)(4 - r^2),$$

$$\frac{dy}{dt} = \frac{dr}{dt}\sin\theta + r\frac{d\theta}{dt}\cos\theta = (r\sin\theta)(4 - r^2).$$

With appropriate multiplication by $\sin\theta$ and $\cos\theta$, this system can be solved for dr/dt and $d\theta/dt$:

$$\frac{dr}{dt} = r(4 - r^2), \qquad \frac{d\theta}{dt} = 0.$$

That the circle of equilibria $r^2 = 4$ is stable can be seen by noticing that $dr/dt > 0$ for $0 < r < 4$ and $dr/dt < 0$ for $r > 4$. A global phase plane portrait for the case $h = 0$ is sketched in Fig. 7.16.

If $h \neq 0$, the coordinates of the equilibrium points satisfy the algebraic equations

$$0 = x(4 - x^2 - y^2) - h, \qquad 0 = y(4 - x^2 - y^2).$$

Therefore $\bar{y} = 0$ and $\bar{x}$ is a root of $x(4 - x^2) - h = 0$. If $|h| < 16/(3\sqrt{3})$, then, as illustrated in Fig. 7.17 where we have graphed $z = x(4 - x^2)$, there are three

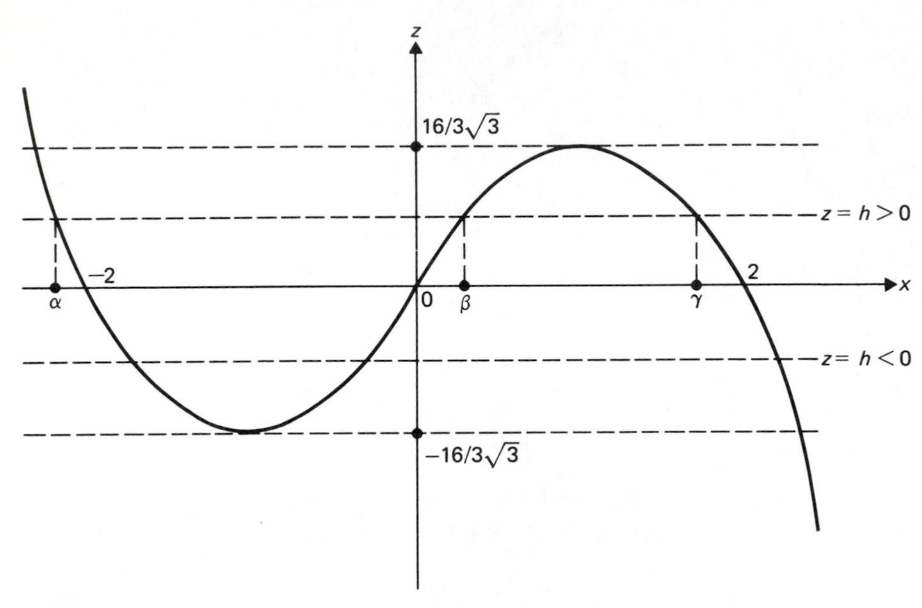

Figure 7.17 Graph of $z = x(4 - x^2)$. (The roots $\bar{x} = \alpha, \beta$, and γ are marked for the case $h > 0$.)

real roots: α, β, γ. If $h > 0$, the roots satisfy

$$\alpha < -2, \quad 0 < \beta < \gamma < 2;$$

whereas if $h < 0$, then

$$-2 < \alpha < \beta < 0, \quad 2 < \gamma.$$

Since $P(x, y) = 4x - x^3 - xy^2 - h$ and $Q(x, y) = 4y - x^2y - y^3$, then

$$P_x(x, 0) = 4 - 3x^2, \qquad P_y(x, 0) = 0,$$

$$Q_x(x, 0) = 0, \qquad Q_y(x, 0) = 4 - x^2.$$

The linear system associated with the equilibrium point $(\bar{x}, 0)$ is

$$\frac{d}{dt}\begin{bmatrix} u \\ v \end{bmatrix} = \begin{bmatrix} 4 - 3\bar{x}^2 & 0 \\ 0 & 4 - \bar{x}^2 \end{bmatrix} \begin{bmatrix} u \\ v \end{bmatrix},$$

with eigenvalues and eigenvectors

$$\lambda_1 = 4 - 3\bar{x}^2, \quad \begin{bmatrix} 1 \\ 0 \end{bmatrix}; \qquad \lambda_2 = 4 - \bar{x}^2, \quad \begin{bmatrix} 0 \\ 1 \end{bmatrix}.$$

Assume that $0 < h < 16/(3\sqrt{3})$ and let $\bar{x}$ take on successively the values α, β, and γ. Because the slope of $z = \bar{x}(4 - \bar{x}^2)$ is $z' = 4 - 3\bar{x}^2$, the value $\bar{x} = \alpha < -2$ corresponds to a point with negative slope, and therefore

$$\lambda_1 = 4 - 3\alpha^2 < 0, \qquad \lambda_2 = 4 - \alpha^2 < 0.$$

We conclude that point $(\alpha, 0)$ is a stable node. The second root, $\bar{x} = \beta$, where $0 < \beta < 2$, corresponds to a point with positive slope; hence,

$$\lambda_1 = 4 - 3\beta^2 > 0, \qquad \lambda_2 = 4 - \beta^2 > 0$$

and $(\beta, 0)$ is an unstable node. For the third root, $\bar{x} = \gamma$, where $0 < \gamma < 2$, the slope is negative, so

$$\lambda_1 = 4 - 3\gamma^2 < 0, \qquad \lambda_2 = 4 - \gamma^2 > 0$$

and $(\gamma, 0)$ is a saddle point.

If $-16/(3\sqrt{3}) < h < 0$, a similar analysis shows that $(\alpha, 0)$ is a saddle point, $(\beta, 0)$ is an unstable node, and $(\gamma, 0)$ is a stable node. Thus $(\beta, 0)$ stays the same regardless of the sign of h, whereas the points $(\alpha, 0)$ and $(\gamma, 0)$ exchange types. This interchange of the stable node and the saddle point is due to the change in sign of λ_2 as h goes from positive to negative.

A sketch of the local phase plane portrait for $0 < h < 16/(3\sqrt{3})$ is shown in Fig. 7.18. A computer generated plot of the phase plane portrait, when $h = 1$, is shown in Fig. 7.19. ■

Figure 7.18 **Local phase plane portrait of $\dot{x} = x(4 - x^2 - y^2) - h$, $\dot{y} = y(4 - x^2 - y^2)$ for $0 < h < 16/(3\sqrt{3})$.**

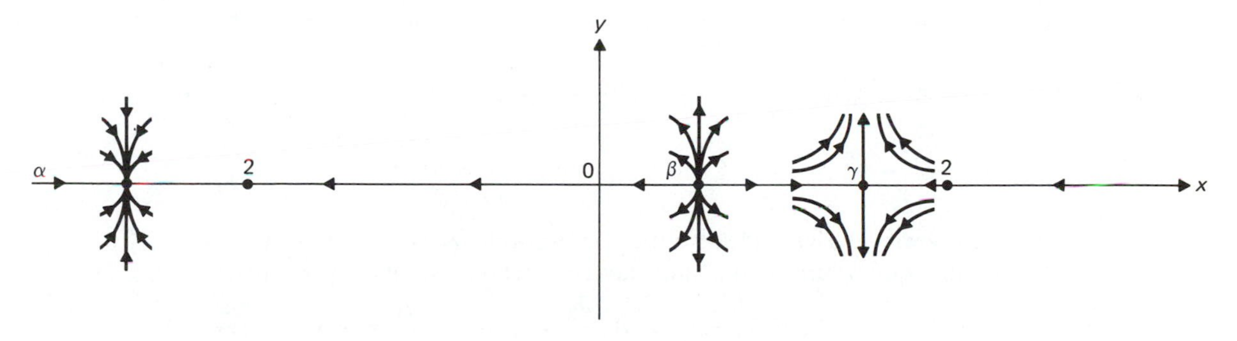

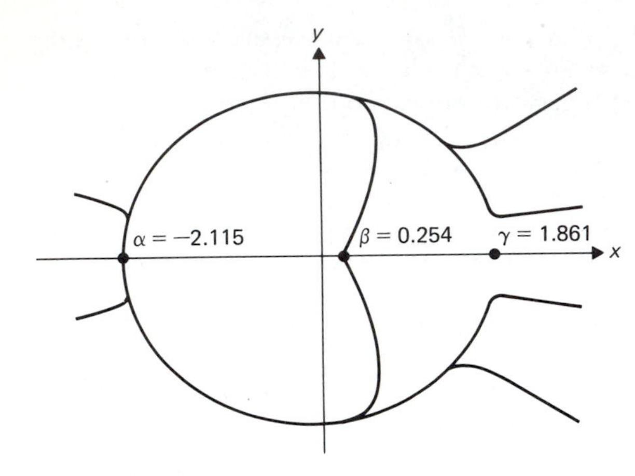

Figure 7.19 Plot generated using RKF45 of $\dot{x} = x(4 - x^2 - y^2) - 1, \dot{y} = y(4 - x^2 - y^2)$.

EXERCISES 7.4

Verify that the given point is an equilibrium of each of the following systems, derive and solve the corresponding linear equations, classify the equilibrium point, and sketch a phase plane portrait near the equilibrium point.

1. $\dfrac{dx}{dt} = 2x + 3y - xy - 6, \quad \dfrac{dy}{dt} = y + x^2 - 3$, given $(\bar{x}, \bar{y}) = (1, 2)$

2. $\dfrac{dx}{dt} = -x + 6y + 2xy - 5, \quad \dfrac{dy}{dt} = 6x + y + x^2 + 4$, given $(\bar{x}, \bar{y}) = (-1, 1)$

3. $\dfrac{dx}{dt} = 6x - 2y + \cos y - 2\pi, \quad \dfrac{dy}{dt} = 2x + y - \dfrac{3\pi}{2} \sin x$, given $(\bar{x}, \bar{y}) = \left(\dfrac{\pi}{2}, \dfrac{\pi}{2}\right)$

4. $\dfrac{dx}{dt} = 3x + 2y + \cos y + 1 - \pi, \quad \dfrac{dy}{dt} = 5x + 5y + \sin x$, given that $(\bar{x}, \bar{y}) = (\pi, -\pi)$

5. $\dfrac{dx}{dt} = 2x + 2y + 5e^{-x} - 5, \quad \dfrac{dy}{dt} = -x - y$, given $(\bar{x}, \bar{y}) = (0, 0)$

6. $\dfrac{dx}{dt} = 3x + 5y + e^x - 1, \quad \dfrac{dy}{dt} = -2x - 2y$, given $(\bar{x}, \bar{y}) = (0, 0)$

Find all equilibrium points, derive and solve the corresponding associated linear equations, classify the equilibrium points, and sketch a local phase plane portrait.

7. $\dfrac{dx}{dt} = x - 4y + 2, \quad \dfrac{dy}{dt} = 4x - 7y - 1$

8. $\dfrac{dx}{dt} = y + 2, \quad \dfrac{dy}{dt} = -x + 2y + 7$

9. $\dfrac{dx}{dt} = x^2 + y, \quad \dfrac{dy}{dt} = -x + y^2$

10. $\dfrac{dx}{dt} = x^2 - 5y, \quad \dfrac{dy}{dt} = x - \sqrt{5}y^2$

11. $\dfrac{dx}{dt} = x - y, \quad \dfrac{dy}{dt} = xy - 1$

12. $\dfrac{dx}{dy} = x + y, \quad \dfrac{dy}{dt} = xy + y$

13. $\dfrac{dx}{dt} = x - y, \quad \dfrac{dy}{dt} = xy + 4y + 3$

14. $\dfrac{dx}{dt} = xy, \quad \dfrac{dy}{dt} = -x + 2y + 1$

15. Sketch a local phase plane portrait for Example 3 in the case $|h| \geq 16/(3\sqrt{3})$.

16. A two-dimensional linear system is said to be degenerate if it has a zero eigenvalue. Show that if it is nontrivial, i.e., the coefficients are not all zero, then it has a line of equilibrium points.

17. Find the general solution and sketch a phase plane portrait of the degenerate system

$$\frac{dx}{dt} = x + y, \qquad \frac{dy}{dt} = x + y.$$

18. Find an integral of the nonlinear system

$$\frac{dx}{dt} = x + y, \qquad \frac{dy}{dt} = x + y - (x + y)^2.$$

19. In the preceding exercise, make the change of dependent variables $z = x + y$, $w = e^x$; derive and solve the resulting nonlinear system.

7.5 ENERGY METHODS FOR SYSTEMS WITH ONE DEGREE OF FREEDOM

A one-degree-of-freedom mechanical system with forces dependent on position can be modeled by a second order differential equation

$$\frac{d^2x}{dt^2} + g(x) = 0, \tag{7.5.1}$$

where x denotes a position coordinate and $-g(x)$ denotes the net force per unit mass. The system is said to be *conservative* since there are no dissipative forces acting. The simple pendulum and the frictionless mass–spring system can be modeled in this way.

The second order differential equation (7.5.1) can be written as a two-dimensional system

$$\frac{dx}{dt} = y, \qquad \frac{dy}{dt} = -g(x). \tag{7.5.2}$$

The equilibrium states are points $(\bar{x}, 0)$ where $g(\bar{x}) = 0$, and they could be analyzed by the methods discussed in the preceding section. But in this section we will analyze the equation in its original form (7.5.1). The significant advantage of retaining the second order equation is that energy methods can be employed to reveal the global geometric structure of its solutions.

Let $E(x, y)$ denote the total energy per unit mass of the system:

$$E(x, y) = \tfrac{1}{2} y^2 + G(x), \tag{7.5.3}$$

where $G(x) = \int^x g(s)\, ds$ denotes the potential energy per unit mass and

$$\frac{1}{2} y^2 = \frac{1}{2}\left(\frac{dx}{dt}\right)^2$$

is the kinetic energy per unit mass. The potential energy function, $G(x)$, is determined up to an additive constant; the constant will be chosen so that the total energy at an equilibrium point is zero. Note that the statement "the total energy is constant for conservative systems" means that the energy function is constant along trajectories. This follows directly since, if $x = x(t)$ is a solution to (7.5.1) and $y = dx(t)/dt$, then

$$\begin{aligned}
\frac{d}{dt} E\left(x(t), \frac{dx}{dt}(t)\right) &= \frac{d}{dt}\left[\frac{1}{2}\left(\frac{dx}{dt}(t)\right)^2 + G(x(t))\right] \\
&= \frac{dx}{dt}(t)\,\frac{d^2x}{dt^2}(t) + \frac{d}{dx} G(x(t))\,\frac{dx}{dt}(t) \\
&= \frac{dx}{dt}(t)\left[\frac{d^2x}{dt^2}(t) + g(x(t))\right] = 0,
\end{aligned} \tag{7.5.4}$$

since $g(x) = G'(x)$. But this implies that

$$E\left(x(t), \frac{dx}{dt}(t)\right) = \text{const.}$$

The *level curves* of the energy surface

$$z = E(x, y) \tag{7.5.5}$$

are the curves $E(x, y) = K = \text{const}$. The relation (7.5.4) shows that they are *integral curves* of the differential equation (7.5.1). They can also be considered as the graphs of the implicit solution

$$\frac{y^2}{2} + G(x) = K, \quad K = \text{const}$$

to the associated first order differential equation

$$\frac{dy}{dx} = \frac{-g(x)}{y} \quad \text{or} \quad g(x)\,dx + y\,dy = 0.$$

The equation $E(x, y) = K$ is called the *energy integral*. The direction of flow on the phase paths corresponding to the integral curves is governed by the two-dimensional system (7.5.2).

EXAMPLE 1 The point (0, 0) is an equilibrium point of the system (7.5.2) corresponding to each of the following:

a) $\dfrac{d^2x}{dt^2} + x = 0$. Here $g(x) = x$ and $G(x) = x^2/2$ satisfies $G'(x) = x$ and $G(0) = 0$, so the total energy is

$$E(x, y) = \frac{y^2}{2} + \frac{x^2}{2}.$$

b) $\dfrac{d^2x}{dt^2} + \sin x = 0$. Here $g(x) = \sin x$ and $G(x) = 1 - \cos x$ satisfies $G'(x) = \sin x$ and $G(0) = 0$. Hence,

$$E(x, y) = \frac{y^2}{2} + 1 - \cos x.$$

c) $\dfrac{d^2x}{dt^2} + x - \dfrac{x^4}{8} = 0$. The point (2, 0) is also an equilibrium point, so we could use

$$G(x) = \frac{x^2}{2} - \frac{x^5}{40} - \frac{6}{5},$$

which satisfies $G'(x) = x - (x^4/8)$ and $G(2) = 0$. In this case,

$$E(x, y) = \frac{y^2}{2} + \frac{x^2}{2} - \frac{x^5}{40} - \frac{6}{5}. \qquad \blacksquare$$

The level curves of the energy surface are constructed by taking the intersection of planes $z = K$ (in the xyz-space) and the surface and then projecting the intersection set onto the xy-plane. This is shown in Fig. 7.20 for Example 1(a), where the energy surface is a paraboloid $z = \frac{1}{2}(x^2 + y^2)$ and the level curves are the circles $x^2 + y^2 = 2K$, corresponding to periodic solutions.

Familiar examples of level curves are the isobars of weather maps and the contour lines of topographic maps.

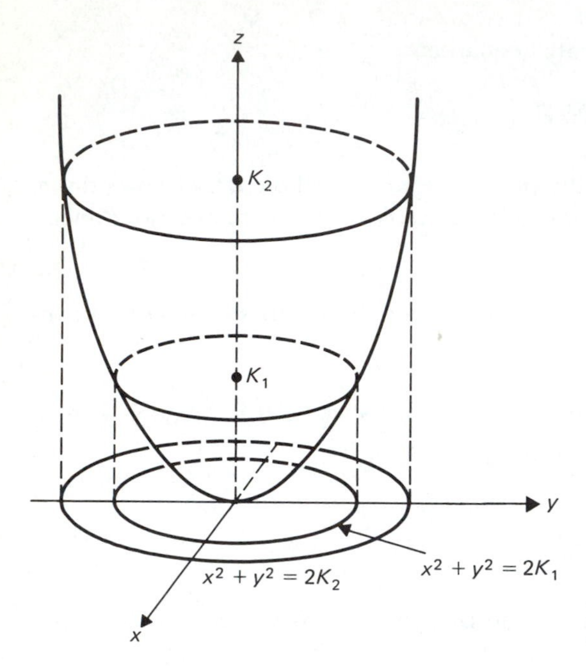

Figure 7.20 Level curves of $z = \frac{1}{2}(x^2 + y^2)$.

We shall exploit the special structure of the energy surfaces by sketching the level curves in the *phase plane* (the xy-plane) with the aid of the graph of the potential energy function $G(x)$ in the *potential plane* (the xz-plane). These two planes can be represented in one diagram by sketching the z-axis of the xz-plane and the y-axis of the xy-plane on the same vertical line (Fig. 7.21). In the potential plane, draw the potential energy curve, $z = G(x)$, and a horizontal line at height K. Since a level curve satisfies

$$E(x, y) = \frac{y^2}{2} + G(x) = K,$$

then

$$y = \pm\sqrt{2[K - G(x)]}. \tag{7.5.6}$$

Clearly, the velocity is real only if $K - G(x)$ is nonnegative.

If, for fixed K, the velocity is real only in certain intervals, then the real solution can exist only on these intervals. Specifically, if $G(x) - K > 0$ only on the interval $a < x < b$, then any solution satisfying $x(t_0) = x_0$ for some x_0 in the interval will satisfy $a < x(t) < b$ for all t. The *integral curves* in the phase plane corresponding to a fixed K are symmetric with respect to the x-axis. They cross the x-axis at any point x_0 where $G(x_0) = K$ and, since this implies that $dy/dx = -g(x)/y = \pm\infty$ at such a point, the curves cross the x-axis at right angles.

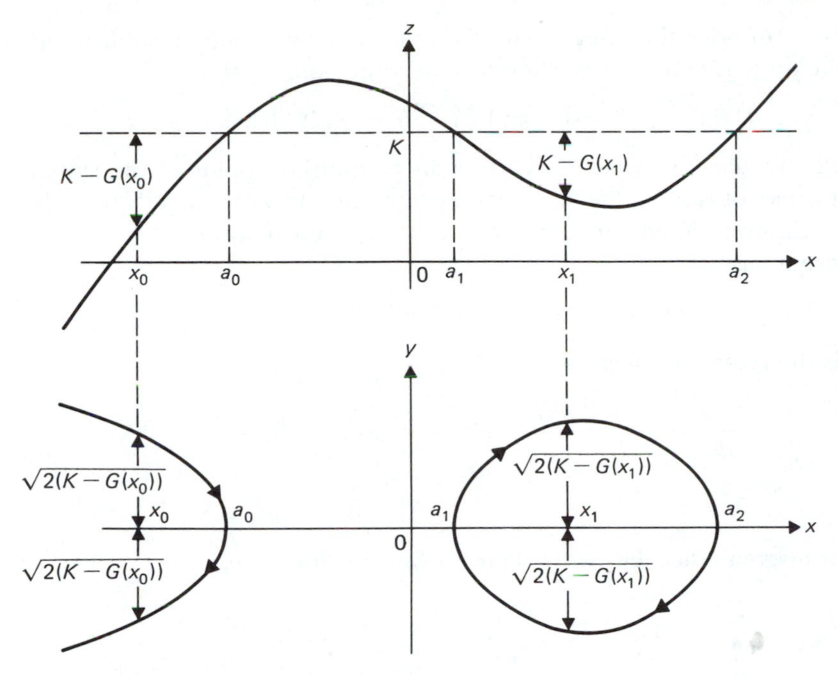

Figure 7.21 The potential plane and the phase plane (in the diagram, $y = \pm\sqrt{2[K - G(x)]}$ is not defined for $a_0 < x < a_1$ and $a_2 < x$).

EXAMPLE 2 Consider $d^2x/dt^2 + 4x = 0$ and $E(x, y) = y^2/2 + 2x^2$.

SOLUTION For a given K_0, the integral curves are given by solving $E(x, y) = K_0$ to obtain $y = \pm\sqrt{2(K_0 - 2x^2)}$. They are defined only for $-\sqrt{K_0/2} \leq x \leq \sqrt{K_0/2}$.By squaring both sides, one obtains the ellipse

$$\frac{x^2}{K_0/2} + \frac{y^2}{2K_0} = 1.$$

It is the locus of all trajectories

$$(x(t), y(t)) = \left(x(t), \frac{dx(t)}{dt}\right), \quad -\infty < t < \infty,$$

for which the energy integral (or total energy) equals K_0. The solutions of the original equation are $x = A\sin(2t - \alpha)$; hence

$$y = \dot{x} = 2A\cos(2t - \alpha),$$

and it is easy to check that the ellipse is the locus of all solutions for which $2A^2 = K_0$. ■

Let us now consider the integral curves resulting from a simple *minimum* of the potential energy function $G(x)$. Therefore at some point $x = \bar{x}$,

$$G'(\bar{x}) = g(\bar{x}) = 0 \quad \text{and} \quad G''(\bar{x}) = g'(\bar{x}) > 0$$

(see Fig. 7.22). In the phase plane, $(\bar{x}, 0)$ is an equilibrium point of the system (7.5.2). But instead of finding the linearized system one can work directly with the second order differential equation. Since $g(\bar{x}) = 0$, then if we let $x = \bar{x} + u$, Taylor's theorem implies

$$g(x) = g(\bar{x} + u) = g'(\bar{x})u + R(u),$$

where $R(u)$ is the remainder term. Hence

$$\frac{d^2x}{dt^2} + g(x) = \frac{d^2u}{dt^2} + g'(\bar{x})u + R(u) = 0,$$

Figure 7.22 **The phase plane diagram when the potential energy function has a local maximum or minimum.**

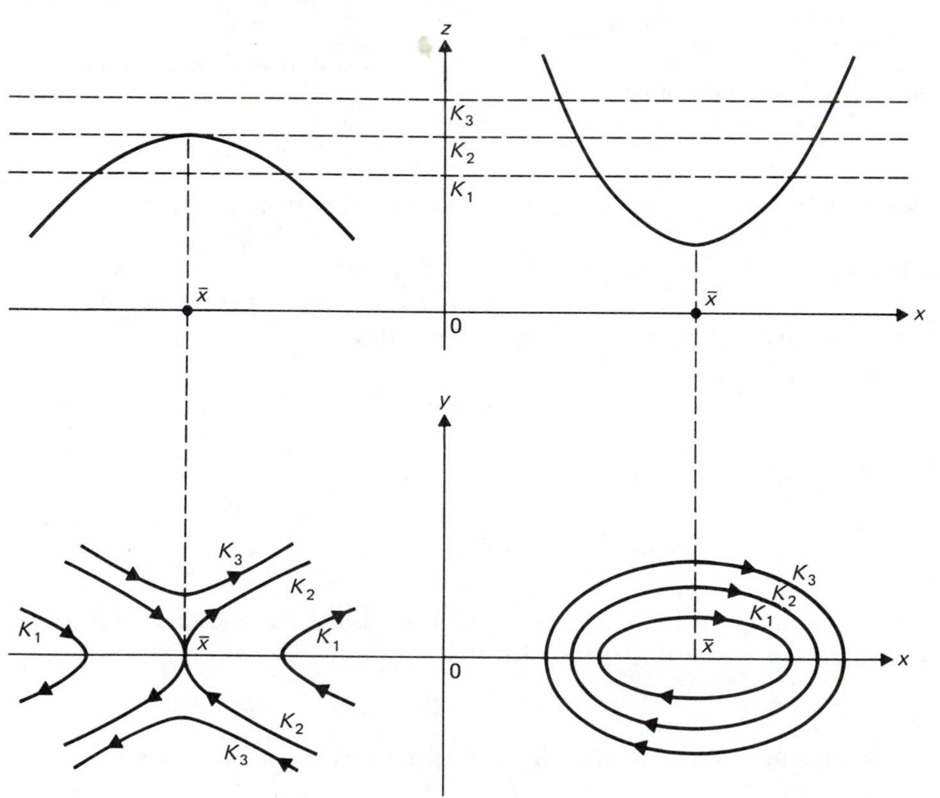

and neglecting $R(u)$ and letting $g'(\bar{x}) = \omega^2 > 0$, we get

$$\frac{d^2u}{dt^2} + \omega^2 u = 0.$$

The roots of the characteristic equation are $\lambda_1 = i\omega$, $\lambda_2 = -i\omega$. In the phase plane of (7.5.2), the point $(\bar{x}, 0)$ is a center and the approximate level curves are ellipses

$$\frac{y^2}{2} + \frac{\omega^2}{2}(x - \bar{x})^2 = K.$$

This implies the following very important fact: In the neighborhood of an equilibrium point $(\bar{x}, 0)$, where $G(\bar{x})$ is a minimum, there is a family of closed level curves. These correspond to periodic solutions of the original differential equation. If, instead, the potential energy function has a simple maximum at some $x = \bar{x}$, then $G'(\bar{x}) = g(\bar{x}) = 0$ and $G''(\bar{x}) = g'(\bar{x}) < 0$. By the same line of reasoning as above, the associated second order linear differential equation is

$$\frac{d^2u}{dt^2} - \omega^2 u = 0, \qquad -\omega^2 = g'(\bar{x}) < 0.$$

The roots of the characteristic equation are $\lambda_2 = -\omega < 0 < \lambda_1 = \omega$; in the phase plane of (7.5.2) the point $(\bar{x}, 0)$ is a saddle point. The approximate level curves are hyperbolas

$$\frac{y^2}{2} - \frac{\omega^2}{2}(x - \bar{x})^2 = K,$$

and straight lines $y = \pm\omega(x - \bar{x})$. This implies that in the phase plane of the original system the equilibrium point $(\bar{x}, 0)$ will be a saddle point.

If the potential energy curve $z = G(x)$ has a horizontal tangent at a simple point of inflection (Fig. 7.23), then at $x = \bar{x}$, we get $G'(\bar{x}) = g(\bar{x}) = 0$, $G''(\bar{x}) = g'(\bar{x}) = 0$, $G'''(\bar{x}) \neq 0$. The degenerate equilibrium state $(\bar{x}, 0)$ is unstable. Its associated linear differential equation, $d^2u/dt^2 = 0$, is of little value, but the approximate potential energy function

$$G(x) = G'''(\bar{x})\frac{(x - \bar{x})^3}{6} = g''(\bar{x})\frac{(x - \bar{x})^3}{6}$$

and approximate energy integral

$$\frac{y^2}{2} + \frac{g''(\bar{x})}{6}(x - \bar{x})^3 = K$$

have a monotonic family of level curves. The curve through $(\bar{x}, 0)$ is distinguished by its *cusp* (see Fig. 7.23).

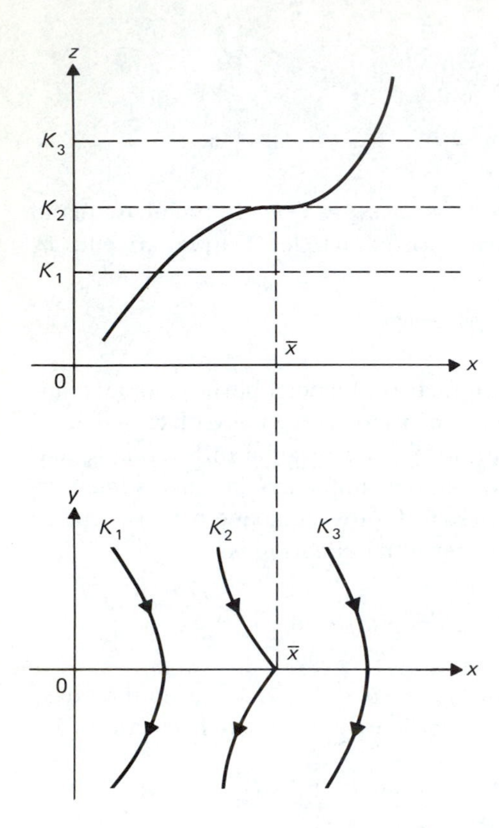

Figure 7.23 **The phase plane diagram when the potential energy function has a point of inflection.**

EXAMPLE 3 Find and classify the equilibrium points of

$$\frac{d^2x}{dt^2} + x - \frac{x^3}{4} = 0.$$

Derive and solve the corresponding linear equations and sketch a local phase plane portrait.

SOLUTION First solve

$$0 = x - \frac{x^3}{4}$$

and obtain $\bar{x} = 0$, $\bar{x} = 2$, $\bar{x} = -2$. Hence, there are three equilibrium points: $(0, 0)$, $(2, 0)$, $(-2, 0)$. A potential energy function $G(x)$ that satisfies $G'(x) = g(x)$ will be of the form

$$G(x) = \frac{1}{2}x^2 - \frac{1}{16}x^4 + C,$$

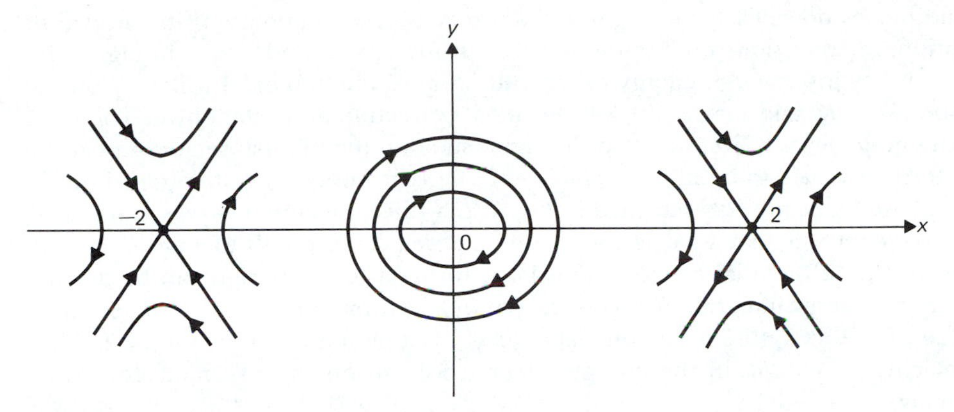

Figure 7.24 Local phase plane portrait of $\ddot{x} + x - x^3/4 = 0$.

and therefore $G'(\bar{x}) = 0$ for $\bar{x} = 0, \pm 2$. Furthermore,

$$G''(\bar{x}) = g'(\bar{x}) = 1 - \frac{3}{4}\bar{x}^2$$

and, consequently, $G''(0) = 1 > 0$ and $G''(\pm 2) = -2 < 0$; hence $x = 0$ is a local minimum and $x = \pm 2$ are local maxima for the potential energy function. We conclude that the points $(0, 0)$ and $(\pm 2, 0)$ in the phase plane are a center and saddle points, respectively.

The linearized second order equations are given by

$$\frac{d^2u}{dt^2} + g'(\bar{x})u = 0,$$

where $\bar{x} = 0, \pm 2$. Consequently at $(0, 0)$ we have

$$\frac{d^2u}{dt^2} + u = 0, \qquad u(t) = A \cos t + B \sin t,$$

and at $(\pm 2, 0)$,

$$\frac{d^2u}{dt^2} - 2u = 0, \qquad u(t) = A \cosh(\sqrt{2}t) + B \sinh(\sqrt{2}t).$$

The local phase plane portrait is sketched in Fig. 7.24. ■

A physically attractive approach to one-degree-of-freedom problems is to associate a bead sliding on the potential curve $z = G(x)$ with a particle moving in the phase plane. The velocity of the bead is given by $v = \pm\sqrt{2[K - G(x)]}$. Suppose that it is released at (x_0, K), which is the point of intersection of the line $z = K$ and the curve $z = G(x)$. The corresponding point in the phase plane is $(x_0, 0)$. If the bead moves downhill to the right and approaches the boundary of its interval of motion, it either slows (in infinite time) to a stop ($K = K_2$ and $x = x_3$ in Fig. 7.25), since it has just enough energy to reach the top of the hill, or, if it pauses on the slope ($K = K_1$ and $x = x_2$), it will reverse its direction and slide down. The bead with energy level K_1 will move to the right, slide up the hill, pause and repeat the pattern. It is said to be in a *potential well.* The corresponding paths traced out in the phase plane are also sketched in Fig. 7.25—they are closed curves.

We conclude that a particle in a potential well corresponds to a periodic solution to the differential equation (7.5.1). A formula for its period can be derived from the energy integral, $E(x, \dot{x}) = K$. On one excursion from $x_{\min}$ (the minimum value of x) to $x_{\max}$ (the maximum value of x) t is a monotonic function of x. After replacing y by dx/dt in the energy integral, we can obtain a formula for dt/dx, namely,

$$\frac{dt}{dx} = \{2[K - G(x)]\}^{-1/2}.$$

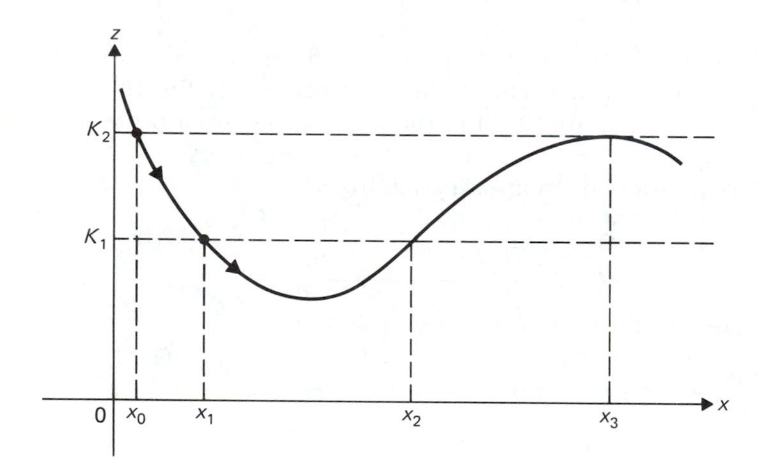

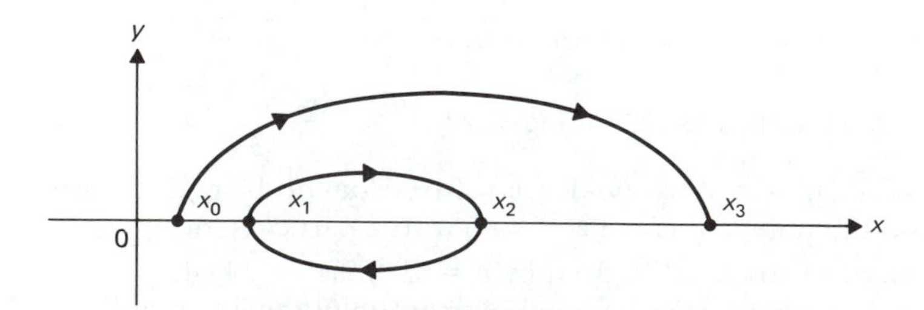

Figure 7.25 **The particle starting at (x_0, 0) with energy level K_2 will reach (x_3, 0) in infinite time. The particle starting at (x_1, 0) with energy level K_1 will reach (x_2, 0), then return to (x_1, 0) in finite time.**

Hence,

$$\frac{T}{2} = \int_{x_{\min}}^{x_{\max}} \{2[K - G(x)]\}^{-1/2}\, dx \tag{7.5.7}$$

is the time needed to traverse from $x_{\min}$ to $x_{\max}$. By symmetry, the return time is the same; hence, T is the *period.* The *frequency* of the motion is $2\pi/T$ and, in general, the period and, consequently, the frequency are functions of the amplitude.

EXAMPLE 4 Consider the example discussed previously:

$$\frac{d^2x}{dt^2} + x - \frac{x^3}{4} = 0, \qquad G(x) = \frac{x^2}{2} - \frac{x^4}{16};$$

the motion will be analyzed for three values of the energy level. Potential and phase plane portraits are sketched in Fig. 7.26 (the reader should glance at them as the discussion proceeds).

a) $K = \dfrac{7}{16}$: $\quad y = \dfrac{dx}{dt} = \pm\sqrt{2\left(\dfrac{7}{16} - \dfrac{x^2}{2} + \dfrac{x^4}{16}\right)}$

$$= \pm\sqrt{\frac{1}{8}(x^2 - 1)(x^2 - 7)}.$$

The bounded motion corresponds to the portion of the level curve that is trapped in the potential well. If the initial conditions are $x(0) = 1$, $\dot{x}(0) = 0$, then in the potential plane this would correspond to the bead being released at $(x, z) = (1, G(1)) = (1, 17/16)$ on the path $z = G(x)$. It will slide down and to the left with negative velocity

$$\frac{dx}{dt} = -\sqrt{\frac{1}{8}(x^2 - 1)(x^2 - 7)}$$

until it reaches $(x, z) = (-1, 17/16)$, at which point its velocity will again be zero. But its acceleration will be

$$\frac{d^2x}{dt^2} = \frac{(-1)^3}{4} - (-1) = \frac{3}{4},$$

so it will move down and to the right with positive velocity

$$\frac{dx}{dt} = \sqrt{\frac{1}{8}(x^2 - 1)(x^2 - 7)}$$

until it reaches $(1, 17/16)$ again. In the phase plane the corresponding particle will start at $(1, 0)$ and move to the left and down until it reaches the

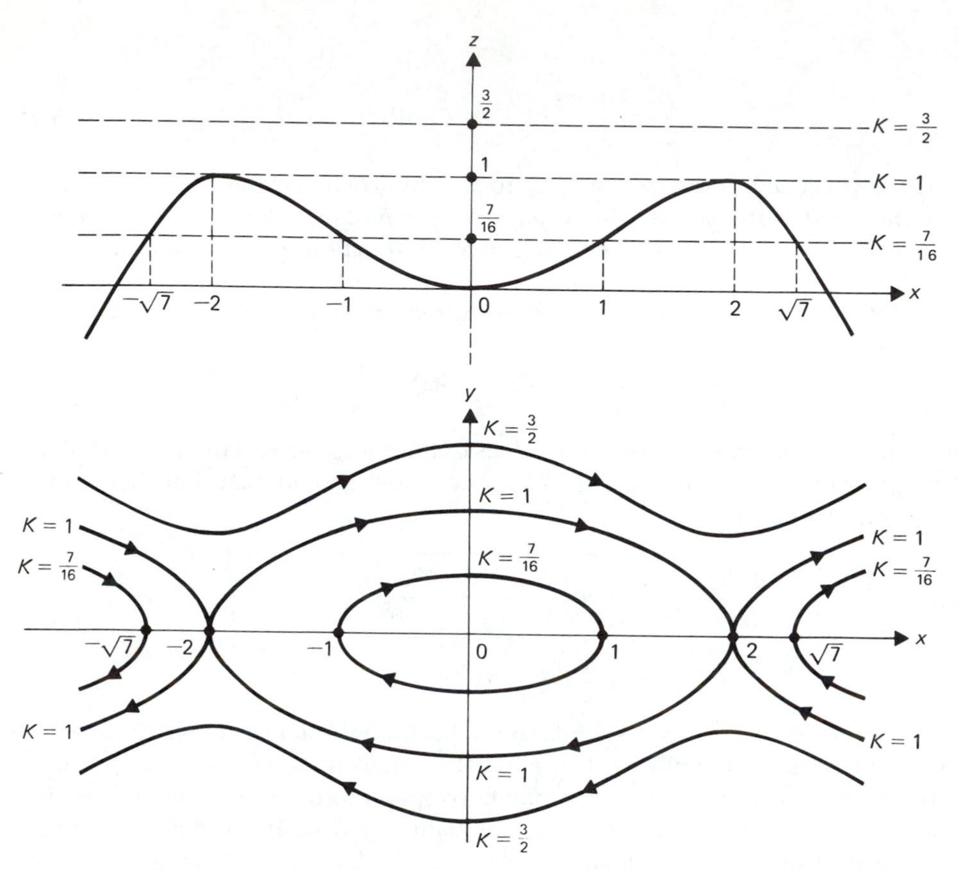

Figure 7.26 The potential and phase plane portraits of $\ddot{x} + x - x^3/4 = 0$ for energy levels $K = 7/16$, 1, and $3/2$.

point $(0, -\sqrt{7/8})$. Then it will move to the left and upward until it reaches the point $(-1, 0)$, then it travels back to $(1, 0)$ in the upper half plane on the mirror image of its path in the lower half plane. Its period could be estimated by approximating the integral

$$T = 2 \int_{-1}^{1} \frac{1}{\sqrt{\frac{1}{8}(x^2 - 1)(x^2 - 7)}}\, dx.$$

The unbounded motion on the right in the potential plane would correspond to the bead sliding up the wire from the right ($x(0) > \sqrt{7}$, $dx/dt < 0$) until it reaches the point $(x, z) = (\sqrt{7}, 7/16)$. But $d^2x/dt^2 > 0$, so it will slide back down with positive velocity. In the phase plane this motion would be represented by a symmetric curve open to the right with vertex at $(\sqrt{7}, 0)$. A similar analysis can be done for the unbounded motion to the left.

b) $K = \frac{3}{2}$: $\quad y = \frac{dx}{dt} = \pm \sqrt{\frac{1}{8}(24 - 8x^2 + x^4)}.$

The discriminant of $q^2 - 8q + 24$ is $8^2 - 4(24) = -32 < 0$, which implies that $24 - 8x^2 + x^4$ will never vanish and will always have the same positive sign. The two level curves correspond to beads moving either to the left or right with enough energy so as to never come to rest. The phase plane path will be a similar motion.

c) $K = 1$: $\quad y = \frac{dx}{dt} = \pm\frac{1}{\sqrt{8}}(x - 2)(x + 2).$

There will be six distinct phase paths. Two of them connect the equilibrium states (saddle points) $(-2, 0)$ and $(2, 0)$ and are called *separatrices.* The other four are unbounded and flow either into or out of the equilibrium states. ■

The last example is a special case of the general *nonlinear spring equation*

$$\frac{d^2x}{dt^2} + \alpha x + \beta x^3 = 0, \quad \alpha > 0, \tag{7.5.8}$$

which, for appropriate values of the parameters, can be used to model a number of distinct physical phenomena. For instance, since $\sin x \cong x - x^3/6$, the choice $\alpha = 1, \beta = -1/6$ gives an approximate nonlinear pendulum equation.

The nonlinear spring equation is a good example of an equation whose solutions change character with a parameter. If $\alpha > 0$ is fixed and $\beta < 0$, there are three equilibrium points: a center at $(0, 0)$ and saddle points at $(\pm[\alpha/(-\beta)]^{1/2}, 0)$. As $\beta \to 0$, the potential well deepens and the saddle points recede to infinity. If $\beta \geq 0$, there is only one equilibrium point, a stable center, and all solutions are periodic (see Fig. 7.27).

In the remainder of this section we shall deal with nonconservative systems, i.e., problems in which damping forces play an essential role. The phase paths of the differential equation

$$\frac{d^2x}{dt^2} + h\left(x, \frac{dx}{dt}\right)\frac{dx}{dt} + g(x) = 0 \tag{7.5.9}$$

are, in general, no longer level curves of an energy function. However, the potential plane sketch is still useful if the horizontal constant-energy lines are replaced

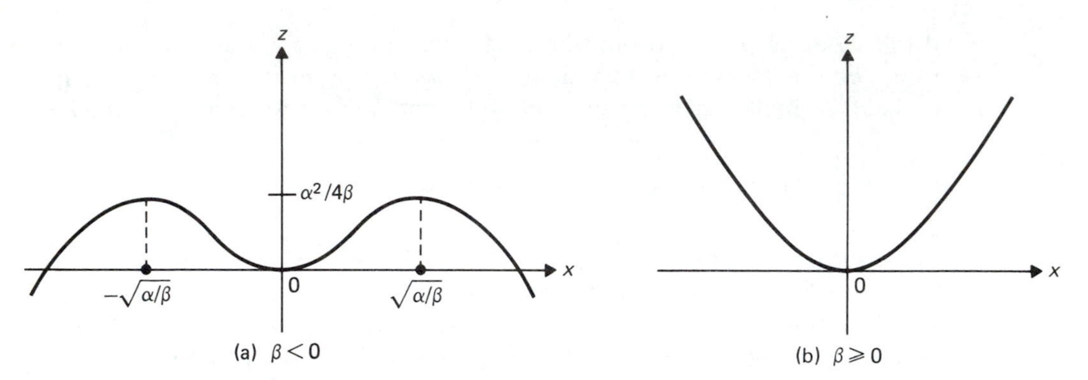

Figure 7.27 **The potential energy function $z = \alpha x^2/2 - \beta x^4/4$, $\alpha > 0$, corresponding to the nonlinear spring equation $\ddot{x} + \alpha x + \beta x^3 = 0$.**

by curves with slope given by

$$\begin{aligned}\frac{d}{dt} E\left(x(t), \frac{dx}{dt}(t)\right) &= \frac{d}{dt}\left[\frac{1}{2}\left(\frac{dx}{dt}(t)\right)^2 + G(x(t))\right] \\ &= \frac{dx}{dt}(t)\left[\frac{d^2x}{dt^2}(t) + g(x(t))\right] \\ &= -h\left(x(t), \frac{dx}{dt}(t)\right)\left[\frac{dx}{dt}(t)\right]^2.\end{aligned} \tag{7.5.10}$$

Letting $y = dx/dt$, we can write this more simply as

$$\frac{d}{dt} E(x, y) = -h(x, y)y^2.$$

For sketching purposes, precise values of dE/dt are not needed and rough estimates are sufficient. This is illustrated with the damped pendulum equation

$$\frac{d^2x}{dt^2} + c\frac{dx}{dt} + \sin x = 0, \quad 0 < c < 2. \tag{7.5.11}$$

For the undamped pendulum ($c = 0$), the total energy is

$$E(x, y) = \frac{y^2}{2} + 1 - \cos x,$$

and (7.5.10) gives for the damped pendulum

$$\frac{d}{dt} E(x, y) = -cy^2,$$

so the total energy is always a decreasing function of time.

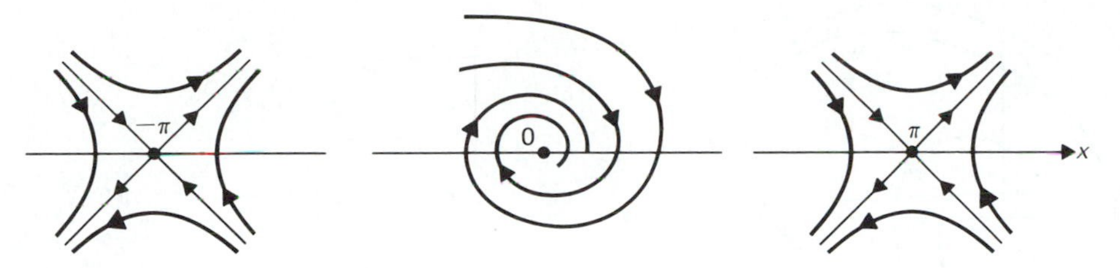

Figure 7.28

By considering the two-dimensional system corresponding to (7.5.11),

$$\dot{x} = y, \quad \dot{y} = -\sin x - cy,$$

one sees that the points (0, 0) and $(n\pi, 0)$, $n = \pm 1, \pm 2, \ldots$, in the phase plane are its equilibrium points. An analysis of the linearized system at each point shows that the points (0, 0), $(2k\pi, 0)$, $k = \pm 1, \pm 2, \ldots$, are stable spiral points, and the points $((2k + 1)\pi, 0)$, $k = 0, \pm 1, \pm 2, \ldots$, are saddle points. A local phase plane portrait is sketched in Fig. 7.28.

In a typical nonconservative system modeled by equation (7.5.9), $h(x, \dot{x})$ is positive and, consequently, dE/dt is negative. Therefore the system is dissipative and any nontrivial motion eventually ceases. However, there is a fascinating phenomenon called *self-sustained oscillations* in which the energy oscillates as it seemingly is pumped into and out of the system even though the modeling differential equation is autonomous. A well-known example is that of a nonlinear electric oscillator that is modeled by the *van der Pol equation*

$$\frac{d^2x}{dt^2} + \epsilon(x^2 - 1)\frac{dx}{dt} + x = 0. \tag{7.5.12}$$

The dependent variable x is proportional to a grid voltage and ϵ is a positive parameter definable in terms of circuit elements. As before, we set $E(x, \dot{x}) = \dot{x}^2/2 + x^2/2$, and (7.4.10) implies that

$$\frac{dE}{dt} = -\epsilon(x^2 - 1)\left(\frac{dx}{dt}\right)^2.$$

Let us assume that $\dot{x} = 0$ only for isolated values of t. If x^2 is greater than one, the energy is decreasing, and if x^2 is less than one, the energy is increasing. A balance between the energy loss and gain would result in self-sustained oscillations. A mathematical proof of this is not elementary, but it has been shown (see [6], pp. 252–256) that for any value of $\epsilon > 0$ the van der Pol equation has an isolated periodic solution, called a *limit cycle,* toward which all other nontrivial solutions tend with increasing time. In the phase plane this would be represented by an isolated closed curve; this is illustrated in Fig. 7.29. If the parameter ϵ is large, a two-

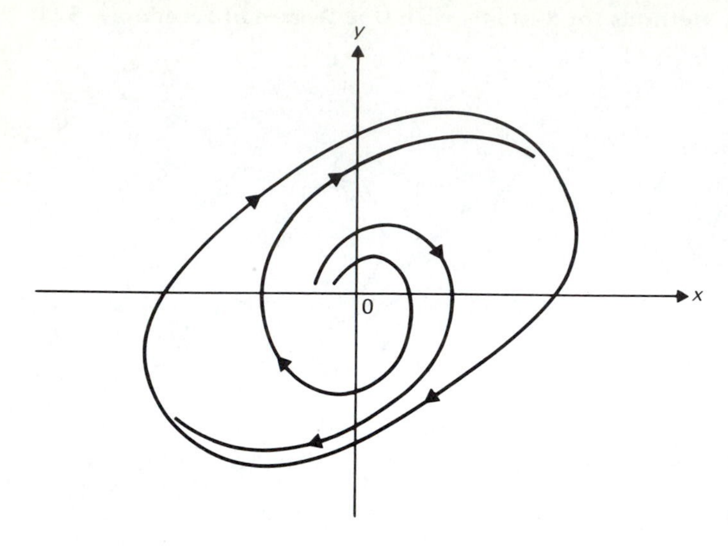

Figure 7.29 A limit cycle of the van der Pol equation when ϵ is small.

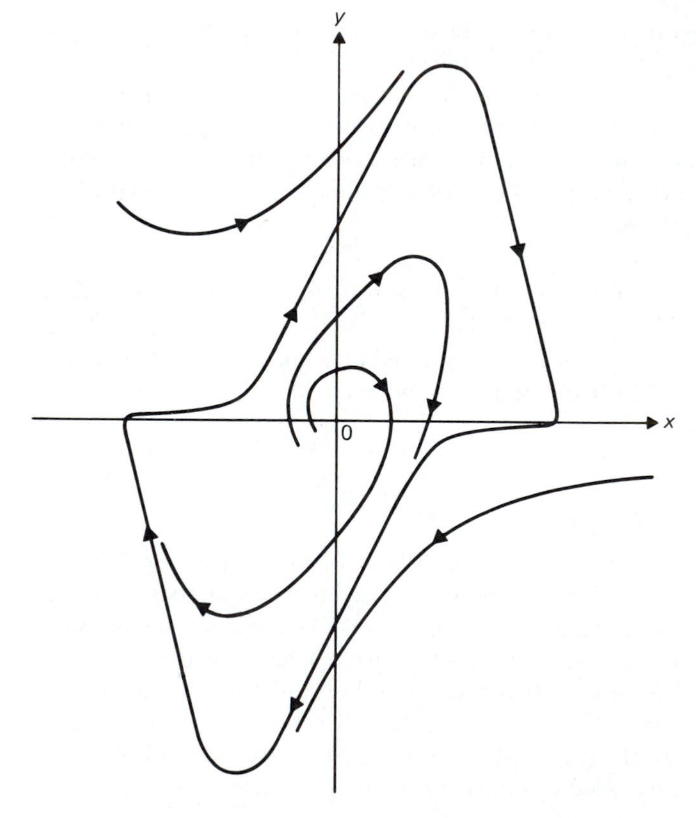

Figure 7.30 A limit cycle of the van der Pol equation when ϵ is large.

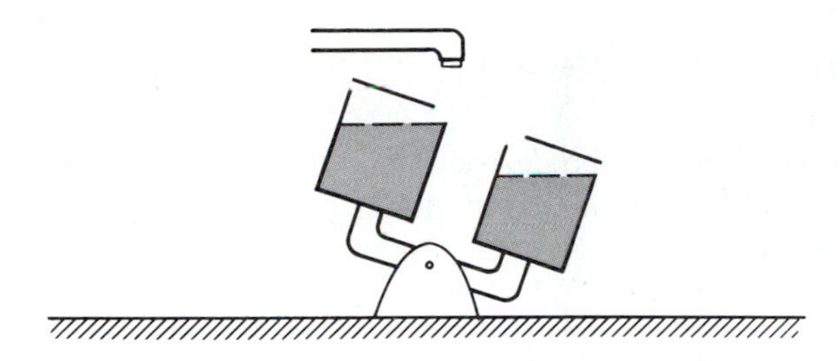

Figure 7.31 Each container has an opening on the top and side. When water flowing from the faucet fills the uppermost container the system inverts its position. Then water will flow out of the first container while the second is being filled.

phase phenomenon called a *relaxation oscillation* is observed. During the first phase energy is slowly stored until a critical threshold is reached when the energy is discharged almost instantaneously (see Fig. 7.30). The corresponding periodic solution would be a very jerky oscillation with abrupt changes in the sign of the amplitude.

The van der Pol equation was first developed in the study of electrical circuits. Self-sustained oscillations also occur in biological organisms, such as the timing of mitosis in slime molds (see [10], p. 184) and in autocatalytic chemical reactions (see [11], p. 146). Relaxation oscillations would typically model physical systems where there are thresholds and abrupt self-correction. An example of a simple system showing this type of behavior is described in Fig. 7.31.

EXERCISES 7.5

For the following differential equations, find and classify the equilibrium points, derive and solve the corresponding linear equations, and sketch *local* phase plane portraits.

1. $\dfrac{d^2x}{dt^2} + x^3 + x^2 - 2x = 0$

2. $\dfrac{d^2x}{dt^2} + x^3 - 2x^2 - 3x = 0$

3. $\dfrac{d^2x}{dt^2} + x - x^3 = 0$

4. $\dfrac{d^2x}{dt^2} + 3x^2 + 5x + 2 = 0$

For the following differential equations, sketch potential plane and global phase plane portraits.

5. $\dfrac{d^2x}{dt^2} + x^3 - \dfrac{x}{4} = 0$

6. $\dfrac{d^2x}{dt^2} + 2x^2 - 5x - 3 = 0$

7. $\dfrac{d^2x}{dt^2} + \sin x - \dfrac{\sqrt{8}}{\pi} x = 0$

8. $\dfrac{d^2x}{dt^2} + \dfrac{\sinh x}{\sinh 1} - x = 0$

9. $\dfrac{d^2x}{dt^2} + \dfrac{x}{x-1} = 0$

10. $\dfrac{d^2x}{dt^2} + \dfrac{x}{x^2-1} = 0$

In the following problems find and classify the equilibrium points, derive and solve the corresponding linear equations, and sketch potential plane and global phase plane portraits. The reader is urged to consult the cited references for more details.

11. According to the general theory of relativity, the path of a particle moving along a timelike geodesic in the gravitational field of a spherically symmetric heavy body is governed by

$$\frac{d^2u}{d\theta^2} + u = 1 + \lambda u^2,$$

where u is inversely proportional to the distance from the heavy body, θ is an angle in the plane of motion, and λ is a parameter proportional to the square of the speed of light, $0 < \lambda \ll 1$. Study the effect of increasing λ on the potential plane and phase plane portraits and observe the disappearance of the potential well at $\lambda = \frac{1}{4}$ (see [12]).

12. The motion of an elastically restrained current-carrying bus bar in the magnetic field of a long fixed parallel wire is governed by

$$m\frac{d^2x}{dt^2} + k\left(x - \frac{\lambda}{L-x}\right) = 0.$$

The parameter λ is proportional to the product of the currents in the two conductors, k is the elastic constant, L is the distance from the fixed wire to a supporting wall, and m is the mass of the bus bar. Study the effect of varying λ on the phase plane portrait (see [2], pp. 50–54).

13. By using carefully selected units, we can model a linear spring–magnet system by

$$\frac{d^2x}{dt^2} + x - \frac{36}{(x-7)^2} = 0.$$

The magnets are assumed to be long enough that the attractive force is inversely proportional to the distance between them (see [8], pp. 268–269). Sketch the phase plane portrait.

14. The Sine–Gordon partial differential equation

$$\frac{\partial^2\phi}{\partial x^2} - \frac{\partial^2\phi}{\partial t^2} = \sin\phi$$

was constructed to illustrate nonlinear wave phenomena. To study the existence of traveling waves, one studies an associated ordinary differential equation obtained as follows: let $\phi(x, t) = f(x - ct)$, where c is the wave velocity, and substitute it into the Sine–Gordon equation. For instance, if we let $z = x - ct$, then

$$\frac{\partial \phi}{\partial t}(x, t) = \frac{\partial}{\partial t} f(x - ct) = \frac{df}{dz}(z)\frac{dz}{dt}$$

$$= -c\frac{df}{dz}, \text{ etc.}$$

Analyze the ordinary differential equation you obtain according to the above instructions. Show that the upper separatrix in the phase plane is the graph of a solution to

$$\frac{df}{dz} = \frac{2}{\sqrt{1 - c^2}} \sin \frac{f}{2}.$$

Solve this equation with initial condition $f(0) = \pi/2$. Its solution is called a "kink" solution (see [13]). Why is $c = 1$ an exceptional value?

Linear damping changes centers into spiral points but saddle points remain saddle points. Sketch potential plane and global phase plane portraits for the following equations. Assume that $0 < c \ll 1$.

15. $\dfrac{d^2x}{dt^2} + c\dfrac{dx}{dt} + x^3 - \dfrac{x}{4} = 0$

16. $\dfrac{d^2x}{dt^2} + c\dfrac{dx}{dt} + 2x^2 - 5x - 3 = 0$

17. $\dfrac{d^2x}{dt^2} + c\dfrac{dx}{dt} + \sin x - \dfrac{\sqrt{8}}{\pi} x = 0$

18. $\dfrac{d^2x}{dt^2} + c\dfrac{dx}{dt} + 3x^2 + 5x + 2 = 0$

19. $\dfrac{d^2x}{dt^2} + c\dfrac{dx}{dt} + x = 1 + \dfrac{1}{8}x^2$

20. $m\dfrac{d^2x}{dt^2} + c\dfrac{dx}{dt} + k\left(x - \dfrac{\lambda}{L - x}\right) = 0$

21. Show that $\ddot{x} + \dot{x}^2 + g(x) = 0$ has an integral,

$$E(x, \dot{x}) = (e^{2x})\frac{\dot{x}^2}{2} + e^{2x}F(x) = \text{const.}$$

with $F'(x) + 2F(x) = g(x)$.

a) Set $g(x) = x$ and show that $(0, 0)$ is a center by showing that the nearby level curves of $E(x, \dot{x})$ are closed.

b) Set $g(x) = \sin x$. Is there a level curve of $E(x, \dot{x})$ joining two unstable equilibrium points? Justify your answer.

c) Introduce a new independent variable s by setting $ds/dt = e^{-x}$ and derive the differential equation

$$x'' + e^{2x}g(x) = 0, \text{ where } (\,)' = d(\,)/ds.$$

Sketch a global phase plane portrait for the integral curves of

$$x'' + xe^{2x} = 0.$$

Are the integral curves the same as those for

$$\ddot{x} + \dot{x}^2 + x = 0?$$

(*Hint:* Are the coordinate systems of the two phase planes the same?)

22. Compute the period of the van der Pol equation

$$\ddot{x} + \epsilon(x^2 - 1)\dot{x} + x = 0$$

with $\epsilon = 0.1$ by integrating it using RK4 on $0 \leq t \leq 30$, $h = 0.1$, $x(0) = 2$, $\dot{x}(0) = 0$. Estimate the period by computing the time between successive crossings of the positive x-axis *after* the integral curve is near the limit cycle. To avoid excessive output, print t, x, $\dot{x}$ only for x near 2 and $\dot{x}$ near 0. Interpolate to find the crossing time. Repeat the computation with $\epsilon = 0.4$, 0.8, 1.6, 2.0, and plot the period versus ϵ.

7.6 MATHEMATICAL MODELS OF TWO POPULATIONS

In Chapter 1 we discussed the mathematical modeling of the growth of one population. The model developed there was based on certain principles that could be satisfied by the per capita rate of growth. This analysis will now be extended to the study of two populations. Much of what we have to say is applicable to models of more than two populations, but the advantage of having the phase plane to describe the qualitative behavior of the models given favors the study of two populations.

Let X and Y represent two populations whose size at time t is given by the functions $x(t)$ and $y(t)$. Both of these will be assumed to be nonnegative (since negative population sizes make no sense), and any time one or both of the components of the vector $(x(t), y(t))$ becomes zero, the study is over, since one or both of the populations has expired. The per capita rates of growth of each population at time t are then $\dot{x}(t)/x(t)$ and $\dot{y}(t)/y(t)$ for populations X and Y, respectively.

The next task is to give plausible conditions that these rates must satisfy. We must first assume that there is some form of interaction between the two populations X and Y, and therefore the per capita rates of growth must depend on both the present size $x(t)$ of X and the present size $y(t)$ of Y. This immediately suggests that there exist functions f and g of two variables such that

$$\frac{\dot{x}(t)}{x(t)} = f(x(t), y(t)), \qquad \frac{\dot{y}(t)}{y(t)} = g(x(t), y(t)),$$

or, equivalently, that the model is described by the autonomous system of two differential equations

$$\dot{x} = x\, f(x, y), \qquad \dot{y} = y\, g(x, y). \tag{7.6.1}$$

There is no inherent biological reason why f or g should not also depend on t, but the obvious advantage of using the phase plane to analyze the qualitative behavior favors an autonomous model.

What should be the nature of f and g? First of all, in the absence of one population, the per capita rate of growth of the remaining population should satisfy whatever principles are decided to be appropriate for the growth of that population without interaction. Since the absence of population Y means $y = 0$, we could have, for instance,

$$f(x, 0) = a > 0, \quad \text{or} \quad f(x, 0) = -a < 0, \quad \text{or} \quad f(x, 0) = r\left(1 - \frac{x}{K}\right), \; r, K > 0,$$

which are exponential growth, exponential decay, and logistic growth of X, respectively. Similarly, $g(0, y)$ should satisfy whatever principles the per capita rate of growth of population Y should follow in the absence of population X.

What should be the nature of f and g when both x and y are not zero? To answer this question one has to specify the type of interaction between the two populations X and Y. There are two general classifications:

Competition models

Both populations are in competition for the same shared resources (e.g., food, water, space). Consequently, an increase in Y will decrease the per capita growth rate of X. Similarly, an increase in X will decrease the per capita growth rate of Y.

Predator–prey models

Population Y is a predator population that has X as its food source, or prey. Consequently, an increase in X will result in an increase in the per capita growth rate of Y, whereas an increase in Y results in a decrease in the per capita growth rate of X.

In the predator–prey case it is sometimes assumed that the only food source for Y is X and that Y will perish in its absence. This would mean that $g(0, y)$ is strictly negative, for example, $g(0, y) = -ay$, $a > 0$, corresponding to exponential decay. While this makes for lovely phase plane portraits, few ecological systems are so simple.

Each class of models will now be discussed in more detail.

COMPETITION MODELS

Since the per capita growth rate of X is $f(x, y)$, the competition assumption suggests that f should decrease as y increases for fixed x and $x > 0$ and $y > 0$. This leads to the requirement that $\partial f/\partial y$ be negative for $x > 0$ and $y > 0$. The same reasoning applied to Y implies that $\partial g/\partial x$ is also negative for $x > 0$ and $y > 0$. Consequently, our model is

$$\dot{x} = x\,f(x, y), \qquad \dot{y} = y\,g(x, y) \tag{7.6.2}$$

$$\frac{\partial f}{\partial y}(x, y) < 0, \quad \frac{\partial g}{\partial x}(x, y) < 0, \quad x > 0, \quad y > 0,$$

where $f(x, 0)$ and $g(0, y)$ are suitable population growth laws for X and Y, respectively.

To analyze a model of the type (7.6.2), one must first find all the equilibrium points in the upper right-hand quadrant $x \geq 0$, $y \geq 0$, and then analyze their stability. These equilibrium points will be of three types, and each will be discussed briefly and some possible phase plane configurations given before analyzing a specific model.

Types of equilibrium points for competition models

1. Point (0, 0). This is always an equilibrium point, and if the system is linearized, one obtains

$$\dot{u} = f(0, 0)\,u, \qquad \dot{v} = g(0, 0)\,v.$$

The point (0, 0) will be asymptotically stable if both $f(0, 0)$ and $g(0, 0)$ are negative. This means that for small initial sizes, the population X will diminish in the absence of Y and the population Y will diminish in the absence of X. Since it is usually assumed that X and Y are growing populations in the absence of competition, $f(0, 0)$ and $g(0, 0)$ are assumed positive and, as a consequence, (0, 0) will be an unstable equilibrium point.

2. Points $(K, 0)$ and/or $(0, J)$, if any, where $f(K, 0) = 0$ and/or $g(0, J) = 0$. The values K and J would represent the natural carrying capacities of the X and Y population, respectively, in the absence of competition. If $(K, 0)$ were asymptotically stable, it would mean that if $x(t)$ were near K and $y(t)$ was small, then competition would favor X and it would survive (with limiting value K) whereas Y would perish. If $(K, 0)$ were unstable, then for X near its carrying capacity and the size of Y small, competition favors Y and it increases in size. This obviously suggests a saddle-point type behavior.

3. Points (x_∞, y_∞), $x_\infty > 0$, $y_\infty > 0$, if any, which are solutions of the two equations

$$f(x, y) = 0, \qquad g(x, y) = 0.$$

If such a point is asymptotically stable, it means that populations X and Y have positive limiting values x_∞ and y_∞, respectively. This is the case of *stable coexistence.* If such a point were unstable or did not exist, it would usually mean that a point of the type $(K, 0)$ or $(J, 0)$ would be asymptotically stable

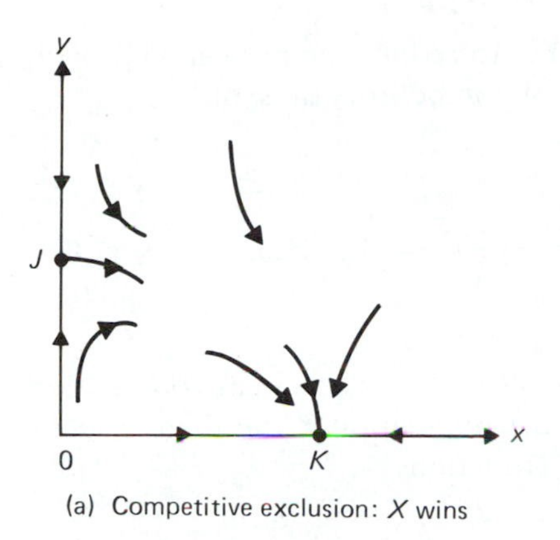

(a) Competitive exclusion: X wins

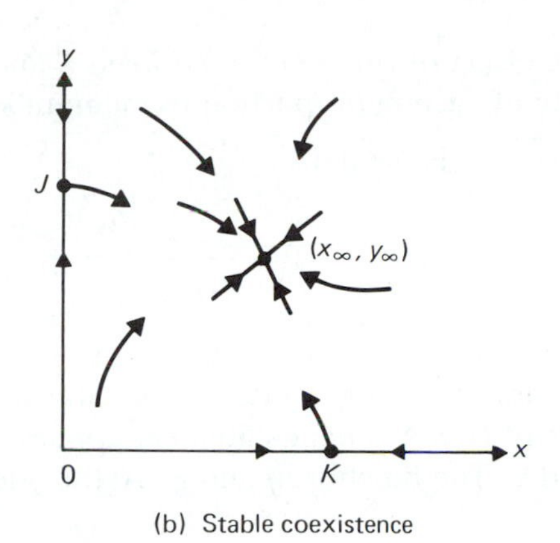

(b) Stable coexistence

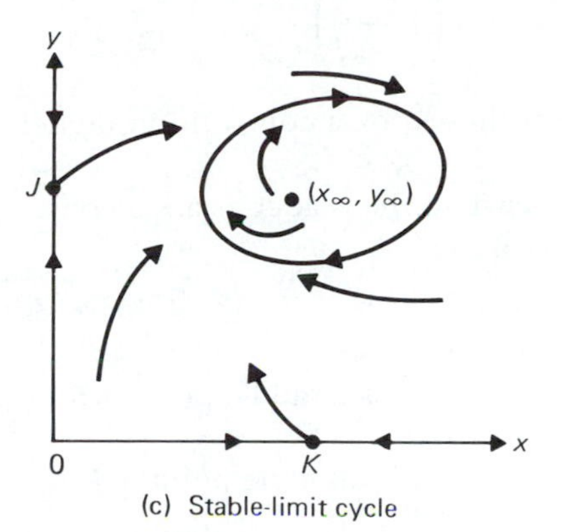

(c) Stable-limit cycle

Figure 7.32

and only one population would survive competition. This is the case of *competitive exclusion.* Another interesting case is where (x_∞, y_∞) is unstable but is enclosed by a closed curve that is a stable limit cycle. Competition will then create a limiting cyclic configuration.

Some possible phase plane portraits for the competition model are shown in Fig. 7.32.

A LOGISTIC COMPETITION MODEL

A simple but frequently used model for competition is where one assumes the following:

1. The growth of populations X and Y in the absence of competition is logistic, and

2. the effect of competition between X and Y is to reduce the per capita growth rate of each in proportion to the number of competitors present.

This leads to the model

$$\dot{x} = xr\left(1 - \frac{x}{K}\right) - \alpha xy, \qquad \dot{y} = ys\left(1 - \frac{y}{J}\right) - \beta xy, \qquad \textbf{(7.6.3)}$$

where r, K, α, s, J, and β are all positive numbers.

The intrinsic growth rate of X is r, its carrying capacity is K, and its competition coefficient is α. A similar statement applies to the population Y and the constants s, J, and β. The functions f and g are the linear functions

$$f(x, y) = r\left(1 - \frac{x}{K}\right) - \alpha y, \qquad g(x, y) = s\left(1 - \frac{y}{J}\right) - \beta x,$$

and since $f_y = -\alpha < 0$, $g_x = -\beta < 0$, the hypotheses for a competition model are satisfied.

It is a straightforward calculation, which we leave to the reader as an exercise, to verify the following properties of the system (7.6.3):

1. Point (0, 0) is an unstable node.
2. Points $(K, 0)$ and $(0, J)$ are equilibrium points and
   a) point $(K, 0)$ is a stable node if $K > s/\beta$, whereas it is a saddle point if $K < s/\beta$;
   b) point $(0, J)$ is a stable node if $J > r/\alpha$, whereas it is a saddle point if $J < r/\alpha$.

One now can graph the lines $f(x, y) = 0$ and $g(x, y) = 0$ and determine that

3. they will have a point of intersection (x_∞, y_∞), where $x_\infty > 0$, $y_\infty > 0$, if
   a) $J > r/\alpha$ and $K > s/\beta$, in which case $(K, 0)$ and $(0, J)$ are stable nodes, or
   b) $J < r/\alpha$ and $K < s/\beta$, in which case $(K, 0)$ and $(0, J)$ are saddle points;
4. they will have no point of intersection in the upper right-hand quadrant if
   a) $J > r/\alpha$ and $K < s/\beta$, which implies that $(0, J)$ is a stable node and $(K, 0)$ is a saddle point. Competition favors the population Y and it will tend toward its natural carrying capacity J, whereas population X will become extinct.
   b) $J < r/\alpha$ and $K > s/\beta$, which gives the same qualitative behavior as in (a), with the roles of X and Y reversed.

A graphical representation of the two cases 4(a) and 4(b), which represent competitive exclusion, is shown in Fig. 7.33.

Finally, we must analyze the case 3 where there is an equilibrium point (x_∞, y_∞) with $x_\infty > 0$, $y_\infty > 0$. If the system (7.5.3) is linearized about the equilib-

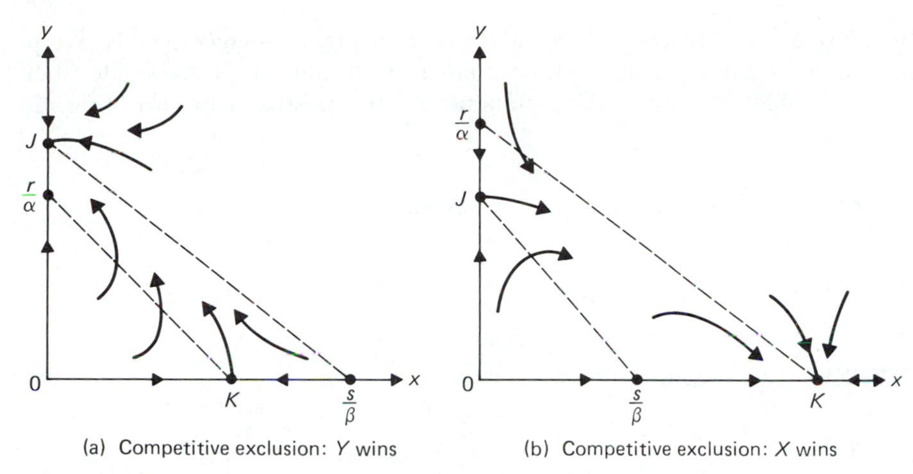

(a) Competitive exclusion: Y wins (b) Competitive exclusion: X wins

Figure 7.33

rium point, then, since (x_∞, y_∞) satisfy the equations

$$r - \frac{r}{K}x_\infty - \alpha y_\infty = 0, \qquad s - \frac{s}{J}y_\infty - \beta x_\infty = 0, \tag{7.6.4}$$

the coefficient matrix of the linearized system will be

$$\begin{bmatrix} \frac{-r}{K}x_\infty & -\alpha x_\infty \\ -\beta y_\infty & \frac{-s}{J}y_\infty \end{bmatrix}.$$

Its characteristic equation is

$$\lambda^2 + \left(\frac{r}{K}x_\infty + \frac{s}{J}y_\infty\right)\lambda + x_\infty y_\infty\left(\frac{rs}{KJ} - \alpha\beta\right) = 0,$$

and since $x_\infty y_\infty > 0$, it will have roots with negative real parts if and only if

$$\frac{r}{K}x_\infty + \frac{s}{J}y_\infty > 0 \quad \text{and} \quad \Delta = \frac{rs}{KJ} - \alpha\beta > 0. \tag{7.6.5}$$

The first condition is always satisfied since all the quantities are positive; a biological interpretation of the condition $\Delta > 0$ will be given below.

Suppose $(K, 0)$ and $(0, J)$ are unstable; then case 3(b) above implies that $\alpha < r/J$ and $\beta < s/K$; hence

$$\alpha\beta < \frac{r}{J} \cdot \frac{s}{K},$$

and therefore $\Delta > 0$. Hence if $(K, 0)$ and $(0, J)$ are unstable, then (x_∞, y_∞) is asymptotically stable. A similar analysis shows that if $(K, 0)$ and $(0, J)$ are stable, then (x_∞, y_∞) is unstable. The phase plane plots for the two possible cases are shown in Fig. 7.34.

EXAMPLE 1 Consider the competitive system

$$\dot{x} = x\left[(0.5) - \frac{1}{400}x - \frac{1}{10^3}y\right], \qquad \dot{y} = y\left[(0.8) - \frac{1}{500}y - \frac{1}{10^3}x\right].$$

SOLUTION This system can be written as

$$\dot{x} = (0.5)x\left[1 - \frac{1}{200}x\right] - \frac{1}{10^3}xy, \qquad \dot{y} = (0.8)y\left[1 - \frac{1}{400}y\right] - \frac{1}{10^3}xy,$$

and therefore

$$r = 0.5, \qquad K = 200, \qquad \alpha = \frac{1}{10^3},$$

$$s = 0.8, \qquad J = 400, \qquad \beta = \frac{1}{10^3}.$$

Figure 7.34

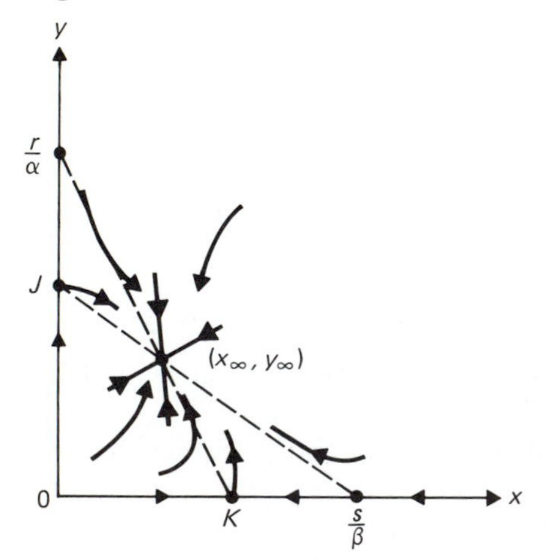

(a) Stable coexistence: (x_∞, y_∞) is a stable node

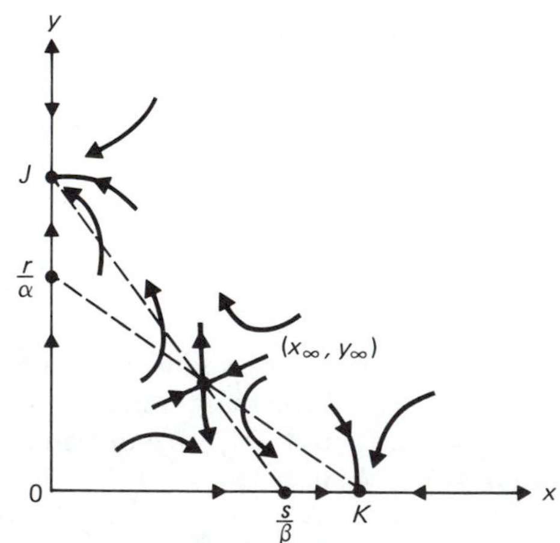

(b) Unstable coexistence: (x_∞, y_∞) is a saddle point and initial conditions will determine which population wins

The point (0, 0) is an unstable node, and since

$$J = 400 < \frac{r}{\alpha} = (0.5)10^3 = 500,$$

$$K = 200 < \frac{s}{\beta} = (0.8)10^3 = 800,$$

the equilibrium points (200, 0) and (0, 400) are saddle points, and there will be a stable equilibrium point (x_∞, y_∞). By solving the pair of equations

$$\frac{1}{400}x + \frac{1}{10^3}y = \frac{1}{2}, \qquad \frac{1}{10^3}x + \frac{1}{500}y = \frac{8}{10}$$

we find that $(x_\infty, y_\infty) = (50, 375)$. The coefficient matrix of the linearized system at (x_∞, y_∞) is

$$\begin{bmatrix} -\dfrac{0.5}{200}(50) & -\dfrac{1}{1000}(50) \\ -\dfrac{1}{1000}(375) & \dfrac{0.8}{400}(375) \end{bmatrix} = \begin{bmatrix} -\dfrac{1}{8} & -\dfrac{1}{20} \\ -\dfrac{3}{8} & -\dfrac{3}{4} \end{bmatrix}.$$

Its characteristic polynomial is $\lambda^2 + \frac{7}{8}\lambda + \frac{3}{40}$ which has distinct negative roots, so the point (50, 375) is a stable node. ■

If the original system of differential equations is written in the form

$$\dot{x} = rx - \frac{r}{K}x^2 - \alpha xy, \qquad \dot{y} = sy - \frac{s}{J}y^2 - \beta xy,$$

the coefficient r/K can be interpreted as a measure of how much population X controls its own growth. The coefficient α can be interpreted as a measure of how much population Y controls X's growth, and similar interpretations can be given for s/J and β. Therefore the condition $\Delta > 0$ is equivalent to

$$\left(\frac{r}{K}\right)\left(\frac{s}{J}\right) > \alpha\beta,$$

which says that stability can occur when species control their own growth stronger than when they try to control their competitor's. Put another way, $\Delta < 0$ means that stability will not occur when species compete too strongly.

PREDATOR–PREY MODELS

In the general classification of types of population models, predator–prey models were characterized by the assumptions that for $x > 0$ and $y > 0$,

1. the per capita rate of growth $f(x, y)$ of population X (the prey) must decrease as the size y of population Y (the predator) increases, and

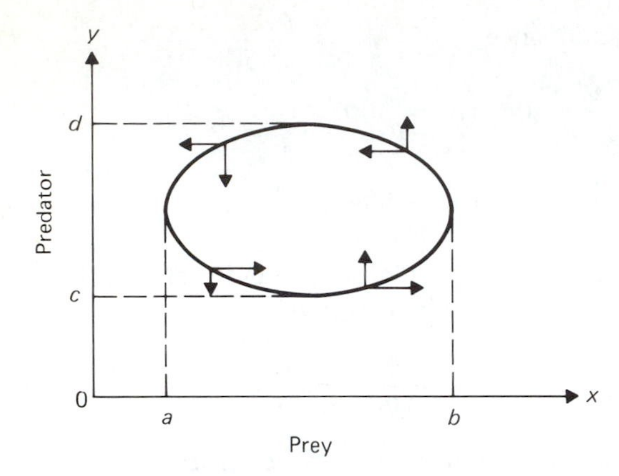

Figure 7.35 A periodic predator–prey model. The prey population size will oscillate between the values a and b; the predator between c and d.

2. the per capita rate of growth $g(x, y)$ of Y must increase as the size x of population X increases.

This leads to the model

$$\dot{x} = x\,f(x, y), \qquad \dot{y} = y\,g(x, y), \tag{7.6.6}$$

$$\frac{\partial f}{\partial y}(x, y) < 0, \qquad \frac{\partial g}{\partial x}(x, y) > 0, \qquad x > 0, \qquad y > 0.$$

The types of equilibrium points for the predator–prey model (7.6.6) are the same as for the competitive model, namely, point (0, 0), possibly points of the type $(K, 0)$ or $(0, J)$, and of the type (x_∞, y_∞), where $x_\infty > 0$, $y_\infty > 0$. By linearizing the system about each of them and studying their asymptotic behavior one tries to put together a phase plane portrait of the system.

But the real interest in predator–prey models arises from the fact that they often exhibit periodic behavior. In the phase plane this would be represented by a closed curve or families of closed curves with an equilibrium point in their interior. If we take such a curve and at points on it put an arrow pointing up or down if Y is increasing or decreasing, and another arrow pointing to the right or left if X is increasing or decreasing, we will obtain the picture given in Fig. 7.35.

There is a certain poetic simplicity in such pictures (rabbits and lynxes living together in harmonious balance but out of phase!), but often such models are based upon simplified assumptions, which makes their biological significance minimal.

A classical predator–prey model is given by the Lotka–Volterra system of equations[1]

$$\dot{x} = x(r - \alpha y), \qquad \dot{y} = y(-s + \beta x), \tag{7.6.7}$$

1. Named after the American biophysicist Alfred J. Lotka (1880–1949) and the Italian mathematician Vito Volterra (1860–1940) who independently developed and analyzed the model.

where r, s, α, and β are positive constants. It is based on the assumptions that

1. the prey X grows exponentially in the absence of predator Y; that is, $f(x, 0) = r > 0$;
2. the predator Y will become extinct exponentially in the absence of prey; that is, $g(0, y) = -s < 0$.

Since $\partial f/\partial y = -\alpha < 0$, $\partial g/\partial x = \beta > 0$, we see that the effect of predation is to reduce or increase the per capita rate of growth in proportion to the numbers of predators or prey present.

The Lotka–Volterra model (7.6.7) has two equilibrium points: $(0, 0)$, which is easily seen to be a saddle point since $r > 0$ and $-s < 0$, and $(x_\infty, y_\infty) = (s/\beta, r/\alpha)$. If the system is linearized about the latter, the coefficient matrix of the linearized system is

$$\begin{bmatrix} 0 & -\alpha s/\beta \\ \beta r/\alpha & 0 \end{bmatrix}.$$

Its characteristic equation is $\lambda^2 + rs = 0$ and since $rs > 0$, the roots are pure imaginary, which says that $(s/\beta, r/\alpha)$ is a center for the linearized system.

As was discussed in a previous section, the case where an equilibrium point of a nonlinear system is a center for the linearized system is an ambiguous case. The nonlinear system may have a family of closed curves (periodic solutions) around the equilibrium point, as in the linear case, or the equilibrium point may be a stable or unstable spiral point.

It is a fact that for the Lotka–Volterra model, given any point (x_0, y_0), $x_0 > 0$, $y_0 > 0$, there is a closed curve, centered at the equilibrium point $(s/\beta, r/\alpha)$ that represents a periodic solution of the system and passes through (x_0, y_0). More simply said, the center structure of the linearization is preserved in the nonlinear model. The proof of this statement depends on the fact that one can easily solve the equivalent first order equation

$$\frac{dy}{dx} = \frac{y(-s + \beta x)}{x(r - \alpha y)}, \qquad y(x_0) = y_0 \tag{7.6.8}$$

since it is separable. The implicitly defined solution will describe the integral curve passing through (x_0, y_0). The proof that the curve is closed is left for the reader as a problem.

The integral curves of the Lotka–Volterra equations are elliptically shaped near the equilibrium point and become more and more distorted as one moves away from it. Graphs of a few of them are shown in Fig. 7.36.

A disturbing consequence of the model is that no matter how small the initial prey population is and how large the initial predator population is, both populations will survive in some sort of cyclic growth curve. Imagine a scenario for the case of 2 rabbits and 4000 lynxes and you will see the consequences of the oversimplified assumptions on which it is based. What if both rabbits were males?

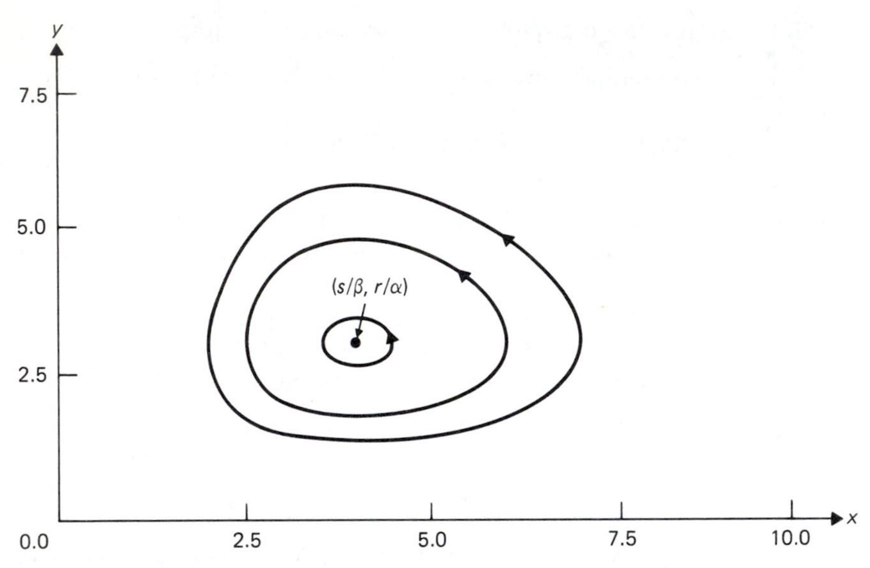

Figure 7.36 A phase plane portrait of the Lotka–Volterra predator–prey system generated with RKF45.

Another problem with the Lotka–Volterra model is its lack of *structural stability* that can be loosely defined as robustness of the phase portrait under small perturbations—a desired property of any good model. To show the lack of structural stability suppose instead that the prey population grows logistically and has a huge carrying capacity $K = r/\epsilon$, where $\epsilon > 0$ is very small. Then the system of equations becomes

$$\begin{aligned} \dot{x} &= xr\left(1 - \frac{x}{K}\right) - \alpha xy = x(r - \epsilon x - \alpha y), \\ \dot{y} &= y(-s + \beta x), \end{aligned} \tag{7.6.9}$$

which are certainly small perturbations of the original Lotka–Volterra equations (7.6.7).

The point (0, 0) is an equilibrium point of (7.6.9) and is still a saddle point. The second equilibrium point is

$$(x_\infty, y_\infty) = \left(\frac{s}{\beta}, \frac{r}{\alpha} - \frac{\epsilon s}{\beta\alpha}\right)$$

and can be made as near to $(s/\beta, r/\alpha)$ as desired by appropriate choice of ϵ. If the perturbed system (7.6.9) is linearized about (x_∞, y_∞), the coefficient matrix of the linearized system is

$$\begin{bmatrix} -\dfrac{\epsilon s}{\beta} & -\dfrac{\alpha s}{\beta} \\ \dfrac{\beta r}{\alpha} - \dfrac{\epsilon s}{\alpha} & 0 \end{bmatrix}.$$

The characteristic equation of the matrix is

$$\lambda^2 + \epsilon \frac{s}{\beta} \lambda + \left(rs - \frac{\epsilon s^2}{\beta} \right) = 0$$

with roots

$$\lambda_1, \lambda_2 = \frac{-\epsilon s}{2\beta} \pm \sqrt{\left(\frac{\epsilon s}{2\beta}\right)^2 - \left(rs - \frac{\epsilon s^2}{\beta}\right)}.$$

One sees that for the linear system and locally for the nonlinear system (7.6.9) the equilibrium point will be

1. a saddle point if $rs - \epsilon s^2/\beta < 0$ (the reader may wish to check that this implies $y_\infty < 0$, which makes no sense);
2. a stable spiral point if $rs - \epsilon s^2/\beta > (\epsilon s/2\beta)^2$;
3. a stable node if $0 < rs - \epsilon s^2/\beta \leq (\epsilon s/2\beta)^2$.

A sketch of the phase plane for some of these possible configurations when $\epsilon > 0$ is in marked contrast to the phase plane portrait of the Lotka–Volterra equations corresponding to $\epsilon = 0$ (see Fig. 7.37). Note that $(K, 0)$ is also an equilibrium

Figure 7.37

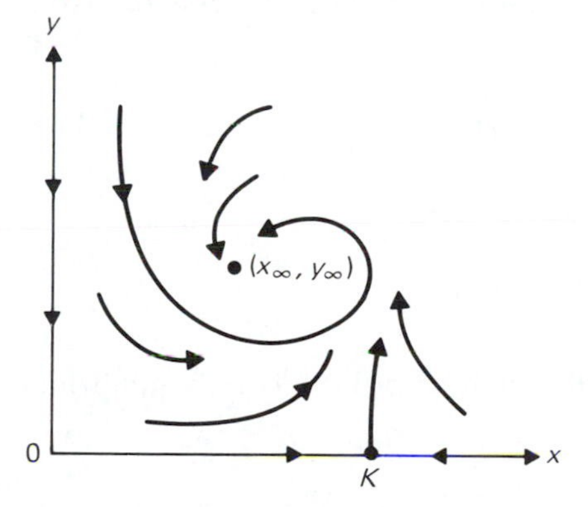

(a) (x_∞, y_∞) is a stable spiral point

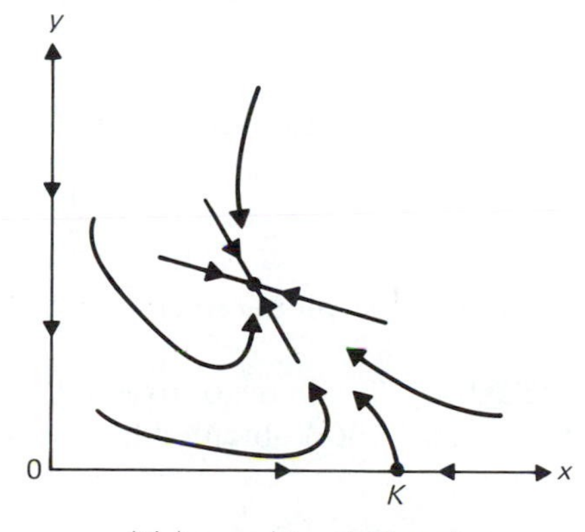

(b) (x_∞, y_∞) is a stable node

point and it will be a saddle point if $K = r/\epsilon > s/\beta$, which will always be the case if ϵ is small.

EXAMPLE 2 Consider the Lotka–Volterra system

$$\dot{x} = x(1 - \frac{1}{5}y) = x - \frac{1}{5}xy,$$

$$\dot{y} = y(-1 + \frac{1}{10}x) = -y + \frac{1}{10}xy.$$

SOLUTION The origin (0, 0) is an equilibrium point and the linearized equations at (0, 0) are

$$\dot{x} = x, \qquad \dot{y} = -y,$$

so it is a saddle point. The other equilibrium point, (10, 5), is obtained by solving

$$1 - \frac{1}{5}y = 0, \qquad -1 + \frac{1}{10}x = 0.$$

The linearized system at (10, 5) is

$$\dot{x} = -2y, \qquad \dot{y} = \frac{1}{2}x,$$

and the coefficient matrix has the characteristic polynomial $\lambda^2 + 1 = 0$ with roots $\pm i$. Since (10, 5) is a center of the nonlinear system as well, the integral curves very near (10, 5) will look like ellipses given by the solutions

$$x(t) = A\cos(t - \theta), \qquad y(t) = \frac{1}{2}A\sin(t - \theta)$$

of the linear system. They will become more distorted as one moves away from (10, 5). ■

EXAMPLE 3 Consider instead the following perturbed variation of Example 2:

$$\dot{x} = x - \epsilon x^2 - \frac{1}{5}xy, \qquad \dot{y} = -y + \frac{1}{10}xy,$$

where $\epsilon > 0$ is a small parameter.

SOLUTION The point (0, 0) is still a saddle point, and the second equilibrium point is $(10, 5 - 50\epsilon)$ obtained by solving

$$1 - \epsilon x - \frac{1}{5}y = 0, \qquad -1 + \frac{1}{10}x = 0.$$

If the system is linearized at the point $(10, 5 - 50\epsilon)$, one obtains the system

$$\dot{x} = -10\epsilon x - 2y, \qquad \dot{y} = \left(\frac{1}{2} - 5\epsilon\right)x.$$

The characteristic equation of the coefficient matrix is

$$\lambda^2 + 10\epsilon\lambda + (1 - 10\epsilon) = 0$$

and its roots are

$$\lambda_1, \lambda_2 = -5\epsilon \pm \sqrt{25\epsilon^2 - (1 - 10\epsilon)}.$$

If $1 - 10\epsilon < 0$ or $\epsilon > \frac{1}{10}$, then the equilibrium point $(10, 5 - 50\epsilon)$ will be a saddle point, but its y-coordinate will be negative. If $25\epsilon^2 < 1 - 10\epsilon$, the equilibrium point will be a stable spiral, and if $0 < 1 - 10\epsilon \leq 25\epsilon^2$, the point will be a stable node. Since $25\epsilon^2 = 1 - 10\epsilon$ for $\epsilon_0 = -\frac{1}{5}(1 - \sqrt{2}) \cong 0.08284$, one can graph the relationship between the value of $\epsilon > 0$ and the nature of the equilibrium point as shown in Fig. 7.38. ■

In general, for both predator–prey and competitive-population models the presence of families of periodic solutions should not be expected. If a periodic solution, i.e., a closed curve in the phase plane does exist, it is more likely to be an isolated one, that is, a limit cycle. It will correspond to a particular set of initial

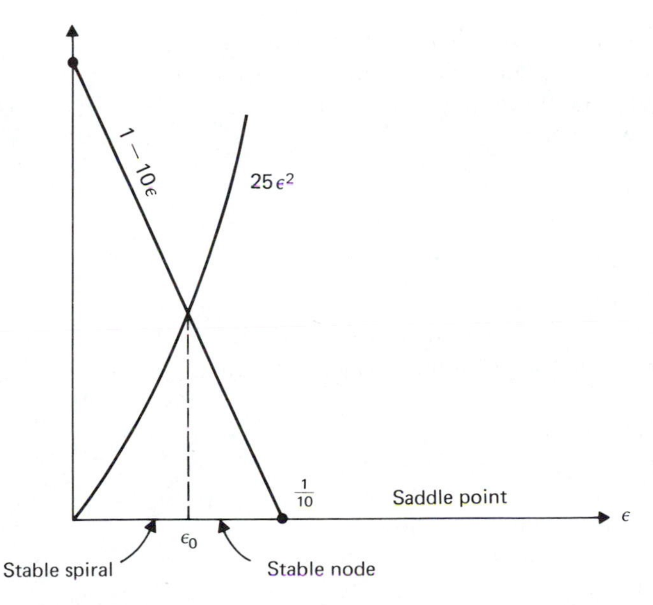

Figure 7.38 A stability diagram for the perturbed Lotka–Volterra system.

conditions or values of certain parameters, and if it is stable, nearby solutions will tend to it as t approaches infinity. The "balance of nature" should in some sense be thought of as a limiting situation rather than a realizable one.

EXERCISES 7.6

1. A simple model of competing populations is

$$\dot{x} = x(r - \alpha y), \qquad \dot{y} = y(s - \beta x),$$

where r, s, α, and β are positive constants. Find the equilibrium points, determine their stability, and sketch a phase plane diagram.

2. In the model given in Exercise 1, suppose that the population Y is being harvested at a constant rate $H > 0$. Such a harvesting could correspond to the effects of disease, hunting, or fishing, or emigration. This leads to the model

$$\dot{x} = x(r - \alpha y), \qquad \dot{y} = y(s - \beta x) - H.$$

For the case of $r = s = 1$, $\alpha = \frac{1}{10}$, $\beta = \frac{1}{5}$, sketch the phase plane for $H = 4$ and $H = 8$. What happens as H approaches 10?

3. For the general logistic competition model (7.6.3) show that
   a) $(0, 0)$ is an unstable node
   b) $(K, 0)$ is a saddle point if $K < s/\beta$
   c) $(0, J)$ is a stable node if $J > r/\alpha$

For the following logistic competition models determine whether there is stable coexistence or competitive exclusion. If the former, determine the nature of the equilibrium point (x_∞, y_∞). If the latter, determine if one population wins the competition. In all cases sketch the phase plane diagram.

4. $\dot{x} = x(2 - \frac{1}{5}x - \frac{1}{6}y)$, $\dot{y} = y(1 - \frac{1}{10}y - \frac{1}{8}x)$
5. $\dot{x} = 2x(1 - \frac{1}{20}x) - \frac{1}{25}xy$, $\dot{y} = 4y(1 - \frac{1}{40}y) - \frac{1}{10}xy$
6. $\dot{x} = x(1 - \frac{1}{20}x - \frac{1}{8}y)$, $\dot{y} = y(1 - \frac{1}{12}y - \frac{1}{16}x)$
7. $\dot{x} = 2x(1 - \frac{1}{100}x) - \frac{1}{40}xy$, $\dot{y} = 10y(1 - \frac{1}{50}y) - \frac{1}{8}xy$
8. A simpler model for competition is where one population grows logistically and the other exponentially in the absence of competition. This leads to the model

$$\dot{x} = xr\left(1 - \frac{x}{K}\right) - \alpha xy, \qquad \dot{y} = sy - \beta xy,$$

where r, K, α, s, and β are all positive numbers. Show that in this case there is no possibility of stable coexistence.

If one population of a competing pair of populations is harvested, the equilibrium structure can be radically changed. For instance, stable coexistence could become competitive exclusion or vice versa. In the following competition models, H is a constant harvest rate. Find the equilibrium points of the system for each given value of H, analyze their stability, and sketch the phase plane.

9. $\dot{x} = x(1 - \frac{1}{10}x - \frac{1}{10}y) - H, \quad \dot{y} = y(1 - \frac{1}{5}y - \frac{1}{8}x), \quad H = 0, \frac{5}{4}, \frac{25}{4}$

10. $\dot{x} = x(1 - \frac{1}{8}x - \frac{1}{10}y) - H, \quad \dot{y} = y(1 - \frac{1}{8}y - \frac{1}{12}x), \quad H = 0, \frac{1}{12}, \frac{1}{2}$

For the following Lotka–Volterra equations find the equilibrium points, then use the code RK4 with step size $h = 0.1$ to approximate the periodic solutions of the given initial value problems. If possible, graph your solutions in the phase plane.

11. $\dot{x} = x(2 - \frac{1}{2}y), \quad \dot{y} = y(-2 + x)$
(i) $x(0) = 1.75, \quad y(0) = 4.00, \quad 0 \le t \le 3.5$
(ii) $x(0) = 2.00, \quad y(0) = 2.00, \quad 0 \le t \le 3.5$

12. $\dot{x} = x(1 - \frac{1}{2}y), \quad \dot{y} = y(-1 + \frac{1}{5}x)$
(i) $x(0) = 5.00, \quad y(0) = 2.25, \quad 0 \le t \le 7.0$
(ii) $x(0) = 3.00, \quad y(0) = 2.00, \quad 0 \le t \le 7.0$

13. To prove that the integral curves of the Lotka–Volterra equations are closed curves, consider the equivalent first order equation:

$$\frac{dy}{dx} = \frac{y(-s + \beta x)}{x(r - \alpha y)}, \qquad y(x_0) = y_0,$$

where r, s, α, and β are positive constants.

a) Using separation of variables, show that the implicit solution of the equation is given by the expression

$$y^r e^{-\alpha y} = C_0 x^{-s} e^{\beta x},$$

where $C_0 > 0$ if $x_0, y_0 > 0$.

b) Consider the function

$$F(y) = y^r e^{-\alpha y}, \quad y \ge 0.$$

Show that $F(0) = 0$, $\lim_{y \to \infty} F(y) = 0$, and that $F(y)$ has a positive maximum for $y = r/\alpha$.

c) Show that the result of (b) implies that the equation

$$F(y) = K, \quad K > 0, \quad K \neq F\left(\frac{r}{\alpha}\right)$$

can have only two positive solutions or no positive solution.

The results of (c) apply as well to the equation

$$G(x) = x^s e^{-\beta x} = K, \quad K > 0, \quad K \neq G\left(\frac{s}{\beta}\right).$$

Recalling that $(s/\beta, r/\alpha)$ is the equilibrium point of the Lotka–Volterra system, show why (a), (b), and (c) imply that the integral curves in the positive xy quadrant are closed curves.

14. For any competition or predator–prey model

$$\dot{x} = x\,f(x, y), \qquad \dot{y} = y\,g(x, y)$$

with an equilibrium point (x_∞, y_∞) with $x_\infty\, y_\infty \neq 0$, show that the matrix of the linearized system at (x_∞, y_∞) is

$$\begin{bmatrix} x_\infty f_x(x_\infty, y_\infty) & x_\infty f_y(x_\infty, y_\infty) \\ y_\infty g_x(x_\infty, y_\infty) & y_\infty g_y(x_\infty, y_\infty) \end{bmatrix}.$$

Find all equilibrium points of the following predator–prey systems and analyze their stability. Sketch the phase plane diagram.

15. $\dot{x} = x\left[2\left(1 - \frac{x}{20}\right) - \frac{1}{4}y\right], \quad \dot{y} = y[-1 + \beta x], \quad \beta = \frac{1}{5} \text{ and } \beta = \frac{1}{40}$

16. $\dot{x} = x\left[1 - \frac{x}{K} - \frac{1}{8}y\right], \quad \dot{y} = y\left[-1 + \frac{1}{16}x\right], \quad K = 12, 18, \text{ and } 20$

17. $\dot{x} = x\left[\frac{1}{2}\left(1 - \frac{x}{20}\right) - \frac{1}{16}y\right], \quad \dot{y} = y\left[\left(1 - \frac{y}{J}\right) + \frac{1}{8}x\right], \quad J = 4 \text{ and } J = 10$

Examples are given below of predator–prey systems that have been suggested by mathematical biologists; the authors have assigned values to certain parameters. For each of the models find all equilibrium points and analyze their stability. (*Hint:* The result of Exercise 14 is useful.)

18. $\dot{x} = x\left[\frac{R}{x} - \frac{y}{x + 10}\right], \quad \dot{y} = \frac{y}{2}\left[\frac{x - 10}{x + 10}\right]$. Let $R = 20$ and $R = 60$.

19. $\dot{x} = x\left[\left(1 - \frac{x}{K}\right) - \frac{y}{x + 10}\right], \quad \dot{y} = \frac{y}{2}\left[\frac{x - 10}{x + 10}\right]$. Let $K = 20$ and $K = 40$.

20. $\dot{x} = x\left[2\left(1 - \frac{x}{K}\right) - \frac{y}{x}(1 - e^{-x/10})\right], \quad \dot{y} = y[e^{-2} - e^{-x/10}]$. Let $K = 25$, $K = 40$, and $K = 60$.

21. In a predator–prey system containing a parameter K, which represents the natural carrying capacity of the prey, enlarging K would correspond to enrichment (e.g., increasing the prey's food supply). This enrichment could, for instance, destabilize an equilibrium point and create oscillations represented by limit cycles.

On the other hand, harvesting the predator will certainly destabilize the system if the harvest rate is too large and could result in extinction of the predator. The following numerical experiment shows the effect on a predator–prey system of enrichment and harvesting.

Consider the system

$$\dot{x} = x\left[2\left(1 - \frac{x}{K}\right) - \frac{y}{x}(1 - e^{x/10})\right],$$

$$\dot{y} = y[e^{-2} - e^{-x/10}] - H,$$

where K is the carrying capacity of the prey and H is the constant rate of harvesting. Use RK4 with $h = 0.2$, $0 \leq t \leq 35$, to compute the solution for the following parameter values and initial conditions. Have your program print out at unit steps of t, that is, $t = 1, 2, \ldots$, and also include a statement in your program to STOP if $y(t) \leq 0$. If possible, graph your results. In all cases use $x(0) = 50$, $y(0) = 10$.

a) $K = 25$; $H = 0$, $H = 0.280$

b) $K = 40$; $H = 0$, $H = 1.10$

c) $K = 60$; $H = 0$, $H = 1.10$, $H = 1.4$

(For reference, see F. Brauer, A. C. Soudack, H. S. Jarosch, *International Journal of Control,* 23 (1976), pp. 553–573.)

SUMMARY

There are no general techniques for finding the solutions of two-dimensional autonomous nonlinear systems

$$\frac{dx}{dt} = P(x, y), \qquad \frac{dy}{dt} = Q(x, y).$$

Such systems can be used to describe a nonlinear mass–spring system, the motion of a pendulum, or the growth of two populations. A general maxim might be that "each nonlinear system is a little world of its own."

However, a very useful approach occurs when the system has an isolated equilibrium point (x_0, y_0) where P and Q both vanish. The solutions near (x_0, y_0) can be studied by creating a system of linear differential equations that approximate the original differential equation near (x_0, y_0). In many cases an analysis of the linear system will determine whether the solutions of the original equation tend toward (x_0, y_0) as t approaches ∞ or move away from (x_0, y_0). This introduces the notion of stability of an equilibrium point, which is the key to the study of almost any nonlinear system.

A second class of nonlinear equations discussed are those describing a mechanical system of one degree of freedom with forces dependent solely on position. They are of the form

$$\frac{d^2x}{dt^2} + g(x) = 0.$$

For instance, the motion of an undamped pendulum is given by such an equation with $g(x) = k \sin x$. Such an equation can be studied graphically by examining the level curves of the total energy of the system

$$E(x, y) = \frac{1}{2}y^2 + G(x),$$

where $G(x)$ is an antiderivative of $g(x)$. The level curves are integral curves of the differential equation, and by analyzing them one can determine the stability of any equilibrium points x_0 where $g(x_0) = 0$ as well as the general behavior of solutions.

The chapter concludes with an analysis of a general system of equations describing the growth of two populations. Such systems can be classified as predator–prey models, wherein one population depends on the other as a food source, or competitive models, wherein both populations are competing for the same shared resources. The models are analyzed by examining the nature of their solutions near equilibrium points.

MISCELLANEOUS EXERCISES

7.1. In Example 4 of Section 7.5, the following expression is given for the period T of the periodic solution corresponding to the energy level $K = \frac{7}{16}$:

$$T = 2 \int_{-1}^{1} \frac{1}{\sqrt{\frac{1}{8}(x^2 - 1)(x^2 - 7)}}\, dx.$$

To approximate it, first let $x = \sin \theta$ and show that

$$T = 4\sqrt{\frac{8}{7}} \int_0^{\pi/2} \left(1 - \frac{1}{7}\sin^2 \theta\right)^{-1/2} d\theta \cong 2\pi \sqrt{\frac{8}{7}}\left(1 + \frac{1}{28} + \frac{9}{3136}\right).$$

(*Hint:* Use the binomial expansion

$$(1 - u)^{-1/2} = 1 + \frac{1}{2}u + \frac{3}{8}u^2 + \cdots$$

to approximate the integrand and obtain an estimate for T.)

7.2. The idea given in the previous problem can be used to approximate for small k the values of the elliptic integral of the first kind

$$K(k) = \int_0^{\pi/2} (1 - k^2 \sin^2 \theta)^{-1/2}\, d\theta.$$

Do this for $k = 0.05$, 0.10, and 0.50, and compare your answers with the values of $K(k)$ from a table.

The following computer problems can be done by modifying the sample program using RK4 given in Chapter 3. The more advanced algorithm RKF45 given in Chapter 8 may also be employed. Study each problem carefully and design your print statements so that you have adequate but not excessive output.

7.3. Employing a generalization of the van der Pol equation, W. S. Krogdahl (*Astrophysical Journal,* 122 (1955), pp. 43–51) has had considerable success in modeling the velocity curves observed in the pulsation of variable stars of the Cepheid type. Denoting the variable radius by $r(\tau)$ and the variable velocity by $v(\tau)$, with τ a scaled time variable, he modeled r and v by

$$r(\tau) = [1 + q(\tau)]^{1/3},$$

$$v(\tau) = r'(\tau) = \frac{q'(\tau)/3}{[1 + q(\tau)]^{2/3}},$$

where $q(\tau) = \lambda Q(\tau)$ and

$$Q'' = -Q + \frac{2}{3}\lambda Q^2 - \frac{14}{27}\lambda^2 Q^3 + \mu(1 - Q^2)Q' + \frac{2}{3}\lambda(1 - \lambda Q)Q'^2;$$

λ and μ are empirical constants. If $\lambda = 0$, the differential equation for $Q(\tau)$ reduces to the van der Pol equation. Use RK4 with step $h = 0.05$ to approximate $r(\tau)$ and $v(\tau)$ for $0 \le \tau \le 10$ when

a) $\mu = 0.01$, $\lambda = 0.1$, $Q(0) = 1.05$, $Q'(0) = -0.05$
b) $\mu = 0.1$, $\lambda = 0.15$, $Q(0) = 0$, $Q'(0) = 1.9$

In each case plot $v(\tau)$ versus $r(\tau)$ in the phase plane and also $v(\tau)$ versus τ. The velocity curves should have the general shape shown in Fig. 7.39.

7.4. The differential equation $y'' + t^2 e^y = 0$, $t > 0$, occurred in a study on non-isothermal flow of a Newtonian fluid between parallel plates. It may be possible to solve it exactly (see *SIAM Review,* Vol. 23, No. 4, Oct. 1981, p. 524). Even without an exact solution, we can observe that $y'' < -t^2$ if y is positive, and therefore y should become and stay negative as t increases. But then $t^2 e^y$ should approach zero and y' should become constant. Test this conjecture by rewriting the differential equation as

$$y_1' = y_2, \qquad y_2' = -t^2 e^{y_1},$$

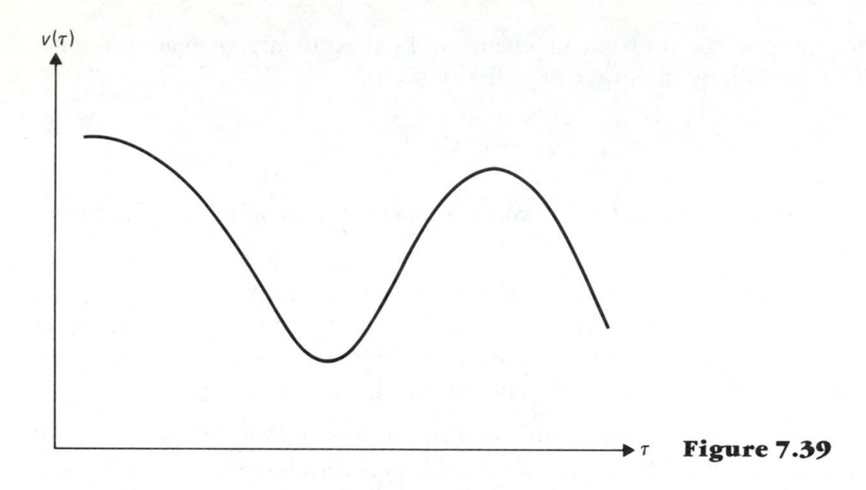

Figure 7.39

and constructing several approximate solutions with different initial values in the square $|y_1| \le 3$, $|y_2| \le 3$ by using RK4 with $h = 0.1$, $0 \le t \le 10$ (or less). Stop your computation if $y_1 < -10$. Plot your results in the y_1y_2-plane. A computer generated plot of several solution curves is given in Fig. 7.40.

7.5. The mass–spring system of Example 3 of Section 5.1 has a mirror image normal-mode solution, $y_1(t) = -y_2(t)$ (see Fig. 7.41) even if the middle linear spring is replaced by a nonlinear spring with spring constant

$$k' = k + \epsilon(y_2 - y_1)^2.$$

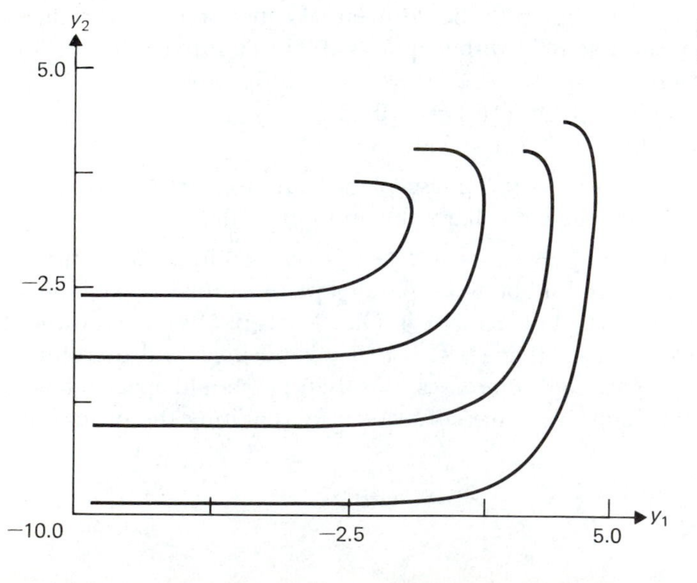

Figure 7.40 Plot of y_1 versus y_2 for Exercise 7.4.

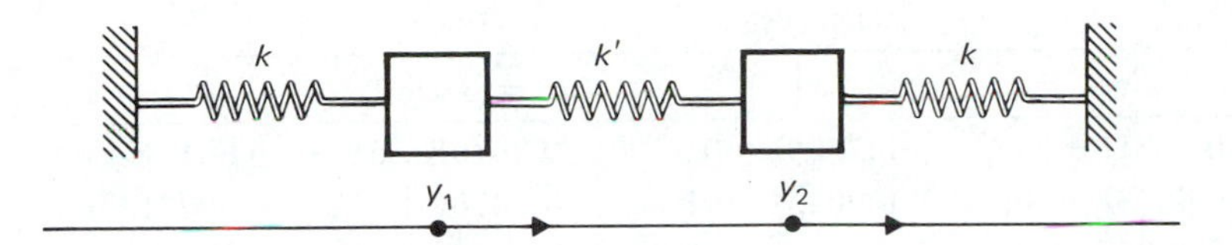

Figure 7.41

If ϵ is positive, the nonlinear spring is said to be *stiff* since k' increases with displacement.

The differential equations governing the motion are

$$\frac{dy_1}{dt} = y_3, \qquad \frac{dy_2}{dt} = y_4,$$

$$\frac{dy_3}{dt} = \frac{-k}{m}y_1 + \frac{k'}{m}(y_2 - y_1), \qquad \frac{dy_4}{dt} = \frac{-k}{m}y_2 - \frac{k'}{m}(y_2 - y_1).$$

To investigate the effect of the nonlinearity, study the mirror image normal mode of vibration with $m = 1$, $k = 1$, $\epsilon = 0.02$, and initial conditions

$$y_1(0) = -y_2(0) = a, \qquad y_3(0) = y_4(0) = 0$$

for several values of a, the initial displacement. The resulting motion is periodic with an amplitude-dependent period. For each choice of a, estimate the period by integrating the differential equations with RK4, $h = 0.025$, and determine the time between successive attainments of the initial configuration. A typical output is given in Table 7.2, and a graph of the period as a function of a is given in Fig. 7.42.

7.6. A biological oscillator subject to a constant positive input b can be modeled by a system of nonlinear differential equations (see [7], p. 67):

$$\dot{x} = x - ay + b, \qquad \dot{y} = x - cy \qquad \text{if } x > 0,\ y \geq 0,$$

$$\dot{x} = (b - ay)u(b - ay), \qquad \dot{y} = -cy \qquad \text{if } x = 0,\ y \geq 0,$$

where $a > c > 0$, $b > 0$, and $u(x)$ is the Heaviside step function: $u(x) = 1$, $x \geq 0$; $u(x) = 0$, $x < 0$. If the equilibrium point in the first quadrant is an unstable spiral point, then there is a limit cycle surrounding it that contains a portion of the positive y-axis.

a) Show that if $1 - c > 0$, $4a > (1 + c)^2$, the equilibrium point is an unstable spiral point.

b) Show that $(0, b/a)$ is on the arc of the limit cycle described by the linear system

$$\dot{x} = x - ay + b, \qquad \dot{y} = x - cy.$$

c) Show that the period of the limit cycle is independent of b.

d) Set $c = \frac{1}{2}$, $a = \frac{25}{16}$, $b = 1$, and use RK4 with $h = 0.1$ to compute the limit cycle.

Table 7.2

a	Time	y_1	y_1'	t^* = period	$y(t^*) \cong a$
0.5	0.05	0.4981E+00	−0.7590E−01	0.3610E+01	0.5019E+00
	3.55	0.4973E+00	0.9043E−01	0.3610E+01	0.5027E+00
	3.57	0.4991E+00	0.5256E−01	0.3610E+01	0.5009E+00
	3.60	0.4999E+00	0.1460E−01	0.3610E+01	0.5001E+00
	3.62	0.4998E+00	−0.2340E−01	0.3610E+01	0.5002E+00
	3.65	0.4988E+00	−0.6135E−01	0.3610E+01	0.5012E+00
	3.67	0.4968E+00	−0.9918E−01	0.3609E+01	0.5033E+00
1.0	3.55	0.9999E+00	0.2295E−01	0.3557E+01	0.1000E+01
	3.57	0.9995E+00	−0.5604E−01	0.3557E+01	0.1000E+01
1.5	3.47	0.1500E+01	0.4252E−01	0.3475E+01	0.1500E+01
	3.48	0.1500E+01	−0.4148E−01	0.3475E+01	0.1500E+01
2.0	3.36	0.2000E+01	0.5080E−01	0.3369E+01	0.2000E+01
	3.37	0.2000E+01	−0.4020E−01	0.3369E+01	0.2000E+01
2.5	3.24	0.2500E+01	0.7333E−01	0.3247E+01	0.2500E+01
	3.25	0.2500E+01	−0.2666E−01	0.3247E+01	0.2500E+01
3.0	3.11	0.3000E+01	0.9147E−01	0.3115E+01	0.3000E+01
	3.12	0.3000E+01	−0.1953E−01	0.3115E+01	0.3000E+01

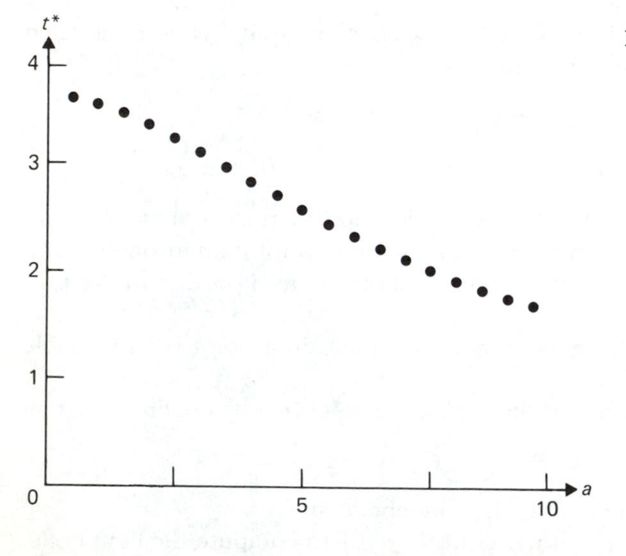

Figure 7.42

Since the approximate curve will penetrate the $x < 0$ half-plane, insert "IF(Y(1).LT.0.0) Y(1) = 0.0" just before the return statement in RK4.

e) The local error is not $O(h^5)$ at the points where the limit cycle enters and leaves the y-axis. Why? This difficulty can be overcome by integrating with different independent variables on different segments of the limit cycle. Write a program to

(i) Integrate with $h = 0.1$

$$\frac{dx}{dt} = x - a + b, \qquad \frac{dy}{dt} = x - cy$$

from $t = 0$, $x = 0$, $y = b/a$ to $t = t_1$, $x = x_1$, $y = y_1$, where $dx/dt \cong -0.5$ at this point;

(ii) Integrate

$$\frac{dy}{dx} = \frac{x - cy}{x - ay + b}, \qquad \frac{dt}{dx} = \frac{1}{x - ay + b}$$

from $x = x_1$, $t = t_1$, $y = y_1$ to $x = 0$, $t = t_2$, $y = y_2$, with $h = -x_1/(\# \text{ of steps}) \cong -0.1$.

(iii) Integrate

$$\frac{dt}{dy} = -\frac{1}{cy}, \qquad \frac{dx}{dy} = 0$$

from $y = y_2$, $t = t_2$, $x = 0$ to $y = b/a$, $t =$ the period, $x = 0$ with $h = (b/a - y_1)/(\# \text{ of steps}) \cong -0.1$.

In your program use three different subroutines for computing the right sides of the differential equations, F1(TI,X,XP), F2(XI,Y,YP), F3(YI,T,TP) with

$$\begin{array}{llll} \text{TI} = t, & \text{X}(1) = x, & \text{X}(2) = y, & \text{NEQN} = 2, \\ \text{XI} = x, & \text{Y}(1) = y, & \text{Y}(2) = t, & \text{NEQN} = 2, \\ \text{YI} = y, & \text{T}(1) = t, & & \text{NEQN} = 1. \end{array}$$

7.7. Show that if $x(t)$ is a solution to

$$\frac{d^2x}{dt^2} + 2 \tan x \sec^2 x = 0, \qquad -\frac{\pi}{2} < x(t) < \frac{\pi}{2}, \qquad x(s) = 0,$$

then

$$\sin\,[x(t)] = \left(\frac{K}{1 + K}\right)^{1/2} \sin[(2 + 2K)^{1/2}(t - s)],$$

and the period is $2\pi/(2 + 2K)^{1/2}$, where

$$\frac{1}{2}\left[\frac{dx}{dt}(t)\right]^2 + \tan^2[x(t)] = K.$$

Hint:

$$\int_0^x \left[\frac{1 + K}{K - \tan^2 w}\right]^{1/2} dw = \arcsin\left[\left(\frac{1 + K}{K}\right)^{1/2} \sin x\right].$$

The potential $G(x) = \tan^2 x$ has been used in a study of particle-channeling effects in crystalline material by J. A. Ellison (see *Physical Review B,* vol. 18, No. 11, Dec. 1978, pp. 5948–5962).

Use RK4 with $h = 0.05$, $0 < t < 10$, to estimate the periods corresponding to $K = \tan[0.05 \times 10^j]$, $j = 0, 1, 2$. Compare your estimated values with the exact values of the periods.

REFERENCES

Classics containing superb examples:

1. N. Minorsky, *Introduction to Nonlinear Mechanics,* J.W. Edwards, Ann Arbor, 1947.
2. J.J. Stoker, *Nonlinear Vibrations in Mechanical and Electrical Systems,* Interscience, New York, 1950.
3. L. Cesari, *Asymptotic Behavior and Stability Problems in Ordinary Differential Equations,* Academic Press, New York, 1963.

Advanced textbooks on the theory of ordinary differential equations:

4. E.A. Coddington and N. Levinson, *Theory of Ordinary Differential Equations,* McGraw-Hill, New York, 1955.
5. W. Hurewicz, *Lectures on Ordinary Differential Equations,* M.I.T. Press, Cambridge, Mass., 1958.

Intermediate textbooks on ordinary differential equations:

6. F. Brauer and J. Nobel, *Qualitative Theory of Ordinary Differential Equations,* W.A. Benjamin, New York, 1969.
7. D.W. Jordan and P. Smith, *Nonlinear Ordinary Differential Equations,* 2nd ed., Clarendon Press, Oxford, 1987.
8. H.K. Wilson, *Ordinary Differential Equations,* Addison-Wesley, Reading, Mass., 1971.

Applications:

9. N. Bogolyubov and Yu.A. Mitropolsky, *Asymptotic Methods in the Theory of Nonlinear Oscillations,* Gordon and Breach, New York, 1961.

10. J. McIntosh and R. McIntosh, *Mathematical Modeling and Computers in Endocrinology,* Springer–Verlag, Berlin, 1980.
11. J.D. Murray, *Nonlinear Differential Equation Models in Biology,* Oxford University Press, New York, 1977.
12. W.M. Smart, *Celestial Mechanics,* Longmans, Green, and Co., New York, 1953.
13. G.B. Whitham, *Linear and Nonlinear Waves,* John Wiley, New York, 1974.

Tables:

14. H.B. Dwight, *Tables of Integrals and other Mathematical Data,* Macmillan, New York, 1947.

8

MORE ON NUMERICAL METHODS

8.1 INTRODUCTION

In Chapter 3 some elementary numerical methods for solving initial value problems for ordinary differential equations were described. There it was mentioned that errors were present in numerical solutions but no attempt was made to identify them or to compensate for their effect. In this chapter these errors will be examined in more detail, and a method will be presented for estimating and controlling them to some extent.

The computer program RKF45, which solves quite general initial value problems for first order systems of differential equations, will also be introduced. This program, based on Runge–Kutta formulas of orders four and five, automatically estimates local error and chooses an appropriate step size h to ensure an accuracy specified by the user. Several numerical examples that illustrate the use of RKF45 are given.

In what follows, the theory and algorithms are developed for the scalar case. The results can be extended to higher order equations and to systems of first order equations, just as they were in Chapter 3.

8.2 ERRORS, LOCAL AND GLOBAL

Errors enter into the numerical solution of initial value problems from two sources. The first is *discretization error;* it is a property of the method being used.

The second is *roundoff error;* it depends on the arithmetic being used. If all calculations were done in exact arithmetic (no roundoff errors), discretization error would be the only error present. In general, one tries to carry enough digits in the calculations (at least one or two more than the accuracy desired) so that roundoff error does not become a problem.

Two types of discretization errors, *local* and *global errors,* should be distinguished. If we are solving the initial value problem

$$y' = f(t, y), \quad a \leq t \leq b, \tag{8.2.1}$$

$$y(a) = y_0, \tag{8.2.2}$$

using a method

$$y_{i+1} = y_i + h\theta(t_i, y_i), \tag{8.2.3}$$

$$t_{i+1} = t_i + h, \tag{8.2.4}$$

with $h = (b - a)/n$, where n is a positive integer, then the global error at $t = t_{i+1}$ is the quantity

$$\text{g.e.} = y(t_{i+1}) - y_{i+1}.$$

This is the difference between the actual solution $y(t_{i+1})$ at $t = t_{i+1}$ and our numerical approximation to it, y_{i+1}, obtained by using the one-step method (8.2.3, 8.2.4). The local error, on the other hand, is the error that is made in trying to follow the solution to the differential equation over just one step of length h, that is, from $t = t_i$ to $t = t_{i+1}$. To make this more precise, let $u(t)$ satisfy the initial value problem

$$u' = f(t, u), \tag{8.2.5}$$

$$u(t_i) = y_i. \tag{8.2.6}$$

Then the local error at $t = t_{i+1}$ is just the quantity

$$\text{l.e.} = u(t_{i+1}) - y_{i+1}. \tag{8.2.7}$$

The local and global errors at $t = t_{i+1}$ are illustrated graphically in Fig. 8.1.

EXAMPLE 1 Suppose the initial value problem

$$y' = 5(y + t), \tag{8.2.8}$$

$$y(0) = -\frac{1}{5} \tag{8.2.9}$$

has been solved by using Euler's method with $h = 0.1$. The results are given in Table 8.1. What are the local and global errors at $t = 10$?

SOLUTION To develop expressions for the local and global errors, first solve the differential equation (8.2.8) subject to the arbitrary initial data $y(a) = A$ to get

$$y(t) = (A + a + \frac{1}{5})e^{5(t-a)} - t - \frac{1}{5};$$

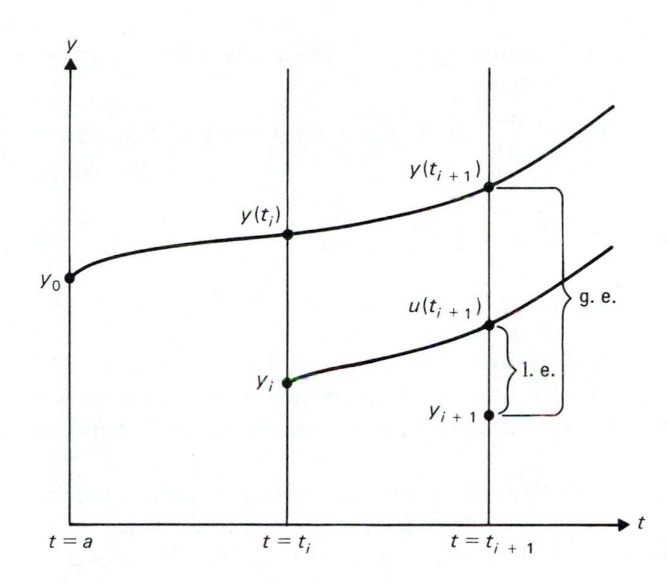

Figure 8.1 Local and global errors at $t = t_{i+1}$.

The solution to (8.2.8, 8.2.9) is then

$$y(t) = -t - \frac{1}{5},$$

whereas the solution $u(t)$ with initial data $u(t_i) = y_i$ is

$$u(t) = (y_i + t_i + \frac{1}{5})e^{5(t-t_i)} - t - \frac{1}{5}.$$

Hence, the local error at $t = t_{i+1}$ is given by

$$\text{l.e.} = [(y_i + t_i + \frac{1}{5})e^{5(t_{i+1}-t_i)} - t_{i+1} - \frac{1}{5}] - y_{i+1},$$

and the global error is given by

$$\text{g.e.} = [-t_{i+1} - \frac{1}{5} - y_{i+1}]$$

Table 8.1

i	t_i	y_i
0	0.00	−0.2000
1	0.10	−0.3000
2	0.20	−0.4000
⋮	⋮	⋮
98	9.80	−5.1085
99	9.90	−2.7632
100	10.00	+0.8053

Referring to the above data, we see that the local and global errors at $t = 10.00$ are

$$\text{l.e.} = [(-2.7632 + 9.90 + 0.2)e^{5(0.1)} - 10.0 - 0.2] - 0.8053 = 1.0910$$

and

$$\text{g.e.} = [-10.0 - 0.2] - 0.8053 = -11.0053. \qquad \blacksquare$$

It is interesting to note that the global error at a given point is not necessarily equal to the sum of all the local errors up to that point. In fact, in Example 1 the global error at $t = 10.0$ is -11.0053; whereas the sum of all local errors from $t = 0.10$ to $t = 10.0$ is -3.2734. The difference between these two quantities is a measure of the robustness of the differential equation itself. For a more detailed discussion of this point, see [1] and [4].

Although we will not develop the details here, something can be said about the effect of roundoff error on a solution of (8.2.1, 8.2.2) by using a method (8.2.3, 8.2.4) of order p. If the error in evaluating $\theta(t_i, y_i)$ is ϵ_i with $|\epsilon_i| \leq \epsilon$ for all i and the error in forming $y_i + h\theta(t_i, y_i)$ is ρ_i with $|\rho_i| \leq \rho$ for all i, then it can be shown (see Chapter 4 of [1]) that

$$\max_{1 \leq i \leq n} |y(t_i) - y_i| \leq |y(t_0) - y_0| e^{L(b-a)} + \frac{e^{L(b-a)}}{L}\left(\frac{ch^p}{2} + \epsilon + \frac{\rho}{h}\right), \qquad \textbf{(8.2.10)}$$

where c and L are constants depending on $y(t)$ and its derivatives. An examination of the estimate shows that the global error at a fixed point in $[a, b]$ has the same qualitative behavior as shown in Fig. 3.3. Values of h too small cause roundoff error problems, while discretization errors become pronounced for h values too large. It seems logical, not only for the sake of efficiency but also to avoid roundoff error difficulties, that the largest value of h consistent with some specified accuracy goal should be used. In the next section we will discuss a method for estimating local errors and devise a procedure for choosing a proper step size h to achieve a given accuracy requirement.

EXAMPLE 2 Solve the initial value problem

$$y' = \frac{y^2 + 2ty}{t^2}, \qquad y(1) = 1, \quad t > 0$$

using the improved Euler method with h values of $5 \times 10^{-1}, 5 \times 10^{-2}, \ldots, 5 \times 10^{-7}$. Tabulate the absolute error in the solution at $t = 1.5$.

SOLUTION First of all, the initial value problem has the solution

$$y(t) = \frac{t^2}{2 - t},$$

Table 8.2

h	$\lvert\text{error}\rvert_{t=1.5}$
5×10^{-1}	1.22222
5×10^{-2}	0.05101
5×10^{-3}	0.00034
5×10^{-4}	0.00043
5×10^{-5}	0.00314
5×10^{-6}	0.06421
5×10^{-7}	1.93206

which, when evaluated at $t = 1.5$, gives 4.5. Using a program written in single precision arithmetic on an IBM 370 computer we obtain Table 8.2. From these data it appears that discretization error is dominant for h values larger than about 5×10^{-3}, while roundoff error tends to become dominant for smaller h values. ■

EXERCISES 8.2

1. Using Euler's method with a step size $h = \frac{1}{4}$, obtain an approximate solution to the initial value problem

$$y' - 10y = t^2, \qquad y(0) = 0$$

at the points $t = \frac{1}{4}, \frac{1}{2}, \frac{3}{4}, 1$. Calculate the local error at each of these points. What is the global error at $t = 1$ and how does it compare with the sum of the four local errors up to that point?

2. Write a computer program to use the improved Euler algorithm to approximate the solution to

$$y' + y = e^{-t}, \qquad y(0) = 0$$

at $t = 1$. Starting with a step size of $h = \frac{1}{2}$, decrease h successively by a factor of $\frac{1}{2}$ until the roundoff error becomes more pronounced than the discretization error.

8.3 ESTIMATING LOCAL ERRORS

Local error is the natural quantity to try to estimate and control. Of course, it is hoped that by controlling local errors, global errors will also be controlled. Fortunately this is the case if the function $f(t, y)$ is smooth, although, as we noted in the last section, the global error at a given point can grow faster than the sum of the local errors up to that point.

A general procedure for estimating local errors is to compare the results of formulas of different orders. To make this idea concrete, suppose we are solving the initial value problem

$$y' = f(t, y), \qquad y(t_0) = y_0,$$

and wish to approximate $y(t_1)$, where $t_1 = t_0 + h$. Assuming that $f(t, y)$ is sufficiently differentiable, we can expand the solution $y(t_1)$ in a Taylor series:

$$\begin{aligned} y(t_1) &= y_0 + hy'(t_0) + \frac{h^2}{2}y''(t_0) + \frac{h^3}{6}y'''(t_0) + \cdots \\ &= y_0 + hf(t_0, y_0) + \frac{h^2}{2}f^{(1)}(t_0, y_0) + \frac{h^3}{6}f^{(2)}(t_0, y_0) + \cdots. \end{aligned} \tag{8.3.1}$$

Suppose $y(t_1)$ is approximated by y_1, which is computed by using the first two terms of the series (Taylor's method of order $p = 1$, the Euler algorithm),

$$y_1 = y_0 + hf(t_0, y_0), \tag{8.3.2}$$

and by $\bar{y}_1$, which is computed using the first three terms (Taylor's method of order $p = 2$),

$$\bar{y}_1 = y_0 + hf(t_0, y_0) + \frac{h^2}{2}f^{(1)}(t_0, y_0). \tag{8.3.3}$$

The difference

$$\bar{y}_1 - y_1 = \frac{h^2}{2}f^{(1)}(t_0, y_0)$$

is an approximation of the local error made in using the Euler algorithm. In fact, if $y(t)$ has at least three continuous derivatives on some interval containing t_0 and t_1, then at $t = t_1$ we get

$$\text{l.e.} = \bar{y}_1 - y_1 + O(h^3). \tag{8.3.4}$$

If h is small, the term $O(h^3)$ will be negligible and the quantity $\bar{y}_1 - y_1$ will be a good approximation to the local error.

This is a general approach for assessing local errors. It can be used with higher order Taylor and Runge–Kutta algorithms. The next example illustrates how the improved Euler algorithm can be used to estimate the local error in the Euler algorithm.

EXAMPLE 1 Solve the initial value problem

$$y' = t + y, \qquad y(0) = 1.0,$$

by using Euler's method with $h = 0.1$. Use the improved Euler algorithm to estimate the local error in the Euler approximation and compare the results with the actual local error.

SOLUTION The Euler and improved Euler methods in this case are

$$y_{i+1} = y_i + h(t_i + y_i)$$

and

$$\bar{y}_{i+1} = y_i + \frac{h}{2}[(t_i + y_i) + (t_i + h + y_i + h(t_i + y_i))].$$

So the estimate of the local error to within $O(h^3)$ is given by

$$r = \bar{y}_{i+1} - y_{i+1} = \frac{h}{2}[h + h(t_i + y_i)]. \tag{8.3.5}$$

The solution to the initial value problem

$$u' = t + u, \qquad u(t_i) = y_i$$

is

$$u(t) = (y_i + t_i + 1)e^{t-t_i} - t - 1,$$

hence the *true* local error at $t = t_{i+1}$ is

$$\begin{aligned} \text{l.e.} &= u(t_{i+1}) - y_{i+1} \\ &= [(y_i + t_i + 1)e^{(t_{i+1}-t_i)} - t_{i+1} - 1] - y_{i+1}. \end{aligned} \tag{8.3.6}$$

Some results are summarized in Table 8.3. ■

Table 8.3

t_i	y_i	Estimated local error (8.3.5)	True local error (8.3.6)
0.0	1.0		
0.1	1.10000	0.10×10^{-1}	0.10×10^{-1}
0.2	1.22000	0.11×10^{-1}	0.11×10^{-1}
0.3	1.36200	0.12×10^{-1}	0.13×10^{-1}
0.4	1.52820	0.13×10^{-1}	0.14×10^{-1}
0.5	1.72102	0.15×10^{-1}	0.15×10^{-1}
0.6	1.94312	0.16×10^{-1}	0.17×10^{-1}
0.7	2.19743	0.18×10^{-1}	0.18×10^{-1}
0.8	2.48718	0.19×10^{-1}	0.20×10^{-1}
0.9	2.81590	0.21×10^{-1}	0.22×10^{-1}
1.0	3.18748	0.24×10^{-1}	0.24×10^{-1}

8.4 A STEP SIZE STRATEGY

We have discussed a procedure for estimating the local error in a typical one step method. The next order of business is to use this estimate to select a step size h, so that the error that is actually committed in taking one step is less than some preassigned value. How this selection is made will be illustrated by using the Euler and the improved Euler algorithms.

Suppose we have just computed y_1 from y_0 by using Euler's method and have also computed an improved Euler approximation $\bar{y}_1$. From the discussion in the previous section it follows that the local error in the Euler approximation can be expressed in the form

$$\text{l.e.} = h^2\tau_0 + O(h^3)$$

for some constant τ_0, and the term $h^2\tau_0$ can be represented within $O(h^3)$ by

$$r = \bar{y}_1 - y_1 = \frac{h}{2}\,[f(t_0 + h, y_0 + hf(t_0, y_0)) - f(t_0, y_0)].$$

(The reader might wish to check these formulas.) Let

$$\text{est.} = \frac{r}{h}$$

be the estimate of local error relative to h. If we had started at $t = t_0$ with a step size of length γh, the local error would have been

$$\text{l.e.} = (\gamma h)^2\tau_0 + O(h^3),$$

and so

$$|\text{l.e.}| \cong |\gamma^2(h^2\tau_0)| = \gamma^2|r| = \gamma^2 h|\text{est.}|.$$

If $\epsilon > 0$ has been selected as an allowable local error per unit change in t, then the allowable error in a step of length γh is $\gamma h\epsilon$. If $|\text{est.}|$ is greater than ϵ, we reject the computation of y_1 and recompute with a smaller step length γh, $0 < \gamma < 1$. Since.

$$|\text{l.e.}| \cong \gamma^2 h|\text{est.}|$$

and we would like $|\text{l.e.}| \cong \gamma h\epsilon$, we have a criterion for choosing γ. For efficiency, we would prefer h has a large as possible. Therefore, we set

$$\gamma h\epsilon = \gamma^2 h|\text{est.}|$$

and

$$\gamma = \frac{\epsilon}{|\text{est.}|}.$$

In practice, a fraction of this γ, usually 0.9, is used to avoid frequent overshoot, which is clearly undesirable. Therefore, if $|\text{est.}| > \epsilon$, reject y_1 and recompute with a new step size. If $|\text{est.}| \leq \epsilon$, accept y_1 and use a new step size for the next step. In either case, use $0.9\epsilon h/|\text{est.}|$ as the new step size.

A simple Fortran program, given in Fig. 8.2, implements the above step size strategy for Euler's method by using a local error estimator based on the improved Euler method. The program solves the initial value problem

$$y' = y, \qquad y(0) = 1,$$

and prints out the solution at $t = 1.0$. The value $\epsilon = 0.01$ has been used as the desired value of the local error per unit change in t.

It should be pointed out that the above program is not "robust," but merely a straightforward implementation of a fairly simple algorithm based on Euler's method. For example, in a sophisticated program to control step size, one should introduce safeguards for the case where the step size h becomes too small for machine precision or when it suddenly becomes large, indicating a possible breakdown in the step size estimate. Besides, lower order schemes, such as Euler's method, are not practical for solving most problems.

8.5 THE SUBROUTINE RKF45

As we pointed out earlier in Chapter 3, Runge–Kutta formulas of order $p = 4$ are probably the most common of the Runge–Kutta type formulas and require four function evaluations per step. If we are going to use the above ideas to approximate the local error in the $p = 4$ method, then a second approximation of order $p \geq 5$ is needed. A $p = 5$ formula would in general require six additional function evaluations per step. These evaluations are not necessarily independent of those required in the $p = 4$ formulas; e.g., in going from $t = t_i$ to $t = t_{i+1}$ both would require the evaluation of $f(t_i, y_i)$. Can a set of $p = 4$ and $p = 5$ formulas that use more points in common be found to minimize the number of function evaluations at each step? E. Fehlberg did about the best that one could hope for and produced a set of Runge–Kutta formulas involving six function evaluations per step, which gives both a fourth and fifth order approximation [7]. The formulas are quite lengthy and will not be given here; for details see [7].

The Fehlberg formulas have been implemented into the highly efficient Fortran subroutine RKF45 by H. A. Watts and L. F. Shampine [8]. A complete listing of the subroutine is given in Appendix 4. Since RKF45 is extensively annotated, no effort will be made here to duplicate the material in those comments. A few remarks about the use of the routine, however, are in order.

Figure 8.2 Sample program for using Euler's method with a local error estimator based on the improved Euler method to solve $y' = y$, $y(0) = 1$.

```
C
C PROGRAM TO SOLVE DY/DT = Y USING EULER'S METHOD WITH
C A LOCAL ERROR ESTIMATOR BASED ON THE IMPROVED EULER
C ALGORITHM.
C
C SET INITIAL VALUES.
C
      F(T,Y) = Y
      T0 = 0.0
      Y0 = 1.0
      EPS = 1.0E-3
      T = T0
      Y = Y0
      TFINAL = 1.0
      H = 0.1
C
C BASIC LOOP (LIMITED TO 1000 STEPS).
C
      DO 20 I = 1,1000
         IF((T+H). GT. TFINAL) H = TFINAL-T
         YE = Y+H*F(T,Y)
         R = 0.5*H*(F(T+H,YE)-F(T,Y))
         EST = R/H
         GAMMA = EPS/ABS(EST)
         HE = 0.9*GAMMA*H
         IF(ABS(EST) .LE. EPS) THEN
            Y = YE
            T = T+H
            H = HE
         ELSE
            H = HE
            Y = Y+H*F(T,Y)
            T = T+H
         ENDIF
         IF(T .GE. TFINAL) THEN
            WRITE(*,5)T,Y
  5         FORMAT(/1X,' T = ',F10.4,3X,' Y = ',F10.5)
            STOP
         ENDIF
 20   CONTINUE
      WRITE(*,10)
 10   FORMAT(1X, 'TFINAL NOT REACHED IN 1000 STEPS')
      STOP
      END
```

OUTPUT

T = 1.0000   Y = 2.71611

The subroutine RKF45 is designed to integrate a system of first order ordinary differential equations of the form

$$\frac{d\mathbf{y}}{dt} = \mathbf{f}(t, \mathbf{y}),$$

where **y** and **f** are vectors of the same length. The *call list* of the routine is

F,NEQN,Y,T,TOUT,RELERR,ABSERR,IFLAG,WORK,IWORK

where the parameters represent:

- F — subroutine F(T,Y,YP) to evaluate the derivatives YP(I) = dY(I)/dt, I = 1,2, . . . ,NEQN, at T;
- NEQN — number of equations to be integrated;
- Y — solution vector at T;
- T — independent variable;
- TOUT — value of independent variable at which the solution is desired;
- RELERR,ABSERR — relative and absolute error tolerances for local error test;
- IFLAG — indicator for status of integration;
- WORK — work vector of dimension at least 3 + 6(NEQN);
- IWORK — work vector of dimension at least 5.

The parameter IFLAG is an important control variable and must be set to 1 before the first call to the subroutine. In the event that the integration reached TOUT with no problems, IFLAG is set to 2 by RKF45 and should be left at that value for continued integration. If a value of IFLAG = 3 is returned by RKF45, the relative error requested (RELERR) was unreasonably small; for instance, the number of working digits might be insufficient for the error requested. The value of RELERR is appropriately increased by the subroutine and integration can be continued. Values of 4 or 7 for IFLAG are warnings that the subroutine has done a significant amount of work or has taken a lot of steps to obtain the accuracy requested. IFLAG values of 5 or 6 indicate that error tolerances must be changed before continuing. IFLAG = 8 indicates improper input parameters. For specifics of each IFLAG value, see the comments at the beginning of the subroutine.

The parameters RELERR and ABSERR, the relative and absolute error tolerances for local error, must be chosen with care. These parameters measure local error by

$$|\mathrm{R}| \le (\mathrm{RELERR})(|\mathrm{Y}|) + \mathrm{ABSERR}$$

for each component of estimated local error R and each component of the solution vector Y. Note that this error test is a pure *relative error test* if ABSERR = 0 and a pure *absolute error test* if RELERR = 0; it is called a *mixed error test* if both ABSERR and RELERR are not zero. To avoid limiting precision difficulties, the subroutine requires RELERR to be larger than an internally computed relative error parameter, which is machine dependent. A pure absolute error test is not permitted and a pure relative error test should be avoided if the solution is expected to vanish for any value of T in the interval of integration. Reasonable values of RELERR and ABSERR, of course, depend on the accuracy desired in the solution, on the particular computer being used, and in a practical problem on the noise level of the data. For the IBM370 series in single precision, the values

RELERR = 1.0E−5 and ABSERR = 1.0E−5

are appropriate for most of the examples in the book.

Perhaps the best way to introduce the subroutine RKF45 is through an example.

EXAMPLE 1 Write a Fortran program that uses RKF45 to solve van der Pol's equation

$$y'' + \epsilon(y^2 - 1)y' + y = 0$$

with the initial conditions

$$y(0) = 1, \qquad y'(0) = 1$$

and the parameter $\epsilon = 1$. This equation arises quite often in nonlinear mechanics.

SOLUTION The first thing that must be done is to change the problem to an equivalent first order system. Let $y_1 = y$, $y_2 = y'$ to get

$$y_1' = y_2, \qquad y_2' = -y_1 - (y_1^2 - 1)y_2, \qquad y_1(0) = 1, \qquad y_2(0) = 1. \qquad \textbf{(8.5.1)}$$

The sample program in Fig. 8.3 solves the problem (8.5.1) and tabulates the solution and its derivative at $t = 0.0, 0.5, \ldots, 10.0$. A phase plane plot (a plot in the yy'-plane) of the solution is shown in Fig. 8.4. ■

In our discussion of numerical methods for solving initial value problems, we have up to this point considered methods of the form (8.2.3, 8.2.4). These methods are called one step methods since they use information at only one point to compute the approximate solution at the next point. There is another class of procedures known as multistep methods, which make use of information of several prior points to compute the approximate solution at the next point. For an excellent discussion of such procedures, as well as for a highly efficient computer program, see [4]. The subroutine RKF45 is adequate for most problems the reader will encounter in practice. Generally speaking, Runge–Kutta type codes are very effective for solving initial value problems where the derivatives are reasonably inexpensive to evaluate and moderate accuracy is required (say, less than 10^{-6}).

Figure 8.3 Sample program and output obtained by using RKF45 to solve van der Pol's equation.

```
C
C  SAMPLE PROGRAM FOR RKF45 — SOLVES VAN DER POL'S EQUATION
C
      SUBROUTINE VDP(T,Y,YP)
      REAL Y(2),YP(2)
      EPS=1.0
      YP(1)=Y(2)
      YP(2)=-Y(1)-EPS*(Y(1)**2-1.0)*Y(2)
      RETURN
      END
C
      EXTERNAL VDP
      REAL WORK(15),Y(2)
      INTEGER IWORK(5)
      T=0.0
      NEQN=2
      Y(1)=1.0
      Y(2)=1.0
      RELERR=1.0E-5
      ABSERR=1.0E-5
      TF=10.0
      DT=0.5
      IFLAG=1
      TOUT=T
C
    1 CALL RKF45(VDP,NEQN,Y,T,TOUT,RELERR,ABSERR,IFLAG,WORK, IWORK)
      WRITE(6,5)T,Y(1),Y(2)
    5 FORMAT(1X,F6.2,2F15.4)
      IF (IFLAG.NE.2) GO TO 6
      TOUT=T+DT
      IF(T.LT.TF) GO TO 1
      STOP
C
    6 WRITE (6,10) IFLAG
   10 FORMAT (1X,'IFLAG =',I2,1X,'; CHECK RKF45 COMMENTS FOR DETAILS)
      STOP
      END
```

OUTPUT

T	Y(1)	Y(2)	T	Y(1)	Y(2)
0.	1.0000	1.0000	5.50	−0.0607	2.1025
0.50	1.3313	0.2800	6.00	1.1871	2.5218
1.00	1.2985	−0.3670	6.50	1.9625	0.5030
1.50	0.9928	−0.8519	7.00	1.9330	−0.4068
2.00	0.4212	−1.4890	7.50	1.6528	−0.6865
2.50	−0.5499	−2.3666	8.00	1.2456	−0.9632
3.00	−1.6348	−1.4855	8.50	0.6485	−1.4914
3.50	−1.9288	0.0799	9.00	−0.3296	−2.4672
4.00	−1.7440	0.5689	9.50	−1.5765	−1.9331
4.50	−1.3920	0.8414	10.00	−2.0083	−0.0342
5.00	−0.8787	1.2581			

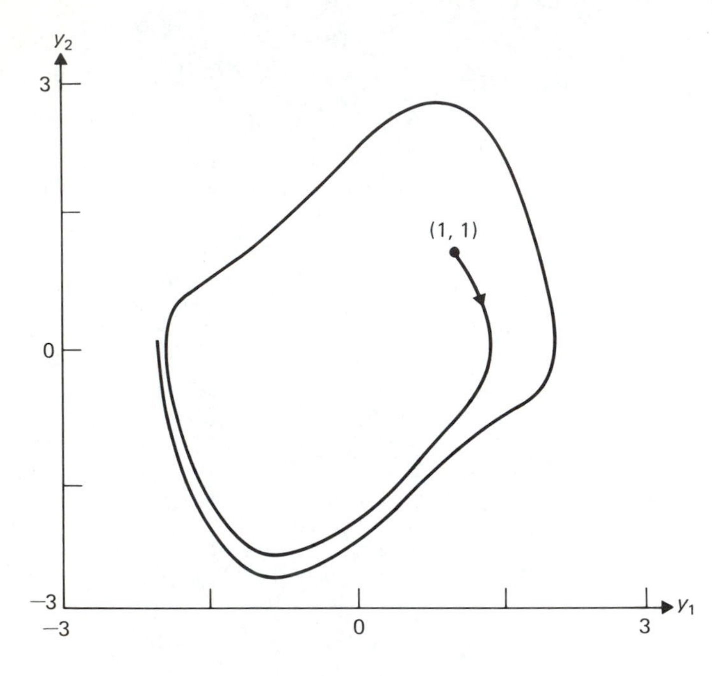

Figure 8.4 Phase plane plot of the solution of van der Pol's equation of Example 1.

8.6 SOME EXAMPLES

In this section we present some reasonably sophisticated examples leading to differential equations for which analytical solutions cannot be found conveniently, if at all. A subroutine like RKF45, which adjusts its step size automatically to control error, is essential for problems of this type, so that the user can have some confidence in the numerical results obtained.

EXAMPLE 1 The orbit of the planet Mercury around the Sun can be represented as a solution to the differential equation

$$\frac{d^2u}{d\theta^2} + u = \frac{\mu}{h^2}(1 + \epsilon u^2), \tag{8.6.1}$$

where $u = 1/r$ and r denotes the distance from the Sun to Mercury. Here θ is an angle in the plane of the orbit, μ is a gravitational constant, h is the angular momentum, and ϵ is a parameter determined by the effects of other planets on Mercury as well as the Sun's oblateness, and a correction required by the general theory of

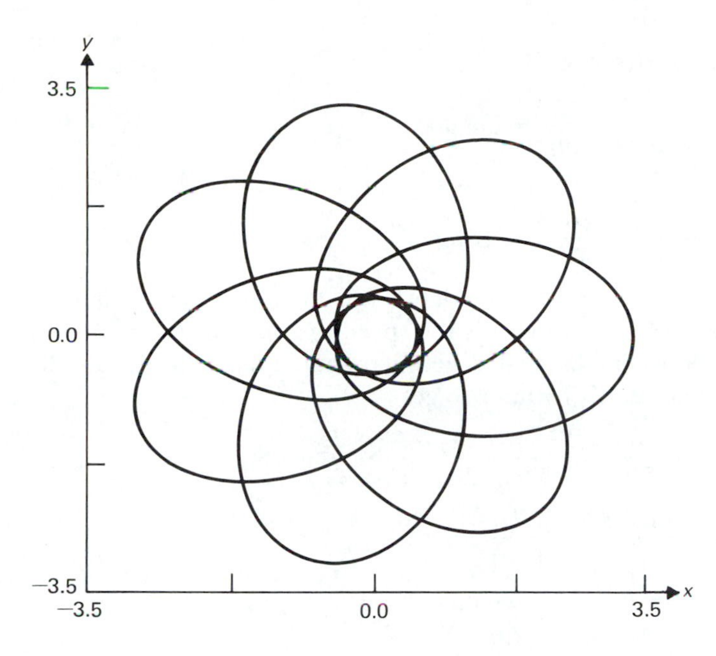

Figure 8.5 Rotating elliptical orbits of Mercury.

relativity. For simplicity, the second order differential equation is replaced by the system

$$\frac{dv}{d\theta} = w, \qquad \frac{dw}{d\theta} = 1 - v + \gamma v^2,$$

where $v = h^2u/\mu$, $\gamma = \epsilon\mu^2/h^4$.

To illustrate the phenomenon of *precession,* we choose $\gamma = 0.1$, $v(0) = 2$, and $w(0) = 0$ (these values do not necessarily correspond to observed values), and integrate the system using RKF45 over several revolutions. As illustrated in Fig. 8.5, where $x = (1/v)\cos\theta$, $y = (1/v)\sin\theta$, Mercury moves on an ellipse that is slowly rotating in the orbital plane. The points of closest approach to the Sun are called perihelia. The *precession* of these points is due to the *perturbing* nonlinearity in the differential equation. The observed *precession of the perihelion of Mercury* could not be explained by Newtonian mechanics and remained a puzzle for many years. The close agreement between observations and the orbit modeled by the differential equation (8.6.1) with ϵ containing a relativistic correction is one of the major experimental confirmations of Einstein's theory of general relativity. ■

EXAMPLE 2[1] According to Fermat's principle applied to the ray theory of sound, sound propagates from one point to another along the curve for which the transit time is the smallest [7, pp. 19, 37]. In water, the speed of sound varies with

1. This example is based on a problem given in [2, p. 173] and on classroom notes written by Stanly Steinberg in collaboration with W. T. Kyner and Cleve Moler (used with permission).

depth. Thus, its path can be described by

$$\frac{dz}{dx} = \tan\theta$$

together with

$$\frac{\cos\theta}{C(z)} = \text{const.}$$

(by Fermat's principle), where z denotes the depth, x is the horizontal distance, θ is the angle from the horizontal, and $C(z)$ is the speed of sound as a function of depth (see Fig. 8.6). To derive a relationship between θ and x, we differentiate the second equation and replace dz/dx by $\tan\theta$ to get

$$\begin{aligned} 0 = \frac{d}{dx}\frac{\cos\theta}{C(z)} &= \frac{-\sin\theta}{C(z)}\frac{d\theta}{dx} - \frac{C'(z)}{C(z)^2}\frac{dz}{dx}\cos\theta \\ &= \frac{-\sin\theta}{C(z)}\frac{d\theta}{dx} - \frac{C'(z)}{C^2(z)}\tan\theta\cos\theta \\ &= \frac{-\sin\theta}{C(z)}\left[\frac{d\theta}{dx} + \frac{C'(z)}{C(z)}\right]. \end{aligned}$$

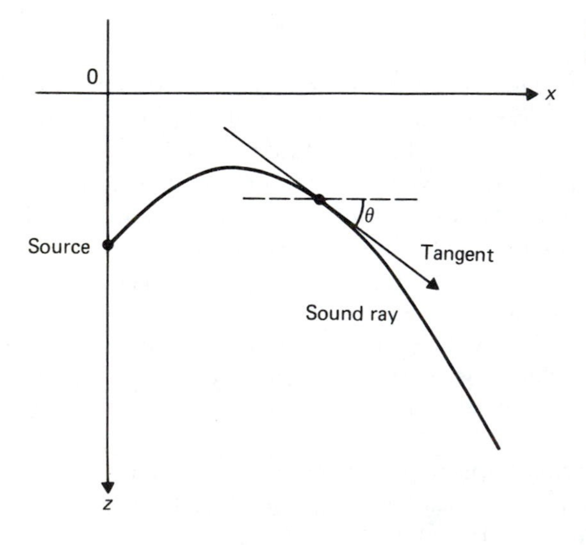

Figure 8.6 The path of a sound ray in water.

It follows that either $\sin\theta = 0$ and the sound propagates horizontally, or the path is governed by the pair of equations

$$\frac{dz}{dx} = \tan\theta, \qquad \frac{d\theta}{dx} = \frac{-C'(z)}{C(z)}.$$

In the ocean, the speed of sound depends on the temperature, salinity, pressure, and other depth-dependent parameters. Because of the difficulty in deriving an analytic expression of $C(z)$, it is common to approximate the speed at various depths with a suitable function. Typical data are given in Table 8.4 [2]. The construction of an empirical function for $C(z)$ must be done carefully since $C'(z)$ is also needed. The following representation due to S. Steinberg works well (see Fig. 8.7):

$$C(z) = 4779 + 0.01668z + \frac{160295}{z + 600}.$$

An inspection of the data reveals that the speed of sound can decrease with depth, attain a minimum value, and then increase. This permits the phenomenon of channeling of sound so that the direction of propagation oscillates about the horizontal and thereby energy is not lost by contact with either the surface or the bottom of the ocean. It is claimed that whales exploit this channeling in order to communicate over large distances. The optimal depth for such channeling is that at which the speed of sound is a minimum. From our data, this depth is nearly 2500 feet.

Table 8.4

Depth (ft)	Speed of sound (ft/sec)
0	5042
500	4995
1000	4948
1500	4887
2000	4868
2500	4863
3000	4865
3500	4869
4000	4875
5000	4875
6000	4887
7000	4905
8000	4918
9000	4933
10000	4949
11000	4973
12000	4991

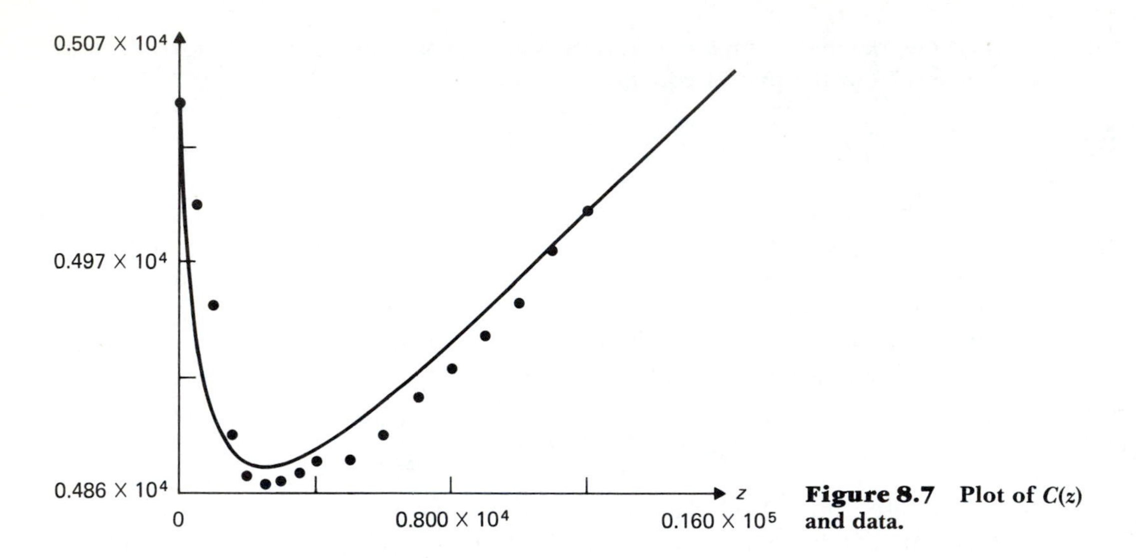

Figure 8.7 Plot of $C(z)$ and data.

Numerical solutions to the differential equations were investigated with RKF45 by assuming the initial conditions: $x = 0$, $z = 2500.016$ (the value where $C(z)$ is a minimum) and $\theta = \pm\alpha$, where the value $\alpha = 0.264$ radians was picked so that the ray grazed the surface of the water. An upper bound was thereby obtained on the cone of rays that can be propagated without energy loss caused by the contact with the surface of the ocean. A graph of the path of one of these bounding rays is given in Fig. 8.8.

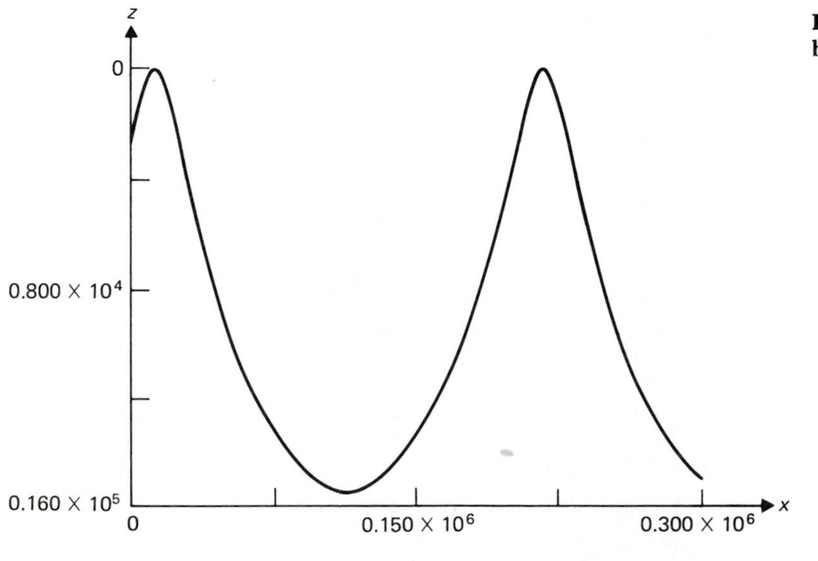

Figure 8.8 The graph of a bounding ray.

■

EXAMPLE 3 In this example, a numerical technique for solving the two-point boundary value problem

$$y'' + p(t)\, y' + q(t)\, y = r(t), \tag{8.6.2}$$

$$y(a) = A, \qquad y(b) = B, \tag{8.6.3}$$

is considered. Geometrically, a function $y(t)$ is sought that satisfies the differential equation (8.6.2) and whose graph passes through the points (a, A) and (b, B). The approach to solving the problem is to try and find the value $y'(a)$, for then we would have an initial value problem and RKF45 could be used directly to solve it. Therefore let $y'(a) = s$, and the task is to find a value $s = s^*$ so that the resulting solution, denoted by $Y(t; s^*)$, satisfies $Y(b; s^*) = B$. The method that is employed is called *shooting* and is described in the following algorithm:

The shooting algorithm. The problem is to find a zero of

$$u(s) = Y(b; s) - B.$$

1. Choose $s = s_1$ and solve the differential equation (8.6.2) with the initial conditions

$$y(a) = A, \qquad y'(a) = s_1;$$

denote the resulting solution by $Y(x; s_1)$. If

$$u(s_1) = Y(b; s_1) - B = 0,$$

set $s^* = s_1$ and stop. The solution is $y(x) = Y(x; s^*)$.

2. Choose $s = s_2 \neq s_1$ and solve (8.6.2) with

$$y(a) = A, \qquad y'(a) = s_2;$$

denote the solution by $y(x; s_2)$. If

$$u(s_2) = Y(b; s_2) - B = 0,$$

set $s^* = s_2$ and stop. The solution is $y(x) = Y(x; s^*)$.

3. Calculate the value of s for which $u(s) = 0$:

$$s^* = s_1 - \frac{s_1 - s_2}{u(s_1) - u(s_2)}\, u(s_1).$$

This can be done because the function $u(s) = Y(b; s) - B$ is a linear function of s and its zero is easy to calculate. The linearity of $u(s)$ follows from the fact that (8.6.2) is a linear equation (see Exercise 2).

4. Solve (8.6.2) with the initial conditions

$$y(a) = A, \qquad y'(a) = s^*$$

to get the desired solution.

The process is illustrated graphically in Fig. 8.9.

To illustrate the algorithm, consider the problem

$$y'' + \frac{2}{t}y' + y = 0 \tag{8.6.4}$$

$$y(1) = 1, \qquad y(2) = 5. \tag{8.6.5}$$

The choice $s_1 = 1$ leads to an initial value problem with initial conditions

$$y(1) = 1, \qquad y'(1) = 1.$$

This is solved at $t = 2$ using RKF45:

$$Y(2; 1) = 1.111622,$$

$$u(1) = 1.111622 - 5 = -3.888378.$$

The choice $s_2 = 2$ leads to the initial conditions

$$y(1) = 1, \qquad y'(1) = 2,$$

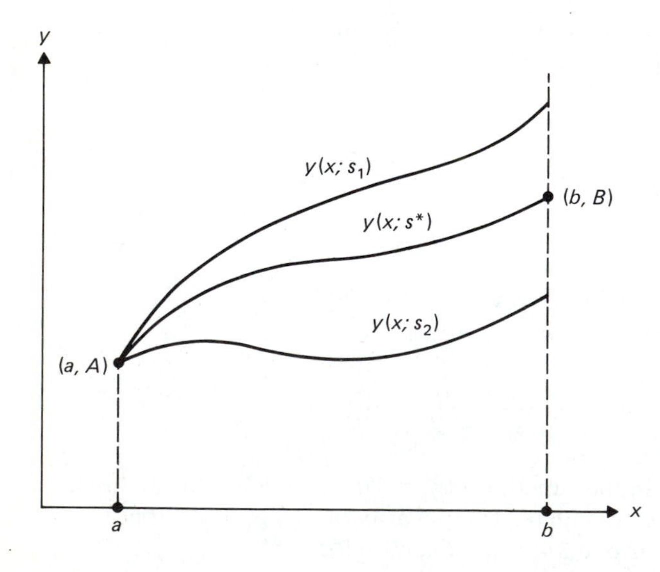

Figure 8.9 Graphic illustration of the shooting algorithm.

Table 8.5

t	$y(t)$	$y'(t)$
1.00	1.000000	10.241848
1.10	1.924833	8.328199
1.20	2.677897	6.784327
1.30	3.290407	5.502938
1.40	3.784888	4.414373
1.50	4.178141	3.472055
1.60	4.483100	2.644097
1.70	4.710023	1.908235
1.80	4.867286	1.248683
1.90	4.961922	0.654103
2.00	4.999996	0.116265

and using RKF45 again gives

$$Y(2;\, 2) = 1.532358,$$

$$u(2) = 1.532358 - 5 = -3.467642.$$

From the formula in step 3 it follows that

$$s^* = 1 - \frac{1 - 2}{(-3.888378) - (-3.467642)}(-3.888378) = 10.241848.$$

The solution of (8.6.4) with initial data

$$y(1) = 1, \qquad y'(1) = 10.241848$$

is shown in Table 8.5. ■

EXERCISES 8.6

1. Problem on propagation of sound in water.

a) Set $B = \dfrac{C(z)}{\cos \theta}$ and show that

$$B^2 = C^2(z) + \left[C(z) \frac{dz}{dx} \right]^2$$

is an integral for the system of differential equations

$$\frac{dz}{dx} = \tan \theta, \qquad \frac{d\theta}{dx} = -\frac{C'(z)}{C(z)}.$$

b) Define a new variable, $y = C(z) \tan \theta$ and show that

$$\frac{dy}{dx} = -C'(z), \qquad \frac{dz}{dx} = \frac{y}{C(z)}.$$

Derive and solve the linear differential equations corresponding to the equilibrium point, $y = 0$, $z = 2500.016$. Use the integral, $B^2 = C^2(z) + y^2$ and sketch a global phase plane portrait of the integral curves in the $z > 0$ half-plane of the nonlinear differential equations.

c) Integrate the y, z differential equations with initial conditions $x = 0$, $z = 2500.016$, $y = \beta$, and by trial and error find β so that the ray corresponding to the solution grazes the surface of the water.

d) If the depth of the water is less than 15,486 feet, parameter β can be selected so that the ray is not reflected by the surface of the water and just grazes the bottom. Assume that the depth is 12,000 feet and find β by trial and error.

e) If two submarines are at a depth of 2500 feet and are 200,000 feet apart, they can communicate by acoustic signals. Find two solutions to the two-point boundary value problem with boundary conditions

$$z(0) = z(200{,}000) = 2500.$$

2. Show that $u(s)$ defined by

$$u(s) = Y(x; s) - B$$

in the shooting algorithm of Example 3 is a linear function of s.

3. Write a computer program to implement the algorithm of Example 3 for solving two-point boundary value problems. Your program should use RKF45 or any other comparable routine for solving initial value problems.

Approximate the solution to Exercises 4–9 at the points indicated.

4. $y'' + y' + xy = 0, \quad y(0) = 1, \quad y(1) = 0; \quad t = 0.1, 0.2, \ldots, 1.0$

5. $y'' + \dfrac{2}{t}y' - \dfrac{2}{t^2}y = \dfrac{\sin t}{t^2}, \quad y(1) = 1, \quad y(2) = 2; \quad t = 1.1, 1.2, \ldots, 2.0$

6. $y'' + 4y = \cos t, \quad y(0) = 0, \quad y\left(\dfrac{\pi}{4}\right) = 0; \quad t = \dfrac{\pi}{32}, \dfrac{\pi}{16}, \dfrac{3\pi}{32}, \ldots, \dfrac{\pi}{4}$

7. $y'' + \sqrt{t}y' + y = e^t, \quad y(0) = 0, \quad y(1) = 0; \quad t = 0.1, 0.2, \ldots, 1.0$

8. $y'' + 4ty' + (4t^2 + 2)y = 0, \quad y(0) = 0, \quad y(1) = \dfrac{1}{e}; \quad t = 0.1, 0.2, \ldots, 1.0$

9. $y'' - \dfrac{2t}{1 - t^2}y' + \dfrac{12}{1 - t^2}y = 0, \quad y(0) = 0, \quad y(0.5) = 4; \quad t = 0.1, 0.2, \ldots, 0.5$

10. Modify the algorithm in Example 3 to solve the problem

$$(t^2 + 1)y'' - 2y = 0, \quad y(0) = 1, \quad y'(1) = 2.$$

Write a computer program to implement your procedure and find $y'(0)$ and $y(1)$.

11. a) Explain why your computer program in Exercise 3 cannot be used to solve the boundary value problem

$$y'' + e^{-y} = 0, \qquad y(0) = 0, \qquad y(1) = 0.$$

b) How would you modify the algorithm of Example 3 so as to obtain an approximate solution to this problem?

SUMMARY

In Chapter 3 a Runge–Kutta method of order 4 was introduced and implemented in the algorithm RK4. For many problems it will give very satisfactory results. It is a fixed step size strategy in that $h = (b - a)/n$ is fixed over the entire interval $[a, b]$ of computation. In general, greater efficiency can be obtained if the step size is adjusted as necessary. For instance, a larger step size would suffice in regions where the solution is smooth; whereas a smaller one would be needed in regions where the solution is oscillatory.

For a variable step size method the size of the step at each calculation is determined by estimating the local relative error at each step. This can be accomplished by applying at each step two algorithms of different order. Then given a preassigned relative error, one can decide whether the step size is satisfactory and proceed to the next computation, or whether to recompute using a smaller step size.

The above concept is incorporated into the subroutine RKF45, described here and in Appendix 3, which uses Runge–Kutta methods of order 4 and 5 to estimate local error. It is extremely accurate, and some examples of nonlinear problems are given for which analytical solutions cannot be found but for which good approximations of solutions over long time intervals are needed. For such problems, a variable step size method such as RKF45 is essential if one is to have confidence in the approximations obtained.

MISCELLANEOUS EXERCISES

8.1. The differential equation

$$\ddot{x} + c\dot{x} + k \sin x = L$$

describes the rotation of a synchronous motor. The dependent variable x is measured from an axis rotating with the electric field, $\ddot{x}$ represents the inertia torque due to the mass of the motor and its connected load, $c\dot{x}$ is due to "slippage," $k \sin x$ is due to the angle between the field of the rotor and armature, and the term L represents the torque of the external load on the motor. An extensive discussion of the mathematical problems arising from the operation of alternating-current motors is given in [11].

One problem of interest is the determination of the *critical damping factor,* the value of c such that there is an integral curve joining two unstable equilibrium states [12]. Remarkably, the same differential equation and the critical damping factor problem occur in celestial mechanics in the study of the spin-orbit coupling of the planet Mercury [13].

Take $k = 1$, $L = 0.1$, $\sin x_j = L$ with

$$x_0 \cong 0.1002, \qquad x_1 \cong 3.041, \qquad x_{-1} \cong -3.242.$$

a) Show that $(0, x_0)$ is a stable equilibrium point and $(0, x_{-1})$, $(0, x_1)$ are the neighboring unstable equilibrium points.

b) Sketch a phase plane portrait including the three equilibrium points with (i) $c = 0$, (ii) $0 < c \ll 0.1$. Note that there is no integral curve joining the two unstable equilibrium points x_{-1} and x_1.

c) Find λ_{-1} and λ_1, $\lambda_1 < 0 < \lambda_{-1}$, so that

$$x(t) = x_{-1} + \delta \exp(\lambda_{-1} t),$$

$$x(t) = x_1 - \delta \exp(\lambda_1 t)$$

are solutions to the linear differential equations corresponding to the two unstable equilibrium points $(x_{-1}, 0)$ and $(x_1, 0)$ with initial data $(x_{-1} + \delta, \delta\lambda_{-1})$ and $(x_1 - \delta, \delta\lambda_1)$. The parameter δ is arbitrary.

d) Let $y = \dot{x}$; then the second order differential equation governing the rotation can be replaced by an equivalent system of differential equations

$$\frac{dx}{dt} = y, \qquad \frac{dy}{dt} = -cy - k \sin x + L\,.$$

The integral curves of this system are also the graphs of the solutions to a corresponding first order differential equation

$$\frac{dy}{dx} = \frac{-cy - k \sin x + L}{y}.$$

Set $c = 0$, $\delta = 0.05$, and use RKF45 to integrate this equation from $x = x_{-1} + \delta$ to $x = x_0$ and then from $x = x_1 - \delta$ to $x = x_0$. Why must δ be positive? Interpolate to get the two values of y at x_0 from the computed values. They will be different since the two curves corresponding to the solutions are not part of the same integral curve. Increase c until the two values of y are the same to three significant figures. (You might use the bisection method of numerical analysis to help you pick reasonable values of c.) The resulting value of c is less than 0.1 and is a good approximation to the critical damping factor. Sketch the corresponding integral curve in the phase plane. (*Remark:* Your programs should be written so that the two integrations are carried out sequentially. Only the two interpolated values of y and their differences should be printed at x_0.)

8.2. A hypothetical reaction exhibiting some of the characteristics of real biochemical oscillators was proposed by I. Prigogine and R. Lefever (a good treatment of this topic is given in [10], Chap. 4). The system of reactions is

$$A \xrightarrow{k_1} X, \qquad B + X \xrightarrow{k_2} Y + D,$$
$$2X + Y \xrightarrow{k_3} 3X, \qquad X \xrightarrow{k_4} E.$$

The reactants A, B, D, and E are kept constant, so $X(t)$ and $Y(t)$ are to be found.

From the law of mass action, we get

$$\frac{dX}{dt} = k_1A - k_2BX + k_3YX^2 - k_4X,$$

$$\frac{dY}{dt} = k_2BX - k_3YX^2.$$

Let

$$s = k_4t, \qquad x = \frac{k_4X}{k_1A}, \qquad y = \frac{k_4Y}{k_1A},$$
$$a = \frac{k_3(k_1A)^2}{k_4^3}, \qquad b = \frac{k_2B}{k_4};$$

then

$$\frac{dx}{ds} = 1 - (b + 1)x + ax^2y, \qquad \frac{dy}{ds} = bx - ax^2y.$$

There is only one equilibrium state, $x = 1$, $y = b/a$, with both x and y positive.

a) Let $x = 1 + u$, $y = v + b/a$; show that

$$\frac{du}{ds} = (b - 1)u + av + g(u, v),$$

$$\frac{dv}{ds} = -bu - av - g(u, v), \quad a > 0, \quad b > 0,$$

where $g(u, v) = u^2(b + av) + 2auv$. Verify that the eigenvalues of the corresponding linear system are

$$\lambda = \frac{b - a - 1 \pm [(b - a - 1)^2 - 4a]^{1/2}}{2}.$$

b) If $b - a < 1$, the equilibrium state is stable, while if $b - a > 1$, it is unstable. The system of differential equations is of considerable interest since, as $b - a$ varies from the stable to the unstable range, a periodic solution (a limit cycle) appears. This is an example of a *Hopf bifurcation.* Set $a = 1$, $b = 3$ and use RKF45 to compute the limit cycle and estimate its period.

c) If a sustained external rhythm is imposed on the internal rhythm of the limit cycle, the internal rhythm is changed until it matches the imposed rhythm. This is called *entrainment.* In the above system, this can be modeled by the differential equations

$$\frac{dx}{ds} = 1 - (b + 1)x + ax^2y,$$

$$\frac{dy}{ds} = bx - ax^2y + c \sin qs,$$

with $a = 1, b = 3, c = 1, q = 1$. Start at $x = 1, y = 2$ and integrate the system using RKF45 on $0 \le s \le 30$ and estimate the period of the solution. Compare it to that of the forcing function and the period of the limit cycle.

REFERENCES

General

1. L.F. Shampine and R.C. Allen, *Numerical Computing: An Introduction,* W.B. Saunders Co., Philadelphia, 1973.
2. G. Forsythe, G. Malcolm, and C.B. Moler, *Computer Methods for Mathematical Computations,* Prentice-Hall, Englewood Cliffs, N.J., 1977.
3. C.W. Gear, *Numerical Initial Value Problems in Ordinary Differential Equations,* Prentice-Hall, Englewood Cliffs, N.J., 1971.

A good discussion of Adams multistep methods is in

4. L.F. Shampine and M.K. Gordon, *Computer Solution of Ordinary Differential Equations: The Initial Value Problem,* W.H. Freeman and Co., San Francisco, 1975.

More advanced discussions of numerical solutions of differential equations and error propagation are in

5. J.W. Daniel and R.E. Moore, *Computation and Theory in Ordinary Differential Equations,* W.H. Freeman and Co., San Francisco, 1970.
6. P. Henrici, *Discrete Variable Methods in Ordinary Differential Equations,* John Wiley and Sons, Inc., New York, 1962.

The Runge–Kutta Fehlberg formulas can be found in

7. E. Fehlberg, "Klassiche Runge–Kutta Formeln vierter und neidrigerer Ordnung mit Schrittweitten-Kontrolle und ihre Anwendung Auf Wärmeleitungsprobleme," *Computing,* 6 (1970), pp. 61–71.

The original report in which RKF45 was developed is

8. L.F. Shampine and H.A. Watts, *Practical Solution of Ordinary Differential Equations by Runge–Kutta Methods,* Sandia Laboratories Report, SAND76–0585.

Applications

9. I.M. Gelfand and S.V. Fomin, *Calculus of Variations,* Prentice-Hall, Englewood Cliffs, N.J., 1963.
10. J.D. Murray, *Nonlinear Differential Equation Models in Biology,* Clarendon Press, Oxford Press, 1977.
11. J.J. Stoker, *Nonlinear Vibrations in Mechanical and Electrical Systems,* Interscience, New York, 1950.
12. W.V. Lyon and H.E. Edgerton, "Transient Torque-Angle Characteristics of Synchronous Machines," *Trans. Amer. Inst. Electr. Eng.*, vol. 49, 1930.
13. T.J. Burns, "On the Rotation of Mercury," *Celestial Mechanics,* vol. 19, 1979.

9

FOURIER SERIES AND PARTIAL DIFFERENTIAL EQUATIONS

9.1 INTRODUCTION

Physical phenomena such as diffusion, fluid dynamics, elasticity, and electromagnetism can be formulated as relationships between unknown functions and their partial derivatives. These relationships are called *partial differential equations.* They differ from *ordinary differential equations* in that there are two or more independent variables. For example, the second order[1] partial differential equation

$$\frac{\partial^2 v}{\partial t^2} = \frac{\partial^2 v}{\partial x^2},$$

the wave equation, is used to model the vibration of a string with t, the time, and x, the position, as the two *independent variables* and v, the displacement, as the *dependent variable.* In this chapter we shall discuss partial differential equations governing wave propagation, diffusion, and potential theory. Initial and boundary conditions appropriate to these problems will be formulated and the qualitative properties of the solutions will be stressed. Since the basic solution technique, the separation of variables method, will employ Fourier series, a brief introduction to this topic is included.

1. The order of a partial differential equation is the order of the highest partial derivative of a dependent variable that appears in the equation.

9.2 THE WAVE EQUATION

DERIVATION OF THE WAVE EQUATION

The physics of musical instruments leads to problems in partial differential equations—for example, the vibrations of violin strings—that can be modeled by the wave equation,

$$\frac{\partial^2 v}{\partial t^2} = c^2 \frac{\partial^2 v}{\partial x^2}.$$

The phrase "can be modeled" is used frequently by mathematicians to motivate the study of mathematics, but it is subject to misuse. The sound produced by a violin is the result of complex interactions of the bow, the strings, and the body. However, fundamental concepts of music such as the superposition of harmonics can be understood by studying the wave equation and its normal modes of vibration. This leads us to the main solution technique of this chapter, the separation of variables method.

We shall take as our starting point the vibrations of a flexible string under tension that has been set into motion in such a way that its points can move only in the vertical direction and show that its displacement satisfies the wave equation. In an actual string, horizontal movement must occur, but for small displacements, the horizontal displacements are negligible compared to the vertical displacements.

We take the x-axis along the undisturbed string and let v denote the vertical displacement of the string. Since horizontal motion is not allowed in our model, v is a function of x, the coordinate that identifies the given points, and t, the time. The function $v(x, t)$ is at least twice differentiable since the flexibility of the string precludes abrupt changes in variables such as curvature that can be represented in terms of first and second derivatives.

The tension in the flexible string is a vector (see Fig. 9.1),

$$\mathbf{T}(x, t) = T_1(x, t)\mathbf{e}_1 + T_2(x, t)\mathbf{e}_2$$

that is parallel to the tangent vector of the string,

$$\mathbf{S}(x, t) = \mathbf{e}_1 + \frac{\partial v}{\partial x}(x, t)\mathbf{e}_2.$$

Therefore $T_1(x, t)\mathbf{S}(x, t) = \mathbf{T}(\mathbf{x}, \mathbf{t})$ and we have

$$T_1(x, t)\frac{\partial v}{\partial x}(x, t) = T_2(x, t).$$

Let us take a segment of string bounded by $x = a$ and $x = b$ and analyze the forces acting on it. Since there is no horizontal displacement, the net horizontal force relative to the arbitrary points a and b must be zero, so

$$-T_1(a, t) = T_1(b, t) = T_0,$$

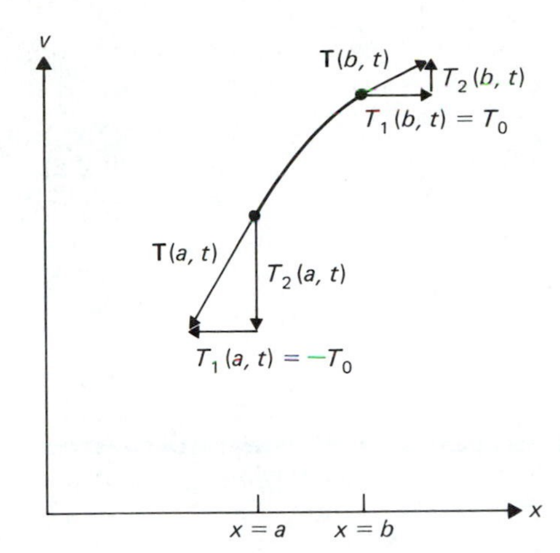

Figure 9.1 **The tension in a vibrating string.**

a constant. The net force in the vertical direction due to the tension is

$$T_2(b, t) + T_2(a, t) = T_1(b, t)\frac{\partial v}{\partial x}(b, t) + T_1(a, t)\frac{\partial v}{\partial x}(a, t)$$

$$= T_0\left(\frac{\partial v}{\partial x}(b, t) - \frac{\partial v}{\partial x}(a, t)\right) = T_0\int_a^b \frac{\partial^2 v}{\partial x^2}(x, t)\, dx.$$

By the extension of Newton's second law to continuous systems, the net force on the segment is equal to the integral with respect to x of the product of $\rho(x)$, the density per unit length, and $v_{tt}(x, t)$, the acceleration,[2]

$$\int_a^b \rho(x)\frac{\partial^2 v}{\partial t^2}(x, t)\, dx = T_0\int_a^b \frac{\partial^2 v}{\partial x^2}(x, t)\, dx.$$

Since a and b are arbitrary and the integrand is continuous, we have

$$\rho(x)\frac{\partial^2 v}{\partial t^2} = T_0\frac{\partial^2 v}{\partial x^2},$$

or

$$\frac{\partial^2 v}{\partial t^2} = c^2\frac{\partial^2 v}{\partial x^2}, \qquad \textbf{(9.2.1)}$$

where $c^2 = T_0/\rho(x)$. We shall assume that the density of the string is constant, hence c^2 is constant.

2. We shall use the two standard notations for the partial derivative, $\frac{\partial v}{\partial t}$ and v_t.

If external forces are present, such as gravity or driving forces, we would have the nonhomogeneous wave equation,

$$\frac{\partial^2 v}{\partial t^2} = c^2 \frac{\partial^2 v}{\partial x^2} + F(x, t). \tag{9.2.2}$$

This equation is discussed in a later section.

EXERCISES 9.2

1. Show that if $c^2 = b^2/a^2$, then $v(x, t) = e^{ax+bt}$ is a solution to $v_{tt} = c^2 v_{xx}$. Find other functions of $ax + bt$ that satisfy the equation.

Use the method of undetermined coefficients to construct solutions to the partial differential equations in Exercises 2–5. In these equations, p, q, and r are given real numbers. State the constraints (if any) on p, q, and r.

2. $v_{tt} - v_{xx} = 6 \sin(px + qt)$ (*Hint:* Try $v(x, t) = A \sin(px + qt)$.)
3. $v_{tt} - v_{xx} = 5 \cosh(px + qt)$
4. $v_{tt} - v_{xx} = px + qt$
5. $v_{tt} - v_{xx} = px^2 + qxt + rt^2$
6. Find $f(t)$ so that $v(x, t) = f(t) \sin x$ is a solution to $v_{tt} - v_{xx} = \cos t \sin x$. (*Hint:* Derive an ordinary differential equation for $f(t)$.) Why is $v(x, t)$ unbounded as $t \to \infty$?
7. Find $f(t)$ so that $v(x, t) = f(t) \sin x$ is a solution to $v_{tt} + v_t - v_{xx} = \cos t \sin x$. Why is $v(x, t)$ bounded for $t \geq 0$?

9.3 NORMAL MODES OF VIBRATION

Let us suppose that a taut flexible string is fastened at both ends, $x = 0$ and $x = L$, displaced from its equilibrium position, and then released. How can we describe its subsequent motion? The displacement of the string, $v(x, t)$, is a solution to the wave equation,

$$\frac{\partial^2 v}{\partial t^2} = c^2 \frac{\partial^2 v}{\partial x^2}.$$

It must satisfy *end* or *boundary conditions*, $v = 0$ at $x = 0$ and $x = L$. Furthermore, it is subject to *initial conditions*, $v(x, 0) = f(x)$, a function that describes its initial displacement from equilibrium, and, for this introductory problem, $v_t(x, 0) = 0$; that is, its initial velocity is zero. A typical presentation of the boundary-initial value problem that governs the motion of a plucked string displays the partial differential equation, the domain of definition of the solution, the boundary conditions, and then the initial conditions:

$$\frac{\partial^2 v}{\partial t^2} = c^2 \frac{\partial^2 v}{\partial x^2}, \qquad 0 < x < L, \quad 0 < t; \tag{9.3.1}$$

$$v(0, t) = 0, \qquad v(L, t) = 0, \qquad 0 < t; \tag{9.3.2}$$

$$v(x, 0) = f(x), \qquad v_t(x, 0) = 0, \qquad 0 < x < L. \tag{9.3.3}$$

The complexity of the solution to the boundary-initial value problem depends on the shape of the initial displacement. The lowest, or *fundamental, tone* of a guitar string is produced by an initial displacement that is modeled by

$$f(x) = A_1 \sin\left(\frac{\pi x}{L}\right).$$

The corresponding displacement is called the *first normal mode of vibration.* It is represented by

$$v_1(x, t) = A_1 \cos(\omega t) \sin\left(\frac{\pi x}{L}\right),$$

where the *fundamental frequency* ω can be found by substituting $v_1(x, t)$ into the wave equation. We find that

$$0 = \frac{\partial^2 v_1}{\partial t^2} - c^2 \frac{\partial^2 v_1}{\partial x^2} = \left(-\omega^2 + \frac{c^2\pi^2}{L^2}\right) A_1 \cos(\omega t) \sin\left(\frac{\pi x}{L}\right).$$

Therefore

$$\omega^2 = \frac{c^2\pi^2}{L^2},$$

and the fundamental frequency is given by

$$\omega = \frac{c\pi}{L}. \tag{9.3.4}$$

The first normal mode was described in 1713 by an English mathematician, Brook Taylor (1685–1731), who also discovered the series expansions known as *Taylor series.* In this mode, the string produces its fundamental tone. Further analysis by Johann Bernoulli (1667–1748), a member of an outstanding family of Swiss mathematicians, resulted in a mathematical derivation of a seventeenth century observation known as *Mersenne's law* that the pitch or frequency of the fundamental tone is proportional to $(T_0/\rho)^{1/2}/L$ (recall that $c = (T_0/\rho)^{1/2}$). A warning is appropriate here. Pitch is a subjective attribute of sound and although it can be identified for simple sounds such as the fundamental tone, complex sounds can be produced with different frequencies but with indistinguishable pitches.

The *overtones* or *harmonics tones* have frequencies that are integral multiples of ω. The corresponding normal modes of vibrations are

$$v_n(x, t) = A_n \cos(n\omega t) \sin\left(\frac{n\pi x}{L}\right), \qquad n = 1, 2, 3, \ldots. \tag{9.3.5}$$

Each normal mode generates a standing wave whose bounding values are $\pm|A_n \sin(n\pi x/L)|$. The fixed points of the vibration, its *nodes*, are the zeros of $\sin(n\pi x/L)$. The number of zeros in the open interval, $0 < x < L$, characterizes the normal modes: the first normal mode has none, the second mode has one, and so on. Several normal modes are sketched in Fig. 9.2.

Since the wave equation is linear and homogeneous, solutions can be constructed by the superposition of normal modes. For example,

$$v(x, t) = \sum_{n=1}^{N} A_n \cos(n\omega t) \sin\left(\frac{n\pi x}{L}\right) \tag{9.3.6}$$

is a solution to the boundary-initial value problem (9.3.1–9.3.3) corresponding to an initial displacement,

$$f(x) = \sum_{n=1}^{N} A_n \sin\left(\frac{n\pi x}{L}\right), \tag{9.3.7}$$

and zero initial velocity.

EXAMPLE 1 Solve the wave equation,

$$\frac{\partial^2 v}{\partial t^2} = c^2 \frac{\partial^2 v}{\partial x^2}, \qquad 0 < x < \pi, \quad 0 < t,$$

with $c = 2$ and with the boundary conditions,

$$v(0, t) = 0, \qquad v(\pi, t) = 0, \qquad 0 < t,$$

and with the initial conditions,

$$v(x, 0) = 10 \sin x + 0.1 \sin 3x, \qquad v_t(x, 0) = 0, \qquad 0 < x < \pi.$$

Plot the solution as a function of x, $0 < x < \pi$, for fixed t with $t = 0, \pi/4, \pi/3$.

SOLUTION For this problem $L = \pi$ and the formula (9.3.7) for the initial displacement is

$$f(x) = A_1 \sin x + A_2 \sin 2x + A_3 \sin 3x + A_4 \sin 4x + \cdots.$$

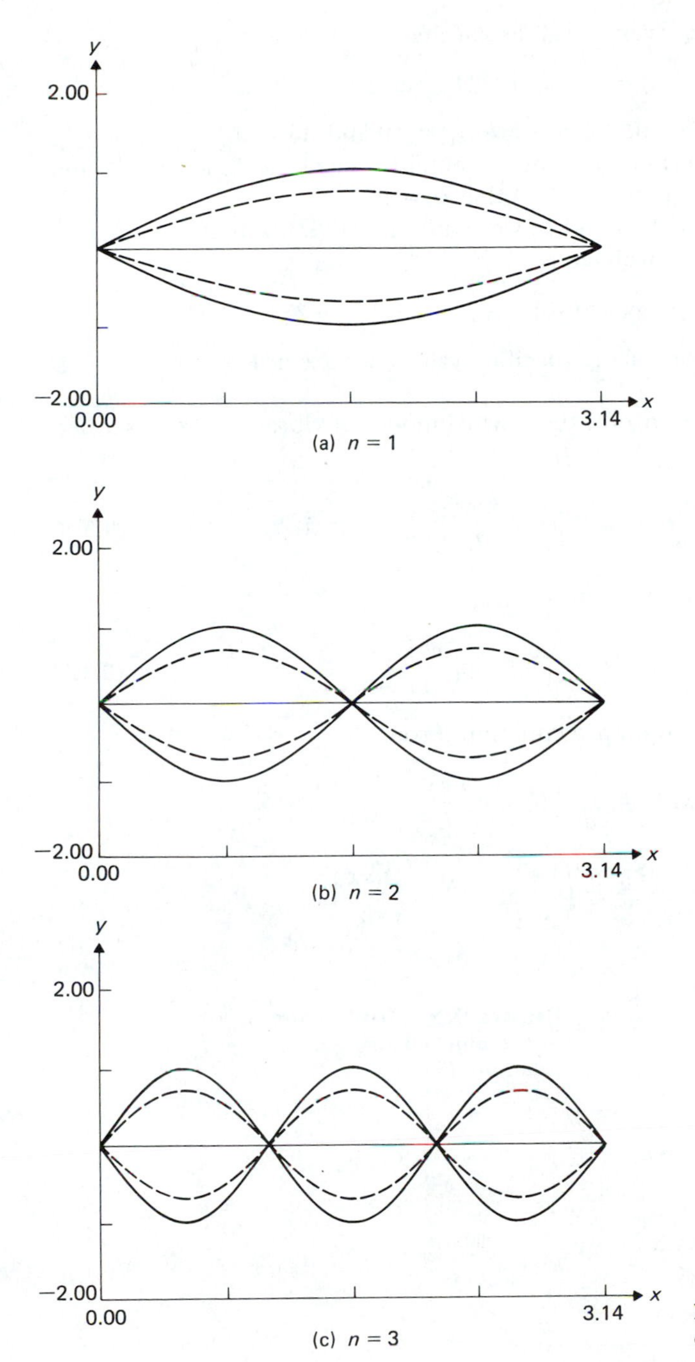

Figure 9.2 Normal modes of a vibrating string.

If we compare this with the given initial displacement,

$$v(x, 0) = 10 \sin x + 0.1 \sin 3x,$$

we see that the only nonzero coefficients are $A_1 = 10$ and $A_3 = 0.1$.

The next step is to determine the fundamental frequency ω. Since $c = 2$ and $L = \pi$, we conclude from equation (9.3.4) that $\omega = 2$.

Knowing the coefficients $\{A_n\}$ and ω, we obtain from (9.3.6) that the solution to this boundary-initial value problem is

$$v(x, t) = 10 \cos 2t \sin x + 0.1 \cos 6t \sin 3x.$$

Sketches of the displacement for the specified values of t are in Fig. 9.3. ■

If the initial velocity is nonzero, the normal modes of vibration are described by

$$v_n(x, t) = \left[A_n \cos(n\omega t) + \frac{B_n}{n\omega} \sin(n\omega t) \right] \sin\left(\frac{n\pi x}{L}\right), \qquad n = 1, 2, 3, \ldots. \quad \textbf{(9.3.8)}$$

This formula can be derived by setting

$$v_n(x, t) = T_n(t) \sin\left(\frac{n\pi x}{L}\right) \quad \textbf{(9.3.9)}$$

and substituting $v_n(x, t)$ into the wave equation. From

$$\frac{\partial^2 v_n}{\partial t^2}(x, t) = \frac{d^2 T_n}{dt^2}(t) \sin\left(\frac{n\pi x}{L}\right),$$

$$\frac{\partial^2 v_n}{\partial x^2}(x, t) = -\left(\frac{n\pi}{L}\right)^2 T_n(t) \sin\left(\frac{n\pi x}{L}\right),$$

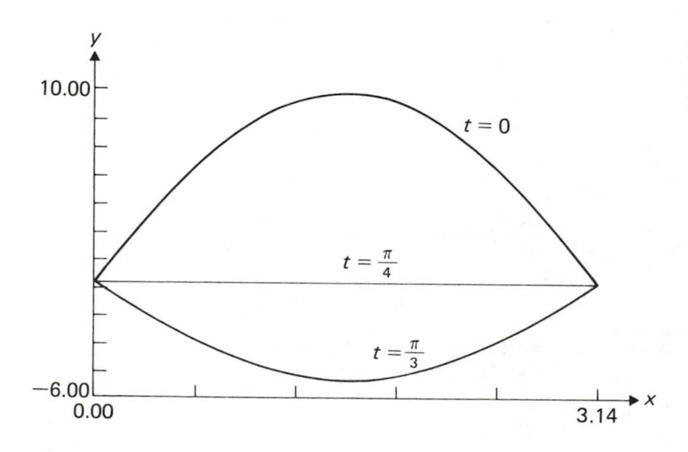

Figure 9.3 Displacements of the plucked string of Example 1.

we have

$$0 = \frac{\partial^2 v_n}{\partial t^2}(x, t) - c^2 \frac{\partial^2 v_n}{\partial x^2}(x, t) = \left\{\frac{d^2 T_n}{dt^2}(t) + \left(\frac{cn\pi}{L}\right)^2 T_n(t)\right\} \sin\left(\frac{n\pi x}{L}\right).$$

This can be valid on the interval $0 < x < L$ only if the coefficient of $\sin(n\pi x/L)$ is zero for all $t > 0$.

It follows that

$$\frac{d^2 T_n}{dt^2}(t) + (n\omega)^2 T_n(t) = 0, \tag{9.3.10}$$

where, as before, $\omega = (c\pi/L)$.

The initial displacement and velocity of the normal mode (9.3.8) generate initial conditions for the ordinary differential equation (9.3.10), namely,

$$\begin{aligned} v_n(x, 0) &= A_n \sin\left(\frac{n\pi x}{L}\right) \text{ implies that } T_n(0) = A_n, \\ \frac{\partial v_n}{\partial t}(x, 0) &= B_n \sin\left(\frac{n\pi x}{L}\right) \text{ implies that } \frac{dT_n}{dt}(0) = B_n. \end{aligned} \tag{9.3.11}$$

Therefore the time-dependent term in (9.3.8) is the solution to the initial value problem (9.3.10), (9.3.11),

$$T_n(t) = A_n \cos(n\omega t) + \frac{B_n}{n\omega} \sin(n\omega t). \tag{9.3.12}$$

As before, a superposition of a finite number of normal modes,

$$v(x, t) = \sum_{n=1}^{N} \left[A_n \cos(n\omega t) + \frac{B_n}{n\omega} \sin(n\omega t) \right] \sin\left(\frac{n\pi x}{L}\right), \tag{9.3.13}$$

is a solution to the wave equation. Its initial displacement and velocity are $v(x, 0) = f(x)$, $v_t(x, 0) = g(x)$, where

$$\begin{aligned} f(x) &= \sum_{n=1}^{N} A_n \sin\left(\frac{n\pi x}{L}\right), \\ g(x) &= \sum_{n=1}^{N} B_n \sin\left(\frac{n\pi x}{L}\right). \end{aligned} \tag{9.3.14}$$

EXAMPLE 2 Solve the wave equation,

$$\frac{\partial^2 v}{\partial t^2} = c^2 \frac{\partial^2 v}{\partial x^2}, \qquad 0 < x < 2, \quad 0 < t,$$

with $c = 1$ and with the boundary conditions,

$$v(0, t) = 0, \qquad v(2, t) = 0, \qquad 0 < t,$$

and initial conditions, $v(x, 0) = f(x)$, $v_t(x, 0) = g(x)$, where

$$f(x) = 6\sin(\pi x),$$

$$g(x) = 2\sin(4\pi x), \qquad 0 < x < 2.$$

Sketch the displacement and velocity on $0 < x < 2$ for fixed t with $t = 0, 1/4, 1/2$.

SOLUTION Since $L = 2$ and $c = 1$, the fundamental frequency, $\omega = \pi/2$ and the nonzero coefficients are $A_2 = 6$ and $B_8 = 2$.

$$v(x, t) = 6\cos(\pi t)\sin(\pi x) + \frac{1}{2\pi}\sin(4\pi t)\sin(4\pi x),$$

$$v_t(x, t) = -6\pi\sin(\pi t)\sin(\pi x) + 2\cos(4\pi t)\sin(4\pi x).$$

Sketches of the displacement and velocity on $0 < x < 2$ for the specified values of t are in Fig. 9.4. ■

Sums of the type (9.3.14) are called trigonometric polynomials. In the above examples, the initial displacement and velocity, $f(x)$ and $g(x)$, have been represented by trigonometric polynomials, and the solutions can be read off directly. Trigonometric polynomials can represent complicated initial displacements and velocities, but the requirement that N be finite is quite restrictive. For example, the function $f(x) = x(\pi - x)$ on $0 < x < \pi$ cannot be represented by a trigonometric polynomial. However, it can be represented by an infinite series,

$$x(\pi - x) = \frac{8}{\pi}\sum_{n=1}^{\infty}\sin^2\left(\frac{n\pi}{2}\right)\frac{\sin nx}{n^3}.$$

This series is an example of a *Fourier series,* the topic of the next two sections.

EXERCISES 9.3

The following boundary-initial value problem is to be solved with the prescribed initial data:

$$v_{tt} = c^2 v_{xx}, \quad 0 < x < \pi, \quad 0 < t;$$

$$v(0, t) = 0, \qquad v(\pi, t) = 0, \quad 0 < t;$$

$$v(x, 0) = f(x), \qquad v_t(x, 0) = g(x), \quad 0 < x < \pi.$$

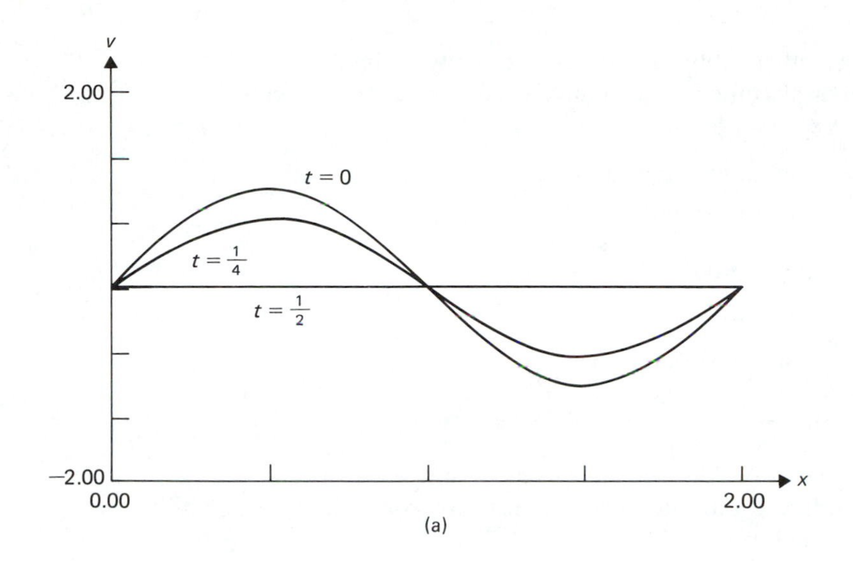

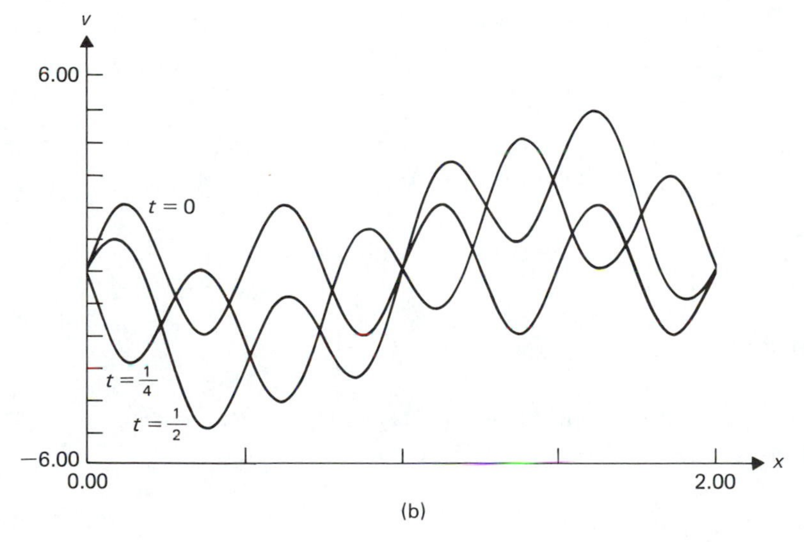

Figure 9.4 Wave profiles of the vibrating string of Example 2.

For Exercises 1–5 sketch the solution as a function of x, $0 < x < \pi$, for $ct = 0$, $\pi/4$, $\pi/2$.

1. $f(x) = 3 \sin 2x$, $g(x) = 0$ with $c = 2$
2. $f(x) = 4 \sin x - 6 \sin 3x$, $g(x) = 0$ with $c = 1$
3. $f(x) = -2 \sin 2x + 2 \sin 4x$, $g(x) = 0$ with $c = 3$
4. $f(x) = -6 \sin 3x + 6 \sin 6x$, $g(x) = 0$ with $c = 2$
5. $f(x) = 2 \sin^3 x$, $g(x) = 0$ with $c = 1$

If a taut string is quickly struck with a rubber hammer, its velocity is changed almost instantaneously, but its displacement is unchanged. Therefore we set $f(x) = 0$, $g(x)$ as prescribed. Solve Exercises 6–10.

6. $f(x) = 0,\ g(x) = 3 \sin x + 4 \sin 2x$ with $c = 3$

7. $f(x) = 0,\ g(x) = 15 \sin 2x - 18 \sin 3x$ with $c = 1$

8. $f(x) = 0,\ g(x) = \sum_{n=1}^{10} (-1)^n \sin nx$ with $c = 2$

9. $f(x) = 0,\ g(x) = \sum_{n=6}^{18} \cos\left(\frac{n\pi}{2}\right) \sin nx$ with $c = 6$

10. $f(x) = 0,\ g(x) = 5 \sin^3 x$ with $c = 1$

Suppose that at $t = 0$ one can measure both the displacement and velocity of a vibrating string. Solve the boundary-initial value problem with the prescribed initial data. Omit terms with zero amplitude.

11. $f(x) = \sin^3 x,\ g(x) = \sin^5 x$ with $c = 3$

12. $f(x) = \sum_{n=2}^{9} \sin\left(\frac{n\pi}{2}\right) \sin nx,\ g(x) = \sum_{n=10}^{14} \cos\left(\frac{n\pi}{2}\right) \sin nx$ with $c = 1$

13. $f(x) = \sum_{n=1}^{5} (2n + 1) \sin(3nx),\ g(x) = \sum_{n=1}^{5} \frac{1}{n} \sin(2nx)$ with $c = 2$

14. $f(x) = \sum_{n=1}^{14} [1 - (-1)^n] \sin nx,\ g(x) = \sum_{n=1}^{14} [1 + (-1)^n] \sin nx$ with $c = 3$

15. $f(x) = \sum_{n=1}^{4} \sin^2\left(\frac{n\pi}{2}\right) \sin(2nx),\ g(x) = \sum_{n=1}^{4} \cos^2\left(\frac{n\pi}{2}\right) \sin nx$ with $c = 1$

If $v(x, t)$ is a solution to the boundary-initial value problem with $f(x)$ and $g(x)$ being twice differentiable functions, the *energy function* is defined by

$$E(t) = \frac{1}{2} \int_0^\pi \{[v_t(x, t)]^2 + c^2[v_x(x, t)]^2\}\, dx. \tag{9.3.15}$$

16. Show that

$$\begin{aligned} \frac{d}{dt} E(t) &= \int_0^\pi [v_t(x, t)v_{tt}(x, t) + c^2 v_x(x, t)v_{xt}(x, t)]\, dx \\ &= c^2 v_x(x, t)v_t(x,t)\Big|_0^\pi + \int_0^\pi [v_t(x, t)v_{tt}(x, t) - c^2 v_t(x, t)v_{xx}(x, t)]\, dx = 0. \end{aligned}$$

(*Hint:* Use integration by parts and note that $v(0, t) = 0$ for $t \geq 0$ implies $v_t(0, t) = 0$ for $t \geq 0$.) Since $dE/dt = 0$, the energy is a constant that is determined by the initial conditions.

For Exercises 17–21, find $E(t)$.

17. Use data of Exercise 3.

18. Use data of Exercise 4.

19. Use data of Exercise 7.

20. Use data of Exercise 10.

21. Use data of Exercise 11.

9.4 FOURIER SERIES

In our study of normal mode solutions to the wave equation, we constructed functions that were finite combinations of trigonometric functions. A related but conceptually different problem is the construction of trigonometric polynomials and trigonometric series that can represent functions on prescribed intervals.

We assume that an *absolutely integrable*[3] function $f(x)$ is defined on the interval $-\pi < x < \pi$ and can be represented as a series,

$$f(x) = \frac{a_0}{2} + \sum_{n=1}^{\infty} (a_n \cos nx + b_n \sin nx), \qquad -\pi < x < \pi. \tag{9.4.1}$$

Such a series, if it exists, is called a *Fourier series* in honor of Jean Baptiste Fourier (1768–1830), the French scientist who made extensive use of them in a study of heat conduction.

The coefficients a_n, b_n are called the Fourier coefficients of $f(x)$ and are given by the *Euler formulas,*

$$\begin{aligned} a_n &= \frac{1}{\pi}\int_{-\pi}^{\pi} f(x)\cos nx\,dx, \\ b_n &= \frac{1}{\pi}\int_{-\pi}^{\pi} f(x)\sin nx\,dx. \end{aligned} \tag{9.4.2}$$

The constant term in (9.4.1) has the factor $\frac{1}{2}$ to avoid a special formula for the calculation of a_0.

The terminology that we have used is standard but confusing. A *trigonometric series* is an infinite series of the form

$$A_0 + \sum_{n=1}^{\infty} (A_n \cos nx + B_n \sin nx).$$

No restriction is placed on the coefficients A_n, B_n, and the series may or may not converge. A *Fourier series* is associated with an absolutely integrable function $f(x)$

3. A function $f(x)$ is absolutely integrable if both $f(x)$ and $|f(x)|$ are integrable.

that is defined on $-\pi < x < \pi$ and has the form (9.4.1) with the coefficients given by the formulas (9.4.2). The term Fourier series is appropriate because Fourier made the first systematic use of them in his study of heat conduction. L. Euler (1707–1783) earlier used such series to solve particular problems. Therefore the Fourier coefficients of a function are computed by means of the Euler formulas. D. Bernoulli (1700–1782) preceded Fourier in his use of these tools to solve vibrating string problems, but his name is not attached to the formulas (9.4.2).

The Fourier series and coefficient formulas are presented here with $-\pi < x < \pi$ as the *fundamental interval.* Other intervals will be considered subsequently. Before deriving the Euler formulas, let us consider an example.

EXAMPLE 1 Compute the Fourier coefficients of $f(x) = x^2 + \pi x$.

SOLUTION From (9.4.2) we have

$$a_0 = \frac{1}{\pi}\int_{-\pi}^{\pi} (x^2 + \pi x)\, dx = \frac{2}{3}\pi^2.$$

For $n \geq 1$,

$$a_n = \frac{1}{\pi}\int_{-\pi}^{\pi} (x^2 + \pi x)\cos(nx)\, dx = \frac{4}{n^2}\cos(n\pi),$$

$$b_n = \frac{1}{\pi}\int_{-\pi}^{\pi} (x^2 + \pi x)\sin(nx)\, dx = -\frac{2\pi}{n}\cos(n\pi).$$ ■

EXAMPLE 2 Compute the Fourier coefficients of

$$f(x) = \begin{cases} \frac{1}{3}(x + \pi) & \text{if } -\pi < x < \frac{\pi}{2}, \\ (\pi - x) & \text{if } \frac{\pi}{2} \leq x < \pi. \end{cases}$$

SOLUTION From (9.4.2) we have

$$\begin{aligned} a_0 &= \frac{1}{\pi}\int_{-\pi}^{\pi} f(x)\, dx \\ &= \frac{1}{3\pi}\int_{-\pi}^{\pi/2} (x + \pi)\, dx + \frac{1}{\pi}\int_{\pi/2}^{\pi} (\pi - x)\, dx \\ &= \frac{\pi}{2}. \end{aligned}$$

For $n \geq 1$,

$$
\begin{aligned}
a_n &= \frac{1}{\pi}\int_{-\pi}^{\pi} f(x)\cos(nx)\,dx \\
&= \frac{1}{3\pi}\int_{-\pi}^{\pi/2} (x+\pi)\cos(nx)\,dx + \frac{1}{\pi}\int_{\pi/2}^{\pi} (\pi - x)\cos(nx)\,dx \\
&= \frac{2}{3n^2\pi}\left[\cos(n\pi) - \cos\left(\frac{n\pi}{2}\right) + \left(\frac{n\pi}{2}\right)\sin\left(\frac{n\pi}{2}\right)\right], \\
b_n &= \frac{1}{\pi}\int_{-\pi}^{\pi} f(x)\sin(nx)\,dx \\
&= \frac{1}{3\pi}\int_{-\pi}^{\pi/2} (x+\pi)\sin(nx)\,dx + \frac{1}{\pi}\int_{\pi/2}^{\pi} (\pi - x)\sin(nx)\,dx \\
&= \frac{2}{3n^2\pi}\left[-3n\pi\cos(n\pi) + \left(\frac{3n\pi}{2}\right)\sin\left(\frac{n\pi}{2}\right) - \sin\left(\frac{n\pi}{2}\right)\right].
\end{aligned}
$$

■

EXAMPLE 3 Compute the Fourier coefficients of $f(x) = (x - \pi/2)\cos(x/2)$ on the fundamental interval, $-\pi < x < \pi$.

SOLUTION From (9.4.2) we have

$$
\begin{aligned}
a_0 &= \frac{1}{\pi}\int_{-\pi}^{\pi}\left(x - \frac{\pi}{2}\right)\cos\left(\frac{x}{2}\right)dx = -2, \\
a_n &= \frac{1}{\pi}\int_{-\pi}^{\pi}\left(x - \frac{\pi}{2}\right)\cos\left(\frac{x}{2}\right)\cos(nx)\,dx \\
&= \frac{1}{2\pi}\int_{-\pi}^{\pi}\left(x - \frac{\pi}{2}\right)\left\{\cos\left[\left(n + \frac{1}{2}\right)x\right] + \cos\left[\left(n - \frac{1}{2}\right)x\right]\right\}dx \\
&= (-1)^n\frac{2}{(4n^2 - 1)}, \\
b_n &= \frac{1}{\pi}\int_{-\pi}^{\pi}\left(x - \frac{\pi}{2}\right)\cos\left(\frac{x}{2}\right)\sin(nx)\,dx \\
&= \frac{1}{2\pi}\int_{-\pi}^{\pi}\left(x - \frac{\pi}{2}\right)\left\{\sin\left[\left(n + \frac{1}{2}\right)x\right] + \sin\left[\left(n - \frac{1}{2}\right)x\right]\right\}dx \\
&= (-1)^{n+1}\frac{32}{(4n^2 - 1)^2}.
\end{aligned}
$$

The formulas for a_n and b_n were simplified with the help of the following:

$$\cos[(n \pm \tfrac{1}{2})\pi] = 0,$$

$$\sin[(n + \tfrac{1}{2})\pi] = (-1)^n,$$

$$\sin[(n - \tfrac{1}{2})\pi] = (-1)^{n+1}. \qquad \blacksquare$$

As a preliminary step in the derivation of the Euler formulas we list the following facts: For positive integers m and n,

$$\int_{-\pi}^{\pi} \cos(mx)\cos(nx)\,dx = \begin{cases} \pi & \text{if } m = n, \\ 0 & \text{if } m \neq n; \end{cases} \tag{9.4.3}$$

$$\int_{-\pi}^{\pi} \sin(mx)\sin(nx)\,dx = \begin{cases} \pi & \text{if } m = n, \\ 0 & \text{if } m \neq n; \end{cases} \tag{9.4.4}$$

$$\int_{-\pi}^{\pi} \sin(mx)\cos(nx)\,dx = 0; \tag{9.4.5}$$

$$\int_{-\pi}^{\pi} \sin(mx)\,dx = \int_{-\pi}^{\pi} \cos(mx)\,dx = 0. \tag{9.4.6}$$

These are called *orthogonality relations* in analogy with the orthogonality properties of certain sets of vectors in finite dimensional vector spaces. They can be derived from trigonometric identities, for example, if we integrate

$$\sin mx \sin nx = \{\cos[(m - n)x] - \cos[(m + n)x]\}/2$$

over $-\pi < x < \pi$, we obtain

$$\int_{-\pi}^{\pi} \sin(mx)\sin(nx)\,dx = \frac{1}{2}\int_{-\pi}^{\pi} \cos[(m - n)x]\,dx - \frac{1}{2}\int_{-\pi}^{\pi} \cos[(m + n)x]\,dx$$

$$= \begin{cases} \pi & \text{if } m = n, \\ 0 & \text{if } m \neq n. \end{cases}$$

The second integral is zero for all choices of positive integers m and n, and the first is equal to π if $m = n$ and is zero if $m \neq n$. Thus, (9.4.4) is valid.

Let us suppose that the series (9.4.1) converges[4] to a known function $f(x)$ on the interval $-\pi < x < \pi$ and that we want to determine a_0. To do this we merely integrate both sides of the equation (9.4.1) from $-\pi$ to π, use (9.4.6), and obtain

$$\int_{-\pi}^{\pi} f(x)\,dx = \int_{-\pi}^{\pi} \frac{a_0}{2}\,dx + \sum_{n=1}^{\infty}\left(a_n \int_{-\pi}^{\pi} \cos(nx)\,dx + b_n \int_{-\pi}^{\pi} \sin(nx)\,dx\right)$$

$$= \pi a_0 + 0.$$

4. A technical note: The series should converge in such a way that term-by-term integration is valid. A sufficient condition is uniform convergence, a concept that is discussed in advanced calculus textbooks.

Therefore

$$a_0 = \frac{1}{\pi} \int_{-\pi}^{\pi} f(x)\, dx.$$

Suppose that we want to determine a_m or b_m for some m. We multiply both sides of the expression (9.4.1) by cos mx or sin mx and integrate from $-\pi$ to π. The orthogonality relations imply that the only nonzero terms on the right side will be

$$a_m \int_{-\pi}^{\pi} \cos^2(mx)\, dx = \pi a_m,$$

or

$$b_m \int_{-\pi}^{\pi} \sin^2(mx)\, dx = \pi b_m.$$

Therefore

$$\int_{-\pi}^{\pi} f(x) \cos(mx)\, dx = a_m \int_{-\pi}^{\pi} \cos^2(mx)\, dx + 0 = \pi a_m,$$

or

$$\int_{-\pi}^{\pi} f(x) \sin(mx)\, dx = b_m \int_{-\pi}^{\pi} \sin^2(mx)\, dx + 0 = \pi b_m,$$

and we have

$$a_m = \frac{1}{\pi} \int_{-\pi}^{\pi} f(x) \cos(mx)\, dx,$$

and

$$b_m = \frac{1}{\pi} \int_{-\pi}^{\pi} f(x) \sin(mx)\, dx,$$

the Euler formulas for the Fourier coefficients of the known function $f(x)$.

The term-by-term integration of the series employed in the derivation of the Euler formula is valid because of our assumptions about the convergence of the series (9.4.1). However, we can define the Fourier coefficients of any absolutely integrable function $f(x)$ by the formulas (9.4.2) and associate with $f(x)$ the series (9.4.1). The notation,

$$f(x) \sim \frac{a_0}{2} + \sum_{n=1}^{\infty} (a_n \cos nx + b_n \sin nx) \qquad \textbf{(9.4.7)}$$

is sometimes used to avoid possible confusion about the convergence of the series.

We also associate with $f(x)$ a sequence of *Fourier approximates,*

$$S_N(x) = \frac{a_0}{2} + \sum_{n=1}^{N} (a_n \cos nx + b_n \sin nx), \quad N = 0, 1, 2, \ldots \qquad \textbf{(9.4.8)}$$

where the coefficients a_n, b_n are given by the Euler formulas (9.4.2). A question of considerable importance is: Do the Fourier approximates converge, in some reasonable way, to $f(x)$ as N becomes infinite? We shall address this question after we have studied several examples.

EXAMPLE 4 Compute the Fourier coefficients of $f(x) = x$ and $g(x) = |x|$ on $-\pi < x < \pi$ and plot the graphs of the functions and their Fourier approximates for $N = 1, 3$.

SOLUTION The function $f(x)$ is an odd function[5] and therefore the coefficients a_n are all zero. (The functions x, $x \cos(nx)$ are odd functions and their integrals on a symmetric interval such as $-\pi < x < \pi$ are zero.) The coefficients b_n are given by

$$b_n = \frac{1}{\pi} \int_{-\pi}^{\pi} x \sin(nx)\, dx = \frac{-2(-1)^n}{n},$$

and the Fourier approximates are

$$S_N(x) = -2 \sum_{n=1}^{N} (-1)^n \frac{\sin nx}{n}.$$

By contrast, the function $g(x) = |x|$ is an even function and therefore the coefficients b_n are all zero. The coefficients a_n are given by

$$a_0 = \frac{1}{\pi} \int_{-\pi}^{\pi} |x|\, dx = \frac{1}{\pi} \int_{-\pi}^{0} (-x)\, dx + \frac{1}{\pi} \int_{0}^{\pi} x\, dx = \frac{2}{\pi} \int_{0}^{\pi} x\, dx = \pi,$$

$$a_n = \frac{1}{\pi} \int_{-\pi}^{\pi} |x| \cos(nx)\, dx = \frac{2}{\pi} \int_{0}^{\pi} x \cos(nx)\, dx = \frac{-2}{\pi n^2} [1 - (-1)^n].$$

The Fourier approximates for $g(x)$ are

$$S_N(x) = \frac{\pi}{2} - \frac{2}{\pi} \sum_{n=1}^{N} [1 - (-1)^n] \frac{\cos nx}{n^2}.$$

The graphs are in Fig. 9.5. ■

5. An odd function $q(x)$ has the property that for all x in its domain of definition, $q(-x) = -q(x)$. The corresponding property for an even function $p(x)$ is $p(x) = p(-x)$. The graph of an odd function is symmetric with respect to the origin. The graph of an even function is symmetric with respect to the vertical axis.

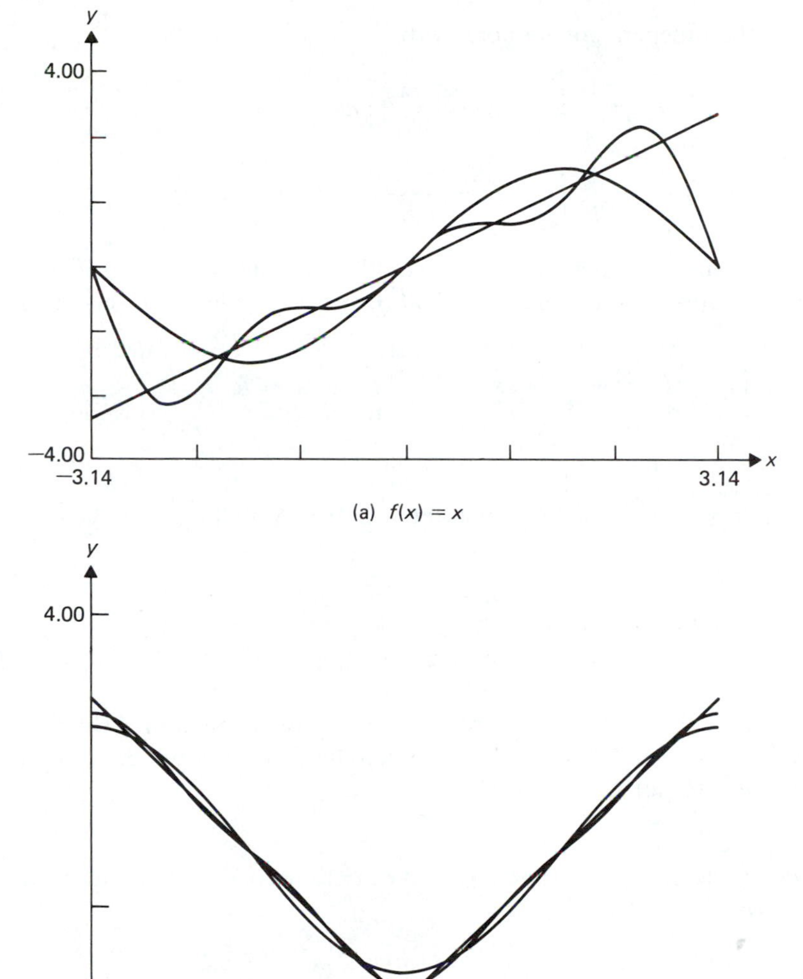

Figure 9.5 Graphs of $f(x) = x$ and $g(x) = |x|$ and their Fourier approximates, $N = 1, 3$, for Example 4.

It is convenient to start the study of Fourier series with functions defined on the interval $-\pi < x < \pi$ because the formulas are simpler than those for the general symmetric interval $-L < x < L$. If we make a change of scale by setting $y = Lx/\pi$, then we find that the Fourier series associated with an absolutely integrable function $f(x)$ defined on the interval $-L < x < L$ is

$$f(x) \sim \frac{a_0}{2} + \sum_{n=1}^{\infty} \left(a_n \cos\left(\frac{n\pi x}{L}\right) + b_n \sin\left(\frac{n\pi x}{L}\right)\right), \tag{9.4.9}$$

(we again use x as the independent variable) with

$$\begin{aligned} a_n &= \frac{1}{L}\int_{-L}^{L} f(x)\cos\left(\frac{n\pi x}{L}\right)dx, \\ b_n &= \frac{1}{L}\int_{-L}^{L} f(x)\sin\left(\frac{n\pi x}{L}\right)dx. \end{aligned} \tag{9.4.10}$$

Instead of employing a change of scale we could have started with (9.4.9) and derived the Euler formulas (9.4.10) with the aid of the orthogonality relations, such as

$$\int_{-L}^{L} \cos\left(\frac{m\pi x}{L}\right)\cos\left(\frac{n\pi x}{L}\right)dx = \begin{cases} L & \text{if } m = n, \\ 0 & \text{if } m \neq n. \end{cases}$$

EXAMPLE 5 Let $f(x)$ be defined on the interval $-3 < x < 3$ by

$$f(x) = \begin{cases} 0 & \text{if } -3 < x < -1, \\ \cosh x & \text{if } -1 \le x \le 1, \\ 0 & \text{if } 1 < x < 3. \end{cases}$$

Compute the Fourier coefficients of $f(x)$ and its Fourier approximates for the fundamental interval $-3 < x < 3$. Plot the function and its Fourier approximates for $N = 1, 3$. Compute $S_N(1)$ for $N = 1, 5, 10$.

SOLUTION Since $f(x)$ is an even function the Fourier coefficients b_n are all zero. The a_n are given by

$$\begin{aligned} a_n &= \frac{1}{3}\int_{-3}^{3} f(x)\cos\left(\frac{n\pi x}{3}\right)dx = \frac{1}{3}\int_{-1}^{1} \cosh(x)\cos\left(\frac{n\pi x}{3}\right)dx \\ &= \frac{6}{(n\pi)^2 + 9}\left[\sinh(1)\cos\left(\frac{n\pi}{3}\right) + \left(\frac{n\pi}{3}\right)\sin\left(\frac{n\pi}{3}\right)\right], \\ S_N(x) &= \frac{1}{3}\sinh(1) + \sum_{n=1}^{N} a_n\cos\left(\frac{n\pi x}{3}\right), \end{aligned}$$

and $S_1(1) \cong 0.71$, $S_5(1) \cong 0.68$, $S_{10}(1) \cong 0.76$. Note that the computed values of $S_N(1)$ are between 0 and $\cosh(1) \cong 1.54$. The graphs are in Fig. 9.6. ■

The function $f(x)$ in Example 5 is a member of the large and important class of functions in applied mathematics that have simple jump discontinuities but are otherwise smooth. They are said to be *piecewise smooth.*

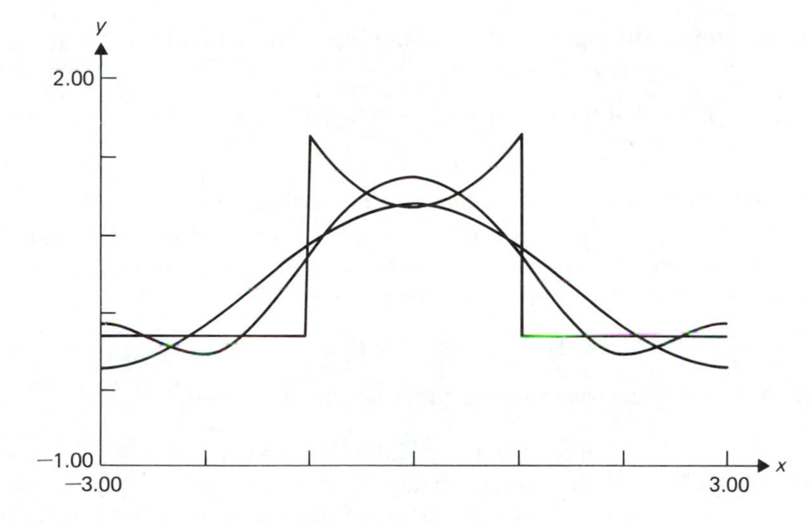

Figure 9.6 Graphs of a function and its Fourier approximates, $N = 1, 3$, for Example 5.

Definition

A function $f(x)$ is *piecewise smooth* on the interval $-L < x < L$ if it is piecewise continuous (that is, continuous on the interval except for a finite number of jump discontinuities) and if it has a first derivative that is also piecewise continuous.

For instance any continuously differentiable function is piecewise smooth, and the continuous function

$$f(x) = |x|, \qquad -\pi < x < \pi,$$

is piecewise smooth because it coincides with the functions

$$f_1(x) = -x, \qquad -\pi < x < 0,$$

$$f_2(x) = x, \qquad 0 < x < \pi,$$

each of which is continuously differentiable. The discontinuous function

$$f(x) = \begin{cases} -1 & \text{if } -\pi < x < 0, \\ 1 & \text{if } 0 \le x < \pi, \end{cases} \tag{9.4.11}$$

is also piecewise smooth since it coincides with the functions

$$f_1(x) = -1, \qquad -\pi < x < 0,$$

$$f_2(x) = 1, \qquad 0 < x < \pi,$$

each of which is continuously differentiable. The class of piecewise smooth functions should suffice for most of the applications the reader will encounter.

Recall the definition of the right- and left-hand limits for a function $f(x)$ at a point, $x = c$:

$$f(c^+) = \lim_{\substack{x \to c \\ x > c}} f(x), \qquad f(c^-) = \lim_{\substack{x \to c \\ x < c}} f(x).$$

Whenever $f(x)$ is continuous at $x = c$ these limits will be equal, whereas at a point of discontinuity of a piecewise smooth function both the limits will exist but not be equal. This is because such a discontinuity will be a simple jump discontinuity. For instance, for the step function (9.4.11) we have at $x = 0$,

$$f(0^+) = 1, \qquad f(0^-) = -1.$$

Finally, note that for a piecewise smooth function the expression

$$\frac{1}{2}[f(x^+) + f(x^-)] = \begin{cases} f(x) & \text{if } x \text{ is a point of continuity,} \\ \text{the average value of the jump at } x & \\ \qquad \text{if } x \text{ is a point of discontinuity.} & \end{cases}$$

For instance, for the function described in Example 5,

$$\frac{1}{2}[f(x^+) + f(x^-)] = \begin{cases} 0 & \text{if } -3 < x < -1, \\ \cosh(-1)/2 & \text{if } x = -1, \\ \cosh(x) & \text{if } -1 < x < 1, \\ \cosh(1)/2 & \text{if } x = 1, \\ 0 & \text{if } 1 < x < 3. \end{cases}$$

We are now ready to state the convergence result for the class of piecewise smooth functions. Its proof can be found in [11, 12].

Theorem

If $f(x)$ is piecewise smooth on $-L < x < L$, then its Fourier approximates converge to $f(x)$ at every point of continuity of $f(x)$, and they converge to the average value $\frac{1}{2}[f(x^+) + f(x^-)]$ at each point of discontinuity. At $x = -L$ and $x = L$, the Fourier approximates converge to $\frac{1}{2}[f(L^-) + f(L^+)]$.

The above result actually holds for a more general class of functions, but as stated it is adequate for most applications.

EXAMPLE 6 Construct the Fourier series of the step function,

$$f(x) = \begin{cases} -1 & \text{if } -\pi < x \leq 0, \\ 1 & \text{if } 0 < x < \pi, \end{cases}$$

and discuss its convergence.

SOLUTION The Fourier coefficients of $f(x)$ are

$$\begin{aligned} a_n &= \frac{1}{\pi}\int_{-\pi}^{\pi} f(x)\cos(nx)\,dx \\ &= \frac{1}{\pi}\int_{-\pi}^{0} -\cos(nx)\,dx + \frac{1}{\pi}\int_{0}^{\pi}\cos(nx)\,dx = 0, \\ b_n &= \frac{1}{\pi}\int_{-\pi}^{\pi} f(x)\sin(nx)\,dx \\ &= \frac{1}{\pi}\int_{-\pi}^{0} -\sin(nx)\,dx + \frac{1}{\pi}\int_{0}^{\pi}\sin(nx)\,dx = \frac{2}{\pi}\frac{[1-(-1)^n]}{n}. \end{aligned}$$

Hence $b_n = 0$ whenever n is even and the Fourier series of $f(x)$ (note the subscript shuffle) is

$$f(x) \sim \frac{4}{\pi}\sum_{n=1}^{\infty}\frac{1}{2n-1}\sin(2n-1)x.$$

By the convergence theorem we conclude that the series converges to $f(x)$ on $-\pi < x < 0$ and on $0 < x < \pi$. It converges to 0, the average value, at $x = 0$ and at the endpoints of the fundamental interval $x = \pm\pi$. ■

The convergence theorem and Example 6 should be sufficient to clarify the association of a piecewise smooth function and its Fourier series. In the sequel, we shall usually employ the symbol = rather than ~ when a Fourier series is displayed.

The Fourier approximates of a piecewise smooth function are finite linear combinations of sines and cosines and therefore are continuously differentiable. According to the convergence theorem, these very smooth functions converge at the points of continuity near a jump discontinuity but abruptly converge to the mean value of the function at the actual point of discontinuity. It should not be surprising that an unusual phenomenon occurs near a point of jump discontinuity. A perfect example of this can be seen in the convergence of the Fourier series of the step function of Example 6. As before, we denote the partial sums of its Fourier series (its Fourier approximates) by $S_N(x)$. Then

$$S_1(x) = \frac{4}{\pi}\sin x, \qquad S_3(x) = \frac{4}{\pi}\left[\sin x + \frac{1}{3}\sin 3x\right],$$

and so on. Plotting these and the step function on the closed interval $-\pi \le x \le \pi$, we obtain the graphs shown in Fig. 9.7. What we observe near the point of discontinuity $x = 0$ and the endpoints $x = \pm\pi$ are oscillations that overshoot the values ± 1 and then converge to those values as N increases. The points at which

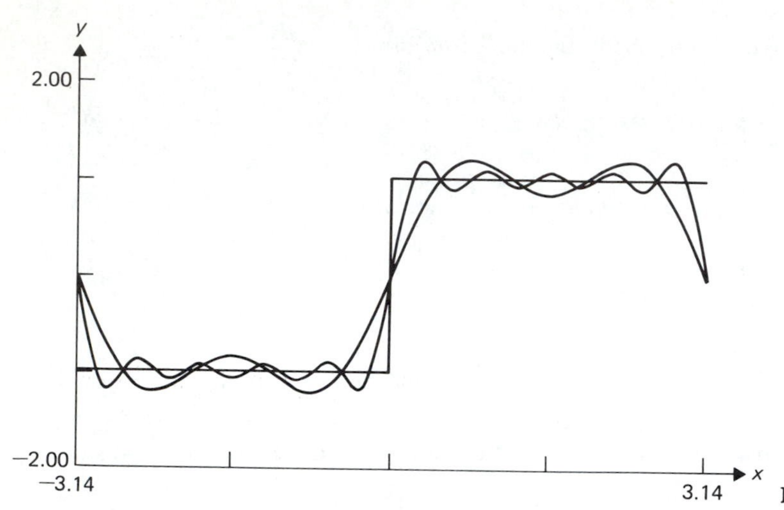

Figure 9.7 Gibbs' phenomenon.

the maximum amplitude of the oscillations occurs move closer to the points of discontinuity as N increases.

It can be shown that the maximum amplitude of $|S_N(x)|$ approaches 1.089490^+ as N approaches infinity. This is an example of Gibbs's phenomenon, named after the eminent American physicist Willard Gibbs (1844–1906). It appears whenever discontinuous functions are approximated by the partial sums of their Fourier series.

EXERCISES 9.4

In Exercises 1–15 construct the Fourier series of the given functions and sketch the functions and their Fourier approximates for $N = 1, 3$. Discuss the convergence of the Fourier series.

1. $f(x) = x^2 - \pi^2, -\pi < x < \pi$

2. $f(x) = x \cos x + x \sin x, -\pi < x < \pi$

3. $f(x) = x + |x|, -\pi < x < \pi$

4. $f(x) = |x|(2x + 1), -\pi < x < \pi$

5. $f(x) = \begin{cases} 2 & \text{if } -\pi < x < 0 \\ 0 & \text{if } 0 \leq x < \pi \end{cases}$

6. $f(x) = \begin{cases} 0 & \text{if } -\pi < x < -\epsilon \\ \dfrac{2}{\epsilon} & \text{if } -\epsilon \leq x \leq \epsilon \\ 0 & \text{if } \epsilon < x < \pi \quad \text{where } \epsilon > 0 \end{cases}$

Take $\epsilon = 0.1$ for the sketches. What happens as $\epsilon \to 0$?

7. $f(x) = \sin\left(\dfrac{x}{2}\right), \ -\pi < x < \pi$

8. $f(x) = \begin{cases} -\dfrac{2}{\pi}(\pi + x) & \text{if } -\pi < x < -\dfrac{\pi}{2} \\ \dfrac{2x}{\pi} & \text{if } -\dfrac{\pi}{2} \leq x \leq \dfrac{\pi}{2} \\ \dfrac{2}{\pi}(\pi - x) & \text{if } \dfrac{\pi}{2} < x < \pi \end{cases}$

(The "triangular" sine function.)

9. $f(x) = e^{x/2}, \ -1 < x < 1$

10. $f(x) = \sinh 3x, \ -2 < x < 2$

11. $f(x) = \begin{cases} \sin\left(\dfrac{\pi x}{2}\right) & \text{if } -4 < x < 2 \\ 0 & \text{if } 2 \leq x < 4 \end{cases}$

12. $f(x) = \begin{cases} \sin(2\pi x) & \text{if } -2 < x < 0 \\ \cos(2\pi x) & \text{if } 0 \leq x < 2 \end{cases}$

13. $f(x) = \begin{cases} 0 & \text{if } -3 < x < 1 \\ \cosh x & \text{if } 1 \leq x < 3 \end{cases}$

14. $f(x) = \begin{cases} e^{x} & \text{if } -3 < x < 0 \\ e^{-x} & \text{if } 0 \leq x < 3 \end{cases}$

15. $f(x) = \begin{cases} \left|\cos\left(\dfrac{\pi x}{3}\right)\right| & \text{if } -6 < x < 0 \\ \left|\sin\left(\dfrac{\pi x}{3}\right)\right| & \text{if } 0 \leq x < 6 \end{cases}$

Set

$$R_N(x) = \frac{1}{2} + \cos x + \cos 2x + \cdots + \cos nx$$
$$= Re\left\{-\frac{1}{2} + \sum_{n=0}^{N} e^{inx}\right\}.$$

Assume the result of Exercise 16 for Exercise 17, etc.

16. Show that

$$R_N(x) = \begin{cases} Re\left\{-\dfrac{1}{2} + \dfrac{1 - e^{i(N+1)x}}{1 - e^{ix}}\right\} & \text{if } e^{ix} \neq 1, \\ \dfrac{1}{2} + N & \text{if } e^{ix} = 1. \end{cases}$$

17. Show that

$$R_N(x) = \frac{\sin[(N + \frac{1}{2})\, x]}{2 \sin\left(\dfrac{x}{2}\right)}.$$

18. Show that

$$S_N(x) = \int_{-\pi}^{\pi} f(s)\{\frac{1}{2} + \cos(s - x) + \cdots + \cos[N(s - x)]\} \, ds.$$

(*Hint:* Use $\cos[n(x - s)] = \cos nx \cos ns + \sin nx \sin ns$ and the Euler formulas for the Fourier coefficients of $f(x)$.)

19. Show that

$$S_N(x) = \frac{1}{\pi}\int_{-\pi}^{\pi} f(s) \frac{\sin[(N + \frac{1}{2})(s - x)]}{2 \sin\left[\dfrac{(s - x)}{2}\right]} \, ds$$

$$= \frac{1}{\pi}\int_{-\pi}^{\pi} f(x + s) \frac{\sin[(N + \frac{1}{2})\, s]}{2 \sin\left(\dfrac{s}{2}\right)} \, ds.$$

What is the value of the integrand of the second integral as $s \to 0$?

20. Write a computer program to evaluate $S_N(x)$ using Simpson's Rule to evaluate the *first* integral of Exercise 19. Take $h = \pi/(30N)$.

In Exercises 21–23 use the computer program of Exercise 20 to evaluate $S_N(x)$, $N = 1, 2, 3$, for $x = 0, \pm\pi/4, \pm\pi/2, \pm\pi$, and compare your results with those obtained from the trigonometric polynomials.

21. Use the given $f(x)$ and the answer to Exercise 1.

22. Use the given $f(x)$ and the answer to Exercise 3.

23. Use the given $f(x)$ and the answer to Exercise 5.

9.5 ODD, EVEN, AND PERIODIC EXTENSIONS, FOURIER SINE AND COSINE SERIES

Our study of Fourier series has been strongly influenced by our motivating example, the vibrations of a taut string of length L with fixed endpoints. In Section 9.3, the displacement and velocity of the string at $t = 0$ are represented by trigonometric polynomials (see 9.3.14),

$$f(x) = \sum_{n=1}^{N} A_n \sin\left(\frac{n\pi x}{L}\right),$$

$$g(x) = \sum_{n=1}^{N} B_n \sin\left(\frac{n\pi x}{L}\right), \qquad 0 < x < L.$$

These trigonometric polynomials are meaningful for all values of x and can be used to extend the domain of definition of $f(x)$ and $g(x)$. The extension of domains of definition of functions is not needed for the vibrating string problem, but it will help us develop tools that are employed in solving a great variety of problems in both ordinary and partial differential equations. In this section we shall define odd, even, and periodic extensions of functions and construct their Fourier series. Only piecewise smooth functions will be considered.

Definition

A function $h(x)$ is said to be an *extension* of a function $f(x)$ if the domain of definition of $h(x)$ includes that of $f(x)$ and for all x in their common domain of definition $h(x) = f(x)$. In particular, if $f(x)$ is defined on $0 < x < L$ and $h(x)$ is defined on $-L < x < L$, then $h(x)$ is an extension of $f(x)$ if for all x in $0 < x < L$, $h(x) = f(x)$. If $h(x)$ is an odd function, then $h(x)$ is said to be an *odd extension* of $f(x)$. If $h(x)$ is an even function, then $h(x)$ is said to be an *even extension* of $f(x)$.

Let us now consider odd and even extensions of functions defined on $0 < x < L$.

Definition

Given the function $f(x)$ defined on the interval $0 < x < L$, its *odd extension* $f_o(x)$ is defined on the interval $-L < x < L$ by

$$f_o(x) = \begin{cases} f(x) & \text{if } 0 < x < L, \\ 0 & \text{if } x = 0, \\ -f(-x) & \text{if } -L < x < 0. \end{cases}$$

Its *even extension* $f_e(x)$ is defined on the interval $-L < x < L$ by

$$f_e(x) = \begin{cases} f(x) & \text{if } 0 < x < L, \\ f(0^+) & \text{if } x = 0, \\ f(-x) & \text{if } -L < x < 0. \end{cases}$$

These formulas may look a little forbidding, but they merely say that the odd extensions may be constructed by reflecting the graph of $f(x)$ through the origin. The even extensions may be constructed by reflecting the graph of $f(x)$ through the vertical axis. Furthermore, the odd and even extensions of piecewise smooth functions are piecewise smooth.

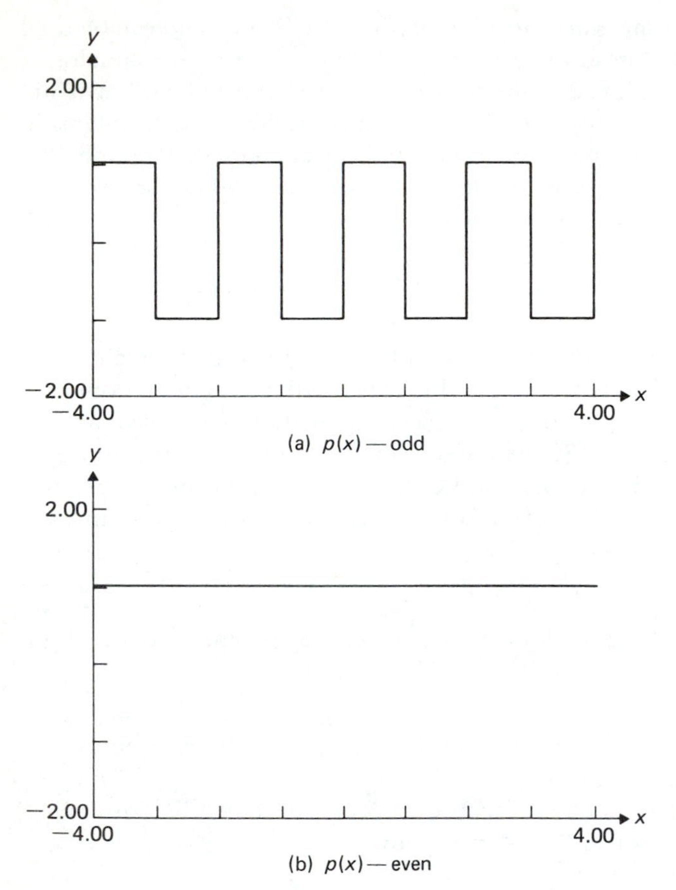

Figure 9.8 The odd and even extensions of $p(x) = 1$, $q(x) = x$ for Example 1.

EXAMPLE 1 Find the even and odd extensions of the functions $p(x) = 1$, $q(x) = x$, $0 < x < 1$. Sketch separate graphs of the functions and their odd and even extensions.

SOLUTION The extensions of $p(x)$ are

$$p_o(x) = \begin{cases} 1 & \text{if } 0 < x < 1, \\ 0 & \text{if } x = 0, \\ -1 & \text{if } -1 < x < 1; \end{cases}$$

$$p_e(x) = \begin{cases} 1 & \text{if } 0 < x < 1, \\ 1 & \text{if } x = 0, \\ 1 & \text{if } -1 < x < 0; \end{cases}$$

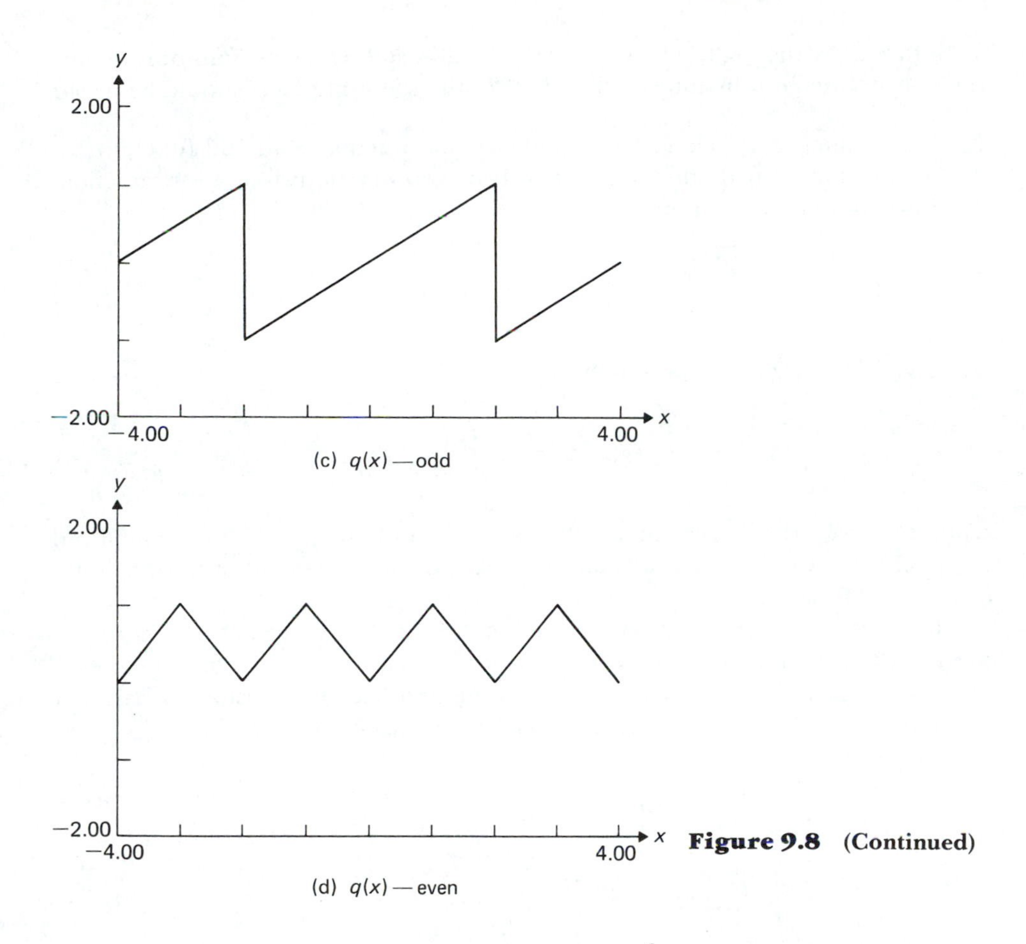

(c) $q(x)$—odd

(d) $q(x)$—even

Figure 9.8 (Continued)

$$q_o(x) = \begin{cases} x & \text{if } 0 < x < 1, \\ 0 & \text{if } x = 0, \\ x & \text{if } -1 < x < 0; \end{cases}$$

$$q_e(x) = \begin{cases} x & \text{if } 0 < x < 1, \\ 0 & \text{if } x = 0, \\ -x & \text{if } -1 < x < 0. \end{cases}$$

Note that we could just as well define

$$p_e(x) = 1, \qquad -1 < x < 1,$$

$$q_o(x) = x, \qquad -1 < x < 1,$$

$$q_e(x) = |x|, \qquad -1 < x < 1.$$

The graphs are in Fig. 9.8. ■

Before constructing the Fourier series of odd and even extensions of functions originally defined on the interval $0 < x < L$, the following facts should be noted:

1. The product of an odd function and an even function is an odd function.
2. The product of two odd functions or two even functions is an even function.
3. If $g(x)$ is an odd function, then

$$\int_{-L}^{L} g(x)\, dx = 0.$$

4. If $g(x)$ is an even function, then

$$\int_{-L}^{L} g(x)\, dx = 2\int_{0}^{L} g(x)\, dx.$$

In addition, we strongly recommend that a good set of integral tables be employed. Our goal here is to learn about Fourier series, not to become idiot savants in the techniques of integration.

If we are given a function $f(x)$ defined on $0 < x < L$, we can construct its odd extension $f_o(x)$ defined on $-L < x < L$ and then construct the Fourier series of $f_o(x)$. Since $f_o(x)\sin(n\pi x/L)$ is an even function, and $f_o(x)$, $f_o(x)\cos(n\pi x/L)$ are odd functions, the Euler formulas for the Fourier coefficients of $f_o(x)$ are

$$\begin{aligned} a_0 &= \frac{1}{L}\int_{-L}^{L} f_o(x)\, dx = 0, \\ a_n &= \frac{1}{L}\int_{-L}^{L} f_o(x)\cos\left(\frac{n\pi x}{L}\right) dx = 0, \\ b_n &= \frac{1}{L}\int_{-L}^{L} f_o(x)\sin\left(\frac{n\pi x}{L}\right) dx = \frac{2}{L}\int_{0}^{L} f(x)\sin\left(\frac{n\pi x}{L}\right) dx. \end{aligned} \tag{9.5.1}$$

In the last integral, $f(x)$ appears because $f(x) = f_o(x)$ on $0 < x < L$.

The Fourier series of $f_o(x)$ is

$$f_o(x) = \sum_{n=1}^{\infty} b_n \sin\left(\frac{n\pi x}{L}\right).$$

It represents an odd function on the interval $-L < x < L$. As shown in (9.5.1), the b_n are computed using the values of the original function $f(x)$. Therefore the Fourier series of the odd extension of $f(x)$ is defined to be the *Fourier sine series of* $f(x)$.

Definition

If $f(x)$ is a piecewise smooth function defined on $0 < x < L$, its *Fourier sine series* is

$$f(x) = \sum_{n=1}^{\infty} b_n \sin\left(\frac{n\pi x}{L}\right), \tag{9.5.2}$$

with

$$b_n = \frac{2}{L}\int_0^L f(x) \sin\left(\frac{n\pi x}{L}\right) dx. \tag{9.5.3}$$

By the convergence theorem of Section 9.4, the Fourier series of a piecewise smooth function $h(x)$ defined on the interval $-L < x < L$ converges to $[h(x^+) + h(x^-)]/2$ on $-L < x < L$ and to $[h(L^-) + h(-L^+)]/2$ at the endpoints, $x = L$ and $x = -L$. Therefore the series (9.5.2) converges to

$$\frac{1}{2}[f_o(x^+) + f_o(x^-)] = \begin{cases} \frac{1}{2}[f(x^+) + f(x^-)] & \text{on } 0 < x < L, \\ 0 & \text{at } x = 0, \\ -\frac{1}{2}[f(-x^-) + f(-x^+)] & \text{on } -L < x < 0. \end{cases} \tag{9.5.4}$$

To justify (9.5.4) we note that if $x = c$, $-L < c < 0$, then

$$f_o(c^+) = \lim_{\substack{x \to c \\ x > c}} f_o(x) = \lim_{\substack{x \to -c \\ x < -c}} -f(-x) = -f(-c^-),$$

$$f_o(c^-) = \lim_{\substack{x \to c \\ x < c}} f_o(x) = \lim_{\substack{x \to -c \\ x > -c}} -f(-x) = -f(-c^+).$$

At $x = 0$ the series converges to

$$\frac{1}{2}[f_o(0^+) + f_o(0^-)] = \frac{1}{2}[f(0^+) - f(0^+)] = 0.$$

At the endpoints $x = L$ and $x = -L$, the sine series converges to 0. This is consistent with the theoretical convergence result because

$$\frac{1}{2}[f_o(L^-) + f_o(-L^+)] = \frac{1}{2}[f(L^-) - f(L^-)] = 0.$$

EXAMPLE 2 Sketch the odd extension of $f(x) = \pi \cos x$, $0 < x < \pi$. Construct the Fourier sine series of $f(x)$ and discuss its convergence on the closed interval $0 \le x \le \pi$.

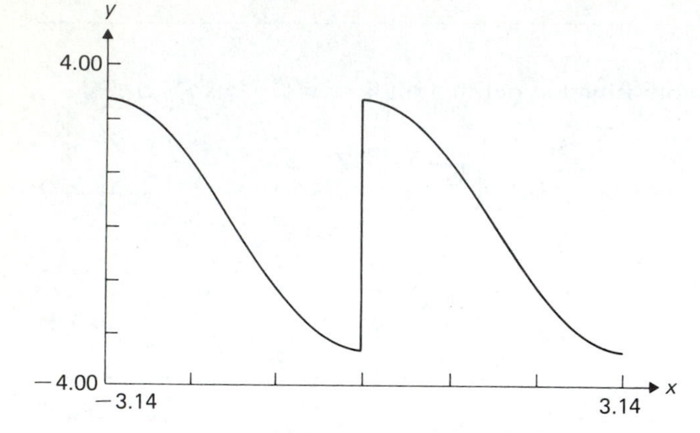

Figure 9.9 The odd extension of $f(x) = \pi \cos x$ for Example 2.

SOLUTION A sketch of $f(x)$ is in Fig. 9.9. By (9.5.2), the Fourier coefficients of $f_o(x)$ are $a_n = 0$ and

$$b_n = 2\int_0^\pi \cos(x)\sin(nx)\,dx = \int_0^\pi [\sin[(n+1)x] + \sin[(n-1)x]]\,dx.$$

Hence $b_1 = 0$, and for $n > 1$,

$$b_n = [1 - (-1)^{n+1}]\left[\frac{2n}{n^2 - 1}\right].$$

The Fourier sine series is

$$f(x) = 2\sum_{n=2}^{\infty} [1 - (-1)^{n+1}]\left[\frac{n}{n^2 - 1}\right] \sin nx.$$

It converges to $f(x) = \pi \cos x$ for x in the open interval $0 < x < \pi$ and to 0 at the endpoints, $x = 0$ and $x = \pi$. ■

The construction of the Fourier series of an even extension of a piecewise smooth function parallels that of an odd extension.

Definition

The *Fourier cosine series* of a piecewise smooth function $f(x)$ defined on $0 < x < L$ is

$$f(x) = \frac{a_0}{2} + \sum_{n=1}^{\infty} a_n \cos\left(\frac{n\pi x}{L}\right), \qquad \textbf{(9.5.5)}$$

with

$$a_n = \frac{2}{L}\int_0^L f(x)\cos\left(\frac{n\pi x}{L}\right)dx. \tag{9.5.6}$$

The Fourier cosine series (9.5.5) represents an even function on the interval $-L < x < L$.

EXAMPLE 3 Sketch the even extension of $f(x) = \pi \sin x$, $0 < x < \pi$. Construct the Fourier cosine series of $f(x)$ and discuss its convergence on the closed interval $0 \leq x \leq \pi$.

SOLUTION A sketch of the even extension of $f(x)$ is in Fig. 9.10. The Fourier coefficients of the cosine series are

$$a_n = \frac{2}{\pi}\int_0^\pi \pi \sin(x)\cos(nx)\,dx.$$

Thus $a_0 = 4$, $a_1 = 0$, and for $n > 1$,

$$a_n = \int_0^\pi [\sin[(n+1)x] + \sin[(n-1)x]]\,dx = \frac{2}{n^2-1}[1 + (-1)^n].$$

The Fourier cosine series of $f(x)$ is

$$f(x) = 2 + 2\sum_{n=2}^{\infty}\left(\frac{1+(-1)^n}{n^2-1}\right)\cos nx.$$

It converges to $f(x) = \pi \sin x$ for x in the closed interval $0 \leq x \leq \pi$. ∎

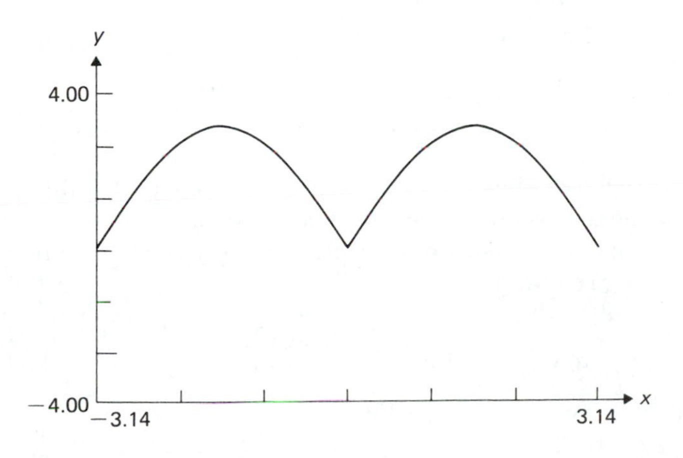

Figure 9.10 The even extension of $f(x) = \pi \sin x$ for Example 3.

Our next topic is the construction of periodic functions from functions defined on a finite interval.

Definition

A function $f(x)$ is a *periodic function* if there exists a positive number T such that for all x, $-\infty < x < \infty$,

$$f(x + T) = f(x). \tag{9.5.7}$$

The *period* of a nonconstant function is the smallest number T such that (9.5.7) is valid.

The trigonometric functions form a familiar class of periodic functions; for example, the trigonometric polynomials,

$$Q_N(x) = A_0 + \sum_{n=1}^{N} A_n \sin nx + B_n \cos nx, \tag{9.5.8}$$

are periodic with period 2π if either A_1 or B_1 is nonzero. Otherwise, $Q_N(x)$ has period $2\pi/k$, where k is the smallest integer such that A_k or B_k is nonzero.

If we start with a piecewise smooth function $f(x)$ that is defined on an open interval, $-L < x < L$, we can construct $f_p(x)$, a periodic extension of $f(x)$, by first requiring that $f_p(L) = f_p(-L)$ and then employing (9.5.7).

Definition

The *periodic extension* $f_p(x)$ of a piecewise smooth function $f(x)$ defined on the open interval $-L < x < L$ is constructed by setting

$$\begin{aligned} f_p(L) &= f_p(-L) = \tfrac{1}{2}[f(L^-) + f(-L^+)], \\ f_p(x) &= f(x), \qquad -L < x < L, \\ f_p(x + 2mL) &= f_p(x), \qquad -L \le x \le L, \quad m = 0, \pm 1, \pm 2, \ldots . \end{aligned} \tag{9.5.9}$$

Extra care is required at the endpoints, $x = \pm L$, since for many partial differential equations problems the data is only known on open intervals.

The Fourier series of the periodic extension of a piecewise smooth function $f(x)$ defined on $-L < x < L$ is given by

$$f_p(x) = \frac{a}{2} + \sum_{n=1}^{\infty} \left[a_n \cos\left(\frac{n\pi x}{L}\right) + b_n \sin\left(\frac{n\pi x}{L}\right) \right],$$

with

$$a_n = \frac{1}{L}\int_{-L}^{L} f(x)\cos\left(\frac{n\pi x}{L}\right) dx,$$

$$b_n = \frac{1}{L}\int_{-L}^{L} f(x)\sin\left(\frac{n\pi x}{L}\right) dx.$$

It converges to $[f_p(x^+) + f_p(x^-)]/2$ for all x, $-\infty < x < \infty$. This follows from the convergence theorem of Section 9.4 and the definition of periodic extension.

EXAMPLE 4 Plot on $-4\pi < x < 4\pi$ the periodic extensions of the following functions that are defined on $-\pi < x < \pi$ and construct their Fourier series,

$$f_1(x) = x,$$

$$f_2(x) = |x|,$$

$$f_3(x) = \begin{cases} 1 & \text{if } 0 < x < \pi, \\ 0 & \text{if } x = 0, \\ -1 & \text{if } -\pi < x < 0, \end{cases}$$

$$f_4(x) = \begin{cases} -\dfrac{1}{2r} & \text{if } \pi - r < x < \pi, \\ 0 & \text{if } r \le x \le \pi - r, \\ \dfrac{1}{2r} & \text{if } -r < x < r, \\ 0 & \text{if } -\pi + r \le x \le -r, \\ -\dfrac{1}{2r} & \text{if } -\pi < x < -\pi + r. \end{cases}$$

SOLUTION The plots are in Fig. 9.11. The Fourier series are

$$f_1(x) = 2\sum_{n=1}^{\infty} (-1)^{n+1}\frac{\sin nx}{n}, \qquad \text{(a saw-tooth wave)}$$

$$f_2(x) = \frac{\pi}{2} - \frac{2}{\pi}\sum_{n=1}^{\infty} [1 - (-1)^n]\frac{\cos nx}{n^2}, \qquad \text{(a triangular wave)}$$

$$f_3(x) = \frac{2}{\pi}\sum_{n=1}^{\infty} [1 - (-1)^n]\frac{\sin nx}{n}, \qquad \text{(a square wave)}$$

$$f_4(x) = \frac{1}{\pi}\sum_{n=1}^{\infty} \frac{\sin nr}{nr}[1 + (-1)^n]\cos nx.$$

■

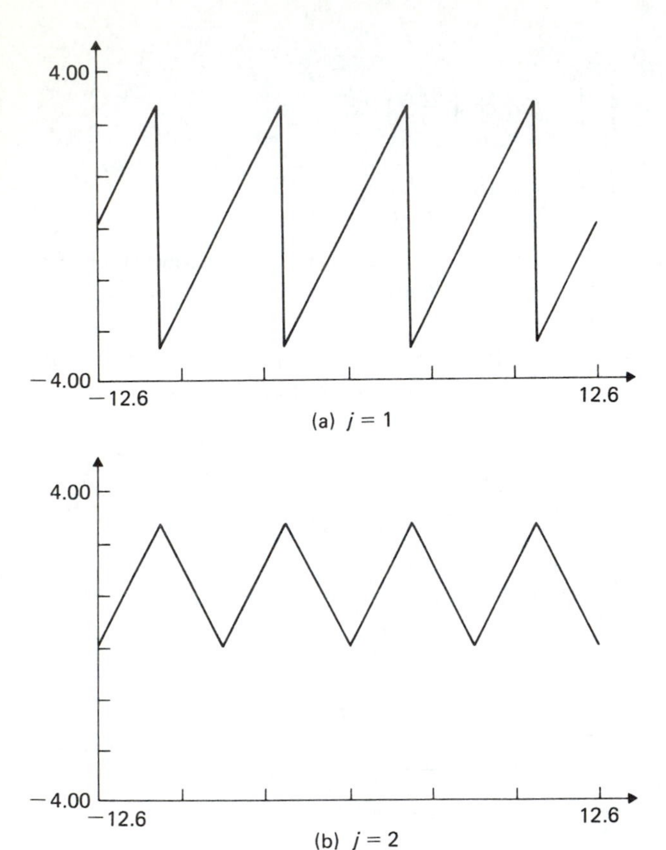

Figure 9.11 **Periodic extensions of the functions of Example 4.**

The periodic functions of Example 4 are important in communication theory. The periodic extension of $f_4(x)$ is a periodic distribution of impulses that, if r is very small, is frequently replaced by a periodic distribution of the Dirac delta functions (see Section 4.9). The mathematical framework that justifies the association of Fourier series with Dirac delta functions is called the theory of generalized functions or distribution theory. It is needed for the understanding of the abstract theory of partial differential equations but not for the problems that we shall be studying.

Our final topic in this section is wave forms that are produced by the rectification of electric signals. (A *rectifier* is a device that conducts only when voltage is applied with one polarity but does not conduct when the polarity is reversed. AC adapters are a type of rectifier.) If $f(t)$ represents an electric signal, then $|f(t)|$ represents a *full-wave rectification* of $f(t)$ and $f^+(t)$ represents a *half-wave rectification,*

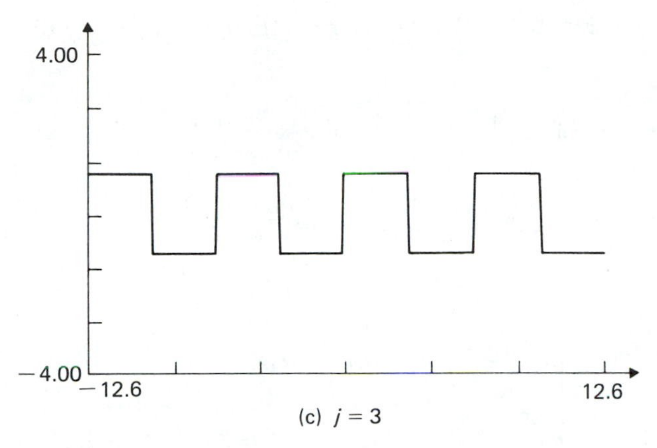

(c) $j = 3$

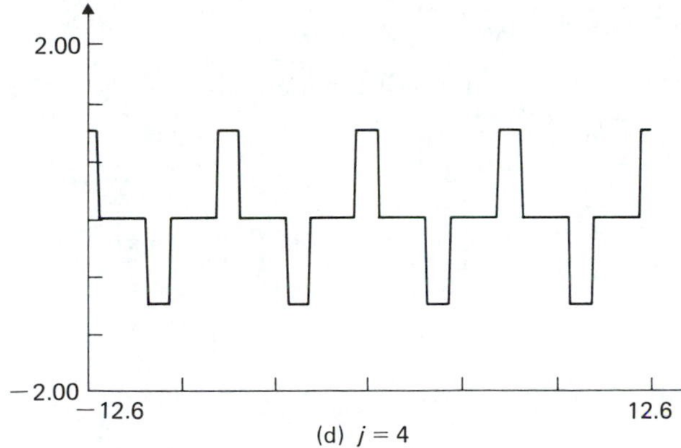

(d) $j = 4$

Figure 9.11 **(Continued)**

where

$$f^+(t) = \begin{cases} f(t) & \text{if } f(t) \geq 0, \\ 0 & \text{if } f(t) < 0. \end{cases}$$

Full and half-wave rectifications are illustrated in the next example.

EXAMPLE 5 Sketch on $-4 < t < 4$ the periodic extensions of $f(t)$, $|f(t)|$, and $f^+(t)$, where

$$f(t) = \begin{cases} t + 2 & \text{if } -2 < t < -1, \\ -t & \text{if } -1 \leq t < 2, \end{cases}$$

and construct their Fourier series.

SOLUTION The graphs are in Fig. 9.12. The Fourier coefficients of $f(t)$ are the values of the integrals,

$$a_n = \frac{1}{2}\int_{-2}^{-1} (t+2)\cos\left(\frac{n\pi t}{2}\right) dt - \frac{1}{2}\int_{-1}^{2} t\cos\left(\frac{n\pi t}{2}\right) dt,$$

$$b_n = \frac{1}{2}\int_{-2}^{-1} (t+2)\sin\left(\frac{n\pi t}{2}\right) dt - \frac{1}{2}\int_{-1}^{2} t\sin\left(\frac{n\pi t}{2}\right) dt,$$

and

$$f(t) = -\frac{1}{4} + \sum_{n=1}^{\infty} \frac{4}{n^2\pi^2}\left[\cos\left(\frac{n\pi}{2}\right) - \cos(n\pi)\right]\cos\left(\frac{n\pi t}{2}\right)$$
$$+ \left[-\frac{4}{n^2\pi^2}\sin\left(\frac{n\pi}{2}\right) + \frac{1}{n\pi}\cos\left(\frac{n\pi}{2}\right)\right]\sin\left(\frac{n\pi t}{2}\right).$$

The Fourier coefficients of $|f(t)|$ are the values of the integrals,

$$a_n = \frac{1}{2}\int_{-2}^{-1} (t+2)\cos\left(\frac{n\pi t}{2}\right) dt - \frac{1}{2}\int_{-1}^{0} t\cos\left(\frac{n\pi t}{2}\right) dt + \frac{1}{2}\int_{0}^{2} t\cos\left(\frac{n\pi t}{2}\right) dt,$$

$$b_n = \frac{1}{2}\int_{-2}^{-1} (t+2)\sin\left(\frac{n\pi t}{2}\right) dt - \frac{1}{2}\int_{-1}^{0} t\sin\left(\frac{n\pi t}{2}\right) dt + \frac{1}{2}\int_{0}^{2} t\sin\left(\frac{n\pi t}{2}\right) dt,$$

and

$$|f(t)| = \frac{3}{4} + \sum_{n=1}^{\infty} \frac{2}{n^2\pi^2}\left[2\cos\left(\frac{n\pi}{2}\right) - 1 - \cos(n\pi)\right]\cos\left(\frac{n\pi t}{2}\right)$$
$$+ \frac{1}{n\pi}\left[2\cos\left(\frac{n\pi}{2}\right) - 2 - 4\cos(n\pi)\right]\sin\left(\frac{n\pi t}{2}\right).$$

The Fourier coefficients of $f^+(t)$ are the values of the integrals,

$$a_n = \frac{1}{2}\int_{-2}^{-1} (t+2)\cos\left(\frac{n\pi t}{2}\right) dt - \frac{1}{2}\int_{-1}^{0} t\cos\left(\frac{n\pi t}{2}\right) dt,$$

$$b_n = \frac{1}{2}\int_{-2}^{-1} (t+2)\sin\left(\frac{n\pi t}{2}\right) dt - \frac{1}{2}\int_{-1}^{0} t\sin\left(\frac{n\pi t}{2}\right) dt,$$

and

$$f^+(t) = \frac{1}{2} + \sum_{n=1}^{\infty} \frac{2}{n^2\pi^2}\left[2\cos\left(\frac{n\pi}{2}\right) - 1 - \cos(n\pi)\right]\cos\left(\frac{n\pi t}{2}\right)$$
$$- \frac{4}{n\pi}\cos(n\pi)\sin\left(\frac{n\pi t}{2}\right).$$

■

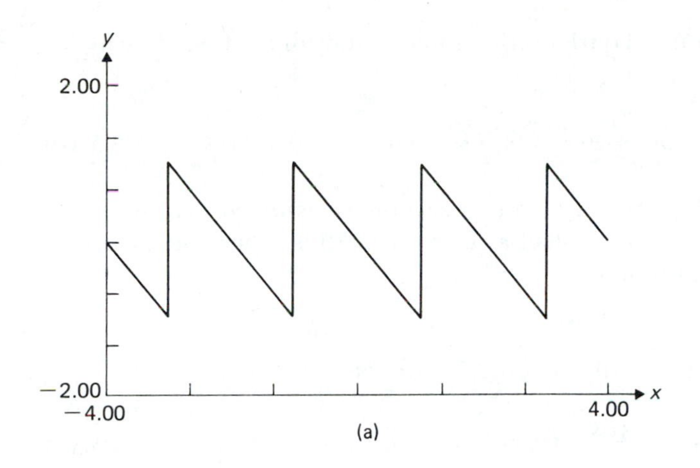

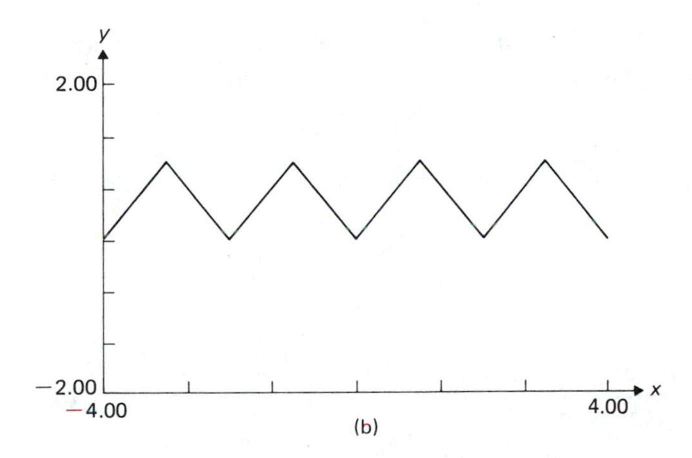

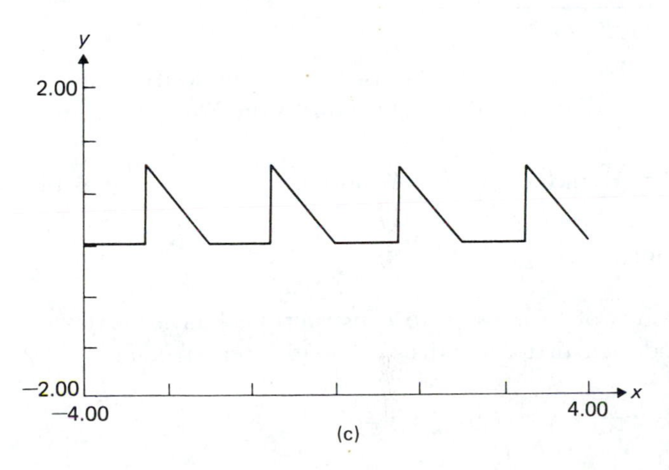

Figure 9.12 Periodic extensions of the function of Example 5.

A significant application of Fourier series is in the study of the forced oscillator equation,

$$m\frac{d^2y}{dt^2} + c\frac{dy}{dt} + ky = F(t), \tag{9.5.10}$$

which was introduced in Section 2.10. In that section we assumed that $F(t)$ was a trigonometric polynomial. Here we shall assume that $F(t)$ is a periodic piecewise smooth function such that for all t,

$$F(t + 2\pi) = F(t).$$

In accordance with the notation of Section 2.10, we set

$$F(t) = P_0 + \sum_{n=1}^{\infty} (P_n \cos nt + R_n \sin nt). \tag{9.5.11}$$

Following our earlier analysis, we can construct a periodic (or steady state) solution to the differential equation (9.5.12) that is represented by the Fourier series,

$$w(t) = A_0 + \sum_{n=1}^{\infty} (A_n \cos nt + B_n \sin nt), \tag{9.5.12}$$

where

$$\begin{aligned} A_0 &= \frac{P_0}{k}, \\ A_n &= \frac{(k - mn^2)P_n - cnR_n}{(k - mn^2)^2 + c^2n^2}, \\ B_n &= \frac{(k - mn^2)R_n + cnP_n}{(k - mn^2)^2 + c^2n^2}, \qquad n \geq 1. \end{aligned} \tag{9.5.13}$$

Although term-by-term differentiation of a Fourier series can destroy convergence, it can be shown that differentiating (9.5.12) produces the Fourier series,

$$w'(t) = \sum_{n=1}^{\infty} (nB_n \cos nt - nA_n \sin nt), \tag{9.5.14}$$

that converges to $w'(t)$ for all t.

EXAMPLE 6 Use the results of Example 4 to construct the Fourier series of the steady state solutions and their first derivatives to the differential equations,

$$\frac{d^2y}{dt^2} + 2\frac{dy}{dt} + y = F_j(t), \qquad j = 1, 3,$$

where $F_j(t)$ is the periodic extension of $f_j(t)$ that is defined in Example 4. Plot the forcing functions and the solution with $-2\pi < t < 2\pi$. Display the infinite series for $w(0)$ and $w'(0)$.

SOLUTION For $j = 1$, $F_1(t)$ is a saw-tooth wave and

$$P_n = 0, \qquad R_n = \frac{2(-1)^{n+1}}{n},$$

$$A_n = \frac{4(-1)^n}{[(1-n^2)^2 + 4n^2]},$$

$$B_n = \frac{(1-n^2)2(-1)^{n+1}}{[(1-n^2)^2 + 4n^2]n},$$

and so

$$w(t) = \sum_{n=1}^{\infty} \frac{2(-1)^n}{[(1-n^2)^2 + 4n^2]n}[2n \cos nt - (1-n^2)\sin nt],$$

$$w'(t) = \sum_{n=1}^{\infty} \frac{2(-1)^{n+1}}{[(1-n^2)^2 + 4n^2]}[(1-n^2)\cos nt + 2n \sin nt],$$

$$w(0) = \sum_{n=1}^{\infty} \frac{4(-1)^n}{[(1-n^2)^2 + 4n^2]},$$

$$w'(0) = \sum_{n=1}^{\infty} \frac{2(-1)^{n+1}(1-n^2)}{[(1-n^2)^2 + 4n^2]}.$$

For $j = 2$, $F_3(t)$ is a square wave and

$$P_n = 0, \qquad R_n = -\frac{2}{\pi}\frac{[1-(-1)^n]}{n^2},$$

$$A_n = \frac{1}{(1-n^2)^2 + 4n^2}\left\{\frac{4}{\pi}\frac{[1-(-1)^n]}{n}\right\},$$

$$B_n = \frac{1}{(1-n^2)^2 + 4n^2}\left\{\frac{-2(1-n^2)}{\pi n^2}[1-(-1)^n]\right\},$$

$$w(t) = \frac{2}{\pi}\sum_{n=1}^{\infty} \frac{1-(-1)^n}{(1-n^2)^2 + 4n^2}\left\{\frac{2\cos nt}{n} - \frac{(1-n^2)}{n^2}\sin nt\right\},$$

$$w'(t) = -\frac{2}{\pi}\sum_{n=1}^{\infty} \frac{1-(-1)^n}{(1-n^2)^2 + 4n^2}\left\{\frac{(1-n^2)}{n}\cos nt + 2\sin nt\right\},$$

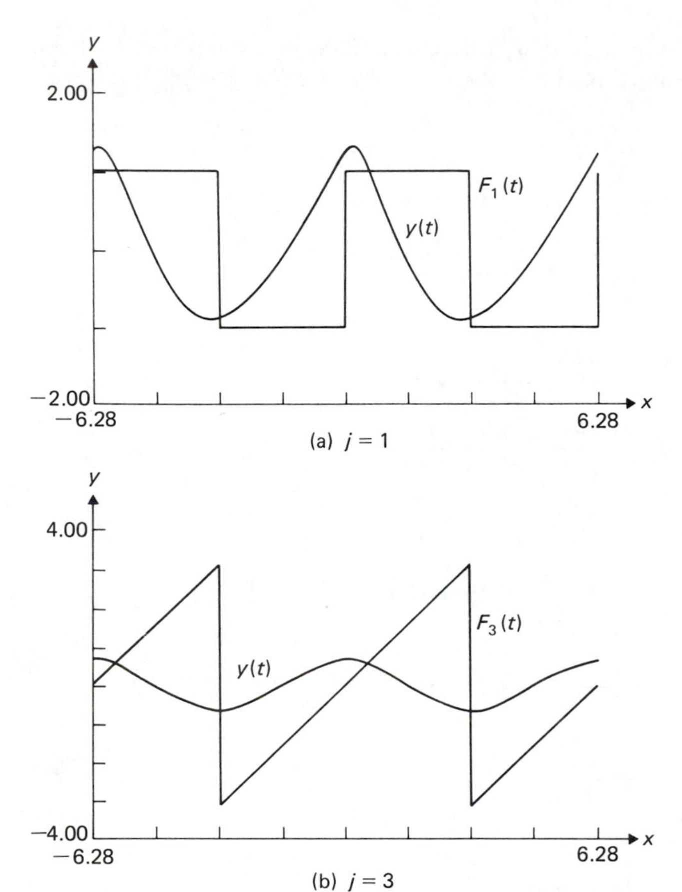

Figure 9.13 The forcing functions and the steady state solutions for Example 6.

$$w(0) = \frac{4}{\pi} \sum_{n=1}^{\infty} \frac{1 - (-1)^n}{(1 - n^2)^2 + 4n^2} \left(\frac{1}{n}\right),$$

$$w'(0) = -\frac{2}{\pi} \sum_{n=1}^{\infty} \frac{1 - (-1)^n}{(1 - n^2)^2 + 4n^2} \left(\frac{1 - n^2}{n}\right).$$

The graphs are in Fig. 9.13. ∎

In the next section, we shall employ cosine series of the form

$$g(x) \sim \sum_{n=1}^{\infty} c_n \cos\left(\frac{n\pi x}{2L}\right), \ n \text{ odd}, \tag{9.5.15}$$

where

$$c_n = \frac{2}{L}\int_0^L g(x)\cos\left(\frac{n\pi x}{2L}\right)dx. \qquad \textbf{(9.5.16)}$$

If the series converges for all x and if $c_1 \neq 0$, then $g(x)$ is an even periodic function with period $4L$. In contrast, a convergent Fourier cosine series (9.5.5) represents an even periodic function with period $2L$. Nevertheless, the series (9.5.15) is usually called a *Fourier cosine series.*

The formulas (9.5.16) that determine the coefficients c_n are not the Euler formulas (9.5.6) of a Fourier cosine series and require a separate justification. Fortunately this is easy because if n and m are positive integers, then

$$\int_0^L \cos\left(\frac{m\pi x}{2L}\right)\cos\left(\frac{n\pi x}{2L}\right)dx = \begin{cases} L/2 & \text{if } m = n, \\ 0 & \text{if } m \neq n. \end{cases}$$

With this information we can parallel the derivation of the Euler formulas to obtain (9.5.16).

EXAMPLE 7 Let $f(x) = \pi x/2,\ 0 < x < \pi$. Compare its Fourier cosine series (9.5.5) with its Fourier cosine series (9.5.15).

SOLUTION

$$a_0 = \int_0^\pi x\,dx = \frac{\pi^2}{2},$$

$$a_n = \int_0^\pi x\cos(nx)\,dx = \frac{1}{n^2}\,[cos(n\pi) - 1],$$

$$c_n = \int_0^\pi x\cos\left(\frac{nx}{2}\right)dx = \frac{4}{n^2}\left[\cos\left(\frac{n\pi}{2}\right) + \left(\frac{n\pi}{2}\right)\sin\left(\frac{n\pi}{2}\right) - 1\right].$$

Therefore,

$$\frac{\pi x}{2} = \frac{\pi^2}{4} + \sum_{n=1}^{\infty}\frac{1}{n^2}\,[\cos(n\pi) - 1]\cos(nx), \qquad 0 < x < \pi,$$

and

$$\frac{\pi x}{2} = 4\sum_{n=1}^{\infty}\frac{1}{n^2}\left[\cos\left(\frac{n\pi}{2}\right) + \left(\frac{n\pi}{2}\right)\sin\left(\frac{n\pi}{2}\right) - 1\right]\cos\left(\frac{nx}{2}\right),\ n \text{ odd}, \quad 0 < x < \pi.$$

■

The main topics of this section are the odd and even extensions of functions defined on an interval, $0 < x < L$, to the symmetric interval, $-L < x < L$; the construction of the corresponding Fourier sine and cosine series; and the periodic extension of functions defined on symmetric intervals. In the next sections, we shall employ Fourier series in the solution of boundary-initial value problems for partial differential equations.

EXERCISES 9.5

Find the Fourier sine and cosine series of the functions (defined on $0 < x < L$) that are given in Exercises 1–10. Sketch the graphs of the odd and even periodic extensions of the functions on $-2L \leq x \leq 2L$.

1. $f(x) = x^2, L = 2$

2. $f(x) = e^x - 1, L = 1$

3. $f(x) = \sin^2 x, L = \pi$

4. $f(x) = \cos^2 x, L = \pi$

5. $f(x) = \sin^3 (3\pi x), L = 1$

6. $f(x) = \cos^3 (3\pi x), L = 1$

7. $f(x) = |\sin^3(3\pi x)|, L = 1$

8. $f(x) = |\cos^3(3\pi x)|, L = 1$

9. $f(x) = [\sin^3(3\pi x)]^+, L = 1$

10. $f(x) = [\cos^3(3\pi x)]^+, L = 1$

In Exercises 11–16, find the steady state solution to

$$\frac{d^2y}{dt^2} + 4\frac{dy}{dt} + 3y = F(t),$$

where $F(t)$ is the given trigonometric polynomial or Fourier series.

11. $F(t) = 3 \sin t + \frac{1}{2} \cos 3t$

12. $F(t) = \sum_{n=1}^{\infty} \frac{1}{n^2} \sin nt$

13. $F(t) = 4 + \sum_{n=1}^{\infty} \frac{1}{n^2} \cos nt$

14. $F(t) =$ the Fourier sine series obtained in Exercise 8 (with $x \to t$).

15. $F(t) =$ the Fourier cosine series obtained in Exercise 7 (with $x \to t$).

16. $F(t) =$ the Fourier sine series obtained in Exercise 10 (with $x \to t$).

In Exercises 17–18, find the Fourier cosine series (9.5.5) and the Fourier cosine series (9.5.15) for the given function.

17. $f(x) = \left| x - \frac{\pi}{2} \right|, 0 < x < \pi$

18. $f(x) = [\sin 3x]^+, 0 < x < \pi$

9.6 THE WAVE EQUATION AND THE SEPARATION-OF-VARIABLES METHOD

The vibrations of a taut string fastened at its endpoints were described in Section 9.3 by the superposition of normal modes of vibration. In this section, we shall use the vibrating string problem to introduce the separation-of-variables method for constructing solutions to boundary-initial value problems of partial differential equations. As we do this, we shall rediscover the normal modes, this time by a mathematical method instead of by observation. The separation-of-variables method is not restricted to wave equation problems; in subsequent sections we shall employ it to solve problems in heat conduction and potential theory. Here, we shall first study unforced vibrations, then forced vibrations, and finally the representation of solutions of the wave equation as the sum of waves traveling to the right and to the left.

UNFORCED VIBRATIONS

The boundary-initial value problem that governs the vibrations of a taut string fastened at its endpoints is

$$\begin{aligned} &\frac{\partial^2 v}{\partial t^2} = c^2 \frac{\partial^2 v}{\partial x^2}, \qquad 0 < x < L, \quad 0 < t, \\ &v(0, t) = v(L, t) = 0, \qquad 0 < t, \\ &v(x, 0) = f(x), \qquad \frac{\partial v}{\partial t}(x, 0) = g(x), \qquad 0 < x < L. \end{aligned} \tag{9.6.1}$$

The separation-of-variables method starts by seeking solutions to the wave equation that are products of functions of the independent variables t and x,

$$v(x, t) = T(t)X(x). \tag{9.6.2}$$

The next step is to substitute (9.6.2) into the wave equation. Since

$$\frac{\partial^2}{\partial t^2}[T(t)X(x)] = T''(t)X(x),$$

$$\frac{\partial^2}{\partial x^2}[T(t)X(x)] = T(t)X''(x),$$

after substituting, we obtain

$$X(x)T''(t) = c^2X''(x)T(t)$$

or

$$\frac{T''(t)}{c^2T(t)} = \frac{X''(x)}{X(x)} \tag{9.6.3}$$

(we can ignore possible zeros of the denominators because the result of the method will be shown to be a solution for all x and t). Because x and t are independent variables, (9.6.3) can be valid only if each ratio is a constant that we shall denote by γ. This observation is the core of the separation-of-variables method. Hence

$$T'' = \gamma c^2 T \quad \text{and} \quad X'' = \gamma X.$$

Because the vibrations of the string are bounded in time, γ must be nonpositive, $\gamma = -\lambda^2$ (otherwise the solutions of the differential equations would be exponential functions), and

$$T'' + \omega^2 T = 0, \qquad X'' + \lambda^2 X = 0, \tag{9.6.4}$$

where $\omega = \lambda c$. The solutions to these ordinary differential equations are

$$T(t) = A\cos(\omega t) + B\sin(\omega t),$$
$$X(x) = D\cos(\lambda x) + E\sin(\lambda x).$$

To proceed further, we impose the boundary conditions $v(0, t) = v(L, t) = 0$ on the function $X(x)$. Clearly $X(0) = 0$ implies $D = 0$, so

$$X(x) = E\sin(\lambda x).$$

If E is zero, the solution is trivial. If $E \neq 0$ and $X(L) = 0$, then

$$\sin(\lambda L) = 0. \tag{9.6.5}$$

This *constraint equation* will be satisfied if and only if

$$\lambda L = n\pi, \qquad n = 0, \pm 1, \pm 2, \ldots$$

or

$$\lambda_n = \frac{n\pi}{L}, \qquad n = 0, \pm 1, \pm 2, \ldots.$$

In this problem, $n = 0$ gives the trivial solution, and the solutions for negative values of n differ only in sign from those for positive values of n. Hence we take

$$\lambda_n = \frac{n\pi}{L}, \qquad n = 1, 2, 3, \ldots. \tag{9.6.6}$$

In summary, we have found that $X(x)$ is a solution to a two point boundary value problem

$$\begin{aligned} X'' + \lambda^2 X &= 0, \qquad 0 < x < L, \\ X(0) &= X(L) = 0. \end{aligned} \tag{9.6.7}$$

The values of λ satisfying (9.6.5) are called *eigenvalues,* and the corresponding solutions,

$$X_n(x) = \sin(\lambda_n x), \tag{9.6.8}$$

are called *eigenfunctions.*

The corresponding functions of t are

$$T_n(t) = A_n \cos(\omega_n t) + B_n \sin(\omega_n t), \tag{9.6.9}$$

where $\omega_n = \lambda_n c$. Therefore we have the normal modes,

$$v_n(x, t) = [A_n \cos(\omega_n t) + B_n \sin(\omega_n t)] \sin(\lambda_n x), \qquad n = 1, 2, 3, \ldots. \tag{9.6.10}$$

Each $v_n(x, t)$ is a solution to the wave equation for all x and t. By the principle of superposition, we have the normal mode solution (or Fourier series solution),

$$v(x, t) = \sum_{n=1}^{\infty} v_n(x, t) = \sum_{n=1}^{\infty} [A_n \cos(\omega_n t) + B_n \sin(\omega_n t)] \sin(\lambda_n x). \tag{9.6.11}$$

(We assume that the convergence is such that the series can be twice differentiated term by term.)

The constants $\{A_n\}$ and $\{B_n\}$ can be determined from the initial data as follows: We expand the initial data in terms of the eigenfunctions

$$\begin{aligned} f(x) &= \sum_{n=1}^{\infty} f_n \sin(\lambda_n x), \\ g(x) &= \sum_{n=1}^{\infty} g_n \sin(\lambda_n x), \end{aligned} \tag{9.6.12}$$

where by the orthogonality of the $\{\sin(\lambda_n x)\}$ (see Section 9.5),

$$\begin{aligned} f_n &= \frac{2}{L} \int_0^L f(x) \sin(\lambda_n x)\, dx, \\ g_n &= \frac{2}{L} \int_0^L g(x) \sin(\lambda_n x)\, dx. \end{aligned}$$

Then we set

$$\begin{aligned} f(x) &= v(x, 0) = \sum_{n=1}^{\infty} A_n \sin(\lambda_n x), \\ g(x) &= \frac{\partial v}{\partial t}(x, 0) = \sum_{n=1}^{\infty} c\lambda_n B_n \sin(\lambda_n x), \end{aligned} \tag{9.6.13}$$

and by matching coefficients of the $\{\sin(\lambda_n x)\}$ conclude that

$$A_n = f_n,$$

$$B_n = \frac{g_n}{c\lambda_n}.$$

EXAMPLE 1 Solve

$$v_{tt} = c^2 v_{xx}, \qquad 0 < x < \pi, \quad 0 < t;$$

$$v(0, t) = 0, \qquad v(\pi, t) = 0, \qquad 0 < t;$$

$$v(x, 0) = 0, \qquad v_t(x, 0) = x(\pi - x), \qquad 0 < x < \pi.$$

SOLUTION Since $L = \pi$, the eigenvalues are $\lambda_n = n$, $n = 1, 2, 3, \ldots$. The normal mode solution (9.6.11) is

$$v(x, t) = \sum_{n=1}^{\infty} [A_n \cos(\omega_n t) + B_n \sin(\omega_n t)] \sin(nx)$$

with

$$\omega_n = nc,$$

$$A_n = 0,$$

$$B_n = \frac{2}{cn\pi} \int_0^{\pi} x(\pi - x) \sin nx \, dx = \frac{8}{cn^4 \pi} \sin^2\left(\frac{n\pi}{2}\right).$$

Hence the normal mode solution (9.6.11) is

$$v(x, t) = \frac{8}{c\pi} \sum_{n=1}^{\infty} \frac{1}{n^4} \sin^2\left(\frac{n\pi}{2}\right) \sin(\omega_n t) \sin(nx).$$ ■

The separation-of-variables method can be employed to solve a great variety of problems. Let us consider a finite string with one end fixed at $x = 0$ so that $v(0, t) = 0$. At $x = L$, the string is fastened to a ring that is free to move up or down a vertical wire. Thus the slope of the string at $x = L$ is zero; that is, $v_x(L, t) = 0$. The corresponding boundary-initial value problem is

$$\frac{\partial^2 v}{\partial t^2} = c^2 \frac{\partial^2 v}{\partial x^2}, \qquad 0 < x < L, \quad 0 < t;$$

$$v(0, t) = 0, \qquad \frac{\partial v}{\partial x}(L, t) = 0, \qquad 0 < t;$$

$$v(x, 0) = f(x), \qquad \frac{\partial v}{\partial t}(x, 0) = g(x), \qquad 0 < x < L.$$

We set $v(x, t) = X(x)T(t)$ and find that

$$T'' + \lambda^2 c^2 T = 0,$$

$$X'' + \lambda^2 X = 0$$

with

$$X(0) = 0, \qquad X'(L) = 0.$$

Therefore

$$X(x) = \sin \lambda x$$

with λ being determined by the constraint equation,

$$X'(L) = \cos(\lambda L) = 0,$$

so

$$\lambda_n = \left(\frac{2n + 1}{2L}\right)\pi,$$

$$X_n(x) = \sin(\lambda_n x), \qquad n = 0, 1, 2, \ldots$$

The nonconstant eigenfunctions, $X_n(x)$, $n \geq 1$, all have period $4L$. The solution is a Fourier sine series,

$$v(x, t) = \sum_{n=0}^{\infty} [A_n \cos(\omega_n t) + B_n \sin(\omega_n t)] \sin \lambda_n x, \tag{9.6.14}$$

where

$$\omega_n = \lambda_n c,$$

$$A_n = f_n = \frac{2}{L}\int_0^L f(x) \sin(\lambda_n x)\, dx,$$

$$\lambda_n c B_n = g_n = \frac{2}{L}\int_0^L g(x) \sin(\lambda_n x)\, dx, \qquad n = 0, 1, 2, \ldots$$

EXAMPLE 2 Solve

$$v_{tt} = c^2 v_{xx}, \qquad 0 < x < 2, \quad 0 < t;$$

$$v(0, t) = 0, \qquad v_x(2, t) = 0, \qquad 0 < t;$$

$$v(x, 0) = x(x - 4), \qquad v_t(x, 0) = 0.$$

SOLUTION The normal mode solution,

$$v(x, t) = \sum_{n=0}^{\infty} [A_n \cos(\omega_n t) + B_n \sin(\omega_n t)] \sin(\lambda_n x),$$

must have

$$\lambda_n = \left(\frac{2n + 1}{4}\right)\pi, \qquad \omega_n = \lambda_n c,$$

$$A_n = \int_0^2 x(x - 4) \sin(\lambda_n x)\, dx = -\frac{2}{\lambda_n^3}, \qquad B_n = 0.$$

Hence

$$v(x, t) = -2 \sum_{n=0}^{\infty} \frac{\cos(\omega_n t)}{\lambda_n^3} \sin(\lambda_n x),$$

where $\lambda_n = (2n + 1)\pi/4$. ■

The separation-of-variables method is summarized in the following algorithm.

Algorithm for solving the finite string problem by the separation-of-variables method—unforced vibrations

Given the homogeneous differential equation,

$$v_{tt} = c^2 v_{xx}, \qquad 0 < x < L, \quad 0 < t,$$

with boundary conditions[6] $v(0, t) = 0$, $v(L, t) = 0$, $0 < t$, and initial conditions, $v(x, 0) = f(x)$, $v_t(x, 0) = g(x)$, $0 < x < L$.

1. Solve

$$X_n'' + \lambda^2 X_n = 0$$

$$X_n(0) = 0, \qquad X_n(L) = 0,^6$$

6. Other boundary conditions may be given as in Example 2. If so, the eigenvalues and eigenfunctions will be different from those of the fixed endpoint problem.

for

$$X_n(x) = \sin(\lambda_n x), \qquad \lambda_n = \frac{n\pi}{L}, \qquad n = 1, 2, 3, \ldots$$

where λ_n satisfies the constraint equation,

$$X_n(L) = \sin(\lambda_n L) = 0.$$

2. Set

$$v(x, t) = \sum_{n=1}^{\infty} T_n(t)X_n(x),$$

$$f(x) = \sum_{n=1}^{\infty} f_n X_n(x),$$

$$g(x) = \sum_{n=1}^{\infty} g_n X_n(x).$$

3. Determine the coefficients in the series for $f(x)$ and $g(x)$ by

$$f_n = \frac{2}{L}\int_0^L f(x)X_n(x)\,dx,$$

$$g_n = \frac{2}{L}\int_0^L g(x)X_n(x)\,dx.$$

4. Set

$$T_n(t) = A_n \cos(\omega_n t) + B_n \sin(\omega_n t),$$

where

$$\omega_n = \lambda_n c,$$

$$A_n = f_n,$$

$$B_n = \frac{g_n}{\omega_n}.$$

5. Substitute the $\{T_n(t)\}$ in the series for $v(x, t)$.

FORCED VIBRATIONS

The problem of a vibrating string subject to distributed forces such as gravity or to the effect of drawing a violin bow across it can be solved using eigenfunction expansions. Consider the following boundary-initial value problem for the inhomogeneous wave equation:

$$\frac{\partial^2 v}{\partial t^2} = c^2 \frac{\partial^2 v}{\partial x^2} + F(x, t), \qquad 0 < x < L, \quad 0 < t;$$

$$v(0, t) = 0, \qquad v(L, t) = 0, \qquad 0 < t;$$

$$v(x, 0) = f(x), \qquad \frac{\partial v}{\partial t}(x, 0) = g(x), \qquad 0 < x < L.$$

We employ the eigenfunctions of the corresponding homogeneous problem (9.6.1),

$$X_n(x) = \sin(\lambda_n x), \qquad \lambda_n = \frac{n\pi}{L}, \qquad n = 1, 2, 3, \ldots$$

and expand all of the functions of the problem in orthogonal series of the $\{X_n(x)\}$:

$$\begin{aligned} v(x, t) &= \sum_{n=1}^{\infty} P_n(t) X_n(x), \\ F(x, t) &= \sum_{n=1}^{\infty} F_n(t) X_n(x), \\ f(x) &= \sum_{n=1}^{\infty} f_n X_n(x), \\ g(x) &= \sum_{n=1}^{\infty} g_n X_n(x), \end{aligned} \tag{9.6.15}$$

where

$$\begin{aligned} F_n(t) &= \frac{2}{L} \int_0^L F(x, t) X_n(x)\, dx, \\ f_n &= \frac{2}{L} \int_0^L f(x) X_n(x)\, dx, \\ g_n &= \frac{2}{L} \int_0^L g(x) X_n(x)\, dx, \end{aligned} \tag{9.6.16}$$

and the $\{P_n(t)\}$ are solutions to an infinite but uncoupled system of ordinary differential equations,

$$\frac{d^2 P_n}{dt^2} + \omega_n^2 P_n = F_n(t), \tag{9.6.17}$$

$$\omega_n = \lambda_n c, \qquad P_n(0) = f_n, \qquad \frac{dP_n}{dt}(0) = g_n, \qquad n = 1, 2, 3, \ldots.$$

The derivation of (9.6.17) is straightforward. We simply substitute the series for $v(x, t)$ and $F(x, t)$ into the nonhomogeneous wave equation and use $X_n'' = -\lambda_n^2 X_n$ to obtain

$$\frac{\partial^2 v}{\partial t^2} - c^2 \frac{\partial^2 v}{\partial x^2} = \sum_{n=1}^{\infty} \left\{ \frac{d^2 P_n(t)}{dt^2} + c^2 \lambda_n^2 P_n(t) \right\} X_n(x)$$

$$= F(x, t) = \sum_{n=1}^{\infty} F_n(t) X_n(x).$$

By matching the coefficients of $X_n(x)$ we have the differential equation of (9.6.17). The initial conditions are found from

$$v(x, 0) = \sum_{n=1}^{\infty} P_n(0) X_n(x) = f(x) = \sum_{n=1}^{\infty} f_n X_n(x),$$

$$\frac{\partial v}{\partial t}(x, 0) = \sum_{n=1}^{\infty} \frac{d}{dt} P_n(0) X_n(x) = g(x) = \sum_{n=1}^{\infty} g_n X_n(x).$$

Again matching the coefficients of the eigenfunctions, $X_n(x)$, we have

$$P_n(0) = f_n, \qquad \frac{dP_n(0)}{dt} = g_n, \qquad n = 1, 2, 3, \ldots .$$

EXAMPLE 3 Solve

$$\frac{\partial^2 v}{\partial t^2} = \frac{\partial^2 v}{\partial x^2} + 9x(\pi - x)\sin 3t, \qquad 0 < x < \pi, \quad 0 < t;$$

$$v(0, t) = 0, \qquad v(\pi, t) = 0, \qquad 0 < t;$$

$$v(x, 0) = 0, \qquad \frac{\partial v}{\partial t}(x, 0) = 3 \sin 3x, \qquad 0 < x < \pi.$$

SOLUTION For this problem, $\lambda_n = n$, $X_n(x) = \sin nx$. From

$$F_n(t) = (9 \sin 3t) \frac{2}{\pi} \int_0^{\pi} x(\pi - x) \sin(nx)\, dx,$$

we have

$$F_n(t) = \frac{72}{n^3 \pi} \sin^2\left(\frac{n\pi}{2}\right) \sin 3t.$$

Clearly,

$$f_n = 0,$$

$$g_n = 0, \qquad n \neq 3, \qquad g_3 = 3.$$

The initial value problems,

$$\frac{d^2P_n}{dt^2} + n^2P_n = \frac{72}{n^3\pi}\sin^2\left(\frac{n\pi}{2}\right)\sin 3t,$$

$$P_n(0) = 0, \qquad \frac{dP_n(0)}{dt} = g_n, \qquad n = 1, 2, 3, \ldots,$$

are easy to solve by the method of undetermined coefficients. The solutions are

$$P_n(t) = \frac{72}{n^3\pi}\sin^2\left(\frac{n\pi}{2}\right)\left[\frac{\sin 3t}{(n^2-9)} - \frac{3\sin nt}{n(n^2-9)}\right], \qquad n \neq 3,$$

$$P_3(t) = \sin 3t.$$

Hence

$$v(x, t) = -\frac{9}{5}\left(\sin 3t - \frac{3}{2}\sin 2t\right)\sin 2x + \sin 3t \sin 3x$$

$$+ \frac{72}{\pi}\sum_{n=4}^{\infty}\frac{\sin^2\left(\frac{n\pi}{2}\right)}{n^3(n^2-9)}\left[\sin 3t - \frac{3}{n}\sin nt\right]\sin nx. \qquad \blacksquare$$

The separation of variables method is summarized in the following algorithm.

Algorithm for solving the finite string problem by the separation-of-variables method—forced vibrations

Given the inhomogeneous differential equation,

$$v_{tt} = c^2v_{xx} + F(x, t), \qquad 0 < x < L, \quad 0 < t,$$

with boundary conditions,[7] $v(0, t) = 0$, $v(L, t) = 0$, $0 < t$ and initial conditions, $v(x, 0) = f(x)$, $v_t(x, 0) = g(x)$, $0 < x < L$.

1. Solve

$$X_n'' + \lambda_n^2X_n = 0,$$

$$X_n(0) = 0, \qquad X_n(L) = 0,^7$$

7. Other boundary conditions may be given as in Example 2. If so, the eigenvalues and eigenfunctions will be different from those of the fixed endpoint problem.

for

$$X_n(x) = \sin(\lambda_n x), \qquad \lambda_n = \frac{n\pi}{L}, \qquad n = 1, 2, 3, \ldots,$$

where λ_n satisfies the constraint equation,

$$X_n(L) = \sin(\lambda_n L) = 0.$$

2. Set

$$v(x, t) = \sum_{n=1}^{\infty} P_n(t)X_n(x), \qquad f(x) = \sum_{n=1}^{\infty} f_n X_n(x),$$

$$F(x, t) = \sum_{n=1}^{\infty} F_n(t)X_n(x), \qquad g(x) = \sum_{n=1}^{\infty} g_n X_n(x).$$

3. Determine the coefficients of the above series by[8]

$$F_n(t) = \frac{2}{L}\int_0^L F(x, t)X_n(x)\,dx,$$

$$f_n = \frac{2}{L}\int_0^L f(x)X_n(x)\,dx,$$

$$g_n = \frac{2}{L}\int_0^L g(x)X_n(x)\,dx,$$

with the $\{P_n(t)\}$ being the solutions to the initial value problems,

$$\frac{d^2P_n}{dt^2} + \omega_n^2 P_n = F_n(t), \qquad \omega_n = \lambda_n c,$$

$$P_n(0) = f_n, \qquad \frac{dP_n}{dt}(0) = g_n, \qquad n = 1, 2, 3, \ldots.$$

4. Substitute the $\{P_n(t)\}$ in the series for $v(x, t)$.

TRAVELING WAVES AND THE D'ALEMBERT FORMULA

On a very long string one can observe waves traveling in both directions without distortion. Let us consider a wave moving to the right with velocity c. If its wave profile at $t = 0$ is described by a function $p(x)$, then its wave profile at later times is described by $p(x - ct)$. A wave moving to the left can be described by a wave $q(x + ct)$. The function

$$v(x, t) = q(x + ct) + p(x - ct) \qquad \textbf{(9.6.18)}$$

8. The formulas for the coefficients may differ with the boundary conditions.

is a mathematical representation of the two traveling waves. In fact, if $q(x)$ and $p(x)$ are twice differentiable, then (9.6.18) is a solution to the wave equation. To verify this, we compute

$$\frac{\partial^2 v(x, t)}{\partial t^2} = c^2 q''(x + ct) + c^2 p''(x - ct),$$

$$\frac{\partial^2 v(x, t)}{\partial x^2} = q''(x + ct) + p''(x - ct),$$

and find that

$$\begin{aligned}\frac{\partial^2 v(x, t)}{\partial t^2} - c^2 \frac{\partial^2 v(x, t)}{\partial x^2} &= c^2 q''(x + ct) + c^2 p''(x - ct) \\ &\quad - c_2 \left[q''(x + ct) + p''(x - ct)\right] \\ &= 0.\end{aligned}$$

The solution (9.6.11), which was derived by the separation-of-variables method, can be decomposed into the sum of traveling waves. From

$$\cos(\lambda_n ct) \sin(\lambda_n x) = \frac{1}{2}\{\sin[\lambda_n(x + ct)] + \sin[\lambda_n(x - ct)]\},$$

$$\sin(\lambda_n ct) \sin(\lambda_n x) = -\frac{1}{2}\{\cos[\lambda_n(x + ct)] - \cos[\lambda_n(x - ct)]\},$$

we obtain

$$\begin{aligned}v(x, t) &= \frac{1}{2}\sum_{n=1}^{\infty} A_n\{\sin[\lambda_n(x + ct)] + \sin[\lambda_n(x - ct)]\} \\ &\quad - \frac{1}{2}\sum_{n=1}^{\infty} B_n\{\cos[\lambda_n(x + ct)] - \cos[\lambda_n(x - ct)]\}.\end{aligned} \tag{9.6.19}$$

This solution to the boundary-initial value problem (9.6.1) can be expressed in terms of odd periodic extensions of the initial data, $f(x)$ and $g(x)$. Let us define $f_o(x)$ and $g_o(x)$ by the expansions (9.6.12) and use the relations (9.6.13) to obtain

$$f_o(x) = \sum_{n=1}^{\infty} f_n \sin(\lambda_n x) = \sum_{n=1}^{\infty} A_n \sin(\lambda_n x),$$

$$g_o(x) = \sum_{n=1}^{\infty} g_n \sin(\lambda_n x) = \sum_{n=1}^{\infty} c\lambda_n B_n \sin(\lambda_n x).$$

Thus

$$\frac{1}{2}[f_o(x + ct) + f_o(x - ct)] = \frac{1}{2}\sum_{n=1}^{\infty} A_n\{\sin[\lambda_n(x + ct)] + \sin[\lambda_n(x - ct)]\}.$$

And from

$$\int^{x} g_o(z)\,dz = -\sum_{n=1}^{\infty} cB_n \cos(\lambda_n x) + \text{const.},$$

we have

$$\frac{1}{2c}\int_{x-ct}^{x+ct} g_o(z)\,dz = -\frac{1}{2}\sum_{n=1}^{\infty} B_n\{\cos[\lambda_n(x+ct)] - \cos[\lambda_n(x-ct)]\}.$$

Therefore, the series expression (9.6.19) for the solution can be expressed as

$$v(x, t) = \frac{1}{2}[f_o(x+ct) + f_o(x-ct)] + \frac{1}{2c}\int_{x-ct}^{x+ct} g_o(z)\,dz. \qquad \textbf{(9.6.20)}$$

This remarkable formula is attributed to J. d'Alembert (1717–1783), who also contributed to the theory of celestial mechanics, fluid mechanics, and the philosophy of mathematics.

D'Alembert's formula can be employed to derive several important properties of solutions to the wave equation. One result is that if the initial velocity, $g(x)$, is zero, then the solution represents two waves, a wave traveling to the right, $\frac{1}{2}f_o(x+ct)$, and a wave traveling to the left, $\frac{1}{2}f_o(x-ct)$. These waves have the same shape as the initial displacement but have one-half its amplitude.

EXAMPLE 4 Solve the boundary-initial value problem

$$v_{tt} = v_{xx}, \qquad 0 < x < 2\pi, \quad 0 < t;$$

$$v(0, t) = 0, \qquad v(2\pi, t) = 0, \qquad 0 < t;$$

$$v(x, 0) = f(x) = \begin{cases} 0 & \text{if } 0 < x < \frac{7}{8}\pi, \\ \cos^2(4x) & \text{if } \frac{7}{8}\pi \le x \le \frac{9}{8}\pi, \\ 0 & \text{if } \frac{9}{8}\pi < x < 2\pi; \end{cases}$$

$$v_t(x, 0) = 0, \qquad 0 < x < 2\pi.$$

Sketch the solution as a function of x for $t = 0, \pi/16, \pi/8, \pi/4, \pi/2, 3\pi/4, 15\pi/16, \pi, 17\pi/16, 5\pi/4, 2\pi, 33\pi/16$. It will be helpful to sketch the odd extension of the wave on adjacent fundamental intervals. They are sometimes referred to as virtual waves. Note how the virtual waves interact with the actual waves so that the endpoint conditions are satisfied.

SOLUTION The graphs are in Fig. 9.14. The solution is

$$v(x, t) = \frac{1}{2}[f_o(x+t) + f_o(x-t)]. \qquad ■$$

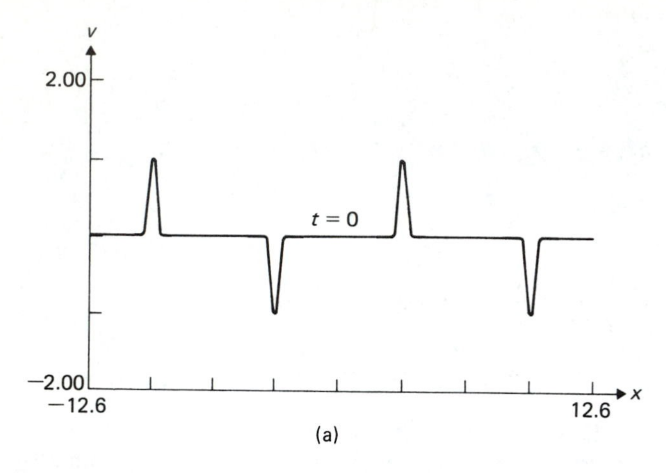

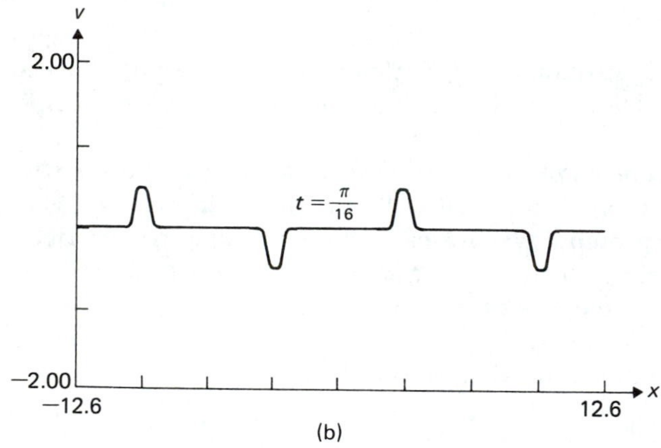

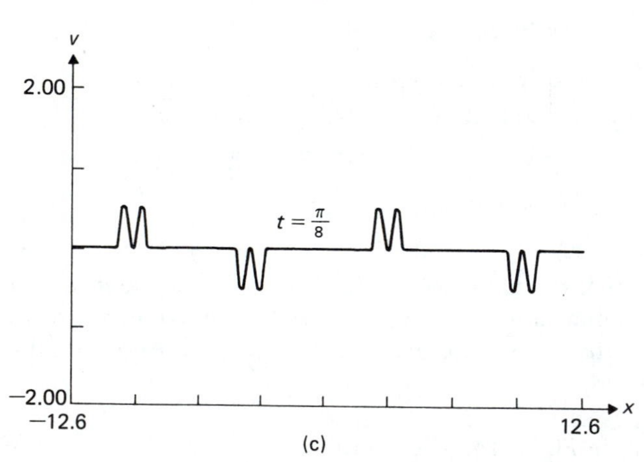

Figure 9.14 Time evolution of a vibrating string for Example 4.

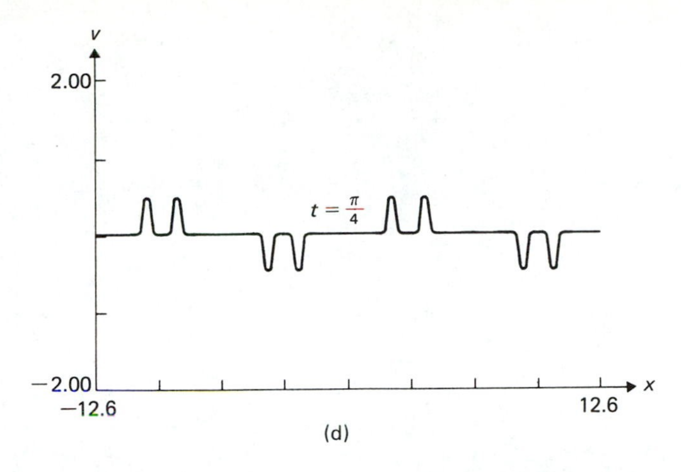

(d)

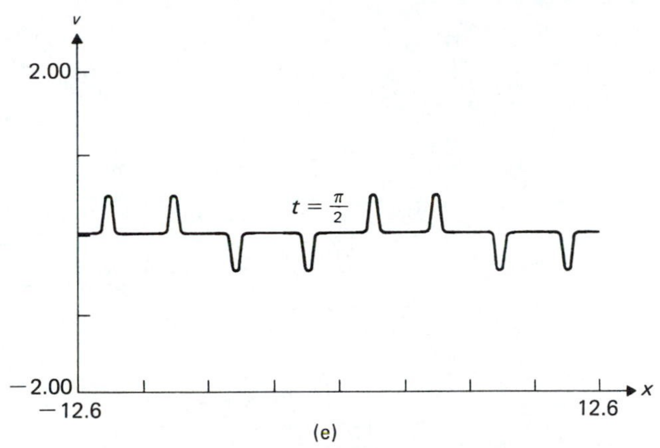

(e)

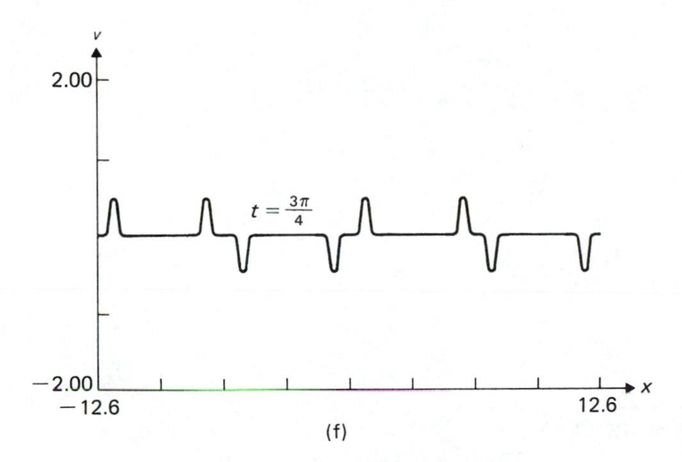

(f)

Figure 9.14 (Continued)

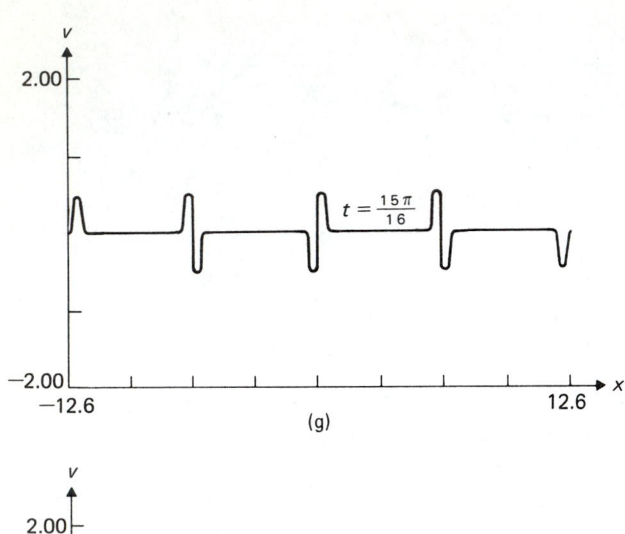

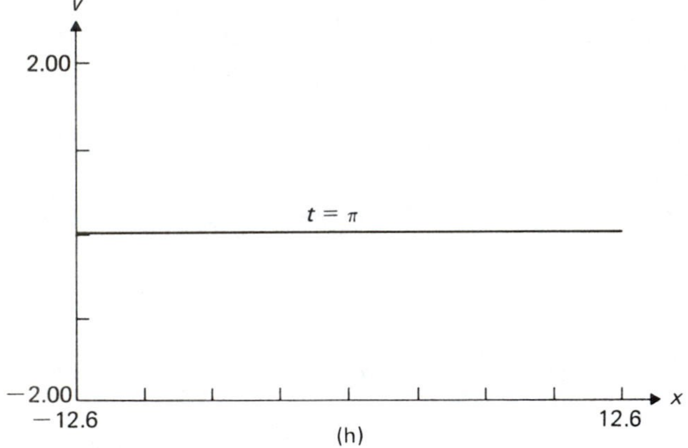

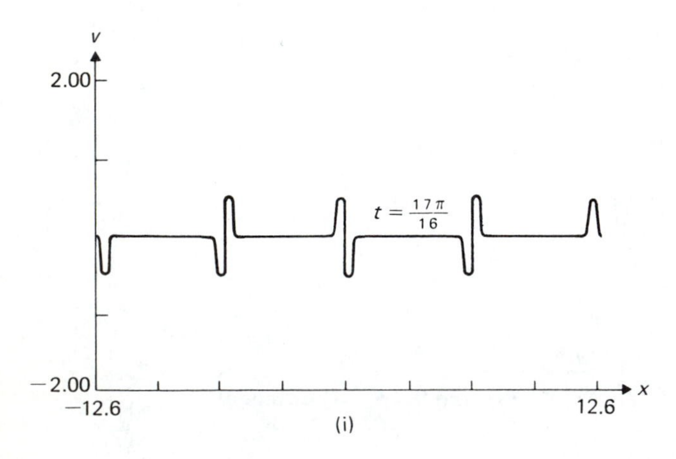

Figure 9.14 (Continued)

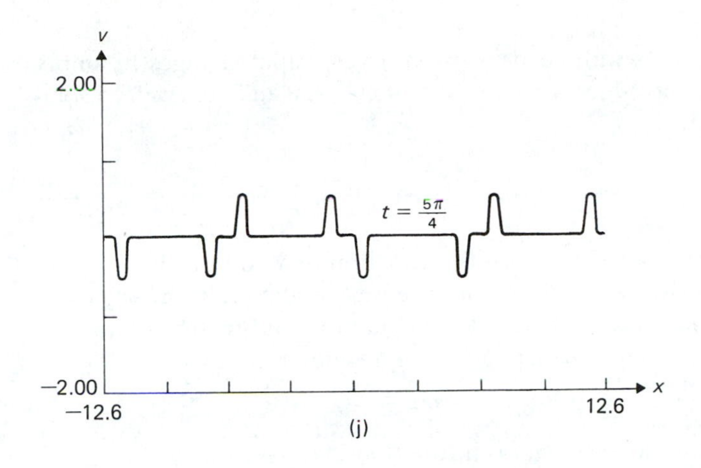

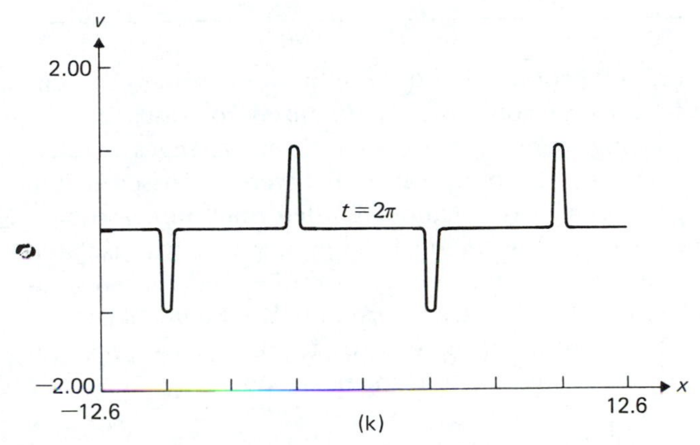

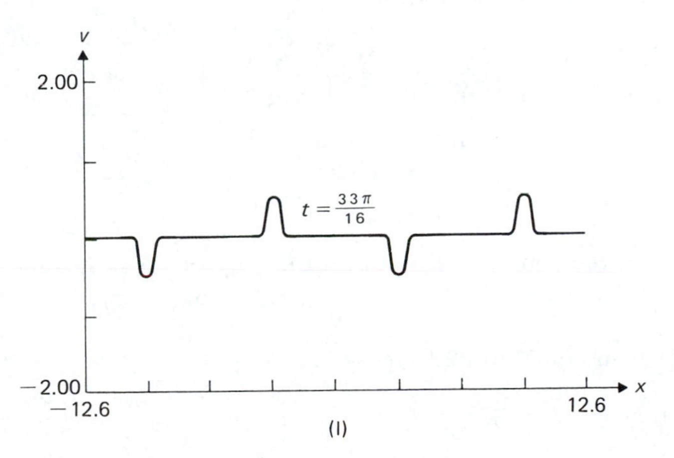

Figure 9.14 **(Continued)**

If we want to study waves with corners or with very rapid changes in amplitudes, it will be helpful to introduce the concepts of classical and generalized solutions to the wave equation.

Definition

A *classical solution* to the wave equation is a function of x and t with continuous second derivatives that satisfies the wave equation in the region of interest. A *generalized solution* to the wave equation is a function of x and t defined in the region of interest that can be represented as

$$v(x, t) = q(x + ct) + p(x - ct),$$

where $q(x)$ and $p(x)$ are piecewise smooth functions.

The normal mode solutions of Section 9.3 and the solutions considered so far in this section have all been classical solutions. Generalized solutions can be employed to represent progressing waves with isolated abrupt changes in their wave profiles such as the square and sawtooth waves that were discussed in the previous section. Of course, an actual string cannot be deformed into a square wave, but such a wave can be considered as an idealization of a wave with isolated wave fronts that have very large but still continuous curvatures. Generalized solutions to the taut string problem with fixed or free endpoints are the limits (except at isolated lines in the xt-plane) of Fourier approximates, each of which is a classical solution to the wave equation. This is illustrated in the next example.

EXAMPLE 5 Construct a generalized solution to

$$\begin{aligned}
v_{tt} &= c^2 v_{xx}, \qquad 0 < x < 4, \quad 0 < t;\\
v(0, t) &= 0, \qquad v(4, t) = 0, \qquad 0 < t;\\
v(x, 0) &= f(x) = \begin{cases} x & \text{if } 0 < x < 1,\\ 0 & \text{if } 1 \le x < 4;\end{cases}\\
v_t(x, 0) &= g(x) = 0, \qquad 0 < x < 4.
\end{aligned}$$

For which values of x and t is the generalized solution the limit of classical solutions? Sketch the solution as a function of x for $ct = 0, \frac{1}{2}, 1, 3, \frac{7}{2}, 4$.

SOLUTION From (9.6.19) and (9.6.20) we have

$$\begin{aligned}
v(x, t) &= \frac{1}{2}[f_o(x + ct) + f_o(x - ct)]\\
&= \frac{1}{2}\sum_{n=1}^{\infty} b_n\{\sin[\lambda_n(x + ct)] + \sin[\lambda_n(x - ct)]\},
\end{aligned}$$

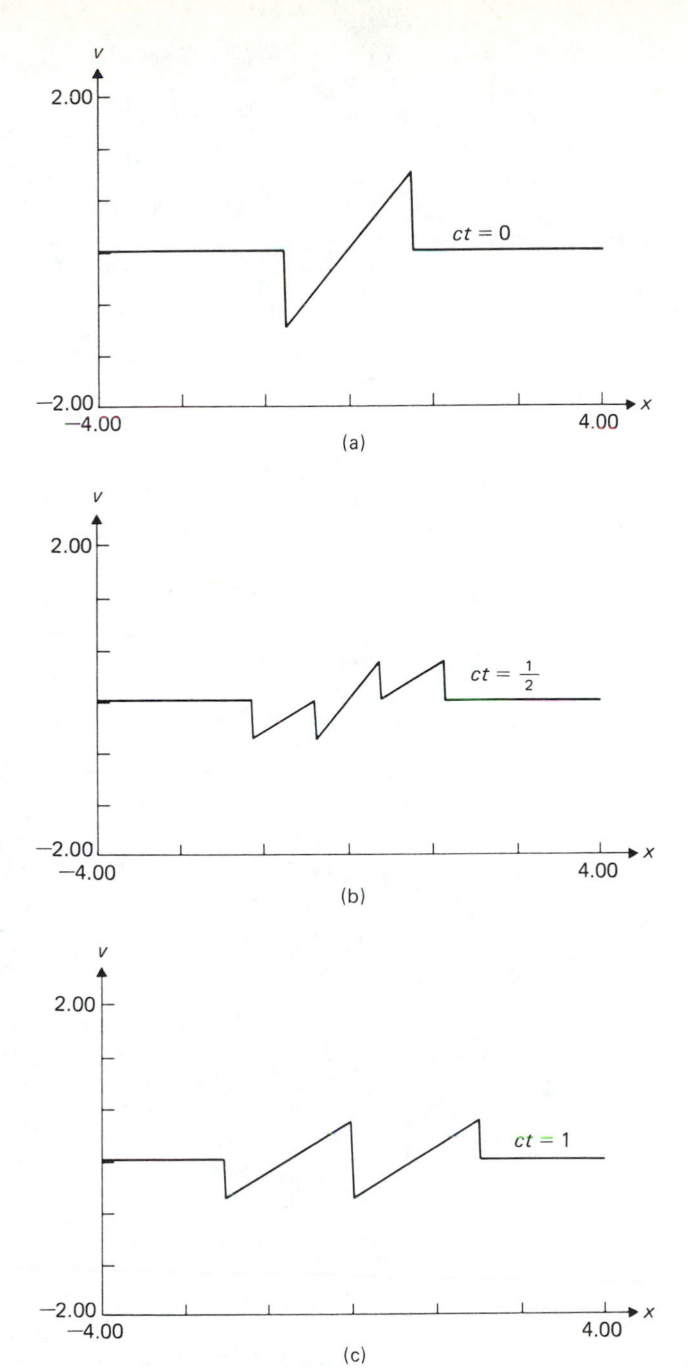

Figure 9.15 Time evolution of a vibrating string for Example 5.

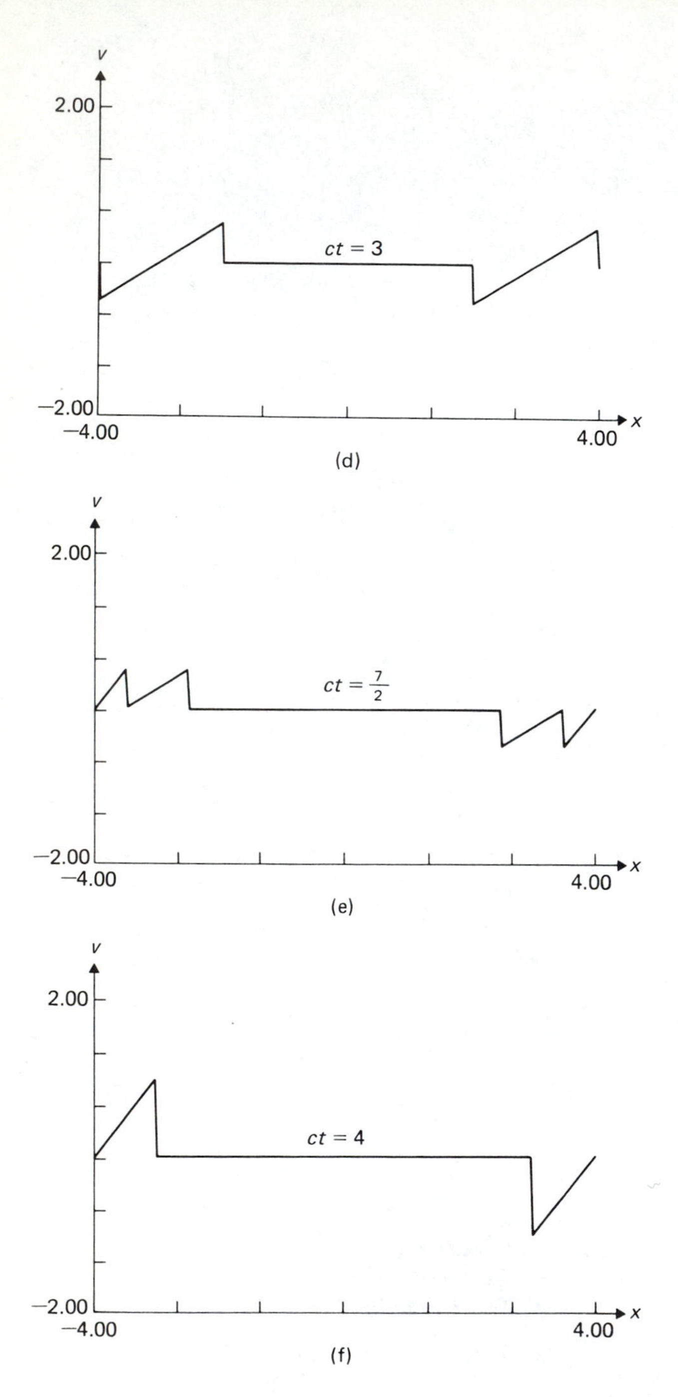

Figure 9.15 (Continued)

where $f_o(x)$ is the odd periodic extension of $f(x)$, $\lambda_n = n\pi/4$, and

$$b_n = \frac{32}{n^2\pi^3}\left[2\sin\left(\frac{n\pi}{4}\right) - \left(\frac{n\pi}{4}\right)\cos\left(\frac{n\pi}{4}\right)\right].$$

The functions $f_o(x + ct)$ and $f_o(x - ct)$ are piecewise smooth; $f_o(x + ct)$ has a jump discontinuity at each point on the line $x + ct = 1$, and $f_o(x - ct)$ has a jump discontinuity at each point on the line $x - ct = 1$. For each N, the Fourier approximate,

$$S_N(x, t) = \frac{1}{2}\sum_{n=1}^{N} b_n\{\sin[\lambda_n(x + ct)] + \sin[\lambda_n(x - ct)]\},$$

is a classical solution to the wave equation and, except on the lines $x + ct = 1$ and $x - ct = 1$,

$$v(x, t) = \lim_{N\to\infty} S_N(x, t).$$

The graphs are in Fig. 9.15. ■

The method of separation of variables was introduced in this section and employed to construct series solutions to several wave equation boundary-initial value problems. We shall use it in the next two sections as we study the equations that are used to model heat conduction and problems in potential theory.

EXERCISES 9.6

For Exercises 1–10 construct the Fourier series solution (9.6.11) to the following boundary-initial value problem with the given initial data.

$$v_{tt} = c^2 v_{xx}, \quad 0 < x < L, \quad 0 < t;$$

$$v(0, t) = 0, \quad v(L, t) = 0, \quad 0 < t;$$

$$v(x, 0) = f(x), \quad v_t(x, 0) = g(x), \quad 0 < x < L.$$

1. $L = \pi$, $f(x) = x(\pi - x)$, $g(x) = 0$

2. $L = \pi$, $f(x) = 0$, $g(x) = x/\pi - \sin(x/2)$

3. $L = \pi$, $f(x) = \sinh x$, $g(x) = \cosh x - 1$

4. $L = \pi$, $f(x) = 4|x - \pi/2|$, $g(x) = 0$

5. $L = \pi$, $f(x) = 0$, $g(x) = e^x$

6. $L = \pi$, $f(x) = |\sin 3x|$, $g(x) = |\sin 5x|$

For Exercises 7–10 in addition to the series solution, find the d'Alembert solution (9.6.20) and sketch it on $0 < x < L$ with $ct = 0, L/4, L/3, L/2, L$.

7. $L = 4, f(x) = \begin{cases} |x - 1| & \text{if } 0 < x < 2 \\ -|x - 3| & \text{if } 2 \leq x < 4 \end{cases}, g(x) = 0$

8. $L = 6, f(x) = \begin{cases} 0 & \text{if } 0 < x < 2 \\ -|x - 3| & \text{if } 2 \leq x \leq 4, \\ 0 & \text{if } 4 < x < 6 \end{cases} g(x) = 0$

9. $L = 2, f(x) = 0, g(x) = |x - 1|$

10. $L = 4, f(x) = 0, g(x) = x - |x - 2|$

For Exercises 11–14 construct the Fourier sine series solution (9.6.14) to the following boundary-initial value problem with the given initial data.

$$v_{tt} = c^2 v_{xx}, \qquad 0 < x < L, \quad 0 < t;$$

$$v(0, t) = 0, \qquad v_x(L, t) = 0, \qquad 0 < t;$$

$$v(x, 0) = f(x), \qquad v_t(x, 0) = g(x), \qquad 0 < x < L.$$

11. $L = 2, f(x) = 6x(x - 2)^2, g(x) = 4 \sin(6\pi x)$

12. $L = 3, f(x) = 4(x - 3)^2 - 36, g(x) = x - 3$

13. $L = \pi, f(x) = \sin^2(2x), g(x) = \cos^2(2x)$

14. $L = \pi, f(x) = \begin{cases} x & 0 < x < \pi/2 \\ (x - \pi)^2 & \pi/2 \leq x < \pi \end{cases}, g(x) = 0$

For Exercises 15–18 construct the Fourier cosine series solution [see (9.5.15)] to the following boundary-initial value problem with the given initial data:

$$v_{tt} = c^2 v_{xx}, \qquad 0 < x < L, \quad 0 < t;$$

$$v_x(0, t) = 0, \qquad v(L, t) = 0, \qquad 0 < t;$$

$$v(x, 0) = f(x), \qquad v_t(x, 0) = g(x), \qquad 0 < x < L.$$

15. $L = 2, f(x) = x \sin(\pi x), g(x) = 0$

16. $L = \pi, f(x) = x^2 - \pi, g(x) = x \cos 2x$

17. $L = 4, f(x) = 0, g(x) = |x - 1|$

18. $L = 1, f(x) = 1 + \cos \pi x, g(x) = 3x^2 - 2$

The phenomenon of resonance can be observed in wave propagation on a finite taut string that is subjected to a periodic impressed force. If the forcing function has a component whose frequency is an integral multiple of the fundamental frequency of the string, the oscillations of the string will grow and the string may snap. This is illustrated in Exercises 19–20.

19. Show that the solution to the following boundary-initial value problem is unbounded and identify the term in the forcing function that induces resonance.

$$\begin{aligned} v_{tt} &= c^2 v_{xx} + F(x, t), \qquad 0 < x < \pi, \quad 0 < t; \\ v(0, t) &= 0, \qquad v(\pi, t) = 0, \qquad 0 < t; \\ v(x, 0) &= 0, \qquad v_t(x, 0) = 0, \qquad 0 < x < \pi, \end{aligned} \tag{9.6.21}$$

where $F(x, t) = 7 \sin[3(x - ct)] + 8 \sin[\pi(x + ct)]$.

20. Solve Exercise 19 with the forcing function replaced by

$$F(x, t) = [7 \cos(3ct) - 3\cos(\pi ct)]\,|x - (\pi/2)|.$$

What would be the effect of changing the length of the string? Discuss the relationship of the length of the string and the frequencies of the forcing function that induce resonance.

The vibrations of a taut string with one end fixed and the other oscillating can be modeled by the following boundary-initial value problem:

$$\begin{aligned} v_{tt} &= c^2 v_{xx}, \qquad 0 < x < L, \quad 0 < t; \\ v(0, t) &= 0, \qquad v(L, t) = h(t), \qquad 0 < t; \\ v(x, 0) &= f(x), \qquad v_t(x, 0) = g(x), \qquad 0 < x < L. \end{aligned} \tag{9.6.22}$$

We shall assume that $h(t)$ has a continuous second derivative, $h(0) = 0$, $h'(0) = 0$, and that $f(x)$ and $g(x)$ are piecewise smooth.

This problem can be solved by a mathematical trick that is borrowed from heat transfer theory. We set

$$v(x, t) = \frac{xh(t)}{L} + w(x, t),$$

then derive and solve a boundary-initial value problem for $w(x, t)$.

21. Show that if $w(x, t)$ is a solution to

$$\begin{aligned} w_{tt} &= c^2 w_{xx} - \frac{xh''(t)}{L}, \qquad 0 < x < L, \quad 0 < t; \\ w(0, t) &= 0, \qquad w(L, t) = 0, \qquad 0 < t; \\ w(x, 0) &= f(x), \qquad w_t(x, 0) = g(x), \qquad 0 < x < L, \end{aligned}$$

then $v(x, t)$ is a solution to (9.6.22).

22. Solve the oscillatory endpoint problem (9.6.22) with $L = \pi$, $c = 1$, $h(t) = 6 \sin^2 3t + 4 \sin^2 4t$, $f(x) = 0$, and $g(x) = 0$. Discuss your results. What would be the effect of changing the length of the string?

23. Solve the oscillatory endpoint problem (9.6.22) with $L = \pi$, $c = 1$, $h(t) = 4(\cos 2t - 1)^2 + 6(\cos 4t - 1)^2$, $f(x) = 4 \sin 2x$, $g(x) = \sin 5x$. Discuss your results.

The wave equation is used to model a variety of physical phenomena. The physics of the following examples is described in books by Coulson [10] and Lamb [11]. Both are excellent introductions to wave motion.

Small-amplitude longitudinal waves in a bar of uniform cross section can be described by the boundary-initial value problem (9.6.1) if $v(x, t)$ denotes a displacement along the bar and $c^2 = \lambda/\rho$, where ρ is the density and λ is an elastic constant. The boundary conditions, $v(0, t) = 0$ and $v(L, t) = 0$, mean that both ends of the bar are rigidly clamped. If the bar is stretched from length L_0 to length L, the bar is in a state of tension and $c^2 = \lambda L^2/\rho_0 L_0^2$, where ρ_0 is the density of the unstretched bar.

24. Show that ω, the fundamental frequency of the rod, is independent of the tension. How does the fundamental frequency of a taut string with fixed endpoints vary with the tension of the string?

Tidal waves in water occur when the wavelengths of the oscillations are much greater than the depth of the water. Coulson shows that tidal waves along a straight horizontal channel can be approximated by solutions to the wave equation, $v_{tt} = c^2 v_{xx}$, where $c^2 = g/h$, g is a gravitational constant, h is the depth of the channel, and v denotes the elevation of the water with respect to the undisturbed surface. Boundary conditions can be omitted if the channel is very long and the region of interest is far from the ends. If so, the domain of definition of $v(x, t)$ is the half-plane, $-\infty < x < \infty$, $0 < t$.

25. Let the initial conditions be $v(x, 0) = f(x)$, $v_t(x, 0) = g(x)$, $-\infty < x < \infty$. Verify that the d'Alembert formula,

$$v(x, t) = \frac{1}{2}[f(x + ct) + f(x - ct)] + \frac{1}{2c}\int_{x-ct}^{x+ct} g(s)\, ds,$$

represents a classical solution to the wave equation if $f(x)$ and $g(x)$ have continuous second derivatives and verify that the initial conditions are satisfied. If $f(x)$ and $g(x)$ are piecewise smooth, the d'Alembert formula represents a generalized solution. Sketch the wave form, $v(x, t)$, on the interval, $-2 \le x \le 8$, with $ct = 0, \frac{1}{2}, 1, \frac{3}{2}, 2, \frac{5}{2}$, and

$$f(x) = \begin{cases} 0 & \text{if } x < 2, \\ \sin 4x & \text{if } 2 \le x \le 4, \\ 0 & \text{if } 4 < x; \end{cases} \qquad g(x) = \begin{cases} 0 & \text{if } x < 2, \\ \sin 8x & \text{if } 2 \le x \le 4, \\ 0 & \text{if } 4 < x. \end{cases}$$

With these initial data, is $v(x, t)$ a classical or generalized solution?

26. If a long horizontal channel has a vertical barrier at $x = 0$, then the corresponding mathematical model for its tidal waves is

$$\begin{aligned} &v_{tt} = c^2 v_{xx}, \qquad 0 < x < \infty, \quad 0 < t; \\ &v(0, t) = 0, \qquad 0 < t; \\ &v(x, 0) = f(x), \qquad v_t(x, 0) = g(x), \qquad 0 < x < \infty. \end{aligned} \tag{9.6.23}$$

This problem is well suited for a study of how the waves are reflected from the vertical barrier ($v(0, t) = 0$). Verify that $v(0, t) = 0$, if the function, $v(x, t)$, is defined by the d'Alembert formula,

$$v(x, t) = \frac{1}{2}[f_o(x + ct) + f_o(x - ct)] + \frac{1}{2c}\int_{x-ct}^{x+ct} g_o(s)\, ds,$$

where $f_o(x)$ and $g_o(x)$ are odd extensions of the piecewise smooth initial data. Sketch the wave profile, $v(x, t)$, on $0 \le x \le 10$ with $ct = 0, 1, \frac{3}{2}, 2, 3, 4$, and

$$f(x) = \begin{cases} 0 & \text{if } 0 < x < 1, \\ |x - 2| & \text{if } 1 \le x \le 3, \\ 0 & \text{if } 3 < x, \end{cases} \qquad g(x) = 0.$$

Sound waves are particular examples of waves in a compressible fluid, such as a gas. Waves there are longitudinal since the oscillatory motion of the particles of the gas in the direction of the wave propagation can be thought of as the propagation of sound. In a tube of constant cross section, the particle displacement, $v(x, t)$ can be approximated by a solution to the wave equation, $v_{tt} = c^2 v_{xx}$, where c is the velocity of sound. The velocity of sound depends on the composition of the gas, the temperature, and the pressure. In ordinary air at 0°C, the velocity of sound $c \cong 332$ m/sec. If the tube is closed at $x = 0$, then $v(0, t) = 0$, since at a fixed boundary the particles cannot move. If the tube is open at $x = 0$, then $v_x(0, t) = 0$.

27. Assume that a very long tube is open at $x = 0$. Verify that d'Alembert's formula,

$$v(x, t) = \frac{1}{2}[f_e(x + ct) + f_e(x - ct)] + \frac{1}{2c}\int_{x-ct}^{x+ct} g_e(s)\, ds,$$

satisfies the boundary condition, $v_x(0, t) = 0$, where $f_e(x)$ and $g_e(x)$ are even extensions of the initial data. Assume that $f(x)$ has a continuous derivative such that $f'(0^+) = 0$ and that $g(x)$ is continuous. Use the same initial data as in Exercise 26 and sketch the wave profile as directed there on $0 \le x \le 4$ with $ct = 0, \frac{1}{2}, 1, \frac{3}{2}, 2, \frac{5}{2}$.

9.7 THE HEAT EQUATION

DERIVATION OF THE HEAT EQUATION

The heat equation,

$$\frac{\partial w}{\partial t} = k\,\frac{\partial^2 w}{\partial x^2}, \tag{9.7.1}$$

can be derived from the principle of conservation of thermal energy and Fourier's law, a relation between the flow of thermal energy and the temperature. For concreteness, we consider the flow of heat through a thin rod of constant cross section with a perfectly insulated lateral surface (see Fig. 9.16). No thermal energy can cross the lateral surface, and so the temperature at any instant is constant in each cross section of the rod. We take the x-axis parallel to the rod and let $w(x, t)$ denote the temperature at time t and position x. The total amount of thermal energy in a segment bounded by $x = a$ and $x = b$ is

$$\int_a^b Ac\rho w(x, t)\, dx,$$

where

A = the cross-sectional area of the rod,

c = the specific heat, the amount of thermal energy needed to increase by one degree the temperature of a unit mass of material, and

$A\rho$ = the mass per unit length.

The principle of conservation of thermal energy requires that the rate of change of thermal energy in a segment of the rod is equal to the thermal energy

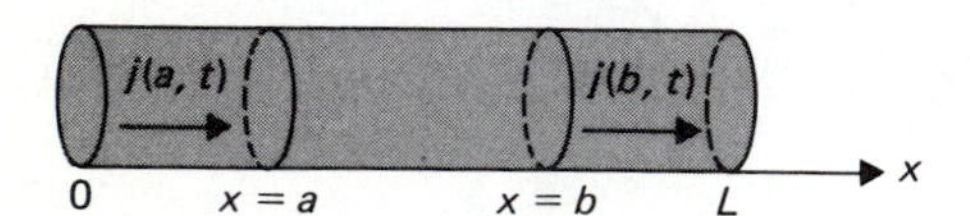

Figure 9.16 Heat flow in a thin rod.

flowing across the boundary per unit time plus the amount generated inside the segment per unit time.

The net transfer at the boundaries is given by

$$Aj(a, t) - Aj(b, t),$$

where $j(x, t)$, the flux, is the transfer of energy per unit area. We also allow the possibility that there are sources of heat within the rod, but that they vary only with x, t and $w(x, t)$. Such sources should contribute

$$AQ(w(x, t), x, t)\, dx$$

units of energy in unit time to the infinitesimal cross section between x and $x + dx$. (The sources of heat could, for instance, be due to chemical reactions occurring within the rod.) The amount of energy generated by the sources per unit time would then be

$$\int_a^b AQ(w(x, t), x, t)\, dx.$$

Conservation of thermal energy implies that

$$\frac{d}{dt}\int_a^b Ac\rho w(x, t)\, dx = Aj(a, t) - Aj(b, t) + \int_a^b AQ(w(x, t), x, t)\, dx.$$

If, as will be true in most problems of physical significance,

$$\frac{d}{dt}\int_a^b Ac\rho w(x, t)\, dx = \int_a^b Ac\rho \frac{\partial}{\partial t} w(x, t)\, dx,$$

and

$$j(a, t) - j(b, t) = -\int_a^b \frac{\partial}{\partial x} j(x, t)\, dx,$$

then

$$\int_a^b A\left\{c\rho \frac{\partial}{\partial t} w(x, t) + \frac{\partial}{\partial x} j(x, t) - Q(w(x, t), x, t)\right\} dx = 0.$$

Since a and b are arbitrary and the integrand is assumed to be continuous, we can conclude that the integrand must be zero. Therefore

$$c\rho \frac{\partial w}{\partial t} + \frac{\partial j}{\partial x} = Q(w, x, t), \qquad \textbf{(9.7.2)}$$

a first order partial differential equation with dependent variables w and j and independent variables x and t. It is called the *conservation equation* or the *equation of continuity*.

The conservation equation is not a complete description of a physical process because a relationship between the temperature w and the flux j must be prescribed. Fourier postulated his law of heat conduction,

$$j = -K\frac{\partial w}{\partial x}, \tag{9.7.3}$$

which implies that thermal energy flows from a hotter region to a cooler region. The positive constant K is called the *thermal conductivity*.

Fourier's law of heat conduction together with the conservation equation gives us a first order system of partial differential equations that can be combined into a single second order partial differential equation,

$$c\rho\frac{\partial w}{\partial t} = \frac{\partial}{\partial x}\left(K\frac{\partial w}{\partial x}\right) + Q(w, x, t). \tag{9.7.4}$$

We shall assume that the thermal coefficients c, ρ, and K are constant. Then the equation (9.7.4) can be written

$$\frac{\partial w}{\partial t} = k\frac{\partial^2 w}{\partial x^2} + F(w, x, t), \tag{9.7.5}$$

the inhomogeneous heat equation, where $k = K/c\rho$ is called the *thermal diffusivity*. For convenience, we have replaced $Q/c\rho$ by F.

THE HOMOGENEOUS HEAT EQUATION

A typical problem that can be modeled by the homogeneous heat equation is the prediction of the flow of the initial thermal energy in a laterally insulated rod after both ends are immersed in an ice bath. This leads us to a boundary-initial value problem,

$$\frac{\partial w}{\partial t} = k\frac{\partial^2 w}{\partial x^2}, \qquad 0 < x < L, \quad 0 < t;$$

$$w(0, t) = 0, \qquad w(L, t) = 0, \qquad 0 < t;$$

$$w(x, 0) = f(x), \qquad 0 < x < L.$$

Because the partial differential equation is of first order in t, only one initial condition is appropriate.

We shall solve this problem by the separation-of-variables method. Set

$$w(x, t) = X(x)T(t)$$

and substitute into the heat equation to get

$$X(x)T'(t) = kX''(x)T(t)$$

or

$$\frac{T'(t)}{kT(t)} = \frac{X''(x)}{X(x)} = -\lambda^2.$$

The separation constant cannot be positive because if it were the temperature would become unbounded as t increases.

Once again we have a two-point boundary value problem for $X(x)$,

$$X'' + \lambda^2 X = 0, \qquad 0 < x < L;$$

$$X(0) = 0, \qquad X(L) = 0,$$

where λ is a solution to the constraint equation,

$$\sin(\lambda L) = 0.$$

Therefore

$$X_n(x) = \sin(\lambda_n x), \qquad \lambda_n = \frac{n\pi}{L}, \qquad n = 1, 2, \ldots.$$

The general solution to

$$T' = -\lambda_n^2 kT$$

is

$$T(t) = A_n \exp(-\lambda_n^2 kt).$$

(Note that the temperature would be unbounded if the separation constant were positive.) Therefore the product solution is

$$w_n(x, t) = A_n \exp(-\lambda_n^2 kt) \sin(\lambda_n x).$$

By the principle of superposition,

$$w(x, t) = \sum_{n=1}^{\infty} A_n \exp(-\lambda_n^2 kt) \sin(\lambda_n x).$$

To determine the coefficients $\{A_n\}$ we set

$$f(x) = \sum_{n=1}^{\infty} f_n \sin(\lambda_n x),$$

$$w(x, 0) = \sum_{n=1}^{\infty} A_n \sin(\lambda_n x)$$

and invoke the initial condition, $w(x, 0) = f(x)$. Therefore

$$A_n = f_n = \frac{2}{L}\int_0^L f(x) \sin(\lambda_n x)\, dx, \qquad n = 1, 2, \ldots.$$

Note that as t gets large, the temperature of the rod approaches the steady state of 0°, the temperature of the two ends.

EXAMPLE 1 Solve

$$\frac{\partial w}{\partial t} = \frac{\partial^2 w}{\partial x^2}, \qquad 0 < x < 1, \quad 0 < t;$$

$$w(0, t) = 0, \qquad w(1, t) = 0, \qquad 0 < t;$$

$$w(x, 0) = f(x) = x\left[1 - u\left(x - \frac{1}{2}\right)\right] + (1 - x)u\left(x - \frac{1}{2}\right), \qquad 0 < x < 1,$$

where $u(x)$ is the Heaviside function.[9] Sketch the approximate solution,

$$\tilde{w}(x, t) = w_1(x, t) + w_2(x, t) + w_3(x, t),$$

as a function of x for $t = 0, 0.01, 0.1$.

SOLUTION The eigenfunctions $\{X_n(x)\}$ for this problem are nonzero solutions to

$$X'' + \lambda^2 X = 0, \qquad 0 < x < 1,$$
$$X(0) = 0, \qquad X(1) = 0,$$

where the corresponding eigenvalues $\{\lambda_n\}$ must satisfy the constraint equation

$$\sin \lambda = 0.$$

Therefore $\lambda_n = n\pi$, and $X_n(x) = \sin(n\pi x)$, $n = 1, 2, 3, \ldots$.

Next we set

$$f(x) = \sum_{n=1}^{\infty} f_n \sin(n\pi x),$$

where

$$\begin{aligned} f_n &= 2\int_0^1 f(x)\sin(n\pi x)\,dx \\ &= 2\int_0^{1/2} x\sin(n\pi x)\,dx + 2\int_{1/2}^1 (1 - x)\sin(n\pi x)\,dx \\ &= \frac{4}{(n\pi)^2}\sin\left(\frac{n\pi}{2}\right). \end{aligned}$$

9. The unit step or Heaviside function $u(x - a)$ is zero if $x < a$ and is unity if $x \geq a$.

Hence

$$w(x, t) = \sum_{n=1}^{\infty} f_n \exp(-n^2\tau^2 t) \sin(n\pi x).$$

Graphs of $\tilde{w}(x, t)$ are in Fig. 9.17. ■

Just as with the wave equation, there are several types of boundary conditions of physical interest. One is the prescription of the temperature w at an end of a rod; another is the prescription of the flux, $j = -Kw_x$. If an end of a rod is insulated, the flux across it is zero, the corresponding boundary condition is $w_x = 0$. In the next example, both ends are insulated.

EXAMPLE 2 Solve

$$w_t = w_{xx}, \qquad 0 < x < 2, \quad 0 < t;$$

$$w_x(0, t) = w_x(2, t) = 0, \qquad 0 < t;$$

$$w(x, 0) = f(x) = 3x(2 - x), \qquad 0 < x < 2.$$

SOLUTION The two-point boundary value problem resulting from the separation-of-variables method,

$$X'' + \lambda^2 X = 0, \qquad 0 < x < 2;$$

$$X'(0) = 0, \qquad X'(2) = 0,$$

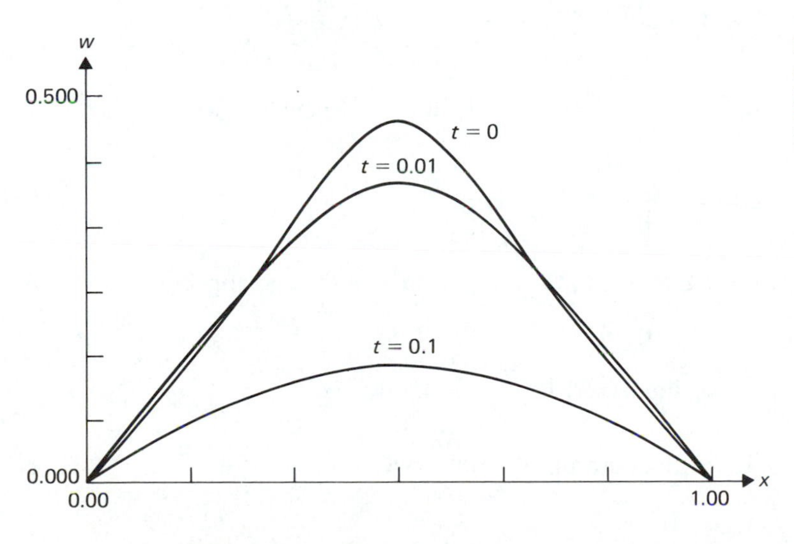

Figure 9.17 An approximate solution to a heat flow problem for Example 1.

has as possible solutions,

$$X_0(x) = A + Bx, \qquad \lambda = 0,$$

$$X(x) = A \cos \lambda x + B \sin \lambda x, \qquad \lambda > 0.$$

It is easy to verify that B must be zero and A nonzero for both cases. For simplicity, we take

$$X_0(x) = 1, \qquad \lambda_0 = 0,$$

$$X_n(x) = \cos \lambda_n x, \qquad n = 1, 2, 3, \ldots .$$

The eigenvalues λ_n for $n \geq 1$ must satisfy the constraint equation

$$X'(2) = -\lambda \sin(2\lambda) = 0.$$

Therefore

$$\lambda_n = \frac{n\pi}{2}, \qquad n = 1, 2, 3, \ldots .$$

Set

$$w(x, t) = f_0 + \sum_{n=1}^{\infty} f_n \exp(-\lambda_n^2 t)\, X_n(x),$$

where

$$f_0 = \frac{1}{2}\int_0^2 f(x)\, dx = \frac{1}{2}\int_0^2 3x(2 - x)\, dx = 2,$$

$$f_n = \int_0^2 f(x) \cos(\lambda_n x)\, dx = \int_0^2 3x(2 - x) \cos(\lambda_n x)\, dx$$

$$= -\frac{24}{n^2\pi^2}\left[\cos(n\pi) + 1\right].$$

In this problem as t gets larger, the temperature of the rod approaches a steady state:

$$\lim_{t\to\infty} w(x, t) = f_0 = \frac{1}{2}\int_0^2 f(x)\, dx,$$

the average of the initial temperature distribution. ■

THE INHOMOGENEOUS HEAT EQUATION

Boundary-initial value problems for the inhomogeneous heat equation can be solved by using eigenfunction expansions. Consider

$$\frac{\partial w}{\partial t} = k\frac{\partial^2 w}{\partial x^2} + F(x, t), \qquad 0 < x < L, \quad 0 < t;$$

$$w(0, t) = 0, \qquad w(L, t) = 0, \qquad 0 < t;$$

$$w(x, 0) = f(x), \qquad 0 < x < L.$$

Following the procedure used to solve boundary-initial value problems for the inhomogeneous wave equation in the preceding section, we employ the eigenfunctions of the corresponding homogeneous equations

$$X_n(x) = \sin(\lambda_n x),$$

$$\lambda_n = \frac{n\pi}{L}, \qquad n = 1, 2, 3 \ldots,$$

and expand all the functions of the problem in orthogonal series of the $\{X_n(x)\}$:

$$w(x, t) = \sum_{n=1}^{\infty} P_n(t)X_n(x),$$

$$F(x, t) = \sum_{n=1}^{\infty} F_n(t)X_n(x),$$

$$f(x) = \sum_{n=1}^{\infty} f_n X_n(x),$$

where

$$F_n(t) = \frac{2}{L}\int_0^L F(x, t)X_n(x)\,dx,$$

$$f_n = \frac{2}{L}\int_0^L f(x)X_n(x)\,dx,$$

and the $\{P_n(t)\}$ are solutions to a first order system of ordinary differential equations

$$\frac{d}{dt}P_n + k\lambda_n^2 P_n = F_n(t),$$

$$P_n(0) = f_n, \qquad n = 1, 2, 3, \ldots.$$

The derivation of these formulas is so similar to the derivation of the corresponding formulas in the previous section that it will be omitted.

EXAMPLE 3 Solve

$$\frac{\partial w}{\partial t} = \frac{\partial^2 w}{\partial x^2} + \left[u\left(x - \frac{\pi}{4}\right) - u\left(x - \frac{3\pi}{4}\right) \right] e^{-t}, \qquad 0 < x < \pi, \quad 0 < t;$$

$$w(0, t) = 0, \qquad w(\pi, t) = 0, \qquad 0 < t < \infty;$$

$$w(x, 0) = x(\pi - x), \qquad 0 < x < \pi,$$

where $u(x)$ is the Heaviside function.

SOLUTION For this problem, $\lambda_n = n$, $X_n(x) = \sin nx$,

$$F_n(t) = \frac{2}{\pi} \int_0^\pi F(x, t) \sin nx \, dx = \frac{2}{\pi} e^{-t} \int_{\pi/4}^{3\pi/4} \sin nx \, dx$$

$$= \frac{2e^{-t}}{n\pi} \left[\cos\left(\frac{n\pi}{4}\right) - \cos\left(\frac{3n\pi}{4}\right) \right] = F_n(0)e^{-t},$$

$$f_n = \frac{2}{\pi} \int_0^\pi x(\pi - x) \sin nx \, dx = \frac{8}{n^4 \pi} \sin^2\left(\frac{n\pi}{2}\right).$$

The initial value problems,

$$\frac{dP_n}{dt} + n^2 P_n = F_n(0)e^{-t}, \qquad P_n(0) = f_n,$$

have solutions,

$$P_1(t) = F_1(0)te^{-t} + f_1 e^{-t},$$

$$P_n(t) = \frac{F_n(0)e^{-t}}{n^2 - 1} + \frac{[f_n - F_n(0)]e^{-n^2 t}}{n^2 - 1}, \qquad n > 1.$$

Therefore

$$w(x, t) = \{F_1(0)te^{-t} + f_1 e^{-t}\} \sin x.$$

$$+ \sum_{n=2}^{\infty} \left\{ \frac{F_n(0)e^{-t}}{n^2 - 1} + \frac{[f_n - F_n(0)]e^{-n^2 t}}{n^2 - 1} \right\} \sin nx. \qquad \blacksquare$$

The method is summarized in the following algorithm.

Algorithm for solving the heat equation by the separation-of-variables method

Given the differential equations

$$w_t = kw_{xx} + F(x, t), \qquad 0 < x < L, \quad 0 < t,$$

with boundary conditions,[10] $w(0, t) = 0$, $w(L, t) = 0$, $0 < t$; and initial condition, $w(x, 0) = f(x)$, $0 < x < L$.

1. Solve

$$X_n'' + \lambda_n^2 X_n = 0,$$
$$X_n(0) = 0, \qquad X_n(L) = 0,^{10}$$

for

$$X_n(x) = \sin \lambda_n x, \qquad \lambda_n = \frac{n\pi}{L}, \qquad n = 1, 2, 3, \ldots .$$

2. Set

$$w(x, t) = \sum_{n=1}^{\infty} P_n(t)X_n(x),$$

$$F(x, t) = \sum_{n=1}^{\infty} F_n(t)X_n(x),$$

$$f(x) = \sum_{n=1}^{\infty} f_n X_n(x).$$

3. Determine the coefficients of the above orthogonal series by[11]

$$F_n(t) = \frac{2}{L}\int_0^L F(x, t)X_n(x)\, dx,$$

$$f_n = \frac{2}{L}\int_0^L f(x)X_n(x)\, dx,$$

and the $\{P_n(t)\}$ are the solutions to the initial value problems:

$$\frac{dP_n}{dt} + k\lambda_n^2 P_n = F_n(t),$$

$$P_n(0) = f_n, \qquad n = 1, 2, 3, \ldots .$$

4. Substitute the $\{P_n(t)\}$ in the series for $w(x, t)$.

---

10. Other boundary conditions may be given as in Example 2. If so, the eigenfunctions will be different from those of the fixed endpoint problem.

11. The formulas for the coefficients may differ with the boundary conditions.

EXERCISES 9.7

For Exercises 1–4 solve the boundary-initial value problem,

$$\begin{aligned} w_t &= kw_{xx}, \quad 0 < x < \pi, \quad 0 < t; \\ w(0, t) &= 0, \quad w(\pi, t) = 0, \quad 0 < t; \\ w(x, 0) &= f(x), \quad 0 < x < \pi, \end{aligned}$$

subject to the given initial data. Plot $v(x, t)$ versus x for $kt = 0, 0.5, 1$.

1. $f(x) = 4 \sin x + 12 \sin 4x$

2. $f(x) = 3 \sin 2x + 15 \sin 5x$

3. $f(x) = 16 \sin^5 x$

4. $f(x) = \cos 3x \sin x$

5. Solve

$$\begin{aligned} w_t &= kw_{xx}, \quad 0 < x < L, \quad 0 < t; \\ w_x(0, t) &= 0, \quad w_x(L, t) = 0, \quad 0 < t; \\ w(x, 0) &= f(x), \quad 0 < x < L. \end{aligned}$$

For Exercises 6–9 solve the boundary-initial value problem of Exercise 5 with the given initial data.

6. $L = 2, f(x) = x^2 - 3x$

7. $L = 4, f(x) = \sinh(x - 2)$

8. $L = 1, f(x) = \cosh x$

9. $L = 2, f(x) = 1 - |x - 1|$

For Exercises 10–11 solve the following inhomogeneous boundary-initial value problem with the given initial data and heat generating function $F(x, t)$,

$$\begin{aligned} w_t &= w_{xx} + F(x, t), \quad 0 < x < \pi, \quad 0 < t; \\ w(0, t) &= 0, \quad w(\pi, t) = 0, \quad 0 < t; \\ w(x, 0) &= f(x), \quad 0 < x < \pi. \end{aligned}$$

10. $F(x, t) = x(\pi - x) \cos t, f(x) = 2 \sin x$

11. $F(x, t) = x(\pi - x) \sin t, f(x) = \sin^3 x$

If nonzero temperature is prescribed at either end of a rod, the temperature $w(x, t)$ can be decomposed into the sum of two functions, one that takes on the boundary data and the other that is a solution to a boundary-initial value problem with an inhomogeneous differential equation and zero boundary data. Assume that $w(x, t)$ is a solution to

$$\begin{aligned} w_t &= kw_{xx}, \quad 0 < x < L, \quad 0 < t; \\ w(0, t) &= h(t), \quad w(L, t) = g(t), \quad 0 < t; \\ w(x, 0) &= f(x), \quad 0 < x < L, \end{aligned} \tag{9.7.6}$$

where $h(t)$ and $g(t)$ have continuous first derivatives, and $f(x)$ is piecewise smooth. Then if we set

$$w(x, t) = \frac{(L - x)}{L} h(t) + \frac{xg(t)}{L} + v(x, t),$$

the function $v(x, t)$ is a solution to the following boundary-initial value problems:

$$\begin{aligned} v_t &= kv_{xx} - h'(t)\frac{(L - x)}{L} - \frac{xg'(t)}{L}, \qquad 0 < x < L, \qquad 0 < t; \\ v(0, t) &= 0, \qquad v(L, t) = 0, \qquad 0 < t; \\ v(x, 0) &= f(x) - \frac{(L - x)h(0)}{L} - \frac{xg(0)}{L}, \qquad 0 < x < L. \end{aligned} \tag{9.7.7}$$

12. Show that if $w(x, t)$ is a solution to (9.7.6), then $v(x, t)$ is a solution to (9.7.7).

For Exercises 13–16, solve (9.7.6) with the prescribed data.

13. $L = \pi$, $k = 0.5$, $h(t) = 2e^{-t}$, $g(t) = 4e^{-2t}$, $f(x) = \sin 2x$
14. $L = \pi$, $k = 0.5$, $h(t) = 3 \cos t$, $g(t) = 4 \sin t$, $f(x) = 0$
15. $L = 1$, $k = 1$, $h(t) = 3 \cos t$, $g(t) = 4 \sin t$, $f(x) = 0$
16. $L = 1$, $k = 1$, $h(t) = 0$, $g(t) = 6e^{-2t}$, $f(x) = 8 \sin(8\pi x)$

The flow of heat along a thin rod with an imperfectly insulated lateral surface can be modeled by the partial differential equation,

$$w_t + sw = kw_{xx}, \tag{9.7.8}$$

where s is a parameter characteristic of the lateral surface of the rod. Experimental data indicate that the effect of the lateral heat loss is exponential damping of the temperature of the rod,

$$w(x, t) = e^{-at}v(x, t). \tag{9.7.9}$$

Substituting this trial solution into (9.7.8), we find that

$$-ae^{-at}v(x, t) + e^{-at}v_t(x, t) + se^{-at}v(x, t) = ke^{-at}v_{xx}(x, t).$$

If we take $a = s$, then we find that $v(x, t)$ will be a solution to the heat equation, $v_t = kv_{xx}$.

17. Use (9.7.9) with $a = s$ and show that if $w(x, t)$ is a solution to

$$\begin{aligned} w_t + sw &= kw_{xx} + F(x, t), \qquad 0 < x < L, \quad 0 < t; \\ w(0, t) &= h(t), \qquad w(L, t) = g(t), \qquad 0 < t; \\ w(x, 0) &= f(x), \qquad 0 < x < L, \end{aligned} \tag{9.7.10}$$

then $v(x, t)$ is a solution to

$$\begin{aligned} v_t &= kv_{xx} + e^{st}F(x, t), \qquad 0 < x < L, \quad 0 < t; \\ v(0, t) &= e^{st}h(t), \qquad v(L, t) = e^{st}g(t), \qquad 0 < t; \\ v(x, 0) &= f(x), \qquad 0 < x < L. \end{aligned} \tag{9.7.11}$$

For Exercises 18–23, solve (9.7.10) with the given data.

18. $L = \pi$, $k = 2$, $s = 2$, $h(t) = 0$, $g(t) = 0$, $f(x) = -4 \sin 3x$, $F(x, t) = 9e^t \sin 4x$

19. $L = \pi$, $k = 1$, $s = 2$, $h(t) = 0$, $g(t) = 0$, $f(x) = (\cos 4x - 1)$, $F(x, t) = -6e^{-t}x(\pi - x)$

20. $L = \pi$, $k = 2$, $s = 1$, $h(t) = 6e^{-t}$, $g(t) = 4e^{-2t}$, $f(x) = \cos x + 5$, $F(x, t) = 0$

21. $L = \pi$, $k = 1$, $s = 2$, $h(t) = -3 \sin t$, $g(t) = 5$, $f(x) = 0$, $F(x, t) = 0$

22. $L = 2$, $k = 1$, $s = 2$, $h(t) = 7 \cos 2t$, $g(t) = 7 \cos 5t$, $f(x) = 7 \cos(5\pi x)$, $F(x, t) = 3 \sin t \sin(\pi x)$

23. $L = 4$, $k = 1$, $s = 1$, $h(t) = -4e^{-2t}$, $g(t) = 0$, $f(x) = \sin(\pi x/4)$, $F(x, t) = -2 \sin 2t \sin(6\pi x)$

9.8 LAPLACE'S EQUATION

DIRICHLET PROBLEM FOR A SQUARE

Laplace's equation,

$$\frac{\partial^2 v}{\partial x^2} + \frac{\partial^2 v}{\partial y^2} = 0, \tag{9.8.1}$$

occurs in a variety of physical and mathematical problems such as in steady-state heat flow, electrostatics, gravitational theory, and hydrodynamics. For example, suppose that the temperature is prescribed on the edges of a square plate of area π^2 (see Fig. 9.18). Then the temperature at any interior point will approach a steady state $v(x, y)$ as time increases. The steady state temperature is a solution to the following boundary value problem:

$$\begin{aligned} \frac{\partial^2 v}{\partial x^2} + \frac{\partial^2 v}{\partial y^2} &= 0, \qquad 0 < x < \pi, \quad 0 < y < \pi, \\ v(x, 0) &= f(x), \qquad 0 < x < \pi, \\ v(x, \pi) &= g(x), \qquad 0 < x < \pi, \end{aligned} \tag{9.8.2}$$

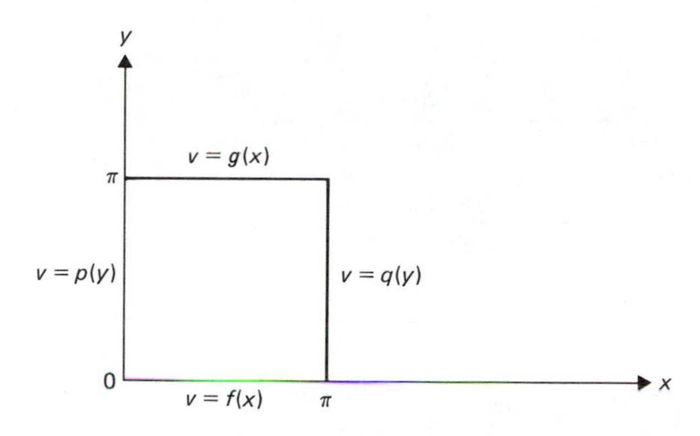

Figure 9.18 The Dirichlet problem for a square.

$$v(0, y) = p(y), \qquad 0 < y < \pi,$$
$$v(\pi, y) = q(y), \qquad 0 < y < \pi,$$

where f, g, p, and q are the prescribed temperatures on the sides of the plate.

The finding of a solution to Laplace's equation in some region that takes on given boundary values is known as a *Dirichlet problem.*[12] The boundary value problem (9.8.2) is an example of an *interior Dirichlet problem* because the region of interest is the interior of the square on which the temperature is given. In an *exterior Dirichlet problem,* the region of interest is the exterior of the boundary curve (or surface).

The interior Dirichlet problem for the square can be solved by the separation-of-variables method by decomposing it into four subproblems and adding their solutions to produce $v(x, y)$. The boundary conditions of the four problems are

1. $v(x, 0) = f(x)$,
   $v(x, \pi) = 0$,
   $v(0, y) = 0$,
   $v(\pi, y) = 0$.

2. $v(x, 0) = 0$,
   $v(x, \pi) = g(x)$,
   $v(0, y) = 0$,
   $v(\pi, y) = 0$.

3. $v(x, 0) = 0$,
   $v(x, \pi) = 0$,
   $v(0, y) = p(y)$,
   $v(\pi, y) = 0$.

4. $v(x, 0) = 0$,
   $v(x, \pi) = 0$,
   $v(0, y) = 0$,
   $v(\pi, y) = q(y)$, $\quad 0 < y < \pi,\ 0 < x < \pi$.

Then

$$v(x, y) = v^1(x, y) + v^2(x, y) + v^3(x, y) + v^4(x, y), \tag{9.8.3}$$

12. P.G.L. Dirichlet (1805–1859) was a German mathematician who made contributions to the theory of numbers, trigonometric series, and differential equations.

where

$$v^1(x, y) = \sum_{n=1}^{\infty} f_n \sin nx \frac{\sinh n(\pi - y)}{\sinh n\pi},$$

$$v^2(x, y) = \sum_{n=1}^{\infty} g_n \sin nx \frac{\sinh ny}{\sinh n\pi},$$

$$v^3(x, y) = \sum_{n=1}^{\infty} p_n \frac{\sinh n(\pi - x)}{\sinh n\pi} \sin ny,$$

$$v^4(x, y) = \sum_{n=1}^{\infty} q_n \frac{\sinh nx}{\sinh n\pi} \sin ny,$$

with

$$f_n = \frac{2}{\pi}\int_0^{\pi} f(x) \sin nx \, dx, \qquad g_n = \frac{2}{\pi}\int_0^{\pi} g(x) \sin nx \, dx,$$

$$p_n = \frac{2}{\pi}\int_0^{\pi} p(y) \sin ny \, dy, \qquad q_n = \frac{2}{\pi}\int_0^{\pi} q(y) \sin ny \, dy.$$

Let us solve problem 3 to obtain $v^3(x, y)$. The construction of the solutions to the other problems is left to the reader. We set $v(x, y) = X(x)Y(y)$, then

$$\frac{X''}{X} + \frac{Y''}{Y} = 0.$$

If γ is the separation constant,

$$\frac{X''}{X} = \gamma, \qquad \frac{Y''}{Y} = -\gamma,$$

then

$$X'' - \gamma X = 0, \qquad Y'' + \gamma Y = 0.$$

Since $v = 0$ at $y = 0$ and $y = \pi$ for problem 3, we have $Y(0) = Y(\pi) = 0$. Consequently, $\gamma = n^2$.

$$Y_n(y) = \sin ny,$$

$$X_n(x) = A_n \cosh nx + B_n \sinh nx,$$

and

$$v^3(x, y) = \sum_{n=1}^{\infty} (A_n \cosh nx + B_n \sinh nx) \sin ny.$$

From the boundary conditions, $v(0, y) = p(y)$, $v(\pi, y) = 0$, we have

$$p(y) = \sum_{n=1}^{\infty} A_n \sin ny,$$

$$0 = \sum_{n=1}^{\infty} (A_n \cosh n\pi + B_n \sinh n\pi) \sin ny.$$

Because

$$p(y) = \sum_{n=1}^{\infty} p_n \sin ny \quad \text{and} \quad 0 = \sum_{n=1}^{\infty} 0 \sin ny,$$

the coefficients A_n and B_n satisfy the algebraic equations,

$$p_n = A_n,$$

$$0 = A_n \cosh n\pi + B_n \sinh n\pi.$$

Therefore

$$X_n(x) = p_n \frac{(\sinh n\pi \cosh nx - \cosh n\pi \sinh nx)}{\sinh n\pi}$$

$$= p_n \frac{\sinh n(\pi - x)}{\sinh n\pi},$$

and

$$v^3(x, y) = \sum_{n=1}^{\infty} p_n \frac{\sinh n(\pi - x)}{\sinh n\pi} \sin ny.$$

EXAMPLE 1 Solve

$$v_{xx} + v_{yy} = 0, \qquad 0 < x < \pi, \quad 0 < y < \pi$$

with boundary conditions

$$v(x, 0) = \sin^2 x, \qquad 0 < x < \pi,$$

$$v(x, \pi) = 0, \qquad 0 < x < \pi,$$

$$v(0, y) = \cos^2 y, \qquad 0 < y < \pi,$$

$$v(\pi, y) = 0, \qquad 0 < y < \pi.$$

SOLUTION Since $g(x)$ and $q(x)$ are identically zero, the substitutions $v^2(x,y)$ and $v^4(x,y)$ are the zero solutions. From the nonzero boundary data $f(x) = \sin^2 x$,

$p(y) = \cos^2 y,$

$$f_n = \frac{2}{\pi}\int_0^\pi \left(\frac{1 - \cos 2x}{2}\right) \sin nx \, dx,$$

$$p_n = \frac{2}{\pi}\int_0^\pi \left(\frac{1 + \cos 2y}{2}\right) \sin ny \, dy.$$

Then

$$f_2 = p_2 = 0,$$

$$f_n = \frac{1}{\pi}(1 - \cos n\pi)\left(\frac{1}{n} - \frac{n}{(n+2)(n-2)}\right), \qquad n \neq 2,$$

$$p_n = \frac{1}{\pi}(1 - \cos n\pi)\left(\frac{1}{n} + \frac{n}{(n+2)(n-2)}\right), \qquad n \neq 2,$$

and

$$v(x, y) = \sum_{n=1}^{\infty} \left\{ f_n \sin nx \frac{\sinh n(\pi - y)}{\sinh n\pi} + p_n \frac{\sinh n(\pi - x)}{\sinh n\pi} \sin ny \right\}. \qquad \blacksquare$$

THE INTERIOR DIRICHLET PROBLEM FOR A CIRCLE

Let us consider the problem of finding a bounded solution to Laplace's equation in the interior of a circle that takes on continuous data on the boundary. It is convenient to use polar coordinates,

$$x = r \cos \theta, \qquad y = r \sin \theta.$$

Using the rules of calculus it can be shown that Laplace's equation (9.8.1) is equivalent to

$$v_{rr} + \frac{1}{r} v_r + \frac{1}{r^2} v_{\theta\theta} = 0. \tag{9.8.4}$$

We shall construct a bounded solution $v(r, \theta)$ in the region, $0 \leq r < 1$, $-\pi \leq \theta \leq \pi$, such that $v = f(\theta)$, a continuous function on the circle, $r = 1$. Both $v(r, \theta)$ and $f(\theta)$ are required to be 2π-periodic in θ, that is,

$$v(r, \theta) = v(r, \theta + 2\pi), \qquad f(\theta) = f(\theta + 2\pi), \quad \text{all } \theta. \tag{9.8.5}$$

Employing the separation-of-variables method we set $v(r, \theta) = R(r)S(\theta)$ and substitute into (9.8.4) to obtain

$$R''S + \frac{1}{r} R'S + \frac{1}{r^2} RS'' = 0,$$

and consequently

$$\frac{r^2R''}{R} + \frac{rR'}{R} = -\frac{S''}{S} = \gamma.$$

From this we have two ordinary differential equations,

$$r^2R'' + rR' - \gamma R = 0,$$
$$S'' + \gamma S = 0.$$

If the separation constant is negative, $\gamma = -\sigma^2$, then the differential equation

$$S'' - \sigma^2 S = 0$$

has no nonzero periodic solutions. Therefore we set $\gamma = \sigma^2$, a nonnegative constant, and conclude that

$$S(\theta) = c_1\theta + c_0 \quad \text{if} \quad \sigma^2 = 0,$$
$$S(\theta) = c_1 \cos \sigma\theta + c_2 \sin \sigma\theta \quad \text{if} \quad \sigma^2 > 0.$$

To satisfy the periodicity condition (9.8.5) we must have $\gamma = \sigma^2 = n^2$, $n = 0, 1, 2, \ldots$. For $n = 0$, we must have $c_1 = 0$ and take $S(\theta) = c_0$. For n positive,

$$S(\theta) = c_1 \cos n\theta + c_2 \sin n\theta.$$

The differential equation for R,

$$r^2R'' + rR' - n^2R = 0,$$

is an Euler differential equation (see Section 2.6). If $n = 0$,

$$R(\theta) = c_1 \ln r + c_0;$$

if n is a positive integer,

$$R(\theta) = c_1r^n + c_2r^{-n}.$$

Consequently,

$$v_0(r, \theta) = a_0 + b_0 \ln r,$$
$$v_n(r, \theta) = (a_n'r^n + b_n'r^{-n})(a_n \cos n\theta + b_n \sin n\theta).$$

For the interior Dirichlet problem, $v(r, \theta)$ is required to be bounded and the terms $\ln r$ and r^{-n} must be discarded since $0 \le r < 1$. It follows that (with $a_n' = 1$)

$$v(r, \theta) = a_0 + \sum_{n=1}^{\infty} r^n(a_n \cos n\theta + b_n \sin n\theta). \tag{9.8.6}$$

The coefficients in (9.8.6) are determined by setting $v(1, \theta) = f(\theta)$. Hence

$$f(\theta) = a_0 + \sum_{n=1}^{\infty} a_n \cos n\theta + b_n \sin n\theta, \tag{9.8.7}$$

with

$$a_0 = \frac{1}{2\pi}\int_{-\pi}^{\pi} f(\phi)\,d\phi, \qquad a_n = \frac{1}{\pi}\int_{-\pi}^{\pi} f(\phi)\cos n\phi\,d\phi,$$

$$b_n = \frac{1}{\pi}\int_{-\pi}^{\pi} f(\phi)\sin n\phi\,d\phi.$$

EXAMPLE 2 Find a bounded solution to the interior Dirichlet problem,

$$v_{rr} + \frac{1}{r}v_r + \frac{1}{r^2}v_{\theta\theta} = 0, \qquad 0 < r < 1, \quad -\pi \le \theta \le \pi,$$

$$v(1, \theta) = \sin 4\theta\left[u\left(\theta + \frac{\pi}{4}\right) - u\left(\theta - \frac{\pi}{4}\right)\right], \qquad -\pi \le \theta \le \pi.$$

SOLUTION In accordance with (9.8.6–9.8.7), we have

$$\begin{aligned} b_n &= \frac{1}{\pi}\int_{-\pi/4}^{\pi/4} \sin 4\phi \sin n\phi\,d\phi \\ &= \frac{1}{\pi}\sin\left(\frac{n\pi}{4}\right)\left[\frac{8}{(4-n)(4+n)}\right], \qquad n \ne 4, \\ b_4 &= \frac{1}{4}, \\ a_0 &= 0, \qquad a_n = 0, \end{aligned}$$

and

$$v(r, \theta) = \sum_{n=1}^{\infty} r^n b_n \sin n\theta. \qquad \blacksquare$$

THE EXTERIOR DIRICHLET PROBLEM FOR A CIRCLE

For the exterior problem where the solution is sought for $r > 1$, the requirement that the solution be bounded implies that $\ln r$ and r^n cannot be used in the expansion of the solution. Therefore we set

$$v(r, \theta) = a_0 + \sum_{n=1}^{\infty} r^{-n}(a_n \cos n\theta + b_n \sin n\theta), \qquad r \ge 1, \quad -\pi \le \theta \le \pi, \tag{9.8.8}$$

and determine the coefficients so that

$$v(1, \theta) = f(\theta), \qquad -\pi \leq \theta \leq \pi.$$

EXAMPLE 3 Find a bounded solution to the exterior Dirichlet problem,

$$v_{rr} + \frac{1}{r} v_r + \frac{1}{r^2} v_{\theta\theta} = 0, \qquad 1 < r, \quad -\pi \leq \theta \leq \pi,$$

$$v(1, \theta) = \pi - |\theta|, \qquad -\pi \leq \theta \leq \pi.$$

SOLUTION In accordance with (9.8.6–9.8.7) with r^n replaced by r^{-n} we have

$$a_0 = \frac{1}{2\pi} \int_{-\pi}^{\pi} (\pi - |\phi|)\, d\phi = \frac{\pi}{2},$$

$$a_n = \frac{1}{\pi} \int_{-\pi}^{\pi} (\pi - |\phi|) \cos n\phi\, d\phi = \frac{2}{\pi n^2}(1 - \cos n\pi),$$

$$b_n = 0,$$

and

$$v(r, \theta) = a_0 + \sum_{n=1}^{\infty} r^{-n} a_n \cos n\theta, \qquad 1 < r, \quad -\pi \leq \theta \leq \pi. \qquad \blacksquare$$

Note that as $r \to \infty$, the solution approaches the average value of the temperature distribution on the boundary of the disk.

EXERCISES 9.8

For Exercises 1–4 solve the boundary value problem

$$v_{xx} + v_{yy} = 0, \qquad 0 < x < \pi, \qquad 0 < y < \pi;$$

$$v(x, 0) = f(x), \qquad v(x, \pi) = g(x), \qquad 0 < x < \pi;$$

$$v(0, y) = p(y), \qquad v(\pi, y) = q(y), \qquad 0 < y < \pi,$$

subject to the given boundary data.

1. $f(x) = g(x) = 3 \sin x + 6 \sin 5x - 10 \sin 6x$, $p(y) = q(y) = 4 \sin^2 y$
2. $f(x) = 2 \sin x + 3 \sin 3x$, $g(x) = 0$, $p(y) = \sin y \cos^2 y$, $q(y) = 6 \sin 6y$
3. $f(x) = \sin^2 y$, $g(x) = \sin^2 3x$, $p(y) = 0$, $q(y) = 0$
4. $f(x) = 0$, $g(x) = 0$, $p(y) = \sin^2 2y$, $q(y) = 6 \sin y \cos y$

For Exercises 5–8 solve both the interior and exterior Dirichlet problems,

$$v_{rr} + \frac{1}{r}v_r + \frac{1}{r^2}v_{\theta\theta} = 0, \qquad -\pi \le \theta \le \pi, \qquad 0 < r < 1 \quad \text{or} \quad 1 < r < \infty,$$

$$v(1, \theta) = f(\theta), \qquad -\pi \le \theta \le \pi,$$

subject to the given boundary data.

5. $f(\theta) = \left(\theta^2 - \frac{\pi}{4}\right)\left[u\left(\theta + \frac{\pi}{4}\right) - u\left(\theta - \frac{\pi}{4}\right)\right]$

6. $f(\theta) = \cos 4\theta \left[u\left(\theta + \frac{\pi}{4}\right) - u\left(\theta - \frac{\pi}{4}\right)\right]$

7. $f(\theta) = e^{-\theta}\sin\theta$

8. $f(\theta) = \cosh\theta\sin^2\theta$

If the potential is given on two concentric circles, $v = h(\theta)$ on $r = 1$ and $v = g(\theta)$ on $r = 2$, then the potential in the annulus, $1 < r < 2$, can be found by setting

$$v(r, \theta) = A_0 + B_0 \ln r + \sum_{n=1}^{\infty} (A_n r^n + B_n r^{-n}) \cos n\theta + (C_n r^n + D_n r^{-n}) \sin n\theta \tag{9.8.9}$$

and determining the coefficients so that the boundary conditions are satisfied. We write

$$\begin{aligned} h(\theta) &= \frac{a_0}{2} + \sum_{n=1}^{\infty} a_n \cos n\theta + b_n \sin n\theta, \\ g(\theta) &= \frac{c_0}{2} + \sum_{n=1}^{\infty} c_n \cos n\theta + d_n \sin n\theta; \end{aligned} \tag{9.8.10}$$

set $v(1, \theta) = h(\theta)$, $v(2, \theta) = g(\theta)$; and match the coefficients of $\cos n\theta$ and $\sin n\theta$ to obtain linear algebraic equations for the A_n, B_n, C_n, D_n.

At $r = 1$, we have

$$A_0 = \frac{a_0}{2}, \qquad A_n + B_n = a_n, \qquad C_n + D_n = b_n, \qquad n \ge 1; \tag{9.8.11}$$

and at $r = 2$, we have

$$A_0 + B_0 \ln 2 = \frac{c_0}{2}, \qquad A_n 2^n + B_n 2^{-n} = c_n, \qquad C_n 2^n + D_n 2^{-n} = d_n, \qquad n \ge 1. \tag{9.8.12}$$

9. Assume that $h(\theta)$ and $g(\theta)$ are continuous 2π-periodic functions with convergent Fourier series and that the coefficients of (9.8.9) satisfy the algebraic equations (9.8.11–9.8.12). Show that $v(r, \theta)$ is a solution to the boundary value problem

$$v_{rr} + \frac{1}{r} v_r + \frac{1}{r^2} v_{\theta\theta} = 0, \qquad 1 < r < 2, \quad -\pi \leq \theta \leq \pi,$$

$$v(1, \theta) = h(\theta), \qquad v(2, \theta) = g(\theta), \qquad -\pi \leq \theta \leq \pi. \tag{9.8.13}$$

This problem is called the Dirichlet problem for an annulus.

For Exercises 10–13, solve (9.8.13) with the given boundary data.

10. $h(\theta) = 3 + 2\cos\theta + 8\sin 3\theta,$
$g(\theta) = 4 - 6\cos 2\theta + 7\sin 4\theta$

11. $h(\theta) = -2 + 5\cos 2\theta - 4\sin\theta,$
$g(\theta) = 2 - 5\cos 2\theta + 6\sin 5\theta$

12. $h(\theta) = \sum_{n=1}^{\infty} \frac{1}{n^2}\cos n\theta,$
$g(\theta) = \sum_{n=1}^{\infty} \frac{1}{n^2}\sin n\theta$

13. $h(\theta) = 4 + \sum_{n=1}^{\infty} \cos(n\pi)\cos n\theta,$
$g(\theta) = \sum_{n=1}^{\infty} \sin\left(\frac{n\pi}{2}\right)\cos n\theta$

9.9 TRANSMISSION LINE EQUATIONS AND SYSTEMS OF PARTIAL DIFFERENTIAL EQUATIONS

DERIVATION OF THE TRANSMISSION LINE EQUATIONS

Because of the interest in transoceanic communication, a problem that stimulated a great deal of mathematical inquiry in the late nineteenth century was the development of a model for the transmission of data in a long, straight, imperfectly insulated (leaky) wire or cable. This problem and its associated partial differential equations, the *telegraph equations* or *transmission line equations*, has a rich history. The name most associated with the problem is that of Lord Kelvin (Sir William Thomson, 1824–1907), who developed a special case of the equations as part of his model for the Atlantic cable. The general equations appeared in papers by G. Kirchhoff (1824–1887) in 1857, O. Heaviside (1850–1925) in 1876, and H. Poincaré (1854–1912) in 1893; a complete analysis of the equations is given in Webster [9, Ch. 4]. In this section we will consider several important cases.

Let Q be the charge density per unit length and J the current—both these quantities depend on t, the time, and x, the distance, along the cable. By the principle of conservation of charge, in a segment of a perfectly insulated cable bounded by $x = a$ and $x = b$ the quantities Q and J satisfy

$$\frac{d}{dt}\int_a^b Q(x, t)\, dx = J(a, t) - J(b, t),$$

or, equivalently,

$$\int_a^b \frac{\partial Q}{\partial t}(x, t)\, dx = -\int_a^b \frac{\partial J}{\partial x}(x, t)\, dx. \tag{9.9.1}$$

If the insulation on the cable is imperfect, some of the current will leak from the cable to the earth, and the amount lost per unit time is proportional to the voltage, that is, V, the potential difference between the cable and the earth. The leakage current from a segment of the length dx is therefore $GVdx$ where G is the conductance per unit length. Subtracting the integral of this quantity from $x = a$ to $x = b$ from the right side of (9.9.1) and combining terms, we have

$$\int_a^b \left\{\frac{\partial Q}{\partial t}(x, t) + \frac{\partial J}{\partial x}(x, t) + GV(x, t)\right\} dx = 0.$$

There are three unknown quantities in the last expression, but we can eliminate one by invoking Faraday's law

$$\frac{\partial Q}{\partial t} = C\frac{\partial V}{\partial t},$$

where C is the capacitance per unit length. Therefore

$$\int_a^b \left\{C\frac{\partial V}{\partial t}(x, t) + \frac{\partial J}{\partial x}(x, t) + GV(x, t)\right\} dx = 0,$$

and because a and b are arbitrary and the integrand is assumed to be continuous, we have

$$C\frac{\partial V}{\partial t} + \frac{\partial J}{\partial x} + GV = 0, \tag{9.9.2}$$

a first order partial differential equation.

The second partial differential equation needed to complete the mathematical model is derived from summing the potential drops along a segment of the cable. By Ohm's law, the voltage drop due to the resistance of the wire is the integral of $RJ(x, t)$, where R is the resistance per unit length. Invoking another result of Faraday, we have that the voltage drop due to the time variation of the current is the integral with respect to x of $L(\partial J/\partial t)(x, t)$, where L is the self-inductance per unit length. Integrating from $x = a$ to $x = b$ we obtain the relation

$$\begin{aligned}\int_a^b \left\{L\frac{\partial J}{\partial t}(x, t) + RJ(x, t)\right\} dx &= V(a, t) - V(b, t) \\ &= -\int_a^b \frac{\partial V}{\partial x}(x, t)\, dx.\end{aligned}$$

By the argument used before we conclude that

$$L\frac{\partial J}{\partial t} + \frac{\partial V}{\partial x} + RJ = 0, \tag{9.9.3}$$

the second partial differential equation of the model. The system of first order partial differential equations

$$\begin{aligned} L\frac{\partial J}{\partial t} + \frac{\partial V}{\partial x} + RJ &= 0, \\ C\frac{\partial V}{\partial t} + \frac{\partial J}{\partial x} + GV &= 0, \end{aligned} \tag{9.9.4}$$

is called the *telegraph equations* or *transmission line equations.*

If the current J and the voltage V are given at $t = 0$,

$$J(x, 0) = j(x), \qquad V(x, 0) = v(x), \qquad -\infty < x < \infty, \tag{9.9.5}$$

where $j(x)$ and $v(x)$ are differentiable functions, then it can be shown [9] that $J(x, t)$ and $V(x, t)$ are determined for all positive t.

The special case of (9.9.4) studied by Lord Kelvin arises if L, the self-inductance, is a negligible quantity. Setting $L = 0$ in the first equation we can solve for J,

$$J = -\frac{1}{R}\frac{\partial V}{\partial x},$$

and substituting this into the second equation we obtain the second order partial differential equation

$$RC\frac{\partial V}{\partial t} + RGV = \frac{\partial^2 V}{\partial x^2}.$$

This equation also models the flow of heat along a rod that is losing energy to its surroundings by radiation, as well as a variety of other diffusion phenomena (see Section 9.7).

Returning to the system (9.9.4), if we assume that neither L nor C is zero, we can write it in the form

$$\begin{aligned} \frac{\partial J}{\partial t} + \frac{1}{L}\frac{\partial V}{\partial x} + \frac{R}{L}J &= 0, \\ \frac{\partial V}{\partial t} + \frac{1}{C}\frac{\partial J}{\partial x} + \frac{G}{C}V &= 0. \end{aligned} \tag{9.9.6}$$

This immediately suggests using vector and matrix notation; so if we let

$$\mathbf{w} = \begin{pmatrix} J \\ V \end{pmatrix}, \qquad \frac{\partial}{\partial t}\mathbf{w} = \begin{pmatrix} \dfrac{\partial J}{\partial t} \\ \dfrac{\partial V}{\partial t} \end{pmatrix}, \qquad \frac{\partial}{\partial x}\mathbf{w} = \begin{pmatrix} \dfrac{\partial J}{\partial x} \\ \dfrac{\partial V}{\partial x} \end{pmatrix},$$

and A and B be the matrices

$$A = \begin{pmatrix} 0 & L^{-1} \\ C^{-1} & 0 \end{pmatrix}, \qquad B = \begin{pmatrix} RL^{-1} & 0 \\ 0 & GC^{-1} \end{pmatrix},$$

we can write (9.9.6) in the convenient form,

$$\frac{\partial}{\partial t}\mathbf{w} + A\frac{\partial}{\partial x}\mathbf{w} + B\mathbf{w} = \mathbf{0}. \tag{9.9.7}$$

The relation (9.9.7) is an example of a *first order linear system of partial differential equations.* We will see that a technique for solving certain such systems is analogous to the eigenvalue–eigenvector method developed in Chapter 5 for solving linear systems of ordinary differential equations.

SOLUTION OF THE NONDISSIPATIVE TRANSMISSION LINE EQUATIONS

If the dissipative effects due to the resistance and leakage of the cable are negligible, then we set $R = 0$, $G = 0$ in (9.9.6) to obtain the system, the nondissipative transmission line equations,

$$\frac{\partial J}{\partial t} + \frac{1}{L}\frac{\partial V}{\partial x} = 0,$$

$$\frac{\partial V}{\partial t} + \frac{1}{C}\frac{\partial J}{\partial x} = 0,$$

or

$$\frac{\partial}{\partial t}\begin{pmatrix} J \\ V \end{pmatrix} + A\frac{\partial}{\partial x}\begin{pmatrix} J \\ V \end{pmatrix} = \begin{pmatrix} 0 \\ 0 \end{pmatrix}, \tag{9.9.8}$$

with

$$A = \begin{pmatrix} 0 & L^{-1} \\ C^{-1} & 0 \end{pmatrix}. \tag{9.9.9}$$

The coefficient matrix A has real, distinct eigenvalues that are roots of its characteristic equation,

$$\det(A - \lambda I) = \lambda^2 - \frac{1}{LC} = 0.$$

For convenience, set $c^2 = 1/LC$, $\gamma^2 = L/C$, then

$$\lambda_1 = c, \qquad \mathbf{b}^1 = \begin{pmatrix} 1 \\ \gamma \end{pmatrix}, \qquad \lambda_2 = -c, \qquad \mathbf{b}^2 = \begin{pmatrix} 1 \\ -\gamma \end{pmatrix}$$

are eigenvalues and linearly independent eigenvectors of A. They will be used in the construction of solutions to the partial differential equations (9.9.8).

Recall how we constructed the solution to the two-dimensional initial value problem,

$$\frac{d}{dt}\mathbf{y} = A\mathbf{y}, \qquad \mathbf{y}(0) = \mathbf{y}^0,$$

where the matrix A had distinct real eigenvalues with corresponding linearly independent eigenvectors $\mathbf{b}^1$ and $\mathbf{b}^2$. We set

$$\mathbf{y}(t) = k_1\mathbf{b}^1 e^{\lambda_1 t} + k_2\mathbf{b}^2 e^{\lambda_2 t}$$

and determined the constants k_1 and k_2 so that

$$\mathbf{y}^0 = k_1\mathbf{b}^1 + k_2\mathbf{b}^2.$$

Motivated by this, we seek a solution to the system of partial differential equations (9.9.8) that can be represented as

$$\begin{pmatrix} J(x, t) \\ V(x, t) \end{pmatrix} = p_1(x, t)\begin{pmatrix} 1 \\ \gamma \end{pmatrix} + p_2(x, t)\begin{pmatrix} 1 \\ -\gamma \end{pmatrix} = p_1(x, t)\mathbf{b}^1 + p_2(x, t)\mathbf{b}^2, \qquad \mathbf{(9.9.10)}$$

where the functions $p_1(x, t)$ and $p_2(x, t)$ are to be determined.

Substituting (9.9.10) into the partial differential equations (9.9.8), we obtain

$$\frac{\partial}{\partial t} p_1(x, t)\mathbf{b}^1 + A\frac{\partial}{\partial x} p_1(x, t)\mathbf{b}^1 + \frac{\partial}{\partial t} p_2(x, t)\mathbf{b}^2 + A\frac{\partial}{\partial x} p_2(x, t)\mathbf{b}^2 = \mathbf{0}. \qquad \mathbf{(9.9.11)}$$

Because

$$A\frac{\partial}{\partial x} p_1(x, t)\mathbf{b}^1 = \frac{\partial}{\partial x} p_1(x, t)A\mathbf{b}^1 = \lambda_1\frac{\partial}{\partial x} p_1(x, t)\mathbf{b}^1,$$

$$A\frac{\partial}{\partial x} p_2(x, t)\mathbf{b}^2 = \frac{\partial}{\partial x} p_2(x, t)A\mathbf{b}^2 = \lambda_2\frac{\partial}{\partial x} p_2(x, t)\mathbf{b}^2,$$

and $\lambda_1 = c$, $\lambda_2 = -c$, equation (9.9.11) is equivalent to

$$\left(\frac{\partial}{\partial t} p_1(x, t) + c\frac{\partial}{\partial x} p_1(x, t)\right)\mathbf{b}^1 + \left(\frac{\partial}{\partial t} p_2(x, t) - c\frac{\partial}{\partial x} p_2(x, t)\right)\mathbf{b}^2 = \mathbf{0}.$$

It follows from the linear independence of $\mathbf{b}^1$ and $\mathbf{b}^2$ that the preceding vector equation is satisfied if and only if the coefficients of $\mathbf{b}^1$ and $\mathbf{b}^2$ are both zero. Therefore we require that

$$\begin{aligned} &\frac{\partial}{\partial t} p_1(x, t) + c\frac{\partial}{\partial x} p_1(x, t) = 0, \\ &\frac{\partial}{\partial t} p_2(x, t) - c\frac{\partial}{\partial x} p_2(x, t) = 0, \qquad -\infty < x < \infty, \quad 0 < t. \end{aligned} \tag{9.9.12}$$

These two uncoupled scalar linear first order partial differential equations, together with the initial data, determine the coefficient functions $p_1(x, t)$ and $p_2(x, t)$ that describe the current $J(x, t)$ and the voltage $V(x, t)$ of the cable.

A telegraph line without resistance or leakage will propagate signals with finite velocity and with no distortion or attenuation. Such signals have the basic property of waves: a recognizable feature moves with finite velocity. Let us consider a signal that propagates with velocity c along the x-axis. At $t = 0$ the shape of the signal, its wave profile, can be described as a function $h(x)$. Because the wave profile propagates without distortion, a photograph taken at a later time t will show the same wave profile as the initial one except that it will be displaced by a distance ct. The signal is then described as $h(x - ct)$. It is readily verified that $p = h(x - ct)$ is a solution to the first equation of (9.9.12). From

$$\frac{\partial}{\partial t} h(x - ct) = h'(x - ct)\frac{\partial}{\partial t}(x - ct) = -ch'(x - ct),$$

$$\frac{\partial}{\partial x} h(x - ct) = h'(x - ct)\frac{\partial}{\partial x}(x - ct) = h'(x - ct),$$

we have

$$\frac{\partial}{\partial t} h(x - ct) + c\frac{\partial}{\partial x} h(x - ct) = -ch'(x - ct) + ch'(x - ct) \equiv 0.$$

A parallel computation shows that $h(x + ct)$ is a solution to the second equation of (9.9.12).

The propagation of waves is illustrated by harmonic waves whose wave profiles are sine or cosine curves; for example, an initial signal may be

$$h(x) = a \cos mx.$$

The maximum value of the wave profile is attained at $x = 0$ and at integral multiples of its wave length, $2\pi/m$. The propagated waves

$$h(x - ct) = a \cos m(x - ct), \qquad h(x + ct) = a \cos m(x + ct)$$

have maxima at $x \pm ct = n(2\pi/m)$, $n = 0, \pm 1, \pm 2, \ldots$. Sketches of these propagating harmonic waves are in Fig. 9.19.

For the transmission line problem we set

$$\begin{pmatrix} J(x, t) \\ V(x, t) \end{pmatrix} = p_1(x - ct)\begin{pmatrix} 1 \\ \gamma \end{pmatrix} + p_2(x + ct)\begin{pmatrix} 1 \\ -\gamma \end{pmatrix}. \tag{9.9.13}$$

If at $t = 0$, $J(x, 0) = j(x)$ and $V(x, 0) = v(x)$, then from (9.9.10), we have

$$j(x) = p_1(x) + p_2(x),$$

$$v(x) = \gamma p_1(x) - \gamma p_2(x).$$

Therefore

$$p_1(x) = \frac{1}{2\gamma}(\gamma j(x) + v(x)),$$

$$p_2(x) = \frac{1}{2\gamma}(\gamma j(x) - v(x)).$$

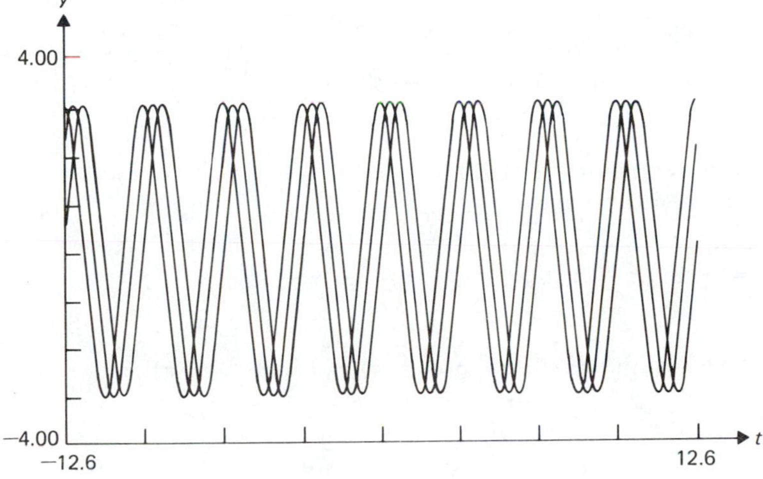

Figure 9.19 Wave profiles of a harmonic wave.

It follows that the current and voltage along the cable are given by

$$\begin{pmatrix} J(x, t) \\ V(x, t) \end{pmatrix} = \frac{1}{2\gamma}[\gamma j(x - ct) + v(x - ct)]\begin{pmatrix} 1 \\ \gamma \end{pmatrix}$$

$$+ \frac{1}{2\gamma}[\gamma j(x + ct) - v(x + ct)]\begin{pmatrix} 1 \\ -\gamma \end{pmatrix},$$

or

$$J(x, t) = \frac{1}{2}[j(x - ct) + j(x + ct)] + \frac{1}{2\gamma}[v(x - ct) - v(x + ct)],$$

$$V(x, t) = \frac{\gamma}{2}[j(x - ct) - j(x + ct)] \tag{9.9.14}$$

$$+ \frac{1}{2}[v(x - ct) + v(x + ct)], \qquad -\infty < x < \infty, \quad 0 < t.$$

We see that both the current $J(x, t)$ and the voltage $V(x, t)$ are superpositions of waves propagating to the right and to the left without distortion. Because of this, the partial differential equations (9.9.8) are called a *two-dimensional system of wave equations.*

EXAMPLE 1 Take $L = 10^{-2}$ henries/ft, $C = 10^{-6}$ farads/ft, and solve equations (9.9.8) with initial data at $t = 0$,[13]

$$j(x) = \sin^2 x[u(x + \pi) - u(x - \pi)],$$

$$v(x) = 0, \qquad -\infty < x < \infty.$$

Plot $J(x, t)$ and $V(x, t)$ for $t = 0, \pi/2, \pi$, and $3\pi/2$.

SOLUTION First we compute $c = (LC)^{-1/2} = 10^4$ ft/sec, $\gamma = (L/C)^{1/2} = 100$ amperes/volt. From (9.9.14), we have

$$J(x, t) = \frac{1}{2}\sin^2(x - ct)[u(x - ct + \pi) - u(x - ct - \pi)]$$

$$+ \frac{1}{2}\sin^2(x + ct)[u(x + ct + \pi) - u(x + ct - \pi)],$$

13. The unit step or Heaviside function $u(x - a)$ is zero if $x < a$ and is unity if $x \geq a$.

$$V(x, t) = \frac{\gamma}{2}\sin^2(x - ct)[u(x - ct + \pi) - u(x - ct - \pi)]$$
$$- \frac{\gamma}{2}\sin^2(x + ct)[u(x + ct + \pi) - u(x + ct - \pi)],$$

with $c = 10^4$ and $\gamma = 100$.

Plots of $J(x, t)$ and $V(x, t)$ are given in Fig. 9.20. ■

SOLUTIONS OF CONSERVATIVE HYPERBOLIC EQUATIONS

The two-dimensional system of wave equations (9.9.8) is an example of a linear system of *conservative hyperbolic* partial differential equations, that is, a system

$$\frac{\partial}{\partial t}\mathbf{w} + A\frac{\partial}{\partial x}\mathbf{w} = \mathbf{0},$$

where $\mathbf{w}$ is an n-dimensional vector and A is an n-by-n real matrix with n real distinct nonzero eigenvalues and n linearly independent eigenvectors.[14]

Definition

A real n-by-n matrix is said to be a *hyperbolic matrix* if it has n real distinct nonzero eigenvalues.

The above discussion leads to the following algorithm.

Algorithm for solving the n-dimensional conservative hyperbolic initial value problem

$$\frac{\partial}{\partial t}\mathbf{w} + A\frac{\partial}{\partial x}\mathbf{w} = \mathbf{0}, \qquad -\infty < x < \infty, \quad 0 < t,$$
$$\mathbf{w}(x, 0) = \mathbf{f}(x), \qquad -\infty < x < \infty. \tag{9.9.15}$$

1. Determine the eigenvalues $\{\lambda_m\}$ and n linearly independent eigenvectors $\{\mathbf{b}^m\}$ of the matrix A.
2. Solve

$$\mathbf{f}(x) = \sum_{m=1}^{n} p_m(x)\mathbf{b}^m \tag{9.9.16}$$

for the functions $\{p_m(x)\}$.

14. It is a result from linear algebra that an $n \times n$ matrix with n distinct eigenvalues has a set of n linearly independent eigenvectors.

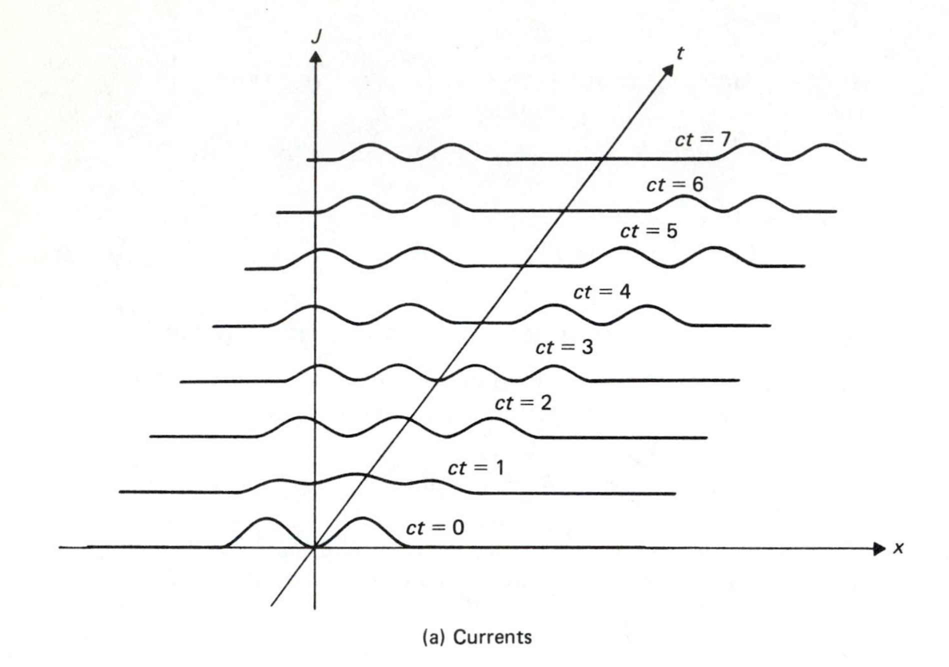

(a) Currents

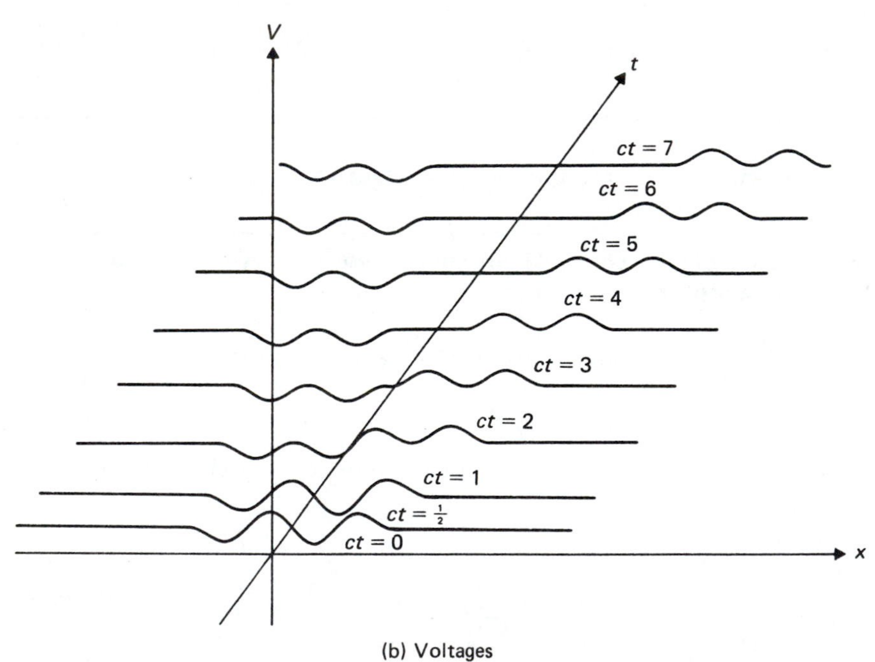

(b) Voltages

Figure 9.20 Wave profiles for Example 1.

3. The solution to the initial value problem is

$$\mathbf{w}(x, t) = \sum_{m=1}^{n} p_m(x - \lambda_m t)\mathbf{b}^m. \tag{9.9.17}$$

Note the important feature that the traveling waves determined by the functions $\{p_m(x)\}$ are not distorted.

EXAMPLE 2 Solve the initial value problem

$$\frac{\partial}{\partial t}\mathbf{w} + A\frac{\partial}{\partial x}\mathbf{w} = \mathbf{0}, \qquad -\infty < x < \infty, \quad 0 < t,$$

with initial data

$$\mathbf{w}(x, 0) = \mathbf{f}(x) = \begin{pmatrix} -6e^{-x^2} \\ xe^{-x^2} \end{pmatrix}, \qquad -\infty < x < \infty,$$

where

$$A = \begin{pmatrix} 2 & 6 \\ -2 & -5 \end{pmatrix}.$$

SOLUTION The eigenvalues and corresponding eigenvectors of A are $\lambda_1 = -1$, $\lambda_2 = -2$,

$$\mathbf{b}^1 = \begin{pmatrix} -2 \\ 1 \end{pmatrix}, \qquad \mathbf{b}^2 = \begin{pmatrix} -3 \\ 2 \end{pmatrix}.$$

We set

$$\begin{pmatrix} w_1(x, t) \\ w_2(x, t) \end{pmatrix} = p_1(x, t)\begin{pmatrix} -2 \\ 1 \end{pmatrix} + p_2(x, t)\begin{pmatrix} -3 \\ 2 \end{pmatrix},$$

then, $p_1(x, t)$ and $p_2(x, t)$ are solutions to

$$\frac{\partial p_1}{\partial t} - \frac{\partial p_1}{\partial x} = 0, \qquad \frac{\partial p_2}{\partial t} - 2\frac{\partial p_2}{\partial x} = 0.$$

Thus, $p_1 = p_1(x + t)$, $p_2 = p_2(x + 2t)$ and

$$\begin{pmatrix} w_1(x, t) \\ w_2(x, t) \end{pmatrix} = p_1(x + t)\begin{pmatrix} -2 \\ 1 \end{pmatrix} + p_2(x + 2t)\begin{pmatrix} -3 \\ 2 \end{pmatrix}.$$

The functions $p_1(x)$ and $p_2(x)$ are to be selected to match the initial data

$$\mathbf{w}(x, 0) = \mathbf{f}(x).$$

Hence,

$$\begin{pmatrix} -6e^{-x^2} \\ xe^{-x^2} \end{pmatrix} = p_1(x)\begin{pmatrix} -2 \\ 1 \end{pmatrix} + p_2(x)\begin{pmatrix} -3 \\ 2 \end{pmatrix},$$

or

$$-6e^{-x^2} = -2p_1(x) - 3p_2(x),$$

$$xe^{-x^2} = p_1(x) + 2p_2(x).$$

It follows that

$$p_1(x) = 12e^{-x^2} - 3xe^{-x^2},$$

$$p_2(x) = -6e^{-x^2} + 2xe^{-x^2},$$

and the solution to the initial value problem is

$$\begin{pmatrix} w_1(x, t) \\ w_2(x, t) \end{pmatrix} = (12 - 3(x + t))e^{-(x+t)^2}\begin{pmatrix} -2 \\ 1 \end{pmatrix} + (-6 + 2(x + 2t))e^{-(x+2t)^2}\begin{pmatrix} -3 \\ 2 \end{pmatrix},$$

or

$$w_1(x, t) = (-24 + 6(x + t))e^{-(x+t)^2} + (18 - 6(x + 2t))e^{-(x+2t)^2},$$

$$w_2(x, t) = (12 - 3(x + t))e^{-(x+t)^2} + (-12 + 4(x + 2t))e^{-(x+2t)^2}.$$

Both $w_1(x, t)$ and $w_2(x, t)$ are the superposition of two waves moving to the left; one with wave speed $c = -1$, the other with wave speed $c = -2$. ■

EXAMPLE 3 Solve the initial value problem

$$\frac{\partial}{\partial t}\mathbf{w} + A\frac{\partial}{\partial x}\mathbf{w} = \mathbf{0}, \qquad -\infty < x < \infty, \quad 0 < t,$$

with initial data

$$\mathbf{w}(x, 0) = \begin{pmatrix} \sin x + \cos x \\ 2\cos x \\ \sin x + 4\cos x \end{pmatrix},$$

where

$$A = \begin{pmatrix} 0 & 1 & 0 \\ 0 & 0 & 1 \\ -2 & 1 & 2 \end{pmatrix}.$$

SOLUTION Taking advantage of the solution to Example 9 of Section 5.2 we have

$$\lambda_1 = 1, \qquad \lambda_2 = -1, \qquad \lambda_3 = 2,$$

and

$$\mathbf{b}^1 = \begin{pmatrix} 1 \\ 1 \\ 1 \end{pmatrix}, \qquad \mathbf{b}^2 = \begin{pmatrix} 1 \\ -1 \\ 1 \end{pmatrix}, \qquad \mathbf{b}^3 = \begin{pmatrix} 1 \\ 2 \\ 4 \end{pmatrix}.$$

Hence from (9.9.16) the $p_m(x)$, $m = 1, 2, 3$, must satisfy

$$p_1(x)\begin{pmatrix} 1 \\ 1 \\ 1 \end{pmatrix} + p_2(x)\begin{pmatrix} 1 \\ -1 \\ 1 \end{pmatrix} + p_3(x)\begin{pmatrix} 1 \\ 2 \\ 4 \end{pmatrix} = \mathbf{w}(x, 0),$$

or

$$\begin{aligned} p_1(x) + p_2(x) + p_3(x) &= \sin x + \cos x, \\ p_1(x) - p_2(x) + 2p_3(x) &= 2\cos x, \\ p_1(x) + p_2(x) + 4p_3(x) &= \sin x + 4\cos x. \end{aligned}$$

It follows that

$$p_1(x) = p_2(x) = \tfrac{1}{2}\sin x, \qquad p_3(x) = \cos x.$$

After substituting in (9.9.17) we find that the solution is

$$\mathbf{w}(x, t) = \frac{1}{2}\sin(x - t)\begin{pmatrix} 1 \\ 1 \\ 1 \end{pmatrix} + \frac{1}{2}\sin(x + t)\begin{pmatrix} 1 \\ -1 \\ 1 \end{pmatrix} + \cos(x - 2t)\begin{pmatrix} 1 \\ 2 \\ 4 \end{pmatrix}. \qquad \blacksquare$$

SOLUTION OF THE TUNED TRANSMISSION LINE EQUATIONS

Let us return to our motivating example, the transmission line equations,

$$\frac{\partial}{\partial t}\mathbf{w} + A\frac{\partial}{\partial x}\mathbf{w} + B\mathbf{w} = \mathbf{0}, \qquad -\infty < x < \infty, \quad 0 < t, \tag{9.9.18}$$

with

$$\mathbf{w} = \begin{pmatrix} J \\ V \end{pmatrix}, \qquad A = \begin{pmatrix} 0 & L^{-1} \\ C^{-1} & 0 \end{pmatrix}, \qquad B = \begin{pmatrix} RL^{-1} & 0 \\ 0 & GC^{-1} \end{pmatrix},$$

and with initial data,

$$J(x, 0) = j(x),$$
$$V(x, 0) = v(x), \qquad -\infty < x < \infty.$$

The *dissipation matrix* B is assumed to be nonzero.

As in the nondissipative case, let us try to construct a solution by setting

$$\begin{pmatrix} J(x, t) \\ V(x, t) \end{pmatrix} = p_1(x, t)\mathbf{b}^1 + p_2(x, t)\mathbf{b}^2, \tag{9.9.19}$$

where $\mathbf{b}^1$ and $\mathbf{b}^2$ are linearly independent eigenvectors of A; for example,

$$\mathbf{b}_1 = \begin{pmatrix} 1 \\ \gamma \end{pmatrix}, \qquad \mathbf{b}_2 = \begin{pmatrix} 1 \\ -\gamma \end{pmatrix}, \qquad \gamma = \sqrt{L/C}\,.$$

Substituting (9.9.19) into (9.9.18) gives

$$\begin{aligned} &\frac{\partial}{\partial t} p_1(x, t)\mathbf{b}^1 + \frac{\partial}{\partial x} p_1(x, t)A\mathbf{b}^1 + p_1(x, t)B\mathbf{b}^1 \\ &\qquad + \frac{\partial}{\partial t} p_2(x, t)\mathbf{b}^2 + \frac{\partial}{\partial x} p_2(x, t)A\mathbf{b}^2 + p_2(x, t)B\mathbf{b}^2 = \mathbf{0}. \end{aligned} \tag{9.9.20}$$

The vectors $B\mathbf{b}^1$ and $B\mathbf{b}^2$ can be written as a linear combination of the linearly independent vectors $\mathbf{b}^1$ and $\mathbf{b}^2$:

$$\begin{aligned} B\mathbf{b}^1 &= \sigma_1\mathbf{b}^1 + \eta_1\mathbf{b}^2, \\ B\mathbf{b}^2 &= \sigma_2\mathbf{b}^1 + \eta_2\mathbf{b}^2. \end{aligned} \tag{9.9.21}$$

We also know that $\mathbf{b}^1$ and $\mathbf{b}^2$ are eigenvectors of A corresponding to the eigenvalues c and $-c$ where $c^2 = 1/LC$:

$$A\mathbf{b}^1 = c\mathbf{b}^1, \qquad A\mathbf{b}^2 = -c\mathbf{b}^2.$$

Therefore the system (9.9.21) can be written as

$$\begin{aligned} &\left(\frac{\partial}{\partial t} p_1(x, t) + c\frac{\partial}{\partial x} p_1(x, t) + \sigma_1 p_1(x, t) + \sigma_2 p_2(x, t)\right)\mathbf{b}^1 \\ &\qquad + \left(\frac{\partial}{\partial t} p_2(x, t) - c\frac{\partial}{\partial x} p_2(x, t) + \eta_1 p_1(x, t) + \eta_2 p_2(x, t)\right)\mathbf{b}^2 = \mathbf{0}. \end{aligned}$$

Because of linear independence, the coefficients of $\mathbf{b}^1$ and $\mathbf{b}^2$ must be zero. We have

$$\begin{aligned} &\frac{\partial}{\partial t} p_1(x, t) + c\frac{\partial}{\partial x} p_1(x, t) + \sigma_1 p_1(x, t) + \sigma_2 p_2(x, t) = 0, \\ &\frac{\partial}{\partial t} p_2(x, t) - c\frac{\partial}{\partial x} p_2(x, t) + \eta_1 p_1(x, t) + \eta_2 p_2(x, t) = 0, \end{aligned} \tag{9.9.22}$$

a coupled system of partial differential equations. It has quite complicated solutions unless σ_2 and η_1, the coupling coefficients, are both zero.

The coefficients of $\mathbf{b}^1$ and $\mathbf{b}^2$ in (9.9.21) are

$$\begin{aligned} \sigma_1 = \eta_2 &= \frac{RL^{-1} + GC^{-1}}{2}, \\ \sigma_2 = \eta_1 &= \frac{RL^{-1} - GC^{-1}}{2}. \end{aligned} \tag{9.9.23}$$

These formulas can be derived by computing

$$B\mathbf{b}^1 = B\begin{pmatrix} 1 \\ \gamma \end{pmatrix} = \begin{pmatrix} RL^{-1} \\ \gamma GC^{-1} \end{pmatrix},$$

$$B\mathbf{b}^2 = B\begin{pmatrix} 1 \\ -\gamma \end{pmatrix} = \begin{pmatrix} RL^{-1} \\ -\gamma GC^{-1} \end{pmatrix},$$

and substituting into (9.9.21). Then the vector equations (9.9.21) are equivalent to four scalar equations,

$$RL^{-1} = \sigma_1 + \eta_1, \qquad RL^{-1} = \sigma_2 + \eta_2,$$

$$\gamma GC^{-1} = \gamma\sigma_1 - \gamma\eta_1, \qquad -\gamma GC^{-1} = \gamma\sigma_2 - \gamma\eta_2.$$

It is easy to verify that the solutions to the algebraic equations are displayed in (9.9.23).

If the parameters of the cable can be adjusted so that $RL^{-1} = GC^{-1}$, then $\sigma_2 = \eta_1 = 0$, and the cable is said to be *tuned*. For a tuned cable, equations (9.9.22) are *uncoupled*. They are

$$\begin{aligned} &\frac{\partial p_1}{\partial t} + c\frac{\partial p_1}{\partial x} + \mu p_1 = 0, \\ &\frac{\partial p_2}{\partial t} - c\frac{\partial p_2}{\partial x} + \mu p_2 = 0, \qquad -\infty < x < \infty, \quad 0 < t, \end{aligned} \tag{9.9.24}$$

where $\mu = 1/\sqrt{LC}$ and μ is the common value of RL^{-1} and GC^{-1}.

The differential equation (9.9.24) can be solved by setting

$$p_1(x, t) = e^{-\mu t} g_1(x, t),$$

$$p_2(x, t) = e^{-\mu t} g_2(x, t).$$

Then we find that $g_1(x, t)$ and $g_2(x, t)$ are solutions to the wave equations,

$$\frac{\partial g_1}{\partial t} + c\frac{\partial g_1}{\partial x} = 0$$

and

$$\frac{\partial g_2}{\partial t} - c\frac{\partial g_2}{\partial x} = 0.$$

It follows that

$$p_1(x, t) = e^{-\mu t} g_1(x - ct),$$
$$p_2(x, t) = e^{-\mu t} g_2(x + ct),$$

and therefore,

$$\begin{pmatrix} J(x, t) \\ V(x, t) \end{pmatrix} = e^{-\mu t} g_1(x - ct) \begin{pmatrix} 1 \\ \gamma \end{pmatrix} + e^{-\mu t} g_2(x + ct) \begin{pmatrix} 1 \\ -\gamma \end{pmatrix}.$$

If the initial data constraints, $J(x, 0) = j(x)$, $V(x, 0) = v(x)$, are imposed, we have

$$\begin{aligned} J(x, t) &= \frac{e^{-\mu t}}{2}[j(x - ct) + j(x + ct)] + \frac{e^{-\mu t}}{2\gamma}[v(x - ct) - v(x + ct)], \\ V(x, t) &= \frac{\gamma e^{-\mu t}}{2}[j(x - ct) - j(x + ct)] + \frac{e^{-\mu t}}{2}[v(x - ct) + v(x + ct)]. \end{aligned} \quad \textbf{(9.9.25)}$$

Both the current $J(x, t)$ and the voltage $V(x, t)$ are superpositions of *damped* waves propagating to the right and to the left without distortion. They are said to be *relatively undistorted.*

EXAMPLE 4 With $c = 3.00 \times 10^8$ m/sec, $\mu = 4 \times 10^{-2}$/sec, $\gamma = 120$ volt/amp, solve the transmission line equations subject to the initial constraints,

$$J(x, 0) = 0,$$
$$V(x, 0) = 100 \tanh^2 x, \qquad -\infty < x < \infty.$$

SOLUTION

$$J(x, t) = \frac{e^{-\mu t}}{2.4}[\tanh^2(x - ct) - \tanh^2(x + ct)],$$

$$V(x, t) = 50e^{-\mu t}[\tanh^2(x - ct) + \tanh^2(x + ct)].$$

Sketches of $J(x, t)$ and $V(x, t)$ are in Fig. 9.21. ■

SYSTEMS OF TUNED HYPERBOLIC EQUATIONS

An n-dimensional hyperbolic system of partial differential equations,

$$\frac{\partial}{\partial t}\mathbf{w} + A\frac{\partial}{\partial x}\mathbf{w} + B\mathbf{w} = \mathbf{0}, \qquad -\infty < x < \infty, \quad 0 < t, \qquad \textbf{(9.9.26)}$$

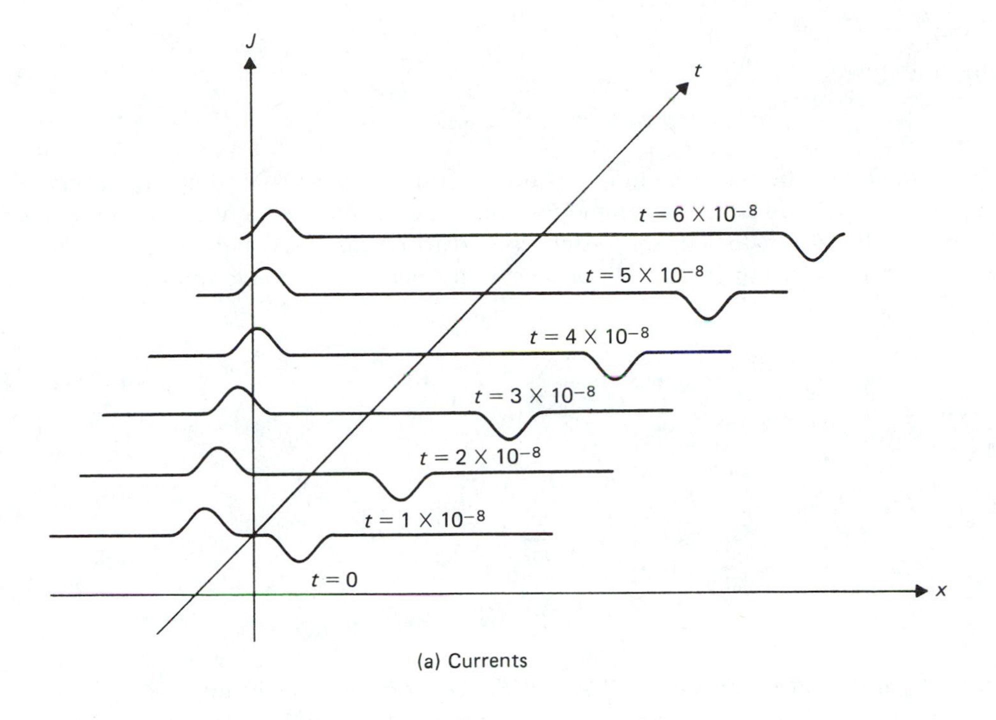

(a) Currents

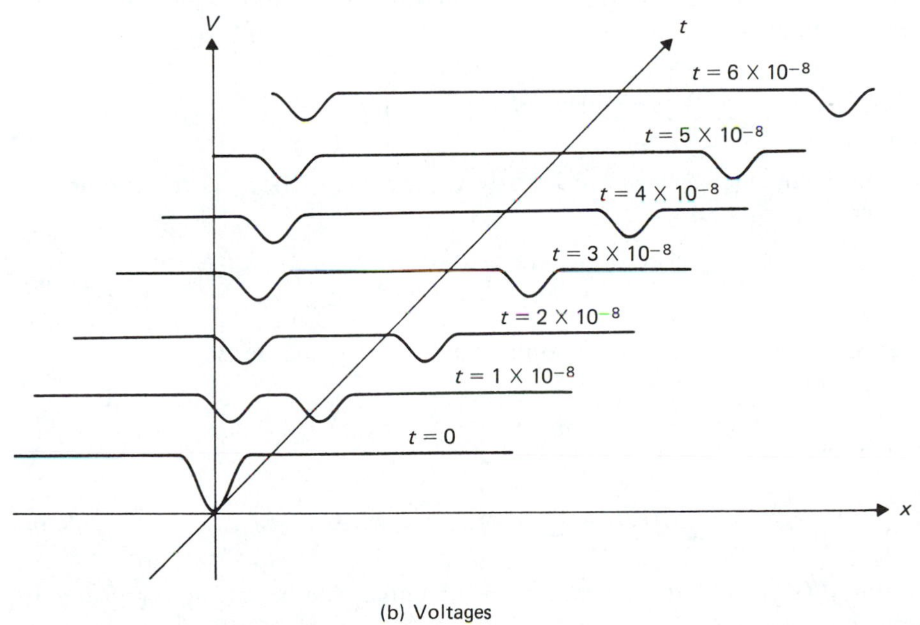

(b) Voltages

Figure 9.21 **Wave profiles for Example 4.**

with initial data

$$\mathbf{w}(x, 0) = \mathbf{f}(x), \qquad -\infty < x < \infty,$$

can be solved by the method employed for the tuned transmission line equations if the matrices A and B have a common set of n linearly independent eigenvectors. If this condition is satisfied, the system is said to be *tuned.* As before, the term *hyperbolic* means that the matrix A has n real, distinct, nonzero eigenvalues.

We set

$$\mathbf{w}(x, t) = \sum_{m=1}^{n} p_m(x, t)\mathbf{b}^m. \tag{9.9.27}$$

Then

$$\begin{aligned} A\frac{\partial}{\partial x}\mathbf{w}(x, t) &= \sum_{m=1}^{n} \lambda_m \frac{\partial}{\partial x} p_m(x, t)\mathbf{b}^m, \\ B\mathbf{w}(x, t) &= \sum_{m=1}^{n} \mu_m p_m(x, t)\mathbf{b}^m, \end{aligned} \tag{9.9.28}$$

where λ_m and μ_m are eigenvalues of A and B, respectively. Substituting the expansion (9.9.27) into the partial differential equation (9.9.26) and using the algebraic relations (9.9.28), we obtain

$$\sum_{m=1}^{n} \left(\frac{\partial p_m}{\partial t} + \lambda_m \frac{\partial p_m}{\partial x} + \mu_m p_m \right) \mathbf{b}^m = \mathbf{0}.$$

Because the eigenvectors are linearly independent, their coefficients in the preceding equation must be zero:

$$\frac{\partial p_m}{\partial t} + \lambda_m \frac{\partial p_m}{\partial x} + \mu_m p_m = 0, \qquad m = 1, 2, \ldots, n. \tag{9.9.29}$$

This set of uncoupled partial differential equations has solutions

$$p_m(x, t) = e^{-\mu_m t} g_m(x - \lambda_m t).$$

It follows that

$$\mathbf{w}(x, t) = \sum_{m=1}^{n} e^{-\mu_m t} g_m(x - \lambda_m t)\mathbf{b}^m, \tag{9.9.30}$$

where the $\{g_m(x)\}$ can be determined from the initial data by solving the system of algebraic equations,

$$\mathbf{f}(x) = \sum_{m=1}^{n} g_m(x)\mathbf{b}^m. \tag{9.9.31}$$

It is useful and important to note that the system of algebraic equations (9.9.31) determining the functions $\{g_m(x)\}$ are exactly the same ones one would obtain in solving the initial value problem with no dissipation ($B = 0$). Therefore the effect of B, under the assumptions given, is to introduce the exponential terms $e^{-\mu_m t}$ in the components of the solution of the nondissipative problem.

EXAMPLE 5 Solve

$$\frac{\partial}{\partial t}\mathbf{w} + A\frac{\partial}{\partial x}\mathbf{w} + B\mathbf{w} = \mathbf{0}, \qquad -\infty < x < \infty, \quad 0 < t,$$

$$\mathbf{w}(x, 0) = \begin{pmatrix} -5e^x \\ 3e^x \end{pmatrix}, \qquad -\infty < x < \infty,$$

where

$$A = \begin{pmatrix} 2 & 6 \\ -2 & -5 \end{pmatrix}, \qquad B = \begin{pmatrix} 7 & 12 \\ -4 & -7 \end{pmatrix}.$$

SOLUTION Taking advantage of the solution to Example 1 of Section 5.3 we have $\lambda_1 = -1$, $\lambda_2 = -2$, and

$$\mathbf{b}^1 = \begin{pmatrix} -2 \\ 1 \end{pmatrix}, \qquad \mathbf{b}^2 = \begin{pmatrix} -3 \\ 2 \end{pmatrix}.$$

To show that $\mathbf{b}^1$ and $\mathbf{b}^2$ are also eigenvectors of B, we calculate

$$B\mathbf{b}^1 = \mathbf{b}^1, \qquad B\mathbf{b}^2 = -\mathbf{b}^2.$$

Hence $\mu_1 = 1$, $\mu_2 = -1$, and the system (9.9.29) is

$$\frac{\partial p_1}{\partial t} - \frac{\partial p_1}{\partial x} + p_1 = 0,$$

$$\frac{\partial p_2}{\partial t} - 2\frac{\partial p_2}{\partial x} - p_2 = 0.$$

It follows that

$$p_1(x, t) = e^{-t}g_1(x + t), \qquad p_2(x, t) = e^t g_2(x + 2t).$$

From the initial data we obtain the algebraic equations for $g_1(x)$ and $g_2(x)$,

$$\begin{pmatrix} -5e^x \\ 3e^x \end{pmatrix} = g_1(x)\begin{pmatrix} -2 \\ 1 \end{pmatrix} + g_2(x)\begin{pmatrix} -3 \\ 2 \end{pmatrix}.$$

Their solution is $g_1(x) = g_2(x) = e^x$. The solution of the initial value problem is

$$\begin{aligned}\mathbf{w}(x, t) &= e^{-t}e^{x+t}\mathbf{b}^1 + e^t e^{x+2t}\mathbf{b}^2 \\ &= e^x\begin{pmatrix} -2 \\ 1 \end{pmatrix} + e^{x+3t}\begin{pmatrix} -3 \\ 2 \end{pmatrix}.\end{aligned}$$

To verify that it is indeed the solution we compute

$$\frac{\partial}{\partial t}\mathbf{w} + A\frac{\partial}{\partial x}\mathbf{w} + B\mathbf{w} = 3e^{x+3t}\mathbf{b}^2 - e^x\mathbf{b}^1 - 2e^{x+3t}\mathbf{b}^2 + e^x\mathbf{b}^1 - e^{x+3t}\mathbf{b}^2 = \mathbf{0},$$

and

$$\mathbf{w}(x, 0) = e^x\mathbf{b}^1 + e^x\mathbf{b}^2 = e^x\begin{pmatrix}-5\\3\end{pmatrix} = \mathbf{f}(x),$$

as required. ■

The procedure discussed above is summarized in the following algorithm:

Algorithm for solving the *n*-dimensional tuned hyperbolic initial value problem

$$\begin{gathered}\frac{\partial}{\partial t}\mathbf{w} + A\frac{\partial}{\partial x}\mathbf{w} + B\mathbf{w} = \mathbf{0}, \qquad -\infty < x < \infty, \quad t < 0,\\ \mathbf{w}(x, 0) = \mathbf{f}(x),\end{gathered} \tag{9.9.32}$$

where A and B are real $n \times n$ matrices with a common set of n linearly independent eigenvectors, and A has distinct, real, nonzero eigenvalues.

1. Determine the eigenvalues $\{\lambda_m\}$ and n linearly independent eigenvectors $\{\mathbf{b}^m\}$ of the matrix A.
2. Find the numbers $\{\mu_m\}$ by computing

$$B\mathbf{b}^m = \mu_m\mathbf{b}^m.$$

3. Solve

$$\mathbf{f}(x) = \sum_{m=1}^{n} g_m(x)\mathbf{b}^m.$$

4. Set

$$\mathbf{w}(x, t) = \sum_{m=1}^{n} e^{-\mu_m t} g_m(x - \lambda_m t)\mathbf{b}^m.$$

The restriction that the eigenvectors of A are also eigenvectors of B has allowed us to uncouple the system (9.9.32). This is certainly a special restriction, and the reader is entitled to know under what conditions on the matrices A and B it holds. A sufficient condition for uncoupling is given in the following lemma.

Lemma

Let A and B be real $n \times n$ matrices satisfying $AB = BA$, and let A have real, distinct, nonzero eigenvalues. Then the eigenvectors of A are also eigenvectors of B.

Proof. Our task is to show that if $A\mathbf{b} = \lambda\mathbf{b}$ for some vector $\mathbf{b} \neq \mathbf{0}$, then there is a real number μ for which $B\mathbf{b} = \mu\mathbf{b}$. Let $\mathbf{d} = B\mathbf{b}$. If $\mathbf{d} = \mathbf{0}$, then

$$\mathbf{d} = \mathbf{0} = 0 \cdot \mathbf{b} = B\mathbf{b},$$

and so $\mathbf{b}$ is an eigenvector of B with associated eigenvalue $\mu = 0$. If $\mathbf{d} \neq \mathbf{0}$, then

$$A\mathbf{d} = AB\mathbf{b} = BA\mathbf{b} = \lambda B\mathbf{b} = \lambda\mathbf{d},$$

and so $\mathbf{d}$ is an eigenvector of A with associated eigenvalue $\lambda \neq 0$. But then $\mathbf{d} = \gamma\mathbf{b}$ for some real number $\gamma \neq 0$, because the hypothesis that the eigenvalues of A are distinct implies that the eigenvectors corresponding to a particular eigenvalue are parallel. It follows that

$$B\mathbf{b} = \mathbf{d} = \gamma\mathbf{b}.$$

Hence $\mathbf{b}$ is an eigenvector of B with associated eigenvalue $\mu = \gamma$, as required. □

EXAMPLE 6 Find a value of s so that $AB(s) = B(s)A$ and then solve the hyperbolic initial value problem

$$\frac{\partial}{\partial t}\mathbf{w} + A\frac{\partial}{\partial x}\mathbf{w} + B\mathbf{w} = \mathbf{0}, \qquad -\infty < x < \infty, \quad 0 < t,$$

where

$$A = \begin{pmatrix} 3 & 1 \\ -3 & 7 \end{pmatrix}, \qquad B = \begin{pmatrix} 1 & s \\ -3 & 5 \end{pmatrix},$$

and

$$\mathbf{w}(x, 0) = \begin{pmatrix} 4\cos 2x \\ 4\sin 2x \end{pmatrix}.$$

SOLUTION We compute

$$AB = \begin{pmatrix} 3 & 1 \\ -3 & 7 \end{pmatrix}\begin{pmatrix} 1 & s \\ -3 & 5 \end{pmatrix} = \begin{pmatrix} 0 & 3s + 5 \\ -24 & -3s + 35 \end{pmatrix},$$

$$BA = \begin{pmatrix} 1 & s \\ -3 & 5 \end{pmatrix}\begin{pmatrix} 3 & 1 \\ -3 & 7 \end{pmatrix} = \begin{pmatrix} 3 - 3s & 1 + 7s \\ -24 & 32 \end{pmatrix}.$$

If $AB = BA$, then by equating entries of the two product matrices, we obtain the equations that determine s,

$$3 - 3s = 0,$$
$$1 + 7s = 3s + 5,$$
$$32 = -3s + 35.$$

They are satisfied by taking $s = 1$ and

$$B = \begin{pmatrix} 1 & 1 \\ -3 & 5 \end{pmatrix}.$$

The eigenvalues of A are $\lambda_1 = 6$ and $\lambda_2 = 4$ with corresponding eigenvectors

$$\mathbf{b}^1 = \begin{pmatrix} 1 \\ 3 \end{pmatrix}, \qquad \mathbf{b}^2 = \begin{pmatrix} 1 \\ 1 \end{pmatrix}.$$

From the computation,

$$B\mathbf{b}^1 = 4\mathbf{b}^1, \qquad B\mathbf{b}^2 = 2\mathbf{b}^2,$$

we conclude that $\mu_1 = 4$ and $\mu_2 = 2$ are eigenvalues of B. Since $\lambda_1 = 6$ and $\lambda_2 = 4$,

$$\mathbf{w}(x, t) = e^{-4t}g_1(x - 6t)\mathbf{b}^1 + e^{-2t}g_2(x - 4t)\mathbf{b}^2.$$

The functions $g_1(x)$ and $g_2(x)$ are determined from the initial conditions,

$$4 \cos 2x = g_1(x) + g_2(x),$$
$$4 \sin 2x = 3g_1(x) + g_2(x).$$

Therefore

$$g_1(x) = 2 \sin 2x - 2 \cos 2x,$$
$$g_2(x) = 6 \cos 2x - 2 \sin 2x,$$

and

$$\begin{pmatrix} w_1(x, t) \\ w_2(x, t) \end{pmatrix} = 2e^{-4t}(\sin 2(x - 6t) - \cos 2(x - 6t)) \begin{pmatrix} 1 \\ 3 \end{pmatrix} + 2e^{-2t}(3 \cos 2(x - 4t) - \sin 2(x - 4t)) \begin{pmatrix} 1 \\ 1 \end{pmatrix}. \qquad \blacksquare$$

In this section we have shown that the technique of representing a vector-valued function as a linear combination of eigenvectors can be employed to solve certain classes of initial value problems for systems of partial differential equations. The hyperbolicity assumption, namely that the eigenvalues of a coefficient matrix be real, distinct, and nonzero, resulted in solutions that propagated as waves.

EXERCISES 9.9

For Exercises 1–4 solve the nondissipative transmission line equations,

$$\frac{\partial J}{\partial t} + \frac{1}{L}\frac{\partial V}{\partial x} = 0, \qquad \frac{\partial V}{\partial t} + \frac{1}{C}\frac{\partial J}{\partial x} = 0, \qquad -\infty < x < \infty, \quad 0 < t,$$

subject to the given initial data with $L = 3.85 \times 10^{-7}$ henrys/m, $C = 2.89 \times 10^{-11}$ farads/m. As in the text, $c^2 = 1/LC$, $\gamma^2 = L/C$. Plot $J(x, t)$ and $V(x, t)$ at the specified times.

1. $J(x, 0) = \tanh x[u(x+1) - u(x-2)]$, $V(x, 0) = 0$, $-\infty < x < \infty$. Plot with $ct = 0, 0.5, 2$.

2. $J(x, 0) = 0$, $V(x, 0) = \gamma \sin 3x[u(x+\pi) - u(x-\pi)]$, $-\infty < x < \infty$. Plot with $ct = 0, \pi, 2\pi, 3\pi$.

3. $J(x, 0) = (x-1)^2(x+2)^2[u(x+1) - u(x-2)]$,
$V(x, 0) = \dfrac{\gamma}{4}(x^2 - 1)^2[u(x+1) - u(x-2)]$, $-\infty < x < \infty$. Plot with $ct = 0, 1.5, 3$.

4. $J(x, 0) = x^2(x-1)^2[u(x) - u(x-1)]$,
$V(x, 0) = -4\gamma x^2 \cos^2(\pi x/2)[u(x) - u(x-1)]$, $-\infty < x < \infty$. Plot with $ct = 0, 0.5, 1.5, 2$.

For Exercises 5–12 solve the conservative hyperbolic system,

$$\frac{\partial}{\partial t}\mathbf{w} + A\frac{\partial}{\partial x}\mathbf{w} = \mathbf{0}, \qquad -\infty < x < \infty, \quad 0 < t,$$

subject to the given initial data and matrix A.

5. $A = \begin{pmatrix} -2 & 6 \\ 0 & 2 \end{pmatrix}$, $\quad \mathbf{f}(x) = \begin{pmatrix} \cosh x \\ \sinh x \end{pmatrix}$, $\quad -\infty < x < \infty$.

6. $A = \begin{pmatrix} 3 & 0 \\ 2 & -1 \end{pmatrix}$, $\quad \mathbf{f}(x) = \begin{pmatrix} \cos x \\ \sin x \end{pmatrix}$, $\quad -\infty < x < \infty$.

7. $A = \begin{pmatrix} 2 & 3 \\ 4 & 3 \end{pmatrix}$, $\quad \mathbf{f}(x) = \begin{pmatrix} x^2 \\ x^4 \end{pmatrix}$, $\quad -\infty < x < \infty$.

8. $A = \begin{pmatrix} 3 & 1 \\ -3 & 7 \end{pmatrix}$, $\quad \mathbf{f}(x) = \begin{pmatrix} x \\ x^3 \end{pmatrix}$, $\quad -\infty < x < \infty$.

9. $A = \begin{pmatrix} 1 & 0 & 0 \\ 1 & 2 & 0 \\ -1 & -1 & -2 \end{pmatrix}$, $\quad \mathbf{f}(x) = \begin{pmatrix} -e^x \\ e^x + 4e^{-x} \\ -e^{-x} \end{pmatrix}$, $\quad -\infty < x < \infty$.

The eigenvalues of A are $\lambda_1 = 1$, $\lambda_2 = 2$, $\lambda_3 = -2$.

10. $A = \begin{pmatrix} 3 & -1 & -1 \\ -12 & 0 & 5 \\ 4 & -2 & -1 \end{pmatrix}, \quad \mathbf{f}(x) = \begin{pmatrix} -3\cos x - \sin x + 1 \\ \cos x + \sin x + 2 \\ -7\cos x - 2\sin x + 2 \end{pmatrix}, \quad -\infty < x < \infty.$

The eigenvalues of A are $\lambda_1 = 1, \lambda_2 = -1, \lambda_3 = 2$.

11. $A = \begin{pmatrix} 5 & -7 & 7 \\ 4 & -3 & 4 \\ 4 & -1 & 2 \end{pmatrix}, \quad \mathbf{f}(x) = \begin{pmatrix} x^2 - x \\ 0 \\ -x^2 \end{pmatrix}, \quad -\infty < x < \infty.$

The eigenvalues of A are $\lambda_1 = 1, \lambda_2 = -2, \lambda_3 = 5$.

12. $A = \begin{pmatrix} 1 & -1 & -1 \\ 0 & 1 & 3 \\ 0 & 3 & 1 \end{pmatrix}, \quad \mathbf{f}(x) = \begin{pmatrix} 2\cosh x \\ -4\sinh x \\ 2\cosh x + 4\sinh x \end{pmatrix}, \quad -\infty < x < \infty.$

The eigenvalues of A are $\lambda_1 = 1, \lambda_2 = -2, \lambda_3 = 4$.

13. Given $G = 1.29 \times 10^{-12}$ ohms/m, $L = 3.85 \times 10^{-7}$ henrys/m, $C = 2.89 \times 10^{-11}$ farads/m, find R so that the transmission line equations are tuned.

$$L\frac{\partial J}{\partial t} + \frac{\partial V}{\partial x} + RJ = 0, \qquad C\frac{\partial V}{\partial t} + \frac{\partial J}{\partial x} + GV = 0, \qquad -\infty < x < \infty, \quad 0 < t.$$

14. Show that the substitutions $t' = \mu t$, $J' = \gamma J$, $V' = V$, $x' = \sigma x$, where $\mu = R/L = G/C$, $\gamma^2 = L/C$, $\sigma = \gamma\mu C$, transforms the tuned transmission line equations into

$$\frac{\partial J'}{\partial t'} + \frac{\partial V'}{\partial x'} + J' = 0, \qquad \frac{\partial V'}{\partial t'} + \frac{\partial J'}{\partial x'} + V' = 0.$$

In Exercises 15–18 use $C = 1$, $L = 1$, $R = 1$, $G = 1$, and solve the tuned transmission line equations subject to the given initial conditions. Plot $J(x, t)$, $V(x, t)$ at the specified times.

15. $J(x, 0) = \cos x$, $V(x, 0) = \sin x$, $-\infty < x < \infty$. Plot with $t = 0, \pi/4, \pi$.

16. $J(x, 0) = e^{-x^2}$, $V(x, 0) = 0$, $-\infty < x < \infty$. Plot with $t = 0, 1, 2$.

17. $J(x, 0) = -(\sinh^2 x)(\sinh^2(x - 1))[u(x) - u(x - 1)]$,
$V(x, 0) = 3(\cosh x - 1)^2 (\cosh x - \cosh 1)^2[u(x) - u(x - 1)]$, $-\infty < x < \infty$.
Plot with $t = 0, 1, 2$.

18. $J(x, 0) = x^2(x - 1)^2[u(x) - u(x - 1)]$,
$V(x, 0) = 5(x^2 - 1)^2[u(x + 1) - u(x - 1)]$, $-\infty < x < \infty$. Plot with $t = 0, 1, 2$.

In Exercises 19–22, find α so that $AB = BA$ and then solve the tuned hyperbolic system,

$$\frac{\partial}{\partial t}\mathbf{w} + A\frac{\partial}{\partial x}\mathbf{w} + B\mathbf{w} = \mathbf{0}, \qquad -\infty < x < \infty, \qquad 0 < t,$$

subject to the given initial conditions and matrices A and B.

19. $A = \begin{pmatrix} 7 & -4 \\ 12 & -7 \end{pmatrix}, \quad B = \begin{pmatrix} 5 & \alpha \\ 6 & \alpha \end{pmatrix}, \quad \mathbf{f}(x) = \begin{pmatrix} x+2 \\ 2x+3 \end{pmatrix}, \quad -\infty < x < \infty.$

20. $A = \begin{pmatrix} 0 & -4 \\ -4 & 0 \end{pmatrix}, \quad B = \begin{pmatrix} \alpha & -1 \\ -1 & \alpha \end{pmatrix}, \quad \mathbf{f}(x) = \begin{pmatrix} 2e^x + e^{2x} \\ -2e^x + e^{2x} \end{pmatrix}, \quad -\infty < x < \infty.$

21. $A = \begin{pmatrix} 1 & 3 \\ 4 & 2 \end{pmatrix}, \quad B = \begin{pmatrix} -5 & 3\alpha \\ 4\alpha & -2 \end{pmatrix}, \quad \mathbf{f}(x) = \begin{pmatrix} 2\sin 2x - 3\sin x \\ -2\sin 2x - 4\sin x \end{pmatrix}, \quad -\infty < x < \infty.$

22. $A = \begin{pmatrix} 1 & 1 \\ 4 & \alpha \end{pmatrix}, \quad B = \begin{pmatrix} 10 & \alpha \\ 4 & 10 \end{pmatrix}, \quad \mathbf{f}(x) = \begin{pmatrix} 5x^2 - 2x^4 \\ 10x^2 + 4x^4 \end{pmatrix}, \quad -\infty < x < \infty.$

In Exercises 23–26, verify that the given matrices A and B have a common set of eigenvectors and solve the corresponding tuned hyperbolic equations subject to the given initial conditions.

23. $A = \begin{pmatrix} -2 & -3 & 3 \\ 0 & -1 & 0 \\ 0 & -2 & 1 \end{pmatrix}, \quad B = \begin{pmatrix} -2 & -2 & 2 \\ 0 & 2 & 0 \\ 0 & 2 & 0 \end{pmatrix}, \quad \mathbf{f}(x) = \begin{pmatrix} 2x + x^3 \\ x^2 \\ 2x + x^2 \end{pmatrix}, \quad -\infty < x < \infty.$

The eigenvalues of A are $\lambda_1 = 1, \lambda_2 = -1, \lambda_3 = -2$.

24. $A = \begin{pmatrix} 15 & -6 & -6 \\ 22 & -9 & -10 \\ 14 & -6 & -5 \end{pmatrix}, \quad B = \begin{pmatrix} 7 & -3 & -3 \\ 13 & -5 & -7 \\ 5 & -3 & -1 \end{pmatrix},$

$\mathbf{f}(x) = \begin{pmatrix} \sin x + \cos x \\ \sin x + 2\cos x \\ \sin x + \cos x \end{pmatrix}, \quad -\infty < x < \infty.$

The eigenvalues of A are $\lambda_1 = 3, \lambda_2 = -3, \lambda_3 = 1$.

25. $A = \begin{pmatrix} -3 & -2 & 6 \\ 7 & 0 & -6 \\ -2 & -2 & 5 \end{pmatrix}, \quad B = \begin{pmatrix} -32 & -8 & 39 \\ -47 & -5 & 51 \\ -38 & -8 & 45 \end{pmatrix},$

$\mathbf{f}(x) = \begin{pmatrix} e^x + e^{-x} \\ e^x + 5e^{-x} \\ e^x + 2e^{-x} \end{pmatrix}, \quad -\infty < x < \infty.$

The eigenvalues of A are $\lambda_1 = 1, \lambda_2 = 2, \lambda_3 = -1$.

26. $A = \begin{pmatrix} -10 & 9 & -6 \\ -24 & 21 & -14 \\ -15 & 13 & -9 \end{pmatrix}, \qquad B = \begin{pmatrix} -43 & 30 & -15 \\ -75 & 52 & -25 \\ -30 & 20 & -8 \end{pmatrix},$

$$\mathbf{f}(x) = \begin{pmatrix} x + 3x^2 + x^3 \\ 2x + 5x^2 + 3x^3 \\ x + 2x^2 + 5x^3 \end{pmatrix}, \qquad -\infty < x < \infty.$$

The eigenvalues of A are $\lambda_1 = 1, \lambda_2 = 2, \lambda_3 = -1$.

SUMMARY

The wave equation was introduced and representative problems were solved by the separation-of-variables method. It was selected as the first topic because its normal mode solutions lead naturally to Fourier series, an essential tool in the separation-of-variables method. In Sections 9.4 and 9.5 the basic ideas and techniques of Fourier series were presented including odd, even, and periodic extensions of functions. Various problems in wave propagation were studied using the separation-of-variables method and the d'Alembert formula. The heat equation and Laplace's equation were studied in Sections 9.7 and 9.8. The last topic covered was first order systems of hyperbolic partial differential equations. They were solved by an extension of the eigenvalue–eigenvector method that was employed in Chapter 5 to solve linear systems of ordinary differential equations.

MISCELLANEOUS EXERCISES

9.1 If $f(x)$ is periodic with period T, show that

a) its derivative $f'(x)$ wherever defined is periodic with period T;

b) the antiderivative

$$F(x) = \int_0^x f(s)\, ds$$

is not periodic with period T unless the first Fourier coefficient a_0 of $f(x)$ is zero; i.e., the function has mean value zero.

In Exercises 9.2–9.7, the fundamental interval is $-\pi < x < \pi$ and the Fourier approximates of an absolutely integrable function $f(x)$ are

$$S_N(x) = \frac{a_0}{2} + \sum_{k=1}^{N} (a_n \cos nx + b_n \sin nx),$$

where the a_n and b_n are computed from the Euler formulas (9.4.2).

9.2

a) Show that if μ is not an integer,

$$\cos \mu x = \frac{2\mu \sin \mu\pi}{\pi}\left(\frac{1}{2\mu^2} + \sum_{n=1}^{\infty} \frac{(-1)^n}{\mu^2 - n^2} \cos nx\right).$$

b) Using the result in (a) show that

$$\cot \pi x = \frac{2\pi}{\pi}\left(\frac{1}{2x^2} + \sum_{n=1}^{\infty} \frac{1}{x^2 - n^2}\right),$$

the so-called resolution of the cotangent into partial fractions.

c) Use the previous results to show that

$$\coth x = \frac{1}{x} + 2x \sum_{n=1}^{\infty} \frac{1}{x^2 + n^2\pi^2}.$$

9.3. Let $f(x) = x^2$, $-\pi < x < \pi$, be extended periodically so as to have period 2π. Show that this implies the relation

$$\frac{\pi^2}{12} = 1 - \frac{1}{2^2} + \frac{1}{3^2} - \frac{1}{4^2} + \cdots = \sum_{n=1}^{\infty} \frac{(-1)^{n+1}}{n^2}.$$

9.4. For the function

$$f(x) = \begin{cases} -1 & \text{if } -\pi < x < 0, \\ 1 & \text{if } 0 < x < \pi, \end{cases}$$

find the following:

a) $\max_{0 \le x \le \pi} S_1(x)$.

b) the values of x in the interval $0 \le x \le \pi$ for which $S_1(x) = 1$.

c) the mean square error

$$\frac{1}{2\pi} \int_{-\pi}^{\pi} [f(x) - S_1(x)]^2 \, dx.$$

9.5. Do the same as in Exercise 9.4 for $S_3(x)$. (You will need the identity $\sin 3x = 3 \sin x - 4 \sin^3 x$ for parts (a) and (b).)

9.6. Let $f(t) = x^2$, $-\pi \leq x \leq \pi$. Find the value of the constant a_0 that minimizes each of the following possible measures of error:

a) $|f(0) - a_0|$

b) $\frac{1}{2\pi} \int_{-\pi}^{\pi} |f(x) - a_0| \, dx$

c) $\frac{1}{2\pi} \int_{-\pi}^{\pi} |f(x) - a_0|^2 \, dx$

d) $\max_{-\pi \leq x \leq \pi} |f(x) - a_0|$

9.7. Extend the function, $f(x) = \pi - x$, $0 < x < \pi$ so that it is even and has period 2π.

a) Show that the Fourier cosine series of $f(x)$ is

$$f(x) = \frac{\pi}{2} + \frac{4}{\pi} \sum_{n=1}^{\infty} \frac{1}{n^2} \sin^2\left(\frac{n\pi}{2}\right) \cos nx.$$

b) Find $|f(0) - S_1(0)|$ and $|f(0) - S_3(0)|$.

c) Find $\max_{-\pi \leq x \leq \pi} S_1(x)$ and $\max_{-\pi \leq x \leq \pi} S_3(x)$ (for the latter you will need the identity $\cos 3x = 4 \cos^3 x - 3 \cos x$).

d) Find the expression for

$$r_N = \frac{1}{2\pi} \int_{-\pi}^{-\pi} |f(x) - S_N(x)|^2 \, dx$$

and evalulate r_1 and r_3.

9.8. Let $f(x)$, $g(x)$, $-\pi < x < \pi$ be real valued functions satisfying

$$\int_{-\pi}^{\pi} [f(x)]^2 \, dx \geq 0, \qquad \int_{-\pi}^{\pi} [g(x)]^2 \, dx > 0.$$

Let $F(\lambda)$ be the nonnegative function

$$F(\lambda) = \int_{-\pi}^{\pi} [f(x) + \lambda g(x)]^2 \, dx.$$

a) Show that $F(\lambda)$ attains a minimum when

$$\lambda = \frac{\int_{-\pi}^{\pi} f(x)g(x) \, dx}{\int_{-\pi}^{\pi} [g(x)]^2 \, dx}.$$

b) By evaluating the minimum obtain the *Cauchy–Schwarz inequality*

$$\left(\int_{-\pi}^{\pi} f(x)g(x) \, dt\right)^2 \leq \int_{-\pi}^{\pi} [f(x)]^2 \, dx \int_{-\pi}^{\pi} [g(x)]^2 \, dx.$$

9.9. The d'Alembert formula

$$v(x, t) = \frac{1}{2}[f(x + ct) + f(x - ct)] + \frac{1}{2c}\int_{x-ct}^{x+ct} g(z)\,dz \qquad \textbf{(9.10.1)}$$

can be written as

$$v(x, t) = \frac{1}{2}[f(x + ct) + f(x - ct)] + \frac{1}{2c}[G(x + ct) - G(x - ct)], \qquad \textbf{(9.10.2)}$$

where

$$G(x) = \int_0^x g(z)\,dz.$$

a) Show that (9.10.1) and (9.10.2) are equivalent.

b) Set

$$f(x) = \frac{a_0}{2} + \sum_{n=1}^{\infty} a_n \cos nx,$$

$$g(x) = \frac{c_0}{2} + \sum_{n=1}^{\infty} c_n \cos nx.$$

Find the Fourier series for $G(x)$.

c) Assume that the series for $f(x)$ and $G(x)$ are term-by-term differentiable. Show that (9.10.2) is a solution to the following boundary-initial value problem:

$$v_{tt} = c^2 v_{xx}, \qquad 0 < x < \pi, \qquad 0 < t;$$

$$v_x(0, t) = 0, \qquad v_x(\pi, t) = 0, \qquad 0 < t;$$

$$v(x, 0) = f(x), \qquad v_t(x, 0) = g(x), \qquad 0 < x < \pi.$$

d) For the taut vibrating string problem, the free endpoint conditions, $v_x(0, t) = 0$, $v_x(\pi, t) = 0$, can mean that the ends of the string are fastened to rings that are free to slide up and down on vertical wires. Show that if

$$\int_0^{\pi} g(z)\,dz > 0,$$

then the solution to (c) is unbounded. Where is the string going?

For Exercises 9.10–9.11 use the d'Alembert formula (9.10.1) to construct a generalized solution.

9.10. Solve the initial value problem

$$v_{tt} = v_{xx}, \qquad -\infty < x < \infty, \quad 0 < t;$$

$$v(x, 0) = \frac{x^3(x-5)^2}{54}[u(x) - u(x-5)];$$

$$v_t(x, 0) = 0, \qquad -\infty < x < \infty,$$

where $u(x)$ is the Heaviside function defined in Section 4.8. Sketch the solution as a function of x for $t = 0, 5/2, 5$.

9.11. Solve the initial value problem

$$v_{tt} = v_{xx}, \qquad -\infty < x < \infty, \quad 0 < t;$$

$$v(x, 0) = 0;$$

$$v_t(x, 0) = 4 \sin^2 x[u(x - \pi) - u(x + \pi)].$$

Sketch $G(x + t)$, $G(x - t)$, and $v(x, t)$ as a function of x for $t = 0, \pi, 2\pi, 4\pi$.

9.12. Show that the Poisson formula (a derivation is in Chapter 8 of [5]),

$$v(x, y) = \frac{y}{\pi}\int_{-\infty}^{\infty} h(z)\frac{1}{(x-z)^2 + y^2}\,dz$$

is a solution to Laplace's equation in the half-plane, $y > 0$, $-\infty < x < \infty$. (Assume that differentiation under the integral sign is valid.)

9.13. Show that (see Chapter 10 of [5])

$$v(x, t) = \int_{-\infty}^{\infty}\left(\frac{1}{4\pi kt}\right)^{1/2} \exp[-(x - z)/4kt]\, f(z)\,dz$$
$$+ \int_0^t\int_{-\infty}^{\infty}\left(\frac{1}{4\pi k(t-s)}\right)^{1/2} \exp[-(x - z)/4k(t - s)]\, F(z, s)\,dz\,ds$$

is a solution to the inhomogeneous heat equation in the half-plane, $t > 0$, $-\infty < x < \infty$. (Assume that differentiation under the integrals is valid.)

The Cartesian coordinates (x,y,z) can be expressed in terms of spherical coordinates by the following relationships:

$$x = \rho \sin\phi \cos\theta, \qquad y = \rho \sin\phi \cos\theta, \qquad z = \rho\cos\phi$$

where

$$\rho \geq 0, \quad 0 \leq \phi \leq \pi, \quad 0 \leq \theta \leq 2\pi.$$

It can be shown that Laplace's equation in spherical coordinates is

$$(r^2 v_r)_r + \frac{1}{\sin\phi}(\sin\phi\, v_\phi)_\phi + \frac{1}{\sin^2\phi} v_{\theta\theta} = 0. \qquad \textbf{(9.10.3)}$$

9.14. Show that a solution to Laplace's equation that is independent of θ and ϕ must have the form

$$v(r) = \frac{c_1}{r} + c_2,$$

where c_1 and c_2 are constants.

9.15. Use the result of Exercise 9.14 and solve the Dirichlet problem:

$$\begin{aligned} (r^2 v_r)_r &= 0, \qquad 1 < r < 2, \\ v(1) &= 10, \qquad v(2) = 20. \end{aligned}$$

9.16. Show that if $R(r)$ is a solution to the Euler differential equation,

$$r^2 R'' + 2rR' - n(n + 1)R = 0, \tag{9.10.4}$$

and $\Phi(\phi)$ is a solution to Legendre's differential equation,

$$[\sin \phi \Phi']' + n(n + 1) \sin \phi \Phi = 0, \tag{9.10.5}$$

then $v(r, \phi) = R(r)\Phi(\phi)$ is a solution to Laplace's equation (9.10.3).

9.17. Let $x = \cos \phi$ and show that the differential equation (9.10.5) is equivalent to

$$(1 - x^2) \frac{d^2 y}{dx^2} - 2x \frac{dy}{dx} + n(n + 1)y = 0, \tag{9.10.6}$$

where $y(x) = \Phi(\phi)$. In Chapter 6, it is shown that $P_n(x)$, the Legendre polynomial of degree n, is a solution to (9.10.6).

9.18. Show that

$$v(r, \phi) = \sum_{n=0}^{N} a_n \left(\frac{r}{b}\right)^n P_n(\cos \phi)$$

is the solution to the axially symmetric interior Dirichlet problem for the sphere:

$$\begin{aligned} (r^2 v_r)_r + \frac{1}{\sin \phi} (\sin \phi v_\phi)_\phi &= 0, \qquad 0 \le \phi \le \frac{\pi}{2}, \quad 0 < r < b, \\ v(b, \phi) &= \sum_{n=0}^{N} a_n P_n(\cos \phi), \qquad 0 \le \phi \le \frac{\pi}{2}, \quad v(0, \phi) < \infty. \end{aligned}$$

9.19. Show that

$$v(r, \phi) = \sum_{n=0}^{N} c_n \left(\frac{b}{r}\right)^{n+1} P_n(\cos \phi)$$

is the solution to the axially symmetric exterior Dirichlet problem for the sphere:

$$(r^2v_r)_r + \frac{1}{\sin\phi}(\sin\phi\, v_\phi)_\phi = 0, \qquad 0 \le \phi \le \frac{\pi}{2}, \quad b < r,$$

$$v(b, \phi) = \sum_{n=0}^{N} c_n P_n(\cos\phi), \qquad 0 \le \phi \le \frac{\pi}{2},$$

$$\lim_{r\to\infty} v(r, \phi) = 0, \qquad 0 \le \phi \le \frac{\pi}{2}.$$

Use the Legendre coefficient formula,

$$a_n = \frac{2n+1}{2}\int_0^\pi h(\phi)P_n(\cos\phi)\sin\phi\, d\phi$$

in the solution of the following exercises.

9.20. Solve the interior Dirichlet problem for the sphere with boundary data

$$v(b, \phi) = 6\cos\phi + 8\cos^2\phi.$$

9.21. Solve the interior Dirichlet problem for the sphere with boundary data

$$v(b, \phi) = 7 + 4\cos\phi - 12\cos(2\phi).$$

9.22. Solve the exterior Dirichlet problem for the sphere with boundary data

$$v(b, \phi) = -\mu + \frac{\mu}{2}J_2(3\cos^2\phi - 1).$$

The solution to this problem represents the gravitational potential of an oblate planet where μ is a gravitational constant and J_2 is the oblateness parameter.

9.23. The equations of motion of a satellite in the equatorial plane ($\phi \equiv \pi/2$) of an axially symmetric oblate planet are

$$\frac{d^2r}{dt^2} - \frac{h^2}{r^3} = -\frac{\partial v}{\partial r}, \qquad \frac{d\theta}{dt} = \frac{h}{r^2},$$

where $v = v(r, \pi/2)$ is the solution to the previous exercise, and h is the angular momentum per unit mass. Set $u = 1/r$, use θ as a new independent variable and derive the differential equation

$$\frac{d^2u}{d\theta^2} + u = \frac{\mu}{h^2}(1 - \epsilon u^2).$$

(This was discussed in Section 6 of Chapter 8.)

REFERENCES

There are many textbooks on ordinary differential equations that include an introduction to partial differential equations. The reader might examine

1. W.E. Boyce and R.C. DiPrima, *Elementary Differential Equations and Boundary Value Problems,* 4th ed., Wiley, New York, 1986.
2. F. Brauer and J.A. Nohel, *Introduction to Differential Equations with Applications,* Harper & Row, New York, 1986.
3. P.D. Ritger and N.D. Rose, *Differential Equations with Applications,* McGraw-Hill, New York, 1968.

Two intermediate textbooks in partial differential equations are

4. S.J. Farlow, *Partial Differential Equations for Scientists and Engineers,* Wiley, New York, 1982.
5. R. Haberman, *Elementary Applied Partial Differential Equations,* Prentice-Hall, Englewood Cliffs, N.J., 1987.

Several more advanced textbooks are

6. C.R. Chester, *Techniques in Partial Differential Equations,* McGraw-Hill, New York, 1971.
7. B. Epstein, *Partial Differential Equations,* McGraw-Hill, New York, 1962.
8. R.L. Street, *Analysis and Solution of Partial Differential Equations,* Brooks/Cole, Monterey, California, 1973.
9. A.G. Webster, *Partial Differential Equations of Mathematical Physics,* Hafner, New York, 1947.

Two excellent introductions to both the physics and mathematics of linear wave motion are

10. C.A. Coulson, *Waves,* Oliver and Boyd, Edinburgh and London, 1961.
11. H. Lamb, *The Dynamical Theory of Sound,* 2nd ed., Dover, New York, 1960.

Two well-written books that treat Fourier series are

12. R.V. Churchill and J.W. Brown, *Fourier Series and Boundary Value Problems,* 3rd ed., McGraw-Hill, New York, 1978.
13. R. Courant, *Differential and Integral Calculus, Volume One,* 2nd ed., Interscience, New York, 1937.

APPENDIX 1: THE FUNDAMENTAL LOCAL EXISTENCE AND UNIQUENESS THEORY

We have had many occasions in the preceding chapters to invoke existence and uniqueness theorems for solutions to differential equations. Such theorems answer questions such as: If I cannot exhibit a solution to a differential equation, how do I know there is one? We shall now prove an existence and uniqueness theorem for first order scalar differential equations and then indicate how the result can be extended to systems and higher order equations.

Mathematics is built on examples, and confusion can be avoided by examining several of them before plunging into a topic that may seem overly complicated and theoretical. The equation

$$\frac{dy}{dt} = y$$

is familiar and nonthreatening. Its solutions, $y(t) = Ae^t$, exist for all t and if initial data are given, there is only one choice of A to be made. Therefore, there is one and only one solution to an initial value problem. On the other hand, the innocuous equation

$$\frac{dy}{dt} = y^2$$

has only one solution that exists for all t, the trivial one, $y(t) = 0$. All others are of the form

$$y(t) = \frac{y_0}{[1 - y_0(t - t_0)]},$$

and for any initial condition $y(t_0) = y_0 \neq 0$ there is a unique solution, but it becomes unbounded for finite t. Finally, the initial value problem

$$\frac{dy}{dt} = y^{1/2}, \qquad y(0) = y_0$$

has no (real) solution if y_0 is negative and has many solutions if y_0 is zero. They are

$$y(t) = \begin{cases} 0 & \text{for } 0 \leq t \leq c, \\ \tfrac{1}{4}(t - c)^2 & \text{for } c \leq t, \end{cases}$$

where c is any positive constant.

It is easy to state reasonable conditions that insure the existence of a solution and prohibit multiple solutions. It will be shown that the initial value problem

$$\frac{dy}{dt} = f(t, y), \qquad y(t_0) = y_0 \tag{1}$$

has one and only one solution if the function $f(t, y)$ is continuously differentiable on a rectangle R: $|t - t_0| \leq a$, $|y - y_0| \leq b$, where a and b are positive numbers. But the problem of the existence of a solution defined for all t or even just for $t_0 \leq t < \infty$ is much more difficult. There are many simple, but still interesting, nonlinear differential equations whose solutions can only be defined on finite time intervals. Therefore, we shall be forced to ask for solutions defined near the initial data and accept the possibility of them becoming unbounded later.

The method that will be used to study the initial value problem (1) is attributed to the French mathematician Emile Picard (1856–1941) and is called *successive approximations*. As its name implies, it is an iterative process that starts with an approximate solution and attempts to improve it again and again. In particular, one starts with an initial approximation $y_0(t) = y_0$, then solves

$$\frac{dy_1}{dt} = f(t, y_0), \qquad y_1(t_0) = y_0$$

by integrating to get

$$y_1(t) = y_0 + \int_{t_0}^{t} f(s, y_0)\, ds$$

as the first iterate. Then one solves

$$\frac{dy_2}{dt} = f(t, y_1(t)), \qquad y_2(t_0) = y_0$$

again by integrating to get

$$y_2(t) = y_0 + \int_{t_0}^{t} f(s, y_1(s))\, ds,$$

for the second iterate.

Given $y_n(t)$, the typical step is to solve

$$\frac{dy_{n+1}}{dt} = f(t, y_n(t)), \qquad y_{n+1}(t_0) = y_0$$

for

$$y_{n+1}(t) = y_0 + \int_{t_0}^{t} f(s, y_n(s))\, ds, \quad n = 0, 1, 2, \ldots. \tag{2}$$

For this method to succeed, the constructed sequence $y_1(t), y_2(t), y_3(t), \ldots$ must be bounded and the graph of each iterate must be contained in the rectangle R. This may not be true. For example, if we define

$$y_{n+1}(t) = 1 + \int_0^t [y_n(s)]^2\, ds, \quad |t| \le 1, \quad |y| \le 1,$$

with $y_0(t) \equiv y_0 = 1$, then

$$y_1(t) = 1 + t,$$

$$y_2(t) = 1 + \int_0^t (1 + s)^2\, ds = 1 + t + t^2 + \frac{t^3}{3},$$

and all subsequent iterates violate the bound $|y| \le 1$. This example, which was easy to construct, corresponds to the initial value problem

$$\frac{dy}{dt} = y^2, \qquad y(0) = 1.$$

As noted earlier, its solution $y(t) = (1 - t)^{-1}$ becomes unbounded, so in fact even if R were the rectangle $|t| \le 1$, $|y| \le 1000$, some $y_n(t)$ would eventually escape from it.

In order to show that the sequence $y_n(t)$ defined in (2) is well behaved, a bound is needed on $f(t, y)$ and, perhaps, a restriction on the interval $|t - t_0| \le a$. Let us assume that

$$|f(t, y)| \le M \quad \text{for} \quad |t - t_0| \le a, \quad |y - y_0| \le b.$$

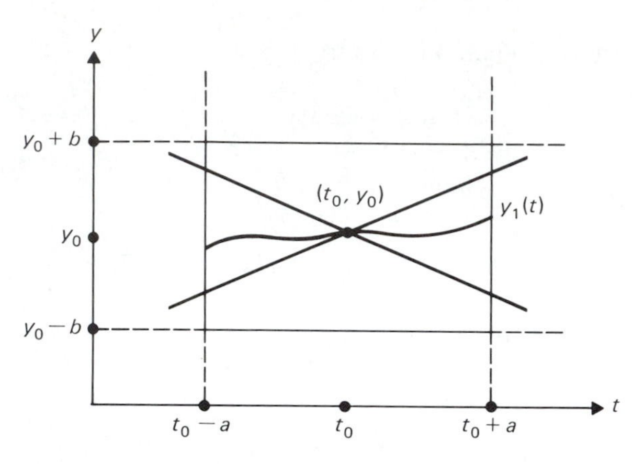

Figure A1.1

(Such a bound exists since $f(t, y)$ is continuous on the closed finite rectangle R.) Then the first iterate $y_1(t)$ satisfies

$$\left|\frac{dy_1}{dt}\right| = |f(t, y_0)| \leq M,$$

and

$$|y_1(t) - y_0| \leq \left|\int_{t_0}^{t} f(s, y_0)\, ds\right| \leq \left|\int_{t_0}^{t} |f(s, y_0)|\, ds\right| \leq M|t - t_0| \leq Ma.$$

Therefore, the graph of $y_1(t)$ must lie between the lines $y = y_0 \pm M(t - t_0)$. If $Ma \leq b$, these lines intersect the vertical boundaries of the rectangle R (see Fig. A1.1). But if $Ma > b$, these lines intersect the horizontal boundaries of R and $y_1(t)$ may leave the rectangle before $t = t_0 + a$ or $t = t_0 - a$ (see Fig. A1.2).

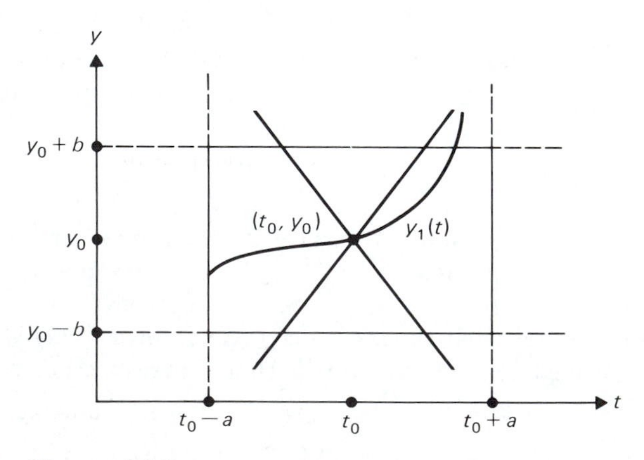

Figure A1.2

If this were to happen, the right side of

$$\frac{dy_2}{dt} = f(t, y_1(t)), \qquad y_2(t_0) = y_0$$

might not be defined after $y_1(t)$ leaves the rectangle, and the iteration process would stop.

Example 1 Show that the method of successive approximations fails for the problem

$$\frac{dy}{dt} = \sqrt{y}, \qquad y(0) = 1,$$

where R is the rectangle $|t| \leq 2$, $|y - 1| \leq 1$.

Starting with $y_0 = 1$, the first iteration gives

$$y_1(t) = 1 + \int_0^t \sqrt{1}\, dt = 1 + t, \quad -2 \leq t \leq 2.$$

But $y_1(-3/2) = -1/2$ is outside the domain of definition of the differential equation, whereas $y_1(3/2) = 5/2$ is outside the rectangle. □

To avoid the difficulty of a poorly defined sequence, the t-interval must be restricted. Let h be the smaller of the numbers a and b/M and require that $|t - t_0| \leq h$. Let R_0 denote the rectangle $|t - t_0| \leq h$, $|y - y_0| \leq b$. Since R_0 is contained in R, $|f(t, y)| \leq M$ for all (t, y) in R_0. Then

$$|y_1(t) - y_0| \leq M|t - t_0| \leq Mh \leq b$$

and $y_2(t)$ is well defined and is subject to the bounds

$$\left|\frac{dy_2(t)}{dt}\right| = |f(t, y_1(t))| \leq M,$$

and for $|t - t_0| \leq h$ we obtain

$$|y_2(t) - y_0| \leq \left|\int_{t_0}^t f(s, y_1(s))\, ds\right| \leq M|t - t_0| \leq Mh \leq b.$$

Therefore the graph of $y_2(t)$ is in R_0 and in fact the functions of the sequence $y_1(t), y_2(t), y_3(t), \ldots$ are subject to the same bounds. This is a consequence of the following lemma.

Lemma 1 Let $f(t, y)$ be a continuous function defined on the rectangle R: $|t - t_0| \le a$, $|y - y_0| \le b$, where a, b are positive numbers, and let $|f(t, y)| \le M$ for all (t, y) in R. If h is the smaller of a and b/M, then the functions $y_n(t)$ given by (2) are defined on the interval $|t - t_0| \le h$, and on this interval

$$|y_n(t) - y_0| \le M|t - t_0| \le Mh \le b, \quad n = 0, 1, 2, \ldots. \qquad (3)$$

Hence, the graph of each $y_n(t)$ is contained in the rectangle R_0: $|t - t_0| \le h$, $|y - y_0| \le b$. Furthermore, each $y_n(t)$ is continuously differentiable and

$$\left|\frac{dy_n}{dt}(t)\right| \le M \quad \text{for} \quad |t - t_0| \le h. \qquad (4)$$

Proof The proof is by induction. First consider $n = 0$. The constant function $y_0(t) \equiv y_0$, defined on the interval $|t - t_0| \le h$, is continuously differentiable and trivially satisfies (3) and (4). Now assume that for $j \le n$, each $y_j(t)$ is continuously differentiable and satisfies (3) and (4). Our task is to show that $y_{n+1}(t)$ has the same properties. By (2), the function $y_{n+1}(t)$ is defined on $|t - t_0| \le h$ and its derivative $dy_{n+1}/dt = f(t, y_n(t))$ is continuous. Hence, $y_{n+1}(t)$ is continuously differentiable and

$$\left|\frac{dy_{n+1}}{dt}(t)\right| = |f(t, y_n(t))| \le M \quad \text{for} \quad |t - t_0| \le h.$$

(The bounds (3) insure that $(t, y_n(t))$ is in R_0 and, therefore, is in R, the rectangle on which $f(t, y)$ is defined and bounded.) It follows from the definition of $y_{n+1}(t)$ that

$$|y_{n+1}(t) - y_0| = \left|\int_{t_0}^{t} f(s, y_n(s))\, ds\right| \le M|t - t_0| \le Mh \le b,$$

and therefore $(t, y_{n+1}(t))$ is in R_0. This completes the proof of the lemma. ∎

So far so good: the sequence of successive approximations $\{y_n(t)\}$ is well defined, properly bounded, and all the graphs are contained in R_0. Picard was a truly accomplished mathematician, so even the most sceptical reader should be willing to grant that the sequence *might* converge. If it does, it is likely that the limit function $y(t)$ is continuously differentiable and that

$$y(t) = \lim_{n\to\infty} y_{n+1}(t) = y_0 + \int_{t_0}^{t} \lim_{n\to\infty} f(s, y_n(s))\, ds = y_0 + \int_{t_0}^{t} f(s, y(s))\, ds.$$

In other words, the hypothetical limit function might be a solution to the *integral* equation

$$y(t) = y_0 + \int_{t_0}^{t} f(s, y(s))\, ds. \tag{5}$$

This has not yet been proved. It has only been asserted that the method of successive approximations leads naturally to the conjecture that there is a solution to (5). But we were seeking a solution to the initial value problem (1) not to an integral equation. Fortunately, the two are equivalent.

Lemma 2 If $y(t)$ is a continuously differentiable solution to the initial value problem (1) for $|t - t_0| \le h$, then $y(t)$ satisfies (5). Conversely, if $y(t)$ satisfies (5) it is a solution of (1).

Proof If $y(t)$ is a continuously differentiable solution of (1), then

$$\frac{dy}{dt}(t) = f(t, y(t)), \quad |t - t_0| \le h.$$

Integrating from t_0 to t, one obtains

$$y(t) - y(t_0) = \int_{t_0}^{t} f(s, y(s))\, ds.$$

Since $y(t_0) = y_0$, this gives (5).

Conversely, if $y(t)$ is a continuously differentiable solution to the integral equation (5), then, from the fundamental theorem of calculus, the derivative of $y(t)$ must equal $f(t, y(t))$, that is,

$$\frac{dy}{dt}(t) = \frac{d}{dt}\left\{y_0 + \int_{t_0}^{t} f(s, y(s))\, ds\right\} = f(t, y(t)).$$

Finally, set $t = t_0$ in (5) to obtain $y(t_0) = t_0$. Hence $y(t)$ is a solution to the initial value problem (1) as required. This completes the proof of Lemma 2. ∎

The next, and hardest, task is to prove that the sequence of successive approximations converges. To do this, the difference between two successive iterates must be estimated. That difference is given by the expression

$$y_{n+1}(t) - y_n(t) = \int_{t_0}^{t} [f(s, y_n(s)) - f(s, y_{n-1}(s))]\, ds,$$

and therefore we must estimate

$$|f(s, y_n(s)) - f(s, y_{n-1}(s))|.$$

By the mean value theorem of calculus, if (t, u) and (t, v) are two points in R, then

$$f(t, u) - f(t, v) = \frac{\partial f}{\partial y}(t, w)(u - v),$$

where w is some number between u and v. Since $\partial f(t, y)/\partial y$ is assumed to be continuous in R, it is bounded there. Hence

$$\left|\frac{\partial f}{\partial y}(t, w)\right| \leq L \text{ for all } (t, w) \text{ in } R,$$

and therefore for any two points (t, u) and (t, v) in R

$$|f(t, u) - f(t, v)| \leq L|u - v|. \tag{6}$$

The inequality (6) is called a *Lipschitz condition,* after the German mathematician Rudolf Lipschitz (1832–1903). With the aid of (5) and (6), we have

$$\begin{aligned} |y_{n+1}(t) - y_n(t)| &\leq \left| \int_{t_0}^{t} |f(s, y_n(s)) - f(s, y_{n-1}(s))| \, ds \right| \\ &\leq L \left| \int_{t_0}^{t} |y_n(s) - y_{n-1}(s)| \, ds \right|, \quad n = 1, 2, 3, \ldots . \end{aligned} \tag{7}$$

The estimate is needed in the proof of the following fundamental local existence theorem.

Theorem Suppose that $f(t, y)$ and $\partial f(t, y)/\partial y$ are continuous on the rectangle R: $|t - t_0| \leq a$, $|y - y_0| \leq b$, where a and b are positive numbers, and that

$$|f(t, y)| \leq M, \quad \left|\frac{\partial f}{\partial y}(t, y)\right| \leq L \quad \text{for all } (t, y) \text{ in } R.$$

Then if h is the smaller of a and b/M, the successive approximations y_n, given by (2), converge to a continuously differentiable solution to the initial value problem (1) for $|t - t_0| \leq h$.

Proof The convergence of y_n will be demonstrated by setting

$$\begin{aligned} y_n(t) = y_0 &+ [y_1(t) - y_0] + [y_2(t) - y_1(t)] \\ &+ \cdots + [y_n(t) - y_{n-1}(t)] \end{aligned} \tag{8}$$

and showing that the corresponding infinite series is dominated by a convergent series of positive constants. To do this, estimates of the difference $|y_{n+1}(t) - y_n(t)|$ are needed. We will prove by induction that

$$|y_{n+1}(t) - y_n(t)| \le ML^n \frac{|t - t_0|^{n+1}}{(n+1)!} \le \frac{ML^n h^{n+1}}{(n+1)!}, \tag{9}$$

where $n = 0, 1, 2, \ldots, |t - t_0| \le h$.

If $n = 0$, then

$$\begin{aligned} |y_1(t) - y_0(t)| &\le \left| \int_{t_0}^{t} f(s, y_0)\, ds \right| \\ &\le \left| \int_{t_0}^{t} |f(s, y_0)|\, ds \right| \le M|t - t_0| \le Mh. \end{aligned}$$

Now assume that (9) is valid for $j \le n$. This implies that

$$|y_n(s) - y_{n-1}(s)| \le ML^{n-1} \frac{|s - t_0|^n}{n!},$$

and then by (7)

$$\begin{aligned} |y_{n+1}(t) - y_n(t)| &\le L \left| \int_{t_0}^{t} ML^{n-1} \frac{|s - t_0|^n}{n!}\, ds \right| \\ &= ML^n \frac{|t - t_0|^{n+1}}{(n+1)!} \le \frac{ML^n h^{n+1}}{(n+1)!}, \quad |t - t_0| \le h, \end{aligned}$$

as required.

It follows that the series (8) is dominated by the series

$$\begin{aligned} |y_0| + Mh + \frac{MLh^2}{2!} + \frac{ML^2h^3}{3!} + \cdots + ML^n \frac{h^{n+1}}{(n+1)!} + \cdots \\ = |y_0| + \frac{M}{L} \left\{ (Lh) + \frac{(Lh)^2}{2!} + \cdots + \frac{(Lh)^{n+1}}{(n+1)!} + \cdots \right\} \\ = |y_0| + \frac{M}{L} \{ e^{Lh} - 1 \}. \end{aligned}$$

Therefore by the Weierstrass M-test (which the reader can find in any advanced calculus text) the series converges uniformly and, consequently, converges to a continuous function $y(t)$. It follows that

$$\lim_{n \to \infty} y_n(t) = y_0 + \lim_{n \to \infty} \sum_{k=1}^{n} [y_k(t) - y_{k-1}(t)] = y(t),$$

and consequently, the successive approximations converge to a continuous function. (The concept of uniform convergence and its consequences are discussed in most advanced calculus texts.)

It will now be shown by deriving an estimate on $|y(t) - y_k(t)|$ that the graph of $y(t)$ is in R_0. By definition,

$$y(t) = y_0 + \sum_{n=0}^{\infty} [y_{n+1}(t) - y_n(t)].$$

Using the identity

$$y_k(t) = y_0 + \sum_{m=0}^{k-1} [y_{m+1}(t) - y_m(t)],$$

we have

$$y(t) - y_k(t) = \sum_{n=k}^{\infty} [y_{n+1}(t) - y_n(t)],$$

and hence by (9)

$$|y(t) - y_k(t)| \leq \sum_{n=k}^{\infty} \frac{ML^n h^{n+1}}{(n+1)!} \leq \frac{M}{L} \frac{(Lh)^{k+1}}{(k+1)!} \sum_{n=0}^{\infty} \frac{(Lh)^n}{n!}$$

or

$$|y(t) - y_k(t)| \leq \frac{M}{L} \frac{(Lh)^{k+1}}{(k+1)!} e^{Lh}, \quad |t - t_0| \leq h. \tag{10}$$

Note that $(Lh)^{k+1}/(k+1)! \to 0$ as $k \to \infty$. It is now easy to show that the graph of $y(t)$ is in R_0. Write

$$|y(t) - y_0| = |y(t) - y_k(t) + y_k(t) - y_0|$$

and use the triangle inequality[1] and the estimates (3), (10) to get

$$\begin{aligned} |y(t) - y_0| &\leq |y(t) - y_k(t)| + |y_k(t) - y_0| \\ &\leq \frac{M}{L} \frac{(Lh)^{k+1}}{(k+1)!} e^{Lh} + b \quad |t - t_0| \leq h. \end{aligned}$$

Letting $k \to \infty$, we obtain

$$|y(t) - y_0| \leq b, \quad |t - t_0| \leq h,$$

and therefore the graph of $y(t)$ is in R_0.

The next step is to show with the aid of the estimates that the continuous function $y(t)$ is a solution to the integral equation (5). By using the

[1] The triangle inequality is used repeatedly in mathematical analysis: $|a + b| \leq |a| + |b|$ for any numbers a, b.

triangle inequality and estimates (7) and (10), we obtain

$$\begin{aligned}
0 &\le \left| y(t) - y_0 - \int_{t_0}^{t} f(s, y(s))\, ds \right| \\
&= \left| y(t) - y_0 - \int_{t_0}^{t} f(s, y_{k-1}(s))\, ds + \int_{t_0}^{t} f(s, y_{k-1}(s))\, ds \right. \\
&\qquad \left. - \int_{t_0}^{t} f(s, y(s))\, ds \right| \\
&\le |y(t) - y_k(t)| + \left| \int_{t_0}^{t} |f(s, y_{k-1}(s)) - f(s, y(s))|\, ds \right| \\
&\le |y(t) - y_k(t)| + L \left| \int_{t_0}^{t} |y_{k-1}(s) - y(s)|\, ds \right| \\
&\le |y(t) - y_k(t)| + L \left| \int_{t_0}^{t} \frac{M}{L} \frac{(Lh)^k e^{Lh}}{k!}\, ds \right| \\
&\le \frac{M}{L} \frac{(Lh)^{k+1}}{(k+1)!} e^{Lh} + M|t - t_0| \frac{(Lh)^k}{k!} e^{Lh} \\
&\le \frac{M}{L} \frac{(Lh)^{k+1}}{k!} e^{Lh} \left(\frac{1}{k+1} + 1 \right).
\end{aligned}$$

This bound tends to zero as k tends to infinity, and hence

$$0 \le \left| y(t) - y_0 - \int_{t_0}^{t} f(s, y(s))\, ds \right| \le 0,$$

so $y(t)$ is a continuous solution of the integral equation as required.

Finally, we note that $f(t, y(t))$ is a continuous function of t, and therefore the right-hand side of the expression

$$y(t) = y_0 + \int_{t_0}^{t} f(s, y(s))\, ds$$

is a continuously differentiable function. It follows that $y(t)$ is a solution to the initial value problem (1). This completes the proof of the theorem. ∎

> **Corollary** (Uniqueness of solutions of the initial value problem) There is only one continuous differentiable solution to the initial value problem (1) on the interval $|t - t_0| \le h$.

Proof Suppose that $y(t)$ and $z(t)$ are two solutions to (1). Then, by Lemma 2, for $|t - t_0| \le h$, we have

$$y(t) = y_0 + \int_{t_0}^{t} f(s, y(s))\, ds$$

and

$$z(t) = y_0 + \int_{t_0}^{t} f(s, z(s))\, ds,$$

from which it follows that

$$|y(t) - z(t)| \leq \left| \int_{t_0}^{t} |f(s, y(s)) - f(s, z(s))|\, ds \right|.$$

Using the Lipschitz condition (6), we have

$$|y(t) - z(t)| \leq L \left| \int_{t_0}^{t} |y(s) - z(s)|\, ds \right|. \tag{11}$$

This inequality is valid for $|t - t_0| \leq h$ or $-h \leq t - t_0 \leq h$. It is convenient to consider separately the intervals $-h \leq t - t_0 \leq 0$ and $0 \leq t - t_0 \leq h$. Assume that $-h \leq t - t_0 \leq 0$ or, equivalently, $t_0 - h \leq t \leq t_0$, and set

$$S(t) = \int_{t}^{t_0} |y(s) - z(s)|\, ds = -\int_{t_0}^{t} |y(s) - z(s)|\, ds$$
$$= \left| \int_{t_0}^{t} |y(s) - z(s)|\, ds \right|.$$

Then by the fundamental theorem of calculus,

$$|y(t) - z(t)| = -\frac{dS}{dt}(t),$$

and hence for $-h \leq t - t_0 \leq 0$, it follows from (11) that

$$-\frac{dS}{dt}(t) \leq LS(t) \qquad \text{or} \qquad 0 \leq \frac{dS}{dt}(t) + LS(t). \tag{12}$$

Guided by our experience with first order differential equations, we multiply the differential inequality (12) by e^{Lt} to obtain

$$0 \leq e^{Lt}\left[\frac{dS}{dt}(t) + LS(t)\right] = \frac{d}{dt}[e^{Lt}S(t)].$$

Since $t \leq t_0$, one can integrate from t to t_0 without destroying the sense of this inequality. Thus,

$$0 \leq \int_{t}^{t_0} \frac{d}{ds}[e^{Ls}S(s)]\, ds = e^{Lt_0}S(t_0) - e^{Lt}S(t)$$

or

$$e^{Lt}S(t) \leq e^{Lt_0}S(t_0).$$

But since by definition $0 \leq S(t)$ and $S(t_0) = 0$, it follows from the last inequality that $S(t) \equiv 0$ for $-h \leq t - t_0 \leq 0$. From the definition of $S(t)$ and (11) it follows that $y(t) = z(t)$ for $-h \leq t - t_0 \leq 0$. The proof for $0 \leq t - t_0 \leq h$ is similar and is left to the reader. ∎

The scalar local existence and uniqueness theorem extends easily to systems of ordinary differential equations. We shall sketch the proof for two-dimensional systems

$$\frac{dy_1}{dt} = f_1(t, y_1, y_2), \qquad \frac{dy_2}{dt} = f_2(t, y_1, y_2)$$

with initial conditions $y_1(t_0) = y_1^0$, $y_2(t_0) = y_2^0$. In vector notation,

$$\frac{d\mathbf{y}}{dt} = \mathbf{f}(t, \mathbf{y}), \qquad \mathbf{y}(t_0) = \mathbf{y}^0, \tag{13}$$

where

$$\mathbf{y} = \begin{bmatrix} y_1 \\ y_2 \end{bmatrix}, \qquad \mathbf{f}(t, \mathbf{y}) = \begin{bmatrix} f_1(t, y_1, y_2) \\ f_2(t, y_1, y_2) \end{bmatrix}.$$

The corresponding integral equations are

$$y_1(t) = y_1^0 + \int_{t_0}^{t} f_1(s, y_1(s), y_2(s))\, ds,$$

$$y_2(t) = y_2^0 + \int_{t_0}^{t} f_2(s, y_1(s), y_2(s))\, ds,$$

or

$$\mathbf{y}(t) = \mathbf{y}^0 + \int_{t_0}^{t} \mathbf{f}(s, \mathbf{y}(s))\, ds.$$

It is more convenient to define the distance between two vectors y and x by

$$\|\mathbf{y} - \mathbf{x}\| = |y_1 - x_1| + |y_2 - x_2|$$

than by the Euclidean distance

$$|\mathbf{y} - \mathbf{x}| = [(y_1 - x_1)^2 + (y_2 - x_2)^2]^{1/2}.$$

The Lipschitz condition is then

$$\|\mathbf{f}(t, \mathbf{y}) - \mathbf{f}(t, \mathbf{x})\| \le L\|\mathbf{y} - \mathbf{x}\|. \tag{14}$$

It can be shown that if $\mathbf{f}(t, \mathbf{y})$ is continuously differentiable in a box R: $|t - t_0| \le a$, $\|\mathbf{y} - \mathbf{y}_0\| \le b$, and if the partial derivatives $\partial f_i(t, y_1, y_2)/\partial y_j$ satisfy

$$\left|\frac{\partial f_i}{\partial y_j}(t, y_1, y_2)\right| \le L,$$

$$i, j = 1, 2, \quad |t - t_0| \le a, \quad |y_1 - y_1^0| + |y_2 - y_2^0| \le b,$$

then (14) is satisfied.

The proof of the existence of a continuously differentiable solution of (13) mimics the earlier proof, with $\|\mathbf{y}\|$ and $\|\mathbf{f}\|$ replacing $|y|$ and $|f|$. The sequence $\{\mathbf{y}^n(t)\}$ is defined as before by letting $y_0(t) \equiv \mathbf{y}^0$ and

$$\mathbf{y}^n(t) = \mathbf{y}^0 + \int_{t_0}^{t} \mathbf{f}(s, \mathbf{y}^{n-1}(s))\, ds, \quad n = 1, 2, \ldots. \tag{15}$$

Then the estimates

$$\left\|\frac{d\mathbf{y}^n}{dt}(t)\right\| \leq M, \qquad \|\mathbf{y}^{n+1}(t) - \mathbf{y}^0\| \leq M|t - t_0|,$$

and

$$\|\mathbf{y}^{n+1}(t) - \mathbf{y}^n(t)\| \leq \frac{ML^n h^{n+1}}{(n+1)!}$$

are obtained and enable one to prove convergence. The uniqueness of the solution follows from the inequality

$$\|\mathbf{y}(t) - \mathbf{z}(t)\| \leq L \left| \int_{t_0}^{t} \|\mathbf{y}(s) - \mathbf{z}(s)\|\, ds \right|.$$

The theorem for N-dimensional systems is the following:

Theorem Suppose that $\mathbf{f}(t, \mathbf{y})$ and $\partial f_i(t, \mathbf{y})/\partial y_j$ $(i, j = 1, 2, \ldots, N)$ are continuous in the box R: $|t - t_0| \leq a$, $\|\mathbf{y} - \mathbf{y}^0\| \leq b$, where a and b are positive numbers, and that

$$\|\mathbf{f}(t, \mathbf{y})\| \leq M, \qquad \left|\frac{\partial f_i}{\partial y_j}(t, \mathbf{y})\right| \leq L \quad (i, j = 1, 2, \ldots, N)$$

for all $(t, \mathbf{y})$ in R. Then if h is the smaller of a and b/M, the successive approximations $\mathbf{y}^n(t)$ given by (15) converge to a continuously differentiable solution to the initial value problem (13) for $|t - t_0| \leq h$.

Corollary There is only one continuously differentiable solution to the initial value problem (13) on the interval $|t - t_0| \leq h$.

The interval of existence estimated by this theorem, $t_0 - h \leq t \leq t_0 + h$, is conservative; it may be much larger. Unfortunately, it is not easy to give a more accurate estimate, except in the important case of systems of linear differential equations, where the interval of existence of solutions is the interval of continuity of the coefficient matrix. We shall now prove this result.

The Linear Existence and Uniqueness Theorem Let $A(t)$ be an $n \times n$ matrix and $\mathbf{f}(t)$ be an n-dimensional vector that are continuous on the interval $a \le t \le b$. Then, if $\mathbf{y}^0$ is an arbitrary n-vector, there exists a unique solution $\mathbf{y}(t)$ to the initial value problem

$$\frac{d\mathbf{y}}{dt} = A(t)\,\mathbf{y} + \mathbf{f}(t), \qquad \mathbf{y}(t_0) = \mathbf{y}^0, \qquad a \le t_0 \le b$$

that is defined on the interval $a \le t \le b$. Furthermore,

$$\mathbf{y}(t) = \Phi(t)\left\{\mathbf{y}^0 + \int_{t_0}^{t} \Phi^{-1}(s)\,\mathbf{f}(s)\,ds\right\},$$

where $\Phi(t)$, a fundamental solution matrix, is a solution to

$$\frac{d\Phi}{dt} = A(t)\,\Phi, \qquad \Phi(t_0) = I, \quad a \le t \le b,$$

and I is the identity matrix (see Chapter 5 for more details).

Proof If $\dfrac{d\Phi}{dt} = A(t)\,\Phi$, $\Phi(t_0) = I$, $a \le t \le b$, and

$$\mathbf{y}(t) = \Phi(t)\left\{\mathbf{y}^0 + \int_{t_0}^{t} \Phi^{-1}(s)\,\mathbf{f}(s)\,ds\right\},$$

then

$$\begin{aligned}\frac{d\mathbf{y}}{dt}(t) &= A(t)\,\Phi(t)\left\{\mathbf{y}^0 + \int_{t_0}^{t} \Phi^{-1}(s)\,\mathbf{f}(s)\,ds\right\} + \Phi(t)\,\frac{d}{dt}\int_{t_0}^{t} \Phi^{-1}(s)\,\mathbf{f}(s)\,ds \\ &= A(t)\,\mathbf{y}(t) + \mathbf{f}(t).\end{aligned}$$

Therefore, it is sufficient to prove the existence of $\Phi(t)$.

Define the sequence of matrices $\{\Phi_k(t)\}$ by

$$\Phi_0(t) = I, \qquad \Phi_{k+1}(t) = I + \int_{t_0}^{t} A(s)\,\Phi_k(s)\,ds,$$
$$a \le t \le b, \quad k = 1, 2, 3, \ldots.$$

Set

$$\Phi_N(t) - I = [\Phi_1(t) - I] + [\Phi_2(t) - \Phi_1(t)] + \cdots + [\Phi_N(t) - \Phi_{N-1}(t)].$$

Our goal is to prove that

$$\lim_{N\to\infty} \Phi_N(t) = \Phi(t), \quad a \le t \le b.$$

It is helpful to employ a *matrix norm* defined by

$$\|C\| = \sum_{i,j=1}^{n} |C_{ij}|,$$

where C_{ij} is a typical entry of the matrix C. The following properties of the norm are needed:

$$\|C + D\| \le \|C\| + \|D\|, \qquad \|CD\| \le \|C\| \cdot \|D\|.$$

(A verification is left to the reader.)

It follows that

$$\|\Phi_N - I\| \le \|\Phi_1(t) - I\| + \|\Phi_2(t) - \Phi_1(t)\| + \cdots + \|\Phi_N(t) - \Phi_{N-1}(t)\|,$$

and

$$\begin{aligned} \|\Phi_1(t) - I\| &= \left\| \int_{t_0}^{t} A(s)\, I\, ds \right\| \\ &= \left\| \int_{t_0}^{t} A(s)\, ds \right\| \le M|t - t_0|, \quad a \le t \le b, \end{aligned}$$

where

$$\|A(t)\| \le M, \quad a \le t \le b.$$

The bound on $\|A(t)\|$ exists since each entry of $A(t)$ is a continuous function on the closed interval $a \le t \le b$. Hence,

$$\begin{aligned} \|\Phi_{k+1}(t) - \Phi_k(t)\| &= \left\| \int_{t_0}^{t} A(s)[\Phi_k(s) - \Phi_{k-1}(s)]\, ds \right\| \\ &\le M \left| \int_{t_0}^{t} \|\Phi_k(s) - \Phi_{k-1}(s)\|\, ds \right| \\ &\le M \left| \int_{t_0}^{t} M^k \frac{(s - t_0)^k}{k!}\, ds \right| = M^{k+1} \frac{|t - t_0|^{k+1}}{(k + 1)!}, \end{aligned}$$

$$a \le t \le b, \quad k = 1, 2, 3, \ldots, N,$$

and

$$\begin{aligned} \|\Phi_N(t) - I\| &\le M|t - t_0| + M^2 \frac{|t - t_0|^2}{2!} + \cdots + M^N \frac{|t - t_0|^N}{N!} \\ &\le M(b - a) + M^2 \frac{(b - a)^2}{2!} + \cdots + M^N \frac{(b - a)^N}{N!}. \end{aligned}$$

We conclude that the series $\Sigma_{k=1}^{\infty} [\Phi_k(t) - \Phi_{k-1}(t)]$ is dominated by a constant convergent series and therefore converges to a continuous function $\Phi(t)$, $a \leq t \leq b$, that satisfies the integral equation

$$\Phi(t) = I + \int_{t_0}^{t} A(s)\, \Phi(s)\, ds,$$

and the equivalent initial value problem

$$\frac{d\Phi}{dt} = A(t)\, \Phi, \qquad \Phi(t_0) = I.$$

Uniqueness follows from the argument in the corollary to the general uniqueness theorem. ∎

The following result is needed in Chapter 2.

Corollary The initial value problem

$$\alpha(t)\, y'' + \beta(t)\, y' + \gamma(t)\, y = g(t), \quad a \leq t \leq b,$$

$$y(t_0) = r, \qquad y'(t_0) = s,$$

where $\alpha(t)$, $\beta(t)$, $\gamma(t)$, $g(t)$ are continuous on $a \leq t \leq b$, $\alpha(t) \neq 0$, has a unique continuous solution defined on $a \leq t \leq b$.

Proof Let $y = y_1$, $y' = y_2$,

$$A(t) = \begin{bmatrix} 0 & 1 \\ -\dfrac{\gamma(t)}{\alpha(t)} & -\dfrac{\beta(t)}{\alpha(t)} \end{bmatrix}$$

$$\mathbf{f}(t) = \begin{bmatrix} 0 \\ \dfrac{g(t)}{\alpha(t)} \end{bmatrix}, \qquad \mathbf{y}^0 = \begin{bmatrix} r \\ s \end{bmatrix}.$$

Then the initial value problem

$$\frac{d\mathbf{y}}{dt} = A(t)\, \mathbf{y} + \mathbf{f}(t), \qquad \mathbf{y}(t_0) = \mathbf{y}^0$$

is equivalent to the second order initial value problem. The result follows from the preceding theorem. ∎

Note that the first component of

$$\mathbf{y}(t) = \Phi(t)\left\{\mathbf{y}^0 + \int_{t_0}^{t} \Phi^{-1}(u)\,\mathbf{f}(u)\,du\right\}$$

can be written as

$$y(t) = rp(t) + sq(t) + z(t),$$

where $z(t)$ is a solution to

$$\alpha(t)\,y'' + \beta(t)\,y' + \gamma(t)\,y = g(t)$$

and $p(t)$, $q(t)$ are solutions to the corresponding homogeneous differential equation.

EXERCISES A.1

Calculate the first three Picard iterates for the following initial value problems.

1. $\dfrac{dy}{dt} = t + y, \quad y(2) = 1$

2. $\dfrac{dy}{dt} = t + y^2, \quad y(-1) = 0$

3. $\dfrac{dx}{dt} = x + y, \quad x(0) = 0, \quad \dfrac{dy}{dt} = y, \quad y(0) = 2$

4. $\dfrac{dx}{dt} = x + 2y,\ x(0) = 2,$

   $\dfrac{dy}{dt} = 3x + 2y, \quad y(0) = 3$

For the following functions, select a rectangle on which a Lipschitz condition is satisfied and find a Lipschitz constant L. Then select a rectangle in which no Lipschitz condition is satisfied.

5. $f(t, y) = t^2y^{1/2}$

6. $f(t, y) = \sin\left(\dfrac{1}{t}\right)\sin\left(\dfrac{1}{y}\right)$

7. $f(t, y) = \dfrac{y}{t^2 + y^2}$

8. $f(t, y) = \arctan\left(\dfrac{y}{t}\right)$

9. Show that the integral equation
$$y(t) = \epsilon\int_0^t e^{-a(t-s)}f(s, y(s))\,ds$$
is equivalent to the initial value problem
$$\frac{dy}{dt} + ay = \epsilon f(t, y), \qquad y(0) = 0.$$

10. Show that the integral equation
$$x(t) = \epsilon\int_0^t \sin(t - s)\,f(s, x(s))\,ds$$
is equivalent to the initial value problem
$$\frac{d^2x}{dt^2} + x = \epsilon f(t, x), \qquad x(0) = 0, \qquad \frac{dx}{dt}(0) = 0.$$

11. Estimate $|y_{n+1}(t) - y_n(t)|$, where $y_0(t) \equiv 0$,
$$y_{n+1}(t) = \epsilon\int_0^t e^{-2(t-s)}[s^2 + y_n^2(s)]\,ds,$$
$$n = 0, 1, 2, \ldots,$$
$$-1 \le t \le 1, \quad -1 \le y \le 1, \quad 0 < \epsilon \le \epsilon_0.$$
It may be necessary to restrict ϵ_0.

12. Estimate $|x_{n+1}(t) - x_n(t)|$, where $x_0(t) \equiv 0$,
$$x_{n+1}(t) = \epsilon\int_0^t \sin(t - s)\,[s^2 + x_n^2(s)]\,ds,$$
$$n = 0, 1, 2, \ldots,$$
$$-1 \le t \le 1, \quad -1 \le x \le 1, \quad 0 < \epsilon \le \epsilon_0.$$
It may be necessary to restrict ϵ_0.

APPENDIX 2: PASCAL PROGRAMS

PROGRAM THAT USES EULER'S METHOD TO SOLVE $y' = ty^{1/3}$, $y(1) = 1$

```
PROGRAM EULER1(OUTPUT);
(*****************************************************************
**   THIS PROGRAM USES THE EULER METHOD TO                       **
**   SOLVE THE INITIAL VALUE PROBLEM:                            **
**                                                               **
**     DY/DT = F(T,Y)     Y(T0) = Y0                             **
*****************************************************************)

VAR I,N : INTEGER;
    T0,Y0,TF,T,Y,H : REAL;

FUNCTION F(T,Y:REAL) : REAL;
    BEGIN
    F: = T*EXP(LN(Y)/3.0)
    END; (* FUNCTION F *)
```

```
BEGIN (* MAIN *)

(* SET INITIAL CONDITIONS *)
T0 := 1.0; Y0 := 1.0;
(* SET FINAL TIME AND NUMBER OF STEPS TO BE TAKEN *)
TF := 2.0; N := 10;
(* EULER LOOP *)
T := T0; Y := Y0;
H := (TF - T0)/N;
WRITELN(' ',T:6:2,Y:10:5);
FOR I := 1 TO N DO
     BEGIN
     Y := Y + H*F(T,Y);
     T := T + H;
     WRITELN(' ',T:6:2,Y:10:5)
     END

END. (* MAIN *)
```

PROGRAM THAT USES THE IMPROVED EULER METHOD TO SOLVE $y' = ty^{1/3}$, $y(1) = 1$

```
PROGRAM EULER2(OUTPUT);
(*****************************************************************
**   THIS PROGRAM USES THE IMPROVED EULER METHOD       **
**   TO SOLVE THE INITIAL VALUE PROBLEM:               **
**                                                     **
**               DY/DT = F(T,Y)     Y(T0) = Y0         **
*****************************************************************)

VAR I,N : INTEGER;
    T0,Y0,TF,T,Y,H,S1,S2 : REAL;

FUNCTION F(T,Y:REAL) : REAL;
    BEGIN
    F: = T*EXP(LN(Y)/3.0)
    END; (* FUNCTION F *)

BEGIN (* MAIN *)
```

```
(* SET INITIAL CONDITIONS *)
T0 := 1.0; Y0 := 1.0;
(* SET FINAL TIME AND NUMBER OF STEPS TO BE TAKEN *)
TF := 2.0; N := 10;
(* IMPROVED EULER LOOP *)
T := T0; Y := Y0;
H: = (TF − T0)/N;
WRITELN(' ',T:6:2,Y:10:5);
FOR I := 1 TO N DO
     BEGIN
     S1 := F(T,Y);
     S2 := F(T+H,Y+H*S1);
     Y := Y + H*(S1 + S2)/2.0;
     T := T + H;
     WRITELN(' ',T:6:2,Y:10:5)
     END

END. (* MAIN *)
```

PROGRAM THAT USES THE IMPROVED EULER METHOD TO SOLVE THE SECOND ORDER LINEAR EQUATION $a(t)y''(t) + b(t)y'(t) + c(t)y(t) = q(t)$

```
PROGRAM EULER2S(OUTPUT);

(*****************************************************************
**   PROGRAM USES THE IMPROVED EULER METHOD TO                  **
**   SOLVE THE INITIAL VALUE PROBLEM:                           **
**                                                              **
**       A(T)*Y''(T) + B(T)*Y'(T) + C(T)*Y(T) = Q(T)            **
**       Y(T0) = Y0,   Y'(T0) = Z0                              **
*****************************************************************)

VAR T0,Y0,Z0,TF,T,Y,Z,H,YBAR,ZBAR,GYZ : REAL;
     N,I : INTEGER;

(*DEFINE FUNCTIONS A(T),B(T),C(T),Q(T),G(Y,Z,T)*)
FUNCTION A(T:REAL) : REAL;
     BEGIN
     A := 1.0
     END;
```

```
FUNCTION B(T:REAL) : REAL;
    BEGIN
    B := 0.0
    END;

FUNCTION C(T:REAL) : REAL;
    BEGIN
    C := -T
    END;

FUNCTION Q(T:REAL) : REAL;
    BEGIN
    Q := 0.0
    END;

FUNCTION G(Y,YP,T:REAL) : REAL;
    BEGIN
    G := (-C(T)*Y-B(T)*YP + Q(T))/A(T)
    END;

BEGIN (* MAIN *)

(* SET INITIAL CONDITIONS, FINAL TIME AND NUMBER OF
STEPS TO BE TAKEN *)

T0 := 0.0; Y0 := 0.35503;
Z0 := 1.0;
TF := 1.0; N:= 10;
(* BASIC LOOP *)
T := T0; Y := Y0;
Z := Z0; H := (TF - T0)/N;
WRITELN(' ',T:6:2,Y:10:5,Z:10:5);
FOR I := 1 TO N DO
    BEGIN
    GYZ := G(Y,Z,T);
    YBAR := Y + H*Z;
    ZBAR := Z + H*GYZ;
    Y := Y + H*(Z + ZBAR)/2.0;
    Z := Z + H*(GYZ + G(YBAR,ZBAR,T+H))/2.0;
    T := T + H;
    WRITELN(' ',T:6:2,Y:10:5,Z:10:5)
    END

END. (*MAIN*)
```

PROGRAM THAT USES EULER'S METHOD WITH AUTOMATIC STEP SIZE ADJUSTMENT TO SOLVE $y' = y$, $y_1(0) = 0$ AT $t = 1$

```
PROGRAM EULER(OUTPUT);

(*****************************************************************
**  PROGRAM TO SOLVE D/DT(Y) = Y USING EULER'S                  **
**  METHOD                                                      **
**  WITH A LOCAL ERROR ESTIMATOR BASED                          **
**  ON THE IMPROVED EULER ALGORITHM.                            **
*****************************************************************)

VAR T0,Y0,EPS,T,Y,TFINAL,H : REAL;
    R,EST,GAMMA,YE,HE : REAL;

FUNCTION F(T,Y:REAL) : REAL;
    BEGIN
    F := Y
    END;

BEGIN (*MAIN*)

T0 := 0.0; Y0 := 1.0;
EPS := 0.001;
T := T0; Y := Y0;
TFINAL := 1.0;
H := 0.1;
REPEAT
    IF (T+H > TFINAL) THEN H := TFINAL-T;
    YE := Y + H*F(T,Y);
    R := 0.5*H*(F(T+H,YE) - F(T,Y));
    EST := R/H;
    GAMMA := EPS/ABS(EST);
    HE := 0.9*GAMMA*H;
    IF (ABS(EST) <= EPS) THEN
      BEGIN
      Y := YE;
      T := T+H;
      H := HE
      END
    ELSE
      BEGIN
      H := HE;
      Y := Y + H*F(T,Y);
```

```
            T := T + H
            END
        UNTIL (T >= TFINAL);
        WRITELN(' T =',T:10:4,' Y =',Y:10:5)

        END. (*MAIN*)
```

SUBROUTINE RK4 AND SAMPLE DRIVER SET-UP TO SOLVE $y_1' = y_2, y_1(0) = 1; y_2' = y_1, y_2(0) = 1$

```
PROGRAM DRIVER (OUTPUT);

(****************************************************************
**   SAMPLE PROGRAM TO ILLUSTRATE PROCEDURE RK4           **
****************************************************************)

TYPE SVECTOR = ARRAY[1..2] OF REAL;

VAR NP,NPRINT,I,NEQN : INTEGER;
    T,H : REAL;
    Y : SVECTOR;

PROCEDURE F(T : REAL; N : INTEGER; Y : SVECTOR;
        VAR YP : SVECTOR);
    BEGIN
    YP[1] := Y[2];
    YP[2] := Y[1]
    END; (* F *)

PROCEDURE RK4(H : REAL; N : INTEGER;
        VAR T : REAL;
        VAR Y : SVECTOR);
(****************************************************************
**   PROCEDURE RK4 IMPLEMENTS THE CLASSICAL 4TH           **
**   ORDER RUNGE-KUTTA FORMULAS OVER                      **
**   ONE STEP OF LENGTH H.                                **
**                                                        **
**   INPUT:                                               **
**                                                        **
**     F — NAME OF A PROCEDURE DEFINING                   **
**     THE DIFFERENTIAL EQUATIONS.                        **
**     F MUST HAVE THE FORM                               **
```

```
**                                                              **
**          PROCEDURE F(T,N,Y,YP);                              **
**          BEGIN                                               **
**            YP(1)  = . . .                                    **
**                     .                                        **
**                     .                                        **
**                     .                                        **
**            YP(N) = . . .                                     **
**          END                                                 **
**                                                              **
**     H — STEP SIZE TO BE USED (H#0.0)                         **
**     N — NUMBER OF COMPONENTS IN SOLUTION VECTOR              **
**     T — VALUE OF THE INDEPENDENT VARIABLE                    **
**     Y — SOLUTION VECTOR OF LENGTH N EVALUATED                **
**     AT T                                                     **
**                                                              **
**   OUTPUT:                                                    **
**     T — NEW VALUE OF INDEPENDENT VARIABLE (INPUT             **
**     VALUE PLUS H)                                            **
**     Y — SOLUTION VECTOR EVALUATED AT T                       **
*****************************************************************)

VAR K1,K2,K3,K4,YTEMP : SVECTOR;
    TEMP : REAL;
    I : INTEGER;

    BEGIN
    F(T,N,Y,K1);
    FOR I := 1 TO N DO
      YTEMP[I] := Y[I] + 0.5*H*K1[I];
    TEMP := T + 0.5*H;
    F(TEMP,N,YTEMP,K2);
    FOR I := 1 TO N DO
      YTEMP[I] := Y[I] + 0.5*H*K2[I];
    F(TEMP,N,YTEMP,K3);
    FOR I := 1 TO N DO
      YTEMP[I] := Y[I] + H*K3[I];
    TEMP := T + H;
    F(TEMP,N,YTEMP,K4);
    FOR I := 1 TO N DO
      Y[I] := Y[I] + H*(K1[I] + 2.0*(K2[I] + K3[I]) +K4[I]) / 6.0;
    T := TEMP;
    END; (* RK4 *)
```

```
BEGIN (* MAIN *)

Y[1] := 1.0; Y[2] := 1.0;
T := 0.0; H := 0.1;
WRITELN(' ',T:10:2,Y[1]:10:2,Y[2]:10:2);
NP := 0; NPRINT := 5;
NEQN := 2;
FOR I := 1 TO 10 DO
     BEGIN
     NP := NP + 1;
     RK4(H,NEQN,T,Y);
     IF (NP >= NPRINT) THEN
        BEGIN
        WRITELN(' ',T:10:2,Y[1]:10:2,Y[2]:10:2);
        NP := 0
        END
     END

END. (*MAIN*)
```

APPENDIX 3: BASIC PROGRAMS

Program that uses Euler's method to solve $y' = ty^{1/3}$, $y(1) = 1$

```
10  ' THIS PROGRAM USES THE EULER METHOD TO SOLVE THE INITIAL
20  '   VALUE PROBLEM:
30  ' DY/DT = F(T,Y)  Y(T0) = Y0
40  '
50  ' SET INITIAL CONDITIONS
60     T0 = 1
70     Y0 = 1
80  ' SET FINAL TIME AND NUMBER OF STEPS
90     TF = 2
100    N = 10
110 ' EULER LOOP
120    T = T0
130    Y = Y0
140    H = (TF - T0)/N
150 PRINT T, Y
160 FOR I = 1 TO N
170      GOSUB 230
180      Y = Y + H * F
```

```
190       T = T + H
200       PRINT T, Y
210    NEXT I
220  END
230  'SUBROUTINE
240  F = T*Y^ (1/3)
245  RETURN
```

Program that uses the improved Euler method to solve $y' = ty^{1/3}$, $y(1) = 1$

```
10   ' PROGRAM IMPROVED EULER
20   ' THIS PROGRAM USES THE IMPROVED EULER METHOD
30   ' TO SOLVE THE INITIAL VALUE PROBLEM:
40   ' DY/DT = F(T,Y)  Y(T0) = Y0
50   'SET INITIAL VALUES
60      T0 = 1
70      Y0 = 1
80   'SET FINAL TIME AND NUMBER OF STEPS
90      TF = 2
100     N = 10
110  'IMPROVED EULER LOOP
120     T = T0
130     Y = Y0
140     H = (TF - T0)/N
150  PRINT T, Y
160  FOR I = 1 T0 N
170          GOSUB 230
180          Y = Y + H*(S1 + S2)/2
190          T = T + H
200          PRINT T, Y
210    NEXT I
220  END
230  'SUBROUTINE
240   S1 = T*Y^ (1/3)
242   S2 = (T+H)*(Y+H*S1)^ (1/3)
245  RETURN
```

Program that uses the improved Euler method to solve the second order linear equation $a(t)y''(t) + b(t)y'(t) + c(t)y(t) = q(t)$

```
10   ' PROGRAM IMPROVED EULER
20   ' THIS PROGRAM USES THE IMPROVED EULER METHOD
30   ' A(T)*Y''(T) + B(T)*Y'(T) + C(T)*Y(T) = Q(T)
40   ' Y(T0) = Y0, Y'(T0) = Z0
50   '
60   'DEFINE FUNCTIONS A(T), B(T), C(T), Q(T), G(Y,T,Z)
70    DEF FNA(T) = 1
```

```
80   DEF FNB(T) = 0
90   DEF FNC(T) = -T
100   DEF FNQ(T) = 0
110   DEF FNG(Y,Z,T) = (-FNC(T)*Y-FNB(T)*Z+FNQ(T))/FNA(T)
120 'SET INITIAL CONDITIONS, FINAL TIME AND NUMBER OF STEPS
130          T0 = 0
140          Y0 = .35503
150          Z0 = 1
160   TF = 1
170   N = 10
180  'BASIC LOOP
190   T = T0
200   Y = Y0
210   Z = Z0
220   H = (TF - T0)/N
230   PRINT T, Y, Z
240   FOR I = 1 TO N
250          YBAR = Y + H*Z
260          GYZ = FNG(Y,Z,T)
270          ZBAR = Z + H*GYZ
280          Y = Y + H*(Z+ZBAR)/2
290          Z = Z + H*(GYZ+FNG(YBAR,ZBAR,T+H))/2
300          T = T + H
310          PRINT T, Y, Z
320   NEXT I
330  END
```

Program that uses Euler's method with automatic step size adjustment to solve $y' = y$, $y_1(0) = 0$ at $t = 1$

```
10  'PROGRAM TO SOLVE DY/DT = Y USING EULER'S METHOD WITH A
20  'LOCAL ERROR ESTIMATOR BASED ON THE IMPROVED EULER ALGORITHM
30  ' DEFINE FUNCTION
40  DEF FNF(T,Y) = Y
60  'SET INITIAL CONDITIONS, FINAL TIME AND NUMBER OF STEPS
70  T0 = 0
80  Y0 = 0
90  EPS = .001
100 T = T0
110 Y = Y0
120 TFINAL = 1
130 H = .1
140  WHILE (T < TFINAL)
150            IF (T+H > TFINAL) THEN H = TFINAL - T
160            YE = Y +H*F(T,Y)
170            R = .5*H*(FNF(T+H,YE) - FNF(T,Y)
```

```
180              EST = R/H
190              GAMMA = EPS/ABS(EST)
200              HE = .9*GAMMA*H
210              IF(ABS(EST) <= EPS) THEN 250
220  ' ELSE
120                   Y = YE
240                   T = T + H
250                   H = HE
260  ' IF TRUE
270                   H = HE
280                   Y = Y +H*FNF(T,Y)
300                   T = T + H
310   WEND
320   PRINT T, Y
330   END
```

RK4 program to solve $y_1' = y_2$, $y_1(0) = 1$; $y_2' = y_1$, $y_2(0) = 1$

```
10  'PROGRAM TO IMPLEMENT 4TH ORDER RUNGE-KUTTA FORMULAS
20  ' OVER ONE STEP OF LENGTH H
30  '
35  OPTION BASE 1
40  DIM K1(2)
50  DIM K2(2)
60  DIM K3(2)
70  DIM K4(2)
80  DIM YTEMP(2)
90  DIM Y(2)
100  Y(1) = 1
110  Y(2) = 1
120  T = 0
130  PRINT T,Y(1),Y(2)
140  H = .1
150  N = 2
160  NP = 0
170  NPRINT = 5
180  FOR J = 1 TO 10
190        NP = NP + 1
200        GOSUB 260
210        IF (NP < NPRINT) THEN 240
220              PRINT T,Y(1),Y(2)
230              NP = 0
240  NEXT J
250  END
260  'SUBROUTINE RK4
270  K1(1) = Y(2)
280  K1(2) = Y(1)
290  FOR I = 1 T0 N
```

```
300          YTEMP(I) = Y(I) + .5*H*K1(I)
310     NEXT I
320     K2(1) = YTEMP(2)
330     K2(2) = YTEMP(1)
240     FOR I = 1 TO N
350          YTEMP(I) = Y(I) + .5*H*K2(I)
260     NEXT I
370     K3(1) = YTEMP(2)
380     K3(2) = YTEMP(1)
390     FOR I = 1 TO N
400          YTEMP(I) = Y(I) + .1*H*K3(I)
410     NEXT I
420     K4(1) = YTEMP(2)
430     K4(2) = YTEMP(1)
440     FOR I = 1 TO N
450          Y(I) = Y(I) + H*(K1(I) + 2*(K2(I) + K3(I)) + K4(I))/6
460     NEXT I
470     T = T + H
480      RETURN
```

APPENDIX 4: LISTING OF SUBROUTINE RKF45

```
      SUBROUTINE RKF45(F,NEQN,Y,T,TOUT,RELERR,ABSERR,IFLAG,WORK,IWORK)
C
C     FEHLBERG FOURTH-FIFTH ORDER RUNGE-KUTTA METHOD
C
C     WRITTEN BY H.A.WATTS AND L.F.SHAMPINE
C
C
C
C    RKF45 IS PRIMARILY DESIGNED TO SOLVE NON-STIFF AND MILDLY STIFF
C    DIFFERENTIAL EQUATIONS WHEN DERIVATIVE EVALUATIONS ARE INEXPENSIVE.
C    RKF45 SHOULD GENERALLY NOT BE USED WHEN THE USER IS DEMANDING
C    HIGH ACCURACY.
C
C ABSTRACT
C
C    SUBROUTINE  RKF45  INTEGRATES A SYSTEM OF NEQN FIRST ORDER
C    ORDINARY DIFFERENTIAL EQUATIONS OF THE FORM
C             DY(I)/DT = F(T,Y(1),Y(2),...,Y(NEQN))
C              WHERE THE Y(I) ARE GIVEN AT T .
C    TYPICALLY THE SUBROUTINE IS USED TO INTEGRATE FROM T TO TOUT BUT IT
C    CAN BE USED AS A ONE-STEP INTEGRATOR TO ADVANCE THE SOLUTION A
C    SINGLE STEP IN THE DIRECTION OF TOUT.  ON RETURN THE PARAMETERS IN
C    THE CALL LIST ARE SET FOR CONTINUING THE INTEGRATION. THE USER HAS
C    ONLY TO CALL RKF45 AGAIN (AND PERHAPS DEFINE A NEW VALUE FOR TOUT).
C    ACTUALLY, RKF45 IS AN INTERFACING ROUTINE WHICH CALLS SUBROUTINE
C    RKFS FOR THE SOLUTION.  RKFS IN TURN CALLS SUBROUTINE  FEHL WHICH
C    COMPUTES AN APPROXIMATE SOLUTION OVER ONE STEP.
C
```

```
C     RKF45  USES THE RUNGE-KUTTA-FEHLBERG (4,5)  METHOD DESCRIBED
C     IN THE REFERENCE
C     E.FEHLBERG , LOW-ORDER CLASSICAL RUNGE-KUTTA FORMULAS WITH STEPSIZE
C                  CONTROL , NASA TR R-315 (ALSO IN COMPUTING,6(1970),
C                  PP. 61-71)
C
C     THE PERFORMANCE OF RKF45 IS ILLUSTRATED IN THE REFERENCE
C     L.F.SHAMPINE,H.A.WATTS,S.DAVENPORT, SOLVING NON-STIFF ORDINARY
C                  DIFFERENTIAL EQUATIONS-THE STATE OF THE ART ,
C                  SIAM REVIEW,18(1976),PP. 376-411
C
C
C     THE PARAMETERS REPRESENT-
C       F -- SUBROUTINE F(T,Y,YP) TO EVALUATE DERIVATIVES YP(I)=DY(I)/DT
C       NEQN -- NUMBER OF EQUATIONS TO BE INTEGRATED
C       Y(*) -- SOLUTION VECTOR AT T
C       T -- INDEPENDENT VARIABLE
C       TOUT -- OUTPUT POINT AT WHICH SOLUTION IS DESIRED
C       RELERR,ABSERR -- RELATIVE AND ABSOLUTE ERROR TOLERANCES FOR LOCAL
C             ERROR TEST. AT EACH STEP THE CODE REQUIRES THAT
C                  ABS(LOCAL ERROR) .LE. RELERR*ABS(Y) + ABSERR
C             FOR EACH COMPONENT OF THE LOCAL ERROR AND SOLUTION VECTORS
C       IFLAG -- INDICATOR FOR STATUS OF INTEGRATION
C       WORK(*) -- ARRAY TO HOLD INFORMATION INTERNAL TO RKF45 WHICH IS
C             NECESSARY FOR SUBSEQUENT CALLS. MUST BE DIMENSIONED
C             AT LEAST  3+6*NEQN
C       IWORK(*) -- INTEGER ARRAY USED TO HOLD INFORMATION INTERNAL TO
C             RKF45 WHICH IS NECESSARY FOR SUBSEQUENT CALLS. MUST BE
C             DIMENSIONED AT LEAST  5
C
C
C   FIRST CALL TO RKF45
C
C     THE USER MUST PROVIDE STORAGE IN HIS CALLING PROGRAM FOR THE ARRAYS
C     IN THE CALL LIST  -      Y(NEQN) , WORK(3+6*NEQN) , IWORK(5)  ,
C     DECLARE F IN AN EXTERNAL STATEMENT, SUPPLY SUBROUTINE F(T,Y,YP) AND
C     INITIALIZE THE FOLLOWING PARAMETERS-
C
C       NEQN -- NUMBER OF EQUATIONS TO BE INTEGRATED.  (NEQN .GE. 1)
C       Y(*) -- VECTOR OF INITIAL CONDITIONS
C       T -- STARTING POINT OF INTEGRATION , MUST BE A VARIABLE
C       TOUT -- OUTPUT POINT AT WHICH SOLUTION IS DESIRED.
C             T=TOUT IS ALLOWED ON THE FIRST CALL ONLY, IN WHICH CASE
C             RKF45 RETURNS WITH IFLAG=2 IF CONTINUATION IS POSSIBLE.
C       RELERR,ABSERR -- RELATIVE AND ABSOLUTE LOCAL ERROR TOLERANCES
C             WHICH MUST BE NON-NEGATIVE. RELERR MUST BE A VARIABLE WHILE
C             ABSERR MAY BE A CONSTANT. THE CODE SHOULD NORMALLY NOT BE
C             USED WITH RELATIVE ERROR CONTROL SMALLER THAN ABOUT 1.E-8 .
C             TO AVOID LIMITING PRECISION DIFFICULTIES THE CODE REQUIRES
C             RELERR TO BE LARGER THAN AN INTERNALLY COMPUTED RELATIVE
C             ERROR PARAMETER WHICH IS MACHINE DEPENDENT. IN PARTICULAR,
C             PURE ABSOLUTE ERROR IS NOT PERMITTED. IF A SMALLER THAN
C             ALLOWABLE VALUE OF RELERR IS ATTEMPTED, RKF45 INCREASES
C             RELERR APPROPRIATELY AND RETURNS CONTROL TO THE USER BEFORE
C             CONTINUING THE INTEGRATION.
C       IFLAG -- +1,-1  INDICATOR TO INITIALIZE THE CODE FOR EACH NEW
C             PROBLEM. NORMAL INPUT IS +1. THE USER SHOULD SET IFLAG=-1
C             ONLY WHEN ONE-STEP INTEGRATOR CONTROL IS ESSENTIAL. IN THIS
C             CASE, RKF45 ATTEMPTS TO ADVANCE THE SOLUTION A SINGLE STEP
C             IN THE DIRECTION OF TOUT EACH TIME IT IS CALLED. SINCE THIS
C             MODE OF OPERATION RESULTS IN EXTRA COMPUTING OVERHEAD, IT
C             SHOULD BE AVOIDED UNLESS NEEDED.
C
C
C   OUTPUT FROM RKF45
C
C       Y(*) -- SOLUTION AT T
```

```
C        T -- LAST POINT REACHED IN INTEGRATION.
C        IFLAG = 2 -- INTEGRATION REACHED TOUT. INDICATES SUCCESSFUL RETURN
C                     AND IS THE NORMAL MODE FOR CONTINUING INTEGRATION.
C              =-2 -- A SINGLE SUCCESSFUL STEP IN THE DIRECTION OF TOUT
C                     HAS BEEN TAKEN. NORMAL MODE FOR CONTINUING
C                     INTEGRATION ONE STEP AT A TIME.
C              = 3 -- INTEGRATION WAS NOT COMPLETED BECAUSE RELATIVE ERROR
C                     TOLERANCE WAS TOO SMALL. RELERR HAS BEEN INCREASED
C                     APPROPRIATELY FOR CONTINUING.
C              = 4 -- INTEGRATION WAS NOT COMPLETED BECAUSE MORE THAN
C                     3000 DERIVATIVE EVALUATIONS WERE NEEDED. THIS
C                     IS APPROXIMATELY 500 STEPS.
C              = 5 -- INTEGRATION WAS NOT COMPLETED BECAUSE SOLUTION
C                     VANISHED MAKING A PURE RELATIVE ERROR TEST
C                     IMPOSSIBLE. MUST USE NON-ZERO ABSERR TO CONTINUE.
C                     USING THE ONE-STEP INTEGRATION MODE FOR ONE STEP
C                     IS A GOOD WAY TO PROCEED.
C              = 6 -- INTEGRATION WAS NOT COMPLETED BECAUSE REQUESTED
C                     ACCURACY COULD NOT BE ACHIEVED USING SMALLEST
C                     ALLOWABLE STEPSIZE. USER MUST INCREASE THE ERROR
C                     TOLERANCE BEFORE CONTINUED INTEGRATION CAN BE
C                     ATTEMPTED.
C              = 7 -- IT IS LIKELY THAT RKF45 IS INEFFICIENT FOR SOLVING
C                     THIS PROBLEM. TOO MUCH OUTPUT IS RESTRICTING THE
C                     NATURAL STEPSIZE CHOICE. USE THE ONE-STEP INTEGRATOR
C                     MODE.
C              = 8 -- INVALID INPUT PARAMETERS
C                     THIS INDICATOR OCCURS IF ANY OF THE FOLLOWING IS
C                     SATISFIED -   NEQN .LE. 0
C                                   T=TOUT  AND  IFLAG .NE. +1 OR -1
C                                   RELERR OR ABSERR .LT. 0.
C                                   IFLAG .EQ. 0  OR  .LT. -2  OR  .GT. 8
C        WORK(*),IWORK(*) -- INFORMATION WHICH IS USUALLY OF NO INTEREST
C                     TO THE USER BUT NECESSARY FOR SUBSEQUENT CALLS.
C                     WORK(1),...,WORK(NEQN) CONTAIN THE FIRST DERIVATIVES
C                     OF THE SOLUTION VECTOR Y AT T. WORK(NEQN+1) CONTAINS
C                     THE STEPSIZE H TO BE ATTEMPTED ON THE NEXT STEP.
C                     IWORK(1) CONTAINS THE DERIVATIVE EVALUATION COUNTER.
C
C
C    SUBSEQUENT CALLS TO RKF45
C
C      SUBROUTINE RKF45 RETURNS WITH ALL INFORMATION NEEDED TO CONTINUE
C      THE INTEGRATION. IF THE INTEGRATION REACHED TOUT, THE USER NEED ONLY
C      DEFINE A NEW TOUT AND CALL RKF45 AGAIN. IN THE ONE-STEP INTEGRATOR
C      MODE (IFLAG=-2) THE USER MUST KEEP IN MIND THAT EACH STEP TAKEN IS
C      IN THE DIRECTION OF THE CURRENT TOUT. UPON REACHING TOUT (INDICATED
C      BY CHANGING IFLAG TO 2),THE USER MUST THEN DEFINE A NEW TOUT AND
C      RESET IFLAG TO -2 TO CONTINUE IN THE ONE-STEP INTEGRATOR MODE.
C
C      IF THE INTEGRATION WAS NOT COMPLETED BUT THE USER STILL WANTS TO
C      CONTINUE (IFLAG=3,4 CASES), HE JUST CALLS RKF45 AGAIN. WITH IFLAG=3
C      THE RELERR PARAMETER HAS BEEN ADJUSTED APPROPRIATELY FOR CONTINUING
C      THE INTEGRATION. IN THE CASE OF IFLAG=4 THE FUNCTION COUNTER WILL
C      BE RESET TO 0 AND ANOTHER 3000 FUNCTION EVALUATIONS ARE ALLOWED.
C
C      HOWEVER,IN THE CASE IFLAG=5, THE USER MUST FIRST ALTER THE ERROR
C      CRITERION TO USE A POSITIVE VALUE OF ABSERR BEFORE INTEGRATION CAN
C      PROCEED. IF HE DOES NOT,EXECUTION IS TERMINATED.
C
C      ALSO,IN THE CASE IFLAG=6, IT IS NECESSARY FOR THE USER TO RESET
C      IFLAG TO 2 (OR -2 WHEN THE ONE-STEP INTEGRATION MODE IS BEING USED)
C      AS WELL AS INCREASING EITHER ABSERR,RELERR OR BOTH BEFORE THE
C      INTEGRATION CAN BE CONTINUED. IF THIS IS NOT DONE, EXECUTION WILL
C      BE TERMINATED. THE OCCURRENCE OF IFLAG=6 INDICATES A TROUBLE SPOT
C      (SOLUTION IS CHANGING RAPIDLY,SINGULARITY MAY BE PRESENT) AND IT
C      OFTEN IS INADVISABLE TO CONTINUE.
```

```
C
C     IF IFLAG=7 IS ENCOUNTERED, THE USER SHOULD USE THE ONE-STEP
C     INTEGRATION MODE WITH THE STEPSIZE DETERMINED BY THE CODE OR
C     CONSIDER SWITCHING TO THE ADAMS CODES DE/STEP,INTRP. IF THE USER
C     INSISTS UPON CONTINUING THE INTEGRATION WITH RKF45, HE MUST RESET
C     IFLAG TO 2 BEFORE CALLING RKF45 AGAIN. OTHERWISE,EXECUTION WILL BE
C     TERMINATED.
C
C     IF IFLAG=8 IS OBTAINED, INTEGRATION CAN NOT BE CONTINUED UNLESS
C     THE INVALID INPUT PARAMETERS ARE CORRECTED.
C
C     IT SHOULD BE NOTED THAT THE ARRAYS WORK,IWORK CONTAIN INFORMATION
C     REQUIRED FOR SUBSEQUENT INTEGRATION. ACCORDINGLY, WORK AND IWORK
C     SHOULD NOT BE ALTERED.
C
C
      INTEGER NEQN,IFLAG,IWORK(5)
      REAL Y(NEQN),T,TOUT,RELERR,ABSERR,WORK(1)
C     IF COMPILER CHECKS SUBSCRIPTS, CHANGE WORK(1) TO WORK(3+6*NEQN)
C
      EXTERNAL F
C
      INTEGER K1,K2,K3,K4,K5,K6,K1M
C
C
C     COMPUTE INDICES FOR THE SPLITTING OF THE WORK ARRAY
C
      K1M=NEQN+1
      K1=K1M+1
      K2=K1+NEQN
      K3=K2+NEQN
      K4=K3+NEQN
      K5=K4+NEQN
      K6=K5+NEQN
C
C     THIS INTERFACING ROUTINE MERELY RELIEVES THE USER OF A LONG
C     CALLING LIST VIA THE SPLITTING APART OF TWO WORKING STORAGE
C     ARRAYS. IF THIS IS NOT COMPATIBLE WITH THE USERS COMPILER,
C     HE MUST USE RKFS DIRECTLY.
C
      CALL RKFS(F,NEQN,Y,T,TOUT,RELERR,ABSERR,IFLAG,WORK(1),WORK(K1M),
     1          WORK(K1),WORK(K2),WORK(K3),WORK(K4),WORK(K5),WORK(K6),
     2          WORK(K6+1),IWORK(1),IWORK(2),IWORK(3),IWORK(4),IWORK(5))
C
      RETURN
      END
      SUBROUTINE RKFS(F,NEQN,Y,T,TOUT,RELERR,ABSERR,IFLAG,YP,H,F1,F2,F3,
     1                F4,F5,SAVRE,SAVAE,NFE,KOP,INIT,JFLAG,KFLAG)
C
C     FEHLBERG FOURTH-FIFTH ORDER RUNGE-KUTTA METHOD
C
C
C     RKFS INTEGRATES A SYSTEM OF FIRST ORDER ORDINARY DIFFERENTIAL
C     EQUATIONS AS DESCRIBED IN THE COMMENTS FOR RKF45 .
C     THE ARRAYS YP,F1,F2,F3,F4,AND F5 (OF DIMENSION AT LEAST NEQN) AND
C     THE VARIABLES H,SAVRE,SAVAE,NFE,KOP,INIT,JFLAG,AND KFLAG ARE USED
C     INTERNALLY BY THE CODE AND APPEAR IN THE CALL LIST TO ELIMINATE
C     LOCAL RETENTION OF VARIABLES BETWEEN CALLS. ACCORDINGLY, THEY
C     SHOULD NOT BE ALTERED. ITEMS OF POSSIBLE INTEREST ARE
C         YP - DERIVATIVE OF SOLUTION VECTOR AT T
C         H  - AN APPROPRIATE STEPSIZE TO BE USED FOR THE NEXT STEP
C         NFE- COUNTER ON THE NUMBER OF DERIVATIVE FUNCTION EVALUATIONS
C
C
      LOGICAL HFAILD,OUTPUT
C
      INTEGER  NEQN,IFLAG,NFE,KOP,INIT,JFLAG,KFLAG
```

```
      REAL  Y(NEQN),T,TOUT,RELERR,ABSERR,H,YP(NEQN),
     1  F1(NEQN),F2(NEQN),F3(NEQN),F4(NEQN),F5(NEQN),SAVRE,
     2  SAVAE
C
      EXTERNAL F
C
      REAL  A,AE,DT,EE,EEOET,ESTTOL,ET,HMIN,REMIN,RER,S,
     1  SCALE,TOL,TOLN,U26,EPSP1,EPS,YPK
C
      INTEGER  K,MAXNFE,MFLAG
C
      REAL  ABS,AMAX1,AMIN1,SIGN
C
C     REMIN IS THE MINIMUM ACCEPTABLE VALUE OF RELERR.  ATTEMPTS
C     TO OBTAIN HIGHER ACCURACY WITH THIS SUBROUTINE ARE USUALLY
C     VERY EXPENSIVE AND OFTEN UNSUCCESSFUL.
C
      DATA REMIN/1.D-12/
C
C
C     THE EXPENSE IS CONTROLLED BY RESTRICTING THE NUMBER
C     OF FUNCTION EVALUATIONS TO BE APPROXIMATELY MAXNFE.
C     AS SET, THIS CORRESPONDS TO ABOUT 500 STEPS.
C
      DATA MAXNFE/3000/
C
C
C     CHECK INPUT PARAMETERS
C
C
      IF (NEQN .LT. 1) GO TO 10
      IF ((RELERR .LT. 0.0E0)  .OR.  (ABSERR .LT. 0.0E0)) GO TO 10
      MFLAG=IABS(IFLAG)
      IF (MFLAG .EQ. 0) GO TO 10
      IF (MFLAG .GT. 8) GO TO 10
      IF (MFLAG .NE. 1) GO TO 20
C
C     FIRST CALL, COMPUTE MACHINE EPSILON
      EPS = 1.0E0
    5 EPS = EPS/2.0E0
      EPSP1 = EPS + 1.0E0
      IF (EPSP1 .GT. 1.0E0) GO TO 5
      U26 = 26.0E0*EPS
      GO TO 50
C
C     INVALID INPUT
   10 IFLAG=8
      RETURN
C
C     CHECK CONTINUATION POSSIBILITIES
C
   20 IF ((T .EQ. TOUT) .AND. (KFLAG .NE. 3)) GO TO 10
      IF (MFLAG .NE. 2) GO TO 25
C
C     IFLAG = +2 OR -2
      IF (KFLAG .EQ. 3) GO TO 45
      IF (INIT .EQ. 0) GO TO 45
      IF (KFLAG .EQ. 4) GO TO 40
      IF ((KFLAG .EQ. 5)  .AND.  (ABSERR .EQ. 0.0E0)) GO TO 30
      IF ((KFLAG .EQ. 6)  .AND.  (RELERR .LE. SAVRE)  .AND.
     1    (ABSERR .LE. SAVAE)) GO TO 30
      GO TO 50
C
C     IFLAG = 3,4,5,6,7 OR 8
   25 IF (IFLAG .EQ. 3) GO TO 45
      IF (IFLAG .EQ. 4) GO TO 40
      IF ((IFLAG .EQ. 5) .AND. (ABSERR .GT. 0.0E0)) GO TO 45
```

```
C
C     INTEGRATION CANNOT BE CONTINUED SINCE USER DID NOT RESPOND TO
C     THE INSTRUCTIONS PERTAINING TO IFLAG=5,6,7 OR 8
   30 STOP
C
C     RESET FUNCTION EVALUATION COUNTER
   40 NFE=0
      IF (MFLAG .EQ. 2) GO TO 50
C
C     RESET FLAG VALUE FROM PREVIOUS CALL
   45 IFLAG=JFLAG
      IF (KFLAG .EQ. 3) MFLAG=IABS(IFLAG)
C
C     SAVE INPUT IFLAG AND SET CONTINUATION FLAG VALUE FOR SUBSEQUENT
C     INPUT CHECKING
   50 JFLAG=IFLAG
      KFLAG=0
C
C     SAVE RELERR AND ABSERR FOR CHECKING INPUT ON SUBSEQUENT CALLS
      SAVRE=RELERR
      SAVAE=ABSERR
C
C     RESTRICT RELATIVE ERROR TOLERANCE TO BE AT LEAST AS LARGE AS
C     2*EPS+REMIN TO AVOID LIMITING PRECISION DIFFICULTIES ARISING
C     FROM IMPOSSIBLE ACCURACY REQUESTS
C
      RER=2.0E0*EPS+REMIN
      IF (RELERR .GE. RER) GO TO 55
C
C     RELATIVE ERROR TOLERANCE TOO SMALL
      RELERR=RER
      IFLAG=3
      KFLAG=3
      RETURN
C
   55 DT=TOUT-T
C
      IF (MFLAG .EQ. 1) GO TO 60
      IF (INIT .EQ. 0) GO TO 65
      GO TO 80
C
C     INITIALIZATION --
C                      SET INITIALIZATION COMPLETION INDICATOR,INIT
C                      SET INDICATOR FOR TOO MANY OUTPUT POINTS,KOP
C                      EVALUATE INITIAL DERIVATIVES
C                      SET COUNTER FOR FUNCTION EVALUATIONS,NFE
C                      ESTIMATE STARTING STEPSIZE
C
   60 INIT=0
      KOP=0
C
      A=T
      CALL F(A,Y,YP)
      NFE=1
      IF (T .NE. TOUT) GO TO 65
      IFLAG=2
      RETURN
C
C
   65 INIT=1
      H=ABS(DT)
      TOLN=0.
      DO 70 K=1,NEQN
        TOL=RELERR*ABS(Y(K))+ABSERR
        IF (TOL .LE. 0.) GO TO 70
        TOLN=TOL
        YPK=ABS(YP(K))
        IF (YPK*H**5 .GT. TOL) H=(TOL/YPK)**0.2E0
```

```
   70 CONTINUE
      IF (TOLN .LE. 0.0E0) H=0.0E0
      H=AMAX1(H,U26*AMAX1(ABS(T),ABS(DT)))
      JFLAG=ISIGN(2,IFLAG)
C
C
C     SET STEPSIZE FOR INTEGRATION IN THE DIRECTION FROM T TO TOUT
C
   80 H=SIGN(H,DT)
C
C     TEST TO SEE IF RKF45 IS BEING SEVERELY IMPACTED BY TOO MANY
C     OUTPUT POINTS
C
      IF (ABS(H) .GE. 2.0E0*ABS(DT)) KOP=KOP+1
      IF (KOP .NE. 100) GO TO 85
C
C     UNNECESSARY FREQUENCY OF OUTPUT
      KOP=0
      IFLAG=7
      RETURN
C
   85 IF (ABS(DT) .GT. U26*ABS(T)) GO TO 95
C
C     IF TOO CLOSE TO OUTPUT POINT,EXTRAPOLATE AND RETURN
C
      DO 90 K=1,NEQN
   90   Y(K)=Y(K)+DT*YP(K)
      A=TOUT
      CALL F(A,Y,YP)
      NFE=NFE+1
      GO TO 300
C
C
C     INITIALIZE OUTPUT POINT INDICATOR
C
   95 OUTPUT= .FALSE.
C
C     TO AVOID PREMATURE UNDERFLOW IN THE ERROR TOLERANCE FUNCTION,
C     SCALE THE ERROR TOLERANCES
C
      SCALE=2.0E0/RELERR
      AE=SCALE*ABSERR
C
C
C     STEP BY STEP INTEGRATION
C
  100 HFAILD= .FALSE.
C
C     SET SMALLEST ALLOWABLE STEPSIZE
C
      HMIN=U26*ABS(T)
C
C     ADJUST STEPSIZE IF NECESSARY TO HIT THE OUTPUT POINT.
C     LOOK AHEAD TWO STEPS TO AVOID DRASTIC CHANGES IN THE STEPSIZE AND
C     THUS LESSEN THE IMPACT OF OUTPUT POINTS ON THE CODE.
C
      DT=TOUT-T
      IF (ABS(DT) .GE. 2.0E0*ABS(H)) GO TO 200
      IF (ABS(DT) .GT. ABS(H)) GO TO 150
C
C     THE NEXT SUCCESSFUL STEP WILL COMPLETE THE INTEGRATION TO THE
C     OUTPUT POINT
C
      OUTPUT= .TRUE.
      H=DT
      GO TO 200
C
  150 H=0.5E0*DT
```

```
C
C
C
C     CORE INTEGRATOR FOR TAKING A SINGLE STEP
C
C     THE TOLERANCES HAVE BEEN SCALED TO AVOID PREMATURE UNDERFLOW IN
C     COMPUTING THE ERROR TOLERANCE FUNCTION ET.
C     TO AVOID PROBLEMS WITH ZERO CROSSINGS,RELATIVE ERROR IS MEASURED
C     USING THE AVERAGE OF THE MAGNITUDES OF THE SOLUTION AT THE
C     BEGINNING AND END OF A STEP.
C     THE ERROR ESTIMATE FORMULA HAS BEEN GROUPED TO CONTROL LOSS OF
C     SIGNIFICANCE.
C     TO DISTINGUISH THE VARIOUS ARGUMENTS, H IS NOT PERMITTED
C     TO BECOME SMALLER THAN 26 UNITS OF ROUNDOFF IN T.
C     PRACTICAL LIMITS ON THE CHANGE IN THE STEPSIZE ARE ENFORCED TO
C     SMOOTH THE STEPSIZE SELECTION PROCESS AND TO AVOID EXCESSIVE
C     CHATTERING ON PROBLEMS HAVING DISCONTINUITIES.
C     TO PREVENT UNNECESSARY FAILURES, THE CODE USES 9/10 THE STEPSIZE
C     IT ESTIMATES WILL SUCCEED.
C     AFTER A STEP FAILURE, THE STEPSIZE IS NOT ALLOWED TO INCREASE FOR
C     THE NEXT ATTEMPTED STEP. THIS MAKES THE CODE MORE EFFICIENT ON
C     PROBLEMS HAVING DISCONTINUITIES AND MORE EFFECTIVE IN GENERAL
C     SINCE LOCAL EXTRAPOLATION IS BEING USED AND EXTRA CAUTION SEEMS
C     WARRANTED.
C
C
C     TEST NUMBER OF DERIVATIVE FUNCTION EVALUATIONS.
C     IF OKAY,TRY TO ADVANCE THE INTEGRATION FROM T TO T+H
C
  200 IF (NFE .LE. MAXNFE) GO TO 220
C
C     TOO MUCH WORK
      IFLAG=4
      KFLAG=4
      RETURN
C
C     ADVANCE AN APPROXIMATE SOLUTION OVER ONE STEP OF LENGTH H
C
  220 CALL FEHL(F,NEQN,Y,T,H,YP,F1,F2,F3,F4,F5,F1)
      NFE=NFE+5
C
C     COMPUTE AND TEST ALLOWABLE TOLERANCES VERSUS LOCAL ERROR ESTIMATES
C     AND REMOVE SCALING OF TOLERANCES. NOTE THAT RELATIVE ERROR IS
C     MEASURED WITH RESPECT TO THE AVERAGE OF THE MAGNITUDES OF THE
C     SOLUTION AT THE BEGINNING AND END OF THE STEP.
C
      EEOET=0.0E0
      DO 250 K=1,NEQN
        ET=ABS(Y(K))+ABS(F1(K))+AE
        IF (ET .GT. 0.0E0) GO TO 240
C
C       INAPPROPRIATE ERROR TOLERANCE
        IFLAG=5
        RETURN
C
  240   EE=ABS((-2090.0E0*YP(K)+(21970.0E0*F3(K)-15048.0E0*F4(K)))+
     1                       (22528.0E0*F2(K)-27360.0E0*F5(K)))
  250   EEOET=AMAX1(EEOET,EE/ET)
C
      ESTTOL=ABS(H)*EEOET*SCALE/752400.0E0
C
      IF (ESTTOL .LE. 1.0E0) GO TO 260
C
C
C     UNSUCCESSFUL STEP
C                       REDUCE THE STEPSIZE , TRY AGAIN
C                       THE DECREASE IS LIMITED TO A FACTOR OF 1/10
```

```
C
      HFAILD= .TRUE.
      OUTPUT= .FALSE.
      S=0.1E0
      IF (ESTTOL .LT. 59049.0E0) S=0.9E0/ESTTOL**0.2E0
      H=S*H
      IF (ABS(H) .GT. HMIN) GO TO 200
C
C     REQUESTED ERROR UNATTAINABLE AT SMALLEST ALLOWABLE STEPSIZE
      IFLAG=6
      KFLAG=6
      RETURN
C
C
C     SUCCESSFUL STEP
C                        STORE SOLUTION AT T+H
C                        AND EVALUATE DERIVATIVES THERE
C
  260 T=T+H
      DO 270 K=1,NEQN
  270   Y(K)=F1(K)
      A=T
      CALL F(A,Y,YP)
      NFE=NFE+1
C
C
C                       CHOOSE NEXT STEPSIZE
C                       THE INCREASE IS LIMITED TO A FACTOR OF 5
C                       IF STEP FAILURE HAS JUST OCCURRED, NEXT
C                          STEPSIZE IS NOT ALLOWED TO INCREASE
C
      S=5.0E0
      IF (ESTTOL .GT. 1.889568D-4) S=0.9E0/ESTTOL**0.2E0
      IF (HFAILD) S=AMIN1(S,1.0E0)
      H=SIGN(AMAX1(S*ABS(H),HMIN),H)
C
C     END OF CORE INTEGRATOR
C
C
C     SHOULD WE TAKE ANOTHER STEP
C
      IF (OUTPUT) GO TO 300
      IF (IFLAG .GT. 0) GO TO 100
C
C
C     INTEGRATION SUCCESSFULLY COMPLETED
C
C     ONE-STEP MODE
      IFLAG=-2
      RETURN
C
C     INTERVAL MODE
  300 T=TOUT
      IFLAG=2
      RETURN
C
      END
      SUBROUTINE FEHL(F,NEQN,Y,T,H,YP,F1,F2,F3,F4,F5,S)
C
C     FEHLBERG FOURTH-FIFTH ORDER RUNGE-KUTTA METHOD
C
C    FEHL INTEGRATES A SYSTEM OF NEQN FIRST ORDER
C    ORDINARY DIFFERENTIAL EQUATIONS OF THE FORM
C             DY(I)/DT=F(T,Y(1),---,Y(NEQN))
C    WHERE THE INITIAL VALUES Y(I) AND THE INITIAL DERIVATIVES
C    YP(I) ARE SPECIFIED AT THE STARTING POINT T. FEHL ADVANCES
C    THE SOLUTION OVER THE FIXED STEP H AND RETURNS
```

```
C    THE FIFTH ORDER (SIXTH ORDER ACCURATE LOCALLY) SOLUTION
C    APPROXIMATION AT T+H IN ARRAY S(I).
C    F1,---,F5 ARE ARRAYS OF DIMENSION NEQN WHICH ARE NEEDED
C    FOR INTERNAL STORAGE.
C    THE FORMULAS HAVE BEEN GROUPED TO CONTROL LOSS OF SIGNIFICANCE.
C    FEHL SHOULD BE CALLED WITH AN H NOT SMALLER THAN 13 UNITS OF
C    ROUNDOFF IN T SO THAT THE VARIOUS INDEPENDENT ARGUMENTS CAN BE
C    DISTINGUISHED.
C
C
      INTEGER  NEQN
      REAL  Y(NEQN),T,H,YP(NEQN),F1(NEQN),F2(NEQN),
     1  F3(NEQN),F4(NEQN),F5(NEQN),S(NEQN)
C
      REAL  CH
      INTEGER  K
C
      CH=H/4.0E0
      DO 221 K=1,NEQN
  221   F5(K)=Y(K)+CH*YP(K)
      CALL F(T+CH,F5,F1)
C
      CH=3.0E0*H/32.0E0
      DO 222 K=1,NEQN
  222   F5(K)=Y(K)+CH*(YP(K)+3.0E0*F1(K))
      CALL F(T+3.0E0*H/8.0E0,F5,F2)
C
      CH=H/2197.0E0
      DO 223 K=1,NEQN
  223   F5(K)=Y(K)+CH*(1932.0E0*YP(K)+(7296.0E0*F2(K)-7200.0E0*F1(K)))
      CALL F(T+12.0E0*H/13.0E0,F5,F3)
C
      CH=H/4104.0E0
      DO 224 K=1,NEQN
  224   F5(K)=Y(K)+CH*((8341.0E0*YP(K)-845.0E0*F3(K))+
     1                             (29440.0E0*F2(K)-32832.0E0*F1(K)))
      CALL F(T+H,F5,F4)
C
      CH=H/20520.0E0
      DO 225 K=1,NEQN
  225   F1(K)=Y(K)+CH*((-6080.0E0*YP(K)+(9295.0E0*F3(K)-
     1        5643.0E0*F4(K)))+(41040.0E0*F1(K)-28352.0E0*F2(K)))
      CALL F(T+H/2.0E0,F1,F5)
C
C     COMPUTE APPROXIMATE SOLUTION AT T+H
C
      CH=H/7618050.0E0
      DO 230 K=1,NEQN
  230   S(K)=Y(K)+CH*((902880.0E0*YP(K)+(3855735.0E0*F3(K)-
     1          1371249.0E0*F4(K)))+(3953664.0E0*F2(K)+
     2          277020.0E0*F5(K)))
C
      RETURN
      END
```

APPENDIX 5: POWER SERIES, COMPLEX NUMBERS, AND EULER'S FORMULA

This appendix contains a brief review of material that is sometimes not stressed in calculus courses but is needed in a study of differential equations, namely, power series, complex numbers, and Euler's formula that represents the complex exponential in terms of real trigonometric functions. A more extensive treatment is given in [1].

Power series are used to represent functions as infinite series and are defined as a series of the form

$$a_0 + a_1 t + a_2 t^2 + \cdots = \sum_{j=0}^{\infty} a_j t^j.$$

They arise when it is desired to approximate a given function $f(t)$ by a sequence of polynomials

$$f_n(t) = a_0 + a_1 t + a_2 t^2 + \cdots + a_n t^n.$$

Such an approximation will, in general, be of value only if t is small. The coefficients $a_0, a_1, a_2, \ldots$ are chosen so that $f_n(0) = f(0)$, and the first n derivatives of $f_n(t)$ are equal to the derivatives of the given function $f(t)$ at $t = 0$. Since $f_n^{(j)}(0) = j!\, a_j$, we have

$$a_0 = f(0),\ a_1 = f'(0),\ a_2 = \frac{f''(0)}{2!},\ a_3 = \frac{f^{(3)}(0)}{3!},\ \ldots,\ a_n = \frac{f^{(n)}(0)}{n!}.$$

If the coefficients a_j are chosen this way, then $f_n(t)$ is called the *Taylor polynomial* of degree n of $f(t)$.

If the function $f(t)$ and its first $n + 1$ derivatives are continuous on an interval containing 0 and t, then it can be shown that

$$f(t) = f_n(t) + R_n(0, t),$$

where the remainder term is given by

$$R_n(0, t) = f^{(n+1)}(\tau)\,\frac{t^{n+1}}{(n + 1)!},\quad \tau \text{ between 0 and } t.$$

The above formula is called *Taylor's formula with remainder*. (A proof of this result can be found in [1], Chap. 16.)

A slight generalization of this result is valuable. Replace t by $s = t + k$ and consider the function

$$f(s) = f(t + k) = q(k),$$

where t is held fixed and k is the independent variable. It follows that

$$q^{(j)}(k) = f^{(j)}(s) \qquad \text{and} \qquad \text{at } k = 0 \quad q^{(j)}(0) = f^{(j)}(t).$$

If we now approximate $f(t + k) = q(k)$ by Taylor polynomials, we obtain

$$\begin{aligned} f(s) &= f(t + k) \\ &= f(t) + f'(t)\,k + f''(t)\,\frac{k^2}{2!} + f^{(3)}(t)\,\frac{k^3}{3!} + \cdots + f^{(n)}(t)\,\frac{k^n}{n!} + R_n(t, k), \end{aligned}$$

where the remainder term is

$$R_n(t, k) = f^{(n+1)}(\tau)\,\frac{k^{n+1}}{(n + 1)!},\quad \tau \text{ between } t \text{ and } t + k.$$

This is most frequently used with $n = 1$ to construct the linear approximation

$$f(t + k) \simeq f(t) + f'(t)k.$$

In Chapters 3 and 7, extension of these formulas to functions of two variables is needed, for example,

$$P(x + u, y + v) \simeq P(x, y) + \frac{\partial P}{\partial x}(x, y)u + \frac{\partial P}{\partial y}(x, y)\,v.$$

This can be obtained from the single-variable approximation by setting

$$q(s) = P(x + su, y + sv)$$

and employing the linear approximation at $s = 0$. Since

$$q(s) \simeq q(0) + q'(0)\, s$$

and since by the chain rule

$$q'(s) = \frac{\partial P}{\partial x}(x + su, y + sv)\, u + \frac{\partial P}{\partial y}(x + su, y + sv)\, v,$$

then

$$q'(0) = \frac{\partial P}{\partial x}(x, y)\, u + \frac{\partial P}{\partial y}(x, y)\, v$$

and the desired formula is obtained letting $s = 1$.

The representation of a function by an infinite series results from taking the limit

$$f(t) = \lim_{n \to \infty} f_n(t) = \sum_{j=0}^{\infty} f^{(j)}(0)\, \frac{t^j}{j!},$$

where the convention $f^{(0)}(t) = f(t)$ is followed. If, instead of approximating $f(t)$ near $t = 0$, we approximate it near $t = a$, the Taylor polynomials and series will contain powers of $(t - a)$:

$$f_n(t) = a_0 + a_1(t - a) + a_2(t - a)^2 + \cdots + a_n(t - a)^n$$

with $a_j = f^{(j)}(a)/j!$, and

$$f(t) = f(a) + f'(a)\,(t - a) + f''(a)\,\frac{(t - a)^2}{2} + \cdots = \sum_{j=0}^{\infty} f^{(j)}(a)\, \frac{(t - a)^j}{j!}.$$

The simplest example of a function represented by an infinite series is offered by the exponential function, $f(t) = e^t$. Since $f^{(j)}(0) = 1$, we obtain $a_j = 1/j!$, and therefore

$$e^t = 1 + t + \frac{t^2}{2!} + \frac{t^3}{3!} + \cdots = \sum_{j=0}^{\infty} \frac{t^j}{j!}.$$

While this expression is valid for all values of t, the approximating polynomials

$$f_n(t) = 1 + t + \frac{t^2}{2!} + \cdots + \frac{t^n}{n!}$$

are only useful for small values of n if t is small (see Fig. A5.1).

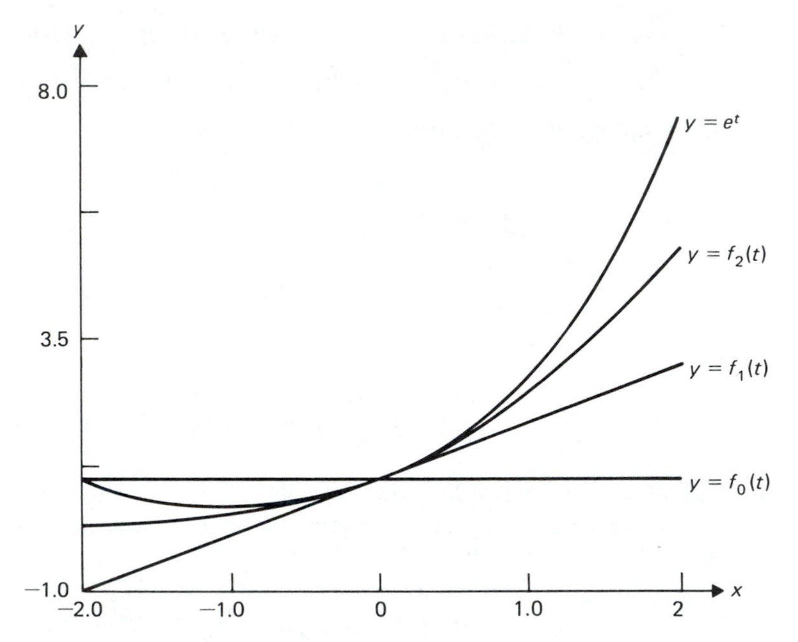

Figure A5.1 Plot of $y = f(t) = e^t$, $y = f_0(t) = 1$, $y = f_1(t) = 1 + t$, $y = f_2(t) = 1 + t + t^2/2$.

Other examples of functions represented by power series are

$$\sin t = t - \frac{t^3}{3!} + \frac{t^5}{5!} - \frac{t^7}{7!} + \cdots = \sum_{j=0}^{\infty} (-1)^j \frac{t^{2j+1}}{(2j+1)!};$$

$$\cos t = 1 - \frac{t^2}{2!} + \frac{t^4}{4!} - \frac{t^6}{6!} + \cdots = \sum_{j=0}^{\infty} (-1)^j \frac{t^{2j}}{(2j)!};$$

$$\sinh t = \frac{e^t - e^{-t}}{2} = \frac{1}{2}\sum_{j=0}^{\infty} \frac{t^j}{j!} - \frac{1}{2}\sum_{j=0}^{\infty} \frac{(-t)^j}{j!} = \sum_{j=0}^{\infty} \frac{t^{2j+1}}{(2j+1)!};$$

$$\cosh t = \frac{e^t + e^{-t}}{2} = \frac{1}{2}\sum_{j=0}^{\infty} \left[\frac{t^j}{j!} + \frac{(-t)^j}{j!}\right] = \sum_{j=0}^{\infty} \frac{t^{2j}}{(2j)!}.$$

The last two series illustrate that, within their common domain of convergence, power series can be added, subtracted, or multiplied by constants, and the result is a convergent power series. Moreover, it is also true that a power series can be differentiated or integrated term by term within its domain of convergence to produce a convergent power series. This permits the construction of power series solutions to differential equations, a topic considered at length in Chapter 6. (A good exposition of operations on power series is in [2], Chap. 9.)

Our next topic is *complex numbers, complex variables* and *functions,* and *power series of complex variables.* Complex numbers are extensions of real numbers that permit the solution of quadratic equations, such as $x^2 + 1 = 0$, and the uniform treatment of linear differential equations as well as the development of powerful mathematical theories, such as two-dimensional fluid flow and potential theory. The complex-number system consists of the complex numbers written as $x + iy$, where x and y are real, together with the rules by which they are added, multiplied, and divided. Rather than giving a list of such rules, we shall take an informal approach and say that the operations are the same as those for real numbers except that $i^2 = -1$. Hence

$$\begin{aligned}(x + iy) + (u + iv) &= (x + u) + i(y + v);\\ (x + iy) \cdot (u + iv) &= xu + iyu + ixv + i^2yv\\ &= (xu - yv) + i(xv + yu).\end{aligned}$$

When convenient, we write $x + iy$ as $x + yi$; then x is called the *real part* and y the *imaginary part* of $x + iy$ (or $x + yi$). The notation

$$x = \text{Re}(x + iy), \quad y = \text{Im}(x + iy)$$

is frequently employed.

The *complex conjugate* of $x + iy$ is defined as $x - iy$. It is customary to write $z^* = x - iy$ to denote the complex conjugate of $z = x + iy$. (The notation $\bar{z} = x - iy$ is also common.) The complex conjugate satisfies the relation

$$zz^* = (x + iy)(x - iy) = x^2 + y^2,$$

which is a real quantity, and if $z = x + iy$, $w = u + iv$, then

$$\frac{z}{w} = \frac{zw^*}{ww^*} = \frac{(x + iy)(u - iv)}{(u + iv)(u - iv)} = \frac{(xu + yv) + i(yu - xv)}{u^2 + v^2}.$$

This is the formula for division of two complex numbers.

The complex number $z = x + iy$ can be represented both as a point in the z plane and as a vector from the origin to the point with coordinates (x, y). Addition of complex numbers then corresponds to the parallelogram rule for addition of vectors. For instance, the addition of $(2 + 3i)$ and $(1 - i)$ can be represented as the result of adding the corresponding vectors (see Fig. A5.2).

The length of a vector corresponding to a complex number is the *absolute value* of the complex number:

$$|z| = |x + iy| = \sqrt{x^2 + y^2} = \sqrt{zz^*}.$$

In terms of polar coordinates

$$x = r \cos \theta, \qquad y = r \sin \theta,$$

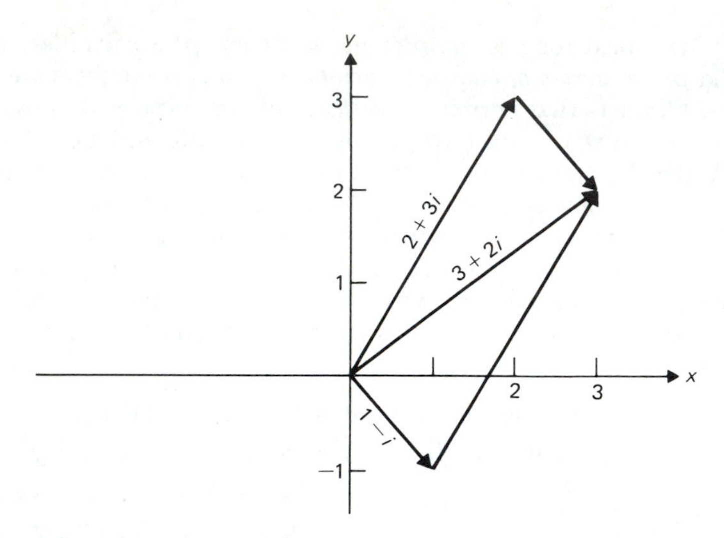

Figure A5.2 Addition of $1 - i$ to $2 + 3i$ to give $3 + 2i$.

where $\theta = \arctan(y/x)$, we have

$$z = x + iy = r(\cos\theta + i\sin\theta)$$

with $r = |z|$. The angle θ is called the *argument* of z. Since

$$\begin{aligned} z_1 z_2 &= r_1(\cos\theta_1 + i\sin\theta_1)\cdot r_2(\cos\theta_2 + i\sin\theta_2) \\ &= r_1 r_2[(\cos\theta_1\cos\theta_2 - \sin\theta_1\sin\theta_2) + i(\cos\theta_1\sin\theta_2 + \sin\theta_1\cos\theta_2)] \\ &= r_1 r_2[\cos(\theta_1 + \theta_2) + i\sin(\theta_1 + \theta_2)], \end{aligned}$$

this gives a geometrical interpretation of multiplication of two complex numbers. The product vector is rotated, so that its argument is the sum of the arguments of the two factors and its length is the product of the two lengths.

Example Let $z_1 = \sqrt{3} + i$, $z_2 = 1 - i$. Compute the lengths and arguments of z_1, z_2, z_1^*, z_2^*, $z_1 z_2$, z_1/z_2, $z_1 + z_2$.

For $z_1 = \sqrt{3} + i$, we have $x_1 = \sqrt{3}$ and $y_1 = 1$; hence

$$|z_1| = \sqrt{(\sqrt{3})^2 + 1^2} = 2 \qquad \text{and} \qquad \theta_1 = \arg z_1 = \arctan\frac{1}{\sqrt{3}} = \frac{\pi}{6}.$$

For $z_2 = 1 - i$, we have $x_2 = 1$, $y_2 = -1$; hence

$$|z_2| = \sqrt{1^2 + (-1)^2} = \sqrt{2} \qquad \text{and} \qquad \theta_2 = \arg z_2 = \arctan\frac{-1}{1} = -\frac{\pi}{4}.$$

Therefore,

$$z_1 = 2\left(\cos\frac{\pi}{6} + i\sin\frac{\pi}{6}\right), \qquad z_2 = \sqrt{2}\left(\cos\frac{\pi}{4} - i\sin\frac{\pi}{4}\right).$$

A similar calculation for $z_1^* = \sqrt{3} - i$ and $z_2^* = 1 + i$ gives $|z_1^*| = 2$, $\arg z_1^* = -\pi/6 = -\arg z_1$, $|z_2^*| = \sqrt{2}$, $\arg z_2^* = \pi/4 = -\arg z_2$; hence

$$z_1^* = 2\left(\cos\frac{\pi}{6} - i\sin\frac{\pi}{6}\right), \qquad z_2^* = \sqrt{2}\left(\cos\frac{\pi}{4} + i\sin\frac{\pi}{4}\right).$$

Furthermore,

$$|z_1z_2| = |z_1||z_2| = 2\sqrt{2}$$

and

$$\arg(z_1z_2) = \arg z_1 + \arg z_2 = \frac{\pi}{6} + \left(-\frac{\pi}{4}\right) = -\frac{\pi}{12},$$

hence

$$z_1z_2 = 2\sqrt{2}\left(\cos\frac{\pi}{12} - i\sin\frac{\pi}{12}\right)$$

is the polar representation of $z_1z_2 = (1 + \sqrt{3}) + (1 - \sqrt{3})i$. For the quotient z_1/z_2 we have

$$\frac{z_1}{z_2} = \frac{z_1z_2^*}{z_2z_2^*} = \frac{z_1z_2^*}{|z_2|^2} = \frac{|z_1|}{|z_2|}[\cos(\theta_1 - \theta_2) + i\sin(\theta_1 - \theta_2)],$$

and therefore

$$\frac{z_1}{z_2} = \frac{2}{\sqrt{2}}\left(\cos\frac{5\pi}{12} + i\sin\frac{5\pi}{12}\right)$$

is the polar representation of

$$\frac{z_1}{z_2} = \frac{1}{2}[(\sqrt{3} - 1) + (1 + \sqrt{3})i].$$

Finally, $z_1 + z_2 = 1 + \sqrt{3}$, hence $|z_1 + z_2| = 1 + \sqrt{3}$ and $\arg(z_1 + z_2) = 0$. □

Exponential functions e^z of a complex variable z are employed in the study of linear differential equations. They can be defined in terms of power series which are limits of Taylor polynomials of a complex variable. Therefore define

$$e^z = \lim_{n\to\infty}\sum_{j=0}^{n}\frac{z^j}{j!} = \sum_{j=0}^{\infty}\frac{z^j}{j!},$$

and compute (the details are messy and are omitted)

$$e^{x+iy} = \sum_{j=0}^{\infty} \frac{(x+iy)^j}{j!} = \left(\sum_{j=0}^{\infty} \frac{x^j}{j!}\right)\left(\sum_{j=0}^{\infty} \frac{(iy)^j}{j!}\right) = e^x e^{iy}$$

and

$$e^{iy} = \sum_{j=0}^{\infty} \frac{(iy)^j}{j!} = \sum_{j=0}^{\infty} (-1)^j \frac{y^{2j}}{(2j)!} + i \sum_{j=0}^{\infty} (-1)^j \frac{y^{2j+1}}{(2j+1)!} = \cos y + i \sin y.$$

In the last computation, the formulas $i^{2j} = (-1)^j$, $i^{2j+1} = i(-1)^j$ were used. The relation

$$e^{iy} = \cos y + i \sin y$$

is called *Euler's formula.* It represents the complex exponential of a purely imaginary variable in terms of real trigonometric functions.

Our final topic is the observation that a *complex solution to a real linear homogeneous equation is equivalent to two real solutions.* For example, $z(t) = e^{it} = \cos t + i \sin t$ is a solution of the linear homogeneous differential equation

$$y'' + y = 0$$

since

$$(e^{it})'' + e^{it} = (i^2 + 1)e^{it} = (-1+1)e^{it} = 0.$$

Furthermore

$$\begin{aligned}(\cos t + i \sin t)'' + (\cos t + i \sin t) &= [(\cos t)'' + \cos t] + i[(\sin t)'' + \sin t] \\ &= [-\cos t + \cos t] + i[-\sin t + \sin t] = 0,\end{aligned}$$

and hence $\cos t = \text{Re}(e^{it})$ and $\sin t = \text{Im}(e^{it})$ are also both solutions.

REFERENCES

1. G. B. Thomas and R. L. Finney, *Calculus and Analytic Geometry,* Addison-Wesley, Reading, 1988.
2. D. V. Widder, *Advanced Calculus,* Prentice-Hall, Englewood Cliffs, 1961.

ANSWERS TO SELECTED EXERCISES

CHAPTER 1

SECTION 1.1

1. Ordinary
t: independent
y: dependent

3. Partial
t, x: independent
u: dependent

5. Ordinary
t: independent
x: dependent

7. 1

9. 2

11. 2

SECTION 1.2

1. $y(t) = ce^{-t}$

3. $y(t) = ce^{-t^3/3}$

5. $y(t) = ce^{\arcsin t}$

7. $y(t) = c(t - 1)^2 e^{-3/(t-1)}$

9. $y(t) = 4e^{-2t}$,
validity: all t

11. $y(t) = 5^{-5/6}(1 - t)^{1/6}(t + 5)^{5/6}$, $-5 < t < 1$

13. $y(t) = 10e^{9-1/3(1+2t)^{3/2}}$, $-\dfrac{1}{2} \leq t$

15. a) $y(t) = 100(\sqrt{0.9})^t$ milligrams
c) Half-life $= -\dfrac{\ln 2}{\ln\sqrt{0.9}}$ hours $\cong 13.157$ hours

17. $\ln 2/\ln \dfrac{5}{4} \cong 3.106$ hours

19. a) $y(t) = 20e^{-t/25}$ lb

21. Approximately 137 minutes
$$T(t) = 65 - 60\left(\frac{11}{12}\right)^{t/20}$$

SECTION 1.3

1. $y(t) = ce^{-t}$

3. $y(t) = ce^{-t^3/3}$

5. $y(t) = ce^{3t}$

7. $y(t) = ce^t - \dfrac{1}{5}(\sin 2t + 2\cos 2t)$

9. $y(t) = (c - \ln t)/t$

11. $y(t) = (c + \sin^3 t)\sin t$

13. $y(t) = 4e^{-2t}$

15. $y(t) = (\sec t + \tan t)/(2 + \sqrt{2})$

17. $y(t) = e^{-t} + 2t + 3$

19. $y(t) = (2 - t)e^{2t}$

21. $y(t) = \dfrac{1}{2}(3e^{t^2} - 1)$

23. $y(t) = \dfrac{1}{4}(\pi + \pi t^2 + (4 + 4t^2)\arctan t)$

25. $y(t) = (t + 1)e^{-t^2}$

27. $y(t) = (\sin t)/t^2$

29. $y(t) = 4e^{2t} + ce^t$

31. $y(t) = (1 + t^2)^{1/2} \ln |t + (1 + t^2)^{1/2}| + c(1 + t^2)^{1/2}$

33. $y(t) = (\ln t + t^2 + c)/t$

35. $y(t) = [-2 + ce^{-t^2}]^{-1/4}$

37. $y(t) = \pm \left[\dfrac{\cos t}{t} + \sin t + \dfrac{c}{t}\right]^{-1/2}$

39. $v(t) = 24\left(1 - \exp\left(-\dfrac{4}{3}t\right)\right)$

$x(t) = 24t - 18\left(1 - \exp\left(-\dfrac{4}{3}t\right)\right)$

a) $v(3) = 24(1 - e^{-4}) \cong 23.56$ ft/sec, $x(3) = 54.3297$ ft
b) If $t \cong 10$, $x \cong 222$

SECTION 1.4

1. $y(t) = \dfrac{3}{2}(2t - 1) + ce^{-2t}$

3. $y(t) = 3 + ce^{-4t}$

5. $y(t) = -\dfrac{6}{5}\sin t - \dfrac{3}{5}\cos t$

7. $y(t) = te^{2t} - e^{2t} + ce^t$

9. $y(t) = \dfrac{1}{25}(-15t\cos t + 30t\sin t + 12\cos t - 9\sin t) + ce^{-2t}$

11. $y(t) = -t^2 + \dfrac{t}{4} - \dfrac{3}{16} + ce^{4t}$

13. $y(t) = \dfrac{1}{4}(2t + 11)e^{2t} + ce^{-2t}$

15. $y(t) = ce^{-2t} + \dfrac{(3t - 1)}{9}e^t$, general solution

$y(t) = \dfrac{37}{9}e^{-2t} + \dfrac{(3t - 1)}{9}e^t$, solution of IVP

17. $y(t) = ce^{-t} + 2t^2 - 4t + 19$
$y(t) = -15e^{-t} + 2t^2 - 4t + 19$

19. $y(t) = (c - 7t)e^{2t}$
$y(t) = (2 - 7t)e^{2t}$

21. $y(t) = ce^{-t} + 2(\sin t + \cos t)$
$y_p(t) = 2(\sin t + \cos t)$

23. $y(t) = ce^{-t} + \dfrac{1}{5}(2\sin 2t + \cos 2t + 5)$

$y_p(t) = \dfrac{1}{5}(2\sin 2t + \cos 2t + 5)$

27.

h	$T(h)$	Error
0.250000	1.717153	−0.349274
0.166667	1.596236	−0.228356
0.125000	1.537458	−0.169579
0.100000	1.502732	−0.134853
0.083333	1.479806	−0.111927
0.071429	1.463540	−0.095661
0.062500	1.451401	−0.083522

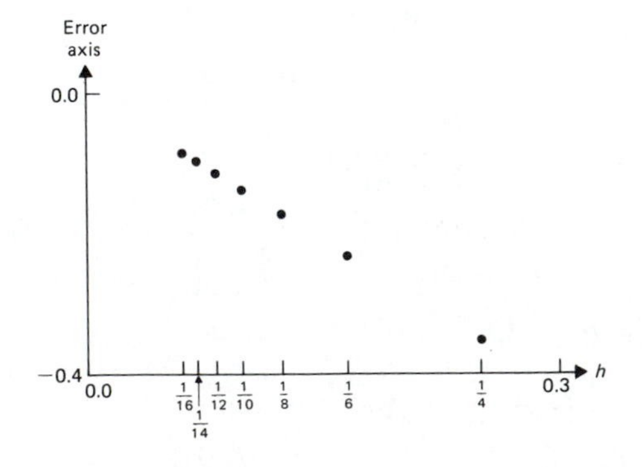

29.

h	$T(h)$	Error
0.250000	−0.272671	0.179220
0.166667	−0.210771	0.117320
0.125000	−0.180650	0.087199
0.100000	−0.162834	0.069383
0.083333	−0.151063	0.057611
0.071429	−0.142705	0.049254
0.062500	−0.136465	0.043014

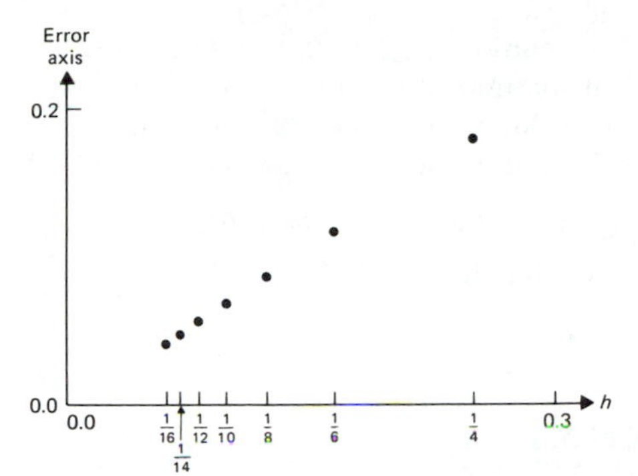

31.

h	$T(h)$	Error
0.250000	0.203198	−0.041956
0.166667	0.188670	−0.027428
0.125000	0.181609	−0.020367
0.100000	0.177437	−0.016195
0.083333	0.174684	−0.013442
0.071429	0.172730	−0.011488
0.062500	0.171272	−0.010030

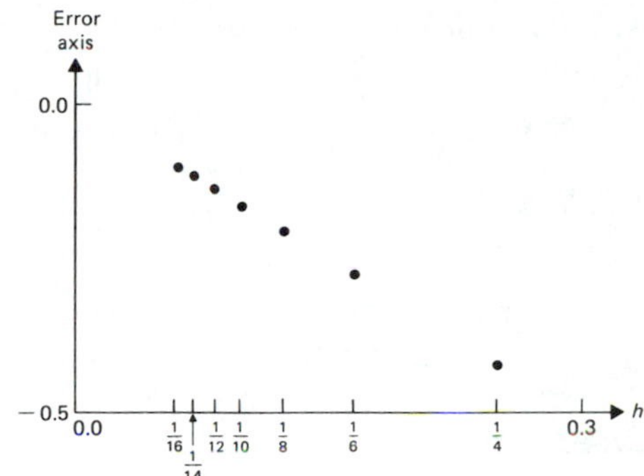

33. $S(\tfrac{1}{4}) = -0.09256$
$S(\tfrac{1}{8}) = -0.09340$

35. $S(\tfrac{1}{4}) = 0.8676$
$S(\tfrac{1}{8}) = 0.9629$

SECTION 1.5

1. Implicit solution: $\dfrac{y^2}{2} = \dfrac{t^3}{3} + C$

Explicit solution: $y(t) = \pm\left(\dfrac{2}{3}t^3 + 2C\right)^{1/2}$

3. Implicit solution: $e^{-y} + e^t + C = 0$
Explicit solution: $y(t) = -\ln|C - e^t|$

5. Implicit solution: $\ln\left|\dfrac{y-2}{y-1}\right| = \ln|Kt|$

Explicit solution: $y(t) = (2 - Kt)/(1 - Kt)$

7. Implicit solution: $y^2 = (t-1)e^t + C$
Explicit solution: $y(t) = \pm[(t-1)e^t + C]^{1/2}$

9. Implicit solution: $\ln|y - 1| + y = -\dfrac{1}{t} + C$

or $y(t) \equiv 1$

11. Implicit solution: $\dfrac{y^2}{2} = \dfrac{3}{2} - \cos t$, all t

Explicit solution: $y = -(3 - 2\cos t)^{1/2}$

13. Implicit solution: $\ln\left|\dfrac{y+1}{y-1}\right| = 2t$

Explicit solution: $y(t) = -(1 - e^{2t})/(1 + e^{2t})$

15. Implicit solution: $(y-1)^2 = t^3 + 2t + 4$ with $y = -1$ or 3 at $t = 0$
Explicit solution: $y(t) = 1 - (t^3 + 2t + 4)^{1/2}$

17. Implicit solution: $\tan y = 2 - 1/t$
Explicit solution: $y(t) = \arctan(2 - 1/t)$, $\lim_{t\to\infty} y(t) = \arctan(2)$

19. Implicit solution: $y^{-1/3} = 2 - \cos t$ if $y(\pi/2) = 1/8$
Explicit solution: $y(t) = (2 - \cos t)^{-3}$, all t
Implicit solution: $y^{-1/3} = 1/2 - \cos t$ if $y(\pi/2) = 8$
Explicit solution: $y(t) = (1/2 - \cos t)^{-3}$, $\pi/3 < t < 5\pi/3$

21. b) $k \cong (3 - \sqrt{3})/16$, $x_{max} \cong 212$ ft

23. $y' = -x/2y$ implies $x^2 + 2y^2 = K$

25. $y' = 4y/x$ implies $y = Kx^4$

27. a) $V = 16\pi h$, area $= 1/144$, $2h^{1/2} = -kt + 2\sqrt{12}$ where $k = 4.8/2304\pi$; hence, $T = 2\sqrt{12}/k$.

SECTION 1.6

1. $-\infty < t < -2$ or $-2 < t < \infty$

3. All t

5. All t

7. $-\infty < t < 2$

9. All t

15. $t_0 = 0$, $y_0 = \pi/2 + n\pi$

17. $y_0 = t_0$

19. $y_0 < 0$

21. c) $-\pi \leq t \leq 0$

23.

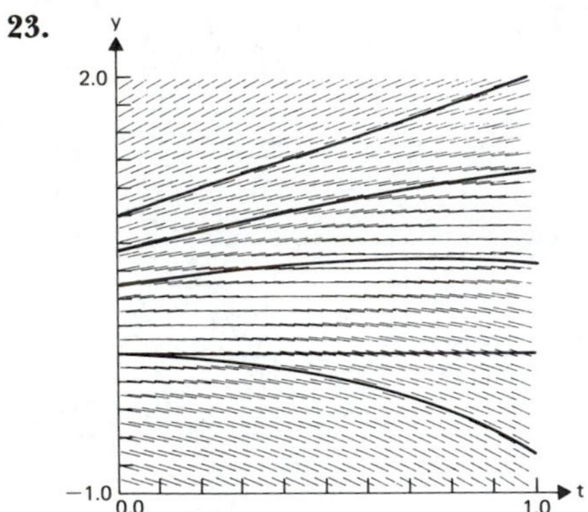

25.

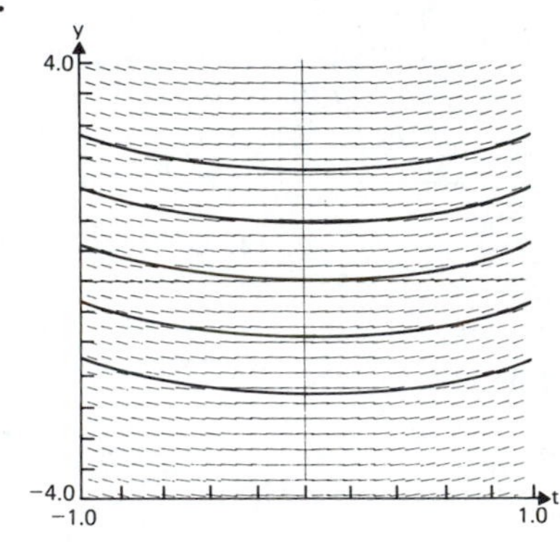

27.

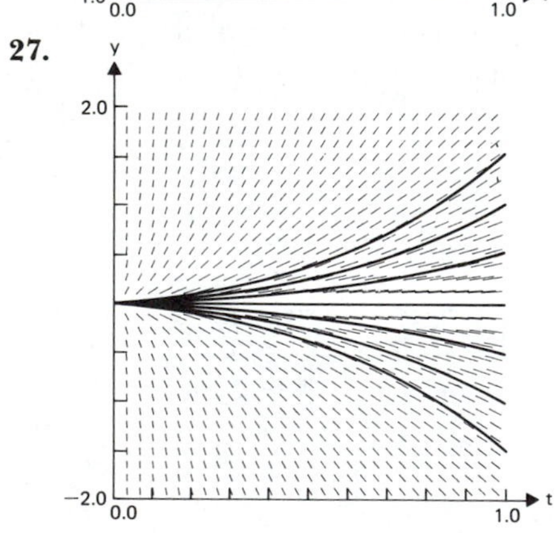

SECTION 1.7

1. a) $h = 1/2$, $y_2 = 1.000$; $h = 1/4$, $y_4 = 1.017$; $h = 1/8$, $y_8 = 1.009$
b) $h = 1/2$, $y_2 = 0.9990$; $h = 1/4$, $y_4 = 1.010$; $h = 1/8$, $y_8 = 1.003$
c) $y(t) = 2/(1 + t^2)$, $y(1) = 1$

3. a) $h = 1/2, y_2 = 2.2500$; $h = 1/4, y_4 = 2.375$; $h = 1/8, y_8 = 2.438$
b) $h = 1/2, y_2 = 2.500$; $h = 1/4, y_4 = 2.500$; $h = 1/8, y_8 = 2.500$
c) $y(t) = t^2/2 + t + 1, y(1) = 2.500$

5. a) $h = 1/2, y_2 = 2.050$; $h = 1/4, y_4 = 2.078$; $h = 1/8, y_8 = 2.092$
b) $h = 1/2, y_2 = 2.1241$; $h = 1/4, y_4 = 2.111$; $h = 1/8, y_8 = 2.107$
c) $y(t) = \sqrt{2}(e^t - 1/2)^{1/2}, y(1) = 2.106$

7.

Exercise	Euler	Improved Euler
1	$h = 0.01807$	$h = 0.1667$
3	$h = 0.01923$	$h = 0.5$
5	$h = 0.01852$	$h = 0.5$

11. $y(t) = 1/1 - t$, so $y(0.5) = 2$.
$y \cong 1.9954$, with $h = 0.025$, $y \cong 1.9988$.
The largest $h = 0.05$.

13. With $h = 0.02564$, $y \cong 5.366$.

15. With $h = 0.03846$, $y \cong -1.4253$.

17. $y(t) = \arctan t$;
$y(1) = \pi/4 \cong 0.7854$;
$y_{(h=1/10)} = 0.7850$

19. $y(t) = t^2$;
$y(2) = 4$;
$y_{(h=1/10)} = 3.987$

21. $y(t) = \ln\left[\frac{t^4}{4} + \frac{3}{4}\right]$;
$y(2) = 1.5581$;
$y_{(h=1/10)} = 1.5580$

23. $y(t) = -\ln(1 + e^{-1} - e^t)$;
$y(0.25) = 2.4787$;
$y^{(h=1/40)} = 2.4781$

25. $y(t) = \frac{19}{16}e^{4t} + \frac{4t-3}{16}$;
$y(0.5) = 8.712$;
$y_{(h=1/80)} = 8.710$

27.

Euler method		Improved Euler method	
h	$[y(1) - y_E]/h$	h	$[y(1) - y_{IE}]/h^2$
1/2	0.236	1/2	−0.091
1/4	0.206	1/4	−0.074
1/8	0.194	1/8	−0.067
1/16	0.189	1/16	−0.064
1/32	0.186	1/32	−0.062
1/64	0.185	1/64	−0.058

31. $Y(T) \cong IE(h) + \frac{1}{3}[IE(h) - IE(2h)]$ with $h = 0.05$.

$Y(T) \cong 1.9954023 + \frac{1}{3}[1.9954023 - 1.9833007]$ so $Y(T) \cong 1.9994362$.

With $h = 0.025$, $Y(T) \cong 1.9987989 + \frac{1}{3}[1.9987989 - 1.9954023]$ so $Y(T) \cong$ 1.9999311. The exact solution is $y(t) = 1/(1 - t)$, so $y(0.5) = 2$.

SECTION 1.8

1. $ty^2 + t^2y = C$

3. $t^2y + y\cos t = C$

5. $\tan t \sin y + ty = C$

7. $y \ln|t| - t^2 \ln|y| = C$

9. $e^x \cos y = C$, $e^x \sin y = K$

11. $\cos x \cosh y = C$, $\sin x \sinh y = K$

13. $\frac{t^2}{y^2} + \frac{1}{t} = C$

15. $2 \ln |ty| + \frac{t}{y} = C$

17. a) Yes c) No e) No

19. $(y + t)(y - 2t)^2 = C$

21. Implicit solution: $2y^2 + ty - t^2 = C$
Explicit solution: $y = (-t \pm [t^2 - 8(t^2 + C)]^{1/2})/4$

23. $\text{Arctan}\left(\frac{y}{t}\right) - \frac{1}{2}\ln\left(1 + \left(\frac{y}{t}\right)^2\right) = \ln t$ or $2 \arctan\left(\frac{y}{t}\right) - \ln(t^2 + y^2) = 0$

25. $\left(\frac{4}{2\sqrt{3} - 9}\right) \ln \left|\frac{t - 2 + \sqrt{3}(y + 1)}{t - 2}\right| - \left(\frac{4}{2\sqrt{3} + 9}\right) \ln \left|\frac{t - 2 - \sqrt{3}(y + 1)}{t - 2}\right| = \ln |t - 2| + C$

27. $y = \frac{x}{2}\left[\left(\frac{c}{x}\right)^k - \left(\frac{x}{c}\right)^k\right]$, $k = \frac{a}{b} < 1$, time $= \frac{c}{b(1 - k^2)}$

29. $y = at$, $x = c - bt$, so $y = k(c - x)$ with $k = a/b$. The value of k determines the slope of the line.

SECTION 1.9

1. $I(t) = 4(1 - e^{-3t})$,
$|I(t) - 4| < 10^{-4}$ if $t > 3.1$.

3. $\tilde{\alpha} = 2.2776762$
$\tilde{\beta} = -.2466857$
$E \cong 9.754$
$L \cong 40.537$

5. $\tilde{\alpha} = 502$
$\tilde{\beta} = -2.17$
$y(16) = 467$ (1979)
$y(17) = 465$ (1980)

7. 0.4203, −0.0633, −0.0768, 0.0258, −0.0761, 0.0418, −0.02006, 0.00259, 0.01184, −0.0003377, using $E = 1.565$, $R = 10$, $L = 31.153$.

SECTION 1.10

3. $P(t) = 1.535 + (6.336)(0.767)/(0.767 + (6.336 - 0.767)e^{-0.023t})$
$P(130) = 6.176$
$P(140) = 6.446$
$P(150) = 6.684$
$P(160) = 6.890$
$P(170) = 7.066$

5. a) $H_C = 4.80$
b) $K_2 = 156.9$
c) $20.0 < t < 20.5$
d) $11.5 < t < 12.0$

7. a) 250 (catch per year), 11 years

11. $X_0 = 3{,}929{,}000$
$r = 0.0313395$
$K = 197{,}272{,}520$
a) $X(130) = 107{,}394{,}310$ (1.6%)
$X(140) = 122{,}397{,}170$ (.6%)
$X(150) = 136{,}317{,}120$ (3.5%)
$X(160) = 148{,}676{,}690$ (1.3%)

SECTION 1.11

1. a) $500 \frac{dC}{dt} = -4C$ b) $t = 125 \ln 10$

3. $t = 200 \ln (99/80)$

5. $C(250) \cong 0.320$ kg/l, $Q(250) \cong 480.5$ kg
$C(500) \cong 0.412$ kg/l, $Q(500) \cong 824.9$ kg

7. $200\dfrac{dC_1}{dt} = -2C_1 + 10;\ 200\dfrac{dC_2}{dt} = -2C_2 + 2C_1$

$C_1(t) = 5(1 - e^{t/100}),\ C_2(t) = 5(1 - e^{-t/100}) - \dfrac{t}{20}e^{-t/100}$

9. $C(t) = 3/100e^{-t/125},\ 0 \le t \le t^*,\ 0.026 = 3/100e^{-t^*/125}$
$C(t) = 0.02 + 0.006e^{-(t-t^*)/125},\ t^* < t$

11. $C(t) = 3/100e^{-t/250},\ 0 \le t \le t^* + 2,\ 0.026 = 3/100e^{-t^*/250}$
$C(t) = 0.02 + (0.026e^{-1/125} - 0.02)\exp\left(-\dfrac{t - t^* - 2}{250}\right),\ t^* + 2 < t$

13. Assume $z > k$: $C_0 > k$ implies that if $C(t) = C_0e^{-t/125}$, $0 \le t \le t^* + \tau$, then the solution oscillates periodically between $ke^{-\tau/125}$ and $z + (k - z)e^{-\tau/125}$; $C_0 \le k$ implies that if $C(t) = z + (C_0 - z)e^{-t/125}$, $0 \le t \le t_* + \tau$, then the above oscillation.

15. a) $\dfrac{d}{dt}[(V_0 + (v_1 - v_2)t)C(t)] = C_1v_1 - C(t)v_2,\ C(0) = 0$
b) $C(t) = c_1t(100 + t)/(2500 + 100t + t^2)$
c) $t = 25(\sqrt{6} - 2)$min

CHAPTER 2

SECTION 2.2

1. b) $10y'' + 20 \cdot 9.8y = 0,\ y(0) = -0.1,\ y'(0) = 0$

3. $4y'' + 10y' + 20 \cdot 9.8y = 0,\ y(0) = 0,\ y'(0) = 2$

5. $my'' + cy' + ky = kA\cos wt$

7. $Q'' + 10Q' + 100Q = 8\sin t,\ Q(0) = 0,\ Q'(0) = 0$

9. $I\dfrac{d^2\theta}{dt^2} = -K\theta$

SECTION 2.3

1. $y_1(0) = 4;\ y_2(0) = -3$

3. $y_1(0) = 2;\ y_2(0) = 5$

5. $y_1(1) = 1;\ y_2(1) = 0$

7. $y_1(\pi) = -(\pi^{-1/2});\ y_2(\pi) = 0$

9. $y(t) = \dfrac{1}{3}\sin 3t$

11. $y(t) = t + 2 + e^{(t-1)}(t - 1)$

13. $y(t) = (1 - 2\ln 2)t^2 + 2t^2\ln t$

SECTION 2.4

1. $y(t) = c_1e^{5t} + c_2e^{4t}$

3. $y(t) = c_1\exp\left[\dfrac{t}{2}(r - \sqrt{r^2 - 16})\right] + c_2\exp\left[\dfrac{t}{2}(r + \sqrt{r^2 - 16}\right]$

5. $y(t) = c_1e^{6t} + c_2e^{-5t}$

7. $y(t) = c_1t + c_2$

9. $y(t) = \dfrac{\sqrt{5}}{10}(e^{\sqrt{5}t} - e^{-\sqrt{5}t})$

11. $y(t) = e^{-t/2+1}(13t/2 - 2)$ **13.** $y(t) = e^{6t}/2 + 7e^{-2t}/2$ **15.** $y(t) = 3e^{3(t-1)}(t - 1)$

17. $y(t) = te^{1-t/4}$ **19.** $y(t) = 3e^{-t/4} - e^{t/4}$ **23.** $y(t) = \cosh\frac{t}{3} - 3\sinh\frac{t}{3}$

SECTION 2.5

1. $y(t) = e^{-3t/2}(k_1 \sin t + k_2 \cos t)$ **3.** $y(t) = e^t(k_1 \sin t + k_2 \cos t)$

5. $y(t) = k_1 \sin(7t/2) + k_2 \cos(7t/2)$ **7.** $y(t) = 2\cos t - 6\sin t$

9. $y(t) = \sqrt{2}\sin 2t$ **11.** $y(t) = \sqrt{2}\cos(t - \pi/4)$

13. $y(t) = \sqrt{2}\cos(2t - \pi/4)$ **17.** $y(t) = 5e^{-t}\cos(3t - \phi)$, where $\cos\phi = 4/5$, $\sin\phi = 3/5$

21. $y^2 + (y')^2/4 = 1^2 + (-1)^2/4 \equiv 5/4$ **25.** $y(t) = \cos t - \sin t$

27. $y(t) = 2e^{-t}[1 + (e^2 - 1)t]$

SECTION 2.6

1. $W = -4$; nonzero for all t **3.** $W = -\omega$; nonzero for all t

5. $W = t^3$; nonzero for $t \neq 0$ **7.** $W = -t^{-1}$; nonzero for $t > 0$

9. $y_1(t) = e^{5t}$, $y_2(t) = te^{5t}$ **11.** $y_1(t) = e^t \sin 2t$, $y_2(t) = e^t \cos 2t$

13. $y_1(t) = \sin 3t$, $y_2(t) = \cos 3t$ **15.** $y(t) = At^{-2} + Bt^{-3}$; yes

17. $y(t) = (A + B\ln t)t^{-3}$; yes **19.** $y(t) = t^{-1}[A\cos(2\ln t) + B\sin(2\ln t)]$; yes

SECTION 2.7

1. $y(t) = -(2t - 1)/4 + C_1e^{-t} + C_2e^{2t}$ **3.** $y(t) = 2\cos t + e^t(C_1 + C_2t)$

5. $y(t) = (t^4 + 2t^3 + 6t^2)/12 + C_1 + C_2t$ **7.** $y(t) = C_1 \sin 2t + C_2 \cos 2t + (2t\cos 2t + t)/4$

9. $y(t) = e^{-t/2}\left[C_1 \sinh\left(\frac{\sqrt{17}}{2}t\right) + C_2 \cosh\left(\frac{\sqrt{17}}{2}t\right)\right] + e^{-t}/4 - e^t/2$

11. $y(t) = 5e^{2t}/12 - 2e^{-t}/3 - (2t - 1)/4$ **13.** $y(t) = 2\cos t + (t - 1)e^t$

15. $y(t) = (\sin 2t)/8 + (2t\cos 2t + t)/4 - (3\pi\cos 2t)/4$

19. $y(t) = C_1t^{(-2+\sqrt{10})} + C_2t^{(-2-\sqrt{10})} - \frac{2}{5}t^{-2} + 2$ **21.** $y(t) = 3(\ln t)^2t^{-2} + C_1t^{-2} + C_2(\ln t)t^{-2}$

SECTION 2.8

1. $y(t) = \frac{e^{3t}}{4} + C_1e^t + C_2e^{-t}$ **3.** $y(t) = (5t - 5)e^{4t} + C_1e^{4t} + C_2e^{3t}$

5. $y(t) = -\frac{1}{32}(\cos 6t - 24t\sin t - 9\cos 2t) + C_1 \sin 2t + C_2 \cos 2t$

7. $y(t) = -e^{-3t}(\ln|t| + 1) + e^{-3t}(C_1 + C_2t)$

9. $y(t) = \dfrac{e^t}{2}\ln(1 + e^{-t}) - \dfrac{e^{3t}}{2}\left[\dfrac{e^{-2t}}{2} - e^{-t} + \ln(1 + e^{-t})\right] + C_1e^t + C_2e^{3t}$

11. $y(t) = -t\displaystyle\int \frac{e^t}{t}\,dt + t^2\int \frac{e^t}{t^2}\,dt + C_1t + C_2t^2$

13. $$y(t) = t^3\left(-\frac{e^t}{t} + \ln t + t + \frac{t^2}{2\cdot 2!} + \frac{t^3}{3\cdot 3!} + \cdots - \frac{\ln t}{3t^3} - \frac{1}{9t^3}\right)$$
$$+\, t^2\left(-\ln t - t - \frac{t^2}{2\cdot 2!} - \frac{t^3}{3\cdot 3!} - \cdots + \frac{\ln t}{2t^2} + \frac{1}{4t^2}\right) + C_1t^3 + C_2t^2$$
$$= -t^2e^t + \frac{\ln t}{6} + \frac{5}{36} + t^3\int \frac{e^t}{t}\,dt - t^2\int \frac{e^t}{t} + C_1t^3 + C_2t^2$$
using $\displaystyle\int \frac{e^{at}}{t}\,dt = \ln t + at + \frac{(at)^3}{2\cdot 2!} + \frac{(at)^3}{3\cdot 3!} + \cdots$

15. $y(t) = \dfrac{2}{9}\sin 3t - \dfrac{2}{9}\cos 3t + \dfrac{t}{3} + \dfrac{2}{9}$

17. $y(t) = -e^{-3t}(\ln|t| + 1) + e^{-3t}(-2e^3 + t(1 + 3e^3))$

19. $y(t) = -t^{-1}\ln t - t^{-1} + 2$

23. $y(t) = \dfrac{t}{2}e^t + C_1t\sin t + C_2t\cos t$

SECTION 2.9

1. $y(t) = \cos 2t$, $p(t) = -2\sin 2t$, not damped

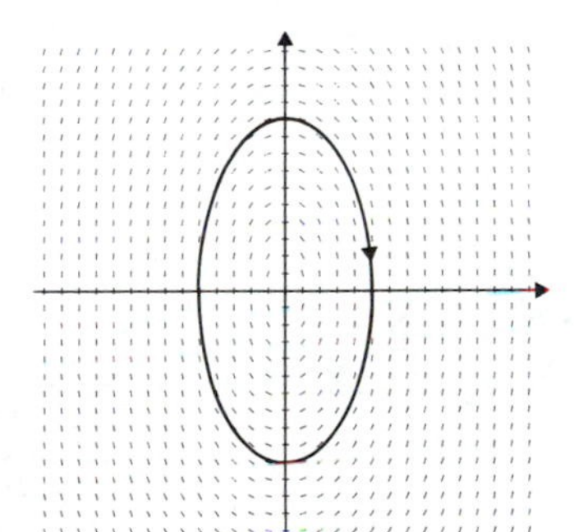

3. $y(t) = \cos t$, $p(t) = -\sin t$, not damped

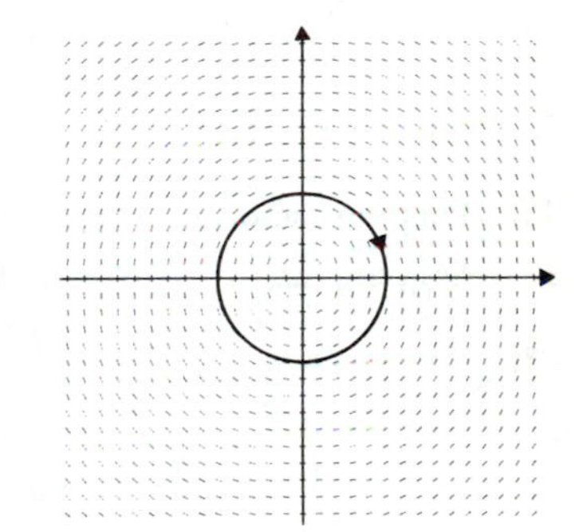

5. $y(t) = e^{-2t}\left(\dfrac{2}{3}\sin 3t + \cos 3t\right)$, $p(t) = e^{-2t}\left(-\dfrac{13}{3}\sin 3t\right)$, underdamped

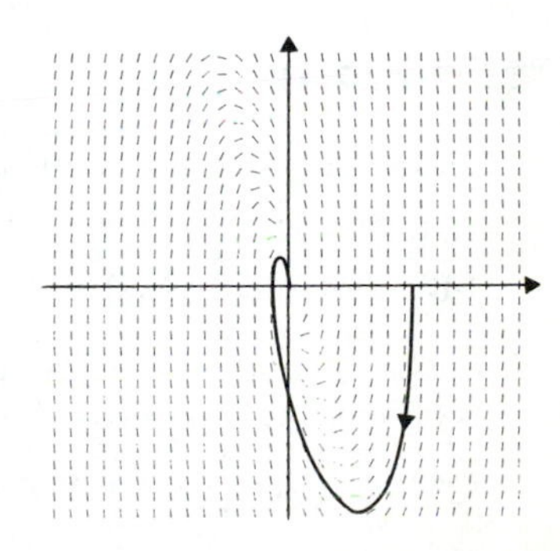

7. $y(t) = (1 + 2t)e^{-2t}$, $p(t) = -4te^{-2t}$, critically damped

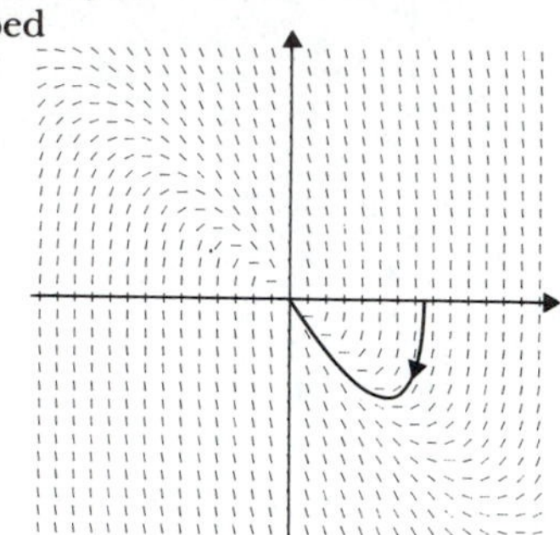

9. $y(t) = (1 + t)e^{-t}$, $p(t) = -te^{-t}$, critically damped

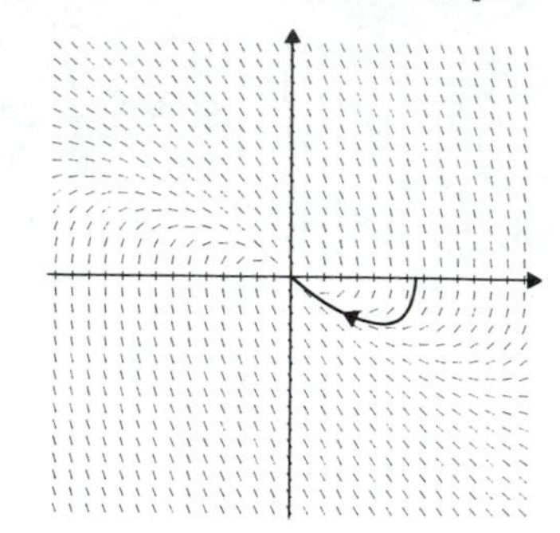

11. $y(t) = 2e^{-t} - e^{-2t}$, $p(t) = 2e^{-2t} - 2e^{-t}$, overdamped

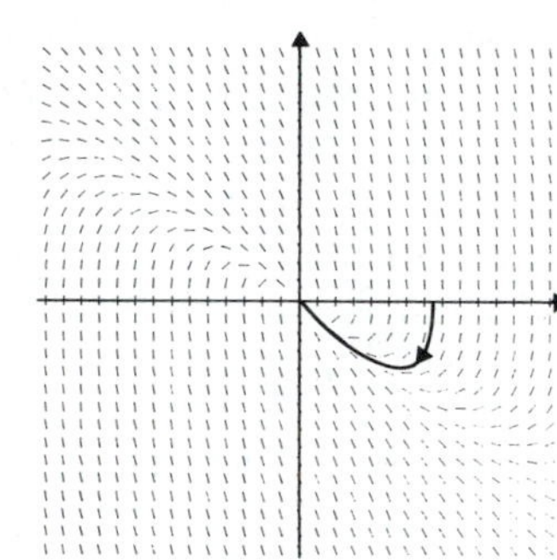

13. Amplitude $\sqrt{5}$, period 2π, $y^2 + (y')^2 = 5$

15. Amplitude 4, period π, $y^2 + \dfrac{(y')^2}{-4} = 16$

17. Damping factor e^{-t}, period 2π, $T = 4 \ln 10 + \ln(2\sqrt{2})$

19. Damping factor $e^{-7/5t}$, period 10π, $T = 25/7 \ln 10$

21. $y(0) = 1$

23. $T = 1/20, 100$

25. $T = -\ln \dfrac{3 - \sqrt{9 - 8\epsilon}}{4}$

27. $T = -\ln\left(\mu - \dfrac{2}{\mu}\right)$ where $\mu = \left(\dfrac{\epsilon}{2} + \dfrac{\sqrt{32 + 27\epsilon^2}}{6\sqrt{3}}\right)^{1/3}$

SECTION 2.10

1. Steady state solution $= (94 \cos 2t + 3 \sin 2t)/145$
transient solution $= e^{-2t}(c_1 \sin 3t + c_2 \cos 3t)$
$Q(2) = \arctan \dfrac{8}{9}$; $K(2) = 145^{-1/2}$

3. Steady state solution $= (15 \sin t - \cos t)/8$
transient solution $= e^{-2t}(c_1 \sin t + c_2 \cos t)$
$Q(1) = \dfrac{\pi}{4}$; $K(1) = 32^{-1/2}$

5. Steady state solution $= (56 \sin 3t - 3 \cos 2t)/370$
transient solution $= c_1e^{-t/3} + c_2e^{-4t}$
$Q(2) = \arctan\left(\dfrac{-13}{4}\right)$; $K(2) = 740^{-1/2}$

7. $Q(3) = 0$; $K(3) = 1/15$

9. $Q\left(\frac{22}{7}\right) = 0;\ K\left(\frac{22}{7}\right) = \left[\left(\frac{22}{7}\right)^2 - \pi^2\right]^{-1}$

11. $c = 1,\ \overline{\omega}^2 = 35/8$ $\qquad c = 5,\ \overline{\omega}^2 = 11/8$ $\qquad c = 8,\ \overline{\omega}^2 = 0$

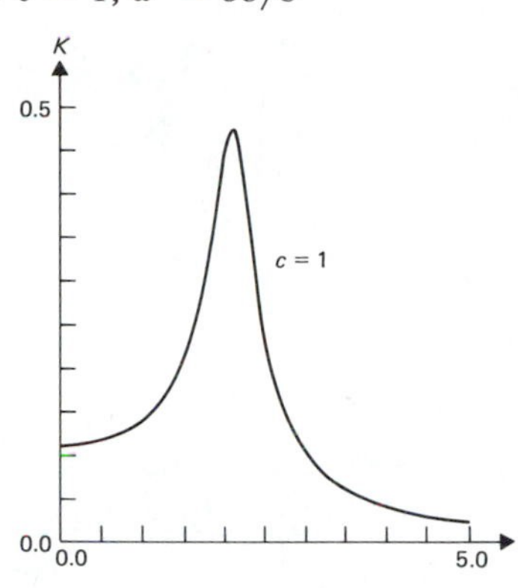

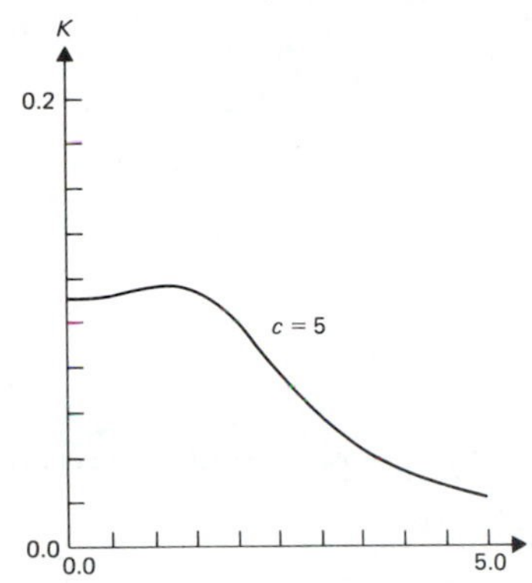

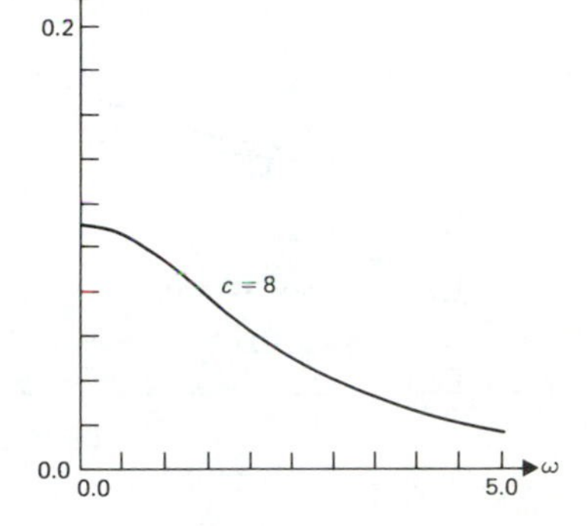

SECTION 2.11

1.

t_i	y_i	$y(t_i)$
0.25	1.0000	1.0314
0.50	1.0625	1.1276
0.75	1.1914	1.2947
1.00	1.3948	1.5431

3.

t_i	y_i	$y(t_i)$
0.25	1.0000	1.0691
0.50	1.1250	1.3163
0.75	1.4258	1.8409
1.00	1.9956	2.8193

5.

t_i	y_i	$y(t_i)$
0.25	0.	0.0282
0.50	0.0625	0.0974
0.75	0.1581	0.1801
1.00	0.2520	0.2477

7.

t_i	y_i	$y(t_i)$
0.25	0.1875	0.1967
0.50	0.3047	0.3161
0.75	0.3779	0.3884
1.00	0.4237	0.4323

9.

t_i	y_i	$y(t_i)$
0.25	0.0313	0.0372
0.50	0.1602	0.1796
0.75	0.4476	0.4954
1.00	0.9936	1.0973

11. Exercises 1, 6

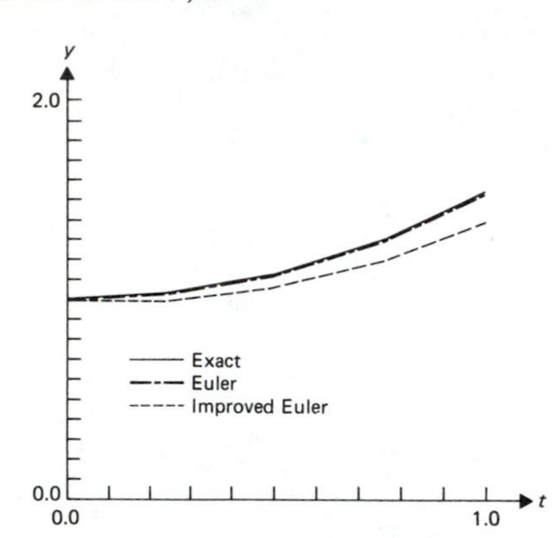

Exercises 4, 9

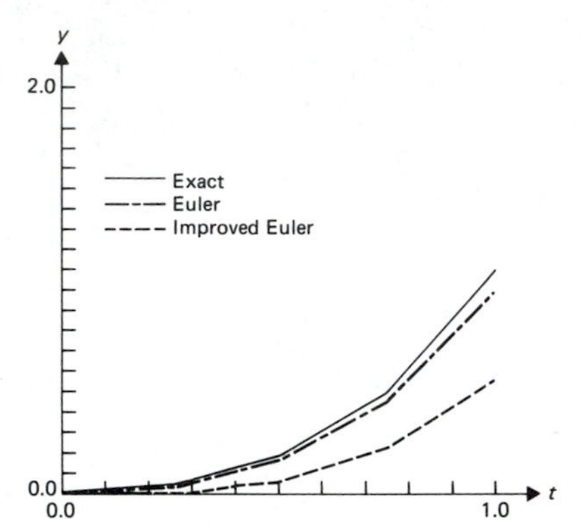

13.

t	y	$y(t)$
0.10	1.1150	1.1148
0.20	1.2587	1.2582
0.30	1.4291	1.4282
0.40	1.6235	1.6220
0.50	1.8386	1.8364
0.60	2.0704	2.0675
0.70	2.3146	2.3110
0.80	2.5663	2.5618
0.90	2.8202	2.8149
1.00	3.0707	3.0647

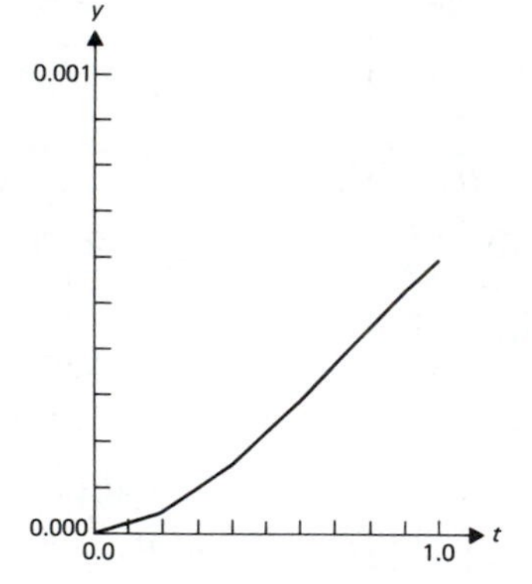

15.

t	y	$y(t)$
0.10	0.0875	0.0883
0.20	0.1550	0.1563
0.30	0.2062	0.2079
0.40	0.2444	0.2462
0.50	0.2719	0.2737
0.60	0.2909	0.2927
0.70	0.3030	0.3047
0.80	0.3097	0.3113
0.90	0.3120	0.3134
1.00	0.3108	0.3121

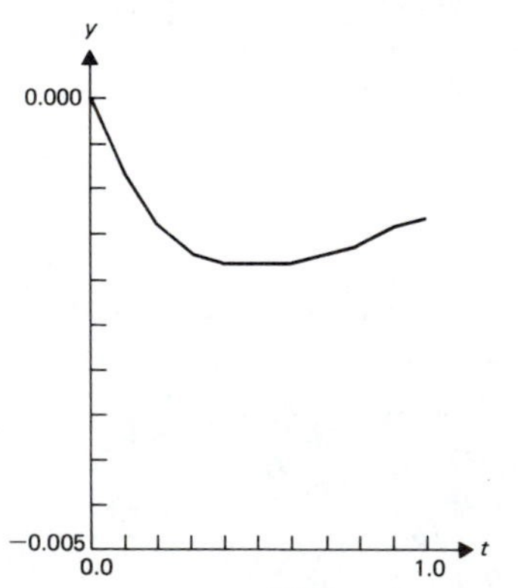

17.

t_i	y_i	y'_i	t_i	y_i	y'_i
0.00	0.35503	1.00000	1.60	1.24900	−0.49276
0.10	0.45503	0.99772	1.70	1.18973	−0.69465
0.20	0.55457	0.98990	1.80	1.11015	−0.89660
0.30	0.65301	0.97455	1.90	1.01050	−1.09346
0.40	0.74949	0.94975	2.00	0.89155	−1.27957
0.50	0.84296	0.91365	2.10	0.75468	−1.44891
0.60	0.93222	0.86454	2.20	0.60187	−1.59522
0.70	1.01588	0.80092	2.30	0.43572	−1.71230
0.80	1.09241	0.72153	2.40	0.25948	−1.79414
0.90	1.16019	0.62543	2.50	0.07696	−1.83529
1.00	1.21752	0.51208	2.60	−0.10754	−1.83106
1.10	1.26264	0.38143	2.70	−0.28924	−1.77784
1.20	1.29383	0.23393	2.80	−0.46312	−1.67341
1.30	1.30946	0.07069	2.90	−0.62398	−1.51715
1.40	1.30802	−0.10659	3.00	−0.76665	−1.31032
1.50	1.28821	−0.29545			

19.

t_i	y_i
0.25	0.031
0.50	0.11
0.75	0.20
1.00	0.24

21.

t_i	y_i
0.20	1.04
0.40	1.20
0.60	1.51
0.80	2.045
1.00	2.91

SECTION 2.12

1. $\omega = e^t$, linearly independent

3. $w = -5e^{2t}$, linearly independent

5. $w = -10$, linearly independent

7. $w = 0$, not linearly independent

9. $y^{(3)} + 4y'' = 0$

11. $y^{(3)} - 12y'' + 48y' - 64y = 0$

13. $y^{(3)} = 0$

15. $y^{(4)} - y'' = 0$

17. $y^{(4)} - 6y^{(3)} + 5y'' - 24y' - 36y = 0$

19. $y(t) = c_1 + c_2 \cos t + c_3 \sin t$

21. $y(t) = c_1e^{p_1t} + c_2e^{p_2t} \sin q_2t + c_3e^{p_2t} \cos q_2t$ where $p_1 = -\dfrac{2}{3} - \dfrac{7}{9\mu} - \mu$, $p_2 = -\dfrac{2}{3} + \dfrac{7}{18\mu} + \dfrac{\mu}{2}$,

$q_2 = \dfrac{7\sqrt{3}}{18\mu} - \dfrac{\sqrt{3}\mu}{2}$, where $\mu = \left(\dfrac{61}{54} - \dfrac{\sqrt{29}}{6}\right)^{1/3}$

23. $y(t) = c_1e^{2t} + c_2e^{3t} + c_3te^{3t}$

25. $y(t) = c_1e^{-t} + c_2te^{-t} + c_3e^t \sin 3t + c_4e^t \cos 3t$

27. $y(t) = -e^t + 2e^{-t} + 3 \sin t - \cos t$

29. $y(t) = 4 \sin t + t \sin t + \cos t - 2t \cos t$

31. $y(t) = \dfrac{19}{3} + \dfrac{1}{2}t + 4t^2 - \dfrac{2}{3}e^t - \dfrac{11}{13}e^{-t}$

33. $y(t) = \dfrac{1}{t^2} - t^4$

35. $y_p(t) = -2 \cos t$

37. $y_p(t) = -2t^2e^{-t}$

39. $y_p(t) = t^4 + 9t^2 - \frac{1}{8}t + \frac{39}{2}$

41. $y_p(t) = e^t + \frac{2}{3}te^{-t}$

43. $y(x) = \frac{1}{180}x^6 - \frac{1}{5}x^5 + 3x^4 + 43.2x^2 - 19.2x^3$
$y(2.7378659) = 69.924129$, maximum deflection

45. $y(x) = \frac{5}{12}x^4 - \frac{5}{3}x^3 + 250x^2$
$y(10) = 12500$, maximum deflection

47. $y(x) = -2x^2 - 21.008565 + 8x + 18.504282e^{-x/2} + 2.504282e^{x/2}$

CHAPTER 3

SECTION 3.3

1. $y(t) = \frac{1}{2}(e^{2t} + 1)$
$y(0.5) = 1.859$, $y_5 = 1.851$

3. $y(t) = e^{-t}$
$y(0.5) = 0.6065$, $y_5 = 0.6071$

5. $y(t) = \frac{2}{t^2 + 1}$
$y(0.5) = 1.600$, $y_5 = 1.595$

7. $y(t) = (-2\cos 2t - \sin 2t + 2e^t)/5$
$y(0.5) = 0.2751$, $y_5 = 0.2744$

9. $y(t) = 2/(2 - e^{-t})$
$y(0.5) = 1.435$, $y_5 = 1.440$

13.

t	$y(p = 1)$	$y(p = 2)$	$y(p = 3)$
0.00	0.0000	0.0000	0.0000
0.10	0.1000	0.1000	0.0993
0.20	0.1980	0.1960	0.1947
0.30	0.2901	0.2844	0.2826
0.40	0.3727	0.3620	0.3599
0.50	0.4429	0.4266	0.4243

SECTION 3.4

3. $\mathbf{y} = \begin{bmatrix} y_1 \\ y_2 \end{bmatrix}, \mathbf{f} = \begin{bmatrix} -4y_1 - y_2 \\ y_1 - 2y_2 \end{bmatrix}, \mathbf{y}_0 = \begin{bmatrix} 1 \\ 0 \end{bmatrix}$

5. $\mathbf{y} = \begin{bmatrix} y_1 \\ y_2 \end{bmatrix}, \mathbf{f} = \begin{bmatrix} y_2 \\ -y_2 + \ln(t) \end{bmatrix}, \mathbf{y}_0 = \begin{bmatrix} 0 \\ -1 \end{bmatrix}$

7. $y(t) = t - 1 + \exp(-t)$

t	y	y(exact)
0.00	0.00000	0.0000
0.10	0.00500	0.004837
0.20	0.01902	0.01873
0.30	0.04121	0.04081

9. $y(t) = t + 3 - \exp(-t)$

	y	y(exact)
0.00	2.000	2.000
0.10	2.195	2.195
0.20	2.381	2.381
0.30	2.559	2.559

11. $y(t) = \exp(t)$

t	$k1$	$k2$	$k3$	$k4$	dy	$y^{(RK)}$	$y^{(EC)}$	y(exact)
0.30	1.221	1.283	1.286	1.350	0.1295	1.350	1.349	1.350

13. $y(t) = \tan(t)$

t	$k1$	$k2$	$k3$	$k4$	dy	$y^{(RK)}$	$y^{(EC)}$	y(exact)
0.30	1.0411	1.0649	1.0655	1.0956	0.1066	0.3093	0.3090	0.3093

15. Set $z = y*y$ and solve for $z(t) = 4 + 5\exp(-t*t)$, then $y(t) = \mathrm{sqrt}(z(t))$.

t	$k1$	$k2$	$k3$	$k4$	dy	$y^{(RK)}$	$y^{(EC)}$	y(exact)
0.30	−0.3238	−0.3989	−0.3975	−0.4683	−0.0397	2.9274	2.9271	2.9274

17.

t	y
0.0	2.000
0.5	3.149
1.0	4.718

19.

t	y
0.0	0.000
0.5	0.4244
1.0	0.5381

21.

t	y
0.0	2.000
0.5	1.600
1.0	1.000

23.

t	y	y'
0.00	1.000	0.0000
0.20	0.8795	−1.029
0.40	0.6424	−1.245
0.60	0.4087	−1.050
0.80	0.2314	−0.7153
1.00	0.1216	−0.3932

25.

t	y	y'
1.00	0.0000	−1.000
2.00	−0.7122	−0.1881
3.00	−0.1537	1.279
4.00	1.547	1.883
5.00	3.095	0.9881
6.00	3.274	−0.6299
7.00	2.088	−1.515
8.00	0.7698	−0.8752
9.00	0.6546	0.6843
10.00	1.957	1.717

27.

t	x	y
0.	0.0000	−1.000
0.10	0.07405	−0.8149
0.20	0.1097	−0.6586
0.30	0.1219	−0.5285
0.40	0.1204	−0.4217
0.50	0.1115	−0.3347
0.60	0.09915	−0.2645
0.70	0.08570	−0.2082
0.80	0.07256	−0.1633
0.90	0.06047	−0.1277
1.00	0.04978	−0.09958

29.

t	x	y
0.	0.0000	1.000
0.20	−0.6964	1.238
0.40	−1.118	1.169
0.60	−1.225	0.8632
0.80	−1.051	0.4288
1.00	−0.6850	−0.01825

31.

t	x	y
0.00	2.000	0.000
0.50	−0.9470	1.816
1.00	−1.749	1.028

CHAPTER 4

SECTION 4.2

1. $\dfrac{1}{s-2}$

3. $\dfrac{1}{s+7}$

5. $\dfrac{1}{s-10}$

7. $\dfrac{e}{s-3}$

11. $L\{1\} = 1/s$, $\mathrm{Re}(s) > 0$, $L\{6\} = 6/s$, $\mathrm{Re}(s) > 0$

SECTION 4.3

1. $y(t) = -3e^{-3t} + 4e^{-2t}$

3. $y(t) = \frac{1}{3} + \frac{2}{3}e^{3t}$

5. $y(t) = e^{3t/2} - e^{t/2}$

7. $y(t) = \frac{5}{2}e^{-t} - \frac{1}{2}e^{t}$

9. $y(t) = 2 - 2e^{-4t}$

11. **a)** $y(t) = 2te^{-t}$
c) $y(t) = bte^{kt}$

12. $y(t) = e^{-t} + 4te^{-t}$

13. $y(t) = 4e^{2t} + te^{2t}$

15. $y(t) = 4e^{-4t} + 4te^{-4t}$

17. $y(t) = 6te^{3-2t}$

18. $y(t) = 2e^{t} - \frac{1}{2}e^{4t} - \frac{1}{2}e^{-2t}$

19. $y(t) = 4e^{t} + 2e^{3t} - 1$

21. $y(t) = \frac{14}{15}e^{-4t} - \frac{4}{3}e^{-t} + \frac{2}{5}e^{t}$

23. $y(t) = 2e^{-2t} + 2te^{-2t} + e^{-t}$

SECTION 4.4

1. $\frac{6}{s^4}$, $\mathrm{Re}(s) > 0$

3. $\frac{3}{s^2 + 9}$, $\mathrm{Re}(s) > 0$

5. $\frac{4}{(s-2)^2 + 1}$, $\mathrm{Re}(s) > 2$

7. $\frac{-4s}{(s^2+1)^2}$, $\mathrm{Re}(s) > 0$

9. $\frac{s+3}{(s+3)^2 - 1}$, $\mathrm{Re}(s) > -2$

11. $\frac{(s-2)^2 - 1}{[(s-2)^2 + 1]^2}$, $\mathrm{Re}(s) > 2$

13. $\frac{72(s^2 - 3)}{(s^2 + 9)^3}$, $\mathrm{Re}(s) > 0$

15. $\frac{3}{s+1} + \frac{8}{s^2 + 4}$, $\mathrm{Re}(s) > 0$

17. $\frac{27}{2}\left[\frac{1}{(s-1)^2 + 9} - \frac{1}{(s+1)^2 + 9}\right]$, $\mathrm{Re}(s) > 1$

19. $2t^2$

21. $5e^{2t}$

23. $4\cos\sqrt{3}t$

25. $\frac{1}{2}e^{t}\sin 2t$

27. $2\sqrt{3}e^{-t}\sin\sqrt{3}t$

29. $3e^{3t}\cos 3t + \frac{10}{3}e^{3t}\sin 3t$

31. $-9te^{2t}$

33. $6e^{3t} + \cos 5t + \frac{4}{5}\sin 5t$

35. $\sin t - t\cos t$

43. $L\{y'(t)\} = sY(s) - 3$
$L\{y''(t)\} = s^2Y(s) - 3s + 2$

45. $L\{y'(t)\} = sY(s) + 1$
$L\{y''(t)\} = s^2Y(s) + s - 1$

47. $L\{y'(t)\} = sY(s) + 2$
$L\{y''(t)\} = s^2Y(s) + 2s + 1$

49. $L\{y'(t)\} = sY(s) - 1$
$L\{y''(t)\} = s^2Y(s) - s - 1$
$L\{y^{(3)}(t)\} = s^3Y(s) - s^2 - s + 1$
$L\{y^{(4)}(t)\} = s^4Y(s) - s^3 - s^2 + s + 1$

51. $L\{y'(t)\} = sY(s) - 2$
$L\{y''(t)\} = s^2Y(s) - 2s - 2$
$L\{y^{(3)}(t)\} = s^3Y(s) - 2s^2 - 2s$
$L\{y^{(4)}(t)\} = s^4Y(s) - 2s^3 - 2s^2 - 1$

53. $L\{y'(t)\} = sY(s) - 1$
$L\{y''(t)\} = s^2Y(s) - s - 2$
$L\{y^{(3)}(t)\} = s^3Y(s) - s^2 - 2s - 3$
$L\{y^{(4)}(t)\} = s^4Y(s) - s^3 - 2s^2 - 3s - 4$

55. Any $\alpha > 0$, $M = 2/\alpha^2$

57. Any $\alpha > 0$, $M = 6 \cdot 201 \cdot \binom{200}{100} \cdot 400!/\alpha^{400}$

59. Any $\alpha > 0$, $M = \exp\left(\frac{4}{27\alpha^2}\right)$

SECTION 4.5

1. $y(t) = \frac{2}{3}\sin 3t + \cos 3t$

3. $y(t) = -\frac{1}{2}\sin t + 2\cos t - \frac{1}{2}t\cos t$

5. $y(t) = 2e^{2t} + 2e^{-2t} - 2$

7. $y(t) = 3e^{-4t} - 4e^{-3t} + 1$

9. $y(t) = \frac{7}{9}e^{t} + \frac{1}{9}e^{4t} - \frac{8}{9}e^{-2t}$

11. $x(t) = 3e^{t}\cosh 3t - e^{t}\sinh 3t$
$y(t) = -3e^{t}\cosh 3t + 9e^{t}\sinh 3t$

13. $x(t) = e^{t}\cosh 3t$
$y(t) = e^{t}\sinh 3t$

15. $x(t) = 4\cos 3t + \sin 3t$
$y(t) = 3\cos 3t + 5\sin 3t$

17. $x(t) = \frac{7}{8}e^{-2t} + \frac{9}{8}e^{6t}$
$y(t) = -\frac{7}{8}e^{-2t} + \frac{15}{8}e^{6t}$

21. $x(t) = \frac{2}{3} - \frac{1}{6}e^{-3t} - \frac{1}{2}e^{-t}$
$y(t) = \frac{1}{3} + \frac{1}{6}e^{-3t} - \frac{1}{2}e^{-t}$

23. $x(t) = e^{t} + \frac{1}{2}e^{2t} - \frac{3}{2}$
$y(t) = 2e^{t} - \frac{1}{2}e^{2t} - \frac{3}{2}$

SECTION 4.6

1. $A_1 = \frac{5}{3}$
$A_2 = \frac{4}{3}$

3. $A_1 = 2$
$A_2 = 1$

5. $A_1 = 1$
$A_2 = 3$
$A_3 = -\frac{2}{3}$

7. $A_1 = 4$
$A_2 = -4$
$A_3 = \sqrt{2}$

9. (Exercise 1) $\frac{4}{3}e^{-t} + \frac{5}{3}e^{2t}$
(Exercise 3) $2e^{-2t} + te^{-2t}$
(Exercise 5) $e^{t} + e^{-t}\left(3\cos 3t - \frac{2}{3}\sin 3t\right)$
(Exercise 7) $4e^{2t} + e^{t}(\sqrt{2}\sin\sqrt{2}t - 4\cos\sqrt{2}t)$

11. $-\frac{2}{3} + \frac{7}{24}e^{-3t} + \frac{3}{8}e^{5t}$

13. $-\frac{2}{5}e^{-t} + \frac{7}{5}\cos 3t - \frac{7}{15}\sin 3t$

15. $e^{2t} - e^{-2t}\left(\cos\sqrt{5}t + \frac{4}{\sqrt{5}}\sin\sqrt{5}t\right)$

17. $\cos t - \cos 3t + \sin 3t$

19. $\frac{1}{2}t^{3}e^{-17t}$

21. $y(t) = \frac{8}{3}e^{-t} - e^{-2t} + \frac{1}{3}e^{2t}$

23. $y(t) = \cos 3t + \frac{1}{3}\sin 3t + \frac{1}{3}t\sin 3t$

25. $y(t) = e^{-t}\left(\frac{8}{25}\cos 2t - \frac{6}{25}\sin 2t\right) + \frac{4}{5}t - \frac{8}{25}$

27. $y(t) = 2e^{-2t} + 4te^{-2t} - \frac{1}{2}t^{2}e^{-2t}$

28. $x(t) = -\frac{5}{32} + \frac{t}{8} + \frac{3}{7}e^{-3t} - \frac{3}{8}e^{-2t} + \frac{23}{224}e^{4t}$
$y(t) = \frac{3}{32} - \frac{3}{8}t - \frac{4}{7}e^{-3t} + \frac{3}{8}e^{-2t} + \frac{23}{224}e^{4t}$

29. $x(t) = t + \frac{1}{5}e^{-t} + \frac{4}{5}\cos 2t + \frac{1}{10}\sin 2t$
$y(t) = t - \frac{1}{2} + \frac{4}{5}e^{-t} + \frac{7}{10}\cos 2t + \frac{9}{10}\sin 2t$

31. $x(t) = e^{-t} + 4te^{-t} - 4t^{2}e^{-t}$
$y(t) = 6te^{-t} - 4t^{2}e^{-t}$

33. $x(t) = 3t \sin 2t + \frac{3}{2} t \cos 2t + \frac{13}{4} \sin 2t$
$y(t) = \cos 2t - 2 \sin 2t - 3t \sin 2t$

SECTION 4.7

1. $\frac{1}{6} t^3$

3. $\sinh t = \frac{1}{2}(e^t - e^{-t})$

5. $\frac{1}{2} - \frac{1}{2} \cos 2t$

7. $\frac{1}{4} e^t - \frac{1}{4} e^{-3t}$

9. $\frac{3}{8} \sin t - \frac{1}{8} \sin 3t$

11. $\frac{1}{4} t - \frac{1}{16} + \frac{1}{16} e^{-4t}$

13. $\frac{1}{8} \sin t - \frac{1}{24} \sin 3t$

15. $\frac{1}{2} \sinh t + \frac{1}{2} \sin t$

17. $\frac{1}{2} t \sinh t$

19. $P(s) = \frac{1}{s^2 - 3s + 10}$
$p(t) = \frac{2}{\sqrt{31}} e^{3t/2} \sin \frac{\sqrt{31}}{2} t$

21. $P(s) = \frac{1}{s^2 - 9}$
$p(t) = \frac{1}{3} \sinh 3t$

23. $P(s) = \frac{1}{s^2 + 4s + 4}$
$p(t) = te^{-2t}$

25. $P(s) = \frac{1}{s^2 + 3s + 3}$
$p(t) = \frac{2}{\sqrt{3}} e^{-3t/2} \sin \frac{\sqrt{3}}{2} t$

27. b) $P(s) = \frac{1}{s^2}(1 + e^{-s} - 2e^{-s/2})$

31. $y(t) = \cos t - \sin t$

33. $y(t) = \frac{1}{2} t \cos t + \frac{1}{2} \sin t$

SECTION 4.8

1. $u(t) - 2u(t - 2) + u(t - 4)$
$\frac{1}{s} - 2\frac{e^{-2s}}{s} + \frac{e^{-4s}}{s}$

3. $tu(t) - 2(t - 1)u(t - 1) + 2(t - 3)u(t - 3) - (t - 4)u(t - 4)$,
$\frac{1}{s^2} - 2\frac{e^{-s}}{s^2} + 2\frac{e^{-3s}}{s^2} - \frac{e^{-4s}}{s^2}$

5. $\sin(t - 2\pi)u(t - 2\pi)$
$\frac{e^{-2\pi s}}{s^2 + 1}$

7. $(\sin t)u(t) + 2 \sin(t - \pi)u(t - \pi) + \sin(t - 2\pi)u(t - 2\pi)$
$\frac{1 + 2e^{-\pi s} + e^{-2\pi s}}{s^2 + 1}$

9. $\frac{e^{-s}}{s} + 4\frac{e^{-2s}}{s}$

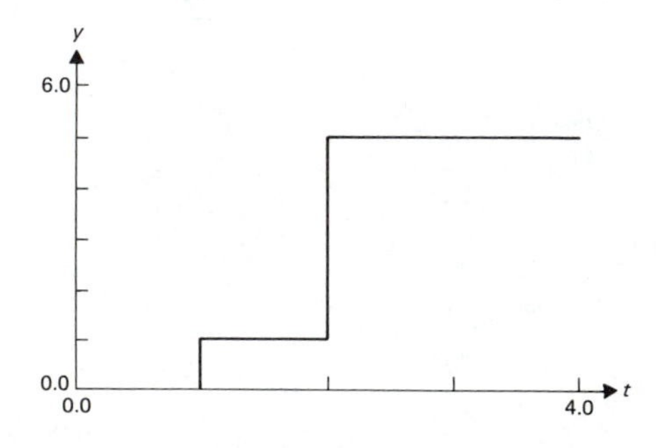

11. $e\frac{e^{-3s}}{s - 1}$

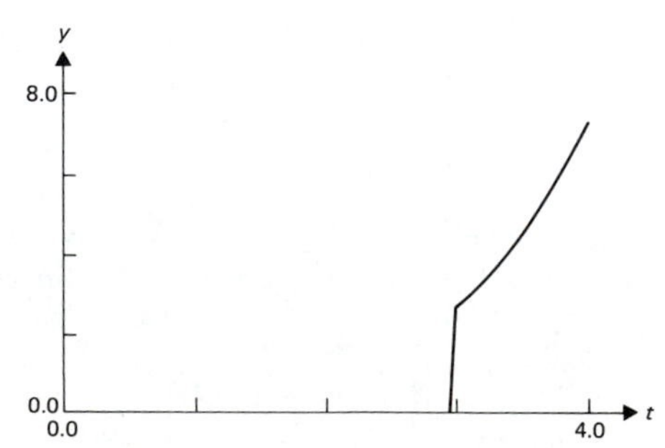

13. $\dfrac{1}{s^2} + \dfrac{e^{-s}}{s^2}$

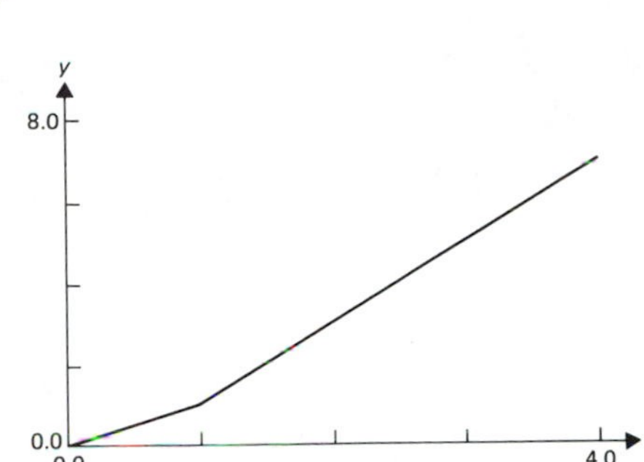

15. $e^{-s}\left(\dfrac{1}{s^2} - \dfrac{1}{s}\right) - e^{-2s}\left(\dfrac{1}{s^2} + \dfrac{1}{s}\right)$

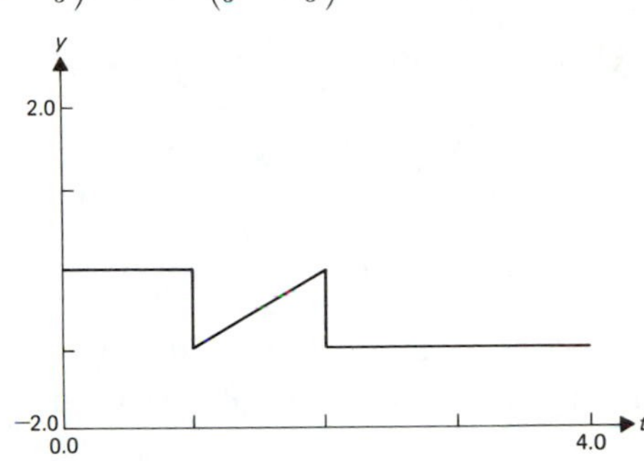

17. $y(t) = f(t) - 2f(t-2)u(t-2) + f(t-4)u(t-4)$, where $f(t) = \dfrac{1}{2} - e^{-t} + \dfrac{1}{2}e^{-2t}$

19. $y(t) = \dfrac{1}{2}[\sin t - (t - 2\pi)\cos t]u(t - 2\pi)$; yes

21. $y(t) = n(t - \sin t) - n\left[\left(t - \dfrac{1}{n}\right) - \sin\left(t - \dfrac{1}{n}\right)\right] u\left(t - \dfrac{1}{n}\right)$; yes

23. $y(t) = \dfrac{1}{2}\sinh 2t - \dfrac{1}{4}(1 - \cosh 2(t-2))u(t-2) + \dfrac{1}{4}(1 - \cosh 2(t-4))u(t-4)$

25. $\dfrac{2(1 - e^{-s})}{s(1 - e^{-2s})} = \dfrac{2}{s(1 + e^{-s})}$

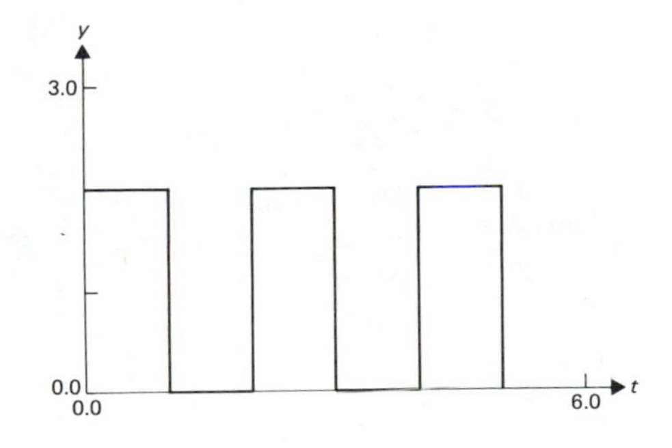

27. $\dfrac{e^{-s/2} - e^{-3s/2}}{s(1 - e^{-2s})} = \dfrac{1}{2s\cosh\dfrac{s}{2}}$

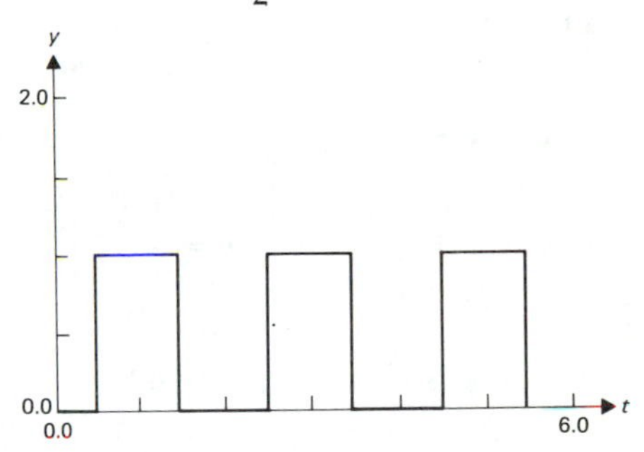

29. $y(t) = \dfrac{2}{3}\sinh 3t + \displaystyle\sum_{n=0}^{\infty} (-1)^n(\cosh 3(t-n) - 1)u(t-n)$

31. $y(t) = 4\cos t + \dfrac{1}{2}\sin t - \dfrac{1}{2}t\cos t + \displaystyle\sum_{n=1}^{\infty} (\sin(t - n\pi) - (t - n\pi)\cos(t - n\pi))u(t - n\pi)$

SECTION 4.9

1. $y(t) = \dfrac{1}{2}(\sin 2t)u(t - \pi) + \cos 2t$

3. $y(t) = e^{-t}\sin 2t + 2e^{-t} + \frac{1}{2}[e^{-(t-2)}\sin 2(t-2)]u(t-2)$

5. $y(t) = \cos 3t + \frac{1}{3}u\left(t - \frac{\pi}{3}\right)\sin 3\left(t - \frac{\pi}{3}\right)$

7. $y(t) = \frac{8}{15}e^{t} - \frac{8}{15}e^{-2t}\cos\frac{3}{2}t - \frac{16}{15}e^{-2t}\sin\frac{3}{2}t + 2e^{-2(t-1)}u(t-1)\sin\frac{3}{2}(t-1)$

11. $y(t) = 2\cos 4t + \frac{1}{4}u\left(t - \frac{\pi}{8}\right)\sin 4\left(t - \frac{\pi}{8}\right)$

13. $y(t) = e^2/2u(t-1)\sinh 2(t-1)$

15. $y(t) = at + a(t-1)u(t-1)$; yes (have used $y(1) = a$)

CHAPTER 5

SECTION 5.1

1.
$$\begin{aligned} y_1(t) &= a\exp(-k_1 t) \\ y_2(t) &= ak_1 t\exp(-k_1 t) + b\exp(-k_1 t) \\ y_3(t) &= C + a(1 - \exp(-k_1 t) - k_1 t\exp(-k_1 t)) \\ &\quad + b(1 - \exp(-k_1 t)) \end{aligned}$$

3. i) $y_1(0) = y_2(0) = a$
$Y_1(s) = Y_2(s) = sa(s^2 + \omega^2)^{-1}$
$y_1(t) = y_2(t) = a\cos\omega t$

ii) $y_1(0) = -y_2(0) = b$
$Y_1(s) = -Y_2(s) = sb(s^2 + 3\omega^2)^{-1}$
$y_1(t) = -y_2(t) = b\cos(\sqrt{3}\,\omega t)$

SECTION 5.2

1. $\mathbf{x} + \mathbf{y} = \begin{bmatrix} 3 \\ 0 \end{bmatrix}$, $\mathbf{x} - \mathbf{y} = \begin{bmatrix} -1 \\ 0 \end{bmatrix}$, $\mathbf{x}\cdot\mathbf{y} = 2$, dependent; $c_1 = 2$, $c_2 = -1$

3. $\mathbf{x} + \mathbf{y} = \begin{bmatrix} 3 \\ 6 \end{bmatrix}$, $\mathbf{x} - \mathbf{y} = \begin{bmatrix} -1 \\ -2 \end{bmatrix}$, $\mathbf{x}\cdot\mathbf{y} = 10$, dependent; $c_1 = 2$, $c_2 = -1$

5. $\mathbf{x} + \mathbf{y} = \begin{bmatrix} 3 \\ 3 \\ 0 \end{bmatrix}$, $\mathbf{x} - \mathbf{y} = \begin{bmatrix} -1 \\ -3 \\ 2 \end{bmatrix}$, $\mathbf{x}\cdot\mathbf{y} = 1$, independent

7. Independent

9. Independent

11. Dependent; $c_1 = -2$, $c_2 = 1$, $c_3 = -1$, $c_4 = 1$

13. $\mathbf{x}'(t) = \begin{bmatrix} 1 \\ 2t \end{bmatrix}$, $\int_0^t \mathbf{x}(s)ds = \begin{bmatrix} t^2/2 \\ t^3/3 \end{bmatrix}$

15. $\mathbf{x}'(t) = \begin{bmatrix} te^t + e^t \\ 2\cos 2t \\ 2 \end{bmatrix}$, $\int_0^t \mathbf{x}(s)\,ds = \begin{bmatrix} te^t - e^t + 1 \\ -\frac{1}{2}\cos 2t + \frac{1}{2} \\ t^2 + t \end{bmatrix}$

17. $\mathbf{x}'(t) = \begin{bmatrix} \cos t - \sin t \\ \cos t + \sin t \end{bmatrix}$, $\int_0^t \mathbf{x}(s)\,ds = \begin{bmatrix} 1 - \cos t + \sin t \\ 1 - \cos t - \sin t \end{bmatrix}$

19. Independent

21. Independent

23. (Exercise 13) $\text{col}(s^{-2}, 2s^{-3})$
(Exercise 15) $\text{col}[(s-1)^{-2}, 2(s^2+4)^{-1}, 2s^{-2}+s^{-1}]$
(Exercise 17) $\text{col}[(s+1)(s^2+1)^{-1}, (1-s)(s^2+1)^{-1}]$

25. c) $AB = \begin{bmatrix} 5 & 5 & 4 \\ 5 & 6 & 4 \\ 5 & 6 & 3 \end{bmatrix}$

27. a) $A^T = \begin{bmatrix} 1 & 3 \\ 2 & 4 \end{bmatrix}$

c) $(A+B)^T = \begin{bmatrix} 6 & 10 \\ 8 & 12 \end{bmatrix}$

e) $(AB)^T = \begin{bmatrix} 19 & 43 \\ 22 & 50 \end{bmatrix}$

29. -2, independent

31. -1, independent

33. 1, independent

37. $\mathbf{x} = \begin{bmatrix} 1 \\ 2 \end{bmatrix}, C = \frac{1}{11}\begin{bmatrix} 5 & 3 \\ -2 & 1 \end{bmatrix}$

39. $\mathbf{x} = \begin{bmatrix} 1 \\ 1 \\ 1 \end{bmatrix}, C = \frac{1}{21}\begin{bmatrix} 8 & -2 & 11 \\ -7 & 7 & -7 \\ 4 & -1 & -5 \end{bmatrix}$

43. $A^{-1} = \begin{bmatrix} 3 & -5 \\ -1 & 2 \end{bmatrix}$

45. $A^{-1} = \begin{bmatrix} 13 & -36 \\ -9 & 25 \end{bmatrix}$

47. $A^{-1} = \begin{bmatrix} 2 & 5 \\ 1 & 3 \end{bmatrix}$

49. $\lambda^2 - 4\lambda + 3 = 0$
$\lambda = 1$, col(1, 1)
$\lambda = 3$, col(1, -1)

51. $\lambda^2 - 3\lambda = 0$
$\lambda = 0$, col(1, -1)
$\lambda = 3$, col(1, 2)

53. $-\lambda^3 + 7\lambda^2 - 11\lambda + 5 = 0$
$\lambda = 1$, col(1, 0, -1), col(0, 1, -2)
$\lambda = 5$, col(1, 1, 1)

55. $A'(t) = \begin{bmatrix} e^t & -e^{-t} \\ 2e^t & -e^t \end{bmatrix}$

57. $A'(t) = e^{2t}\begin{bmatrix} 2\cos t - \sin t & 4\sin t + 2\cos t \\ 6\cos t - 3\sin t & 2\sin t + \cos t \end{bmatrix}$

59. $A'(t) = \begin{bmatrix} 0 & -2\sin t & 5\cos t \\ 2e^t & -2\sin t & 5\cos t \\ e^t & -3\sin t & -5\cos t \end{bmatrix}$

65. (Exercise 55) $\begin{bmatrix} \dfrac{1}{s-1} & \dfrac{-1}{s+1} \\ \dfrac{2}{s-1} & \dfrac{-1}{s-1} \end{bmatrix}$

(Exercise 57) $\dfrac{1}{(s-2)^2+1}\begin{bmatrix} s-2 & 2 \\ 3(s-2) & 1 \end{bmatrix}$

(Exercise 59) $\begin{bmatrix} 0 & \dfrac{-2s}{s^2+1} & \dfrac{5}{s^2+1} \\ \dfrac{2}{s-1} & \dfrac{-2s}{s^2+1} & \dfrac{5}{s^2+1} \\ \dfrac{1}{s-1} & \dfrac{-3s}{s^2+1} & \dfrac{-5}{s^2+1} \end{bmatrix}$

SECTION 5.3

1. $\mathbf{y}(t) = c_1 \begin{bmatrix} 1 \\ 2 \end{bmatrix} e^{-t} + c_2 \begin{bmatrix} 2 \\ 1 \end{bmatrix} e^{2t}$

3. $\mathbf{y}(t) = c_1 \begin{bmatrix} 1 \\ \dfrac{1+\sqrt{5}}{2} \end{bmatrix} e^{(-5+\sqrt{5})t/2} + c_2 \begin{bmatrix} 1 \\ 1 - \dfrac{\sqrt{5}}{2} \end{bmatrix} e^{(-5-\sqrt{5})t/2}$

5. $\mathbf{y}(t) = c_1 \begin{bmatrix} 1 \\ 1 \end{bmatrix} e^{-t} + c_2 \begin{bmatrix} 1 \\ -1 \end{bmatrix} e^{-5t}$

7. $\mathbf{y}(t) = \begin{bmatrix} 2e^{-t} - e^t \\ e^{-t} - e^t \end{bmatrix}$

9. $\mathbf{y}(t) = \dfrac{1}{5} \begin{bmatrix} 9e^t - 4e^{-4t} \\ 3e^t + 2e^{-t} \end{bmatrix}$

11. $\mathbf{y}(t) = \dfrac{1}{2} \begin{bmatrix} 3e^{-t} - e^t \\ 9e^{-t} - e^t \end{bmatrix}$

13. $\mathbf{y}(t) = e^{-t} \left\{ c_1 \begin{bmatrix} -2 \sin 2t \\ \cos 2t \end{bmatrix} + c_2 \begin{bmatrix} 2 \cos 2t \\ \sin 2t \end{bmatrix} \right\}$

15. $\mathbf{y}(t) = c_1 e^{3t} \begin{bmatrix} 3 \cos 3t + \sin 3t \\ 5 \sin 3t \end{bmatrix} + c_2 e^{3t} \begin{bmatrix} 2 \sin 3t \\ \sin 3t - 3 \cos 3t \end{bmatrix}$

17. $\mathbf{y}(t) = c_1 e^t \begin{bmatrix} 5 \cos t \\ -2 \cos t - \sin t \end{bmatrix} + c_2 e^t \begin{bmatrix} 5 \sin t \\ \cos t - 2 \sin t \end{bmatrix}$

19. $\mathbf{y}(t) = e^{-t} \left\{ \dfrac{1}{2} \begin{bmatrix} 2 \cos t \\ -5 \cos t - \sin t \end{bmatrix} + \dfrac{5}{2} \begin{bmatrix} 2 \sin t \\ -5 \sin t + \cos t \end{bmatrix} \right\}$

21. $\mathbf{y}(t) = e^{2(t-\pi)} \left\{ (-2) \begin{bmatrix} \sin 2t \\ -\dfrac{\cos 2t}{2} \end{bmatrix} + (1) \begin{bmatrix} \cos 2t \\ \dfrac{\sin 2t}{2} \end{bmatrix} \right\}$

23. $\mathbf{y}(t) = (-1) \begin{bmatrix} \cos t \\ -\sin t \end{bmatrix} + (-1) \begin{bmatrix} \sin t \\ \cos t \end{bmatrix}$

27. $\mathbf{y}(t) = c_1 \begin{bmatrix} 1 \\ 1 \end{bmatrix} e^{3t} + c_2 \left\{ \begin{bmatrix} 0 \\ 1 \end{bmatrix} + t \begin{bmatrix} 1 \\ 1 \end{bmatrix} \right\} e^{3t}$

29. $\mathbf{y}(t) = c_1 \begin{bmatrix} 2 \\ 1 \end{bmatrix} e^t + c_2 \left\{ \begin{bmatrix} 1 \\ 0 \end{bmatrix} + t \begin{bmatrix} 2 \\ 1 \end{bmatrix} \right\} e^t$

31. $\mathbf{y}(t) = c_1 \begin{bmatrix} 1 \\ -2 \end{bmatrix} e^{4t} + c_2 \left\{ \begin{bmatrix} 0 \\ -1 \end{bmatrix} + t \begin{bmatrix} 1 \\ -2 \end{bmatrix} \right\} e^{4t}$

33. $\mathbf{y}(t) = e^{-3t} \begin{bmatrix} 8t + 3 \\ 8t + 1 \end{bmatrix}$

35. $\mathbf{y}(t) = e^{-3t} \begin{bmatrix} 1 - t \\ t \end{bmatrix}$

37. $\mathbf{y}(t) = e^{-2(t-1)} \begin{bmatrix} 7 - 7t \\ 8 - 7t \end{bmatrix}$

39. a) $\mathbf{y}(t) = \operatorname{col}(2e^{4t}, -3(e^{4t} - e^{-2t}), -3(e^{4t} + e^{-2t}))$
b) $\mathbf{y}(t) = e^t \operatorname{col}(2 \cos 2t, \sin 2t, 3 \sin 2t)$

41. $\mathbf{y}(t) = e^{3t} \operatorname{col}(0, 1, 0)$

43. $\mathbf{y}(t) = \operatorname{col}(5 \sin t, 5 \cos t, -2 \cos t + \sin t)$

SECTION 5.4

3. $\Phi(t) = \dfrac{e^{-t}}{6} \begin{bmatrix} 4e^{3t} + 2e^{-3t} & e^{3t} - e^{-3t} \\ 8e^{3t} - 8e^{-3t} & 2e^{3t} + 4e^{-3t} \end{bmatrix}$

5. $\Phi(t) = e^{2t} \begin{bmatrix} 1 + 3 & -3t \\ +3t & 1 - 3t \end{bmatrix}$

7. $\Phi(t) = \begin{bmatrix} \cos 2t + \frac{1}{2}\sin 2t & -\frac{1}{2}\sin 2t \\ \frac{5}{2}\sin 2t & \cos 2t - \frac{1}{2}\sin 2t \end{bmatrix}$

9. $\Phi(t) = e^{t}\begin{bmatrix} \cos 2t & -\frac{1}{2}\sin 2t \\ 2\sin 2t & \cos 2t \end{bmatrix}$

11. $\Phi(t) = \begin{bmatrix} e^{-t} + 2te^{-t} & -2te^{-t} \\ 2te^{-t} & e^{-t} - 2te^{-t} \end{bmatrix}$

13. $\Phi(t) = e^{2t}\begin{bmatrix} \cos t + \sin t & \sin t \\ -2\sin t & \cos t - \sin t \end{bmatrix}$

15. $\Phi(t) = e^{4t}\begin{bmatrix} 1 + t & -t \\ t & 1 - t \end{bmatrix}$

17. $\Phi(t) = e^{3t}\begin{bmatrix} 1 & t \\ 0 & 1 \end{bmatrix}$

19. $\Phi(t) = e^{t}\begin{bmatrix} \cos\sqrt{3}\,t & \sqrt{3}/3\,\sin\sqrt{3}\,t \\ -\sqrt{3}\sin\sqrt{3}\,t & \cos\sqrt{3}\,t \end{bmatrix}$

21. $\Phi(t) = e^{-3t}\begin{bmatrix} 1 - t & -t \\ t & 1 + t \end{bmatrix}$

$\mathbf{y}(t) = e^{-3t}\begin{bmatrix} 2 - 3t \\ 1 + 3t \end{bmatrix}$

$\mathbf{y}(t) = e^{-3(t-1)}\begin{bmatrix} 4 - 3t \\ 3t - 1 \end{bmatrix}$

23. $\Phi(t) = e^{t}\begin{bmatrix} \cos\sqrt{6}t + \frac{2}{\sqrt{6}}\sin\sqrt{6}t & -\frac{2}{\sqrt{6}}\sin\sqrt{6}t \\ \frac{5}{\sqrt{6}}\sin\sqrt{6}t & \cos\sqrt{6}t - \frac{2}{\sqrt{6}}\sin\sqrt{6}t \end{bmatrix}$

$\mathbf{y}(t) = e^{t}\begin{bmatrix} \cos\sqrt{6}\,t + \frac{4}{\sqrt{6}}\sin\sqrt{6}t \\ \frac{7}{\sqrt{6}}\sin\sqrt{6}t - \cos\sqrt{6}t \end{bmatrix}$

$\mathbf{y}(t) = e^{t-1}\begin{bmatrix} \cos\sqrt{6}(t-1) \\ \cos\sqrt{6}(t-1) + \frac{3}{\sqrt{6}}\sin\sqrt{6}(t-1) \end{bmatrix}$

25. $\Phi(t) = \frac{1}{2}\begin{bmatrix} 3e^{-2t} - e^{-4t} & 3e^{-2t} - 3e^{-4t} \\ e^{-4t} - e^{-2t} & 3e^{-4t} - e^{-2t} \end{bmatrix}$

$\mathbf{y}(t) = e^{-4t}\begin{bmatrix} -2 \\ 2 \end{bmatrix}$

$\mathbf{y}(t) = \begin{bmatrix} -80e^{-4t} + 24e^{-2t} \\ 80e^{-4t} - 8e^{-2t} \end{bmatrix}$

27. $\Phi(t) = \begin{bmatrix} e^{t} & 0 & 0 \\ 13\sin t + 11\cos t - 11e^{t} & 3\sin t + \cos t & -5\sin t \\ 10\sin t + 4\cos t - 4e^{t} & 2\sin t & -3\sin t + \cos t \end{bmatrix}$

$\mathbf{y}(t) = \begin{bmatrix} 3e^{t} \\ 37\sin t + 34\cos t - 33e^{t} \\ 29\sin t + 13\cos t - 12e^{t} \end{bmatrix}$

$\mathbf{y}(t) = \begin{bmatrix} 2e^{t-\pi} \\ (60e^{-\pi} - 18)\sin t + (20e^{-\pi} - 21)\cos t - 22e^{t-\pi} \\ (40e^{-\pi} - 15)\sin t - 9\cos t - 8e^{t-\pi} \end{bmatrix}$

33. $\mathbf{y}(t) = \begin{bmatrix} -\frac{3}{4} - \frac{7}{8}e^{-4t} + \frac{21}{18}e^{4t} \\ -\frac{5}{8} + \frac{21}{16}e^{-4t} + \frac{21}{16}e^{4t} \end{bmatrix}$ **35.** $\mathbf{y}(t) = \begin{bmatrix} -2 + e^{3t}(3\sin t - \cos t) \\ -1 + e^{3t}(4\sin t + 2\cos t) \end{bmatrix}$

37. a) $\mathbf{y}'(t) = \begin{bmatrix} -\frac{10}{200} & \frac{5}{200} \\ \frac{10}{200} & -\frac{10}{200} \end{bmatrix} \mathbf{y}(t),\ \mathbf{y}(0) = \begin{bmatrix} 0 \\ y_0 \end{bmatrix}$

b) $\mathbf{\Phi}(t) = \begin{bmatrix} e^{-t/20}\cosh\frac{\sqrt{2}}{40}t & \frac{1}{\sqrt{2}}e^{-t/20}\sinh\frac{\sqrt{2}}{40}t \\ \frac{2}{\sqrt{2}}e^{-t/20}\sinh\frac{\sqrt{2}}{40}t & e^{-t/20}\cosh\frac{\sqrt{2}}{40}t \end{bmatrix}$

39. b) $I(t) = \frac{2}{5} - \frac{1}{2}e^{-2t}(I(0) + 10Q_2(0) - 1) + \frac{1}{10}e^{-10t/3}(15I(0) + 50Q_2(0) - 9)$ amperes

$Q_2(t) = \frac{3}{50} + \frac{1}{20}e^{-2t}(3I(0) + 30Q_2(0) - 3) - \frac{1}{100}e^{-10t/3}(15I(0) + 50Q_2(0) - 9)$coulombs

Steady state values: $Q_2 = \frac{3}{50}$ coulombs, $I = \frac{2}{5}$ amperes

SECTION 5.5

1. $\mathbf{y}(t) = \begin{bmatrix} -\frac{7}{8} - \frac{37}{24}e^{-4t} + \frac{41}{12}e^{2t} \\ -3 + \frac{37}{6}e^{-4t} + \frac{41}{6}e^{2t} \end{bmatrix}$ **3.** $\mathbf{y}(t) = \begin{bmatrix} -1 + \frac{1}{3}e^{-t} + \frac{7}{6}e^{2t} + te^{2t} \\ -e^{-t} + e^{2t} \end{bmatrix}$

5. $\mathbf{y}(t) = \begin{bmatrix} -\frac{1}{8} + \frac{1}{4}e^{2t} - \frac{1}{8}e^{4t} \\ -\frac{1}{8} + \frac{3}{4}e^{2t} + \frac{3}{8}e^{4t} \end{bmatrix}$ **7.** $\mathbf{y}(t) = \begin{bmatrix} \frac{5}{4}e^{t} + 2e^{2t} - \frac{1}{4}e^{5t} \\ -\frac{5}{4}e^{t} - 2e^{2t} - \frac{3}{4}e^{5t} \end{bmatrix}$

9. $\mathbf{y}(t) = \begin{bmatrix} -\frac{9}{2} - 5\cos t + 10\sin t + \frac{1}{2}e^{t-\pi}\cos t - \frac{27}{2}e^{t-\pi}\sin t \\ \frac{3}{2} - \cos t - 3\sin t + \frac{5}{2}e^{t-\pi}\cos t + \frac{11}{2}e^{t-\pi}\sin t \end{bmatrix}$

11. $\mathbf{y}(t) = e^{-t}\begin{bmatrix} -\frac{8}{25} \\ \frac{2}{25} \end{bmatrix} + e^{t}\begin{bmatrix} -\frac{1}{9} \\ -\frac{4}{9} \end{bmatrix}$

$$+ e^{4(t-2)}\begin{bmatrix} t-1 & 2-t \\ t-2 & 3-t \end{bmatrix}\left\{\begin{bmatrix} -3 \\ 2 \end{bmatrix} + e^{-2}\begin{bmatrix} \frac{8}{25} \\ -\frac{2}{25} \end{bmatrix} + e^{2}\begin{bmatrix} \frac{1}{9} \\ \frac{4}{9} \end{bmatrix}\right\}$$

13. $\hat{\Phi}(s) = \begin{bmatrix} \dfrac{1}{s-3} & 0 \\ 0 & \dfrac{1}{s-1} \end{bmatrix}$

$\Phi(t) = \begin{bmatrix} e^{3t} & 0 \\ 0 & e^{t} \end{bmatrix}$

15. $\hat{\Phi}(s) = \dfrac{1}{s^2+9}\begin{bmatrix} s+1 & 5 \\ -2 & s-1 \end{bmatrix}$

$\Phi(t) = \begin{bmatrix} \cos 3t + \dfrac{1}{3}\sin 3t & \dfrac{5}{3}\sin 3t \\ -\dfrac{2}{3}\sin 3t & \cos 3t - \dfrac{1}{3}\sin 3t \end{bmatrix}$

17. $\hat{\Phi}(s) = \begin{bmatrix} \dfrac{1}{s-1} & 0 & 0 \\ 0 & \dfrac{1}{s-2} & \dfrac{1}{(s-2)^2} \\ 0 & 0 & \dfrac{1}{s-2} \end{bmatrix}$

$\Phi(t) = \begin{bmatrix} e^{t} & 0 & 0 \\ 0 & e^{2t} & te^{2t} \\ 0 & 0 & e^{2t} \end{bmatrix}$

19. $x(t) = 28 - 8e^{-3t/100}\cosh\dfrac{\sqrt{26}}{200}t - \dfrac{83}{\sqrt{26}}e^{-3t/100}\sinh\dfrac{\sqrt{26}}{200}t$ lb

$y(t) = 20 - 15e^{-3t/100}\cosh\dfrac{\sqrt{26}}{200}t - \dfrac{25}{\sqrt{26}}e^{-3t/100}\sinh\dfrac{\sqrt{26}}{200}t$ lb

Steady state solution: $x(t) = 28$ lb, $y(t) = 20$ lb

21. $I(t) = \dfrac{1}{545}(232\sin t - 26\cos t) - c_1e^{-2t} + c_2e^{-10t/3}$ amperes

$Q_2(t) = \dfrac{1}{2725}(102\sin t - 96\cos t) + \dfrac{3}{10}c_1e^{-2t} - \dfrac{1}{10}c_2e^{-10t/3}$ coulombs

Steady state solution: $Q_2(t) = \dfrac{1}{2725}(102\sin t - 96\cos t)$ coulombs,

$I(t) = \dfrac{1}{545}(232\sin t - 26\cos t)$ amperes

CHAPTER 6

SECTION 6.2

3. $a_2 = a_3 = 0,\ a_n = \dfrac{a_{n-4}}{n(n-1)},\ n = 4, 5, \ldots$

$y(x) = a_0\left(1 - \dfrac{1}{4\cdot 3}x^4 + \dfrac{1}{8\cdot 7\cdot 4\cdot 3}x^8 + \cdots\right) + a_1\left(x - \dfrac{1}{5\cdot 4}x^5 + \dfrac{1}{9\cdot 8\cdot 5\cdot 4}x^9 + \cdots\right)$

5. $a_0 = -2,\ a_1 = 2,\ a_2 = \dfrac{a_0}{2\cdot 1},\ a_3 = \dfrac{a_1}{3\cdot 2},\ a_n = \dfrac{a_{n-4} + a_{n-2}}{n(n-1)},\ n = 4, 5, \ldots$

$y(x) = -2 + 2x - x^2 + \dfrac{1}{3}x^3 - \dfrac{1}{4}x^4 + \dfrac{7}{5\cdot 4\cdot 3}x^5 + \cdots$

7. $y(x) = 1 - x - \dfrac{1}{2}x^2 + \dfrac{1}{12}x^4 + \dfrac{1}{24}x^5 + \cdots$

9. $y(x) = a_0 + a_1(x-1) - a_0(x-1)^2 - \frac{a_1}{2}(x-1)^3 + \frac{a_0}{3}(x-1)^4 + \cdots$

11. $y(x) = a_0 + a_1(x-1) - \frac{a_1}{2}(x-1)^2 + \frac{(2a_1 + a_0)}{6}(x-1)^3 - \frac{a_1}{12}(x-1)^4 + \cdots$

13. $y(x) = a_0 + a_1\left(x - \frac{\pi}{2}\right) - \frac{a_0}{2}\left(x - \frac{\pi}{2}\right)^2 - \frac{a_1}{6}\left(x - \frac{\pi}{2}\right)^3 + \frac{a_0}{12}\left(x - \frac{\pi}{2}\right)^4 + \cdots$

15. $y(x) = -x + \frac{x^5}{20} - \frac{252}{9!}x^9 + \frac{27720}{13!}x^{13} + \cdots$

17. $y(x) = x - \frac{2}{3!}x^3 + \frac{9}{5!}x^5 - \frac{75}{7!}x^7 + \cdots$

19. $y(x) = x + \frac{x^2}{2} + \frac{2}{3!}x^3 + \frac{2}{4!}x^4 + \frac{3}{5!}x^5 + \cdots$

SECTION 6.3

1. $r^2 + \frac{1}{2}r - \frac{1}{2} = 0,\ r_1 = \frac{1}{2},\ r_2 = -1$

$y_1(x) = x^{1/2} \sum_0^\infty a_n x^n$

$y_2(x) = x^{-1} \sum_0^\infty b_n x^n$

3. $r^2 - \frac{1}{2}r = 0,\ r_1 = \frac{1}{2},\ r_2 = 0$

$y_1(x) = x^{1/2} \sum_0^\infty a_n x^n$

$y_2(x) = \sum_0^\infty b_n x^n$

5. $r^2 + 4r + 4 = 0,\ r_1 = r_2 = -2$

$y_1(x) = x^{-2} \sum_0^\infty a_n x^n$

$y_2(x) = y_1(x)\beta \ln x + x^{-2} \sum_0^\infty b_n x^n$

7. $r^2 + \frac{1}{2}r - \frac{3}{16} = 0,\ r_1 = \frac{1}{4},\ r_2 = -\frac{3}{4}$

$y_1(x) = x^{1/4} \sum_0^\infty a_n x^n$

$y_2(x) = x^{-3/4} \sum_0^\infty b_n x^n$ or

$y_2(x) = y_1(x)\beta \ln x + x^{-3/4} \sum_0^\infty b_n x^n$

9. $a = -1,\ r^2 + \frac{1}{2}r = 0,\ r_1 = 0,\ r_2 = -\frac{1}{2}$

$y_1(x) = \sum_0^\infty a_n (x+1)^n$

$y_2(x) = (x+1)^{-1/2} \sum_0^\infty b_n (x+1)^n$

11. $a = 2,\ r^2 - \frac{5}{2}r + \frac{3}{2} = 0,\ r_1 = \frac{3}{2},\ r_2 = 1$

$y_1(x) = (x-2)^{3/2} \sum_0^\infty a_n (x-2)^n$

$y_2(x) = (x-2) \sum_0^\infty b_n (x-2)^n.$

A second singular point is $a = 0,\ r^2 + \frac{1}{2}r = 0,$

$r_1 = 0,\ r_2 = -\frac{1}{2}$

$y_1(x) = \sum_0^\infty a_n x^n$

$y_2(x) = x^{-1/2} \sum_0^\infty b_n x^n$

13. $r_1 = \frac{1}{2}: a_{n+1} = \dfrac{a_n}{(n+1)(2n+3)}$

$$y_1(x) = a_0x^{1/2}\left(1 + \frac{1}{3}x + \frac{1}{30}x^2 + \frac{1}{160}x^3 + \cdots\right)$$

$$= a_0x^{1/2}\left(1 + \sum_{n=1}^{\infty} \frac{1}{n!(2n+1)(2n-1)\cdots 5\cdot 3}x^n\right)$$

$r_2 = 0: b_{n+1} = \dfrac{b_n}{(n+1)(2n+1)}$

$$y_2(x) = b_0\left(1 + x + \frac{1}{6}x^2 + \frac{1}{90}x^3 + \cdots\right)$$

$$= b_0\left(1 + \sum_{n=1}^{\infty} \frac{1}{n!(2n-1)(2n-3)\cdots 3\cdot 1}x^n\right)$$

15. $r_1 = \frac{1}{2}: a_{n+1} = \dfrac{-a_n}{2(n+1)(2n+3)}$

$$y_1(x) = a_0x^{1/2}\left(1 - \frac{1}{6}x + \frac{1}{120}x^2 - \frac{1}{5040}x^3 + \cdots\right)$$

$$= a_0x^{1/2}\left(1 + \sum_{n=1}^{\infty} \frac{(-1)^n}{2^n n!(2n+1)(2n-1)\cdots 5\cdot 3}x^n\right)$$

$r_2 = 0: b_{n+1} = \dfrac{-b_n}{2(n+1)(2n+1)}$

$$y_2(x) = b_0\left(1 - \frac{1}{2}x + \frac{1}{24}x^2 - \frac{1}{720}x^3 + \cdots\right)$$

$$= b_0\left(1 + \sum_{n=1}^{\infty} \frac{(-1)^n}{2^n n!(2n-1)(2n-3)\cdots 3\cdot 1}x^n\right)$$

17. $r_1 = 3: y_1(x) = a_0x^3\left(1 - \dfrac{5}{36}x^2 + \dfrac{19}{1404}x^4 + \cdots\right)$

$r_2 = \frac{1}{2}: y_2(x) = b_0x^{1/2}\left(1 + \dfrac{5}{24}x^2 - \dfrac{119}{3456}x^4 + \cdots\right)$

19. $r_1 = 0$, $r_2 = -1$: When r_2 is used, a_0 and a_1 are arbitrary, $a_2 = 0$, and

$$a_{n+3} = \frac{-a_n}{(n+3)(n+2)},\ n \geq 0$$

$$y(x) = a_0x^{-1}\left(1 - \frac{1}{6}x^3 + \frac{1}{180}x^6 - \frac{1}{12960}x^9 + \cdots\right) + a_1x^{-1}\left(x - \frac{1}{12}x^4 + \frac{1}{504}x^7 - \frac{1}{45360}x^{10} + \cdots\right)$$

$$= a_0y_0(x) + a_1y_2(x)$$

21. $r_1 = 5$, $r_2 = 0$: When r_2 is used, a_0 and a_5 are arbitrary, $a_4 = 0$, and

$$a_{n+1} = \frac{n^2+4}{(n+1)(n-4)}a_n,\ n = 0, 1, 2, \text{ and } n \geq 5$$

$$y(x) = a_0\left(1 - x + \frac{5}{6}x^2 - \frac{10}{9}x^3\right)$$

$$+ a_5x^3\left(1 + \frac{29}{6}x + \frac{290}{21}x^2 + \frac{15370}{378}x^3 + \cdots\right)$$

22. $r_1 = 0$, $r_2 = -4$: a_0 and a_4 arbitrary, $a_n = 0$ for $n \geq 7$,

$$y(x) = a_0(x-1)^4[1 + 4(x-1) + 5(x-1)^2] + a_4\left[1 + \frac{4}{5}(x-1) + \frac{1}{5}(x-1)^2\right].$$

(It is useful to let $z = x - 1$, express x and $4 + x$ in terms of z, and let $y(z) = z^{-4}\sum_0^\infty a_n z^n$.)

SECTION 6.4

1. $y_2(x) = 1/3e^{2x}\sin 3x$

3. $y_2(x) = x\int^x \frac{1}{r^2}e^{r^2/2}\,dr$

5. $y_2(x) = x^3$

7. $y_2(x) = -(1 + x\tan x)$

9. $$\frac{1}{P(x)^2} = \frac{1}{1 + 2x + \frac{3}{2}x^2 + \frac{13}{18}x^3 + \cdots} = 1 - 2x + \frac{5}{2}x^2 - \frac{49}{18}x^3 + \cdots$$

11. $P(x)Q(x) = 1 - \frac{1}{4}x^2 + \frac{7}{36}x^3 - \frac{7}{144}x^4 + \cdots$

13. $P(x)Q^2(x) = 1 - x + \frac{1}{4}x^2 + \frac{5}{18}x^3 + \cdots$

15. $$\int x^2\left(1 - x + \frac{1}{4}x^2 + \cdots\right)dx = \frac{1}{3}x^3 - \frac{1}{4}x^4 + \frac{1}{20}x^5 + \cdots$$

17. $$\int \frac{1}{x^3}\left(1 + \frac{2}{3}x^2 + \frac{2}{9}x^4 + \cdots\right)dx = -\frac{1}{2x^2} + \frac{2}{3}\ln x + \frac{1}{9}x^2 + \cdots$$

19. $r_1 = r_2 = 0$: a_0 is arbitrary, $a_1 = 0$, $a_{n+2} = a_n/(n+2)^2$

$$y_1(x) = a_0\left(1 + \frac{1}{2^2}x^2 + \frac{1}{4^2\cdot 2^2}x^4 + \frac{1}{6^2\cdot 4^2\cdot 2^2}x^6 + \cdots\right) = a_0\sum_{n=1}^\infty \frac{1}{2^{2n}(n!)^2}x^{2n}$$

$$y_2(x) = y_1(x)\int \frac{1}{x}\left(1 - \frac{1}{2}x^2 + \frac{5}{32}x^4 + \cdots\right)dx$$

$$= y_1(x)\left[\ln x - \frac{1}{4}x^2 + \frac{5}{128}x^4 + \cdots\right]$$

21. $r_1 = 1$, $r_2 = -1$: a_0 is arbitrary, $a_{n+1} = -a_n/[(n+1)(n+3)]$

$$y_1(x) = a_0x\left(1 - \frac{1}{3}x + \frac{1}{24}x^2 - \frac{1}{360}x^3 + \cdots\right) = 2a_0x\sum_{n=0}^\infty \frac{(-1)^n}{(n+2)!n!}x^n$$

$$y_2(x) = y_1(x)\int \frac{1}{x^3}\left(1 + \frac{2}{3}x + \frac{1}{4}x^2 + \frac{19}{270}x^3 + \cdots\right)dx$$

$$= y_1(x)\left(-\frac{1}{2x^2} - \frac{2}{3x} + \frac{1}{4}\ln x + \frac{19}{270}x + \cdots\right)$$

23. $r_1 = r_2 = 2$: a_n is arbitrary, $a_{a+1} = -4a_n/(n+1)^2$

$$y_1(x) = a_0x^2\left(1 - 4x + 4x^2 - \frac{16}{9}x^3 + \frac{4}{9}x^4 + \cdots\right) = a_0x^2\sum_{n=0}^\infty \frac{(-1)^n4^n}{(n!)^2}x^n$$

$$y_2(x) = y_1(x)\int \frac{1}{x}\left(1 + 8x + 40x^2 + \frac{1472}{9}x^3 + \frac{5416}{9}x^4 + \cdots\right)dx$$

$$= y_1(x)\left(\ln x + 8x + 20x^2 + \frac{1472}{27}x^3 + \frac{1354}{9}x^4 + \cdots\right)$$

SECTION 6.5

1. $\Gamma(5.185) = 31.8304$, $\Gamma(-3.185) = -0.60453$
3. With 20 terms, the error is 41.275; with 30 terms, the error is $3.7805 \cdot 10^{-3}$ and since $e^{-10} \cong 4.5399 \cdot 10^{-5}$, this is still unacceptable.
5. $J_{-3/2}(0.587) = -\left[\dfrac{1}{0.587}J_{-1/2}(0.587) + J_{1/2}(0.587)\right] = -2.0539$

11. Write as $x^2y'' + 4x^3y = 0$ with solutions $y\,(x) = c_1x^{1/2}J_{1/3}\left(\dfrac{4}{3}x^{3/2}\right) + c_2x^{1/2}J_{-1/3}\left(\dfrac{4}{3}x^{3/2}\right)$

13. $y(x) = c_1x^{3/4}J_{1/4}\left(\dfrac{\sqrt{2}}{2}x\right) + c_2x^{3/4}J_{-1/4}\left(\dfrac{\sqrt{2}}{2}x\right)$

19. $H_0(x) = 1$
$H_1(x) = 2x$
$H_2(x) = 4x^2 - 2$
$H_3(x) = 8x^3 - 12x$
$H_4(x) = 16x^4 - 48x^2 + 12$

CHAPTER 7

SECTION 7.1

1. Integrate the expression for $\overline{\text{KE}}$ by parts and use the fact that $x(0) = x(T)$ for any periodic motion.
3. Observe that for $0 < a < \pi$ and $0 \le x \le a$,

$$\int_0^\alpha \left(\sin^2\frac{\alpha}{2} - \sin^2\frac{x}{2}\right)^{-1/2} dx > \int_0^\alpha \left(1 - \sin^2\frac{x}{2}\right)^{-1/2} dx$$

7. Since $\dot{\eta}^2 + \sin^2\eta = 2E$, $E = \dfrac{1}{2}$ will do it. The equation $\dot{\eta} = \pm\cos\eta$ is separable and solving gives

$$\eta(t) = 2\arctan(Ke^{\pm t}) - \frac{\pi}{2},\; K = \text{const}$$

$$\dot{\eta}(t) = \pm 2Ke^{\pm t}/(1 + K^2e^{\pm 2t})$$

SECTION 7.2

1. $(1, -1)$
3. $(2, 4)$, $(-1, 1)$
5. $(2, 1)$, $(2, -1)$, $(-2, 1)$, $(-2, -1)$
7. $(2, 3)$; $x(t) = 2 - \dfrac{1}{2}e^{5t}(u_0 + v_0) + \dfrac{1}{2}e^{3t}(3u_0 + v_0)$

$y(t) = 3 + \dfrac{3}{2}e^{5t}(u_0 + v_0) - \dfrac{1}{2}e^{3t}(3u_0 + v_0)$, (u_0, v_0) near $(0, 0)$

9. $(0, 0)$; $x(t) = u_0\cos t + v_0\sin t$
$y(t) = -u_0\sin t + v_0\cos t$, (u_0, v_0) near $(0, 0)$

11. $(2, -4)$; $x(t) = 2 - e^{2t}\left[u_0 \cosh \sqrt{7}t + \frac{1}{\sqrt{7}}(2u_0 + v_0)\sinh \sqrt{7}t\right]$

$$y(t) = -4 + e^{2t}\left[v_0 \cosh \sqrt{7}t + \frac{1}{\sqrt{7}}(3u_0 - 2v_0)\sinh \sqrt{7}t\right], \ (u_0, v_0) \text{ near } (0, 0)$$

$(-1, -1)$; $x(t) = -1 + u_0 e^{-t}\cos \sqrt{2}t - \frac{(u_0 - v_0)}{\sqrt{2}} e^{-t} \sin \sqrt{2}t$

$$y(t) = -1 + v_0 e^{-t}\cos \sqrt{2}t - \frac{(3u_0 - v_0)}{\sqrt{2}} e^{-t} \sin \sqrt{2}t, \ (u_0, v_0) \text{ near } (0, 0).$$

13. Note that $x \sin y = (\rho \cos \phi) \sin(\rho \sin \phi)$

$$= \rho \cos \phi \left[\rho \sin \phi - \frac{(\rho \sin \phi)^3}{3!} + \cdots\right]$$

$$= \rho^2 \left[\cos \phi \sin \phi - \frac{\rho^2 \cos \phi\, (\sin \phi)^3}{3!} + \cdots\right]$$

and similarly for the term $y \sin x$. Also, let $\begin{bmatrix} -4 & 5 \\ 1 & 4 \end{bmatrix} = -21 \neq 0$.

15. $u(t) = 10^{-4}e^{-t}(\cos 3t - \sin 3t)$
$v(t) = 10^{-4}e^{-t}(\sin 3t + \cos 3t)$

SECTION 7.3

1. Saddle point, $\lambda = 7, -5$; portrait similar to Fig. 7.8.

$$\begin{pmatrix} x(t) \\ y(t) \end{pmatrix} = c_1 \begin{pmatrix} 2 \\ 1 \end{pmatrix} e^{7t} + c_2 \begin{pmatrix} 2 \\ -1 \end{pmatrix} e^{-5t}$$

3. Stable node, $\lambda = -1, -2$; portrait similar to Fig. 7.6.

$$\begin{pmatrix} x(t) \\ y(t) \end{pmatrix} = c_1 \begin{pmatrix} 1 \\ 1 \end{pmatrix} e^{t} + c_2 \begin{pmatrix} 3 \\ 2 \end{pmatrix} e^{-2t}$$

5. Unstable spiral, $\lambda = 1 \pm 2i$; portrait similar to Fig. 7.12(b).

$$\begin{pmatrix} x(t) \\ y(t) \end{pmatrix} = c_1 e^{t} \begin{pmatrix} 2 \cos 2t \\ \sin 2t \end{pmatrix} + c_2 e^{t} \begin{pmatrix} 2 \sin 2t \\ -\cos 2t \end{pmatrix}$$

7. Saddle point, $\lambda = 2, -1$; portrait similar to Fig. 7.8.

$$\begin{pmatrix} x(t) \\ y(t) \end{pmatrix} = c_1 \begin{pmatrix} 1 \\ 4 \end{pmatrix} e^{2t} + c_2 \begin{pmatrix} 1 \\ 1 \end{pmatrix} e^{-t}$$

9. Stable node, $\lambda = -1$, double root; portrait similar to Fig. 7.10.

$$\begin{pmatrix} x(t) \\ y(t) \end{pmatrix} = c_1 \begin{pmatrix} 2 \\ 1 \end{pmatrix} e^{-t} + c_2 \begin{pmatrix} -1 + 2t \\ t \end{pmatrix} e^{-t}$$

11. Unstable node, $\lambda = 2$, double root; portrait similar to Fig. 7.11.

$$\begin{pmatrix} x(t) \\ y(t) \end{pmatrix} = c_1 \begin{pmatrix} 1 \\ -1 \end{pmatrix} e^{2t} + c_2 \begin{pmatrix} -1 + t \\ -t \end{pmatrix} e^{2t}$$

13. Center, $\lambda = \pm 4i$; portrait similar to Fig. 7.12(c).

$$\begin{pmatrix} x(t) \\ y(t) \end{pmatrix} = c_1 \begin{pmatrix} 2\cos 4t \\ \sin 4t \end{pmatrix} + c_2 \begin{pmatrix} -2\sin 4t \\ \cos 4t \end{pmatrix}$$

15. Asymptotically stable node, $\lambda = -2$, double root; portrait similar to Fig. 7.9.

$$\begin{pmatrix} x(t) \\ y(t) \end{pmatrix} = c_1 \begin{pmatrix} 1 \\ 0 \end{pmatrix} e^{-2t} + c_2 \begin{pmatrix} 0 \\ 1 \end{pmatrix} e^{-2t}$$

17. Saddle point, $\lambda = 1, -1$; portrait similar to Fig. 7.8.

$$\begin{pmatrix} x(t) \\ y(t) \end{pmatrix} = c_1 \begin{pmatrix} 4 \\ -1 \end{pmatrix} e^{t} + c_2 \begin{pmatrix} 2 \\ -1 \end{pmatrix} e^{-t}$$

19. Unstable spiral, $\lambda = 2 \pm 2i$, portrait similar to mirror image of Fig. 7.12(b).

$$\begin{pmatrix} x(t) \\ y(t) \end{pmatrix} = c_1 e^{2t} \begin{pmatrix} \cos 2t \\ \cos 2t + 2\sin 2t \end{pmatrix} + c_2 e^{2t} \begin{pmatrix} -\sin 2t \\ 2\cos 2t - \sin 2t \end{pmatrix}$$

SECTION 7.4

1. Saddle point, $\lambda = (1 \pm \sqrt{17})/2$; portrait similar to Fig. 7.8

3. Unstable node, $\lambda = 3, 4$; portrait similar to Fig. 7.7

5. Stable spiral, $\lambda = -2 \pm i$; portrait similar to Fig. 7.12(a)

7. $(2, 1)$, stable node, $\lambda = -3$, double root

$x(t) = 2 + u_0 e^{-3t} + 4te^{-3t}(u_0 - v_0)$

$y(t) = 1 + v_0 e^{-3t} + 4te^{-3t}(u_0 - v_0)$, (u_0, v_0) near $(0, 0)$

9. $(0, 0)$, center, $\lambda = \pm i$

$x(t) = u_0 \cos t + v_0 \sin t$

$y(t) = -u_0 \sin t + v_0 \cos t$, (u_0, v_0) near $(0, 0)$

$(1, -1)$, saddle point, $\lambda = \pm\sqrt{3}$, $x(t) = 1 + u_0 \cosh \sqrt{3}t + \dfrac{(2u_0 + v_0)}{\sqrt{3}} \sinh \sqrt{3}t$

$y(t) = -1 + v_0 \cosh \sqrt{3}t - \dfrac{(u_0 + 2v_0)}{\sqrt{3}} \sinh \sqrt{3}t$, (u_0, v_0) near $(0, 0)$

11. $(1, 1)$, unstable spiral, $\lambda = 1 \pm i$

$x(t) = 1 + e^t(u_0 \cos t - v_0 \sin t)$

$y(t) = 1 + e^t(u_0 \sin t + v_0 \cos t)$, (u_0, v_0) near $(0, 0)$

$(-1, -1)$, saddle point, $\lambda = \pm\sqrt{2}$

$x(t) = -1 + u_0 \cosh \sqrt{2}t + \dfrac{(u_0 - v_0)}{\sqrt{2}} \sinh \sqrt{2}t$

$y(t) = -1 + v_0 \cosh \sqrt{2}t - \dfrac{(u_0 + v_0)}{\sqrt{2}} \sinh \sqrt{2}t$, (u_0, v_0) near $(0, 0)$

13. $(-1, -1)$, unstable node, $\lambda = 2 \pm \sqrt{2}$,

$x(t) = -1 + \dfrac{e^{2t}}{\sqrt{2}} [u_0 \sqrt{2} \cosh \sqrt{2}t - (u_0 + v_0) \sinh \sqrt{2}t]$,

$y(t) = -1 + \dfrac{e^{2t}}{\sqrt{2}} [v_0 \sqrt{2} \cosh \sqrt{2}t - (u_0 - v_0) \sinh \sqrt{2}t]$, (u_0, v_0) near $(0, 0)$

$(-3, -3)$, saddle point, $\lambda = 1 \pm \sqrt{3}$,

$x(t) = -3 + e^t \left[u_0 \cosh \sqrt{3}t - \dfrac{v_0}{\sqrt{3}} \sinh \sqrt{3}t \right]$,

$y(t) = -3 + e^t[v_0 \cosh \sqrt{3}t - u_0\sqrt{3} \sinh \sqrt{3}t]$, (u_0, v_0) near $(0, 0)$

15.

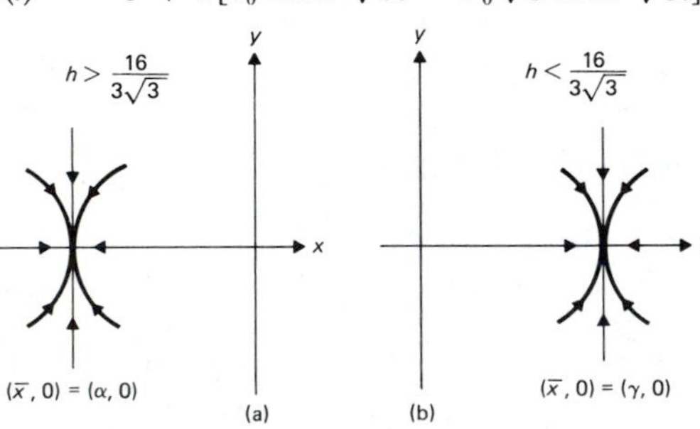

17. $x(t) = Ae^{2t} - B$
$y(t) = Ae^{2t} + B$
A, B arbitrary constants

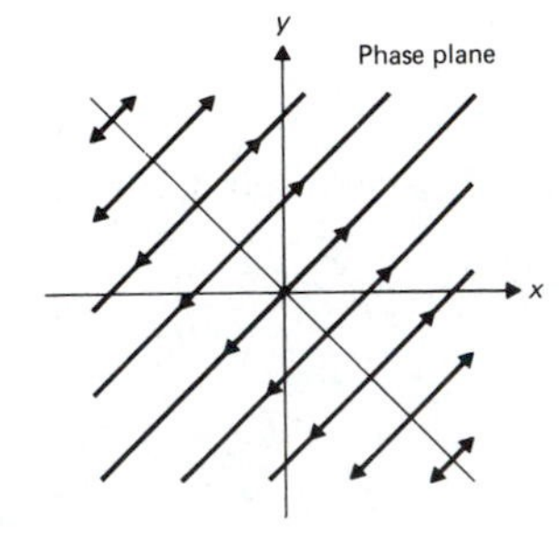

19. $dz/dt = 2z - z^2 = 2z(1 - z/2)$(compare with logistic equation of Chapter 1), $dw/dt = wz$,

$z(t) = \dfrac{2z_0}{z_0 + (2 - z_0)e^{-2t}} = \dfrac{2z_0e^{2t}}{(2 - z_0) + z_0e^{2t}}$ if $z(0) = z_0$,

$w(t) = w_0 \exp \left[\int_0^t z(s)\, ds \right] = \dfrac{w_0}{2} [(2 - z_0) + z_0e^{2t}]$ if $w(0) = w_0$.

There is a line of equilibrium points ($z = 0$) and an isolated equilibrium point $(2, 0)$ which is a saddle point.

SECTION 7.5

1. $(0, 0)$, a saddle point

$x(t) = u_0\cosh \sqrt{2}t + \dfrac{v_0}{\sqrt{2}} \sinh \sqrt{2}t$

$y(t) = v_0 \cosh \sqrt{2}t + \sqrt{2}\, u_0 \sinh \sqrt{2}t$, (u_0, v_0) near $(0, 0)$; $(1, 0)$, a center

$x(t) = 1 + u_0 \cos \sqrt{3}t + \dfrac{v_0}{\sqrt{3}} \sin \sqrt{3}t$

$y(t) = v_0 \cos \sqrt{3}t - \sqrt{3}u_0 \sin \sqrt{3}t$, (u_0, v_0) near $(0, 0)$; $(-2, 0)$, a center

$x(t) = -2 + u_0 \cos \sqrt{6}t + \dfrac{v_0}{\sqrt{6}} \sin \sqrt{6}t$

$y(t) = v_0 \cos \sqrt{6}t - \sqrt{6}u_0 \sin \sqrt{6}t$, (u_0, v_0) near $(0, 0)$

3. $(0, 0)$, a center

$x(t) = u_0 \cos t + v_0 \sin t$

$y(t) = v_0 \cos t - u_0 \sin t$, (u_0, v_0) near $(0, 0)$; $(1, 0)$, a saddle point

$x(t) = 1 + u_0 \cosh \sqrt{2}t + \dfrac{v_0}{\sqrt{2}} \sinh \sqrt{2}t$

$y(t) = v_0 \cosh \sqrt{2}t + \sqrt{2}u_0 \sinh \sqrt{2}t$, (u_0, v_0) near $(0, 0)$;

5.

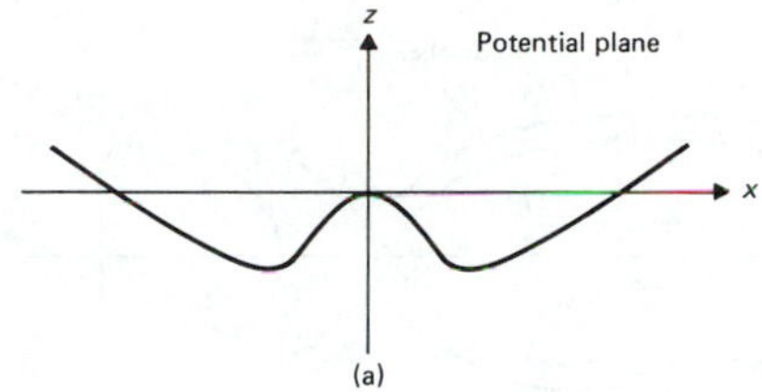

(a)

y

Phase plane

x

(b)

7.

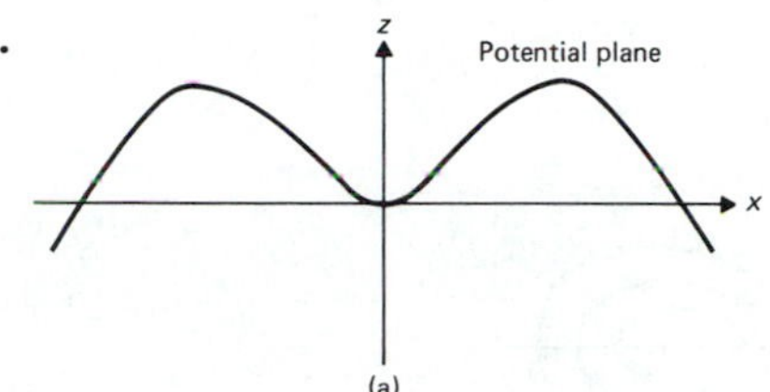

(a)

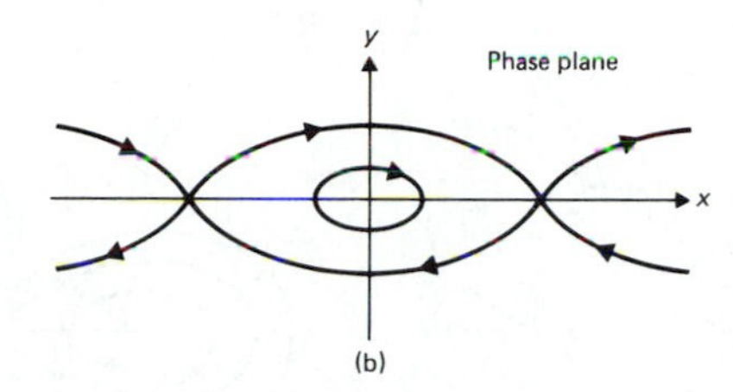

(b)

9.

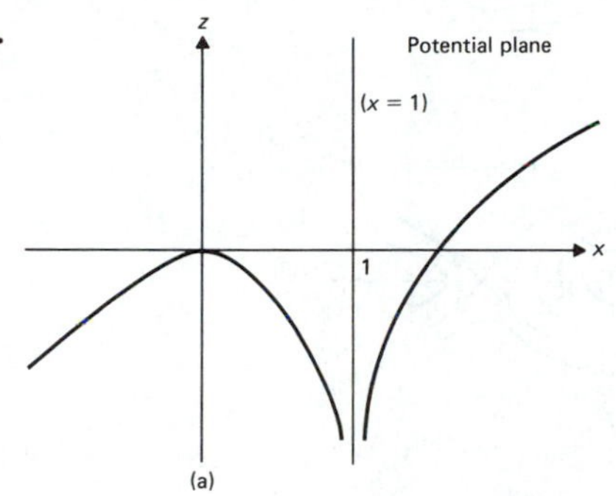

(a)

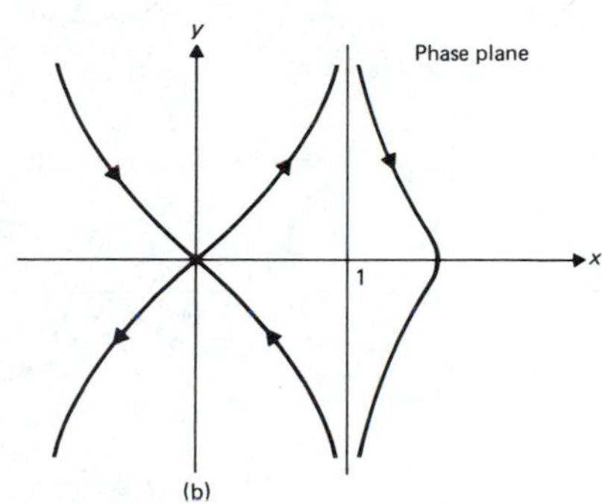

(b)

11. The potential function $F(u) = \frac{u^2}{2} - u - \lambda \frac{u^3}{3}$ has a local minimum at $u = \frac{1}{2\lambda}[1 - \sqrt{1 - 4\lambda}]$ and a local maximum at $u = \frac{1}{2\lambda}[1 + \sqrt{1 - 4\lambda}]$ for $0 < \lambda < \frac{1}{4}$. For $\lambda > \frac{1}{4}$ it is a strictly decreasing function.

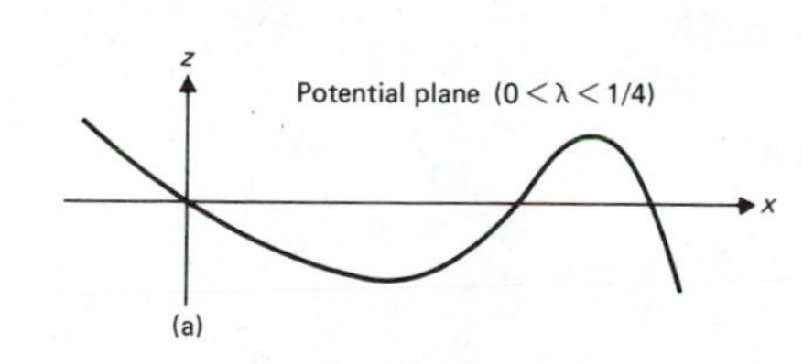

(a)

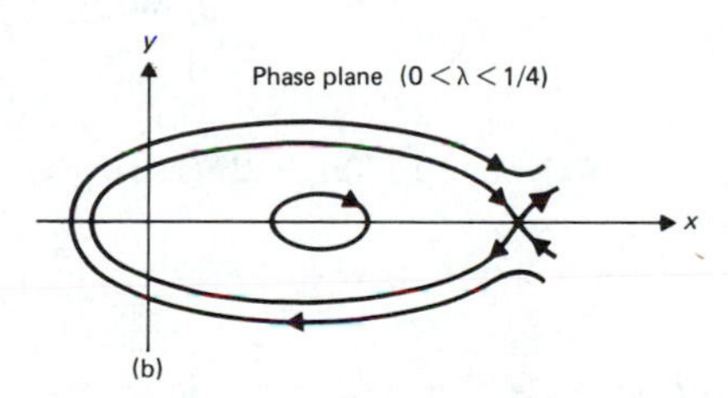

(b)

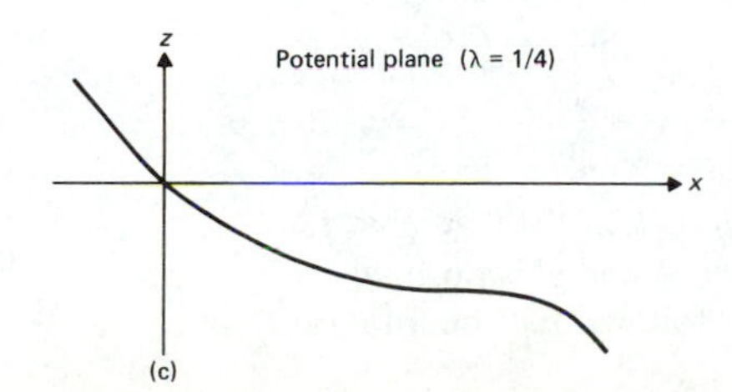

(c)

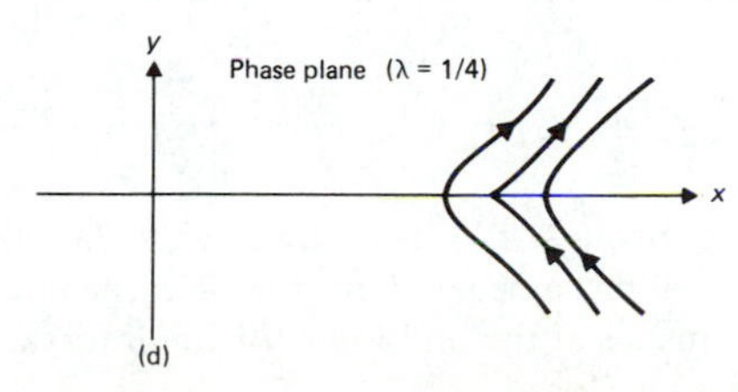

(d)

13. 

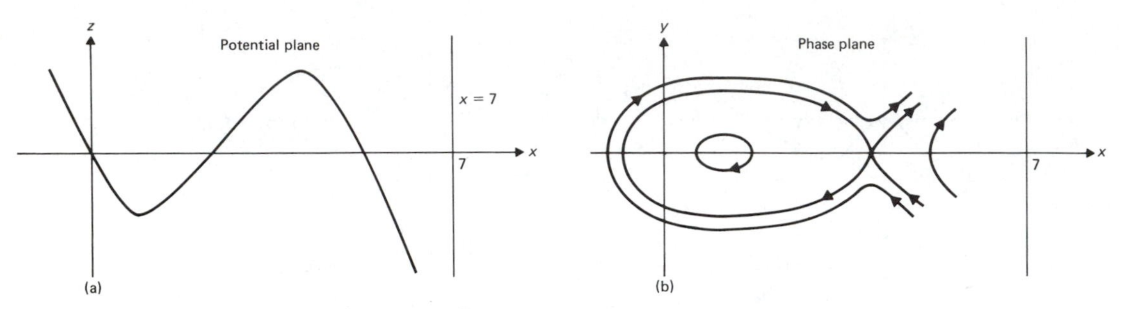

15.

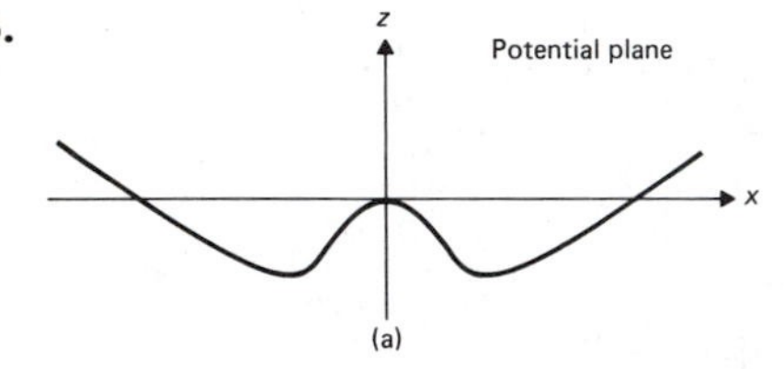

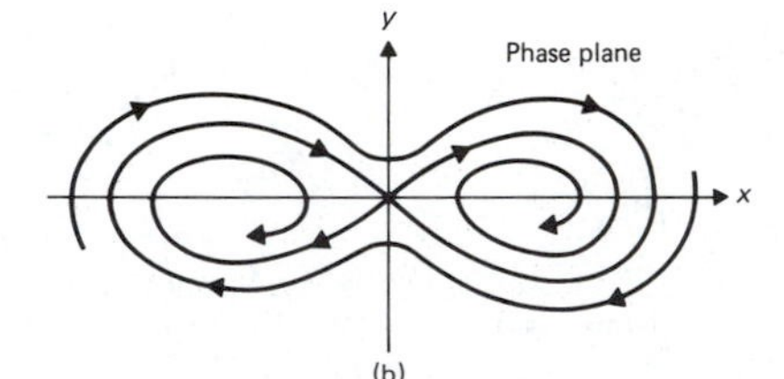

17.

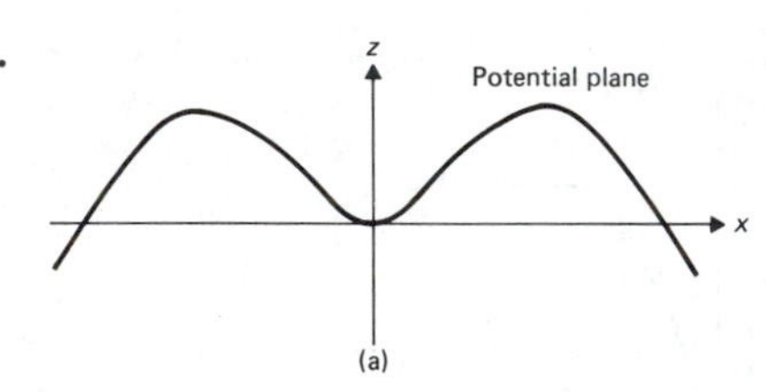

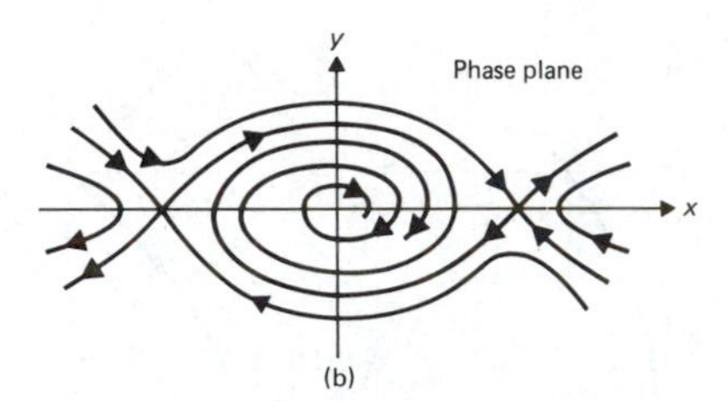

19.

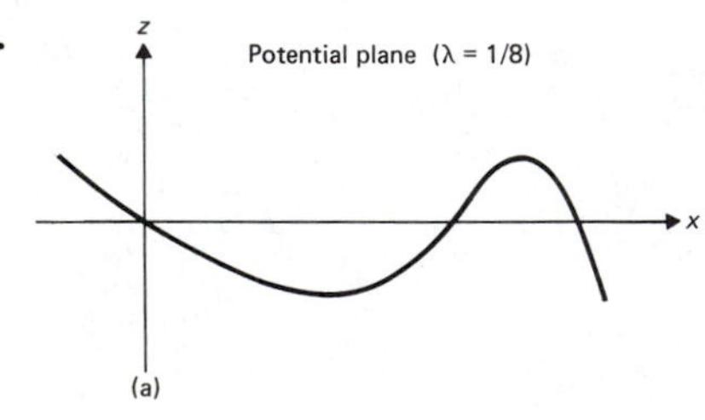

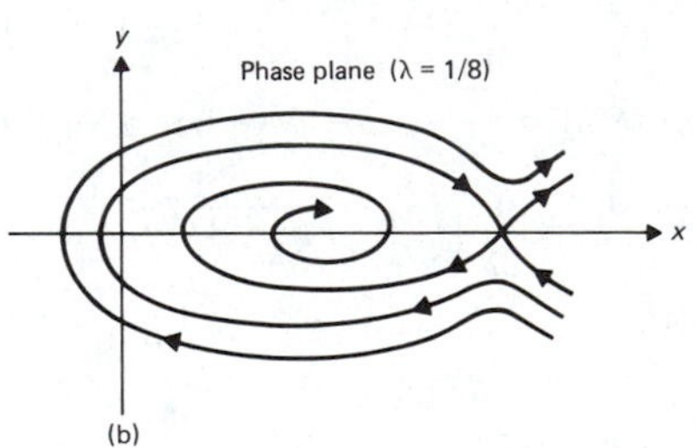

21. If $E(x, \dot{x}) = e^{2x}\dot{x}^2/2 + e^{2x}F(x)$, where $F'(x) + 2F(x) = g(x)$, then

$$\frac{dE(x, \dot{x})}{dt} = e^{2x}\dot{x}[\ddot{x} + \dot{x}^2 + 2F(x) + F'(x)] = 0.$$

a) If $g(x) = x$, then $F(x) = Ae^{-2x} + x/2 - \frac{1}{4}$. Set $A = \frac{1}{4}$, then $F(0) = 0$ and

$$E(x, y) = \frac{e^{2x}y^2}{2} + h(x) = k, \quad \text{where} \quad h(x) = \frac{e^{2x}(2x - 1)}{4} + \frac{1}{4}.$$

The integral curve through $(b, 0)$ is given by $y^2 = 2e^{-2x}[h(b) - h(x)]$. If $0 < b \ll 1$, then $h(x) - h(b) = 0$ has a negative root $(x = a)$ near $x = -b$ and y^2 is positive on $a < x < b$ and vanishes at the endpoints of the interval. It follows that the integral curve through $(b, 0)$ is closed.

b) $E(x, y) = \dfrac{e^{2x}y^2}{2} + \dfrac{e^{2x}(2 \sin x - \cos x)}{5} = k$. Since the points $(\pi, 0)$ and $(-\pi, 0)$ determine different values of k, they lie on different integral curves.

c) Let $G(x)$ be an antiderivative of $e^{2x}g(x)$; then $G(x) = e^{2x}F(x)$. Then, if $y = dx/dt$, $z = dx/ds$, $E(x, y) = e^{2x}y^2/2 + G(x) = k$ and $E^*(x, z) = z^2/2 + G(x) = k^*$ are the corresponding integrals. Therefore, the integral curves are different.

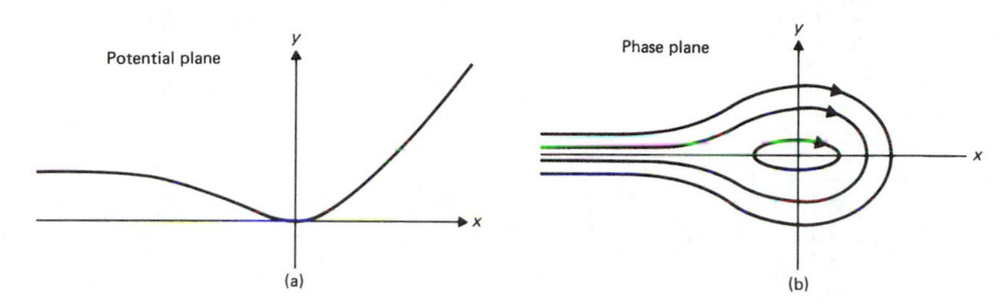

SECTION 7.6

1. a) (0, 0), unstable node b) $(s/\beta, r/\alpha)$, saddle point

5. $r = 2, K = 20, \alpha = 1/25, s = 2, J = 40, \beta = 1/10$
$K = s/\beta$ implies that $(K, 0)$ is a stable node (double root). $J < r/\alpha$ implies that $(0.J)$ is a saddle point. Hence there is competitive exclusion with Y becoming extinct (see Fig. 7.33).

7. $r = 2, K = 100, \alpha = 1/40, s = 10, J = 50, \beta = 1/8$
$K > s/\beta$ implies that $(K, 0)$ is a stable node. $J < r/\alpha$ implies that $(0, J)$ is a saddle point. Hence there is competitive exclusion with Y becoming extinct (see Fig. 7.33).

9. $H = 0$: competitive exclusion; $H = 5/4$: (10/3, 35/12) is a saddle point and unstable coexistence; $H = 25/4$: extinction.

11. a) i)

t	x	y
0.	1.750	4.000
0.50	1.851	3.591
1.00	2.085	3.525
1.50	2.264	3.876
2.00	2.186	4.391
2.50	1.926	4.522
3.00	1.759	4.147
3.50	1.802	3.684

ii)

t	x	y
0.	2.000	2.000
0.50	3.137	2.646
1.00	3.291	5.334
1.50	1.748	6.944
2.00	1.048	4.918
2.50	1.079	3.016
3.00	1.575	2.112
3.50	2.569	2.147

15. $\beta = 1/3$: (5, 6) is a stable spiral point; $\beta = 1/40$: no equilibrium points in the interior of the first quadrant.

17. (20, 0) is a saddle point. $(0, J)$ is a saddle point if $J = 4$, a stable node if $J = 10$.

19. $K = 20$: equilibrium points are (0, 0), a saddle point; (20, 0), a saddle point; (10, 10), a stable spiral point.
$K = 40$: equilibrium points are (0, 0), a saddle point; (40, 0), a saddle point; (10, 15), an unstable spiral point.

21. $H = 0$

t	x	y
30.00	19.91	23.08
31.00	19.98	23.06
32.00	20.04	23.07
33.00	20.07	23.08
34.00	20.07	23.11
35.00	20.06	23.13

$H = 1.1$

t	x	y
30.00	38.73	2.082
31.00	39.17	1.170
32.00	39.67	0.1471
33.00	40.21	−1.002
34.00	40.80	−2.294
35.00	41.45	−3.753

CHAPTER 8

SECTION 8.2

1.

t_i	y_i	**Local error**
0.25	0.0	0.01111
0.50	0.0156	0.1088
0.75	0.1172	0.4507
1.00	0.5508	1.6472

The actual solution is $y(t) = \dfrac{e^{10t}}{500} - \dfrac{50t^2 + 10t + 1}{500}$, so $y(1) = 43.9309$. The global error at $t = 1$ is 43.3801; the sum of local errors is 2.2178. Notice that even though the sum of local errors is small, the global error is not.

2.

h	**Error**	h	**Error**
$\frac{1}{2}$	0.027198	$\frac{1}{64}$	0.000016
$\frac{1}{4}$	0.005096	$\frac{1}{128}$	0.000006
$\frac{1}{8}$	0.001104	$\frac{1}{256}$	0.000005
$\frac{1}{16}$	0.000257	$\frac{1}{512}$	0.000009
$\frac{1}{32}$	0.000063	$\frac{1}{1024}$	0.000019

SECTION 8.6

1. c) The value $\beta = -1317.057$ will produce a ray that comes close to the surface of the water ($z = 0$). The output below illustrates the situation.

x	y	z
0.0	−0.1317E+04	0.2500E+04
⋮	⋮	⋮
0.9000E+04	−0.8795E+03	0.2583E+03
0.1200E+05	0.1578E+03	0.3165E+01
0.1500E+05	0.1005E+04	0.3984E+03
⋮	⋮	⋮
0.1110E+06	0.4590E+02	0.1547E+05
0.1140E+06	−0.2278E+01	0.1549E+05
0.1170E+06	−0.5046E+02	0.1547E+05

3. A program to solve the boundary value problem is

```
C     PROGRAM TO IMPLEMENT THE SHOOTING ALGORITHM, EXAMPLE 3
C     OF CHAPTER 8, TO SOLVE A LINEAR TWO-POINT BOUNDARY
C     VALUE PROBLEM
      SUBROUTINE F(T,Y,YP)
      REAL Y(2), YP(2)
      YP(1) = Y(2)
      YP(2) = -(2.0/T)*Y(2)+(2.0/T**2)*Y(1)+SIN(T)/T**2
      RETURN
      END
C
      EXTERNAL F
      REAL WORK(15), Y(2)
      INTEGER IWORK(5)
      A=1.0
      B=2.0
      CAPA=1.0
      CAPB=2.0
      NEQN=2
      RELERR=1.0E-5
      ABSERR=1.0E-5
C
C     SOLVE WITH Y(A)=CAPA, Y'(A)=S1
C
      S1=1.0
      T=A
      Y(1)=CAPA
      Y(2)=S1
      IFLAG=1
      TOUT=B
      CALL RKF45(F,NEQN,Y,T,TOUT,RELERR,ABSERR,IFLAG,WORK,IWORK)
```

(Continued)

```
      IF(IFLAG.NE.2)GO TO 10
      US1=Y(1)-CAPB
      IF(US1.NE.0.0)GO TO 1
      SSTAR=S1
      GO TO 3
C
C     SOLVE WITH Y(A)=CAPA,Y'(A)=S2
C
    1 S2=2.0
      T=A
      Y(1)=CAPA
      Y(2)=S2
      IFLAG=1
      TOUT=B
      CALL RKF45(F,NEQN,Y,T,TOUT,RELERR,ABSERR,IFLAG,WORK,IWORK)
      IF(IFLAG.NE.2)GO TO 10
      US2=Y(1)-CAPB
      IF(US2.NE.0.0)GO TO 2
      SSTAR=S2
      GO TO 3
C
C     SOLVE WITH Y(A)=A, Y'(A)=SSTAR
C
    2 SSTAR=S1-(S1-S2)*US1/(US1-US2)
    3 T=A
      Y(1)=CAPA
      Y(2)=SSTAR
      IFLAG=1
      TOUT=A
      DT=0.1
    4 CALL RKF45(F,NEQN,Y,T,TOUT,RELERR,ABSERR,IFLAG,WORK,IWORK)
      WRITE(6,5)T,Y(1),Y(2)
    5 FORMAT(1X,F6.2,2F15.4)
      IF (IFLAG.NE.2)GO TO 10
      TOUT=T+DT
      IF(T.LT.B)GO TO 4
      STOP
C
   10 WRITE(6,20)IFLAG
   20 FORMAT(1X, 'IFLAG =', I2, '; CHECK RKF45 COMMENTS FOR DETAILS')
      STOP
      END
```

5.

t	y	y'
1.00	1.0000	0.6659
1.10	1.0733	0.7932
1.20	1.1574	0.8846
1.30	1.2495	0.9523
1.40	1.3473	1.0033
1.50	1.4497	1.0424
1.60	1.5555	1.0725
1.70	1.6640	1.0959
1.80	1.7745	1.1140
1.90	1.8867	1.1279
2.00	2.0000	1.1383

7.

t	y	y'
0.00	0.0000	−0.8291
0.10	−0.0770	−0.7040
0.20	−0.1400	−0.5527
0.30	−0.1869	−0.3845
0.40	−0.2165	−0.2048
0.50	−0.2277	−0.0182
0.60	−0.2200	0.1719
0.70	−0.1933	0.3628
0.80	−0.1475	0.5520
0.90	−0.0830	0.7382
1.00	−0.0000	0.9202

9.

t	y	y'
0.00	0.0000	13.7143
0.10	1.3486	13.0286
0.20	2.5600	10.9714
0.30	3.4971	7.5429
0.40	4.0229	2.7429
0.50	4.0000	−3.4286

11. a) A fundamental assumption in our shooting algorithm is that the differential equation is linear. This makes the determination of s^* easy because $u(s) = y(b; s) - B$ is a linear function of s (see Exercise 2.). The differential equation $y'' + e^{-y} = 0$ is not linear and hence the function $u(s)$ is not, in general, a linear function of s.

b) One procedure that might be used to compute a zero of $u(s)$ to some specified accuracy is the secant method:

(i) Choose $s = s_1$ and solve the differential equation with initial conditions

$$y(a) = A,$$
$$y'(a) = s_1.$$

Form $u(s_1) = y(b; s_1) - B$.

(ii) Choose $s_2 \neq s_1$ and solve the differential equation with initial conditions

$$y(a) = A,$$
$$y'(a) = s_2.$$

(iii) Calculate (the secant method)

$$s_3 = s_2 - u(s_2)\frac{s_2 - s_1}{u(s_2) - u(s_1)}$$

and solve the differential equation with initial conditions

$$y(a) = A,$$
$$y'(a) = s_3.$$

Form $u(s_3) = y(b; s_3) - B$, and so forth.

The process is continued until $|u(s_i)|$ and $|s_i - s_{i-1}|$ are sufficiently small where

$$s_i = s_{i-1} - u(s_{i-1}) \frac{s_{i-1} - s_{i-2}}{u(s_{i-1}) - u(s_{i-2})}.$$

Of course if any s_i results in $u(s_i) = 0$, we are finished. The usual precautions should be taken when using the secant method to find a zero of a nonlinear function (see Ref. [1] for more details).

CHAPTER 9

SECTION 9.2

1. $v_{tt} = b^2 e^{aX+bt}$, $c^2 v_{xx} = \frac{b^2}{a^2}(a^2 e^{ax+bt}) = v_{tt}$. Any function $f(ax + bt)$ that is twice differentiable will be a solution.

3. $V(x, t) = \frac{5}{q^2 - p^2} \cosh(px + qt)$

5. $V(x, t) = At^4 + Bxt^3 + Cx^3t + Dx^4$ where $A = r/12$, $D = -p/12$, and $B - C = q/6$

7. $f(t)$ must satisfy $f'' + f' + f = \cos t$. Solving gives $f(t) = Ae^{-t/2} \sin\left(\frac{\sqrt{3}}{2}\right)t + Be^{-t/2} \cos\left(\frac{\sqrt{3}}{2}\right)t + \sin t$, which is bounded for $t \geq 0$. Presence of v_t term ensures boundedness.

SECTION 9.3

1. $v(x, t) = 3 \cos 4t \sin 2x$

3. $v(x, t) = -2 \cos 6t \sin 2x + 2 \cos 12t \sin 4x$

5. $v(x, t) = \frac{3}{2} \cos t \sin x - \frac{1}{2} \cos 3t \sin 3x$

7. $v(x, t) = \frac{15}{2} \sin 2t \sin 2x - 6 \sin 3t \sin 3x$

9. $v(x, t) = \sum_{n=6}^{18} \frac{B_n}{6n} \sin 6nt \sin nx$ with $B_n = \cos \frac{n\pi}{2}$. Hence $B_6 = -1$, $B_7 = 0$, $B_8 = 1$, $B_9 = 0$, $B_{10} = -1$, etc.

11. $v(x, t) = \left(\frac{3}{4} \cos 3t + \frac{5}{24} \sin 3t\right) \sin x - \left(\frac{1}{4} \cos 9t + \frac{5}{144} \sin 9t\right) \sin 3x + \frac{1}{240} \sin 15t \sin 5x$

13. $v(x, t) = \sum_{n=1}^{5} (2n + 1) \cos 6nt \sin 3nx + \sum_{n=1}^{5} \frac{1}{4n^3} \sin 4nt \sin 2nx$

15. $v(x, t) = \left(\cos 2t + \frac{1}{2} \sin 2t\right) \sin 2x + \frac{1}{4} \sin 4t \sin 4x + \cos 6t \sin 6x$

17. 180π (*Hint:* You may wish to show that $E(t) = \frac{\pi}{4}\sum_{n=1}^{N}(n^2\omega^2A_n^2 + B_n^2)$.)

19. $\frac{549\pi}{4}$

21. $\frac{1359\pi}{512}$

SECTION 9.4

1. $f(x) = x^2 - \pi^2 = -\frac{2}{3}\pi^2 + 4\sum_{n=1}^{\infty}\frac{(-1)^n}{n^2}\cos nx$. (Note: $\cos n\pi = (-1)^n$.) Converges to $f(x)$ on $-\pi \le x \le \pi$.)

3. $a_0 = \pi$, $a_n = 0$, n even, $a_n = -\frac{4}{n^2\pi}$, n odd, $b_n = -\frac{2}{n}(-1)^n$

$f(x) = x + |x| = \frac{\pi}{2} - \sum_{n=1}^{\infty}\frac{4}{(2n-1)^2\pi}\cos(2n-1)x + \frac{2}{n}(-1)^n \sin nx$. Converges to $f(x)$ on $-\pi < x < \pi$ and to π at $x = \pm\pi$.

5. $a_0 = 2$, $a_n = 0$, $n \ne 0$, $b_n = \frac{2}{n\pi}[\cos n\pi - 1]$. Hence $b_n = -\frac{4}{n\pi}$, n odd, $b_n = 0$, n even.

$f(x) = 1 - \frac{4}{\pi}\sum_{n=1}^{\infty}\frac{1}{(2n-1)}\sin(2n-1)x$. Converges to 1 at $x = 0, \pm\pi$ and to $f(x)$ on $-\pi < x < 0$ and $0 < x < \pi$.

7. $f(x) = -\frac{8}{\pi}\sum_{n=1}^{\infty}\frac{(-1)^n n}{4n^2-1}\sin nx$. Converges to $f(x)$ on $-\pi < x < \pi$ and to 0 at $x =$

9. $f(x) = \sinh\left(\frac{1}{2}\right)\left\{2 + 4\sum_{n=1}^{\infty}\frac{(-1)^n}{4n^2\pi^2+1}[\cos n\pi x - 2n\pi \sin n\pi x]\right\}$. Converges to $f(x)$ on $-1 < x < 1$ and to $\sinh\left(\frac{1}{2}\right)$ at $x = \pm 1$.

11. $a_0 = \frac{1}{\pi}$, $a_2 = 0$, $b_2 = \frac{3}{4}$, $a_n = -\frac{2}{\pi(n^2-4)}\left[\cos n\pi + \cos\frac{n\pi}{2}\right]$, $b_n = \frac{2}{\pi(n^2-4)}\sin\frac{n\pi}{2}$.
Converges to $f(x)$ on $-4 \le x \le 4$.

13. $a_0 = \frac{1}{3}[\sinh(3) - \sinh(1)]$

$$a_n = \frac{1}{n^2\pi^2+9}\left[3\sinh(3)\cos n\pi - 3\sinh(1)\cos\frac{n\pi}{3} - n\pi\cosh(1)\sin\frac{n\pi}{3}\right]$$

$$b_n = \frac{1}{n^2\pi^2+9}\left[n\pi\cosh(1)\cos\frac{n\pi}{3} - n\pi\cosh(3)\cos n\pi - 3\sinh(1)\sin\frac{n\pi}{3}\right]$$

Converges to $f(x)$ on $-3 < x < 1$ and $1 < x < 3$, to $\frac{1}{2}\cosh(1)$ at $x = 1$, and to $\frac{1}{2}\cosh(3)$ at $x = \pm 3$.

15. $a_0 = \dfrac{4}{\pi}$, $a_2 = 0$, $b_2 = 0$; otherwise

$$a_n = \frac{-2}{\pi(n^2 - 4)}\left[1 + \cos n\pi - 2\cos\frac{n\pi}{2} + 2\cos\frac{n\pi}{4} + 2\cos\frac{3n\pi}{4}\right],$$

$$b_n = \frac{-2}{\pi(n^2 - 4)}\left[n(\cos n\pi - 1) - \sin\frac{n\pi}{2} + 2\sin\frac{n\pi}{4} + 2\sin\frac{3n\pi}{4}\right].$$

Converges to $f(x)$ on $-6 < x < 0$ and $0 < x < 6$, to $\dfrac{1}{2}$ at $x = 0$, $x = \pm 6$.

21.

x	Program $SN(x)$	Direct
$N = 1$		
−3.1416	−2.5797	−2.5797
−1.5708	−6.5797	−6.5797
−0.7854	−9.4082	−9.4082
0.0000	−10.5797	−10.5797
0.7854	−9.4082	−9.4082
1.5708	−6.5797	−6.5797
3.1416	−2.5797	−2.5797
$N = 2$		
−3.1416	−1.5797	−1.5797
−1.5708	−7.5797	−7.5797
−0.7854	−9.4082	−9.4082
0.0000	−9.5797	−9.5797
0.7854	−9.4082	−9.4082
1.5708	−7.5797	−7.5797
3.1416	−1.5797	−1.5797
$N = 3$		
−3.1416	−1.1353	−1.1353
−1.5708	−7.5797	−7.5797
−0.7854	−9.0939	−9.0939
0.0000	−10.0242	−10.0242
0.7854	−9.0939	−9.0939
1.5708	−7.5797	−7.5797
3.1416	−1.1353	−1.1353

23.

x	Program $SN(x)$	Direct
$N = 1$		
−3.1416	0.9972	1.0000
−1.5708	2.2760	2.2732
−0.7854	1.9070	1.9003
0.0000	1.0083	1.0000
0.7854	0.1064	0.0997
1.5708	−0.2705	−0.2732
3.1416	0.9972	1.0000
$N = 2$		
−3.1416	0.9986	1.0000
−1.5708	2.2746	1.8488
−0.7854	1.8970	2.2004
0.0000	0.9931	1.0000
0.7854	0.0963	−0.2004
1.5708	−0.2719	0.1512
3.1416	0.9986	1.0000
$N = 3$		
−3.1416	0.9991	1.0000
−1.5708	1.8479	2.1035
−0.7854	2.2013	2.0204
0.0000	1.0065	1.0000
0.7854	−0.1995	−0.0204
1.5708	0.1502	−0.1035
3.1416	0.9991	1.0000

SECTION 9.5

1. $b_n = \dfrac{16}{n^3\pi^3}[\cos n\pi - 1] - \dfrac{8}{n\pi}\cos \pi$. Since $\cos n\pi = (-1)^n$, $b_n = \dfrac{8}{n\pi} - \dfrac{32}{n^2\pi^2}$, n odd; $b_n = \dfrac{8}{n\pi}$, n even. $a_0 = \dfrac{8}{3}$, $a_n = \dfrac{16}{n^2\pi^2}\cos n\pi = \dfrac{16}{n^2\pi^2}(-1)^n$.

3. $b_2 = 0,\ b_n = \dfrac{\cos n\pi - 1}{\pi}\left[\dfrac{4}{n(n^2-4)}\right],\ n \neq 2$, or $b_n = \dfrac{-8}{\pi n(n^2-4)},\ n$ odd, $b_n = 0,\ n$ even.

(Note: $\cos(n \pm 2)\pi = \cos n\pi$). $a_0 = 1,\ a_2 = -\dfrac{1}{2},\ a_n = 0,\ n \neq 0, 2$.

5. $b_3 = \dfrac{3}{4},\ b_9 = -\dfrac{1}{4},\ b_n = 0,\ n \neq 3, 9$

$a_3 = a_9 = 0,\ a_n = \dfrac{1 + \cos n\pi}{\pi}\left[\dfrac{324}{(n^2-9)(n^2-81)}\right],\ n \neq 3, 9$, or $a_n = 0,\ n$ odd,

$a_n = \dfrac{648}{\pi(n^2-9)(n^2-81)},\ n$ even. (Note: $\cos(n \pm 3)\pi = \cos(n \pm 9)\pi = -\cos n\pi$.)

7. Use the fact that $2\displaystyle\int_0^1 |\sin^3 3\pi x| \sin n\pi x\, dx = 2\int_0^1 \sin^3 3\pi x \sin n\pi x\, dx - 4\int_{1/3}^{2/3} \sin^3 3\pi x \sin n\pi x\, dx$,

the results of Ex. 5, and the change of variables $r = \pi x$ to simplify calculations.

$b_3 = \dfrac{1}{4},\ b_9 = -\dfrac{1}{12},\ b_n = \dfrac{1}{\pi}\left[\dfrac{648}{(n^2-81)(n^2-9)}\right]\left(\sin\dfrac{2n\pi}{3} + \sin\dfrac{n\pi}{3}\right)$

$a_3 = a_9 = 0,\ a_n = \dfrac{324}{\pi(n^2-9)(n^2-81)}\left[1 + \cos n\pi + 2\left(\cos\dfrac{2n\pi}{3} + \cos\dfrac{n\pi}{3}\right)\right],\ n \neq 3, 9$

9. $b_3 = \dfrac{1}{2},\ b_9 = -\dfrac{1}{6},\ b_n = \dfrac{324}{\pi(n^2-81)(n^2-9)}\left[\sin\dfrac{2n\pi}{3} + \sin\dfrac{n\pi}{3}\right]$

$a_3 = a_9 = 0,\ a_n = \dfrac{324}{\pi(n^2-81)(n^2-9)}\left[1 + \cos n\pi + \cos\dfrac{2n\pi}{3} + \cos\dfrac{n\pi}{3}\right],\ n \neq 3, 9$

11. $w(t) = -\dfrac{3}{5}\cos t + \dfrac{3}{10}\sin t - \dfrac{1}{60}\cos 3t + \dfrac{1}{30}\sin 3t$

13. $w(t) = \dfrac{4}{3} + \displaystyle\sum_{n=1}^{\infty} \dfrac{(3-n^2)\cos nt + 4n \sin nt}{n^2[(3-n^2)^2 + 16n^2]}$

15. $w(t) = \dfrac{a_0}{6} + \displaystyle\sum_{n=1}^{\infty} a_n\left[\dfrac{(3-n^2)\cos nt + 4n\sin nt}{(3-n^2)^2 + 16n^2}\right]$

17. $\left|x - \dfrac{\pi}{2}\right| = \dfrac{\pi}{4} + \dfrac{2}{\pi}\displaystyle\sum_{n=1}^{\infty}\dfrac{1}{n^2}\left[1 + \cos n\pi - 2\cos\dfrac{n\pi}{2}\right]\cos nx$. Note that the bracketed term is zero except when $n = 2, 6, 10 \ldots, 2(2n+1)$, so the series could be written as $\left|x - \dfrac{\pi}{2}\right| = \dfrac{\pi}{4} + \dfrac{2}{\pi}\displaystyle\sum_{n=0}^{\infty}\dfrac{\cos 2(2n+1)x}{(2n+1)^2}$.

$\left|x - \dfrac{\pi}{2}\right| = \dfrac{2}{\pi}\displaystyle\sum_{n=1}^{\infty}\dfrac{1}{n^2}\left[4 - 8\cos\dfrac{n\pi}{4} + \pi n \sin\dfrac{n\pi}{2}\right]\cos\dfrac{n}{2}x,\ n$ odd

SECTION 9.6

1. $A_n = \dfrac{4}{\pi n^2}(1 - \cos n\pi) = \dfrac{8}{\pi n^2}\cos^2\dfrac{n\pi}{2},\ B_n = 0,\ v(x, t) = \displaystyle\sum_{n=0}^{\infty} A_n \cos nct \sin nx$

3. $A_n = \dfrac{-2n}{\pi(n^2+1)} \sinh \pi \cos n\pi = (-1)^{n+1} \dfrac{2n}{\pi(n^2+1)} \sinh \pi$

$$ncB_n = \frac{2n}{\pi(n^2+1)}[1 - \cosh \pi \cos n\pi] + \frac{2}{n\pi}[\cos n\pi - 1]$$

$$= \frac{2n}{\pi(n^2+1)}[1 + (-1)^{n+1} \cosh \pi] - \frac{2}{n\pi} \cos^2 \frac{n\pi}{2}$$

$$v(x, t) = \sum_{n=0}^{\infty} [A_n \cos nct + B_n \sin nct] \sin nx$$

5. $A_n = 0$, $B_n = \dfrac{2}{c\pi(n^2+1)}[1 + (-1)^{n+1} e^{\pi}]$, $v(x, t) = \displaystyle\sum_{n=1}^{\infty} B_n \sin nct \sin nx$

7. $A_n = \left[\dfrac{2}{n\pi} - \dfrac{16}{n^2\pi^2} \sin \dfrac{n\pi}{4} - \dfrac{4}{n\pi} \cos \dfrac{n\pi}{2} + \dfrac{16}{n^2\pi^2} \sin \dfrac{3n\pi}{4} - \dfrac{2}{n\pi} \cos n\pi\right]$, $B_n = 0$,

$v(x, t) = \displaystyle\sum_{n=1}^{\infty} A_n \cos c \frac{n\pi}{4} t \sin \frac{n\pi}{4} x$. Also $v(x, t) = \dfrac{1}{2}[f_0(x + ct) + f_0(x - ct)]$ where $f_0(x)$ is

$$f_0(x) = \begin{cases} |x+3| & \text{if } -4 < x < -2 \\ -|x+1| & \text{if } -2 < x < 0 \\ f(x) & \text{if } 0 < x < 4, \; f_0(x+8) = f_0(x). \end{cases}$$

9. $A_n = 0$, $B_n = \dfrac{2}{n\pi c}\left[\dfrac{2}{n\pi} - \dfrac{8}{n^2\pi^2} \sin \dfrac{n\pi}{2} - \dfrac{2}{n\pi} \cos n\pi\right]$, $v(x, t) = \displaystyle\sum_{n=1}^{\infty} B_n \sin \frac{n\pi c}{2} t \sin \frac{n\pi}{2} x$.

Also $v(x, t) = \dfrac{1}{2c}\displaystyle\int_{x-ct}^{x+ct} g_0(z)dz$ where

$$g_0(x) = \begin{cases} -|x+1| & \text{if } -2 < x < 0 \\ |x-1| & \text{if } 0 < x < 2, \; g_0(x+4) = g_0(x). \end{cases}$$

11. $\lambda_n = \dfrac{2n+1}{4}\pi$. (Note: $\cos 2\lambda_n = 0$.)

$A_n = \dfrac{48}{\lambda_n^3} - \dfrac{36}{\lambda_n^4} \sin 2\lambda_n$, $\lambda_n c B_n = \dfrac{24\pi \sin 2\lambda_n}{\lambda_n^2 - 36\pi^2}$

13. $\lambda_n = \dfrac{2n+1}{2}$, $A_n = \dfrac{-16}{\pi\lambda_n(\lambda_n^2 - 16)}$, $\lambda_n c B_n = \dfrac{2\lambda_n^2 - 16}{\pi\lambda_n(\lambda_n^2 - 16)}$

15. $\lambda_n = \dfrac{(2n+1)\pi}{4}$, $A_n = \dfrac{1}{2}\sin(2\lambda_n)\left[\dfrac{1}{(\lambda_n + \pi)^2} - \dfrac{1}{(\lambda_n - \pi)^2}\right] = \dfrac{8 \cos n\pi}{\pi^2}\left[\dfrac{1}{(2n+5)^2} - \dfrac{1}{(2n-3)^2}\right]$,

$B_n = 0$, $v(x, t) = \displaystyle\sum_{n=0}^{\infty} A_n \cos c\lambda_n t \cos \lambda_n x$

17. $\lambda_n = \dfrac{(2n+1)\pi}{8}$, $A_n = 0$, $\lambda_n c B_n = \dfrac{1}{2}\left[\dfrac{1 - 2\cos \lambda_n}{\lambda_n^2} + \dfrac{3 \sin 4\lambda_n}{\lambda_n}\right]$,

$$v(x, t) = \sum_{n=0}^{\infty} B_n \sin c\lambda_n t \cos \lambda_n x$$

19. The term $7 \sin[3(x - ct]$ induces resonance. $F_3(t) = 7 \cos 3ct$, $f_3 = g_3 = 0$; hence $P_3(t) = \dfrac{7}{6c} t \sin 3ct$

23. $F(x, t) = \dfrac{-32x}{\pi}[\cos 2t + 5 \cos 4t - 5 \cos 8t]$, $F_n(t) = \dfrac{64}{n\pi}[\cos 2t + 5 \cos 4t - 6 \cos 8t] \cos n\pi$, $f_2 = 4, f_n = 0, n \neq 2, g_5 = 1, g_n = 0, n \neq 5$. For $n \neq 2, 4, 5, 8$

$$P_n(t) = \frac{64}{n\pi}\left[\frac{\cos 2t}{n^2 - 4} + \frac{5 \cos 4t}{n^2 - 16} - \frac{6 \cos 8t}{n^2 - 64}\right] \cos n\pi - \frac{64}{n\pi}\left[\frac{1}{n^2 - 4} + \frac{5}{n^2 - 16} - \frac{6}{n^2 - 64}\right] \cos n\pi \cos nt$$

$$P_2(t) = \frac{8}{\pi} t \sin 2t + \frac{32}{\pi}\left[-\frac{5 \cos 4t}{12} + \frac{\cos 8t}{10}\right] - \frac{32}{\pi}\left[-\frac{5}{12} + \frac{1}{10} - \frac{\pi}{8}\right] \cos 2t$$

$$P_4(t) = \frac{10}{\pi} t \sin 4t + \frac{16}{\pi}\left[\frac{\cos 2t}{12} + \frac{\cos 8t}{8}\right] - \frac{16}{\pi}\left[\frac{1}{12} + \frac{1}{8}\right] \cos 4t$$

$$P_5(t) = -\frac{64}{5\pi}\left[\frac{\cos 2t}{21} + \frac{5 \cos 4t}{9} + \frac{2 \cos 8t}{13}\right] + \frac{64}{5\pi}\left[\frac{1}{21} + \frac{5}{9} + \frac{2}{13}\right] \cos 5t + \frac{1}{5} \sin 5t$$

$$P_8(t) = -\frac{3}{\pi} t \sin 8t + \frac{8}{\pi}\left[\frac{\cos 2t}{60} + \frac{5 \cos 4t}{48}\right] - \frac{8}{\pi}\left[\frac{1}{60} + \frac{1}{48}\right] \cos 8t$$

$$w(x, t) = \sum_{n=1}^{\infty} P_n(t) \sin nx$$

$$v(x, t) = \frac{x}{\pi}[4(\cos 2t - 1)^2 + 6(\cos 4t - 1)^2] + w(x, t)$$

SECTION 9.7

1. $w(x, t) = 4e^{-kt} \sin x + 12e^{-16kt} \sin 4x$

3. $w(x, t) = 10e^{-kt} \sin x - 5e^{-9kt} \sin 3x + e^{-25kt} \sin 5x$

5. $w(x, t) = \dfrac{c_0}{2} + \displaystyle\sum_{n=1}^{\infty} c_n e^{-\lambda_n^2 kt} \cos \lambda_n x$, $\lambda_n = \dfrac{n\pi}{L}$, $c_n = \dfrac{2}{L}\displaystyle\int_0^L f(x) \cos(\lambda_n x)\, dx$

7. $w(x, t) = (8 \cosh 2) \displaystyle\sum_{n=1}^{\infty} \frac{\cos n\pi - 1}{16 + n^2\pi^2} \exp(-n^2\pi^2 kt/16) \cos \frac{n\pi}{4} x$

9. $w(x, t) = \dfrac{1}{2} + 4 \displaystyle\sum_{n=1}^{\infty} \frac{1}{n^2\pi^2}\left[2 \cos \frac{n\pi}{2} - \cos n\pi - 1\right] \exp(-n^2\pi^2 kt/4) \cos \frac{n\pi}{2} x$

11. $f_1 = \frac{3}{4}$, $f_3 = -\frac{1}{4}$, $F_n = 0$, $n \neq 1, 3$; $f_n(t) = \frac{4}{\pi n^3}(1 - \cos n\pi)\sin t$; hence $F_n(t) = \frac{8}{\pi n^3}\sin t$, n odd, $F_n(t) = 0$, n even.

$P_n(t) = \frac{8}{\pi n^3(1 + n^4)}[e^{-n^2 t} - \cos t + n^2 \sin t]$, n, odd, $n \neq 1, 3$; $P_n(t) = 0$, n even.

$P_1(t) = \left(\frac{3}{4} + \frac{4}{\pi}\right)e^{-t} + \frac{4}{\pi}\sin t - \frac{4}{\pi}\cos t$

$P_3(t) = \left(\frac{4}{1107\pi} - \frac{1}{4}\right)e^{-9t} + \frac{4}{123\pi}\sin t - \frac{4}{1107\pi}\cos t$

$w(x, t) = \sum_{n=1}^{\infty} P_n(t)\sin nx$

13. $F(x, t) = \frac{2}{\pi}e^{-t}(\pi - x) + \frac{8}{\pi}e^{-2t}x$, $f(x) = \sin 2x - \frac{2}{\pi}(\pi + x)$, $f_n = \frac{-12}{n\pi}$, n odd; $f_2 = 1 + \frac{2}{\pi}$,

$f_n = \frac{4}{n\pi}$, n even, $n \neq 2$, $F_n(t) = \frac{4}{n\pi}e^{-t} - \frac{16}{n\pi}e^{-2t}\cos n\pi$.

$P_n(t) = A_n e^{-(n2/2)t} + \frac{8}{\pi n(n^2 - 2)}e^{-t} - \frac{32\cos n\pi}{\pi n(n^2 - 4)}e^{-2t}$, $n \neq 2$ where A_n is chosen so that $P_n(0) = \frac{4}{\pi n}(2\cos n\pi - 1)$.

$P_2(t) = e^{-2t} + \frac{2}{\pi}e^{-t} - \frac{8}{\pi}te^{-2t}$

$v(x, t) = \sum_{n=1}^{\infty} P_n(t)\sin nx$

$w(x, t) = \frac{2}{\pi}e^{-t}(\pi - x) + \frac{4}{\pi}e^{-2t}x + v(x, t)$

15. $F(x, t) = 3(1 - x)\sin t - 4x\cos t$, $f(x) = -3(1 - x)$, $f_n = -\frac{6}{n\pi}$, $F_n(t) = \frac{6}{n\pi}\sin t + \frac{8}{n\pi}\cos n\pi\cos t$.

$P_n(t) = A_n \exp(-n^2\pi^2 t) + \frac{2(3n^2\pi^2 + 4\cos n\pi)}{n\pi(1 + n^4\pi^4)}\sin t + \frac{2(3 + 4n^2\pi^2\cos n\pi)}{n\pi(1 + n^4\pi^4)}\cos t$ where A_n is chosen so that $P_n(0) = \frac{-6}{n\pi}$

$v(x, t) = \sum_{n=1}^{\infty} P_n(t)\sin n\pi x$

$w(x, t) = 3(1 - x)\cos t + 4x\sin t + v(x, t)$

19. $e^{2t}F(x, t) = -6e^t x(\pi - x)$, $f(x) = \cos 4x - 1$, $f_n = \frac{16}{\pi}\frac{1}{n(n^2 - 4)}$, n odd, $f_n = 0$, n even. $F_n(t) = \frac{-48}{\pi n^2}e^t$, n odd, $F_n(t) = 0$, n even.

$P_n(t) = A_n e^{-n2t} - \frac{48}{\pi n^2(n^3 + 1)}e^t$ where A_n is chosen so that $P_n(0) = \frac{16}{\pi}\frac{1}{n(n^3 - 4)}$, n odd; $P_n(t) = 0$, n even.

$v(x, t) = \sum_{n=1}^{\infty} P_n(t)\sin nx$,

$w(x, t) = e^{-2t}v(x, t)$

21. $w(x, t) = e^{-2t}v(x, t)$ where $v(x, t)$ satisfies $v_t = v_{xx}$, $0 < x < \pi$, $t > 0$, $v(0, t) = -3e^{2t}\sin t$, $v(\pi, t) = 5e^{2t}$, $v(x, 0) = 0$.

From Exercise 12, $v(x, t) = -3\,\dfrac{\pi - x}{\pi}\,e^{2t}\sin t + \dfrac{5x}{\pi}e^{2t} + u(x, t)$ where $u(x, t)$ satisfies

$$u_t = u_{xx} + [6e^{2t}\sin t + 3e^{2t}\cos t]\,\frac{\pi - x}{\pi} - 10e^{2t}\,\frac{x}{\pi}, \qquad u(0, t) = u(\pi, t) = 0, \qquad u(x, 0) = -\frac{5x}{\pi}.$$

Hence

$$f_n = \frac{10}{n\pi}\cos n\pi, \qquad F_n(t) = \frac{2}{n\pi}[6e^{2t}\sin t + 3e^{2t}\cos t] + \frac{20}{n\pi}e^{2t}\cos n\pi,$$

$$P_n(t) = A_n\exp(-n^2t) + \frac{6}{n\pi[(n^2+2)^2+1]}[(2n^2+5)e^{2t}\sin t + (n^2+4)e^{2t}\cos t] + \frac{20\cos n\pi}{n\pi(n^2+2)}e^{2t}$$

where A_n is chosen so that

$$P_n(0) = \frac{10}{n\pi}\cos n\pi, \qquad u(x, t) = \sum_{n=1}^{\infty} P_n(t)\sin nx.$$

23. $w(x, t) = e^{-t}v(x, t)$ where $v(x, t)$ satisfies

$$v_t = v_{xx} - 2e^t\sin 6\pi x,\ 0 < x < 4,\ t > 0, \qquad v(0, t) = -4e^{-t}, \qquad v(4, t) = 0, \qquad v(x, 0) = \sin\frac{\pi}{4}x.$$

From Exercise 12, $v(x, t) = -(4 - x)e^{-t} + u(x, t)$ where $u(x, t)$ satisfies

$$u_t = u_{xx} - 2e^t\sin 6\pi x - (4 - x)e^{-t}, \qquad u(0, t) = u(4, t) = 0, \qquad u(x, 0) = \sin\frac{\pi}{4}x + (4 - x).$$

Hence $f_1 = 1 + \dfrac{8}{\pi}$, $f_n = \dfrac{8}{n\pi}$, $n \neq 1$; $F_n(t) = -\dfrac{8}{n\pi}e^{-t}$, $n \neq 24$.

$$F_{24}(t) = -2e^t - \frac{1}{3\pi}e^{-t}, \qquad P_n(t) = A_{24}e^{-(n2\pi 2/16)t} + \frac{128}{n\pi(16 - n^2\pi^2)}e^{-t}$$

where A_n is chosen so that $P_n(0) = f_n$, $n \neq 24$.

$$P_{24}(t) = A_{24}e^{-36\pi 2t} - \frac{2}{36\pi^2 + 1}e^t + \frac{1}{3\pi(1 - 36\pi^2)}e^{-t}$$

where A_{24} is chosen so that $P_{24}(0) = \dfrac{1}{3\pi}$.

$$u(x, t) = \sum_{n=1}^{\infty} P_n(t)\sin\frac{n\pi}{4}x$$

SECTION 9.8

1. $f_1 = g_1 = 3$, $f_5 = g_5 = 6$, $f_6 = g_6 = -10$, all other $f_n = g_n = 0$;
$p_2 = q_2 = 0$, $p_n = q_n = \dfrac{16(\cos n\pi - 1)}{\pi n(n^2 - 4)}$, $n \neq 2$

3. $f_2 = 0, f_n = \dfrac{4(\cos n\pi - 1)}{\pi n(n^2 - 4)}$, $n \neq 2$; $g_6 = 0$, $g_n = \dfrac{36(\cos n\pi - 1)}{\pi n(n^2 - 36)}$, $n \neq 6$, $p_n = q_n = 0$, all n

5. $a_0 = \dfrac{\pi}{16}\left(\dfrac{\pi}{12} - 1\right)$; $a_n = \dfrac{2}{\pi}\left[\dfrac{\pi}{2n^2}\cos\dfrac{n\pi}{4} + \left(\dfrac{\pi^3}{16n} - \dfrac{2}{n^3} - \dfrac{\pi}{4n}\right)\sin\dfrac{n\pi}{4}\right]$, $n \neq 0$

7. $a_0 = -\dfrac{1}{2\pi}\sinh \pi$; $a_n = \dfrac{1}{\pi}\left[\dfrac{n-1}{(n-1)^2+1} - \dfrac{n+1}{(n+1)^2+1}\right]\sinh \pi \cos n\pi$

11. $A_0 = -2$, $B_0 = \dfrac{4}{\ln 2}$, $A_2 = -\dfrac{5}{3}$, $B_2 = \dfrac{20}{3}$, $A_n = B_n = 0$, $n \neq 0, 2$

$C_1 = \dfrac{4}{3}$, $D_1 = -\dfrac{16}{3}$, $C_5 = \dfrac{192}{1023}$, $D_5 = \dfrac{-192}{1023}$, $C_n = D_n = 0$, $n \neq 1, 5$

13. $A_0 = 4$, $B_0 = \dfrac{-4}{\ln 2}$, $A_n = \dfrac{2^{-n}\cos n\pi - \sin\dfrac{n\pi}{2}}{2^{-n} - 2^n}$, $B_n = \dfrac{\sin\dfrac{n\pi}{2} - 2^n\cos n\pi}{2^{-n} - 2^n}$, $C_n = D_n = 0$, $n \geq 1$

SECTION 9.9

1. $J(x, t) = 0.5[j(x - ct) + j(x + ct)]$, $V(x, t) = \dfrac{\gamma}{2}[j(x - ct) - j(x + ct)]$, where $j(x) = \tanh x[u(x + 1) - u(x - 2)]$.

3. $J(x, t) = 0.5[j(x - ct) + j(x + ct)] + (0.5/\gamma)[v(x - ct) - v(x + ct)]$,
$V(x, t) = 0.5\gamma[j(x - ct) - j(x + ct)] + 0.5[v(x - ct) + v(x + ct)]$, where
$j(x) = (x - 1)^2(x + 2)^2[u(x + 1) - u(x - 2)]$, $V(x) = 0.25\gamma(x^2 - 1)^2[u(x + 1) - u(x - 2)]$.

5. $\mathbf{w}(x, t) = 0.25[(5e^{-x-2t} - e^{x+2t})\mathbf{b}' + (e^{x-2t} - e^{-x+2t})\mathbf{b}^2]$ where $\mathbf{b}_1 = \text{col}(1, 0)$, $\mathbf{b}^2 = \text{col}(3, 2)$.

7. $\mathbf{w}(x, t) = [(x - 6t)^2 + (x - 6t)^4]/7\mathbf{b}' + [4(x + t)^2 - 3(x + t)^4]/7\mathbf{b}^2$ where $\mathbf{b}^1 = \text{col}(3, 4)$, $\mathbf{b}^2 = \text{col}(1, -1)$.

9. $w_1(x, t) = -\exp(x - t)$, $w_2(x, t) = \exp(x - t) + 4\exp(2t - x)$, $w_3(x, t) = -\exp(2t - x)$

11. $\mathbf{w}(x, t) = (x - t)\mathbf{b}^1 + (x + 2t)^2\mathbf{b}^2 + (5t - x)\mathbf{b}^3$ where $\mathbf{b}^1 = \text{col}(0, 1, 1)$, $\mathbf{b}^2 = \text{col}(1, 0, -1)$, $\mathbf{b}^3 = \text{col}(1, 1, 1)$.

13. $R = 1.72 \times 10^{-8}$ ohms/m

15. $J(x, t) = 0.5\exp(-t)[\cos(x - t) + \cos(x + t) + \sin(x - t) - \sin(x + t)]$
$V(x, t) = 0.5\exp(-t)[\cos(x - t) - \cos(x + t) + \sin(x - t) + \sin(x + t)]$

17. $J(x, t) = 0.5\exp(-t)[j(x - t) + j(x + t) + v(x - t) - v(x + t)]$,
$V(x, t) = 0.5\exp(-t)[j(x - t) - j(x + t) + v(x - t) + v(x + t)]$, where
$j(x) = -(\sinh^2 x)(\sinh^2(x - 1))[u(x) - u(x - 1)]$,
$V(x) = 3(\cosh x - 1)^2(\cosh x - \cosh 1)^2[u(x) - u(x - 1)]$.

19. $\alpha = -2$, $\mathbf{w}(x, t) = \exp(-2t)\mathbf{b}^1 + \exp(-t)(x + t)\mathbf{b}^2$ where $\mathbf{b}^1 = \text{col}(2, 3)$, $\mathbf{b}^2 = \text{col}(1, 2)$.

21. $\alpha = 3$, $\mathbf{w}(x, t) = -\exp(-7t)\sin(x - 5t)\mathbf{b}^1 + 2\exp(14t)\sin(2x - 10t)\mathbf{b}^2$ where $\mathbf{b}^1 = \text{col}(3, 4)$, $\mathbf{b}^2 = \text{col}(1, -1)$.

23. $\mathbf{w}(x, t) = 2(x - t)\mathbf{b}^1 + e^{-2t}(x + t)^2\mathbf{b}^2 + e^{2t}(x + 2t)^3\mathbf{b}^3$ where $\mathbf{b}^1 = \text{col}(1, 0, 1)$, $\mathbf{b}^2 = \text{col}(0, 1, 1)$, $\mathbf{b}^3 = \text{col}(1, 0, 0)$.

25. $\mathbf{w}(x, t) = e^t e^{(x-t)}\mathbf{b}^1 + e^{-6t}e^{-(x+t)}\mathbf{b}^3 + 0\mathbf{b}^2$ where $\mathbf{b}^1 = \text{col}(1, 1, 1)$, $\mathbf{b}^3 = \text{col}(1, 5, 2)$, $\mathbf{b}^2 = \text{col}(2, 1, 2)$.

INDEX